ANALYSIS OF COSMETIC PRODUCTS

ANALYSIS OF COSMETIC PRODUCTS

SECOND EDITION

Edited by

AMPARO SALVADOR and ALBERTO CHISVERT

University of Valencia, Valencia, Spain

Elsevier
Radarweg 29, PO Box 211, 1000 AE Amsterdam, Netherlands
The Boulevard, Langford Lane, Kidlington, Oxford OX5 1GB, United Kingdom
50 Hampshire Street, 5th Floor, Cambridge, MA 02139, United States

Copyright © 2018 Elsevier B.V. All rights reserved.

No part of this publication may be reproduced or transmitted in any form or by any means, electronic or mechanical, including photocopying, recording, or any information storage and retrieval system, without permission in writing from the publisher. Details on how to seek permission, further information about the Publisher's permissions policies and our arrangements with organizations such as the Copyright Clearance Center and the Copyright Licensing Agency, can be found at our website: www.elsevier.com/permissions.

This book and the individual contributions contained in it are protected under copyright by the Publisher (other than as may be noted herein).

Notices
Knowledge and best practice in this field are constantly changing. As new research and experience broaden our understanding, changes in research methods, professional practices, or medical treatment may become necessary.

Practitioners and researchers must always rely on their own experience and knowledge in evaluating and using any information, methods, compounds, or experiments described herein. In using such information or methods they should be mindful of their own safety and the safety of others, including parties for whom they have a professional responsibility.

To the fullest extent of the law, neither the Publisher nor the authors, contributors, or editors, assume any liability for any injury and/or damage to persons or property as a matter of products liability, negligence or otherwise, or from any use or operation of any methods, products, instructions, or ideas contained in the material herein.

Library of Congress Cataloging-in-Publication Data
A catalog record for this book is available from the Library of Congress

British Library Cataloguing-in-Publication Data
A catalogue record for this book is available from the British Library

ISBN: 978-0-444-63508-2

For information on all Elsevier publications visit our website at
https://www.elsevier.com/books-and-journals

Publisher: John Fedor
Acquisition Editor: Kathryn Morrissey
Editorial Project Manager: Amy Clark
Production Project Manager: Maria Bernard
Cover Designer: Victoria Pearson

Typeset by TNQ Books and Journals

To our fathers
To Alberto Jr and Carla

CONTENTS

List of Contributors — xiii
Foreword — xvii
Preface — xix
Acknowledgements — xxiii

Part I: General Concepts and Cosmetic Legislation on Cosmetic Products

1. General Concepts: Current Legislation on Cosmetics in Various Countries — 3
Stefano Dorato

General Concepts — 3
Current Legislation on Cosmetics in Various Countries — 5
The International Cooperation on Cosmetics Regulation Role in Global Alignment — 19
Safety and Efficacy of Cosmetics: A Brief Survey of Other Significant Worldwide Markets — 20
References — 34

2. Quality Control of Cosmetic Products: Specific Legislation on Ingredients — 39
Eliseo F. González Abellán and Desirée Martínez Pérez

Quality Control of Cosmetic Products — 39
Specific Legislation on Ingredients — 41
Summary — 52
References — 52

Part II: Analytical Methods for Monitoring Ingredients and Quality Control of Cosmetics Products

3. Selection From Bibliographic Resources of an Analytical Method for Cosmetic Products: Validation of a Method — 57
Juan G. March, Amparo Salvador and Alberto Chisvert

Introduction — 57
Validation of a Method for Cosmetic Analysis — 58
Selection of the Appropriate Method in Cosmetic Analysis — 61
References — 65

4. General Review of Official Methods of Analysis of Cosmetics — 67
Gerd Mildau

Introduction — 67
History of Official Methods for Cosmetics — 68
Cascade Approach for Selecting Appropriate Methods — 70

New Approach: Standardization of Methods for Cosmetics	71
International Standards on Analytical Methods for Cosmetics	75
References	82

5. Ultraviolet Filters in Cosmetics: Regulatory Aspects and Analytical Methods — 85
Alberto Chisvert and Amparo Salvador

Introduction	85
Regulatory Aspects	87
Analytical Methods for Ultraviolet Filters in Cosmetic Products	93
References	104

6. Tanning and Whitening Agents in Cosmetics: Regulatory Aspects and Analytical Methods — 107
Alberto Chisvert, Juan L. Benedé and Amparo Salvador

Introduction	107
Regulatory Aspects	111
Analytical Methods for Tanning and Whitening Agents Determination in Cosmetic Products	112
References	119

7. Colouring Agents in Cosmetics: Regulatory Aspects and Analytical Methods — 123
Adrian Weisz, Stanley R. Milstein, Alan L. Scher and Nancy M. Hepp

Introduction	123
Regulatory Aspects of Colouring Agents in Cosmetic Products	123
Determination of Colouring Agents	135
Acknowledgements	154
References	154

8. Hair Dyes in Cosmetics: Regulatory Aspects and Analytical Methods — 159
Alberto Chisvert, Pablo Miralles and Amparo Salvador

Introduction	159
Regulatory Aspects	163
Analytical Methods for Hair Dye Determination in Cosmetic Products	166
References	172

9. Preservatives in Cosmetics: Regulatory Aspects and Analytical Methods — 175
Gerardo Alvarez-Rivera, Maria Llompart, Marta Lores and Carmen Garcia–Jares

Introduction	175
Analytical Methods	185
References	221

10. Perfumes in Cosmetics: Regulatory Aspects and Analytical Methods — 225
Alberto Chisvert, Marina López-Nogueroles, Pablo Miralles and Amparo Salvador

Introduction	225
Regulatory Aspects and Analytical Methods for Fragrance Ingredients in Cosmetics	229
Determination of Fragrance Ingredients in Cosmetics	231
References	246

11. Surfactants in Cosmetics: Regulatory Aspects and Analytical Methods — 249
M. Carmen Prieto-Blanco, María Fernández-Amado, Purificación López-Mahía, Soledad Muniategui-Lorenzo and Darío Prada-Rodríguez

Introduction	249
Regulatory Aspects of Surfactants in Cosmetic Products	249
Analytical Methods for the Determination of Surfactants in Cosmetics and Raw Materials	252
Determination of Residual Products From Surfactants	272
References	283
Further Reading	287

12. Nanomaterials in Cosmetics: Regulatory Aspects — 289
Diana M. Bowman, Nathaniel D. May and Andrew D. Maynard

Introduction	289
Nanomaterials in Cosmetics	289
What's New to Regulate, and Implications for Regulation	291
Regulatory Developments for Nano-Based Cosmetics	293
Conclusion	300
References	300

13. Green Cosmetic Ingredients and Processes — 303
Carla Villa

Introduction	303
Green Cosmetic Ingredients and Processes	303
References	327
Further Reading	330

14. Main Chemical Contaminants in Cosmetics: Regulatory Aspects and Analytical Methods — 331
Isela Lavilla, Noelia Cabaleiro and Carlos Bendicho

Introduction	331
Overview of Regulations Concerning Chemical Contaminants in Cosmetics	336
Analytical Methods for the Detection of Main Chemical Contaminants in Cosmetics and Raw Materials	339

x   Contents

Future Prospects	376
References	377
Further Reading	383

Part III: Analytical Methods for Monitoring of Cosmetic Ingredients in Biomedical and Environmental Studies

15. Human Biomonitoring of Select Ingredients in Cosmetics   387
Rajendiran Karthikraj and Kurunthachalam Kannan

Introduction	387
Biomonitoring as an Approach to Assessing Exposure	389
Matrix of Choice and Analytical Methods in Biomonitoring	390
Conclusions	418
References	430

16. Environmental Monitoring of Cosmetic Ingredients   435
Alberto Chisvert, Dimosthenis Giokas, Juan L. Benedé and Amparo Salvador

Introduction	435
UV Filters	436
Synthetic Musks	472
Preservatives	502
Insect Repellents	534
References	540

Part IV: Safety Evaluation of Cosmetic Products

17. Alternative Methods to Animal Testing in Safety Evaluation of Cosmetic Products   551
Octavio Díez-Sales, Amparo Nácher, Matilde Merino and Virginia Merino

Introduction	551
Acute Toxicity	553
Skin Corrosion/Irritation	554
Eye Irritation Tests	558
Skin Sensitization	564
Skin Absorption Studies	566
Repeated-Dose Toxicity	569
Genotoxicity/Mutagenicity	570
Ultraviolet-Induced Toxic Effects	571
Carcinogenicity	573
Reproductive and Developmental Toxicity	574
Toxicokinetics Studies	575
Conclusions	576
References	577

18. Microbiological Quality in Cosmetics — 585
Gabriel A. March, Maria C. Garcia-Loygorri, José M. Eiros, Miguel A. Bratos and Raúl Ortiz de Lejarazu

Introduction	585
Impact on Human Health of Cosmetic Products Contaminated by Microorganisms	586
Legislation on Cosmetic Microbiology	586
Microbiological Analysis of Cosmetic Products	587
Recall of Cosmetics Contaminated by Microorganisms From the Market	592
Bacterial Resistance in Cosmetics	593
Conclusions	595
References	595

Index — *599*

LIST OF CONTRIBUTORS

Gerardo Alvarez-Rivera
Universidade de Santiago de Compostela, Santiago de Compostela, Spain

Carlos Bendicho
University of Vigo, Vigo, Spain

Juan L. Benedé
University of Valencia, Valencia, Spain

Diana M. Bowman
Arizona State University, Phoenix, AZ, United States

Miguel A. Bratos
University Clinic Hospital of Valladolid, Valladolid, Spain; University of Valladolid, Valladolid, Spain

Noelia Cabaleiro
University of Vigo, Vigo, Spain

Alberto Chisvert
University of Valencia, Valencia, Spain

Octavio Díez-Sales
University of Valencia, Valencia, Spain

Stefano Dorato
Cosmetica Italia, Milano, Italy

José M. Eiros
University of Valladolid, Valladolid, Spain

María Fernández-Amado
Universidade da Coruña, A Coruña, Spain

Carmen Garcia–Jares
Universidade de Santiago de Compostela, Santiago de Compostela, Spain

Maria C. Garcia-Loygorri
Medina del Campo Hospital, Medina del Campo, Spain

Dimosthenis Giokas
University of Ioannina, Ioannina, Greece

Eliseo F. González Abellán
Regional Government of Comunidad Valenciana, Valencia, Spain

Nancy M. Hepp
U.S. Food and Drug Administration, College Park, MD, United States

Kurunthachalam Kannan
Wadsworth Center, Albany, NY, United States; University of New York at Albany, Albany, NY, United States

Rajendiran Karthikraj
Wadsworth Center, Albany, NY, United States

Isela Lavilla
University of Vigo, Vigo, Spain

Maria Llompart
Universidade de Santiago de Compostela, Santiago de Compostela, Spain

Purificación López-Mahía
Universidade da Coruña, A Coruña, Spain

Marina López-Nogueroles
University of Valencia, Valencia, Spain; Instituto de Investigación Sanitaria La Fe (IIS La Fe), Valencia, Spain

Marta Lores
Universidade de Santiago de Compostela, Santiago de Compostela, Spain

Gabriel A. March
University Clinic Hospital of Valladolid, Valladolid, Spain

Juan G. March
University of Balearic Islands, Palma de Mallorca, Spain

Desirée Martínez Pérez
Regional Government of Comunidad Valenciana, Valencia, Spain

Nathaniel D. May
Arizona State University, Phoenix, AZ, United States

Andrew D. Maynard
Arizona State University, Tempe, AZ, United States

Virginia Merino
University of Valencia, Valencia, Spain

Matilde Merino
University of Valencia, Valencia, Spain

Gerd Mildau
Chemisches und Veterinäruntersuchungsamt (CVUA), Karlsruhe, Germany

Stanley R. Milstein
U.S. Food and Drug Administration, College Park, MD, United States

Pablo Miralles
University of Valencia, Valencia, Spain

Soledad Muniategui-Lorenzo
Universidade da Coruña, A Coruña, Spain

Amparo Nácher
University of Valencia, Valencia, Spain

Raúl Ortiz de Lejarazu
University Clinic Hospital of Valladolid, Valladolid, Spain; University of Valladolid, Valladolid, Spain

Darío Prada-Rodríguez
Universidade da Coruña, A Coruña, Spain

M. Carmen Prieto-Blanco
Universidade da Coruña, A Coruña, Spain

Amparo Salvador
University of Valencia, Valencia, Spain

Alan L. Scher (deceased)
U.S. Food and Drug Administration, College Park, MD, United States

Carla Villa
University of Genova, Genova, Italy

Adrian Weisz
U.S. Food and Drug Administration, College Park, MD, United States

FOREWORD

Cosmetic products are usually complex mixtures of chemicals formulated to be applied to the human body to improve its external presence or to provide protection against environmental challenges such as UV radiation. They may also contain compounds that help to stabilize the product and inadvertent contaminants. This complexity demands the development of increasingly powerful and accurate analytical methods for quality control and quality assurance based on a stringent European or other international regulatory framework, to minimize harmful effects on human health. This second edition of *Analysis of Cosmetic Products* by Professors Salvador and Chisvert thoroughly covers state-of-the-art developments, especially since 2006, in terms of analytical method development and validation against official methods of analysis. The methods are applied to the main ingredients in cosmetics, including sunscreens, colourants, tanning and whitening agents, product preservatives, hair dyes, surfactants and perfumes, with extension to novel measurands, such as carbon, metallic and liquid nanomaterials, and unwanted organic and metallic contaminants and impurities at trace level concentrations. This book comprehensively surveys sorptive and liquid-phase microextraction approaches with 'green' chemical credentials and solvent-free approaches under microwave and ultrasound irradiation, supercritical and subcritical extraction conditions and dielectric heating for synthetic processes allied to modern chromatographic and mass spectrometric detection techniques. Particular emphasis is thus given to the link between method development, including sample pre-treatment, and regulatory aspects for the diversity of cosmetic products. This is the core of the monograph. Based on the current interest in human and environmental exposure to multiple, potentially toxic, chemicals found in cosmetics—called emerging contaminants—this monograph has greatly expanded its coverage of (bio)analytical methods for biomonitoring the accumulation of potential oestrogenic species in human and biota tissues and also in environmental compartments.

This monograph has been written by an international, multi-disciplinary team of authors from the cosmetics industry, public offices of cosmetics and colours and medical research institutes and by analytical chemistry and pharmaceutical researchers from academia. It is indispensable for bolstering basic knowledge of cosmetic science for the novice while providing up-to-date information on legislation issues and advanced analytical procedures to practitioners. In fact we believe that this book is currently the major compilation of cutting-edge analytical methodology for personal care chemists, (bio)analytical chemistry researchers, technicians and stakeholders interested in trends in cosmetic analysis, alternative manufacturing procedures, quality control/quality assurance of products and risk assessment of unwanted ingredients. It serves as an authoritative text for Bachelor, MSc and PhD students of analytical chemistry, cosmetic science and related disciplines.

Manuel Miro
FI-TRACE Group, Department of Chemistry,
University of the Balearic Islands, Palma, Spain

Alan Townshend
Department of Chemistry, University of Hull,
United Kingdom

PREFACE

Both editors of this book are analytical chemists who have been working together for 20 years on the development and validation of analytical methods for cosmetic products, and particularly in the field of sunscreen products.

In this multi-authored book we share our experience in this field with our readers and give them advice, with the help of the other authors participating in this book, on the often difficult choice of suitable analytical methods for production monitoring and quality control of cosmetic products according to their composition.

To do this, we have aimed to provide extensive and varied information on the topic enabling the reader to gain insight into the aspects related to the world of cosmetics from the viewpoint of analytical chemistry. We have divided the book into four parts:

PART I

This part is composed of two chapters that provide definitions and general concepts regarding cosmetic products (Chapter 1) as well as the current legislation in various countries and specific legislation on ingredients (Chapter 2).

PART II

The central body of this book is dedicated to the analytical methods for monitoring and controlling the quality of cosmetic products.

The fundamental objective of this part of the book is to put at the reader's disposal scientific reviews, carried out by experts in analytical chemistry, of existing methods published in the scientific bibliography for different types of analytes and/or cosmetic samples, their use, their potential, etc., also indicating the sources where one can find the original article, with the idea that the reader will be the one to choose the method that he or she considers most suitable to tackle a specific analytical problem.

To begin, the book offers the reader information about how to select an adequate analytical method (Chapter 3) and it details the official methods for cosmetic product analysis and explains how to access the corresponding documents (Chapter 4).

Next a detailed revision is given of the published analytical procedures specific to the different analytes and cosmetic samples, including sample preparation, analytical techniques to be used, particular methodologies, etc. The first chapters are devoted to sun protection (Chapter 5, on UV filters) and other aspects related to skin bronzing (Chapter 6, on tanning and whitening ingredients). Two chapters are devoted to general colouring agents (Chapter 7) and more specifically to hair dyes (Chapter 8). Next, three other types of ingredients very useful in most cosmetic preparations are treated, namely preservatives (Chapter 9), perfumes (Chapter 10), and surfactants (Chapter 11). Three new topics (which were not treated in the first edition of the book) have been included because of their increasing interest: nanomaterials (Chapter 12), 'green' cosmetic ingredients (Chapter 13), and main chemical contaminants (Chapter 14).

PART III

This part is devoted to the human (Chapter 15) and environmental (Chapter 16) monitoring of cosmetic ingredients. These two important issues have been included in this second edition, as cosmetic ingredients can be percutaneously absorbed and they also can be found as residues in the environment, with subsequent possible negative consequences.

PART IV

Finally, in this part the safety evaluation of cosmetic products is treated, i.e., alternative methods to the use of animals for cosmetic product evaluation are considered (Chapter 17), and last we include a new chapter devoted to the microbiological quality of cosmetics (Chapter 18).

We sincerely hope that this book is useful for both scientists and technologists specializing in cosmetic research, manufacturing, and quality control, as well as for students of cosmetic science and related topics, in addition to others who are connected with the field of cosmetics (practitioners, consultants, etc.).

We are at the entire disposal of our readers, and will be delighted to answer any questions about the topics described here, insofar as we are able, via electronic mail.

Amparo Salvador Carreño **Alberto Chisvert Sanía**

University of Valencia (Spain)
October 2017

STRUCTURE OF THE BOOK

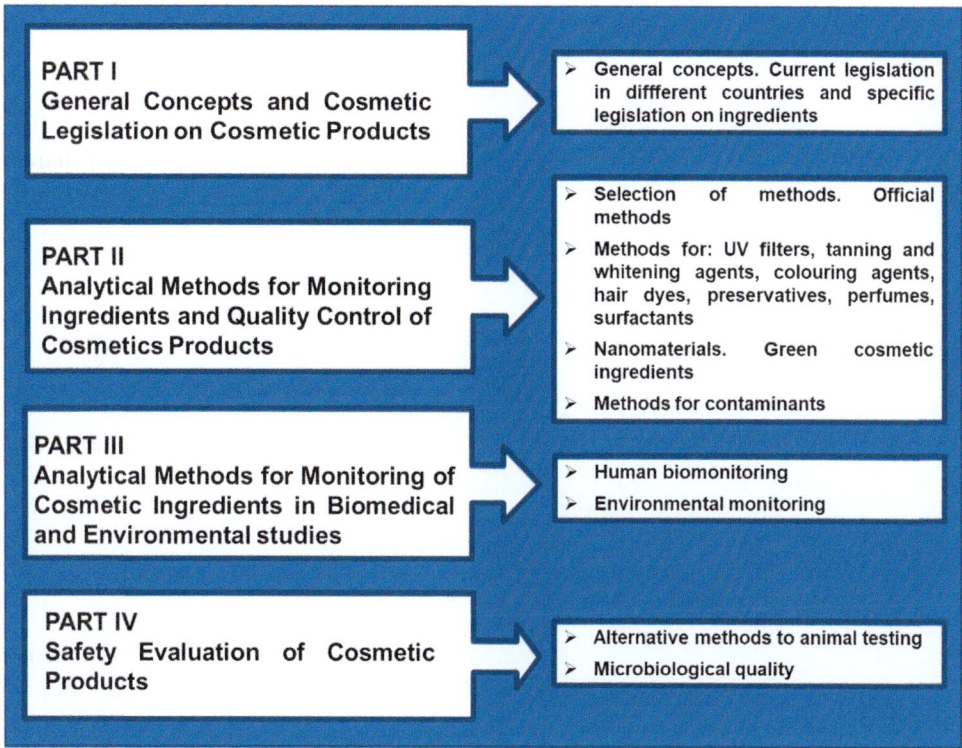

ACKNOWLEDGEMENTS

We would like to express our most sincere thanks to Elsevier for again placing their trust in us to carry out this new edition of our book.

We are grateful to Beth Campbell for her initial proposal to do this second edition of the book and for her help to begin it. We also thank Katy Morrissey, Jill Cetel, and of course Amy Clark for her ongoing assistance and patience with us. Thanks also to Sandhya Narayanan and Maria Bernard for their aid in the last stage. A very good team of professionals!

We thank the authors from the various universities and international institutions for their participation. They have done a splendid job and we would like to give them special thanks for their patience in bearing with our many revisions and adaptations of the texts until finally managing to complete the book.

We appreciate the valuable help of our proofreader (and dear colleague), Dr. Marina López Nogueroles. She has worked with us to improve the texts, for which we thank her wholeheartedly.

Our thanks to Dr. Joan March for his help. He has assisted us, with patience and thoroughness, in the difficult task of revising the references of the various chapters.

We would like to thank the *Subdirección General de Proyectos de Investigación* of the Spanish *Ministerio de Economía y Competitividad* for financing our research projects in the field of cosmetic product analysis, which has enabled us to continue this line of research for so many years.

Last but not least, we would like to thank our respective families (especially Joan and Rosa), and also our friends, and ask them to forgive us for the lost leisure time that we should have spent with them and that was 'stolen' by the preparation of this second edition (especially Alberto Jr. and Carla), but we hope they will share the happiness we feel when at last we hold the finished project in our hands.

Amparo Salvador Carreño   **Alberto Chisvert Sanía**

University of Valencia (Spain)
October 2017

PART I

General Concepts and Cosmetic Legislation on Cosmetic Products

CHAPTER 1

General Concepts: Current Legislation on Cosmetics in Various Countries

Stefano Dorato
Cosmetica Italia, Milano, Italy

GENERAL CONCEPTS

The cosmetics industry is a real global one and its main markets are the European Union (EU), the United States of America, the People's Republic of China, Brazil and Japan, with approximate values of 77, 64, 41, 24 and 22 billion euros, respectively, in 2016, according to data compiled by Cosmetics Europe, *i.e.* the European Personal Care Association (Cosmetics Europe, 2017).

To ensure safety and efficacy, cosmetic products are regulated and controlled worldwide. However, the alignment of laws dealing with cosmetics is far from being achieved and regulatory frameworks vary greatly between countries, making it practically impossible for a global industry to sell the same product on all markets and generating difficulties for trading. In the first decade of the 21st century the main markets were regulated by different models: an approach applying a vertical, risk-based directive in the EU; a liberal one in the United States based on regulations dating back to the late 1930s, having a narrow definition of cosmetics with few restrictions on the ingredients; a complex pre-market approval in the People's Republic of China and an EU-like law but with significant deviations in Brazil. In 2000, Japanese cosmetic deregulation moved regulatory compliance responsibility from health authorities to manufacturers, bringing the law closer to that of the EU.

Starting in the 1990s the EU Cosmetics Directive (EU Directive) framework became an international model, easy to understand and more or less easy to apply, representing a dynamic legislation that allows for continuous adaptation to technical progress, leading to relevant restrictions and/or warnings and to rapid inclusion of new regulated molecules (e.g., preservatives, colours, UV filters). As an example of global convergence, the Association of Southeast Asia Nations (ASEAN), the Andean Community (Comunidad Andina) and Mercosur/Mercosul aligned, with some divergence, their local cosmetic regulations to the EU Directive. Notwithstanding the international appeal and resulting success of the EU Directive, EU regulators considered, at the beginning of the new century, that it was the right time to recast the cosmetics law and developed the new EU Cosmetics Regulation (EU Regulation), an improved up-to-date regulatory model published in 2009 and entered fully into force in 2013.

Looking at the major markets, there are broadly similar regulatory elements which fuel expectations for a promising future global alignment: manufacturer responsibility for product safety, notification process or only voluntary requirements, cosmetic good manufacturing practices (GMP) guidelines, management of ingredients through positive and negative lists, use of International Nomenclature of Cosmetic Ingredients (INCI) names, basic labelling requirements and safety and efficacy information to be developed and kept available by manufacturers.

Nonetheless at the international level the main distinction in regulatory approaches resides in the so-called in-market control versus pre-market approval. The key features of the in-market control system are the following:

1. Time-consuming and expensive pre-market licensing/registration of products is avoided.
2. The company responsible for placing the cosmetic on the market is also responsible for regulatory compliance and for the safety and efficacy of the product, and must be able to justify this on authority investigation.
3. The appointed competent authority is notified of data concerning the company.
4. Competent national authorities have effective in-market control by monitoring the market to check compliance with regulations. Market surveillance is performed through random control on the market (in retail shops), by routine checks on compliance with regulations or by systematic control of given product categories, in the event of a problem arising, such as a consumer complaint.
5. In case of non-compliance there can be penalties of varying strength, and if the manufacturer fails to prove the cosmetic is safe for use the product will be withdrawn from the market.
6. It encourages companies to ensure product safety and therefore it offers better health protection benefits.
7. Administrations must be aligned and compatible, not identical.
8. It has one safety standard for consumers.
9. It increases the competitiveness of the market, ensuring that new products will be available to consumers more quickly.
10. It guarantees equal treatment of all companies.

On the other hand, the key features of pre-market approval are:

1. Registration time delays the launch of cosmetics that change quickly in line with developments in technology and/or fashion.
2. It does not prevent fraud or stop illegal products from reaching the market, so it does not necessarily increase product safety.
3. In any event, in-market surveillance is needed to monitor the system, implying a duplication of enforcement procedures.

4. It restricts the availability of new products on the market, discouraging innovation and investments.
5. Time-consuming, and somewhat uncertain, procedures and high registration fees imply higher costs for companies and consequently more expensive products.

CURRENT LEGISLATION ON COSMETICS IN VARIOUS COUNTRIES
The European Union Cosmetics Regulation

For 37 years, and until July 2013, the manufacture and marketing of cosmetics in the EU was regulated by the globally well-known EU Directive (EU Council Directive 76/768/EEC on the approximation of the laws of the Member States relating to cosmetic products, adopted on 27 July 1976 and published in the *Official Journal of the European Community* L 262 on 27 September 1976), which underwent seven council and parliament amendments (i.e., changes in articles) and more than 50 commission adaptations to technical progress (i.e., changes in the annexes regulating banned or restricted substances, cosmetic colourants, preservatives and UV filters).

The EU Directive represented a well-established, autonomous sectorial legislation, based on the principle of risk assessment and ensuring a high level of protection of human health, having a clear scope based on a clear definition. EU Member States, however, had to implement each adaptation in their own national laws, a process that proved to be slow, sometimes not perfect and prone to the introduction of requirements against the proper functioning of the internal market.

Therefore, regulators, through the 'recast' process, a necessary regulatory cleaning up and streamlining exercise, passed the EU Regulation, Regulation (EC) No. 1223/2009 of the European Parliament and of the Council of 30 November 2009 on cosmetic products, published in the *Official Journal of the European Union* L 342 on 22 December 2009.

The EU Regulation entered fully into force on 11 July 2013 and represents the logical evolution towards a comprehensive harmonization across the EU Member States: a regulation is a single piece of legislation instantly and directly enforced across the whole territory of the EU with no room for diverging transposition (in content and/or in time) by Member States.

The fundamental principles of the EU cosmetics legislation remain unchanged: a wide definition of a cosmetic with no intermediary category between cosmetics and pharmaceuticals, product safety is an obligation, adequate and non-misleading information to be provided to consumers, responsibility for compliance of the person placing a cosmetic on the EU market, no pre-market registration or certification of cosmetics but a simple notification to an EU centralized portal prior to placement on the market, in-market controls of compliance by competent authorities of Member States, regulation of specific substances through ingredient lists. Adaptation to technical progress of the annexes will continue as before by the commission but using the legal instrument of the EU Regulation, directly enforced in the Member States.

The principal changes introduced by the EU Regulation are related to the setting up of the Cosmetic Product Safety Report, new rules on nanomaterials and CMR (carcinogenic, mutagenic or toxic to reproduction) substances, EU-wide product notification prior to placement on the market (Cosmetic Product Notification Portal—CPNP), enhancement of in-market control of compliance by authorities, a set of definitions, obligations of responsible person and distributor, the role of standards and criteria for claims.

Table 1.1 illustrates the layout of the EU Regulation.

The European Commission retains overall responsibility for cosmetics legislation within the EU, while a competent authority, which enforces the legislation and will work with other authorities in areas where cooperation is necessary to ensure proper application of the EU Regulation, is designated in each of the 28 Member States (Austria, Belgium, Bulgaria, Croatia, Cyprus, Czech Republic, Denmark, Estonia, Finland, France, Germany, Greece, Hungary, Ireland, Italy, Latvia, Lithuania, Luxemburg, Malta, the Netherlands, Poland, Portugal, Romania, Slovak Republic, Slovenia, Spain, Sweden, United Kingdom).

Each Member State appoints national inspectors, responsible for the in-market surveillance system, who may monitor compliance, visit local manufacturing sites and any retail outlet to see what is sold, take products from the marketplace to official laboratories for compliance testing, act against those companies that do not comply and that may present a threat to consumers and the market (e.g., counterfeit cosmetics) and have access to product information via direct consultation or through another Member State authority under the principle of mutual recognition. The cooperation system among

Table 1.1 The layout of EU Regulation 1223/2009 on cosmetic products

Chapter	Title	Articles
I	Scope, definitions	1 and 2
II	Safety, responsible person, free movement	3–9
III	Safety assessment, Product Information File, sampling and analysis, notification	10–13
IV	Restrictions on certain substances	14–17
V	Animal testing	18
VI	Consumer information	19–21
VII	Market surveillance	22–24
VIII	Non-compliance, safeguard clause	25–27
IX	Administrative cooperation	28–30
X	Implementing measures, final provisions	31–40
	Annexes	I–X

competent national authorities focuses on mutual recognition, enforcement in the event of non-compliant products and administrative cooperation.

The definition in Article 2.a of a cosmetic product is unchanged:

cosmetic product means any substance or mixture intended to be placed in contact with the external parts of the human body (epidermis, hair system, nails, lips and external genital organs) or with the teeth and the mucous membranes of the oral cavity with a view exclusively or mainly to cleaning them, perfuming them, changing their appearance, protecting them, keeping them in good condition or correcting body odours.

A set of explicit definitions, however, to codify the existing understanding is listed in Article 2 and in the Preamble to the Annexes, e.g., substance, mixture, making available on the market, placing on the market, manufacturer, importer, distributor, end user, undesirable effect, serious undesirable effect, withdrawal, frame formulation, preservative, colourant, nanomaterial, hair products, rinse-off products, professional use. In the EU Directive, Annex I provided a non-exhaustive, illustrative list of the products to be considered as cosmetics; in contrast, in the EU Regulation such a list is present only in the Whereas 7 of the introduction. The definition of a cosmetic identifies the site of application (epidermis, hair system, nails, lips, external genital organs, teeth, oral cavity mucous membranes) and the intended functions (cleaning, perfuming, changing the appearance, protecting, keeping in good condition, correcting body odours). By using the phrase '*exclusively or mainly*' the EU Regulation foresees that cosmetics may have some functions other than those specified in Article 2, as long as one of the six aforementioned functions is predominant. In contrast to the United States and Japan, within the EU regulatory frameworks products are mutually exclusive; therefore, there are no intermediate product categories among cosmetics and pharmaceuticals or foodstuffs or medical devices (RPA, 2004). The European Court of Justice (ECJ) repeatedly states, in several case laws, that a product can be only either a cosmetic or another type of product; it cannot be both at the same time (the so-called non-cumulation rule) (ECJ, 1991a,b, 2005). To decide among the borderline products the EU Commission has developed different guidelines to address assessment criteria for specific borderline cases between cosmetics and medicinal products, medical devices, biocides, toys, food and chemicals, even if the final and only decision resides in the ECJ.

The responsible person in the EU is outlined in Article 4:

1. *Only cosmetic products for which a legal or natural person is designated within the Community as 'responsible person' shall be placed on the market.*
2. *For each cosmetic product placed on the market, the responsible person shall ensure compliance with the relevant obligations set out in this Regulation.*

The obligations of the responsible person and the distributor can be found in Articles 5 and 6.

The responsible person (in most instances the manufacturer or the importer) has a central role in the application of the EU Regulation, retaining the best qualifications and know-how to assess the safety of the cosmetic and to avoid any risk to human health and to decide any necessary and appropriate corrective measures with the competent authorities concerned, to manage the communication along the supply chain.

Safety of cosmetics is primarily addressed in Article 3:

A cosmetic product made available on the market shall be safe for human health when used under normal or reasonably foreseeable conditions of use, taking account, in particular, of the following:
(a) presentation including conformity with Directive 87/357/EEC;
(b) labelling;
(c) instructions for use and disposal;
(d) any other indication or information provided by the responsible person defined in Article 4.
The provision of warnings shall not exempt persons defined in Articles 2 and 4 from compliance with the other requirements laid down in this Regulation.

Articles 10 and 11 indicate further obligations of responsible persons towards safety and efficacy by requiring that a cosmetic product must undergo a safety assessment and that a Cosmetic Product Safety Report is set up in accordance with Annex I of the EU Regulation. An appropriate weight-of-evidence approach must be used in the safety assessment for reviewing data from all existing sources, while the cosmetic safety assessment shall be carried out by a person in possession of a diploma or other evidence of formal qualifications (e.g., a degree in pharmacy, toxicology, medicine or a similar discipline).

The Product Information File, which has to be readily accessible in electronic or other format at the address indicated on the label to the competent authority of the Member State in which the file is kept, shall contain the following information and data, which shall be updated as necessary: a description of the cosmetic; the Cosmetic Product Safety Report, a description of the method of manufacturing and a statement on compliance with GMP, proof of the effect claimed for the cosmetic product and data on any animal testing performed by the manufacturer, his agents or suppliers, relating to the development or safety assessment of the cosmetic product or its ingredients, including any animal testing performed to meet the legislative or regulatory requirements of third countries.

Compliance with GMP is mandatory under Article 8, but certification is not required and the choice of GMP set of rules is voluntary. Compliance is presumed if the manufacture is in accordance with the relevant harmonized standards, the references of which have been published in the *Official Journal of the European Union*. CEN 22716:2007—Cosmetics—Good Manufacturing Practices has been published, but GMP conformity can also be demonstrated in other ways, e.g., via industry-recognized standards and codes.

The EU Regulation has also introduced a harmonized EU online notification system: one electronic EU notification portal (the aforementioned CPNP) replacing former, non-harmonized national ones to facilitate in-market control and post-marketing surveillance. All cosmetic products must be notified at commission level to the CPNP at the time of first placement on the market and the process includes also the notification of frame formulations accessible to poison control centres. Article 13 lists the information which must be submitted by the responsible person to the commission.

Obligations are also imposed on distributors if they translate, on their own initiative, any element of the labelling of a cosmetic to comply with national law.

The in-market control role of national competent authorities is strengthened by new requirements regarding identification within the supply chain (traceability), the possibility of authorities asking for the formulation in cases of 'serious doubts' about safety, corrective measures (withdrawal included) in cases of non-compliance and provisions on good administrative cooperation among EU Member States.

The approach to ingredient management has not been changed in the EU Regulation and the choice of safe ingredients is the responsibility of the manufacturer and its safety assessor, even though the regulators have confirmed the need to keep EU-wide, harmonized restrictions through two negative lists (banned and restricted substances) and three positive ones (preservatives, colourant and UV filters):

- banned substances listed in Annex II
- substances which cosmetics must not contain except subject to the restrictions laid down in Annex III
- colourants allowed in Annex IV
- preservatives listed in Annex VI
- UV filters listed in Annex VII

However, the EU Regulation also distinguishes two categories of ingredients in Articles 15 and 16, respectively: CMR substances and nanomaterials.

CMR substances are prohibited because of their chemical hazard classification, but a derogation for cosmetic use may apply following a positive safety evaluation by the EU Scientific Committee on Consumer Safety (SCCS), in the case of a CMR 2, or by fulfilling a series of conditions, if a substance is classified as CMR 1A or 1B, and by being evaluated and found safe by the SCCS, in particular in view of exposure to cosmetics and taking into consideration the overall exposure from other sources, taking particular account of vulnerable population groups.

Nanomaterials defined according to Article 2.k, *'"nanomaterial" means an insoluble or biopersistant and intentionally manufactured material with one or more external dimensions, or an internal structure, on the scale from 1 to 100 nm'*, are subject to a special pre-notification regime. Notification of products containing nanomaterials needs to be made 6 months prior to placement on the market, with full safety information. In the event that the commission has concerns regarding the safety of a nanomaterial, it will request the SCCS to give its opinion on the safety of such nanomaterial for use in the relevant categories of cosmetic products and on the reasonably foreseeable exposure conditions. The SCCS shall deliver its opinion within 6 months. This notification does not apply for nanomaterials which are already authorized in a positive list or listed in Annex III. Moreover, for the sake of consumer information, in the ingredient list the INCI name of a nanomaterial shall be followed by the word 'nano' in brackets, e.g., titanium dioxide (nano).

Introducing the labelling of cosmetics, Article 19.1 states that a cosmetic should bear, in indelible, easily legible and visible lettering:

- the name or registered name and the address of the responsible person, which might be abbreviated (if several addresses are indicated, the one at which the responsible person makes readily available the Product Information File shall be highlighted);
- the country of origin, which applies only to extra-EU cosmetics;
- the nominal content by weight (g) or by volume (mL) (derogations apply);
- particular precautions to be observed in use, and at least those listed in Annexes III–VI and any special precautionary information on cosmetic products for professional use;
- the batch number;
- the function of the cosmetic product, unless it is clear from its presentation;
- the list of ingredients in descending order of weight using the INCI names;
- the date of minimum durability (using the words 'best used before the end of') or the following symbol:

The date of minimum durability shall be clearly expressed and shall consist of either the month and year or the day, month and year, in that order. Indication of the date of minimum durability shall not be mandatory for cosmetic products with a minimum durability of more than 30 months. For such products, there shall be an indication of the period of time after opening (the so-called PaO—Period after Opening), for which the product is safe and can be used without any harm to the consumer. This information shall be indicated, except where the concept of durability after opening is not relevant, by the symbol shown in point 2 of Annex VII followed by the period (in months and/or years, e.g., 12 M):

In some cases, in which it is impossible for practical reasons to label the information, derogations apply. For example, the following symbol in Annex VII.1 might appear on the container and the packaging as a reference to enclosed or attached information:

Some information, notably but not exclusively warnings and functions, must be labelled in the language(s) of the Member States in which the product is made available to the end user.

The EU Regulation, in contrast to the former EU Directive, tackles, in Article 20, the issue of product claims. Claims convey explicitly or implicitly cosmetic product characteristics or functions and are marketing tools which are essential to help consumers choose a product, to foster competition and to promote innovation.

Article 20.1 states:

In the labelling, making available on the market and advertising of cosmetic products, texts, names, trademarks, pictures and figurative or other signs shall not be used to imply that these products have characteristics or functions which they do not have.

This aims principally to protect end users of cosmetic products from misleading claims. This same article requests the commission to

adopt a list of common criteria for claims which may be used in respect of cosmetic products ..., taking into account the provisions of Directive 2005/29/EC.

(Note: Directive 2005/29/EC is the horizontal legislation on unfair commercial practices.) As a consequence, Commission Regulation (EU) No. 655/2013 of 10 July 2013, published in the *Official Journal of the European Union* L 190 on 11 July 2013, laid down common criteria for the justification of claims used in relation to cosmetic products. At the same time the Guidelines to Commission Regulation (EU) No. 655/2013 were released too, to provide guidance on how to apply each of the common criteria and to focus on best practice regarding efficacy testing for evidence support. In July 2017 an updated *Technical Document on Cosmetic Claims—Agreed by the Sub-working Group on Claims (version of 3 July 2017)*, which supersedes the former guideline, was released by the EU Commission. It also contains two new annexes providing guidance for the application of common criteria to 'free from' claims and to the specific claim "hypoallergenic".

The six common criteria (legal compliance, truthfulness, evidential support, honesty, fairness and informed decision-making) govern claims by defined principles, not by regulating specific wording, and are mandatory and legally binding for products qualifying as cosmetics. They apply to any claim within the scope of the EU Regulation, irrespective of the medium or type of marketing tool used, the product functions claimed and the target audience.

A new element also introduced by the EU Regulation in Article 23 is the harmonized communication of confirmed serious undesirable effects (SUEs) linked to the use of a cosmetic by responsible persons and distributors to national competent authorities where SUEs occur. An SUE is defined as an undesirable effect which results in temporary or permanent functional incapacity, disability, hospitalization, congenital anomalies or immediate vital risk or death.

Replacement of animal testing for cosmetics by alternative methods has long been a high priority in the EU and provisions to ban animal testing for these purposes were introduced into the legislation already in 1993. The management of animal testing is enclosed in Article 18 and the last ban entered into force on 11 March 2013, extending the prohibition also to toxicological tests concerning repeated-dose toxicity, reproductive toxicity and toxicokinetics, notwithstanding the fact that there are no alternatives yet available and no specific timeline can be estimated because of core scientific challenges.

Last but not least it is worth citing the Commission Recommendation of 22 September 2006 on the efficacy of sunscreen products and the claims made related to them, which gives guidance on the efficacy and testing of sunscreen products, the claims (text, names, trademarks, pictures or other signs) which should be made or not in relation to sunscreens and the warnings and instructions for use. Claims indicating UVB and UVA protection should be made only if the protection equals or exceeds the levels set out in the recommendation, while the UVB efficacy of sunscreen products should be indicated on the label by reference to categories and numbers, the variety of which is limited to facilitate comparison between products without reducing the choice for the consumer. A standardized UVA logo for products offering a minimum UVA protection has been issued by the EU trade association Cosmetics Europe. The recommendation, which is due to be updated, is a non-binding text but the industry has committed to follow it.

The United States' Regulatory Approach to the Safety and Efficacy of Cosmetics

Cosmetics in the United States are defined according to the Food, Drugs and Cosmetics Act (FD&C Act) dating back to 1938 and, in practice, unchanged, excluding the amendments to colour additives. Cosmetics manufactured in or imported into the United States must comply with the provisions of the FD&C Act, the Fair Packaging and Labeling Act (FP&L Act), dated 1967, and the regulations published under the authority of these laws. The complete regulations published by the Food and Drug Administration (FDA) can be found in Title 21 of the Code of Federal Regulations (21 CFR).

The FD&C Act defines cosmetics as

articles intended to be rubbed, poured, sprinkled, or sprayed on, introduced into, or otherwise applied to the human body ... for cleansing, beautifying, promoting attractiveness, or altering the appearance.

Included in this definition are skin care products, fragrance preparations, manicuring products, lipsticks, eye and facial make-up preparations, bath preparations, hair preparations (non-colouring), hair colouring preparations, oral hygiene products, baby products, personal cleanliness products, shaving products, tanning products and deodorants. Soaps consisting primarily of an alkali salt of a fatty acid and making no label claim other than cleansing of the human body are not considered cosmetics under the law.

The same FD&C Act defines drugs as

... articles intended for use in the diagnosis, cure, mitigation, treatment, or prevention of disease, and articles (other than food) intended to affect the structure or any function of the body of man or other animals ....

Over-the-counter (OTC) drugs are those which can be purchased without a doctor's prescription.

In the United States some products considered as cosmetics in the EU are classified as OTC drugs. The FDA reviews these products to establish single monographs under which the drugs are generally recognized as safe and effective, and not misbranded. The OTC monograph dictates the appropriate active ingredients and their allowed concentrations and combinations, uses and permissible claims, warnings and directions for use to be included on the label and testing requirements for some product categories like sunscreens. A product covered by a final monograph must conform to the terms of the monograph. In this sense, specific regulations of various OTC drugs were published, e.g., anticaries (21 CFR Part 355) and oral health care products (21 CFR Part 356), antiperspirants (21 CFR Part 350), antidandruff and antiseborrheic products (21 CFR Part 358) and sunscreens (21 CFR Part 352). The OTC monograph process is an ongoing one, also for final monographs; therefore manufacturers should periodically consult either the Federal Register or *The Rulemaking History for Drug Products: Drug Category List* at the FDA's website. Many products may qualify both as cosmetics and as OTC drugs when they have two intended uses. For instance, a shampoo is a cosmetic, because its intended use is to cleanse the hair. An antidandruff treatment is a drug, because its intended use is to treat dandruff. Consequently, an antidandruff shampoo is both a cosmetic and a drug. Fluoride-containing toothpastes, deodorants that are also antiperspirants and moisturizers and make-up products with sunscreen claims are other examples of cosmetic/drug combinations.

In many cases, assigning a product to the category of cosmetic or drug depends on the concept of 'intended use'. Claims stated on the product label, in advertising, on the Internet or in any promotional material; consumer perception and some ingredients having a well-known therapeutic use may cause a product to qualify as a drug, even if marketed as if it were a cosmetic. Claims regarding treatment or prevention of disease, or implying that the product otherwise affects the structure or functions of the human body, make a cosmetic a drug. The FDA enforces control on cosmetic claims through warning letters sent to companies for cosmetics making drug claims, but the FDA also works closely with the US Customs and Border Protection to monitor imports at the time of entry through US customs. Products that do not comply with FDA laws and regulations (e.g., labelling requirements, claims, microbial contamination, illegal colours, banned ingredients) are subject to refusal of admission into the United States and must be brought into compliance (if practicable), destroyed or re-exported.

The FDA regulates the safety of cosmetics, establishes labelling requirements and is also responsible for enforcing laws about cosmetics, and the specific department is the Office of Colors and Cosmetics within the Center for Food Safety and Applied Nutrition. Issues related to drugs are handled by the Center for Drug Evaluation and Research.

Such laws are enforced by the FDA by means of checking and investigating products, inspecting establishments where products are manufactured or held, determining what toxicological and/or other testing the firm has conducted, if any, to substantiate product safety, and seizure of adulterated (e.g., cosmetics which are not properly manufactured or do not meet their specifications or are contaminated) or misbranded cosmetics (e.g., when cosmetics labelling is untruthful, misleading and inadequate to ensure that the cosmetics are safe and effective for their intended uses).

The FDA does not require pre-market approval of cosmetics, neither does it ask firms to register manufacturing premises nor make available safety data or other information before marketing. Manufacturers or distributors of cosmetics may submit online information about establishment and cosmetic product ingredients to the agency voluntarily under the Voluntary Cosmetic Registration Program. However, cosmetic firms are responsible for marketing safe, properly labelled products, without using banned ingredients, and adhering to limits on restricted ingredients. Other than by government agencies, the safety of cosmetic ingredients is evaluated and reviewed by the Cosmetic Ingredient Review, an independent panel of scientific experts set up by the Personal Care Products Council (formerly known as the Cosmetic, Toiletry and Fragrance Association), looking at all available data and assessing the safety of ingredients used in cosmetics.

No obligation to report adverse event information is in place but the FDA manages MedWatch, a system for consumers, medical professionals and manufacturers to report, also online, bad experiences associated with FDA–regulated products, including cosmetics. The Center for Food Safety and Applied Nutrition (CFSAN) manages a database (CAERS; CFSAN Adverse Event Reporting System) incorporating information on adverse events associated with various consumer products, cosmetics included.

The regulatory requirements for OTC drugs are more extensive, as, for example, the FD&C Act requires that all establishments that manufacture, package or label a drug register electronically (Electronic Submissions Gateway) with the FDA at the beginning of the activity, and then every year, and update their lists of all manufactured drugs twice annually. Additionally, OTC drugs must be manufactured in accordance with current drug GMP regulations (21 CFR 210 and 211). While there are no compulsory GMP requirements for cosmetics, GMP yet remains an important factor in ensuring that a cosmetic is neither adulterated nor misbranded. Registered establishments are subject to regular inspections by the FDA, even overseas.

Cosmetic manufacturers may use essentially any raw material as a cosmetic ingredient, as long as it is not considered as only a drug active ingredient, and market the product without approval. The law regulates only colour additives specifically approved for cosmetics and a few prohibited and restricted ingredients (e.g., bithionol, mercury compounds, vinyl chloride, halogenated salicylanilides). Cosmetics should not be contaminated with nitrosamines, 1,4-dioxane or pesticide residues, whose potential presence is regularly checked by the FDA. Regarding nanomaterials the FDA issued in April 2012 the Guidance for industry—Safety of nanomaterials in cosmetic products. To date the FDA has not established a definition of a nanomaterial, but the guide refers to the range of 1–100 nm as a first reference point.

The Microbead-Free Waters Act of 2015, amending Section 301 of the FD&C Act, will ban the manufacture (2017) and the marketing (2018) of rinse-off cosmetics, toothpaste included, containing plastic microbeads (*any solid plastic particle that is less than 5 mm and is intended for use to exfoliate or cleanse the human body*). On the other hand, OTC drug active ingredients are strictly regulated: only the substances present in the positive list of each monograph can be used, and they must conform to the maximum and minimum permitted concentrations. It is extremely difficult to add a new active ingredient to these positive lists, because it must first be used in a therapeutic formula authorized by the cumbersome procedure called 'New Drug Application'.

Cosmetics marketed in the United States must comply with the labelling regulations of the FD&C Act and the FP&L Act. The information must be in the English language and if the label contains any description in a foreign language, all words, statements and other information required by US law must also appear in the foreign language.

The principal display panel (PDP), i.e., '*the part of a label that is most likely to be displayed, presented, shown or examined under customary conditions of display for retail sale*', shall bear a statement identifying the product and an accurate statement of the net quantity in terms of weight, measure, numerical count or a combination of numerical count and weight or measure. Weight and measure must be expressed using US (e.g., ounces or fluid ounces) and metric units. The net content declaration must be distinct, placed at the bottom of the PDP in line generally parallel to the base on which the package rests. The law prescribes also that the type size must be proportionate to the size of the container. Other information required on the label are the name and place of business of the manufacturer, packer or distributor; the country of origin; appropriate warnings and directions for use (warnings for coal-tar hair dyeing products, aerosols, feminine deodorant sprays, foam baths and products whose safety has not been proven are compulsory) and a declaration of the name of each ingredient in descending order of predominance. The ingredients are identified by using the FDA-established name, such as the INCI names or the names adopted by regulation by the FDA or generally recognized technical names. Ingredients exempt from public disclosure may be stated as 'and other ingredients'. Cosmetics not habitually distributed for retail sale (e.g., hair preparations or make-up products used by professionals on customers at their establishments) are exempt from ingredient listing provided they are not also sold to consumers at professional establishments or workplaces for their consumption at home.

Labels for cosmetics which are also drugs must first identify the drug ingredient(s) as '*active ingredients(s)*' before listing the cosmetic ingredients. They must also mention a use-by date if their validity is less than 3 years, and warnings or instructions for use as required by each specific monograph. It is worth noting that OTC product labelling was standardized to make it easy to read for consumers. The so-called Drug Facts Box has been in use since May 2002 except for OTC products which are not yet regulated by a final ruling.

State laws (e.g., on volatile organic compounds—VOCs) demand compliance beyond federal rules. Above all California has different laws directly related to cosmetics, like the California Safe Cosmetics Act, which requires the manufacturer, packer and/or distributor named on the product label to provide to the California Safe Cosmetics Program of the California Department of Public Health a list of all cosmetic products that contain any ingredients known or suspected to cause cancer, birth defects or other reproductive harm. Since 1987 the California Office of Environmental Health Hazard Assessment has managed a list, called Proposition 65, of naturally occurring and synthetic chemicals known to cause cancer or birth defects or other reproductive harm. Businesses, including cosmetics businesses, are required to provide a clear and reasonable warning before knowingly and intentionally exposing anyone to a listed chemical.

Regulatory Features in Marketing Cosmetics in Japan

The Pharmaceutical and Food Safety Bureau inside the Ministry of Health, Labour and Welfare (MHLW) is the regulatory body in charge of cosmetics, monitoring compliance to the Pharmaceutical Affairs Law (PAL), which was first adopted in 1943 but drastically revised and changed for cosmetics in 2001. The PAL regulates drugs, quasi-drugs, cosmetics and medical devices to guarantee their quality, safety and efficacy. The PAL, notwithstanding the fact that it contains provisions affecting all categories, specifies that a product can fall within, and comply with, the definition and requirements of only one category.

Effective 1 April 2001, new regulations, such as MHLW Notification No. 331/2000, changed the cosmetics system in Japan. This so-called 'deregulation' implied a radical change from the past: the abolition of the pre-market approval or licensing system for each cosmetic product, meaning that, as in the EU and the United States, the manufacturer/seller is now responsible for ensuring that any marketed cosmetic is safe and to substantiate its harmlessness.

Generally speaking, in Japan there are cosmetics and quasi-drugs. Cosmetics are defined in Article 2, Paragraph 3 of the PAL as:

... any article intended to be used by means of rubbing, sprinkling or by similar application to the human body for cleaning, beautifying, promoting attractiveness, altering the appearance of the human body and for keeping the skin and hair healthy provided that the action of the article on the human body is mild.

Cosmetics are classified into six different categories: perfume and eau de cologne, make-up, skin care products, hair care products, special purpose (e.g., sunscreen) and soap.

Under the PAL, quasi-drugs (Article 2, Paragraph 2) are defined as

... articles which have the following purposes and exert mild actions on the human body: prevention of nausea or other discomfort, or prevention of foul breath or body odour; prevention of prickly heat, sore and the like; prevention of hair loss, promotion of hair growth, or depilation; killing or repellence of rats, flies, mosquitoes, fleas, etc., for maintaining the health of man or animals, and exert mild action on the human body but are not intended for use in the diagnosis, cure or prevention of disease or to affect the structure or any function of the human or animal body and are not equipment or instruments.

In this sense, the products designated as quasi-drugs by the MHLW are, as an example, hair dyes, permanent waving products, hair growers, depilatories, deodorants, medicated cosmetics, insect repellents, sanitary cotton products and others.

Following the abolition of the former licensing system, manufacturers and/or importers take full responsibility for the safety of a cosmetic. Importers assume all quality assurance and product liability for cosmetics and need to check their compliance with the PAL, then a series of notifications will be filed, providing also results of testing and inspection verifying that the cosmetic does not contain any banned substance.

Since 1 June 2009, the revised PAL has required the importer to obtain a cosmetics manufacturing and sales license from the competent prefectural pharmaceutical affairs division, and the license is to be renewed every 5 years. To obtain a license to market cosmetics, a company must observe GQP (good quality practice) and GVP (good vigilance practice) in accordance with the standards specified by MHLW ordinances in terms of quality control and post-marketing safety management. To comply with GQP, the company must have a product quality manager responsible for keeping records on product quality, for proper manufacturing as well as shipment, and, if necessary, for product recall. Even if there are no mandatory GMPs, voluntary guidelines have been issued by the Japanese cosmetic trade association (JCIA—Japanese Cosmetic Industry Association). To comply with GVP, a safety control manager is necessary to monitor post-marketing product safety control and keep all necessary records. Furthermore, the company must have a general marketing business controller, who oversees proper marketing practices and supervises the GQP and GVP managers. If the license holder becomes aware of any information indicating that one of the imported cosmetic products may have a harmful effect, they must report that fact to the MHLW within 30 days.

On the other hand, quasi-drugs need to be pre-market registered, and product registration might take 6 months or more. Product formula, manufacturing method, application method and claimed effects are checked on this occasion. In both cases, notification and registration, only local residents can apply.

Cosmetics must fulfil the following labelling requirements (in Japanese): name and address of the approved and licensed company, brand name and function, batch code, full ingredient list in descending order (only on outer container) using the INCI names transliterated into Japanese, nominal content using metric units of weight or volume, country of origin, expiry date only for cosmetics designated by the MHLW, warning statements (prescribed by law or voluntary), instructions for use and instructions for storage (if required). The JCIA has compiled a Japanese version of the INCI names for listing all ingredient names on the label. If a new label name needs to be devised, then a request can be filed with the association.

Regarding the scope and efficacy of cosmetics, reference has to be made to Notification No. 1339/2000, which lists 55 functions for a cosmetic product, e.g., clean the scalp and hair, make the skin strong, mask unpleasant odours, give the skin lustre, make the skin smooth, moisturize the scalp and hair, improve the skin after shaving, keep moisture in the scalp and hair, prevent sunburn, make hair easier to comb, give a pleasant fragrance, protect nails, treat dandruff and scalp itch, prevent lip roughness, improve skin texture, keep the skin healthy, moisturize the skin, protect the skin, whiten teeth, clean the oral cavity (toothpastes), prevent cavities (toothpastes used in brushing).

Following deregulation, ingredient use is regulated by positive and negative lists, such as banned substances, restricted cosmetic ingredients, positive lists of preservatives and UV filters for skin protection. The use of coal-tar colouring agents (which correspond to synthetic colourants) has been regulated since 1966 by several ordinances. The responsibility of ensuring the safety of a cosmetic resides with cosmetic manufacturers and importers.

On the other hand, all active ingredients used in quasi-drugs, as well as excipients, must be in the positive list, a situation similar to US OTC monographs. Authorization of new ingredients by Japanese authorities might prove extremely expensive and time consuming.

The PAL prohibits any form of misleading, exaggerated or false labelling that deceives consumers about the nature or quality of a product, while the Fair Trade Commission is the authority on applying advertising codes for cosmetics, quasi-drugs, drugs and medical devices. The JCIA has also published guidelines for the fair advertising practice of cosmetics. Sunscreens are self-regulated and the sun protection factor (SPF) claimed must be between 2 and 50+ values. ISO Standard 24444 for UVB and ISO Standard 24442 for UVA are applied.

Table 1.2 gives a brief summary showing a comparison of the five main cosmetics regulations.

Table 1.2 Comparison of the main features of the Brazil, EU, Japan, People's Republic of China and United States cosmetic regulations

Item	Brazil	EU	Japan	PRC	USA
Product notification	Yes (distinct for Grau 1 and Grau 2)	Yes	Yes	Yes (registration)	No (voluntary)
Pre-market approval	No	No	No (if cosmetics)	Yes	No (if cosmetics)
Product safety: manufacturer responsibility	Yes	Yes	Yes	Yes	Yes
List of banned/restricted ingredients	Yes	Yes	Yes	Yes	Yes
Positive list for colours	Yes	Yes	Yes	Yes	Yes
Positive list for preservatives	Yes	Yes	Yes	Yes	No
Positive list for UV filters	Yes	Yes	Yes	Yes	Yes (but as OTC)
Use of INCI names	Yes	Yes	Yes (in Japanese)	Yes (in Chinese)	Yes
Product categories	Grau 1 or Grau 2 cosmetics	Cosmetics	Cosmetics; some are classified as quasi-drugs	Generic and special cosmetics	Cosmetics; some are classified as OTCs

EU, European Union; *INCI*, International Nomenclature of Cosmetic Ingredients; *OTC*, over-the-counter drug; *PRC*, People's Republic of China; *USA*, United States of America.

THE INTERNATIONAL COOPERATION ON COSMETICS REGULATION ROLE IN GLOBAL ALIGNMENT

The International Cooperation on Cosmetics Regulation (ICCR) is a voluntary group organized since 2007 by health regulators of Brazil (accepted in 2014), Canada, the EU, Japan and the United States dealing with cosmetics. In contrast to previous regulator forums, the cosmetics industry associations of these countries are allowed to participate and give opinions in a particular session of the annual meeting and in specific working groups, making the ICCR a formal industry/regulators dialogue. The purpose of the ICCR is to discuss common issues on cosmetics regulation, remove regulatory obstacles and minimize international trade barriers while keeping the highest level of global consumer protection.

Themes up to now tackled by the ICCR have been common cosmetics GMP, criteria to identify nanomaterials and nanotechnologies for cosmetics regulatory purposes, promotion of the validation of alternatives to animal testing, clarification of state-of-the-art science (e.g., in silico/quantitative structure–activity relationship models), common and general principles relevant to the safety assessment of cosmetic ingredients, UV protection test methods/UV filters, recommendations for trace substances in cosmetics, review of available risk assessment methodologies for allergens, product preservation, microbiological limits and the interrelationship between the ICCR and the International Organization for Standardization (ISO).

The ICCR also set up a dedicated website to help promote its achievements through online publication of reports. The reports, documents and infographics available relate to alternatives to animal testing, nanotechnology and nanomaterials, trace elements, safety assessment, allergens and product preservation.

Throughout the years the ICCR has attracted the interest of other countries and industry associations: People's Republic of China representatives have attended yearly meetings since 2012; Thailand was an observer in 2015, as was Saudi Arabia in 2016; whilst in 2017 at the 11th annual meeting in Brazil representatives from Argentina, Chile, Colombia, the Republic of Korea, South Africa and Taiwan participated as observers.

SAFETY AND EFFICACY OF COSMETICS: A BRIEF SURVEY OF OTHER SIGNIFICANT WORLDWIDE MARKETS

Association of Southeast Asia Nations

The ASEAN was instituted in 1967 as a political and economic organization and consists of 10 Member States: Brunei Darussalam, Cambodia, Indonesia, Laos, Malaysia, Myanmar, Philippines, Singapore, Thailand and Vietnam.

To eliminate restrictions on the trade of cosmetics and to facilitate the free movement of cosmetics in the ASEAN region, an agreement signed by ASEAN Economic Ministers on 2 September 2003 established that, beginning on 1 January 2008, a harmonized regulatory scheme modelled on EU Cosmetic Directive 76/768/EEC up to the sixth council amendment (Directive 93/35/EEC) had to be implemented by Member States. All cosmetics meeting the requirements of the ASEAN Cosmetics Directive should have equal and immediate access to the market and be able to circulate freely throughout ASEAN countries. It still has to be noted that, while a high degree of alignment has been reached, to date some Member States have not yet fully implemented the ASEAN Cosmetics Directive or continue to ask for requirements not included in it (e.g., notification number printed on label, mandatory expiry date, legalized letter of representation, label artwork or picture to be uploaded).

The contents of the ASEAN Cosmetic Directive are the following: general provisions, definition and scope, safety requirements, ingredient listings, labelling requirements, product claims, product information, methods of analysis, institutional arrangements, special cases and implementation. Common Technical Documents have also been developed dealing with an illustrative list of cosmetic products, labelling requirements, claims guidelines and GMP guidelines.

The definition of cosmetics is the same as in the EU regulations and there is only one category of products. The company responsible for placing a cosmetic on the market must be in a position to demonstrate its safety and has to complete a Product Information File for each cosmetic, which is signed by a duly qualified safety assessor. This technical and safety information about the cosmetic (e.g., qualitative and quantitative formulation, ingredients and cosmetic specifications, summary of manufacturing method in compliance with GMP, safety assessment, existing data on adverse effects, supporting data for efficacy and claims) must be readily accessible to the regulatory authority at the address on the product label.

There is no pre-market registration, but there is a notification procedure using the Product Notification Form prescribed by the regulatory authority. A cosmetic can be marketed after notification has been sent to the competent authority and acknowledgement has been received. Member countries will make an effort to ensure that notifications will receive acknowledgement within three working days. The notification has to be presented to the competent authorities of each Member State in which the product will be marketed. Notification fee and number and validity period vary among the ASEAN Member States.

All cosmetics put on the market must be manufactured according to the ASEAN Cosmetic GMP Guideline or approved equivalent, i.e., ISO 22716:2007, but there is no requirement for a GMP certificate and companies have to certify compliance in the notification form.

The essentials of cosmetic labelling are product name and function, name and address of company responsible for placing the product on the local market (not one for all ASEAN countries), country of manufacture, net content by weight or volume, batch number, expiry date and/or manufacturing date, full ingredient listing, conditions of use and warnings as required. The expiry date is mandatory for cosmetic products with durability less than 30 months and shall consist of month and year or of day, month and year.

Ingredients must be declared using INCI names, in descending order, but those in concentrations of <1% may be listed in any order.

Ingredients are regulated by distinct annexes: Annex II lists banned ingredients, whereas Annex III contains restricted ingredients; Annexes IV, VI and VII are positive lists of colours, preservatives and UV filters, respectively.

ASEAN Member States have the responsibility to ensure that formulation, manufacturing, labelling and advertising of marketed cosmetics comply with the requirements by way of inspection of manufacturing, storage and sale premises; assessment of the Product Information File and monitoring and analysis of products on the market.

Claims must be supported by scientific evidence as reported in the claims guidelines; they should not be misleading and should be aligned with international practice. Products are classified as a cosmetic or drug based on composition and suggested use as indicated by presentation, label, text, trademarks, pictures and advertising.

Whenever there is a reasonable suspicion (e.g., there is evidence to suggest a causal relationship) that the cosmetic might be the cause of a serious adverse event, regardless of the source of the account, it must be reported by the company to the regulatory authority of the ASEAN Member State in which the adverse event occurred. Non-serious adverse events are not required to be reported.

Australia

Following the introduction of a new regulatory framework in September 2007, according to the Australian Industrial Chemicals (Notification and Assessment) Act of 1989 and the Cosmetics Standard, 2007, a cosmetic product means:

A substance or preparation intended for placement in contact with any external part of the human body, including: the mucous membranes of the oral cavity and the teeth; with a view to: altering the odours of the body; or changing its appearance; or cleansing it; or maintaining it in good condition; or perfuming it; or protecting it.

The main categories of cosmetics are face and nail products, hair care and hairdressing products, oral hygiene, perfumes, personal hygiene and skin care.

Product-specific requirements detailed in the National Industrial Chemicals Notification and Assessment Scheme (NICNAS) Cosmetics Standard apply to certain cosmetic categories: tinted bases or foundations (liquids, pastes, powders) with sunscreen; products intended for application to the lips with sunscreen; moisturizing products with sunscreen for dermal application, including but not limited to anti-wrinkle, anti-ageing and skin whitening products; sunbathing products (e.g., oils, creams or gels, including products for tanning without sun and after-sun care products) with SPF ≥ 4 and ≤ 15; antibacterial skin products (must be presented only as being active against bacteria); anti-acne products (must be presented as controlling or preventing acne only through cleansing, moisturizing, exfoliating or drying the skin); products for care of the teeth and the mouth (such as dentifrices, mouthwashes and breath fresheners, but the product must not claim benefits in relation to other diseases or ailments such as gum disease, oral disease or periodontal condition) and antidandruff products (must be presented as controlling or preventing dandruff only through cleansing, moisturizing, exfoliating or drying the scalp).

While cosmetics are controlled by the NICNAS inside the Department of Health, the Therapeutic Goods Administration (TGA), also inside the Department of Health, regulates therapeutic products within the definition outlined by the Therapeutic Goods Act, 1989. In general a therapeutic good can be described as a product for use in humans intended for 'therapeutic use', which includes influencing, inhibiting or modifying a physiological process or preventing, diagnosing, curing or alleviating a disease, ailment or defect. Examples of therapeutic goods are primary sunscreens with SPF ≥4 and secondary sunscreens as moisturizers with SPF >15. In addition, a cosmetic may be classified as a medicine depending on its ingredients, its route of administration and if therapeutic claims are made on its label or in advertising.

Cosmetics and their manufacturers must register with NICNAS and pay a yearly fee as well as a charge for the value of the cosmetics introduced into the Australian market, and therapeutic goods and their manufacturers must be included in the Australian Register of Therapeutic Goods (ARTG) and manufacturers must be licensed to demonstrate GMP compliance.

The Australian Consumer Law introduced mandatory labelling standards to provide consumers with specified information about goods. The Trade Practices (Consumer Product Information Standards) (Cosmetics) Amendment Regulation of 1998 establishes the labelling of cosmetics under the authority of the Australian Competition and Consumer Commission (ACCC), which is responsible for enforcing mandatory consumer product safety and information standards.

Moreover, the Trade Practices Regulations and other regulations such as the Trade Measurement Act of 1990 require the following to be mentioned, in English and in clear, distinct and legible lettering: list of ingredients using INCI names, name and address of the packer or distributor, manufacturer or person responsible for placing the product on the market, product name and function, net quantity of contents, warnings, instructions for use and country of origin. The Poison Standard of June 2017 (otherwise known as *Standard for the uniform labeling of drugs and poisons*) provides for mandatory 'warning' labelling of containers when certain scheduled substances have been added to the cosmetic.

Labelling of therapeutic goods includes information such as the names of the active ingredients according to the Australian Approved Names nomenclature and their concentration, name and Australian address of the product sponsor, batch number, expiry date, storage conditions, warnings and the TGA listing or registration number.

Regarding cosmetic ingredients, including fragrances, these must all be included in the Australian Inventory of Chemical Substances; if not, they have to be assessed under the NICNAS. Other sources of regulatory information should be consulted to check which chemicals must not be used in cosmetics, e.g., the Poison Standard of June 2017 and the Hazardous Substances Information System (chemicals used in the workplace).

As an example, the labelling and advertising of therapeutic sunscreens included in the ARTG must comply with the relevant requirements of each of the following: The Labelling Order, Therapeutic Goods Order No. 69, The Therapeutic Goods Advertising Code, Australian/New Zealand Standard AS/NZS 2604:2012 Sunscreen products—Evaluation and classification and the current edition of *Required Advisory Statements for Medicine Labels*.

The ACCC conducts random surveys of retail outlets throughout Australia to detect products that do not comply with the Trade Practices and Consumer Affairs legislation.

It also investigates allegations by consumers and suppliers about non-complying goods and frequently seeks the immediate withdrawal of defective goods from sale, as well as the recall of the goods for corrective advertising.

A cosmetic will be considered defective '*if its safety is not such as persons generally are entitled to expect*'. Generally it is the manufacturer or importer or the person responsible for placing the products on the Australian market that is liable. However, in instances in which other suppliers, such as retailers, cannot identify the manufacturer, they may be deemed liable for damages.

The Uniform Recall Procedure for Therapeutic Goods defines the action to be taken by health authorities and sponsors when therapeutic goods for human use, for reasons relating to their quality, safety and efficacy, are to be removed from supply or use, or subjected to corrective action. As of this writing, the Australian government is revising the NICNAS, which by 1 July 2018 should be replaced by the Australian Industrial Chemicals Introduction Scheme, which could introduce restrictions on animal testing, e.g., a ban on the use of new animal test data for cosmetic ingredients.

Brazil

Brazil is a member of Mercosur/Mercosul (Mercado Común del Sur, founded by Argentina, Brazil, Paraguay and Uruguay, joined in 2012 by Venezuela and in 2015 by Bolivia), which adopted in the mid-1990s, as a general reference for its Member States, EU Cosmetics Directive 76/768//EEC. The Brazilian Health Regulatory Agency (ANVISA) is a federal organization that enforces regulations on manufacturing, importation and sales of cosmetics, perfumes and toiletries other than drugs, medical devices, foods and home care products.

Cosmetics are defined as:

> ... preparations whose ingredients (natural or synthetic) are to be used externally in various parts of the human body such as skin, hair, nails, lips, genitals, teeth, membranes of the oral cavity, with the only objective to clean them, odorize them, change their appearance, or correct body odours, or protect and/or keep the body in good conditions.

They are divided into four categories (hygiene products, cosmetics, fragrances and baby products) and two risk groups [Grau 1 and Grau 2].

Grau 1 cosmetics include those with a minimal safety risk, having simple formulation and characteristics (e.g., soaps, shampoos, hair conditioners, make-up, mild creams and lotions, deodorants, shaving products, etc.). Grau 2 cosmetics might present a potential risk or have specific indications and characteristics demanding detailed data on safety and efficacy (e.g., antidandruff preparations, anticaries toothpastes, antiperspirants, anti-wrinkle products, sunscreens, hair dyes, hair bleaching products, permanent waving products, nail hardeners, all baby products).

The registration procedure differs for the two risk groups. Grau 1 cosmetics may be simply notified (which involves nevertheless sending a small file to ANVISA) online at least 30 days before marketing. By contrast, Grau 2 cosmetics must be registered and are subject to a specific tax, which value depends on the company turnover, and the registration process may require 60–90 days, if ANVISA does not require the company to submit any additional data. Grau 2 cosmetics also need to delay marketing until the registration number is published in the Brazilian Official Gazette. Notification and registration are valid for 5 years, and then can be renewed 6 months before expiring if no significant changes in the required information have occurred.

All cosmetics labelling must provide the consumer with clear, precise and easily legible information in Portuguese: name and function of the product; trademark; registration number; batch number; expiry date; net content; country of origin; name and address of the manufacturer, importer or person responsible for putting the product on the market; specific instructions for use and warnings required for certain categories of cosmetics and the ingredient list using INCI names (untranslated).

Ingredients are regulated through negative and positive lists and are regularly updated. However, discrepancies exist in comparison with the annexes of the EU law and a local scientific body, the Câmara Técnica de Cosméticos (Technical Group on Cosmetics), may introduce different restrictions or bans.

Manufacturers and importers must have supporting data attesting to the quality, safety and efficacy of cosmetics, and a complete file (including, e.g., formula, chemico-physical and microbiological specifications, manufacturing process, stability data, label artwork, packaging information, safety and efficacy data) on each finished product must be kept at the disposal of the controlling authorities for both notified and registered cosmetic products.

Companies must develop a 'cosmetovigilance' system and keep specified records, which must be made available to controlling authorities upon request.

Canada

Health Canada is the authority responsible for cosmetics, drugs, nonprescription drugs and natural health products.

The Food and Drugs Act, which includes the Cosmetic Regulations, defines a cosmetic as:

> … any substance or mixture of substances manufactured, sold or represented for use in cleansing, improving or altering the complexion, skin, hair or teeth, and include deodorants and perfumes. Soap is considered to be included.

Cosmetics therefore include, for example, lipsticks, mascara, eye shadows, nail polish, shampoos, non-fluoride toothpaste, conditioners, soaps, moisturizers, cleansers, hair dyes, hair permanents and depilatories.

Products such as sunscreens, acne treatments, antidandruff shampoos, anticaries toothpastes and antiperspirants, however, are considered to be non-prescription drugs subject to Category IV monographs. These regulatory instruments indicate, notably, the definition of the non-prescription drug, the approved active ingredients and permitted combinations, the maximum and minimum permitted concentrations and the labelling requirements (statement of identity, indications for use, warnings and directions).

Any product that has a therapeutic claim or that contains certain ingredients not permitted in cosmetics is considered to be an OTC drug and is handled by the Therapeutic Products Program, classified as drug and assigned a Drug Identification Number.

The Natural Health Products Regulations rule on products containing natural therapeutic ingredients. Some products previously classified as cosmetics or non-prescription drugs might be regulated as natural health products (NHPs) if the active ingredient has a natural (animal, plant or mineral) origin.

Cosmetics are subject to the provisions of both the Food and Drugs Act and its regulations (regarding composition, safety, labelling, advertising) and the Consumer Packaging and Labelling Act and its regulations (regarding bilingual labelling, deceptive packaging, net quantity declaration in metric units).

Manufacturers and importers of cosmetics are required to submit online the Cosmetic Notification Form to Health Canada within 10 days of the first sale of a cosmetic in Canada. Cosmetic notification is not a product evaluation or approval procedure or acceptance by the authorities, but rather the manufacturer has the responsibility of ensuring that a cosmetic meets the requirements of acts and regulations.

Products falling into Category IV monographs and NHPs are subject to pre-market approval and registration.

Manufacturers of cosmetics must print certain information on the label of each product, that includes the identity and function of the product, the net quantity in metric units, the name of the manufacturer or distributor and the address of the principal place of business, country of origin for imported products and any warnings or cautions necessary for the safe use of the product, especially mandatory ones such as those required for coal-tar hair dyes, etc. The statement of identity of the product and the net content must be placed in the PDP. Ingredient labelling is mandatory for all cosmetics sold in Canada, and Health Canada has identified the INCI system required in the EU as the reference standard for ingredient labelling on the basis that it is applied in most countries and that the EU system made more use of Latin, the international scientific language. However, if a cosmetic is sold in Canada, the EU and the United States, trivial names need to listed in English, Latin and French, e.g., water/aqua/eau.

The labelling of cosmetics, with the exception of the manufacturer's name and address, must be in both English and French. Special labelling is required under the Quebec Official Languages Act for products sold in that province.

Non-prescription drugs must follow stricter mandatory labelling, including, for example, batch number, expiry date, registration number, percentage of active ingredients, claims, instructions for use and warnings according to the relevant monograph. Special and composite rules also apply for the labelling of an NHP complex.

A list of restricted or banned ingredients is to be found in the Cosmetic Regulations. Excluding colours used in the area of the eye, there is no approved or banned list of colours. Health Canada also makes available an ingredients 'Hot List', a non-exhaustive list of restricted raw materials updated annually, which closely follows the annexes of the EU Regulation. If an ingredient on the notification form also appears in this Hot List the manufacturer will be warned, depending on the ingredient, to take action, including possible removal of the ingredient or concentration reduction or registration of the product as a drug, etc.

Category IV monographs identify the ingredients recognized as safe and effective for a specific purpose, e.g., a list of UV filters.

The acts and the regulations also set out the safety requisites for all cosmetics sold in Canada, prohibiting the sale of a cosmetic that is either prepared under unsanitary conditions or unsafe when used as directed. It is the manufacturer's responsibility to ensure that the cosmetics are safe for their intended use.

The Guidelines for Cosmetic Advertising and Labelling Claims have been published to help manufacturers in choosing acceptable and not misleading claims for cosmetics.

Kingdom of Saudi Arabia

Cosmetics in the Kingdom of Saudi Arabia are regulated and overseen by the Saudi Food & Drug Authority (SFDA). Following the adoption of the GCC Standardization Organization (GSO) *1943–2016 Safety Requirements of Cosmetics and Personal Care Products* there was a marked alignment with EU Regulation 1223/2009. However, in contrast to the EU law, perfumery products based on ethanol were classified into three categories according to the essential oil content.

Products that possess a therapeutic effect and claim medical properties or contain pharmaceutical substances are not considered to be cosmetics.

To clear customs and enter the Saudi market, cosmetics and perfumery products shall be accompanied by a Certificate of Conformity obtained in the country of export through pre-shipment testing procedures (e.g., pH, microbiology, heavy metals/1,4-dioxane/formaldehyde/triclosan/phthalates content) and inspections performed by an SFDA-recognized and -approved certification body, and based on mandatory Saudi Arabian Standards Organization (SASO) standards applied equally to both imported and domestically produced products.

Since April 2015 cosmetics need to be entered in eCosma, the cosmetic product online notification system established by the SFDA and divided into different sections, including clearance, domestic manufacture licensing, warehouse licensing and product (simple, composite and kit) notification. Depending on the kind of request it takes from 15 to 30 days for the SFDA to review the product and give a decision.

Labelling requirements may differ by product type according to each applicable standard; however, the following information should be legibly and indelibly marked: name of the product and brand name (in Arabic or in both Arabic and English); name and address of manufacturer or name and address of the importer; expiry date if less than 30 months or symbol of PaO if expiry date is >30 months; list of ingredients according to INCI names; function of product unless it is indicated by virtue of its presentation, in Arabic or in both Arabic and English; warnings, instructions for use and storage in Arabic or in both Arabic and English; manufacturing date or batch number; country of origin and net content if >5 mL or >5 g. Perfumery products must be labelled with the percentage of ethanol and 'for external use only' in Arabic.

Banned, restricted and regulated ingredient lists are aligned to the *EU Regulation*. In compliance with Islamic law, cosmetics shall contain no pork or its derivatives, ethanol must be denatured and should not exceed 90%, containers containing ethanol should not exceed 250 mL and a non-removable pump is mandatory.

Manufacturers or exporters are requested to provide evidence that their finished products are safe for consumer use based upon the toxicological profile of the ingredients, their chemical structure and their exposure level. The toxicological evaluation needs to be signed by an authorized/recognized safety assessor and should contain a series of data similar to that included in the EU Product Information File.

No medical or unacceptable claims are allowed and claims should comply with the definition of cosmetic products; moreover, pictures or references inconsistent with the prevailing social customs, traditions and religious values (e.g., reference to alcoholic beverages, portrait of women) cannot not be used. Since late 2016, the management of claims has been performed by applying the GSO 2528-2016 standard named *Cosmetic Products—Technical Regulation of Cosmetic and Personal Care Product Claims,* which is based on the principles of EU Regulation 655/2013 but also contains a list of unacceptable cosmetic claims.

People's Republic of China

The China Food and Drug Administration (CFDA) is the competent authority managing the cosmetics sector, applying rules contained in a number of regulations and technical standards and guidelines. The General Administration for Quality Supervision, Inspection and Quarantine (AQSIQ) is involved in the inspection of imported cosmetics upon arrival. It is worth taking note of the profound revision of the cosmetics laws and guidelines which should be completed by 2018.

The definition of a cosmetic is close to that outlined in the former EU Directive, but cosmetics are differentiated into general cosmetics and special purpose cosmetics. General cosmetics are those which are used for hair care, skin care, fragrance and make-up. Special purpose cosmetics include antiperspirants, hair dyes, sunscreens, hair waving products, depilatories, body slimming/breast enhancement products, deodorants and skin whiteners. Soaps are not classified as cosmetics, while oral care products are not regulated in cosmetic categories but through Chinese National Standards.

A time-consuming, costly and occasionally unpredictable pre-market approval from the CFDA is compulsory and can be carried out only by a Chinese legal entity. The requirements are slightly different for imported products compared to locally made products. Indeed local general cosmetics require only a file to be submitted to a provincial bureau, and manufacturing can start before license approval.

Conversely, for both kinds of imported cosmetics as well as Chinese-made special purpose cosmetics, a full dossier has to be submitted to the CFDA for review. It is to be noted that, starting in 2015, foreign general cosmetics online registration to the local FDA has been gradually introduced. National designated laboratories carry out the pre-approval testing and check hygienic and chemical properties (e.g., heavy metals, methanol, pH), microbiological quality, toxicology (selective testing based on claim and formula) and efficacy (e.g., sunscreens). It is important that consistency (e.g., in product name) is kept throughout the application because any discrepancy will result in a delay in the application process.

In the registration process care must be applied also in checking the ingredients present in the formula. In June 2014 the CFDA issued a list of more than 8700 ingredients (last updated in December 2015) which can be used in cosmetics, including those regulated via the annexes, because they are 'known' to the authorities since they were previously reported as used in cosmetics put on the Chinese market. If the formula to be registered contains an ingredient which is not on this list, the ingredient must be first registered with the Chinese authorities.

General labelling of cosmetics is also addressed by a specific Chinese standard (GB 5296.3-2008). Information required on the label, in Chinese, includes identity and function of the product, net quantity in metric units, name of the manufacturer/importer or distributor and the address of the principal place of business, date of manufacture plus shelf life or batch number plus expiry date, health licence number (for imported cosmetics), any warnings or cautions necessary for the safe use of the product, instructions for storage (if applicable), country of origin and ingredient list (INCI names have to be transliterated into Chinese).

Ingredients are mainly regulated through the *Safety and Technical Standard for Cosmetics—2015,* which includes lists of banned, allowed and restricted ingredients similar to the EU Regulation.

Safety is assessed during the registration process by means of both the documents presented by the manufacturer/importer and local chemical, physical, microbiological, toxicological and clinical testing. Concerning safety testing on animals, on June 2014 the CFDA, in its plan to modernize its cosmetics regulatory framework, began to phase out the requirement that new, locally manufactured general cosmetic products be tested on animals.

Cosmetics advertising is managed through different across-the-board laws, the latest published in April 2015, which established an advertising code of conduct, for example, a ban on the use of therapeutic and medical words and misleading pictures and depictions. The State Administration for Industry and Commerce monitors the advertising activities of commercial cosmetic products.

Republic of Korea

The Ministry of Food and Drug Safety (MFDS) regulates cosmetic production and marketing through the Cosmetic Act and its amendments, which identify three categories of cosmetics: cosmetics, functional cosmetics or cosmeceuticals and quasi-drugs. Cosmetics are defined as

items with mild action on the human body for the purpose of cleaning, beautifying, adding to the attractiveness, altering the appearance, or keeping or promoting the skin or hair in good condition.

Functional cosmetics, for which a positive list of active ingredients exists, include skin whiteners, sunscreens (primary and secondary), suntanning products, anti-wrinkle (minimizing the appearance of lines in the face and body) products, oxidation hair dyes, hair bleaches, hair lotions, chemical depilatories and products reducing the dryness of sensitive skin. Examples of quasi-drugs are acne products, mouthwashes, toothpastes, tooth whiteners and personal hygiene products.

There is no pre-market registration for cosmetics but manufacturers and importers have to assess and guarantee the safety and efficacy of the products. A notification for imported products is to be submitted electronically to the Korean Pharmaceutical Trade Association (KPTA) along with the free sale certificate and the notarized formula, and a declaration related to a measure adopted against possible bovine spongiform encephalopathy/transmissible spongiform encephalopathy contamination must be presented by importers. On the other hand registration for functional cosmetics or quasi-drugs is quite complex and costly and might take up to 6 or 10 months, respectively.

Korean is the language for labelling requirements, which include product name or function of the product, name and address of the manufacturer/importer or of the company responsible for marketing the product, use-by date after opening (the EU PaO symbol is used) or best-before date and date of manufacture, batch number, net content in volume or weight, country of origin, ingredients list, the statement 'functional cosmetic' (if applicable), retail price, instructions for use and storage (if needed) and warnings concerning restricted substances or voluntary warnings.

Ingredients for cosmetics are regulated through lists of banned ingredients and positive lists for preservatives, colours and UV filters. Positive lists of active ingredients are available for functional cosmetics as well as for quasi-drugs.

The manufacturer or the importer is responsible for the safety and efficacy of cosmetics marketed in the Republic of Korea. Cosmetics must not present a risk to human health owing to inadequate ingredients, microbial or heavy metal contamination, unhealthy manufacturing practices or insufficient packaging. Standards for manufacture and quality control of cosmetics including GMP according to ISO 22716 are applied.

Guidelines on cosmetic claims have been published by local authorities. In cosmetics advertising it is considered misleading to make any reference to therapeutic properties, public acknowledgement or recommendation by a health professional or suggestion leading an end user to believe that a domestic product is a foreign product or vice versa. Comparative advertising shall be made in an objective manner, while it is prohibited to use words like 'the best' or ' the most outstanding'.

Russian Federation

Since July 2012 cosmetics have been regulated by the Technical Regulation on the Safety of Perfumery and Cosmetic Products, TR CU 009/2011 (as amended on December 2015), adopted by the Customs Union in September 2011. The Eurasian Customs Union was established in 2010 by Belarus, Kazakhstan and Russia and enlarged in 2015 to Armenia and Kyrgyzstan. According to the agreement of the Customs Union, cosmetics proved to be compliant with the requirements of the Technical Regulation developed by the Customs Union have free access to territory of the Member States.

The Technical Regulation is modelled on the EU Regulation, so the same definition for cosmetics applies, and it is structured in various articles and the following annexes:

- *Annexes 1 to 5*: related to banned, restricted or allowed substances
- *Annex 6*: pH Value Requirements
- *Annex 7*: Microbiological Safety
- *Annex 8*: Toxicological Requirements
- *Annex 9*: Clinical Laboratory Requirements
- *Annex 10*: Clinical Requirements for Oral Hygiene Products
- *Annex 11*: The Symbol Indicating Additional Information on Perfumery and Cosmetics
- *Annex 12*: List of Perfumery and Cosmetics Subject to State Registration

In contrast to the EU Regulation, pre-market approval prior to sale in Russia is compulsory through two different procedures: state registration or conformity declaration.

The state registration procedure is compulsory for the products listed in Annex 12 (e.g., skin whitening, cosmetics for children, hair dyes and hair bleaching products, hair waving and straighteners, products containing nanomaterials, products for dental and oral hygiene containing fluorine). State registration must be applied at the Federal Consumer Protection and Public Welfare Service (RosPotrebNadzor).

A conformity declaration, released by an authorized Russian certification body, must be obtained for all other cosmetics. The manufacturer or importer duly registered in Russia must present the following documents: list of ingredients and concentrations of restricted substances, chemico-physical specifications, testing (performed in Russia) results confirming that the products comply with safety requirements contained in the Technical Regulation (e.g., pH, microbiological parameters, heavy metal limits), label or artwork, manufacturer's GMP declaration or quality management certificate, efficacy data.

Labelling requirements, mostly in Russian, are name, brand name (may be in Latin letters) and product function, if not self-evident; name and address of the manufacturer (may be in Latin letters); country of origin; name and registered address of the Russian authorized representative; net content; manufacturing date and shelf life or expiry date; storage conditions; colour if a make-up; baby product (if applicable); special precautions, directions for use and warnings (if applicable); batch number; ingredients list (using INCI names in Latin letters); ingredients which are present in the form of nanomaterials must be marked with the word нано in brackets.

The Conformity Mark has to be put on each individual cosmetic packaging as proof that the product is certified:

If requested information cannot be listed on the label, the following symbol can be used:

GMP, pH limits, microbiological criteria, heavy metals limits and toxicological, clinical and laboratory parameters are set out in the Technical Regulation or in its annexes.

The lists of prohibited, restricted and allowed ingredients are greatly overlapping with the EU Regulation lists but lack the same timing of update.

Alcohol-containing perfumery and cosmetic products must also follow specific restrictions, registration and the authorities must be notified of their import (Federal Service for Surveillance on Consumer Rights Protection and Human Wellbeing) as part of the Customs Import Declaration.

South Africa

South African cosmetics regulations represent the only example of cosmetics self-regulation in the world: Self-Regulatory Code of Practice by the South African National Standards. Another source of information for manufacturing and marketing compliance is the *Cosmetics, Toiletry, and Fragrance Association of South Africa (CTFA-SA) Compendium*,

4th edition. The Foodstuffs, Cosmetics and Disinfectants Act, the Foodstuffs, Cosmetics and Amendment Act and the Consumer Protection Act are legislations also affecting cosmetics. ISO standards related to cosmetics have also been adopted to complete an appropriate regulatory system for cosmetics in South Africa.

The definition of a cosmetic product, with illustrative list by category of cosmetics included, is close to the EU definition except for the last sentence which has been added to this definition: '*[…] except where such cleaning, perfuming, protection, changing, keeping or correcting is for the purpose of treating or preventing disease*'.

Guidance on secondary antibacterial/antifungal function, cosmetic cellulite products, hair care products, skin lighteners and sunscreens has also been published.

No pre-market notification is required but cosmetics must be in compliance with the aforesaid guidelines.

The following information must be on the label in English and possibly one further official national language: name and address of the manufacturer or of the person responsible for placing it on the market, product name and function if not obvious from the presentation, net quantity of contents, list of ingredients (INCI names allowed), warnings and directions for use, storage (if needed), date of minimum durability if less than 30 months, batch number, country of origin.

Special labelling and testing are required for sunscreens according to the Standard SANS 1557 published by the South African Bureau of Standards or equivalent, such as the ISO standards.

Making reference to the EU regulatory system, various positive and negative lists control the use of cosmetic ingredients: banned substance, restricted ingredients in cosmetics and cosmetic oral hygiene products, restricted list of hair dyes and allowed colourants, preservatives and UV filters.

Manufacturers must ensure the safety and efficacy of cosmetics marketed in South Africa. GMP and microbiology testing requirements are in place.

The Advertising Standards Authority of South Africa, which receives complaints from the public or companies from time to time on unsubstantiated claims, use of non-permitted ingredients or sale of a substandard product, can check cosmetics. In these cases, the non-compliant companies have been asked to withdraw their products.

Switzerland

On 1 May 2017 a new federal law on cosmetics entered into force and it is quite fully aligned with the EU Regulation. The *Federal act on foodstuff and commodities,* the *Ordinance on foodstuff and commodities* and the *Ordinances on cosmetics* and *aerosols* need to be consulted for one to have a complete understanding of the applied legislation. The Federal Office for Public Health has published a guidance on the borderline between cosmetics, drugs and biocides: *Critères de delimitation des produits cosmetiques par rapport aux produit thérapeutiques et aux produits biocides.*

Neither the Federal Office for Public Health nor the local cantonal authorities require a mandatory notification of cosmetics, but the latter are charged with market control, product sampling and analysis. Labelling is also in line with the provisions of the EU Regulation and should be in at least one the official languages (French, German and Italian). It is not compulsory to have a Swiss address on the label if there is a responsible person in the EU. The symbols of the PaO and the hourglass and the symbol to be used where it is impossible for practical reasons to label specific information are included in the labelling requirements and are equal to those used in the EU. Ingredients are basically allowed, restricted or banned according to the annexes of the EU Regulation, with minor divergences, and will be automatically adapted to the EU updates. No provision regarding the ban on animal testing is in force; however, the Federal Food Safety and Veterinary Office has not permitted any animal testing for cosmetic ingredients since 2007, with only one exception. De facto animal testing is banned under the federal law on animal welfare. As in the EU, manufacturers or importers or the person responsible for putting the product on the market must assure the safety and efficacy of the cosmetic concerned by completing a Product Information File for each cosmetic, signed by a duly qualified safety assessor. The files may not be kept in Switzerland, but local cantonal authorities may ask companies for proof of safety assessment. Claims are regulated by making reference to EU Regulation 655/2013. As of this writing, Switzerland is the only country in Western Europe imposing a tax on the VOC content of cosmetics through the *Ordinance on the Incentive Tax on VOC*.

REFERENCES

Association of South East Asia Nations

http://asean.org.

Agreement on the ASEAN Harmonized Cosmetic Regulatory Scheme, Phnom Penh, September 2, 2003. http://www.asean.org/?static_post=agreement-on-the-asean-harmonized-cosmetic-regulatory-scheme-phnom-penh-2-september-2003.

ASEAN Cosmetics Directive. http://aseancosmetics.org/information-center/asean-cosmetic-directive/.

ASEAN Cosmetics Directive Technical Documents. http://aseancosmetics.org/asean- -cosmetics-directive/technical-documents/.

Australia

Australia Competition and Consumer Commission – Product Safety Australia. http://www.productsafety.gov.au/content/index.phtml/itemId/971652; http://www.productsafety.gov.au/content/index.phtml/itemId/971654/fromitemId/971652.

Australia Inventory of Chemical Substances. https://www.nicnas.gov.au/regulation-and-compliance/aics.

Australian Register of Therapeutic Goods. https://www.tga.gov.au/australian-register-therapeutic-goods.

Australian Trade Practices (Consumer Product Information Standards) (Cosmetics) Regulations 1991 as amended. https://www.legislation.gov.au/Details/F2008C00244.

Cosmetics Standard, 2007. https://www.nicnas.gov.au/cosmetics-and-soaps/cosmetics-standard-and-sunscreens.

Hazardous Substance Information System. http://hsis.safeworkaustralia.gov.au/.

Industrial Chemical (Notification and Assessment), 1989. https://www.legislation.gov.au//Details/C2016C00816.

NICNAS Cosmetic Requirements. https://www.nicnas.gov.au/cosmetics-and-soaps.

Poisons Standard, June 2017. https://www.tga.gov.au/publication/poisons-standard-susmp.

The Australian/New Zealand Standard AS/NZS 2604:2012 Sunscreen Products–Evaluation and Classification. https://infostore.saiglobal.com/en-gb/Search/Standard/?searchTerm=As/nzs%202604:2012&productFamily=STANDARD.

Therapeutic Goods Act 1989 as amended. https://www.tga.gov.au/legislation-legislative-instruments.

Therapeutic Goods Administration. https://www.tga.gov.au/, https://www.tga.gov.au/regulation-basics; https://www.tga.gov.au/sunscreens.

Brazil

All Information and Regulation About Cosmetics (in Portoguese) are available at the following website. http://portal.anvisa.gov.br/wps/content/Anvisa+Portal/Anvisa/Inicio/Cosmeticos.

ANVISA. http://portal.anvisa.gov.br/wps/portal/anvisa/home.

Cosmetic Registration. http://portal.anvisa.gov.br/wps/content/Anvisa+Portal/Anvisa/Inicio/Cosmeticos/Assuntos+de+Interesse/Regularizacao+de+Produtos.

Canada

Health Canada /Santé Canada All information and Regulation on cosmetics are available at the following website: https://www.canada.ca/en/health-canada/services/consumer-product-safety/cosmetics.html; https://www.canada.ca/en/health-canada/services/consumer-product-safety/cosmetics/regulatory-information.html; https://www.canada.ca/en/health-canada/services/consumer-product-safety/cosmetics/notification-cosmetics.html; https://www.canada.ca/en/health-canada/services/consumer-product-safety/cosmetics/cosmetic-ingredient-hotlist-prohibited-restricted-ingredients/hotlist.html.

Information about drug products are accessible at the following website: https://www.canada.ca/en/health-canada/services/drugs-health-products.html.

Information about natural health products are accessible at the following website: https://www.canada.ca/en/health-canada/corporate/about-health-canada/branches-agencies/health-products-food-branch/natural-non-prescription-health-products-directorate.html.

Guidelines for the Nonprescription and Cosmetic Industry Regarding Non-Therapeutic Advertising and Labelling Claims (Guidelines). http://www.adstandards.com/en/Clearance/Cosmetics/Guidelines-for-the-Nonprescription-and-Cosmetic-Industry.pdf.

European Union

All Information about Legislation, Reference Documents, Scientific and Technical Assessment, Cosmetic Ingredient Database, Cosmetic Notification, Borderline Products, Ban on Animal Testing are Available at the Following Website. http://ec.europa.eu/growth/sectors/cosmetics/_en.

Commission Recommendation of 22 September 2006 on the Efficacy of Sunscreen Products and the Claims Made Related to Them. http://ec.europa.eu/growth/sectors/cosmetics/products/sunscreen/index_en.htm.

Commission Regulation (EU) No 655/2013 Laying Down Common Criteria for the Justification of Claims Used in Relation to Cosmetic Products. http://eur-lex.europa.eu/legal-content/EN/TXT/?uri=CELEX:32013R0655.

Cosmetics Europe, 2016. Annual Report. https://www.cosmeticseurope.eu/library/6.

Council Directive of 27 July 1976 on the Approximation of the Laws of the Member States Relating to Cosmetic Products. http://eur-lex.europa.eu/legal-content/EN/TXT/?qid=1459262299959&uri=CELEX:31976L0768.

Council Directive 87/357/EEC of 25 June 1987 on the Approximation of the Laws of the Member States Concerning Products Which, Appearing to be Other Than They are, Endanger the Health or Safety of Consumers. http://eur-lex.europa.eu/legal-content/EN/TXT/uri=CELEX%3A31987L0357.

Directive 2005/29/EC of the European Parliament and of the Council of 11 May 2005 Concerning Unfair Business-to-Consumer Commercial Practices in the Internal Market. http://eur-lex.europa.eu/legal-content/EN/TXT/uri=CELEX%3A32005L0029.

ECJ–European Court of Justice, 1991a. C-369/88 of 21.03.1991 "Delattre", ECR 1991 I-1487, para. 12.

ECJ–European Court of Justice, 1991b. C-112/89 of 16.04.1991, "Upjohn", ECR 1991 I-1703.
ECJ–European Court of Justice, 2005. C-193/2005 of 6.08.2005 in joined cases C-211/03, C-299/03, C-316/03 to C 318/03, "HLH Warenvertriebs GmbH, Orthica BC v Federal Republic of Germany", para. 44 and 45. http://ec.europa.eu/DocsRoom/documents/13175/attachments/1/translations/en/renditions/pdfhttp://curia.europa.eu/juris/recherche.jsf?cid=167559.
EN ISO 22716:2007, 2007. Cosmetics – Good Manufacturing Practices (GMP) – Guidelines on Good Manufacturing Practices. https://standards.cen.eu/dyn/www/f?p=204:110:0::::FSP_PROJECT,FSP_ORG_ID:26991,679535&cs=1E0A2BDA6418D4772BD7751A37DD668C2.
Guidelines on Borderline Products. http://ec.europa.eu/growth/sectors/cosmetics/products/borderline-products/index_en.htm.
Regulation (EC) No.1223/2009 of the European Parliament and of the Council of 30 November 2009 on Cosmetic Products. http://eur-lex.europa.eu/legal-content/EN/TXT/HTML/?uri=CELEX:02009R1223-20150416&from=EN.
RPA–Risk and Policy Analysts, 2004. Comparative Study on Cosmetics Legislation in the EU and Other Principal Markets with Special Attention to so-called Borderline Products, RPA, Norfolk, UK. http://rpaltd.co.uk/projects/comparative-study-on-cosmetics-legislation-in-the-eu-and-other-principle-markets-with-special-attention-to-so-called-borderline-products.
Scientific Committee on Consumer Safety Website. http://ec.europa.eu/health/scientific_committees/consumer_safety/index_en.htm.
Technical document on cosmetic claims - Agreed by the Sub-Working Group on Claims - (version of 3 July 2017). https://ec.europa.eu/docsroom/documents/24847.

ICCR
www.iccrnet.org.

Japan
Approval and Licensing for Drugs, Quasi-Drugs and Cosmetics. http://www.mhlw.go.jp/english/wp/wp-hw2/part2/p3_0034.pdf.
ISO 24442:2011, 2011. Cosmetics – Sun Protection Test Methods – In Vivo Determination of Sunscreen UVA Protection. http://www.iso.org/iso_catalogue/catalogue_tc/catalogue_detail.htm?csnumber=46521.
ISO 24444:2010, 2010. Cosmetics – Sun Protection Test Methods – In Vivo Determination of the Sun Protection Factor (SPF). http://www.iso.org/iso/catalogue_detail.htm?csnumber=46523.
Ministry of Health, Welfare and Labour. http://www.mhlw.go.jp/english; http://www.mhlw.go.jp/english/topics/cosmetics/index.html; http://www.mhlw.go.jp/english/org/policy/p13-14.html.
Ministry of Health and Welfare of Japan, Notification No.1339, December 28, 2000.
Japanese Cosmetic Industry Association, Application for Japanese Labelling Names. http://www.jcia.org/n/en; http://www.jcia.org/n/jcia/e/.
Standards for Cosmetics MHW–Ministry of Health and Welfare, 2000. Notification 331/2000. http://www.mhlw.go.jp/file/06-Seisakujouhou-11120000-Iyakushokuhinkyoku/0000032704.pdf.

Kingdom of Saudi Arabia
Saudi Arabia Standards Organization (SASO). http://www.saso.gov.sa/en/pages/default.aspx.
Saudi Food and Drug Authority (SFDA). http://www.sfda.gov.sa; https://ecosma.sfda.gov.sa/home.aspx?enc=9Zo/OcjXLUTAdEFdDlLN8A==.

People's Republic of China
Administration of Quality Supervision, Inspection and Quarantine (AQSIQ). http://english.aqsiq.gov.cn/.
China Food and Drug Administration (CFDA). http://eng.sfda.gov.cn/WS03/CL0755/; http://eng.sfda.gov.cn/WS03/CL0772/; http://www.sfda.gov.cn/directory/web/WS01/images/MjAxNcTqtdoyNji6xbmruOa4vbz+LnBkZg==.pdf; http://www.sfda.gov.cn/WS01/CL1272/140365.html.

Republic of Korea

Korean Cosmetic Act. http://www.moleg.go.kr/english/korLawEng;jsessionid=uBgnlOEoJ1C6kPgmdt4i7PoreMROZOHJvh1sV9az30A3La7OeKRxUb8imPtKqCis.moleg_a2_servlet_engine2?pstSeq=58340&pageIndex=2; http://www.mfds.go.kr/eng/eng/index.do?nMenuCode=16&searchKeyCode=131&page=1&mode=view&boardSeq=67029.

Korean Pharmaceutical Trade Association (KPTA). http://www.kpta.or.kr/eng/main/main.asp.

Russian Federation

Technical Regulation (TR) on the Safety of Cosmetic Products for Customs Union (TR CU 009/2011), 2011. http://customsunioncertificate.com/wp-content/uploads/2014/06/TR-TC-009-Perfumery-Cosmetics-ENG.pdf; http://www.rospotrebnadzor.ru/en/; http://www.gost-r.info/news-2012-09-13.php; http://ww.fsa.gov.ru; http://www.eurasiancommission.org/en/Pages/default.aspx.

South Africa

Advertising Standards Authority of South Africa. http://www.asasa.org.za/.

South African Bureau of Standards. https://www.sabs.co.za/.

South African Cosmetics, Toiletry, and Fragrance Association, Cosmetics Compendium. http://www.ctfa.co.za/compendium; http://www.ctfa.co.za/understanding-cosmetic-product-labelling.

The Foodstuffs, Cosmetics and Disinfectants Act. http://www.nda.agric.za/vetweb/Legislation/Other%20acts/Act%20-%20Foodstuffs,%20Cosmetics%20and%20Disinfectants%20Act-54%20of%201972.pdf.

Switzerland

Regulations Available (in French, German, Italian). Available at: https://www.admin.ch/opc/it/official-compilation/2017/249.pdf; https://www.admin.ch/opc/it/official-compilation/2017/283.pdf; https://www.admin.ch/opc/it/official-compilation/2017/1597.pdf; https://www.admin.ch/opc/it/official-compilation/2017/1633.pdf; https://www.blv.admin.ch/2010_final_FR.pdf.

Regulation on VOC. http://www.admin.ch/ch/i/rs/8/814.018.it.pdf.

United States

California Proposition 65. http://oehha.ca.gov/; http://oehha.ca.gov/prop65/background/p65plain.html.

California Safe Cosmetics Program. http://www.cdph.ca.gov/programs/cosmetics/Pages/default.aspx.

Electronic Code of Federal Regulations, Title 21. https://www.accessdat.fda.gov/scripts/cdrh/cfdocs/cfcfr/cfrsearch.cfm.

Federal and State VOC Regulations. http://www.epa.gov/indoor-air-quality-iaq/volatile-organic-compounds-impact-indoor-air-quality; http://www.issa.com/data/moxiestorage/regulatory_education/voc_limits_summary_8-12-15.pdf; http://www.arb.ca.gov/consprod/regs/regs.htm.

Guidance for Industry: Safety of Nanomaterials in Cosmetic Products. http://www.fda.gov/Cosmetics/GuidanceRegulation/GuidanceDocuments/ucm300886.htm.

Regulations and Guidelines are Available at the Specific FDA Websites. https://www.fda.gov/aboutfda/centersoffices/officeoffoods/cfsan/; https://www.fda.gov/Cosmetics/; http://www.fda.gov/Drugs/DevelopmentApprovalProcess/DevelopmentResources/Over-the-CounterOTCDrugs/StatusofOTCRulemakings/default.htm; https://www.fda.gov/aboutfdacentersoffices/officeofmedicalproductsandtobacco/cder/.

The Fair Packaging and Labeling Act, 1967, 15 U.S.C. 1451-1461. https://www.ftc.gov/enforcement/rules/rulemaking-regulatory-reform-proceedings/fair-packaging-labeling-act.

Personal Care Product Council. http://www.personalcarecouncil.org/.

CHAPTER 2

Quality Control of Cosmetic Products: Specific Legislation on Ingredients

Eliseo F. González Abellán, Desirée Martínez Pérez
Regional Government of Comunidad Valenciana, Valencia, Spain

QUALITY CONTROL OF COSMETIC PRODUCTS

As was mentioned in the first edition of this book (Fernández de Córdoba and González Abellán, 2007), legislation concerning cosmetic products in the main markets worldwide, such as the European Union (EU), the United States of America, Japan and South Korea, demands the assurance of three very important features, namely safety, efficacy and quality of cosmetic products, as is the case for pharmaceuticals or foods.

General aspects of current legislation in the different countries, including label information requirements, were dealt with in Chapter 1.

As indicated in Chapter 1, manufacturers must have enough data available to assure cosmetic product safety under the normal conditions of use. Data can either be obtained specifically on the finished products or be deduced from the properties of their ingredients. Moreover, data can be obtained through different studies (toxicology, sensitivity, allergic reactions, etc.), some of which are commented on in Chapter 17. Sometimes, surveillance of cosmetics in use can be requested to detect possible side effects.

Likewise, as mentioned in Chapter 1, manufacturers must have enough data available to demonstrate cosmetic efficacy (fulfilling that claimed on the label). These can be obtained through different studies (measurements on subjects after using the product: hydration, roughness, firmness, elasticity of the skin, other measurements such as the determination of the sun protection factor of a product, etc.) as can be seen in the guidelines published by Cosmetic Europe (2008).

Both safety and efficacy have to be considered under the following conditions:
- the final product must accord with the composition designed by the manufacturer and be in a perfect state;
- the cosmetic has to be applied by the user under the normal given conditions.

Another very important feature of cosmetic products is their quality and this requires thorough control.

Sometimes it is not easy to differentiate between quality and safety problems. Both could cause adverse effects on users; however, the origin is different. Experience shows that quality problems affect specific batches, which have to be withdrawn from the market, whereas if there is a safety problem it affects all the batches. This is because in the latter case, product use has proven to have toxicological effects on users, thus it is a design failure of the product in question, which must then be completely withdrawn from the market.

Often, quality failures are so evident (for example, separation of phases, rarefaction of the fat phases, etc.) that users will realize that they should not apply the product. On the other hand, some quality failures do not cause adverse effects on users. For example, an error in the label does not usually cause adverse effects, although it may in certain cases; for instance, the wrong sun protection factor labelled on a sunscreen product could give rise to solar erythema in users who would trustingly overexpose themselves to the sun.

Differences between quality and safety failures are shown in Fig. 2.1 using the following example: Let us suppose that some dermatological infections have occurred and the authorities are searching for the origin. Several cases can be given that could have caused adverse effects on the user, but the origins of the defect are different:

Case 1: The amount of antimicrobial preservatives in the cosmetic formulation has not been calculated properly in the product design, and the cosmetic is not preserved well enough. This is an example of a safety problem; the manufacturer must modify the formulation, and all the batches that were put onto the market will have to be withdrawn.

Case 2: The product should contain a specific amount of antimicrobial preservatives (according to its formulation), but owing to a production failure several batches were

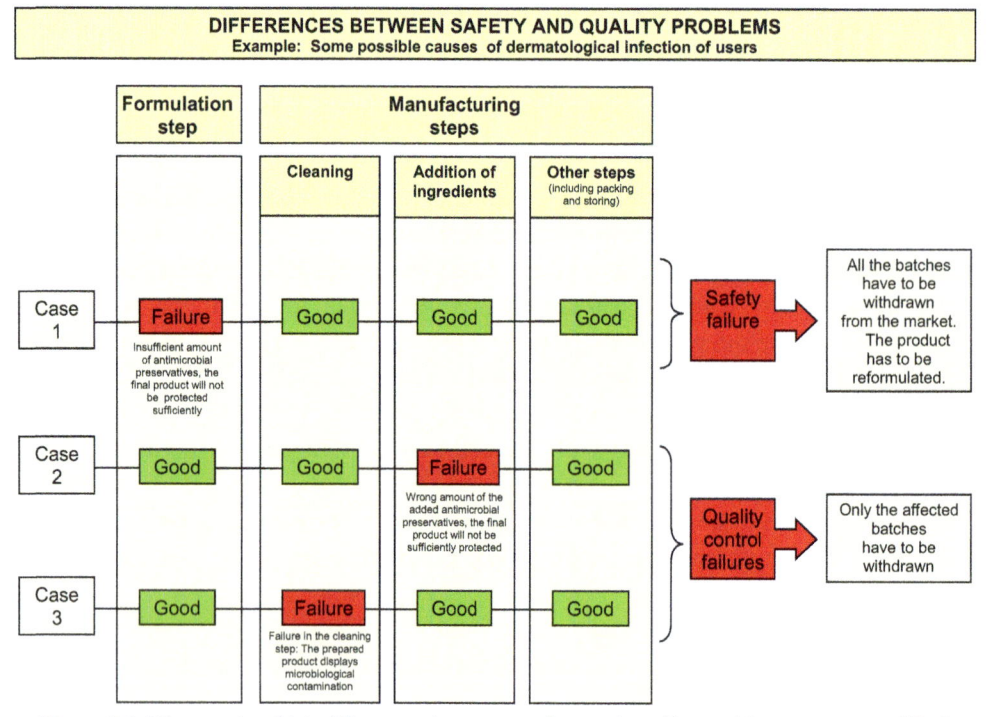

Figure 2.1 Diagram in which differences between safety and quality problems are exemplified.

produced without the correct dose. This is an example of a quality problem; the cosmetic product is well formulated but the operator in charge made a mistake and added an insufficient amount of preservatives in several batches. Only the affected batches will have to be withdrawn from the market.

Case 3: The product has the correct amount of preservatives but there was a failure during the cleaning steps of production and several batches were damaged. This is also a quality problem; failure to maintain proper standards of cleanliness in the factory plant gave rise to a microbiological increase, causing contamination. Only the contaminated batches have to be withdrawn.

Proper quality control of the manufacturing process or of the final product would avoid the aforementioned quality problems. To this end, quality control requires the manufacturing laboratory (or an external laboratory) to do the following:

- The lab must use appropriate chemical, physicochemical, biological or microbiological analytical procedures to control production. Precision and accuracy of the applied procedures have to be known. These methods must be modified according to new scientific research and advances.
- The stability and good preservation of the final product must be assured through the necessary assays. The expiry date of the product must also be considered.

Some countries have established specific practices for the manufacture of cosmetic products, usually named 'good manufacturing practices', such as those proposed by the International Organization for Standardization (ISO 22716:2007) to avoid possible problems or errors in each and every step of the manufacturing process. By following these rules, one will obtain a final product with the expected quality. The final product must have a constant and specific qualitative and quantitative composition.

Moreover, authorities can carry out analytical controls of commercial products or raw materials, packing, preservation conditions, etc., to assure the quality of the finished product.

SPECIFIC LEGISLATION ON INGREDIENTS

One of the main aspects to be considered in quality control of cosmetic products concerns the substances they contain, i.e., cosmetic ingredients. Different authorities have used the ingredients as a way to put in place more direct or indirect controls on cosmetic products, thus enabling them to regulate and manage the market.

In more developed countries, different strategies have been adopted to classify cosmetic products, thus enabling gradual requirements to be established in terms of different aspects such as:

- legal requirements, for example, to prohibit narcotic and/or psychotropic substances, whose trade is regulated by international treaties, or to classify new or toxic substances that could be allowed/restricted in the formulation of cosmetic products;
- public health requirements, such as sunscreen products to prevent sunburn, pediculicidal products, etc.;

- products considered in principle as cosmetics, but which could exert a marked pharmacological effect;
- toxicological aspects.

Moreover, the competent authorities also establish requirements with regard to labelling the ingredient composition, warnings about the presence of certain ingredients, etc.

The aim of this chapter is to give a specific overview of the requirements imposed by the main legislators on cosmetics (EU, United States, Japan and South Korea) in terms of allowed, restricted or prohibited ingredients and their labelling requirements and other special requests.

Readers can also find more detailed information on specific ingredients and/or groups of ingredients (UV filters, tanning and whitening agents, colouring agents, hair dyes, preservatives, perfumes, allergenic fragrances, surfactants and other) in the following chapters of this book.

International Nomenclature of Cosmetic Ingredients

The use of cosmetic products is on the increase around the world. As described in Chapter 1, there are different organs regulating the manufacture of this type of product in different countries. An increasing number of new cosmetic products are appearing on the market with new or improved properties, implying the use of new chemical substances. This means that a great many substances are employed in cosmetic products worldwide. To avoid language barriers that may promote problems of free trade, and may also confuse consumers, it is necessary to harmonize the nomenclature of the substances employed in cosmetics.

The former Cosmetic, Toiletry and Fragrance Association (CTFA), currently The Personal Care Products Council, was a pioneer in trying to harmonize cosmetic nomenclature following the guidelines of the US Food and Drug Administration (FDA). In a survey carried out on cosmetic companies in the late 1960s, they realized that the same chemical was identified by different trade and chemical names. They then created a committee comprising industrial experts in the fields of chemistry, cosmetic science and technology, as well as members of the American Medical Association, the US Adopted Names Council and the FDA. In 1973 the CTFA published the first edition of the *CTFA Cosmetic Ingredient Dictionary* (Estrin, 1973) in which the substances employed in cosmetic products were described by their CTFA Adopted Names. Afterwards, the FDA cited this dictionary as the primary source of nomenclature for cosmetic product labelling. Later, in 1993, because of the strong repercussions it had in different countries around the world, the designation was changed from CTFA Adopted Names to International Nomenclature of Cosmetic Ingredients (INCI), as it is known nowadays (Gottschalck and McEwen, 2006).

This nomenclature was officially adopted by other legislations on cosmetic products throughout the world, i.e., in the EU and Japanese frameworks, in 1996 and 2001, respectively, although a few discrepancies can be observed in the cases of colouring agents, botanical extracts, and a few trivial names. Nevertheless, the Personal Care Products Council is working closely with Cosmetics Europe and with the Japan Cosmetic Industry Association to harmonize these final discrepancies.

The INCI names may be assigned only by the Personal Care Products Council's International Nomenclature Committee (INC). To insert a new substance into the dictionary, an application needs to be addressed to the Personal Care Products Council, which, after a preliminary review, will be submit it to the INC; then an INCI name is assigned based on the chemical structure and composition and is published in the next edition of the dictionary and in the INCI Application website.

Now, the *International Cosmetic Ingredient Dictionary and Handbook* is in its 16th edition (Nikitakis and Lange, 2016), and incorporates more than 22,600 ingredients. Nevertheless, it should be emphasized that this dictionary does not represent a positive list of the cosmetic ingredients that appear here. The inclusion of any chemical means only that this chemical is or was sold for use in cosmetic products, and does not imply that the substance is safe for use as a cosmetic ingredient, nor does it indicate that its use as a cosmetic ingredient complies with the laws concerning cosmetic products. On the other hand, the absence of a chemical substance from this list does not imply that this substance may not be used in cosmetic products. In this sense, when a cosmetic product is going to be marketed in a certain domain, manufacturers have to consult the specific legislation in force on cosmetic products of the country.

As already mentioned, the use of a harmonized nomenclature not only helps free trade, but also makes it easier for consumers and the medical community to act when a dermatological problem arises.

Next, we will summarize the legislation concerning cosmetic ingredients set out by the three main legislative bodies worldwide.

Cosmetic Ingredients in the European Union Framework

Regulation (EC) 1223/2009 of the European Parliament and of the Council of 30 November 2009 on cosmetic products (henceforth EU Regulation) established that the European Commission should adapt the annexes of the regulation (lists of prohibited substances, restricted substances, colourants, preservative substances and UV filters) to technical and scientific progress. In addition, the same regulation defines a cosmetic ingredient as '*any substance or mixture intentionally used in the cosmetic product during the process of manufacturing*'. Nevertheless, according to Article 19, impurities in the raw materials are not considered ingredients, nor are the subsidiary technical materials used in the preparation but not present in the final product. Perfume and aromatic compositions and their raw materials shall be referred to by the terms *parfum* or *aroma*; moreover, the presence of substances, the mention of which is required under the column *Other* in Annex III, shall be indicated in the label of the product.

According to Article 33 of the EU Regulation, the Commission should compile a glossary of common cosmetic ingredient names to ensure uniform labelling and to facilitate the identification of the cosmetic ingredients. This glossary is compiled in the database named CosIng (Cosing, website). The following fields can be found:
- INCI name
- international nonproprietary name
- name in the European pharmacopoeia (where applicable)
- Chemical Abstracts Service number
- European Inventory of Existing Commercial Chemical Substances number or European List of Notified Chemical Substances name (where applicable), International Union of Pure and Applied Chemistry name
- restrictions
- usual function as used in cosmetic products

Nevertheless, it should be emphasized that this list does not constitute a positive list of the substances authorized for use in cosmetic products that appear here.

The lists that reflect regulatory aspects in the EU framework are, as already mentioned in Chapter 1, the various annexes of the EU Regulation.
- Annex II is a negative list of substances (or families of substances) that are banned for use in the composition of cosmetic products.
- Annex III gives ingredients which cosmetic products may only contain subject to the restrictions and conditions established therein.
- Annexes IV, V and VI are positive lists of colouring agents, preservatives and UV filters, respectively, permitted for use in cosmetic products within the limits and under the conditions therein.

These lists are not closed, and are permanently updated according to the data provided by the Scientific Committee on Consumer Safety (SCCS) (formerly known as the Scientific Committee of Cosmetic and Non-food Products Intended for Consumers), in response to technical progress and/or concerns about the impacts of particular ingredients on safety, taking the final decision on the addition or removal of substances to or from the lists by the Commission and the Member States.

Table 2.1 shows the number of substances considered in these annexes at the time this chapter was written.

Moreover, the EU Regulation, in its Article 15, prohibits the use in cosmetic products of the substances classified as carcinogenic, mutagenic or toxic for reproduction, of categories 1A, 1B and 2, under Part 3 of Annex VI of the Regulation (EC) 1272/2008, which regulates the classification, packaging and labelling of dangerous substances placed on the market in the Member States of the EU. Nevertheless, a substance classified in category 2 may be used in cosmetics if the substance has been evaluated by the SCCS and found acceptable for use in cosmetic products.

Table 2.1 Number of substances or groups of substances considered in Annexes II to VI of the EU Regulation (Data as of 4 September 2016)

Annex	Type of substance	Prohibited substances	Restricted substances
II	General	1378	
III	General		287
IV	Colourants allowed		153
V	Preservatives allowed		59
VI	UV filters allowed		28

Despite all these lists regulating the substances prohibited or restricted in cosmetics, it is worth mentioning that confusion could arise among manufacturers, because there are substances that are not listed as such, but are included in any one of the listed families of substances. This is the case, for example, of cocaine (a narcotic drug), which is not listed as such but is included under Entry 306 of Annex II, which lists '*Narcotics, natural and synthetic: All substances listed in Tables I and II of the single Convention on narcotic drugs signed in New York on 30 March 1961*'. This example shows the importance of knowing legislation concerning cosmetics appropriately.

On the other hand, considerations on packaging and labelling of cosmetic products were discussed in Chapter 1; however, those concerning cosmetic ingredients are described here in depth.

Taking into account Article 19 of the EU Regulation, Member States have to take all measures necessary to ensure that the cosmetic products marketed in their territory are labelled with a list, preceded by the word *Ingredients*, of ingredients in descending order of weight at the time they were added. Those ingredients whose concentration is less than 1% may be listed in any order after those in concentrations of more than 1%, and colouring agents may be listed in any order after all the other ingredients, in accordance with the colour index if appropriate. Moreover, in the event of decorative cosmetic products marketed in several colour shades, all colouring agents used in the range may be listed, provided that the words *may contain* or the symbol '+/−' is added. In addition, all warnings for any ingredients used that are described in Annexes III, IV, V and VI of the EU Regulation must also appear on the label of the product. Where that is impossible for practical reasons, an enclosed leaflet, label, tape or card must contain the ingredients to which the consumer is referred either by abbreviated information or by the symbol given in Annex VII, which must appear on the packaging.

In the event of perfume and aromatic compositions and their raw materials, as indicated earlier, they have to be referred to by the word *parfum* or *aroma*. However, the presence of aromatic substances, for which there is a special mention according to Annex III of the EU Regulation, must be declared in the labelling list irrespective of their function in the product. This is the case for the 26 potentially allergenic fragrance compounds, which according to the aforementioned annex must be indicated when their content in the finished product is higher than 0.001% in leave-on products and 0.01% in rinse-off products.

Cosmetic Ingredients in the United States

In the US framework, regulations published by the FDA concerning cosmetic products can be found in Title 21 of the Code of Federal Regulations (CFR) Parts 701, 710, 720 and 740, which states that '... *"ingredient" means any single chemical entity or mixture used as a component in the manufacture of a cosmetic product*'.

Title 21 CFR Part 701.3 about cosmetic labelling establishes that each cosmetic package has to bear a list, in descending order of predominance, naming each ingredient, except that fragrance or flavour may be listed as *fragrance* or *flavor*. If it is not possible to declare this on the package for practical reasons, the declaration may appear on a firmly affixed tag, tape or card. Similar to the EU Regulation, a permitted alternative is to list those ingredients, other than colouring agents, present at a concentration greater than 1%, in descending order of predominance, followed by those (other than colouring agents) present at a concentration of not more than 1% without respect to order of predominance. However, all these could also be joined together and listed in order of predominance, and finally followed by colouring agents, without respect to the order of predominance. In the event of shaded products or products with similar composition and intended for the same use, colouring agents may be included in the label even they are not in the cosmetic, provided the phrase *may contain* followed by the colouring agent name is written. The term *and other ingredients* at the end of the ingredient declaration will replace the name of ingredients that the FDA has authorized the company to exclude from the label for confidentiality purposes, according to 21 CFR Part 720.8.

In the event that there is a current or anticipated shortage of a cosmetic ingredient, alternative ingredients may be used. These must be declared either immediately after the normally used ingredient it replaces with the word *or*, or following the declaration of all normally used ingredients after the phrase *may also contain*.

The incidental ingredients that could be present in a cosmetic product at insignificant levels and that have no technical or functional effect in the cosmetic need not be declared on the label. This is the case of substances that have no technical or functional effect in the cosmetic but are present because they have been incorporated into the cosmetic as an ingredient of another cosmetic ingredient. Nor is it necessary to declare on the label substances that are added to a cosmetic during manufacture for technical and functional effects, but are removed before the cosmetic product is packaged in its finished form, or are converted into the same substances as those constituents of declared ingredients, without significantly increasing the concentration of these constituents, or are present in the finished cosmetic at insignificant levels and do not have any technical or functional effect on the cosmetic.

When a cosmetic product is also considered an over-the-counter (OTC) drug product (see Chapter 1), the active ingredients are to be listed first of all, after the phrase *Active ingredients*, and the quantity of each one must also be declared. The rest of the ingredients will be listed next, after the phrase *Inactive ingredients* according the rules listed earlier.

The cosmetic ingredients have to be identified in the declaration of ingredients by the name specified in 21 CFR Part 701.30, where only eight ingredients are listed (a few chlorofluorocarbon derivatives and ethyl esters of hydrolysed animal protein). In the event of not-listed ingredients the source to be employed will be the Personal Care Products Council's *International Cosmetic Ingredient Dictionary and Handbook*. In the unusual case that an ingredient does not appear in this database, other sources such as the US Pharmacopeia, the National Formulary, the Food Chemicals Codex and finally the United States Adopted Name will be consulted in this order of preference. If the ingredient does not appear in any of the aforementioned databases, the name generally recognized by consumers will be used, or finally the chemical or other technical name or description.

However, none of the aforementioned sources constitutes a list of substances allowed for use in cosmetic formulations. As described in Chapter 1, the FDA lists only a small number of strictly regulated or prohibited ingredients, which are summarized in Table 2.2, and also have positive lists for colouring ingredients.

One of these colouring ingredient lists (21 CFR Part 74) shows those colouring agents subject to batch certification of composition and purity by the FDA. These colouring agents are synthetic organic chemicals, and are usually referred to as 'coal-tar' colouring agents, owing to their original source in the 19th century. On the other hand, the other positive list (21 CFR Part 73) comprises colourants obtained primarily from mineral, plant or animal sources. These last colour additives are exempt from batch certification and although they are free of such testing, manufacturers must assure that each colouring agent complies with the identity, specifications, labelling requirements, use and restrictions of colouring agent regulations. Nevertheless, with the exception of coal-tar hair dyes, all colour additives, whether they are subject to certification or not, must be approved by the FDA for their intended use, otherwise the cosmetics containing them will be considered adulterated. So manufacturers need to check the aforementioned lists to determine whether a colouring agent is approved for an intended use and whether it is subject to certification requirements.

Moreover, as mentioned in Chapter 1, various products considered cosmetics in the EU are considered OTC drugs in the United States. These include antiperspirants, sunscreens, anticaries toothpastes and antidandruff and antiseborrheic products, among others. For each of these product types, the FDA has published a monograph containing positive lists of the active ingredients that can be employed (FDA, Title 21).

Table 2.2 Prohibited or restricted substances in cosmetic products by the FDA (in alphabetical order)

Substance	Restriction/prohibition	Cause	FD&C Act
Bithionol	Prohibited	Photo-contact sensitization	21 CFR 700.11
Cattle material (prohibited)	Cosmetics may not be manufactured from, be processed with, or otherwise contain prohibited cattle materials. These materials include specified risk materials, material from non-ambulatory cattle, material from cattle not inspected and passed, or mechanically separated beef. Prohibited cattle materials do not include tallow that contains no more than 0.15% insoluble impurities, tallow derivatives, and hides and hide-derived products, and milk and milk products	To protect against bovine spongiform encephalopathy, also known as 'mad cow disease'	21 CFR 700.27
Chlorofluorocarbon propellants	Prohibited in aerosol products		21 CFR 700.23
Chloroform	Prohibited	Animal carcinogenicity and likely hazard to human health; the regulation makes an exception for residual amounts from its use as a processing solvent during manufacture, or as a by-product from the synthesis of an ingredient	21 CFR 700.18
Halogenated salicylanilides (di-, tri-, metabromsalan and tetrachlorosalicylanilide)	Prohibited	They may cause serious skin disorders	21 CFR 700.15
Hexachlorophene	It may be used only when no other preservative has been shown to be as effective; the concentration in a cosmetic may not exceed 0.1%, and it may not be used in cosmetics that are applied to mucous membranes	Toxic effect and ability to penetrate human skin	21 CFR 250.250

Mercury compounds	65 mg/kg of metallic mercury in eye area cosmetics when no other effective and safe preservative is available for use; <1 mg/kg of metallic mercury in other area cosmetics when unavoidable under conditions of good manufacturing practice	Absorption through the skin on topical application and tendency to accumulate in the body; may cause allergic reactions, skin irritation or neurotoxic manifestations	21 CFR 700.13
Methylene chloride	Prohibited	Animal carcinogenicity and likely hazard to human health	21 CFR 700.19
Sunscreens	Use of the term 'sunscreen' or similar sun protection wording in a product's labelling generally causes the product to be subject to regulation as a drug or a drug/cosmetic, depending on the claims; however, sunscreen ingredients may also be used in some cosmetic products to protect the products' colour; the labelling must also state why the sunscreen ingredient is used, for example, 'Contains a sunscreen to protect product color'; if this explanation is not present, the product may be subject to regulation as a drug		21 CFR 700.35
Vinyl chloride	Prohibited in aerosol products	Carcinogenicity and other problems	21 CFR 700.14
Zirconium-containing complexes	Prohibited in aerosol products	Toxic effect on lungs, including the formation of granulomas	21 CFR 700.16

FD&C Act, Federal Food, Drug, and Cosmetic Act.

Thus, regarding cosmetic products as such, excepting colouring agents, any substance could be used as a cosmetic ingredient except those few listed in Table 2.2, but under the responsibility of the manufacturer. Qualitative and quantitative formulas have to be available in case of FDA inspection. Nevertheless, manufacturers can send the FDA data about cosmetic product ingredients voluntarily under the Voluntary Cosmetic Registration Program.

Nevertheless, there are different cosmetic and fragrance trade associations that have recommended eliminating or limiting maximum levels of various ingredients taking into account the health issue. For example, the Cosmetic Ingredient Review (CIR) Expert Panel, an independent panel of scientific experts established by the Personal Care Products Council, with support of the FDA, thoroughly reviews and assesses the safety of numerous ingredients used in cosmetics and publishes the results yearly in open, peer-reviewed scientific literature (CIR website). According to its Annual Reports, the CIR has classified ingredients according their safety profile. So, it lists substances as safe at certain concentrations of use, substances as safe with certain restrictions and substances whose safety is not well documented and are classified as unsafe. Its webpage gives a detailed list of all findings. The output of the CIR has no legal weight, because the final decision belongs to the FDA.

In a similar way, the International Fragrance Association establishes usage guidelines for cosmetic ingredients related to fragrance products (IFRA website). In its Code of Practice, available online, can be found recommendations for avoiding many ingredients, but once again the final decision belongs to the FDA.

Other substances to be considered are those not added intentionally, but formed by the reactions between different ingredients, whether during manufacture or storage, as is the case of nitrosamines. These hazardous substances can be formed by the reaction of amines with nitrosating agents [such as sodium nitrite or preservatives like 2-bromo-2-nitropropane-1,3-diol, 5-bromo-5-nitro-1,3-dioxane or tris(hydroxymethyl) nitromethane]. The FDA expressed its concern about the contamination of cosmetics with nitrosamines in a Federal Register notice dated 10 April 1979, which stated that cosmetics containing nitrosamines may be considered adulterated and subject to enforcement action.

Japan

Until recently, Japan had a positive list (the Comprehensive Licensing Standards) system under which each ingredient used in a cosmetic formulation had to be pre-approved by the Ministry of Health, Labour and Welfare (MHLW). Since a deregulation process in April 2001 (see Chapter 1), and with regard to cosmetic ingredients, the establishment of negative and positive lists like in the EU has been carried out by the MHLW in the Standards for Cosmetics notification (MHLW, 2000). In this sense, Japan adopted a list of around 30 prohibited ingredients (Appendix 1 of the aforementioned notification), a list containing approximately 20 restricted ingredients (Appendix 2), a positive list of more than 40 preservatives (Appendix 3) and a positive list of nearly 30 UV filters (Appendix 4).

Nevertheless, a positive list of colouring agents has been regulated since 1966 (MHLW, 1966).

According to the Standards for Cosmetics, '*cosmetics shall not contain any medical drug ingredients (excluding those used only as additives and those listed in Appendix 2 through 4), or any ingredients that do not meet the Standards for Biological Materials (Ministry of Health, Labour and Welfare Notification No. 210 of 2003), or any of the materials listed in Appendix 1*'. Moreover, cosmetic products will not be allowed to contain any of the ingredients listed in Appendix 2 at amounts higher than those specified in the aforementioned list, and no other preservative included in Appendix 3 nor other UV filters included in Appendix 4 under the conditions established therein.

Moreover, full ingredient labelling must be provided for cosmetics using the INCI names translated or transliterated into Japanese, in descending order of predominance.

On the other hand, as has been seen in Chapter 1, different products considered cosmetics in the EU are considered quasi-drugs in Japan, for example, hair dyes, permanent waving products, depilatories, deodorants and antidandruff shampoos, among others. For this type of product, the deregulation process of cosmetics in 2001 and the changes it involved are not applied, and a pre-market approval is needed from the MHLW. Moreover, registration of all ingredients used in product manufacture, as well as product safety data which specify the active ingredients, usage and dosage, indications or effects, is also required. Full lists of approved quasi-drug ingredients are not published, although the MHLW has published lists of ingredients approved for use in certain categories, such as hair dyes, permanent waving agents, medicated toothpaste and bath preparations. A full ingredient listing is not required for quasi-drugs; however, the MHLW has listed 138 ingredients that must be indicated on the label.

South Korea

In South Korea, cosmetics products and their ingredients are regulated by the Korea Food and Drugs Administration (KFDA). The Korean Cosmetic Act created it in 1999 and it incorporates a set of rules that encompass the fields of manufacturing, importing and selling cosmetics products.

According the Korean Cosmetic Act, it is necessary that the ingredients that form part of the cosmetic product are appointed or are noted as a cosmetic ingredient by the Commission of the KFDA, otherwise the Commission of the KDFA should evaluate the specifications, the standards and the security of the cosmetic ingredient.

With the purpose of assurance of manufacturing and security, Article 8 of the Korean Cosmetic Act defines the creation of the Security Standards. In these standards are the lists of prohibited ingredients and of ingredients with restrictions and positive lists of preservatives, colourants and UV factors. In the same document we find the specifications and assays for the more common raw materials that are used in cosmetics products.

SUMMARY

The main similarity between the four main legislations regarding cosmetic products, i.e., those in force in the EU, the United States, Japan and South Korea concerning cosmetic ingredients, lies in the fact that pre-market approval is unnecessary, and thus, full responsibility for the safety of products, which depends mainly on the ingredients, falls on the manufacturer. Meanwhile, authorities carry out in-market surveillance to check compliance with each one of the legislations.

Another similar point is that INCI nomenclature is widely used in regulatory systems, avoiding consumer confusion and improving the marketing of the products between domains.

In terms of the major difference between these four regulatory systems we can see that while the EU, Japan and South Korea maintain lists of prohibited and restricted substances, together with positive lists for colouring agents, preservatives and UV filters, in the United States there are no positive lists for cosmetic ingredients (except for colouring agents) and the list for prohibited ingredients is somewhat shorter than in the other three legislations. Moreover, the European, Japanese and South Korean lists are not identical and some ingredients that are prohibited or restricted in one market are permitted in another. Furthermore, in some cases the authorized content for the same permitted ingredient differs from one legislation to another.

Finally, different products considered cosmetics in the EU are considered OTC drugs in the United States, and similarly, in Japan there is another category that is named quasi-drugs. These product categories need pre-market approval of the competent authority and obey different regulations compared to cosmetics. In addition, in the case of OTC drugs in the United States, the FDA has published different monographs which compile specific data regarding this type of product with regard to the ingredients allowed and their maximum contents. In contrast, the Japanese MHLW discloses ingredients only for hair dyes, permanent waving agents, medicated toothpaste and bath products. Consequently, it may be difficult for manufacturers to identify which ingredients are approved for which uses.

REFERENCES

CIR – Cosmetic Ingredient Review. http://www.cir-safety.org.
CosIng – Inventory of Ingredients. https://ec.europa.eu/growth/sectors/cosmetics/cosing/inventory_es.
Cosmetics Europe, 2008. Guidelines for the Evaluation of the Efficacy of Cosmetic Products. www.cosmeticseurope.eu/publications-cosmetics-europe-association/guidelines.html?view=item&id=23.
Estrin, N.F. (Ed.), 1973. Cosmetic Ingredient Dictionary. CTFA – Cosmetic, Toiletry, and Fragrance Association, Washington, DC.

FDA – Food and Drug Administration. Code of Federal Regulations, Title 21, Parts 70–82 for Colorants; Parts 330–360 for OTC drugs; Parts 700–740 for Cosmetics. http://www.accessdata.fda.gov/scripts/cdrh/cfdocs/cfcfr/cfrsearch.cfm.

Fernández de Córdoba Manent, B., González Abellán, E.F., 2007. Quality control of cosmetic products. Specific legislation on ingredients. In: Salvador, A., Chisvert, A. (Eds.), Analysis of Cosmetic Products, first ed. Elsevier.

Gottschalck, T.E., McEwen, G.N. (Eds.), 2006. International Cosmetic Ingredient Dictionary and Handbook, tenth ed. CTFA – Cosmetic, Toiletry and Fragrance Association, Washington, DC.

IFRA – International Fragrance Association. http://www.ifraorg.org.

ISO 22716:2007, 2007. Cosmetics – Good Manufacturing Practices (GMP) – Guidelines on Good Manufacturing Practices. https://www.iso.org/obp/ui/#iso:std:iso:22716:ed-1:v2:en.

MHLW – Ministry of Health and Welfare, 1966. Ordinance No. 30/1966: Ordinance to Regulate Coal-Tar Colors Permitted for Use in Drugs, Quasi-drugs and Cosmetics (As Amended by Ordinance No. 55/1972 and Ordinance No. 126/2003).

MHLW – Ministry of Health and Welfare, 2000. Notification No. 331/2000, Standards for Cosmetics. http://www.mhlw.go.jp/english/topics/cosmetics/index.html.

Nikitakis, J., Lange, B. (Eds.), 2016. International Cosmetic Ingredient Dictionary and Handbook, sixteenth ed. The Personal Care Products Council, Washington, DC.

Regulation (EC) No 1272/2008 of the European Parliament and of the Council of 16 December 2008 on Classification, Labelling and Packaging of Substances and Mixtures, Amending and Repealing Directives 67/548/EEC and 1999/45/EC, and Amending Regulation (EC) No 1907/2006.

Regulation (EC) No 1223/2009 of the European Parliament and of the Council of 30 November 2009 on Cosmetic Products.

PART II

Analytical Methods for Monitoring Ingredients and Quality Control of Cosmetics Products

CHAPTER 3

Selection From Bibliographic Resources of an Analytical Method for Cosmetic Products: Validation of a Method

Juan G. March[1], Amparo Salvador[2], Alberto Chisvert[2]
[1]University of Balearic Islands, Palma de Mallorca, Spain; [2]University of Valencia, Valencia, Spain

INTRODUCTION

As already mentioned in Chapters 1 and 2 of this book, the regulations of various countries deal with both the currently banned ingredients and those restricted to certain applications and/or that should be used at concentration levels below the stated maximum value. Taking this into account, and to avoid risks to the user of the cosmetic product, analytical methods are required to enable:

- the detection of the possible presence of prohibited substances and, if appropriate, their content in the cosmetic product;
- the determination of the content of the ingredients with restricted concentration in the cosmetic product, to verify that the limits established in the legislation are not exceeded.

In addition, it is desirable to verify that the finished product responds to the formulation initially designed, to ensure the quality and expected efficiency of the commercialized product. Therefore, analytical methods are also necessary to enable:

- the determination of the content of the authorized ingredients in the cosmetic product.

Despite the high number of substances that should be analytically controlled for these purposes, the number of official methods of analysis for cosmetic products is small. Because of this, companies in the cosmetics industry often find difficulties in selecting an appropriate method to carry out the required analytical determination for quality control using the available instrumentation. Similar problems can occur at official laboratories responsible for carrying out health inspections, which need validated analytical methods to carry out their supervisory tasks and detect errors, contamination and/or possible fraud.

Available reference methods from the corresponding agencies are itemized and specified in Chapter 4 of this book.

The implementation of protocols from the scientific literature may be advisable in some cases, especially if no reference methods have been officially adopted for the target analyte. To select the appropriate method it is important to take into account, depending on the analyte, the required information from the analytical determination and/or the available techniques in a concrete laboratory. A great number of published papers on methodologies or protocols for the chemical analysis of cosmetics products can be found in the scientific literature. The published scientific literature on analytical methods for the different types of cosmetic ingredients or contaminants is described in Chapters 5–16 of this book.

Complementarily, this chapter is focused on a general strategy to select a method from the literature using flexible web tools.

VALIDATION OF A METHOD FOR COSMETIC ANALYSIS

To analytically control a cosmetic ingredient an appropriate method should be selected taking into account the ingredient, the type of sample (matrix) and its expected level of concentration. Once an appropriate method is identified from the literature and adapted to the conditions of the laboratory (i.e., specific analytical equipment and matrix of the samples), a validation of the method shall be performed before the initial use in routine analysis to demonstrate that the chosen method produces results that are fit for the purpose. If any parameter does not meet the expected values a further development of the method and revalidation will be required.

In this sense, the Institute for Reference Materials and Measurements of the Joint Research Centre published an interesting guideline for the selection of appropriate methods for cosmetic analysis (Vincent, 2015). It is based on a review of analytical methods and the selection of the most suitable method, taking into account its validation (verification) status (i.e., official or harmonized methods, methods that have passed a ring trial, methods that have passed at least a single-laboratory validation process). In case the selected method does not fully match the actual scope (different analyte, matrix composition or analyte concentration) an additional validation should be performed. Finally, if there is not a validated method available, it will be necessary to develop and validate a new method.

Main Analytical Features to Consider in the Selection of a Method

A validation study of an analytical method should involve the establishment of the following parameters: limit of detection (LOD), limit of quantification (LOQ), linearity, precision (repeatability, intermediate precision and reproducibility), selectivity, robustness and, of course, accuracy. Unfortunately, slightly different definitions of these concepts can be found, and also different approaches can be followed for their quantification. A brief description of the mentioned parameters is given in the following paragraphs. Further information, textbooks on practical considerations and detailed calculations can be found in the literature.

Limit of Detection

This is an indicative value of the lowest concentration of analyte that can be detected with an established level of confidence.

There are several approaches to determine the LOD. It can be evaluated by multiplying the standard deviation of replicate measurements of the blank (s_B) (or a sample containing a low amount of analyte) by an appropriate factor based on statistical considerations. So, the LOD is often reported as $3.3\,s_B$ (sometimes rounded to $3\,s_B$).

For chromatographic methods, the LOD can also be determined by comparing the height of peaks from samples containing a known low concentration of analyte and the baseline noise. Then, the LOD can be defined as the concentration that provides a peak three times higher than the baseline noise level.

Limit of Quantification

This is an indicative value of the lowest concentration of analyte that can be quantitatively determined with reasonable precision and accuracy.

The LOQ is often reported as $10\,s_B$ (s_B, as indicated earlier, is the standard deviation of the blank or a low concentration sample) or, for chromatographic methods, the concentration that provides a peak 10 times higher than the baseline noise level.

Linearity

This is the capability (within a range) to obtain analytical signals that are directly proportional to the concentration of the analyte in the sample.

Linearity can be evaluated from the observation of the scatter plot of the instrument response (or analytical signal) versus the analyte concentration. A good linearity will provide a regression line with a coefficient of determination (R^2) close to 1.

Another frequently used graph to assess linearity is the plot of the residuals of each point as a function of analyte concentration. For linear ranges, this graph should be similarly scattered between positive and negative values.

Precision

This is defined as the agreement between the results of a set obtained from different aliquots of the same homogeneous sample. The sample used for these measurements must be stable and representative of the samples used in routine testing in terms of matrix composition and analyte concentration. Each repeated measurement should be obtained by application of the whole method, including sample treatment if any.

Precision can be expressed as the standard deviation or the relative standard deviation (normally expressed as %) of the obtained results. Precision can be established at three levels, i.e., repeatability, intermediate precision and reproducibility. The repeatability is the precision achieved under identical operating conditions in a short lapse of time. The intermediate precision is the precision achieved within a laboratory but under different operating conditions that should emulate, as much as possible, the routine working conditions, e.g., different operator, different instrument, different day, and so on; the intermediate precision informs on the precision in regular use. The reproducibility refers to the precision obtained from replicate measurements performed in different laboratories. So, the evaluation of the reproducibility involves an inter-laboratory study.

Selectivity

This is the ability to measure the target analyte in the presence of other components that could be potentially present in the analytical sample as matrix components, impurities or degradation products. Frequently, this information is not entirely available. In such case, proportional errors that are dependent on analyte concentration can be estimated by the analysis of real samples (containing these potentially interfering substances), using standard addition calibration and comparison of results with those obtained by conventional external calibration. More systematic selectivity studies for specific potentially interfering substances can be done by the analysis of synthetic samples prepared with known concentrations of both the analyte and the potentially interfering target substance; this allows us to detect both errors, dependent and not dependent on analyte concentration.

Accuracy

This is defined as the agreement between the assumed true value and the found value.

The accuracy is normally measured in terms of bias, which is the difference between the average of the obtained results and the accepted true value. Often, bias is expressed as a percentage (also named **relative error**) of the true value. Also, accuracy can be measured as the ratio between the found value and the true value, normally expressed as a percentage (named **recovery**).

When a reference sample with a certified (true) value is not available, the accuracy can be established using a sample that was previously analysed by a reference method, a synthetic sample with a known concentration or an analyte-free sample spiked with an appropriate and known amount of analyte. When working with spiked samples, it is very important to ensure that chemical equilibrium between the spiked amount and the sample matrix constituents is reached before any treatment is started. In all cases, the concentration of the measured sample solutions should include values close to the LOQ, to the centre of the linear range and to the upper level of the calibration graph.

Robustness

This is the capacity of the method to remain unaffected by small variations in the operating conditions (experimental variables) within a realistic range, such as the pH of a used buffer, temperature, reagent concentration, injection volume, and so on.

For the determination of the robustness a number of selected operational conditions are deliberately varied within an established range, and the influence of such variations on the results is evaluated. The method is considered robust with respect to an operational condition when the variation in the results is within a previously established range. So, the robustness test identifies the method conditions that significantly affect the results when the method is under regular usage. The robustness test can be done by considering either each variable separately or several variables at the same time according to the planned experimental design.

Additional Features for Selecting a Method

In addition to the main analytical features, the viability of an analytical method may also depend on other practical characteristics, especially if the method will be used for periodic production controls that require a considerable number of samples and/or lots to be analysed. These other additional features are:
- the cost and equipment availability: these features are especially important for small companies, so it is wise to select the analytical method depending on the available laboratory instrumentation;
- the sample throughput: tedious sample handling or time-consuming methods are characteristics of relevance, because they indirectly affect the cost;
- the safety: the ideal analytical method should be safe for both the operator and the environment, especially if a large number of samples will be processed. Analytical methods of the so-called green chemistry, replacing or reducing toxic reagents as much as possible, are desirable.

SELECTION OF THE APPROPRIATE METHOD IN COSMETIC ANALYSIS

Once the analytical problem is defined (i.e., the cosmetic product and the analyte for which the chemical analysis is required), the selection of the most appropriate analytical method can be made from the information in the different chapters of this book and additionally from databases.

Using the Information Provided by This Book

The publications on cosmetic analysis in the different chapters of this book have been comprehensively reviewed by the authors, experts in this field. Readers can find extensive and diverse information in addition to the corresponding bibliographic references in:
- Chapter 4: official methods on cosmetic analysis
- Chapters 5–14: other published methods (each chapter is devoted to a type of ingredient: UV filters, whitening agents, colouring agents, hair dyes, preservatives, perfumes and related substances, surfactants, possible contaminants, etc.)

If official methods are not available for the concrete analytical problem, the reported validation data of the selected published method must be considered. If the validation status of the method is not enough and/or does not fully match the actual scope, a validation study before its implementation in an analytical laboratory for routine monitoring is advisable.

Using the Information Provided by Databases

Several database search engines are appropriate to explore the available literature on cosmetic analysis. Scientific bases such as ScienceDirect (Elsevier), Scopus (Elsevier), SciFinder Scholar (American Chemical Society) or Web of Science (Thomson Reuters) are usually preferred but other databases such as the freely available Google Scholar can be also useful. The success of the obtained results strongly depends on the search strategy. The right searching scheme depends on the searching engine and the databases.

Next, a short introduction to the use of the mentioned web tools is described.

ScienceDirect

The **Advanced Search** is recommended to find references on a particular topic using this search machine. For this, the search words that are specific (or closely related) to the inquiry have to be elected and typed into two text boxes titled as **Search for**. These two boxes can be combined using Boolean connectors (**AND, OR** and **AND NOT**) from the drop-down menu. **AND** is used to narrow the search to references included in both text boxes, **OR** expands the search to terms in either text box, **AND NOT** is used to exclude references containing certain terms from the output. When several terms are typed in the same box all of them are searched and combined with the connector AND. Optionally, the search can be limited to **Journals, Books, Reference Works** (e.g., Encyclopaedia of Analytical Sciences) or **Images** (figures or videos). Moreover, the Science field can be selected from a drop-down list (for analysis of cosmetic products, the items **Chemistry** and **Pharmacology, Toxicology and Pharmaceutical Sciences** are appropriate) and the publication year range for the documents can also be specified. Each text box can be associated with a **Field**, such as **Abstract, Title, Keywords,** among others.

Search tips should be considered to make the search successful. Briefly, stop words (such as by, in, of, etc.) are not searchable and, consequently, they can be omitted. Typing the singular form of a word will also search for the corresponding plural and possessive forms. These functionalities can be turned off by enclosing words or phrases in double quotation marks (" ") or braces ({ }), in which case the search is focused on the exact marked text. Wild card characters, such as an asterisk '*' and a question mark '?', can be used to find variations of the search words; '*' replaces any number of characters at the end of the word and '?' replaces a single character in a word.

The displayed search results can be, subsequently, refined according to **Publication Year**, **Publication Title** (i.e., journal name), **Topic** and/or **Content Type** (i.e., journal, book or reference work).

Example

A bibliographic search focused on the determination of preservatives in cosmetics by chromatography but excluding gas chromatography can be done by combining two text boxes titled as **Search for** with the Boolean connector **AND NOT** and typing *determination of preservatives in cosmetics by chromatography* in the upper box, and *gas chromatography* in the other one. The same search results would be obtained by omitting the stop words, i.e., typing *determination preservatives cosmetics chromatography*. Also, the search result is not dependent on the order of the typed words; thus, the same search result would be obtained by typing *preservatives determination cosmetics chromatography* or *chromatography gas*.

Nevertheless, a clearly different search result would be obtained if the search scheme is *chromatographic determination of preservatives in cosmetics* AND NOT *gas chromatography*. To extend the search to chromatographic and chromatography, *chromatograph** should be typed as a search term, or better *chromatogra**, which will search for all terms with the same root (as chromatography, chromatographic, chromatogram or chromatograph).

In addition, when a search includes terms that differ by one character (or more), '?' should be used, e.g., for adsorption and absorption, *a?sorption* should be typed as the search term.

Scopus

The advanced search is also recommended. Once the search terms have been chosen, they have to be combined with the Boolean operators (**AND, OR** and **AND NOT**) and typed in the **Enter query string** box. When no operator is typed, AND is automatically inserted between terms. When several operators are used in a search, they are processed according to the order **OR, AND** and **AND NOT**. Search tips are similar to those for ScienceDirect. Thus, when entering singular nouns, the corresponding plurals are also searched; the search ignores stop words unless they are in brackets or double quotation marks; to search for an exact phrase, enclose it in braces '{ }'; if several terms must appear adjacent to one another, enclose the terms in double quotation marks (" "); to search variations of a word, the wild cards asterisk '*' and question mark '?' can be used to replace multiple characters or a single character in a word, respectively.

The obtained list of references can be refined by either limiting or excluding several search criteria, such as **Year** of publication, **Author Name**, **Subject Area** (chemistry and pharmacology, toxicology and pharmaceutics are recommended when searching for analysis of cosmetics), **Document Type** (article, review, book or conference paper), **Source Title**, **Keywords**, **Affiliation** of the authors, **Country**, **Source Type** and **Language**.

Example

The search outlined before can be done using Scopus by typing *determination of preservatives in cosmetics by chromatography* AND NOT *"gas chromatography"* (or merely *determination preservatives cosmetics chromatography* AND NOT *"gas chromatography"*) in the **Enter query string** box. Notice that if *gas chromatography* is not in double quotation marks, similar but different search results will be obtained. When double quotation marks are not used, the machine searches for gas AND chromatography; if the terms are marked, the engine selects only the documents in which gas and chromatography appear adjacently.

SciFinder Scholar

The option named **Research Topic** is convenient to find references on analysis of cosmetic products. For this, type the query into the text box in plain English, i.e., two or three concepts shared with a preposition, conjunction and other simple parts of speech. Synonyms can also be included in parentheses next to the topic. SciFinder understands which terms are research topics and how to relate the terms. Optionally, to narrow the answer, the dialogue box **Advanced Search** can be run by adding more specific search criteria such as publication year, type of document (e.g., book, journal, patent, review, …), language, author name and/or company name.

Then, a **Research Topic Candidates** list is displayed. Each candidate is a concrete combination of the concepts that SciFinder recognized from the query. So, stop words in the query affect the number of candidates. One or more candidates can be selected from this list and the corresponding bibliographic references obtained (**Get References**). Moreover, a narrower search can be obtained by means of the **Refine** function, using an additional **Research Topic** or other concepts also accessible from the **Advanced Search** already mentioned.

Web of Science

The search strategy of this database search engine is based on combining typed significant words (topics), as a search term, in the corresponding text boxes using the Boolean connectors **AND, OR** and **NOT**. The number of text boxes is not limited. In addition, the content of each text box can be associated with the desired **Search Field**, such as topic, title, author, journal or publication year, among others.

Search term rules are similar to those of ScienceDirect or Scopus. Briefly, typing singular nouns will also search for the corresponding plural, typing multiple words in a text box will search for all words (connected with AND), searches in double quotation marks will search for the exact typed text, '*' can be used to replace any number of characters at the end of the word, '?' replaces a single character and '$' denotes an eventual presence of a character (e.g., *colo$r* searches for color and colour).

The obtained answer (list of references) can be refined according to several criteria, such as **Topic**, **Category** (e.g., chemistry analytical), **Document Type** (e.g., article), **Author, Journal, Publication Year, Language** or **Open Access** availability, among others.

Google Scholar

References on cosmetics analysis can also be successfully found using Google Scholar. The **Advanced Scholar Search** allows carrying out an inquiry by typing words or phrases in the corresponding search boxes. The answer can be narrowed by excluding references containing some specific words. Also, the search can be refined according to **Author, Journal Name** and **Date Interval.**

REFERENCES

Vincent, U., 2015. Guidelines for: 1-selecting and/or validating analytical methods for cosmetics 2-recommending standardization steps for analytical methods for cosmetics. Joint Research Center (JRC), Belgium.

Official Methods for Chemical Analysis of Cosmetic Products
See Chapter 4 of this book.

Other Chemical Analytical Methods for Cosmetic Products
See Chapters 5–14 of this book.

Some Useful Databases for Searching Methods
Google Scholar. Google. https://scholar.google.es.
ScienceDirect. Elsevier. http://www.sciencedirect.com.
SciFinder-Chemical Abstracts Service. American Chemical Society. http://www.cas.org/products/scifinder.
Scopus. Elsevier. https://www.elsevier.com/solutions/scopus.
Web of Science. Thomson Reuters. http://wokinfo.com.

CHAPTER 4

General Review of Official Methods of Analysis of Cosmetics

Gerd Mildau
Chemisches und Veterinäruntersuchungsamt (CVUA), Karlsruhe, Germany

INTRODUCTION

The analysis of cosmetics constitutes a challenge mainly because of the large variety of ingredients and formulations leading to huge matrix complexity and variability. Official methods of analysis of cosmetics are necessary to guarantee the compliance of these products on the market. This chapter will focus mostly on the European market and its legislation, but also discusses the international level in terms of International Organization for Standardization (ISO) standards.

The new European Union (EU) regulation on cosmetic products (Regulation EC 1223/2009) imposes clear and detailed rules throughout the EU to achieve free circulation of products into the internal market for cosmetic products while ensuring a high level of protection of public health. According to the good manufacturing practices (GMP) described in ISO 22716:2007 manufacturers are responsible for ensuring the safety of the products they put on the market and for selecting the most appropriate method for their production, quality control, storage and shipment. The competent authorities shall ensure the analytical control of the products within the frame of market surveillance.

To ensure the uniform application and control of restrictions on substances, sampling and analysis of cosmetic products, they shall be performed in a reliable, reproducible and standardized manner in accordance with Article 12 of the EU Regulation. Prohibited and/or restricted substances are listed in different annexes, i.e., Annex II (prohibited substances), Annex III (restricted substances for intentional use), Annex IV (list of permitted colourants), Annex V (list of permitted preservatives) and Annex VI (list of permitted ultraviolet filters). Article 12 stipulates that the reliability and reproducibility of analytical methods for those substances shall be presumed if the methods used are in accordance with the relevant harmonized standards. Harmonized standards are defined in Article 2 No. 1 (j) as standards adopted by one of the European standardization bodies listed in Annex I of Directive 98/34/EC laying down a procedure for the provision of information in the field of technical standards, e.g., the European Committee for Standardization (CEN).

As stated on the CEN website: '*CEN, the European Committee for Standardization, is an association that brings together the National Standardization Bodies of 34 European countries. CEN is one of three European Standardization Organizations ... that have been officially recognized by the European Union and by the European Free Trade Association (EFTA) as being responsible for developing and defining voluntary standards at European level*'. CEN is the only one of these three organizations devoted to analytical methods, as the other two organizations deal with electrotechnical (European Committee for Electrotechnical Standardization) and telecommunication (European Telecommunications Standards Institute) standards.

Therefore, the creation of CEN standards for analytical methods to identify and, where appropriate, quantify substances (ingredients or traces) in cosmetics is deemed necessary. These harmonized standards are considered officially recognized methods for the analysis of cosmetic products to be used as a reference for all stakeholders (control laboratories of the competent authorities and laboratories of responsible persons in terms of Article 4 of the EU Regulation, e.g., laboratories of manufacturers or ordered private laboratories) if they are referenced in the *Official Journal of the European Union* as described in Article 12.

HISTORY OF OFFICIAL METHODS FOR COSMETICS

The EU Regulation replaced Council Directive 76/768/EC on cosmetics (known as the Cosmetics Directive), which was adopted in 1976 and had been substantially revised on numerous occasions. The Cosmetics Directive did not deal in a concrete manner with harmonized methods. Article 8 of this directive merely stated that the European Commission was responsible for the necessary analytical methods. On this point the European Commission published between 1980 and 1996 a total of seven directives with 36 official methods in the *Official Journal of the European Union*. They looked in more detail at the sampling and specific analytical methods for cosmetic products. All seven directives had been published in a volume of the European Commission (European Commission, 2000). In the foreword the methods are described as European official methods: '*A certain number of methods of analysis have already been validated at European level and accepted as official methods and described in seven Commission Directives on the approximation of the laws of the Member States relating to methods of analysis necessary for checking the composition of cosmetic products. This means that official testing of cosmetic products by laboratories of any kind (national, control, etc.) have to be carried out in accordance with the European official methods described in these Directives*'.

One example is Commission Directive 80/1335/EEC, which sets out the general criteria for the sampling and analysis of substances, among others oxalic acid in hair care products or chloroform in toothpaste. These methods were developed in the 1980s and 1990s in national working groups (in Germany, for instance, in the German 35 Act on Food and Articles in Everyday Use Working Group), in the Cosmetic Products Working Group of the Member States or under the aegis of the Joint Research Centre (JRC) of the European Commission (JRC website). Eventually the working group was disbanded in 2003 because the European Commission had recognized that standardization work can be undertaken more effectively on the European CEN level or internationally on the ISO level (New Approach). New Approach means that technical specifications—in contrast to official methods—are not mandatory and maintain their status of voluntary standards. But the standards offer a guarantee of quality. The 36 published official methods are listed in Table 4.1, and they were reviewed in the first edition of this book (Salvador and Chisvert, 2007). In most cases they no longer reflect the state of the art, and the competent Platform of European Market Surveillance Authorities in Cosmetics (expert group of the European Commission, composed of representatives from Member States) actually plans to replace or update them.

Table 4.1 The 36 official methods of analysis according to Article 8 (1) of Council Directive 76/768/EEC

Commission directive	Official method
80/1335/EEC	1. Sampling of cosmetic products and the laboratory preparation of test samples 2. Identification and determination of free sodium and potassium hydroxides 3. Identification and determination of oxalic acid and alkaline salts in hair care products 4. Determination of chloroform in toothpastes 5. Determination of zinc 6. Identification and determination of phenolsulphonic acid
82/434/EEC	1. Identification of oxidizing agents and determination of hydrogen peroxide in hair care products 2. Identification and semi-quantitative determination of certain oxidation colourants in hair dyes 3. Identification and determination of nitrite 4. Identification and determination of free formaldehyde 5. Determination of resorcinol in shampoos and hair lotion 6. Determination of methanol in relation to ethanol or propan-2-ol
83/514/EEC	1. Determination of dichloromethane and 1,1,1-trichloroethane 2. Identification and determination of quinolin-8-ol and bis(8-hydroxyquinolinium) sulphate 3. Determination of ammonia 4. Identification and determination of nitromethane 5. Identification and determination of mercaptoacetic acid in hair-waving, hair-straightening and depilatory products 6. Identification and determination of hexachlorophene 7. Determination of tosylchloramide sodium 8. Determination of total fluorine in dental creams 9. Identification and determination of organomercury compounds 10. Determination of alkali and alkaline earth sulphides

Continued

Table 4.1 The 36 official methods of analysis according to Article 8 (1) of Council Directive 76/768/EEC—cont'd

Commission directive	Official method
85/490/EEC	1. Identification and determination of glycerol 1-(4-aminobenzoate) 2. Determination of chlorobutanol 3. Identification and determination of quinine 4. Identification and determination of inorganic sulphites and hydrogen sulphites 5. Identification and determination of chlorates of the alkali metals, and identification and determination of sodium iodate
93/73/EEC	1. Identification and determination of silver nitrate 2. Identification and determination of selenium disulphide in antidandruff shampoos 3. Determination of soluble barium and soluble strontium in pigments in the form of salts or lakes 4. Identification and determination of benzyl alcohol 5. Identification of zirconium, and determination of zirconium, aluminium and chlorine in non-aerosol antiperspirants 6. Identification and determination of hexamidine, dibromohexamidine, dibromopropamidine and chlorhexidine
95/32/EEC	1. Identification and determination of benzoic acid, 4-hydroxybenzoic acid, sorbic acid, salicylic acid and propionic acid 2. Identification and determination of hydroquinone, hydroquinone monomethyl ether, hydroquinone monoethyl ether and hydroquinone monobenzyl ether
96/45/EEC	1. Identification and determination of 2-phenoxyethanol; 1-phenoxypropan-2-ol; methyl, ethyl, propyl, butyl and benzyl 4-hydroxybenzoate

CASCADE APPROACH FOR SELECTING APPROPRIATE METHODS

To ensure that testing results related to official controls are sufficiently robust and reliable, the analysis should be performed in accordance with the principles laid down in ISO 17025:2005. The need for formal accreditation to ISO 17025:2005 or for audits against this standard will depend on circumstances and will not always be appropriate, e.g., for laboratories of manufacturers. The manufacturers are responsible for the selection of the most appropriate method to perform the production, quality control, storage and shipment of their cosmetic products according to GMP. At this point it must, of course, be stressed that a manufacturer can routinely check as a rule the proof or non-proof of certain substances in the formulation within the framework of cosmetic GMP (e.g., via weighing protocols). Of course, analytical methods are also used in internal quality assurance. What is needed here are reliable, reproducible methods. They need not, however, necessarily be internationally standardized methods. In-house methods can also be appropriate.

Table 4.2 Cascade approach for appropriate methods to perform the analysis of cosmetics extracted from Joint Research Centre guidelines (Vincent, 2015)

Step	Action
1. Review of appropriate analytical methods	• Use international or EU standard, if possible • If not, consult other valuable sources
2. Selection of the method based on its validation status	Priority order: 1. Community method ('official method') 2. CEN European standard (EN) or ISO/CEN standard (EN ISO) 3. Other internationally harmonized protocols 4. Ring-trial-validated method 5. Single-laboratory-validated method
3. Measurand outside of method scope	If a suitable method exists but has a non–fully matching scope (e.g., different cosmetic matrix and/or concentration): • To extend the scope through an additional validation of each result
4. No validated method available	If no (validated) method exists: • To develop and validate an appropriate method

A useful tool to choose appropriate methods and develop European standards is the set of guidelines published by the JRC (Vincent, 2015), where it is stated that '*This document describes a pragmatic approach not only for validation of methods necessary for market surveillance of cosmetics but also for further standardisation of the methods if deemed necessary*'.

The JRC recommends first applying a 'cascade approach' for the best available method for the daily analysis of cosmetics for both official labs and manufacturers' labs. This approach is similar to the one described in Regulation EC 882/2004 on official controls performed to ensure the verification of compliance with feed and food law, animal health and animal welfare rules. It must be criticized that this cascade approach focuses again on the priority of 'community methods' (described in Table 4.1), which still exist despite their unfitness as already mentioned earlier (see Table 4.2).

Details of standardization will be explained in the following section. In terms of step 4 ('No validated method available'), the development and validation of an in-house method is required (see Table 4.3). To reach this aim some experience in analysing cosmetic products is helpful. The JRC guidelines recommend using the criteria which are listed in the aforementioned official controls regulation.

NEW APPROACH: STANDARDIZATION OF METHODS FOR COSMETICS

In contrast to the background of the official methods in Europe (see Table 4.1), Article 12 of the new EU Regulation refers, from the angle of a new approach, to existing and future standards for the sampling and analysis of cosmetic products which are elaborated

Table 4.3 Characterization of methods of analysis extracted from Regulation EC 882/2004, Annex III

Analytical properties to be considered	Accuracy, applicability (matrix and concentration range), limit of detection, limit of determination, precision, repeatability, reproducibility, recovery, selectivity, sensitivity, linearity, measurement uncertainty; also other criteria may be required
Precision criteria	The precision values shall either: • Be obtained from a collaborative trial (with an internationally recognized protocol) or • Be based on criteria compliance tests (using established performance criteria for analytical methods) The repeatability and reproducibility values shall be expressed in an internationally recognized form The results from the collaborative trial shall be published or freely available
Applicability	Methods of analysis which are applicable uniformly to various groups of commodities should be given preference over methods which apply only to individual commodities
Validation	If the method can be validated only within a single laboratory: • Use IUPAC harmonized guidelines or • Use criteria compliance tests
Layout	Methods should be edited in the standard layout recommended by the International Organization for Standardization

and published by European standardization authorities. These are harmonized test methods and standard operating procedures which are to be made available to both official cosmetic surveillance authorities and cosmetic manufacturers for the in-house control of their products. In the EU Regulation it is stated that sampling and analysis should be carried out in a reproducible and standardized manner to ensure the uniform application and control of the restrictions on substances.

To meet the requirement of uniform standardization within the intendment of EU Regulation—and also with a view to Europe-wide harmonized, reliable surveillance—efficient and, if possible, unbureaucratic standards must be available for the sampling and analysis of cosmetic products. They should correspond to the best available analytical methods, which means that they have to be efficiently updated.

In 2009 the CEN technical committee (TC) 392 Cosmetics was set up on the European level. This TC consists of three working groups (WGs), which are involved in the standardization of the following methods:

- CEN/TC 392/WG 1: analytical methods (active ingredients, restricted or prohibited substances);
- CEN/TC 392/WG 2: microbiological methods;
- CEN/TC 392/WG 4: proof of efficacy, e.g., ultraviolet protection (mirror body from ISO/TC 217/WG 7).

An excerpt from the business plan of CEN/TC 392 Cosmetics (CEN/TC 392, 2016) reads as follows:

The purpose of the CEN/TC 392 is to develop appropriate standards in the field of cosmetics to the final benefit of consumer health and well being. However, it is recognized that certain products (substances or mixtures), although applied to the body for permanent decorative purposes, do not fall under the definition of cosmetic products [amended by the author: tattoo and permanent make-up inks]. The task of the WGs is to identify
- *which successfully collaboratively tested methods are available;*
- *which of these give comparable results;*
- *which of the available methods, if possible one of the comparable ones, is the most appropriate for adoption.*

The scope of WG 1, analytical methods, is the following: '*The purpose of the CEN/TC 392/WG 1 is to develop appropriate standards for analytical methods to identify, determine and, where appropriate, quantify ingredients or traces in cosmetics. The analysis of cosmetic products shall be performed in a reliable and reproducible manner*'.

On the ISO level (ISO/TC 217, 2016) the WG on analytical methods in the field of cosmetics was organized within ISO/TC 217/WG 3 (ISO website). Furthermore, in ISO/TC 217/WG 1 fundamental work took place to standardize microbiological methods for cosmetics, whereas methods for the efficacy of sunscreen products were dealt with within ISO/TC 217/WG 7.

European official laboratories are associated with the European Network of Official Cosmetics Control Laboratories (OCCLs), whose activities are coordinated by the European Directorate for the Quality of Medicines & HealthCare (EDQM). As stated on the EDQM website: '*Considering that only few official methods exist in European legislation on cosmetic products for a great number of prohibited or restricted substances, a need for increasing the number of available methods is evident. Following the provisions in Article 12 of Regulation (EC) No 1223/2009 for "reliable and reproducible" methods to be used in cosmetics testing, the European Network of Official Cosmetics Control Laboratories (OCCLs) has developed two analytical methods to support control laboratories in their testing activities: Determination of hydrogen peroxide in cosmetic products and Determination of free formaldehyde in cosmetic products. The OCCL Network activities are coordinated by the EDQM*'.

According to the aforementioned JRC guidelines, '*a method proposed for standardization should fulfil all requirements of this guideline to reach the status of an official CEN method in terms of Article 12 of Regulation (EC) No 1223/2009, both as Technical Specification (TS) and as European Standard (EN)*'.

Standardization of analytical methods is a resource-intensive process and it is therefore recommended by the JRC to initiate it only if clear needs have been identified and if the method proposed for standardization fulfils all the requirements listed in Table 4.4.

The JRC guidelines describe three ways to produce robust and reliable standards, whereas reliability is defined there as a combination of the two parameters precision and accuracy. These ways are described in Table 4.5.

Table 4.4 Requirements for standardization of a method, extracted from Joint Research Centre guidelines (Vincent, 2015)

Need	The method is necessary: • On the basis of a health risk priority (proposed by competent authorities) • For analysis of cosmetics on a frequency-of-use basis • To fulfil legal requirements indispensable for the harmonization of measurements among different laboratories/countries
Features	The method is • Robust • Reliable (taking into account technological advances, instrumentation and cost efficiency)

Table 4.5 Three types of standardization extracted from Joint Research Centre guidelines (Vincent, 2015)

Peer-Review Process (e.g., OCCL)	Three to less than eight laboratories having successfully participated in a verification study using the same method, on the same analyte(s) and in the same matrix
Technical Specification TS (CEN)	Three to less than eight laboratories having successfully participated in a ring trial using the same method, on the same analyte(s) and in the same matrix
European standard EN (CEN)	At least eight laboratories having successfully participated in a ring trial using the same method, on the same analyte(s) and in the same matrix

CEN, European Committee for Standardization; *OCCL*, Official Cosmetics Control Laboratory.

The JRC defines the prerequisites for a ring trial (collaborative study) as follows:

within-laboratory standard deviation (repeatability) and between-laboratory standard deviation (reproducibility). It does not aim at establishing e.g. limits of detection (LOD) and quantification (LOQ), specificity or accuracy, which are usually assessed in the laboratory of the method developer. Key pre-requisites for a collaborative trial are that the trueness of the method (e.g. the recovery rate) and its robustness (e.g. using factorial design approaches) are established during single-laboratory validation. The classical design of a ring trial for the validation of quantitative methods is given in ISO 5725-1:1994 and in the IUPAC harmonised protocol and summarized hereafter:
- *At least five different materials (same analyte and different matrix/concentration combinations)*
- *Blind samples are sent out in replicates*
- *At least eight laboratories deliver valid results,*
- *Outlier treatment (Cochran – within lab variation, Grubbs – between lab variation),*
- *Performance characteristics are calculated by Analysis of Variance of the retained results after rejection of outliers.*

In terms of a verification study for the peer review process, according to the JRC there are no concrete requirements. But they have naturally to be defined before the peer review procedure by the responsible organization.

ISO standards are commonly recognized via the Vienna Agreement between ISO and CEN (Vienna Agreement, 1991): '*The Vienna Agreement, signed in 1991, was drawn up with the aim of preventing duplication of effort and reducing time when preparing standards. As a result, new standards projects are jointly planned between CEN and ISO. Wherever appropriate priority is given to cooperation with ISO provided that international standards meet European legislative and market requirements and that non-European global players also implement these standards. The Vienna Agreement allows expertise to be focused and used in an efficient way to the benefit of international standardization*'.

INTERNATIONAL STANDARDS ON ANALYTICAL METHODS FOR COSMETICS

Table 4.6 shows the published standards on analytical methods for cosmetics by ISO, CEN and OCCLs, and also the draft standards currently under consideration. Summaries in terms of the scope and principles of these methods are given thereafter.

Published Standards

Detection and Determination of N-*nitrosodiethanolamine in Cosmetics by High-Performance Liquid Chromatography, Post-Column Photolysis and Derivatization (ISO 10130:2009)*

Because of their carcinogenic potential *N*-nitrosamines are banned in cosmetics (Annex II No. 410 Regulation EC 1223/2009). Among *N*-nitrosamines, *N*-nitrosodiethanolamine (NDELA) has been recognized as a potential contaminant of cosmetics. Traces of NDELA are allowed in cosmetics only in technically unavoidable amounts providing that the product is safe (compliance with Article 17 of Regulation EC 1223/2009, which stipulates that 'the non-intended presence of a small quantity of a prohibited substance, stemming from impurities of natural or synthetic ingredients, the manufacturing process, storage, migration from packaging, which is technically unavoidable in good manufacturing practice, shall be permitted provided that such presence is in conformity with Article 3'.). Article 17 shows that technical unavoidability of prohibited substances depends on different factors. Prerequisite to consider the appropriate amount is a reliable, reproducible and very sensitive method to determine such traces of NDELA in different cosmetic matrices.

This ISO standard uses high-performance liquid chromatography (LC) with a reversed-phase C_{18} column coupled with a post-column photolysis unit, to convert NDELA to nitrite, and then coupled with a derivatization unit to derivatize nitrite into an azo dye, which is measured spectrophotometrically. Clean-up is performed using either solid-phase extraction with a C_{18} cartridge or liquid–liquid extraction using dichloromethane when the samples are not dispersible in water.

Table 4.6 Published (or under development) standards (CEN, ISO) and peer review methods (OCCL)

Method	ISO/CEN (EN)/OCCL	Published
Nitrosamines: Detection and determination of NDELA in cosmetics by HPLC, post-column photolysis and derivatization	ISO 10130	2009
Nitrosamines: Detection and determination of NDELA in cosmetics by HPLC–MS/MS	ISO 15819	2014
Detection and quantitative determination of diethanolamine by GC/MS	ISO/PRF TR 18818	Under development
Measurement of traces of heavy metals in cosmetic finished products using ICP–MS	ISO/AWI 21392	Under development
	prEN ISO 21392	Under drafting
Quantification of suspected fragrance allergens in consumer products, step 1: GC analysis of ready-to-inject sample	EN 16274	2012
Quantitative determination of zinc pyrithione, piroctone olamine and climbazole in surfactant-containing cosmetic antidandruff products	EN 16342	2013a
Determination of 3-iodo-2-propynyl butylcarbamate in cosmetic preparations, LC/MS methods	EN 16343	2013b
Screening for UV filters in cosmetic products and quantitative determination of 10 UV filters by HPLC	EN 16344	2013c
GC/MS method for the identification and assay of 12 phthalates in cosmetic samples ready for analytical injection	EN 16521	2014
HPLC/UV method for the identification and quantitative determination in cosmetic products of the 22 organic UV filters in use in the European Union	prEN	Under drafting
HPLC/UV method for the identification and assay of hydroquinone, ethers of hydroquinone and corticosteroids in skin whitening cosmetic products	FprEN 16956	Under approval
HPLC for the determination of free formaldehyde in cosmetic products using 2,4-DNPH for derivatization	OCCL	2016a
HPLC for the determination of hydrogen peroxide present in or released by tooth whitening or bleaching products	OCCL	2016b

CEN, European Committee for Standardization; *2,4-DNPH*, 2,4-dinitrophenylhydrazine; *GC*, gas chromatography; *HPLC*, high-performance liquid chromatography; *ICP*, inductively coupled plasma; *ISO*, International Organization for Standardization; *LC*, liquid chromatography; *MS/MS*, tandem mass spectrometry; *NDELA*, N-nitrosodiethanolamine; *OCCL*, Official Cosmetics Control Laboratory.

Detection and Determination of N-nitrosodiethanolamine in Cosmetics by High-Performance Liquid Chromatography–Tandem Mass Spectrometry (ISO 15819:2014)

Similar to the aforementioned standard, in this one a clean-up is performed also using solid-phase extraction or liquid–liquid extraction when the samples are not dispersible in water. LC with a reversed-phase C_{18} column is also used, but in this case it is coupled to tandem mass spectrometry. Extraction of the nitrosamine NDELA in cosmetic samples is carried out with water in the presence of deuterated d_8-NDELA used as an internal standard. Characteristic ions 135 (M+H), 104 and 74 (fragment ions) identify the presence of NDELA in the sample. Quantification is made by using the ratio of the intensity of fragment ions 104 (from NDELA) and 111 (from d_8-NDELA).

Quantification of Suspected Fragrance Allergens in Consumer Products, Step 1: Gas Chromatography Analysis of Ready-to-Inject Sample (EN 16274:2012)

This standard was elaborated by CEN/TC 347 (Allergens), WG 4 (CEN/TC 347/WG 4). Whether there will be a second step in terms of analysis of fragrance allergens in emulsions after appropriate sample preparation is still open.

Twenty-six fragrance substances were restricted in the cosmetics regulation, with labelling requirements if they exceeded the following limit values: 100 mg/kg in rinse-off products, such as shower gels, and 10 mg/kg in leave-on products, such as perfumes. Twenty-four of those 26 restricted substances are volatile enough to be determined by gas chromatography (GC) using two GC capillary columns with different polarities.

The method deals only with leave-on products such as perfumes, which can be directly injected into GC systems without further sample preparation steps.

The detection takes place by mass spectrometry (MS) acquired in selected-ion monitoring (SIM) mode using characteristic fragments. Confirmation of the presence of allergens is done in SCAN mode. Typically a mass range from 35 to 350 mass units allows a structural confirmation of all allergen analytes. For every substance three characteristic ions are used.

Quantification is performed by SIM using 1,4-dibromobenzene and 4,4'-dibromobiphenyl as internal standards. The final result depends on the agreement of the different ion ratios obtained for both injections according to specific requirements.

Quantitative Determination of Zinc Pyrithione, Piroctone Olamine and Climbazole in Surfactant-Containing Cosmetic Antidandruff Products (EN 16342:2013a)

Zinc pyrithione, piroctone olamine and climbazole are permitted as preservatives in cosmetics (Annex V of Regulation EC 1223/2009). These substances are also used widely as active antimicrobial agents in antidandruff shampoos, but there is a limit value only for zinc pyrithione according to Annex III. They mainly inhibit the development of microorganisms, especially specific fungus species, on the human scalp. The antidandruff agents are extracted from the cosmetic sample matrix using dichloromethane and methanol within a volumetric flask and an extraction-promoting temperature-controlled ultrasonic bath (10 min at 35°C–40°C). The membrane-filtered extract is directly analysed using reversed-phase LC with UV detection. For the gradient elution two eluents are used: oxalic acid/EDTA in water (pH 4.0) and acetonitrile. This gradient delivers a sufficient separation performance of the analytes within the matrix. The quantitative determination is made using the external standard method of calibration.

The reliability of the method was demonstrated in a ring test of 10 laboratories where good recovery, repeatability and reproducibility were calculated. The method also mentions the determination of ketoconazole and ciclopirox olamine as further active antidandruff agents; however, ketoconazole is banned as a carcinogenic, mutagenic or toxic to reproduction substance. But for these two substances there exist no ring-test data within the standard.

Determination of 3-Iodo-2-propynyl Butylcarbamate in Cosmetic Preparations: Liquid Chromatography–Mass Spectrometry Methods (EN 16,343:2013b)

Two hundred milligrams of the sample are dissolved in a mixture of tetrahydrofuran (THF)/methanol (5 min ultrasonic bath at room temperature), filtered through a membrane filter and analysed using LC coupled to MS or to tandem MS. For poorly soluble or suspendable matrices 2-propanol instead of THF is recommended as well as stirring with a magnetic stirrer for 30 min prior to the treatment in the ultrasonic bath. LC is carried out using a reversed-phase column and a gradient eluent (formic acid/water/methanol).

MS detection is possible with two techniques:
- SIM mode with mass traces: m/z 282 $[M+H]^+$ and m/z 304 $[M+Na]^+$; evaluation is based on the total ion current (from the sum of the two masses);
- multiple-reaction monitoring mode with protonated molecule ion: m/z 282 $[M+H]^+$; fragment ion 1, 57; fragment ion 2, 165; evaluation is based on the most sensitive fragment ion.

The analyte may form adducts with sodium ions and therefore false low results can be obtained if larger amounts of sodium ions are present in the sample. Therefore a verification of out-of-specification results via standard addition is recommended.

This method has been successfully tested in an inter-laboratory test for a shower gel and a skin cream with a total of 13 participants. Statistical data in terms of repeatability and reproducibility are listed together with data in terms of recovery.

Screening for UV Filters in Cosmetic Products and Quantitative Determination of 10 UV Filters by High-Performance Liquid Chromatography (EN 16344:2013c)

With this standard method 21 UV filters are prepared in three different calibration solutions according to their polarity (i.e., the most polar filters in methanol, the less polar ones in a methanol/acetone mixture, and the non-polar ones in THF). Then 100 mg of a sample is weighed in a flask and dissolved in the acetone/methanol mixture (15 min at

50°C in an ultrasonic bath). Specific UV filter combinations may require modified solvents to improve their recovery.

All 21 UV filters can be determined within one chromatographic run despite their different polarities. The LC analysis is performed by means of ion-pair elution reversed-phase chromatography. The gradient consists of an aqueous solution of lauryl trimethyl ammonium bromide (LTAB) and ammonium bromide and EDTA in water and an organic phase of LTAB and ammonium bromide in ethanol. The peak areas are measured at a wavelength of 300 or 350 nm, respectively. The UV spectra are compared with the reference spectra in a database.

The reliability (recovery) and reproducibility of the method has been tested in an inter-laboratory test with 10 participants. In three samples the reproducibility and repeatability of 10 UV filters was successfully determined. Sample 1 was a sunscreen milk with sun protection factor (SPF) 40; sample 2, a sunscreen spray with SPF 50, and sample 3, a sunscreen oil-in-water product specifically made for the ring test.

Gas Chromatography/Mass Spectrometry Method for the Identification and Assay of 12 Phthalates in Cosmetic Samples Ready for Analytical Injection (EN 16521:2014)

About 80% of all phthalates manufactured are used as is without sacrificing strength or durability. Phthalates are present in cosmetic products like perfumes and toiletries both as traces from packaging materials ('plasticizers' to make plastics flexible) and as ingredients (e.g., diethyl and dimethyl phthalate are used as solvents and perfume fixatives or diethyl phthalate can be used as to denature alcohol). Their presence as a contaminant could also be due to the manufacturing process or raw materials used. Cosmetic samples are analysed directly or after a previous dilution in ethanol. Twelve phthalates were separated on a cross-linked 5% phenyl/95% dimethylpolysiloxane capillary column (30 m × 0.25 μm film thickness) using a temperature gradient. The detection was carried out on a GC/MS system with electron impact ionization mode. Phthalate quantification was performed by external calibration using an internal standard (dibromodiphenyl) or by the standard addition. The limit of quantification using the described sample preparation was set at 5 ppm (1:10 dilution with ethanol). This method does not include requirements for the preparation of samples in cosmetic matrices for which direct injection in GC is not feasible.

High-Performance Liquid Chromatography Method for the Determination of Free Formaldehyde in Cosmetic Products Using 2,4-Dinitrophenylhydrazine for Derivatization (OCCL, 2016)

This procedure describes a method for the determination of free formaldehyde in cosmetic products for concentrations ranging from 75 to 1200 mg/kg. Free formaldehyde is extracted using THF/water (9:1) as a solvent. The sample extract and the standard solutions are derivatized at pH≈1 using 2,4-dinitrophenylhydrazine. The derivative is stabilized at pH 1.5–2.5 by using phosphate buffer. The yellow derivative is analysed using LC with a reversed-phase stationary phase (C_8) and UV detection at 354 nm. The identity of formaldehyde is confirmed by comparing UV spectra with reference samples.

High-Performance Liquid Chromatography Method for the Determination of Hydrogen Peroxide Present in or Released by Tooth Whitening or Bleaching Products (OCCL, 2016b)

This procedure describes a method for the quantification of hydrogen peroxide present in or released by tooth whitening products for concentrations ranging from 0.06% to 10% (m/m) in the finished product. An LC/UV method based on oxidation of triphenylphosphine into triphenylphosphine oxide is used for the determination of hydrogen peroxide. In most tooth whitening products, the active ingredient is hydrogen peroxide or a substance able to generate hydrogen peroxide (carbamide peroxide, sodium perborate, calcium peroxide).

Outcome—Draft Standards

Different methods are in development at the time of writing as ISO and/or CEN standards. They are described next.

Detection and Quantitative Determination of Diethanolamine by Gas Chromatography/Mass Spectrometry (ISO/PRF TR 18818, Under Approval)

This project is arriving at the final step before its publication as an ISO standard. It describes the determination of diethanolamine (DEA) in cosmetic products and raw materials. The method is based on GC/MS. DEA is extracted with anhydrous ethanol from the sample matrix by ultrasonic extraction, and then centrifuged. The extract is then ready for injection into the GC/MS instrument. An external standard method using SIM mode is used for quantitative analysis, whereas confirmation is performed by its mass spectrum.

High-Performance Liquid Chromatography–UV Method for the Identification of Hydroquinone, Ethers of Hydroquinone and Corticosteroids in Skin Whitening Cosmetic Products (FprEN 16956, Under Approval)

In 2016 there was a ring trial organized to get results in terms of reproducibility for this method. The assessment of the ring trial is still in preparation at the time of writing.

Hydroquinone is no longer used in cosmetic products for skin whitening and depigmentation of dermal spots or imperfections. Because of its cytotoxic effects its use has been regulated. Hydroquinone and three of its ethers [hydroquinone monomethyl ether (MME), hydroquinone monoethyl ether (MEE) and hydroquinone monobenzyl ether (MBE)] have been regulated since 11 July 2013 by EC Regulation 1223/2009. Nowadays, its use is prohibited in skin whitening cosmetics. Depigmentation is a side effect of topical steroids; in this way corticosteroids might be used as whitening agents in products illegally sold as cosmetics. Corticosteroids most commonly found in products illegally sold as cosmetics are clobetasol propionate, betamethasone dipropionate, fluocinonide and fluocinolone acetonide. Corticosteroids are prohibited in cosmetic products. All these substances work on the same principle as hydroquinone, which mainly consists of inhibition of melanin.

Cosmetic Directive 95/32/EC gives an analytical method for the assay of hydroquinone and MME, MEE and MBE in cosmetic products for lightening the skin. To update and extend this official method for the identification and assay of corticosteroids in cosmetic products, this standard describes an LC/UV method for the identification and assay of hydroquinone, three ethers of hydroquinone and four corticosteroids most frequently found in illegally sold skin whitening cosmetic products: clobetasol propionate, betamethasone dipropionate, fluocinonide and fluocinolone acetonide. This standard also proposes LC/UV methods for the identification of 38 corticosteroids that may be found in skin whitening products. Indeed, as corticosteroids could be deliberately introduced into skin whitening cosmetics, despite the fact that they are forbidden to use, the identification of the presence of one of these illicit compounds could be enough for a market survey control. This standard is not dedicated to artificial nail products or soaps.

Liquid Chromatography–UV Method for the Identification and Quantitative Determination in Cosmetic Products of the 22 Organic UV Filters in Use in the European Union (prEN, Under Drafting)

In 2016 there was organized by Spain a ring trial to get results in terms of reproducibility for this method. The assessment of the ring trial is still in preparation at this writing.

With the aim of ensuring efficacy (in terms of the SPF) and compliance with the EU Regulation, this CEN draft describes an analytical method based on LC with UV/visible spectrometry (UV/VIS) detection for the detection and quantitative determination of 22 organic UV filters from among the 26 compounds listed in Annex VI of the EU Regulation, excluding titanium dioxide (because of its inorganic nature) and benzylidene camphor sulphonic acid, polyacrylamidomethyl benzylidene camphor and 3-benzylidene camphor (which are protected by patents and are no longer used in cosmetics). The target 22 compounds are grouped into three groups according to their nature, i.e., the 'water-soluble' (comprising six of these compounds), the 'fat-soluble' (comprising 15 of them) and the 'polymeric' (comprising just polysilicone-15).

In this method, the cosmetic sample is weighed and solved in ethanol, and then, aliquots are taken and diluted with a mixture of ethanol/acetate buffer, ethanol or THF for the determination of the water-soluble group, the fat-soluble group or the polymeric, respectively. These three different groups are measured in the LC/UV/VIS system by employing different chromatographic conditions, i.e., a reversed-phase column and a gradient mobile phase of ethanol/acetate buffer for the water-soluble group, a reversed-phase column and a gradient mobile phase of ethanol/aqueous formic acid containing hydroxypropyl-β-cyclodextrin for the fat-soluble group or a size-exclusion column and isocratic mobile phase of THF for the polymeric.

This method has been validated for emulsion-based cosmetic products, lip balms, lotions and waters.

Measurement of Traces of Heavy Metals in Cosmetic Finished Products Using Inductively Coupled Plasma–Mass Spectrometry Technique (ISO/AWI 21392 Under Development, prEN ISO 21392, Under Drafting)

This standard, devoted to the determination of traces of banned elements in cosmetics, is under development between ISO and CEN via the Vienna Agreement. Specifically, the elements are antimony, arsenic, barium, lead, cadmium and nickel, and inductively coupled plasma–MS after pressure digestion is employed.

The homogenized sample is digested with mineral acids in a sealed vessel under elevated temperature and pressure. Pressure digestion is performed with microwave-assisted heating at a temperature of 200°C. Depending on their type and composition, not all cosmetic preparations can be dissolved residue free. To obtain comparable results it is necessary to precisely comply with the conditions provided in this method. The digestion solution is atomized and the aerosol is transferred to an inductively coupled argon plasma where the elements are ionized. The ions are transferred into a mass spectrometer where they are separated according to their mass/charge ratio and determined with the help of a detector system. To avoid or minimize interference with the mass of the elements antimony, arsenic, barium, lead, cadmium and nickel, the mass spectrometer has to be suitably equipped. Therefore a mass spectrometer which has the potential to remove or reduce interference (e.g., correction equation cell technique, resolution above 3000) must be used.

REFERENCES

CEN Website. http://www.cen.eu.
CEN/TC 392, 2016. https://standards.cen.eu/dyn/www/f?p=204:7:0::::FSP_ORG_ID:679535&cs=1847A554BB84699437FBAD7EEA797BCB9.
Commission Directive 80/1335/EEC on the Approximation of the Laws of the Member States Relating to Methods of Analysis Necessary for Checking the Composition of Cosmetic Products, Amended by Commission Directive 87/143/EEC.
Council Directive 76/768/EEC of 27 July 1976 on the Approximation of the Laws of the Member States Relating to Cosmetic Products.
Directive 98/34/EC of the European Parliament and of the Council of 22 June 1998 Laying Down a Procedure for the Provision of Information in the Field of Technical Standards and Regulations.
EDQM Website. https://www.edqm.eu/en/cosmetics-testing.
EN 16274, 2012. Methods for Analysis of Allergens – Quantification of Suspected Fragrance Allergens in Consumer Products – Step 1: GC Analysis of Ready-to-Inject Sample.
EN 16342, 2013a. Cosmetics – Analysis of Cosmetic Products – Quantitative Determination of Zinc Pyrithione, Piroctone Olamine and Climbazole in Surfactant Containing Cosmetic Anti-Dandruff Products.
EN 16343, 2013b. Cosmetics – Analysis of Cosmetic Products – Determination of 3-iodo-2-propynyl butylcarbamate (IPBC) in Cosmetic Preparations, LC-MS Methods.

EN 16344, 2013c. Cosmetics – Analysis of Cosmetic Products – Screening for UV-filters in Cosmetic Products and Quantitative Determination of 10 UV-Filters by HPLC.

EN 16521, 2014. Cosmetics – Analytical Methods – GC/MS Method for the Identification and Assay of 12 Phthalates in Cosmetic Samples Ready for Analytical Injection.

European Commission, 2000. The Rules Governing Cosmetic Products in the European Union Vol. 2, Methods of Analysis, European Commission, Bruxelles.

FprEN 16956 Cosmetics – Analytical Methods – HPLC/UV Method for the Identification and Assay of Hydroquinone, Ethers of Hydroquinone and Corticosteroids in Skin Whitening Cosmetic Products.

ISO 5725–1, 1994. Accuracy (Trueness and Precision) of Measurement Methods and Results – Part 1: General Principles and Definitions.

ISO 10130, 2009. Cosmetics – Analytical Methods – Nitrosamines: Detection and Determination of N-nitrosodiethanolamine (NDELA) in Cosmetics by HPLC, Post-Column Photolysis and Derivatisation.

ISO 15819, 2014. Cosmetics – Analytical Methods – Nitrosamines: Detection and Determination of N-nitrosodiethanolamine (NDELA) in Cosmetics by HPLC-MS-MS.

ISO 17025, 2005. General Requirements for the Competence of Testing and Calibration Laboratories.

ISO 22716, 2007. Cosmetics – Good Manufacturing Practices (GMP) – Guidelines on Good Manufacturing Practices.

ISO Website. http://www.iso.org.

ISO/PRF TR 18818 Cosmetics – Analytical Method – Detection and Quantitative Determination of Diethanolamine (DEA) by GC/MS.

ISO/TC 217, 2016. http://www.iso.org/iso/home/standards_development/list_of_iso_technical_committees/iso_technical_committee.htm?commid=54974.

JRC Website. https://ec.europa.eu/jrc/en.

OCCL Publication, 2016a. Determination of Free Formaldehyde in Cosmetic Products. https://www.edqm.eu/en/cosmetics-testing.

OCCL Publication, 2016b. Determination of Hydrogen Peroxide in Cosmetic Products. https://www.edqm.eu/en/cosmetics-testing.

prEN Cosmetics – Analytical Methods – HPLC/UV Method for the Identification and Quantitative Determination in Cosmetic Products of the 22 Organic UV Filters in Use in the EU.

prEN ISO 21392 Cosmetics – Analytical Methods – Measurement of Traces of Heavy Metals in Cosmetic Finished Products Using ICP/MS Technique.

Regulation (EC) No 1223/2009 of the European Parliament and of the Council of 30 November 2009 on Cosmetic Products, and Its Successive Amendments. http://eur-lex.europa.eu/legal-content/EN/ALL/?uri=CELEX%3A32009R1223.

Regulation EC 882/2004 of the European Parliament and of the Council of 29 April 2004 on Official Controls Performed to Ensure the Verification of Compliance with Feed and Food Law, Animal Health and Animal Welfare Rules.

Salvador, A., Chisvert, A. (Eds.), 2007. Analysis of Cosmetic Products. Elsevier, Amsterdam.

Vienna Agreement, 1991. Cooperation between CEN and ISO. http://www.cencenelec.eu/intcoop/StandardizationOrg/Pages/default.aspx.

Vincent, U., 2015. JCR Guidelines for 1-Selecting and/or Validating Analytical Methods for Cosmetics and 2-Recommending Standardization Steps for Analytical Methods for Cosmetics. European Commission, Joint Research Centre, Institute for Reference Materials and Measurements, Bruxelles.

CHAPTER 5

Ultraviolet Filters in Cosmetics: Regulatory Aspects and Analytical Methods

Alberto Chisvert, Amparo Salvador
University of Valencia, Valencia, Spain

INTRODUCTION

It is well known that in the past decades there has been a progressive increase in the UV radiation reaching the earth's surface owing to the damage in the stratospheric ozone layer. Consequently, there has been an increase in the number of disorders caused by this deleterious radiation, such as melanoma and non-melanoma skin cancers, skin photo-ageing, systemic immune suppression and various eye diseases (e.g., cataracts and pterygium). Fortunately, the Montreal Protocol has limited, and is now probably reversing, this damage. Despite all the harmful effects and the awareness campaigns promoted by public health bodies from different countries, exposure to solar radiation has not diminished but has increased in recent years. Comparing habits of the so-called first world countries since 1965, people now spend more of their free time exposed to solar radiation. Moreover, protective clothes are not used properly and thus skin is now more exposed. In addition, people usually prefer their skin tanned, because tanning is associated with a beauty standard (Lucas et al., 2015).

In response to this, the so-called UV filters have been used as active ingredients in the formulation of cosmetic products to protect human skin from the deleterious radiation of sunlight. First, they were formulated into beach products (i.e., sunscreen products), and later in products for practicing outdoor sports, especially winter sports, but nowadays UV filters are incorporated into all kinds of daily use cosmetics, such as moisturizing creams, anti-ageing creams, aftershave products, lipsticks, makeup, and even shampoos and hair masks (Chisvert and Salvador, 2007). Different cosmetic matrices require UV filters with different physico-chemical properties, especially to ensure chemical compatibility (i.e., no reaction with other matrix compounds) but also physical stability (i.e., easily soluble or dispersible). Therefore, different chemical compounds are used as UV filters, which are mainly classified into two groups, according to their nature (Shaath, 2016; Daly et al., 2016):

- *inorganic UV filters* (also called physical UV filters), which principally work by reflecting and scattering the UV radiation, although they can also absorb it;

- *organic UV filters* (also called chemical UV filters), which mainly work by absorbing the UV light, although a new generation of particulated organic UV filters, which also scatter the light, is now also being used.

The physical UV filters are generally metallic oxides and provide generally higher protection than the chemical ones. Typically, titanium dioxide and zinc oxide have been used for years in their micronized forms, but nowadays, nanosized titanium dioxide and zinc oxide are being used. This increases the reflection/scattering power while diminishing the uncomfortable white layer they formed on the user's skin (Singh and Nanda, 2014). They are not soluble in any media, but they are dispersed and properly stabilized.

Chemical UV filters are organic compounds with high molar absorptivity in the UV range. These compounds usually possess single or multiple aromatic structures, sometimes conjugated with carbon–carbon double bonds and/or carbonyl moieties. They are usually soluble in fatty matrices, although some of them contain ionizable moieties, such as sulphonic (—SO$_3$H) or carboxylate (—COOH), which enables their solubility in water. They are classified into different families according to their chemical structure: benzophenone derivatives, *p*-aminobenzoic acid derivatives, salicylates, cinnamates, benzotriazole derivatives, benzimidazole derivatives, camphor derivatives, triazine derivatives, and others, whereas they can be classified as UVA or UVB filters depending on the radiation they attenuate (Chisvert and Salvador, 2007).

To achieve a broad-spectrum protection, UV filters are usually combined into cosmetic formulations (Baki and Alexander, 2015; Cole, 2016). The sun protection factor (commonly referred to as SPF) is the parameter used to show the efficacy of a sunscreen product or any other cosmetic product containing UV filters (ISO 24444, 2010). It is a ratio calculated from the energy (i.e., exposure time) required to induce a minimum erythemal response with and without the product applied to the skin of human volunteers. The SPF provides an estimation of how long the coated skin can be exposed to the sun before erythema appears and depends on the nature and concentration of the UV filter contained in the cosmetic product. It should be said that those cosmetic products claiming sun protection have to be labelled with the SPF value to show their expected efficacy.

Many concerns about UV filters' safety, however, have been raised in past years. Organic UV filters were the first studied since it was proved that they were absorbed through the skin, further metabolized and eventually bioaccumulated and/or excreted (Chisvert et al., 2012). These percutaneous absorption processes may result in various adverse health effects (i.e., allergic contact dermatitis) and other more serious systemic effects, such as carcinogenic and estrogenic activity (Schlumpf et al., 2001; Schreurs et al., 2005; Gomez et al., 2005; Krause et al., 2012; Gilbert et al., 2013; Stiefel and Schwack, 2015). Later, inorganic UV filters were also in the spotlight because of their nanometric size (Shi et al., 2013; Nohynek and Dufour, 2012; Gilbert et al., 2013). Thus, in an attempt to achieve an optimum compromise between adequate protection and minimal side effects, those compounds that can be used as UV filters in cosmetics as well as their maximum allowed concentrations have been regulated by the legislation in force in each country.

Additionally, it is worth mentioning that UV filters have reached the aquatic environment by direct sources (e.g., sunbathing or swimming) and/or indirect sources (wastewater treatment plants, showering or domestic washing), thus accumulating in sea, lake or river waters, where they can cause a negative impact on the flora and fauna (Brausch and Rand, 2011; Tóvar-Sánchez et al., 2013; Molins-Delgado et al., 2016). For all of this, UV filters have been labelled as emergent contaminants.

For all the aforementioned reasons, analytical methods are needed to control the concentration of UV filters and thus to ensure the efficacy and the safety of cosmetic products that contain them as ingredients. Moreover, analytical methods for analysing biological fluids and tissues are of great interest to study the distribution in the human body, metabolism and excretion processes of these products. Finally, analytical methods to control their impact in the environment are also welcomed.

The aim of this chapter is to familiarize the reader with the different compounds used as UV filters and the legislation regulating these compounds, and especially to review the analytical methods for UV filter determination in cosmetic products since 2006. Those articles published earlier (i.e., before 2006) were reviewed in the first edition of this book (Salvador and Chisvert, 2007). Those articles dealing with the determination of UV filters in biological fluids and tissues or in environmental samples escape the aim of this chapter, but interested readers are kindly invited to read Chapters 15 and 16, respectively, which deal with the determination of cosmetic ingredients, including UV filters, in these types of samples.

REGULATORY ASPECTS

As was already mentioned, to guarantee consumers' health, the compounds that can be used as UV filters in cosmetics, and their maximum allowed concentrations, are regulated by legislation in each country. As mentioned in Part I of this book, the regulatory systems for cosmetic products differ from one part of the world to the other, and the allowed UV filters differ consequently.

Ultraviolet Filters in the European Union

In the European Union (EU), sunscreens or any other product containing UV filters are considered cosmetics, and are regulated under the EU Cosmetics Regulation (Regulation EC 1223/2009). According to this regulation, UV filters are defined as '*substances which are exclusively or mainly intended to protect the skin against certain UV radiation by absorbing, reflecting or scattering UV radiation*'. No comment about the protection of the cosmetic product from light is made (Chisvert and Salvador, 2007) as was made in the previous EU Cosmetics Directive (i.e., Directive 76/768/EEC).

The EU Cosmetics Regulation lists in Annex VI not only the compounds that can be used as UV filters in cosmetic products, but also their maximum allowed concentration and other conditions of use. In this regard, as of this writing, there are 27 compounds listed in this annex that can be used as UV filters in cosmetics (Table 5.1).

Among them, titanium dioxide and zinc oxide are of inorganic nature, whereas the other 25 are organic UV filters, the chemical structures of which are shown in Fig. 5.1.

Of the organic UV filters, seven of them (i.e., CBM, PBSA, TDSA, BCSA, P25, BZ4 and PDTA; see Table 5.1 for full names) have a hydrophilic character, whereas the remaining 18 present lipophilic properties and are the most commonly employed. Nevertheless, it should be noted that the manufacture of BCSA and PBC is protected under L'Oréal patents and they are no longer manufactured nor used in cosmetic products.

Annex VI is subject to changes in accordance with the knowledge on safety and efficacy for each compound. In this regard, the UV filters are reviewed periodically by the Scientific Committee on Consumer Safety (SCCS; formerly known as the Scientific Committee on Consumer Products), which studies and evaluates the data provided by different sources, such as hospitals, industries and research centres. The recommendations can be freely accessed throughout its website (SCCS, 2017). After the study is reported, the European Commission together with the Member States adopts the appropriate actions. In this sense, some compounds initially allowed have been eliminated as a consequence of data showing them not to be safe for human health or because of their low capacity to afford protection from sunlight or poor photostability. The last UV filter prohibited in the EU as of this writing was 3-benzylidene camphor, because of safety reasons (Regulation EC, 2015/1298). In contrast, some new compounds have been incorporated into the list because they have been shown to have high UV light absorption potential, great stability and few side effects. An example of this is the 2016 approval of zinc oxide (Regulation EC, 2016/621).

Ultraviolet Filters Outside the European Union

In many other countries, such as Japan, China, South Africa, the countries of the Association of Southeast Asian Nations and the Southern Common Market in South America, sunscreens are also regulated as cosmetics, whereas in the United States, Canada, Australia and New Zealand they are regulated as drugs (Stiefel and Schwack, 2015). While in some cases the list of UV filters converges with that of the EU, in others it is far different (Table 5.1), and only eight UV filters (i.e., BZ3, BZ4, BMDM, EHDP, EHMC, EHS, HMS, OC, PBSA; see Table 5.1 for full names) can be considered allowed worldwide. A more detailed discussion can be found in Stiefel and Schwack (2015) and in Daly et al. (2016).

Table 5.1 Ultraviolet filters allowed in different legislative frameworks and their maximum authorized contents (in alphabetical order)

INCI name	Key[a]	EU[b]	USA[c]	Japan[d]	ASEAN[e]	Mercosur[f]
1-(3,4-Dimethoxyphenyl)-4,4-dimethyl-1,3-pentanedione	DDP			7		
3-Benzylidene camphor	3BC				2	2
4-(2-β-Glucopyranosiloxy)propoxy-2-hydroxybenzophenone	GPHB			5		
4-Methylbenzylidene camphor	MBC	4			4	4
Benzophenone-1	BZ1			10		
Benzophenone-2	BZ2			10		
Benzophenone-3	BZ3	10	6	5	10	10
Benzophenone-4 and/or its sodium salt (benzophenone-5)	BZ4	5/5[i]	10/−	10/10	5/5[i]	10/5[i]
Benzophenone-6	BZ6			10		
Benzophenone-8	BZ8		3			3
Benzophenone-9	BZ9			10		
Benzylidene camphor sulphonic acid	BCSA	6[h]			6[h]	6[h]
Bis-ethylhexyloxyphenol methoxyphenyl triazine	BEMT	10		3	10	10
Butyl methoxydibenzoylmethane	BMDM	5	3	10	5	5
Camphor benzalkonium methosulfate	CBM	6			6	6
Cinoxate	CNX		3	5		3
Diethylamino hydroxybenzoyl hexyl benzoate	DHHB	10		10	10	10
Diethylhexyl butamido triazone	DEBT	10			10	10
Diisopropyl methyl cinnamate	DPMC			10		
Disodium phenyl dibenzimidazole tetrasulfonate	PDTA	10[i]			10[i]	10[i]
Drometrizole trisiloxane	DTS	15		15	15	15
Ethylhexyl dimethoxybenzylidene dioxoimidazolidine propionate	EHDDP			3		
Ethylhexyl dimethyl PABA	EHDP	8	8	10	8	8

Continued

Table 5.1 Ultraviolet filters allowed in different legislative frameworks and their maximum authorized contents (in alphabetical order)—cont'd

INCI name	Key[a]	EU[b]	USA[c]	Japan[d]	ASEAN[e]	Mercosur[f]
Ethylhexyl methoxycinnamate	EHMC	10	7.5	20	10	10
Ethylhexyl salicylate	EHS	5	5	10	5	5
Ethylhexyl triazone	EHT	5		5	5	5
Ferulic acid	FA			10		
Glyceryl ethylhexanoate dimethoxycinnamate	GEDC			10		
Homosalate	HMS	10	15	10	10	15
Isoamyl dimethyl PABA	IDP			10		
Isoamyl p-methoxycinnamate	IMC	10			10	10
Isopentyl trimethoxycinnamate trisiloxane	ITT			7.5		
Isopropyl methoxycinnamate	IPMC			10[k]		
Methyl anthranilate	MA		5		5	5
Methylene bis-benzotriazolyl tetramethylbutylphenol	MBBT	10		10	10	10
Octocrylene	OC	10[i]	10	10	10[i]	10[i]
PABA	PABA		15	4[j]		15
PEG-25 PABA	P25	10			10	10
Phenylbenzimidazole sulphonic acid	PBSA	8[g]	4	3	8[g]	8[g]
Polyacrylamidomethyl benzylidene camphor	PBC	6			6	6
Polysilicone-15	P15	10		10	10	10
TEA-salicylate	TS		12			12
Terephthalylidene dicamphor sulphonic acid	TDSA	10[h]		10	10[h]	10[h]
Titanium dioxide	TiO$_2$	25	25		25	25
Tris-biphenyl triazine	TBT	10			10	10
Zinc oxide	ZnO	25	25		25	25

INCI, International Nomenclature of Cosmetic Ingredients; PABA, para-aminobenzoic acid.
[a]Key system adopted by authors.
[b]Regulation (EC) 1223/2009.
[c]Food and Drug Administration, Code of Federal Regulations (FDA, 2017).
[d]Ministry of Health and Welfare Notification (MHLW Japan, 2000).
[e]Association of Southeast Asian Nations Cosmetic Directive (ASEAN, 2003).
[f]Mercosur (Southern Common Market) Regulation (Mercosur, 2015).
[g]And its potassium, sodium and triethanolamine salts (expressed as acid).
[h]And its salts (expressed as acid).
[i]Expressed as acid.
[j]And its esters.
[k]It contains 3%–9% methyl diisopropyl cinnamate and 15%–21% ethyl diisopropyl cinnamate.

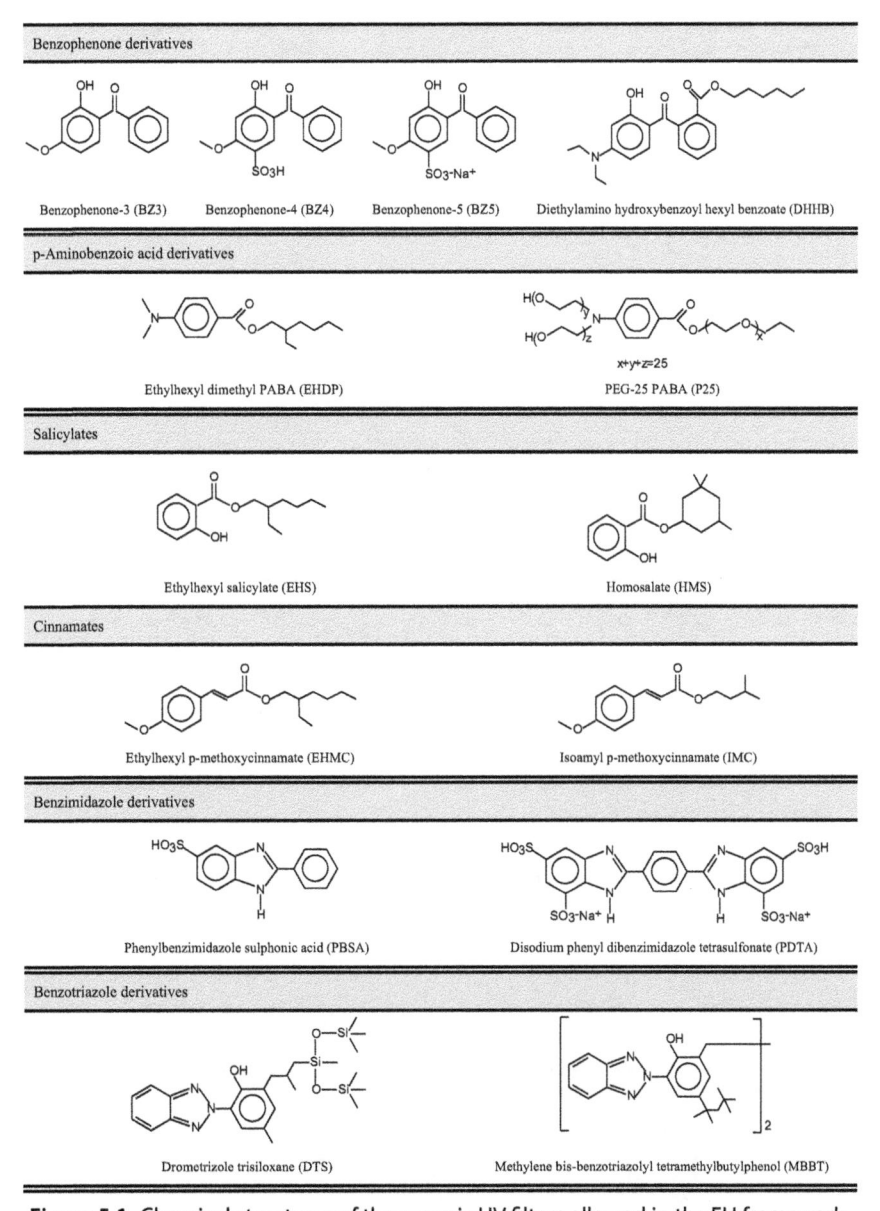

Figure 5.1 Chemical structures of the organic UV filters allowed in the EU framework.

Camphor derivatives

Benzylidene camphor sulphonic acid (BCSA)

Camphor benzalkonium methosulfate (CBM)

4-Methylbenzylidene camphor (MBC)

Polyacrylamidomethyl benzylidene camphor (PBC)

Terephthalylidene dicamphor sulphonic acid (TDSA)

Triazine derivatives

$R_1=R_2$
R_3
Bis-ethylhexyloxyphenol methoxyphenyl triazine (BEMT)

$R_1=R_2=R_3$
Ethylhexyl triazona (EHT)

$R_1=R_2$
R_3
Diethylhexyl butamido triazone (DEBT)

$R_1=R_2=R_3$
Tris-biphenyl triazine (TBT)

Others

Butyl methoxydibenzoylmethane (BMDM)

Octocrylene (OC)

$R = CH_3$ approx. 92.5 %

$R =$ approx. 6.0 %

$R =$ approx. 1.5 %

Polysilicone-15 (P15)

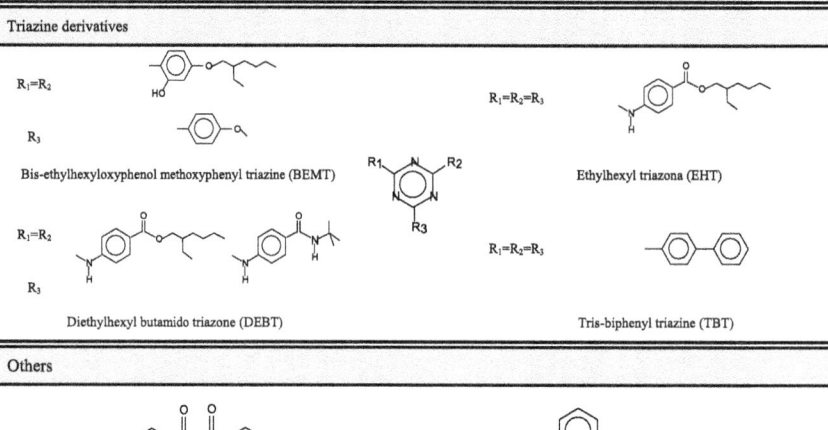

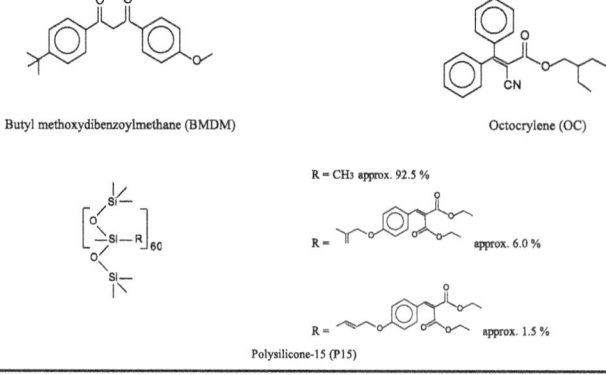

Figure 5.1, cont'd

ANALYTICAL METHODS FOR ULTRAVIOLET FILTERS IN COSMETIC PRODUCTS

All the aforementioned restrictions established for UV filters need to be accompanied by efficient and reliable analytical methods to control for them.

An interesting procedure, based on a paper by Hauri et al. (2003), was approved as the EU standard (EN 16344:2013) for the determination of 10 organic UV filters.

With the aim of complementing the aforementioned standard and implementing a broad-spectrum analytical method to improve and facilitate quality control in the cosmetics industry, a new proposal covering the determination of the 22 organic UV filters in use in the EU at the time of writing is being approved as an EU standard (prEN 17156, WI 00392030, 2017) (see Chapter 4). This standard is based on papers published by the authors of this chapter (Salvador and Chisvert, 2005; Chisvert et al., 2013), which will be discussed later. In summary, the target 22 compounds are grouped into three categories according to their solubility, i.e., the 'water-soluble' (comprising 6 of these compounds), the 'fat-soluble' (comprising 15 of them) and the 'polymeric' (comprising just polysilicone-15), which are measured by liquid chromatography (LC) with UV/VIS spectrometry under different chromatographic conditions.

The analytical literature covers many other papers dealing with the determination of UV filters in cosmetic products. A detailed and exhaustive review of those methods published up to 2006 can be found in the chapter devoted to UV filters in the first edition of this book (Chisvert and Salvador, 2007). At that time, a detailed study of those published papers indicated two main shortcomings: most of them did not deal with a high number of UV filters and some of the methods were not particularly suitable for routine production control as they required laborious sample preparation procedures, with an excessive amount of analysis time and/or use of toxic organic solvents. Since 2006, the demand for reliable and broad-spectrum analytical methods to improve and facilitate quality control in the cosmetics industry seems to have been heard. Yet it is worth mentioning the methods published by Schneider et al. (1996), Rastogi (2002), Hauri et al. (2003, 2004), Schakel et al. (2004) and Salvador and Chisvert (2005), by which a broad spectrum of UV filters were determined.

A detailed review of the papers published since the last edition has been carried out here. Table 5.2 presents the experimental details and some interesting remarks on published papers dealing with the determination of organic UV filters, whereas Table 5.3 collects those published papers dealing with inorganic UV filters.

Determination of Organic Ultraviolet Filters

The fact that UV filters are usually combined in cosmetic formulations makes it difficult to determine them by direct measurement without a previous separation step, so chromatography is usually used. Among the chromatographic techniques, LC is, by far, the

94   Analysis of Cosmetic Products

Table 5.2 Papers published from January 2006 to December 2016 concerning organic ultraviolet filter determination in cosmetic products (in chronological order)

Authors	Target UV filters[a]	Type of matrix[b]	Sample preparation[c]	Analytical technique[c]	Remarks[d]
De Orsi et al. (2006)	BEMT, BMDM, BZ3, BZ4, DHHB, EHMC, EHT, MBC, MBBT, PBSA, OC	EM, OI	Sample is dispersed in methanol:acetic acid (aq) or water plus dimethylacetamide:isopropanol, depending on the analytes, sonicated (10 min) and filtered	LC–UV/VIS, C_{18}, C_8 or amide C_{16} columns at 35°C, and gradient acetonitrile:perchloric acid (aq), isocratic methanol:acetonitrile or isocratic methanol:perchloric acid (aq) as mobile phase, respectively	LOD: 0.005%–0.1% R.: 94%–103%
Cardoso et al. (2007)	BZ3, EHMC, MBC	EM	Sample is dispersed in methanol by sonication (10 min), centrifuged and diluted	SWV, mercury hanging electrode and Britton–Robinson buffer containing cetyl trimethylammonium bromide	LOD: no data R.: 97%–100%
Dencausse et al. (2008)	BEMT, BMDM, BZ3, EHMC	EM	Sample is solved in ethanol	LC–UV/VIS, C_{18} column, gradient tetrahydrofuran:acetonitrile:acetic acid (aq) as mobile phase	LOD: no data R.: 99%–105%
Wu et al. (2008)	BZ1, BZ2, BZ3	EM	Sample is dispersed in methanol by sonication (30 min), centrifuged, filtered and subjected to CPE (Triton X-114 as surfactant, 50°C, 20 min incubation time) and centrifuged; surfactant-rich phase is diluted with ethanol	MEKC–UV/VIS, uncoated fused-silica capillary, borate buffer containing sodium dodecyl sulphate as running buffer	LOD: 0.0007%–0.001% R.: 89%–102% EF: 20

Lee et al. (2008)	BEMT, EHMC, EHS, EHT, IMC, MBBT	EM, FO, LO, LS, MU	Sample is dispersed in dimethyl formamide:methanol by sonication (10 min) and filtered	LC–UV/VIS, C$_{18}$ column at room temperature, gradient ethanol:water as mobile phase	LOD: no data R: 99%–102% UPLC shorter runtime than LC
Imamovic et al. (2009)	BMDM	EM	Sample is dispersed in acetonitrile containing polysorbate-80, then magnetically stirred (60 min, 60°C) and diluted	LC–UV/VIS, C$_{18}$ column at room temperature, gradient ethanol:water as mobile phase	LOD: 0.06% R: no data
Balaguer et al. (2009)	BEMT, BMDM, BZ3, DEBT, EHMC, EHS, HMS, IMC, MBBT, MBC, OC, PBSA	EM, LO, LS, PE	Sample is dispersed in ethanol by sonication (15 min) and filtered	LC–UV/VIS, C$_{18}$ monolithic column at 30°C, gradient ethanol:acetic acid (aq) as mobile phase	LOD: 0.02%–0.3% R: 93%–123%
Nyeborg et al. (2010)	BEMT, BMDM, BZ3, DEBT, DHHB, EHMC, EHS, EHT, IMC, MBBT, OC, PBSA	EM, LS, MU	Sample is dispersed in ethanol by sonication (30 min) and filtered	LC–UV/VIS, C$_{18}$ column at 40°C, gradient ethanol:phosphoric acid (aq) as mobile phase	LOD: 0.002%–0.08% R: 98%–101%
Wharton et al. (2011)	BMDM, BZ3, EHDP, EHMC, EHS, MBC, OC	ST		LC–UV/VIS, C$_{18}$ column at room temperature, gradient ethanol:acetic acid (aq) as mobile phase DART–MS by dipping a glass rod	No real samples were analysed
Haunschmidt et al. (2011)	BMDM, BZ3, EHDP, EHMC, EHS, HMS, MBC, OC	EM, LO, LS, MU, OI	Sample is dispersed in methanol by sonication (10 min) and diluted		LOD: 0.0002%–0.05% R: 71%–120%

Continued

Table 5.2 Papers published from January 2006 to December 2016 concerning organic ultraviolet filter determination in cosmetic products (in chronological order)—cont'd

Authors	Target UV filters[a]	Type of matrix[b]	Sample preparation[c]	Analytical technique[c]	Remarks[d]
Liu and Wu (2011)	BZ3, BZ4, EHMC, EHS, HMS, IMC, MBBT, MBC, OC, PABA, PBSA	EM, FO, LO, LS, OI	Sample is mixed with methanol: tetrahydrofuran:perchloric acid (aq) or tetrahydrofuran, depending on the sample, sonicated (30 min), diluted and filtered	LC–UV/VIS, C$_{18}$ column, gradient methanol:tetrahydrofuran:perchloric acid (aq) as mobile phase	LOD: 0.02%–0.05% R: 98%–102%
Pinto et al. (2011)	BZ3, EHMC, EHS	EM, GE, LO	Sample is dispersed in ethanol, diluted and filtered	LC–UV/VIS, C$_{18}$ column at room temperature, isocratic methanol:acetic acid (aq) as mobile phase	LOD: 0.3%–0.4% R: 87%–106%
Kim et al. (2011)	BMDM, BZ3, BZ8, EHDP, EHMC, EHS, IMC, MBC, OC	EM	Sample is dispersed in methanol:ethyl acetate by sonication (10 min), diluted and filtered	LC–UV/VIS, C$_{18}$ column at 30°C, gradient acetonitrile:acetic acid (aq) as mobile phase	LOD: 0.05%–0.3% R: 98%–102%
Yang et al. (2011)	BZ3, BZ4, EHDP, EHMC, EHS	EM, FO, LO, LS	Sample is mixed, dispersed in hexane by sonication (5 min) and centrifuged; supernatant is filtered, dried, reconstituted in water and subjected to dHF-LPME (toluene as extraction solvent, 30 min extraction time)	LC–UV/VIS, C$_{18}$ column at 40°C, isocratic methanol:water as mobile phase	LOD: 0.001%–0.1% R: 98%–102% EF: 24–57
Sobanska and Pyzowski (2012)	EHT	EM, LO	Sample is dispersed in methanol by shacking (60 min)	TLC-DS, silica plate, cyclohexane:diethyl ether as mobile phase	LOD: 0.15% R: 95%–105%
Junior et al. (2012)	OC	EM	Sample is dispersed in ethanol by sonication (10 min), centrifuged and diluted	SWV, mercury hanging electrode and Britton–Robinson buffer containing ethanol	LOD: no data R: 9100%–106%

Kapalavavi et al. (2012)	BMDM, EHS, OC, PBSA	EM	Sample is dispersed in methanol by vortex agitation and filtered	HTLC, C$_{18}$ column at 150°C–200°C, isocratic methanol:water as mobile phase SBWC, C$_{18}$ column at 250°C, isocratic water as mobile phase	LOD: no data R: 90%–113%
Kim et al. (2012)	BMDM, BZ3, BZ8, DEBT, DTS, EHDP, EHMC, EHT, IMC, MA, MBBT, MBC, OC, PABA, PBSA, TDSA	EM, LO	Sample is mixed with methanol or tetrahydrofuran, depending on the analyte, sonicated and filtered	LC–UV/VIS, C$_{18}$ column at 40°C, gradient methanol:phosphate buffer as mobile phase	LOD: no data R: 95%–117%
Yousef-Agha et al. (2013)	BZ3, EHMC, EHS, OC	EM	Sample is dispersed in methanol/isopropanol by sonication and filtered	LC–UV/VIS, C$_{18}$ column at room temperature, isocratic methanol:water as mobile phase	LOD: no data R: no data
Almeida et al. (2013)	BZ1, BZ3 + (BZ, HBZ)	EM, LO	Sample is diluted, subjected to BAμE (modified pyrrolidone or activated carbon as sorbent, 4 or 16h as extraction time), back-extracted in acetonitrile:methanol, evaporated and reconstituted in methanol	LC–UV/VIS, C$_{18}$ column at 25°C, isocratic methanol:water	LOD: 0.001%–0.6% R: no data
Chisvert et al. (2013)	BEMT, BMDM, BZ3, DEBT, DHHB, DTS, EHDP, EHMC, EHS, EHT, HMS, IMC, MBC, MBBT, OC	EM, LO, LS	Sample is dispersed in ethanol by sonication (2min) and filtered	LC–UV/VIS, C$_{18}$ column at 60°C, gradient ethanol:formic acid (aq) containing hydroxypropyl-β-cyclodextrin as mobile phase	LOD: 0.001%–0.02% R: 97%–104%

Continued

Table 5.2 Papers published from January 2006 to December 2016 concerning organic ultraviolet filter determination in cosmetic products (in chronological order)—cont'd

Authors	Target UV filters[a]	Type of matrix[b]	Sample preparation[c]	Analytical technique[c]	Remarks[d]
Kale et al. (2014)	BMDM, BZ3, EHMC	EM	Sample is dispersed in methanol by sonication (30 min) and filtered	TLC–DS, silica or C_{18} plate, cyclohexane:n-hexane:acetone:diethyl ether or acetonitrile:water, respectively, as mobile phase	LOD: no data R.: 93%–102%
Chang et al. (2015)	BMDM, BZ3, BZ4, DEBT, DTS, EHDP, EHMC, EHS, EHT, HMS, MBC, OC, PABA, PBSA	EM	Sample is dispersed in ethanol by sonication (30 min) and filtered	LC–UV/VIS, C_{18} column at 28°C, gradient ethanol:phosphate buffer	LOD: 0.002%–0.2% R.: 91%–110% The pair EHMC–BMDM was not satisfactorily separated
Lesellier et al. (2015)	BMDM, BZ3, DEBT, DHHB, EHMC, EHS, EHT, OC	EM	Sample is dissolved in methanol	SFC–UV/VIS, 2EP column at 35°C, 15 MPa, gradient CO_2:methanol:ethanol as mobile phase	LOD: no data R.: no data Only one sample was analysed
Li et al. (2015)	BZ1, BZ3 + (BZ)	LO	Sample is filtered, diluted in water and subjected to DSPE (MIL-101 as sorbent, 40 min extraction time), back-extracted with methanol:acetonitrile, evaporated and reconstituted in methanol	LC–UV/VIS, C_{18} column at 30°C, gradient methanol:phosphoric acid (aq) as mobile phase	LOD: 0.0000009%–0.0000012% (3S/N criterion) R.: 94%–105%

López-Gazpio et al. (2015)	BZ3, BZ4	GE, EM, SH, PE	Sample is dispersed in methanol:water by sonication (5–10 min), diluted and filtered	MEKC-UV/VIS, uncoated fused-silica capillary at 25°C, borate buffer containing sodium dodecyl phosphate as running buffer	LOD: no data R.: 90%–107%
Sobanska et al. (2015)	BZ4	SH	Sample is dissolved and diluted with phosphate buffer	TLC-UV/VIS, silica plate, ethyl acetate:ethanol:water as mobile phase	LOD: no data R.: 98%–103%
Hsiao et al. (2015) and Lin et al. (2015)	BMDM, BZ1, BZ2, BZ3, BZ4, BZ6, BZ8, EHMC, OC	EM, LO	Sample is dispersed in methanol by sonication (15 min), centrifuged, filtered and diluted with Tris buffer	MEKC-UV/VIS, uncoated fused-silica capillary at 25°C, Tris buffer containing sodium dodecyl sulphate and γ-cyclodextrin as mobile phase	LOD: No data R.: 95%–105%
Wharton et al. (2015)	BMDM, BZ3, EHDP, EHMC, EHS, HMS, IMC, MBC, OC, PBSA	EM	Sample is dispersed in ethanol by sonication, diluted and filtered	LC-UV/VIS, C_{18} column at 35°C, isocratic ethanol:acetic acid (aq)	LOD: 0.025%–0.5% R.: No data
Vila et al. (2015)	BZ1, BZ2, BZ3, BZ4, BZ6, BZ8, EHDP, EHMC, EHS, HMS, IMC, MA, MBC, OC + (EC)	EM, LS, NP	Sample is mixed with Na_2SO_4/Florisil, subjected to PLE (acetonitrile as extraction solvent, 10 min, 90°C, 1500 psi), diluted with ethyl acetate and derivatized with acetic anhydride (100°C, 60 min)	GC-MS/MS, silphenylene polymer column, helium as carrier	LOD: 0.0000003%–0.00006% R.: 37%–110%

Continued

Table 5.2 Papers published from January 2006 to December 2016 concerning organic ultraviolet filter determination in cosmetic products (in chronological order)—cont'd

Authors	Target UV filters[a]	Type of matrix[b]	Sample preparation[c]	Analytical technique[c]	Remarks[d]
Ma et al. (2015)	BZ1, BZ3 + (BZ, HBZ)	EM	Sample is dispersed in methanol by sonication (10 min) and centrifuged; supernatant is diluted and then subjected to DLLME ([C6MIM][PF6] as extraction solvent and acetonitrile/Triton X-114 as disperser solvent); finally the sedimented phase is diluted	MEKC–UV/VIS, uncoated fused-silica capillary at 25°C, borate buffer containing sodium dodecyl phosphate as running buffer	LOD: no data R: 80%–118% EF: 25.3–40.5
Meng et al. (2016)	BMDM, BZ1, BZ2, BZ3, BZ6, BZ8, DHHB, EHS, HMS, IMC, MBC, PBSA + (BZ, BZ10, BZ12)	EM, LO, LS	Sample is mixed with tetrahydrofuran containing ammonia (aq), vortexed, diluted with methanol, sonicated (10 min) and ultracentrifuged; supernatant is evaporated, reconstituted in methanol and filtered	LC–MS/MS, C_{18} column at 30°C, gradient methanol:formic acid (aq) for ESI$^+$ and gradient methanol:ammonia (aq) for ESI$^-$	LOD: 0.0002%–0.002% R: 87%–103%
Vila et al. (2016)	BMDM, BZ1, BZ2, BZ3, BZ4, BZ6, BZ8, EHDP, EHMC, IMC, MA, MBC, OC, PBSA + (EC)	GE, EM, LO, LS, MU, NP	Sample is mixed with Na_2SO_4/sand, subjected to PLE (methanol:acetone as extraction solvent, 10 min, 90°C, 1500 psi) and diluted	LC–MS/MS, C_{18} column at 30°C, gradient methanol:formic acid/ammonia (aq)	LOD: 0.00000004%–0.000003% R: 64%–112%

[a]See Table 5.1 for abbreviation key, except for those in parentheses: *BZ*, benzophenone; *BZ10*, benzophenone-10; *BZ12*, benzophenone-12; *EC*, etocrylene; *HBZ*, hidroxybenzphenone.

[b]*EM*, emulsion; *FO*, foundation; *GE*, gel; *LO*, lotion; *LS*, lipstick; *MU*, makeup; *NP*, nail polish; *OI*, oil; *PE*, perfume; *SH*, shampoo; *ST*, standard solution.

[c]*BAμE*, bar adsorptive microextraction; *CPE*, cloud point extraction; *DART*, direct analysis in real time; *DLLME*, dispersive liquid–liquid microextraction; *dHF–LPME*, dynamic hollow fibre–liquid phase microextraction; *DS*, densitometry; *DSPE*, dispersive solid-phase extraction; *EF*, enrichment factor; *ESI*, electrospray ionization; *GC*, gas chromatography; *HTLC*, high-temperature liquid chromatography; *LC*, liquid chromatography; *MEKC*, micellar electrokinetic chromatography; *MS*, mass spectrometry; *MS/MS*, tandem mass spectrometry; *PLE*, pressurized liquid extraction; *SBWC*, subcritical water chromatography; *SFC*, supercritical fluid chromatography; *SWC*, subcritical water chromatography; *SWV*, square-wave voltammetry; *TLC*, thin-layer chromatography; *UPLC*, ultrahigh-performance liquid chromatography; *UV/VIS*, ultraviolet/visible spectrometry.

[d]*LOD*, limit of detection (referred to sample, not to sample solution); *R*, recovery (referred to sample, not to sample solution).

Table 5.3 Papers published from January 2006 to December 2016 concerning inorganic ultraviolet filter determination in cosmetic products (in chronological order)

Authors	Target UV filters[a]	Type of matrix[b]	Sample preparation[c]	Analytical technique[c]	Remarks[d]
Kim et al. (2006)	TiO_2	EM, FO, MU	Sample is calcined and residue is dissolved in hot sulphuric acid/ammonium sulphate; Ti(IV) is reduced with metallic aluminium	Redox titration, ferric ammonium sulphate as titrant and potassium thiocyanate as indicator	LOD: no data R: 96%–105%
Melquiades et al. (2008)	TiO_2	EM	No sample preparation	pXRF	LOD: no data R: no data
Zacharidis and Sahanidou (2009)	TiO_2, ZnO and other metallic elements	EM	Sample is acid digested (2 h, pressurized vessels) and diluted, or it is emulsified in a slurry (Triton X-100 as surfactant)	ICP-AES	LOD: 0.00002% (3 s criterion) R: 95%–98%
Oh et al. (2010)	TiO_2	EM	No sample preparation	Raman	LOD: no data R: no data
Menneveux et al. (2015)	TiO_2	EM	No sample preparation	LIBS	LOD: no data R: no data
Dan et al. (2015)	TiO_2	EM	Sample is dispersed in Triton X-100 (aq) by sonication/vortex and diluted	SP-ICP-MS	LOD: no data R: no data
Bairi et al. (2016)	TiO_2, ZnO	EM	No sample preparation	pXRF	LOD: 0.01%–0.06% (3 s criterion) R: 93%–136%
Benítez-Martínez et al. (2016)	TiO_2	EM	Sample is subsequently sonicated by adding water, methanol and hexane, then centrifuged; the aqueous phase is cleaned up with hexane, evaporated and reconstituted in methanol	Quenched fluorimetry of N-doped graphene quantum dots	LOD: 0.00014 (3 s_a criterion) R: 85%–106%

[a] See Table 5.1 for abbreviation key.
[b] EM, emulsion; FO, foundation; MU, makeup.
[c] AES, atomic emission spectrometry; ICP, inductively coupled plasma; LIBS, laser-induced breakdown spectroscopy; MS, mass spectrometry; pXRF, portable X-ray fluorescence; SP, single particle.
[d] LOD, limit of detection (referred to sample, not to sample solution); R, recovery (referred to sample, not to sample solution).

most commonly employed, as UV filters have relatively high boiling points, especially the water-soluble ones. In this sense, gas chromatography (GC) was used on just one occasion (Vila et al., 2015), being necessary to perform a derivatization step to form more volatile compounds.

Regarding LC, the reversed phase is usually employed, mainly with traditional C_{18} stationary phases, although other types such as C_8, amide C_{16} (De Orsi et al., 2006) or C_{18} monolithic (Balaguer et al., 2009) have been used. Although classical hydro-organic mixtures with acetonitrile, methanol or tetrahydrofuran as organic solvents have been used, more recently authors have used ethanol–water mixtures (Lee et al., 2008; Balaguer et al., 2009; Nyeborg et al., 2010; Wharton et al., 2011, 2015; Chisvert et al., 2013; Chang et al., 2015), thus contributing to the development of more environmentally green analytical methods. The use of ethanol–water mixtures for the determination of UV filters had already been proposed by the authors of this chapter for the analysis of various cosmetic products (Chisvert et al., 2001a,b; Chisvert and Salvador, 2002; Salvador et al., 2003, 2005; Salvador and Chisvert, 2005).

The high capability of these compounds to absorb UV radiation makes the UV/VIS spectrometry detector the most suitable for LC analysis, achieving limits of detection (LODs) below 0.1% in most cases. This is more than enough to carry out quality control in the cosmetics industry, because UV filters are added at much higher concentrations in the final product. More sophisticated (and expensive) detectors, such as tandem mass spectrometry, that could achieve lower LODs (i.e., <0.001%) have been less used (Meng et al., 2016; Vila et al., 2016), maybe because UV filters are used as major ingredients and such LODs are unnecessary.

In general terms, the pairs IMC–MBC, MBC–DHHB, BMDM–HMS and BMDM–EHMC (see Table 5.1 for full names) present difficulties in their separation by LC. There is only one published method whereby all these mixtures are considered and satisfactorily resolved (Chisvert et al., 2013), using 2-hydroxypropyl-β-cyclodextrin as the mobile phase modifier. This method is part of the aforementioned EU standard under study (prEN 17156 WI 00392030, 2017) (see Chapter 4).

In addition to LC and GC, there are a few articles using other separation techniques, such as micellar electrokinetic chromatography (MEKC) (Wu et al., 2008; López-Gazpio et al., 2015; Hsiao et al., 2015; Lin et al., 2015; Ma et al., 2015), thin-layer chromatography (Sobanska and Pyzowski, 2012; Kale et al., 2014; Sobanska et al., 2015), high-temperature LC and subcritical water chromatography (Kapalavavi et al., 2012) and supercritical fluid chromatography (Lesellier et al., 2015).

There are also a very few articles without a previous separation step. First, Cardoso et al. (2007), and later Junior et al. (2012), used square-wave voltammetry, but just three and one UV filter(s), respectively, were determined.

Finally, it is worth emphasizing the article published by Haunschmidt et al. (2011), who carried out the semiquantitative determination of eight UV filters by direct analysis in real time—mass spectrometry. They just diluted the sample and measured it using a time-of-flight mass spectrometer, thus obtaining a mixed spectrum containing the exact mass of all the target compounds.

Regarding sample preparation, most of the cases consisted just in adding an appropriate solvent (usually methanol, ethanol or acetonitrile) to the cosmetic sample and then sonicating it for several minutes (from less than 2 to 30 min, depending on the method, see Table 5.2) to dissolve the sample totally or at least leach the target compounds (e.g., difficult-to-dissolve samples such as makeups or foundations containing insoluble metallic oxides). In another few cases, magnetic stirring (Imamovic et al., 2009), vortexing (Kapalavavi et al., 2012) or mechanical shaking (Sobanska and Pyzowski, 2012) was used, and in just a couple of cases, pressurized liquid extraction was proposed (Vila et al., 2015, 2016).

The use of sample preconcentration techniques is not really necessary in this case, because UV filters are major ingredients. Nevertheless, some authors propose the use of microextraction techniques, such as cloud-point extraction (Wu et al., 2008), dynamic hollow fibre–liquid phase microextraction (Yang et al., 2011), bar adsorptive microextraction (Almeida et al., 2013), dispersive solid-phase microextraction (Li et al., 2015) and dispersive liquid–liquid microextraction (Ma et al., 2015). Hsiao et al. (2015) and Lin et al. (2015) used MEKC and achieved online analyte enrichment by working with the analyte focusing by micelle collapse stacking approach.

Determination of Inorganic Ultraviolet Filters

As can be seen in Table 5.3, many fewer are the published articles devoted to the quantitative determination of inorganic UV filters compared to those devoted to the determination of organic UV filters and summarized earlier. It should be said that all the papers focused on particle size determination are excluded from this chapter. Nevertheless, interested readers are kindly encouraged to read Chapter 12 devoted to nanomaterials.

Kim et al. (2006) proposed a redox titration for TiO_2 determination, based on calcination of the sample, residue dissolution in hot sulphuric acid/ammonium sulphate, reduction of Ti(IV) to Ti(III) with metallic aluminium and then titration with Fe(III).

The use of portable X-ray fluorescence was proposed for TiO_2 direct determination (Melquiades et al., 2008) in a manner similar to what Kawaguchi et al. (1996) previously did. Later, Bairi et al. (2016) used a similar approach to determine both TiO_2 and ZnO.

Oh et al. (2010) and Menneveux et al. (2015) also carried out the direct determination of TiO_2 without further sample preparation by using Raman spectroscopy and laser-induced breakdown spectroscopy, respectively.

Zacharidis and Sahanidou (2009) used inductively coupled plasma–atomic emission spectrometry for the determination of both TiO_2 and ZnO and other metallic elements, either with acid digestion of the sample in closed pressurized vessels or using an emulsified slurry with a non-ionic surfactant, similar to Salvador et al. (2000). A similar approach was followed by Dan et al. (2015) for the determination of TiO$_2$ by single-particle inductively coupled plasma–mass spectrometry.

Finally, Benítez-Martínez et al. (2016) determined TiO$_2$ indirectly by means of the quenching effect that it produced in an N-doped graphene quantum dots solution used as a fluorescent probe.

REFERENCES

Almeida, C., Stepkowska, A., Alegre, A., Nogueira, J.M.F., 2013. J. Chromatogr. A 1311, 1.
ASEAN Cosmetic Directive, 2003.
Bairi, V.G., Lim, J.-H., Quevedo, I.R., Mudalige, T.K., Linder, S.W., 2016. Spectrochim. Acta Part B 116, 21.
Baki, G., Alexander, K.S., 2015. Introduction to Cosmetic Formulation and Technology. Wiley.
Balaguer, A., Talamantes, S., Duran, E., Cuadrado, P., 2009. LC GC Eur. 22, 562.
Benítez-Martínez, S., López-Lorente, A.I., Valcárcel, M., 2016. Microchim. Acta 183, 781.
Brausch, J.M., Rand, G.M., 2011. Chemosphere 82, 1518.
Cardoso, J.C., Armondes, B.M.L., de Araújo, T.A., Raposo Jr., J.L., Poppi, N.R., Ferreira, V.Z., 2007. Microchem. J. 85, 301.
Chang, N.I., Yoo, M.Y., Lee, S.H., 2015. Int. J. Cosmet. Sci. 37, 175.
Chisvert, A., Salvador, A., 2002. J. Chromatogr. A 977, 277.
Chisvert, A., Salvador, A., 2007. UV filters in sunscreens and other cosmetics. In: Salvador, A., Chisvert, A. (Eds.), Regulatory Aspects and Analytical Methods. Elsevier, Amsterdam.
Chisvert, A., Pascual-Martí, M.C., Salvador, A., 2001a. Fresenius' J. Anal. Chem. 369, 638.
Chisvert, A., Pascual-Martí, M.C., Salvador, A., 2001b. J. Chromatogr. A 921, 207.
Chisvert, A., León-González, Z., Tarazona, I., Salvador, A., Giokas, D., 2012. Anal. Chim. Acta 752, 11.
Chisvert, A., Tarazona, I., Salvador, A., 2013. Anal. Chim. Acta 790, 61.
Cole, C., 2016. Sunscreen formulation: optimizing efficacy of UVB and UVA protection. In: Wang, S.Q., Lim, H.W. (Eds.), Principles and Practice of Photoprotection. Springer.
Commission Regulation (EC) 2015/1298 of 28 July 2015 Amending Annexes II and VI to Regulation (EC) No 1223/2009 of the European Parliament and of the Council on Cosmetic Products. http://eur-lex.europa.eu/legal-content/EN/TXT/?uri=CELEX:02009R1223-20160812&from=EN.
Commission Regulation (EU) 2016/621 of 21 April 2016 Amending Annex VI to Regulation (EC) No 1223/2009 of the European Parliament and of the Council on Cosmetic Products. http://eur-lex.europa.eu/legal-content/EN/ALL/?uri=CELEX%3A32016R0621.
Daly, S., Ouyang, H., Maitra, P., 2016. Chemistry of sunscreens. In: Wang, S.Q., Lim, H.W. (Eds.), Principles and Practice of Photoprotection. Springer.
Dan, Y., Shi, H., Stephan, C., Liang, X., 2015. Microchem. J. 122, 119.
De Orsi, D., Giannini, G., Gabliardi, L., Porrá, R., Berri, S., Bolasco, A., Carpani, I., Tonelli, D., 2006. Chromatographia 64, 509.
Dencausse, I., Galland, A., Clamou, J.L., Basso, J., 2008. Int. J. Cosmet. Sci. 30, 373.
Directive 76/768/CEE of the Council of 27 July 1976 on the Approximation of the Laws of the Member States Relating to Cosmetic Products, and Its Successive Amendments and Adaptations.
EN 16344, 2013. Cosmetics – Analysis of Cosmetic Products – Screening for UV-filters in Cosmetic Products and Quantitative Determination of 10 UV-Filters by HPLC. https://standards.cen.eu.
FDA, 2017. Code of Federal Regulations, Title 21: Food and Drugs, Chapter I, Subchapter D, Part 352 – Sunscreen Drug Products for Over-the-Counter Human Use. http://www.accessdata.fda.gov/SCRIPTS/cdrh/cfdocs/cfCFR/CFRSearch.cfm.
Gilbert, E., Pirot, F., Berthlle, V., Roussel, L., Falson, F., Padois, K., 2013. Int. J. Cosmet. Sci. 35, 208.
Gomez, E., Pillon, A., Fenet, H., Rosain, D., Duchesne, M.J., Nicolas, J.C., Balaguer, P., Casellas, C., 2005. J. Toxicol. Environ. Health A 68, 239.
Haunschmidt, M., Buchberger, W., Klampfl, C.W., Hertsens, R., 2011. Anal. Methods 3, 99.
Hauri, U., Lütolf, B., Hohl, C., 2003. Mitt. Leb. Hyg. 94, 80.

Hauri, U., Lütolf, B., Schlegel, U., Hohl, C., 2004. Mitt. Leb. Hyg. 95, 147.
Hsiao, W.-Y., Jiang, S.-J., Feng, C.-H., Wang, S.-W., Chen, Y.L., 2015. J. Chromatogr. A 1383, 175.
Imamovic, B., Sober, M., Becic, E., 2009. Int. J. Cosmet. Sci. 31, 383.
ISO 24444, 2010. Cosmetics – Sun Protection Test Methods – In Vivo Determination of the Sun Protection Factor (SPF).
Junior, J.B.G., Araujo, T.A., Trindade, M.A.G., Ferreira, V.S., 2012. Int. J. Cosmet. Sci. 34, 91.
Kale, S., Kulkarni, K., Ugale, P., Jadav, K., 2014. Int. J. Pharm. Pharm. Sci. 6, 391.
Kapalavavi, B., Marple, R., Gamsky, C., Yang, Y., 2012. Int. J. Cosmet. Sci. 34, 169.
Kawaguchi, A., Ishida, M., Saitoh, I., 1996. Spectrosc. Lett. 29, 345.
Kim, Y.S., Kim, B.-M., Park, S.-C., Jeong, H.-J., Chang, I.S., 2006. J. Cosmet. Sci. 57, 377.
Kim, K., Mueller, J., Park, Y.-B., Jung, H.-R., Kang, S.-H., Yoon, M.-H., Lee, J.-B., 2011. J. Chromatogr. Sci. 49, 554.
Kim, D., Kim, S., Kim, S.-A., Choi, M., Kwon, K.-J., Kim, M., Kim, D.-S., Kim, S.-H., Choi, B.-K., 2012. J. Cosmet. Sci. 63, 103.
Krause, M., Klit, A., Jensen, M.B., Søeborg, T., Frederiksen, H., Schlumpf, M., Lichtensteiger, W., Skakkebaek, N.E., Drzewiecki, K.T., 2012. Int. J. Androl. 35, 424.
Lee, S.-M., Jeong, H.-J., Chang, I.S., 2008. J. Cosmet. Sci. 59, 469.
Lesellier, E., Mith, D., Dubrulle, I., 2015. J. Chromatogr. A 1423, 158.
Li, N., Zhu, Q., Yang, Y., Huang, J., Dang, X., Chen, H., 2015. Talanta 132, 713.
Lin, Y.-H., Lu, C.-Y., Jiang, S.-J., Hsiao, W.-Y., Cheng, H.-L., Chen, Y.-L., 2015. Electrophoresis 36, 2396.
Liu, T., Wu, D., 2011. Int. J. Cosmet. Sci. 33, 408.
López-Gazpio, J., García-Ramos, R., Millán, E., 2015. Electrophoresis 36, 1064.
Lucas, R.M., Norval, M., Neale, R.E., Young, A.R., de Grujil, F.R., Takizawa, Y., var del Leun, J.C., 2015. Photochem. Photobiol. Sci. 14, 53.
Ma, T., Li, Z., Niu, Q., Li, Y., Zhou, W., 2015. Electrophoresis 36, 2530.
Melquiades, F.L., Ferreira, D.D., Appoloni, C.R., Lopes, F., Lonni, A.G., Oliveira, F.M., Duarte, J.C., 2008. Anal. Chim. Acta 613, 135.
Meng, X., Ma, Q., Bai, H., Wang, Z., Han, C., Wang, C., 2016. Int. J. Cosmet. Sci. 38, 1.
Menneveux, J., Wang, F., Lu, S., Bai, X., Motto-Ros, V., Gilon, N., Chen, Y., Yu, J., 2015. Spectrochim. Acta Part B 109, 9.
Mercosur, 2015. Reglamento técnico Mercosur sobre lista de filtros ultravioletas permitidos para productos de higiene personal, cosméticos y perfumes, MERCOSUR/GMC/RES. N°44/15.
MHLW Japan, 2000. Standard for Cosmetics, No.331.
Molins-Delgado, D., Gago-Ferrero, P., Diaz-Cruz, M.S., Barcelo, D., 2016. Environ. Res. 145, 126.
Nohynek, G., Dufour, E.K., 2012. Arch. Toxicol. 86, 1063.
Nyeborg, M., Pissavini, M., Lemasson, Y., Doucet, O., 2010. Int. J. Cosmet. Sci. 32, 47.
Oh, C., Yoon, S., Kim, E., Han, J., Chung, H., Jeong, H.-J., 2010. J. Pharm. Biomed. Anal. 53, 762.
Pinto, F.M., Fonseca, Y.M., Vicentini, F.T.M.C., Vieira, M.J., Henriques, M.P., 2011. Quim. Nova 34, 879.
prEN 17156 (WI 00392030), CEN/TC 392-Cosmetics, 2017. Analytical methods - HPLC/UV method for the identification and quantitative determination in cosmetic products of the 22 organic UV filters in use in the EU. https://standards.cen.eu.
Rastogi, S.C., 2002. Contact Dermat. 46, 348.
Regulation (EC) 1223/2009 of the European Parliament and of the Council of 30 November 2009 on Cosmetic Products, and its successive amendments. http://eur-lex.europa.eu/legal-content/EN/ALL/?uri=CELEX%3A32009R1223.
Salvador, A., Chisvert, A., 2005. Anal. Chim. Acta 537, 15.
Salvador, A., Chisvert, A. (Eds.), 2007. Analysis of Cosmetic Products. Elsevier.
Salvador, A., Pascual-Martí, M.C., Adell, J.R., Requeni, A., March, J.G., 2000. J. Pharm. Biomed. Anal. 22, 301.
Salvador, A., de la Ossa, M.D., Chisvert, A., 2003. Int. J. Cosmet. Sci. 25, 97.
Salvador, A., Chisvert, A., Jaime, M.A., 2005. J. Sep. Sci. 28, 2319.
SCCS, 2017. Scientific Committee on Consumer Safety's Website. https://ec.europa.eu/health/scientific_committees/consumer_safety/opinions_en.

Schakel, D.J., Kalsbeek, D., Boer, K., 2004. J. Chromatogr. A 1049, 127.
Schlumpf, M., Cotton, B., Conscience, M., Haller, V., Steinmann, B., Lichtensteiger, W., 2001. Environ. Health Perspect. 109, 239.
Schneider, P., Bringhen, A., Gonzenbach, H., 1996. Drug Cosmet. Ind. 159, 32.
Schreurs, R.H.M.M., Sonneveld, E., Jansen, J.H.J., van der Burg, B., 2005. Toxicol. Sci. 83, 264.
Shaath, N., 2016. The chemistry of UV filters. In: Wang, S.Q., Lim, H.W. (Eds.), Principles and Practice of Photoprotection. Springer.
Shi, H., Magaye, R., Castranova, V., Zhao, J., 2013. Part. Fibre Toxicol. 10, 15.
Singh, R., Nanda, A., 2014. Int. J. Cosmet. Sci. 36, 273.
Sobanska, A.W., Pyzowski, J., 2012. Sci. World J. 1.
Sobanska, A.W., Kalebasiak, K., Pyzowski, J., Brzezinska, E., 2015. J. Anal. Methods Chem. 1.
Stiefel, C., Schwack, W., 2015. Int. J. Cosmet. Sci. 37, 2.
Tovar-Sánchez, A., Sánchez-Quiles, D., Basterretxea, G., Benedé, J.L., Chisvert, A., Salvador, A., Moreno-Garrido, I., Blasco, J., 2013. PLoS One 8, e65451.
Vila, M., Lamas, J.P., García-Jares, C., Dagnac, T., Llompart, M., 2015. J. Chromatogr. A 1405, 12.
Vila, M., Facorro, R., Lamas, J.P., García-Jares, C., Dagnac, T., Llompart, M., 2016. Anal. Methods 8, 2016.
Wharton, M., Geary, M., O'Connor, N., Murphy, B., 2011. Int. J. Cosmet. Sci. 33, 164.
Wharton, M., Geary, M., O'Connor, N., Curtin, L., Ketcher, K., 2015. J. Chromatogr. Sci. 53, 1289.
Wu, Y.-W., Jiang, Y.-Y., Liu, J.-F., Xiong, K., 2008. Electrophoresis 29, 819.
Yang, H.Y., Li, H.F., Masahito, I., Lin, J.-M., Guo, G.S., Ding, M.Y., 2011. Sci. China 54, 1627.
Yousef, A.N., Haidar, S., Al-Khayat, 2013. Int. J. Pharm. Sci. Rev. Res. 23, 254.
Zacharidis, G.A., Sahanidou, E., 2009. J. Pharm. Biomed. Anal. 50, 342.

CHAPTER 6

Tanning and Whitening Agents in Cosmetics: Regulatory Aspects and Analytical Methods

Alberto Chisvert, Juan L. Benedé, Amparo Salvador
University of Valencia, Valencia, Spain

INTRODUCTION

Tanning and whitening cosmetic products are closely related products despite their opposite effects (i.e., employed for tanning or for bleaching the skin, respectively). Leaving aside pharmaceutical products to treat severe skin disorders, such as vitiligo (hypopigmentation) or melasma (hyperpigmentation), tanning and whitening cosmetic products are used to darken or to lighten the original tone of our skin for aesthetic reasons closely related with cultural and socioeconomical status or simply for fashion (Herrmann et al., 2015; Naidoo et al., 2016).

Before we describe all of these products, the natural process associated with the synthesis of melanin should be described to better understand how they work. Melanin is the natural pigment that appears as a protective measure when skin is exposed to the sun. This process is called melanogenesis and involves enzymatically catalysed and chemical reactions. Briefly, the biosynthesis of melanin is initiated with the first step of tyrosine oxidation to dopaquinone, catalysed by tyrosinase. This key enzyme for the melanogenesis process is located in the membrane of melanosomes, which are vesicles inside specialized cells called melanocytes located at the basal epidermis layer (Gillbro and Olsson, 2011; Burger et al., 2016). It should be emphasized that this first step is the rate-limiting step of the overall process (Chang, 2009). From dopaquinone synthesis, the melanin biosynthesis diverges into two different pathways depending on whether there is cysteine or glutathione present (Burger et al., 2016). This results in two types of melanin depending on the synthesis pathway, eumelanin (dark brown–black) and pheomelanin (light red–yellow), and the relative amounts of them are responsible for the constitutive skin pigmentation. Once melanin is synthesized, melanocytes transport the mature melanosomes containing melanin through their dendrites to surrounding keratinocytes (Desmedt et al., 2016), where it is accumulated and protects us from the sun (Burger et al., 2016).

Tanning Products

Among the different tanning products available on the market, two groups should be differentiated, i.e., sunless tanning products and sun tanning accelerators.

The sunless tanning products, also called self-tanners, produce a tanning effect without sun, and thus they avoid the harmful side effects associated with solar radiation (see Chapter 5). Their active ingredients interact with amino acids at the stratum corneum level to form substances called melanoidins (Balogh et al., 2011). These orange to brown polymeric compounds artificially formed should not be confused with melanin. Despite this simple staining reaction, this artificial tanning has been shown to provide a slight protection against UVA radiation in animals and humans (Brown, 2001). It should be noted that the tan provided by these products is not removed by simple washing but it does not last more than 5–7 days, as the skin cells are continuously being shed (Pantini et al., 2007). These products can be spread by the users themselves, but usually they are sprayed by tanning booths to achieve a more homogeneous tone. The most popular active ingredient is dihydroxyacetone, commonly known by its acronym DHA. Erythrulose is a compound chemically very similar to DHA that in fact works similar to DHA and, despite having better tanning properties at high concentrations, it has been used as a self-tanning active ingredient to a much lesser extent (Lenz et al., 2010). There are other active ingredients with self-tanning properties, but they are under study and under patents.

Regarding tanning accelerators, they increase tanning when sunbathing outdoors, or also indoors after sun exposure. Although different compounds have been shown to accelerate skin pigmentation in vitro via different mechanisms (Brown, 2001), the marketed cosmetic products are mainly based on tyrosine as the active ingredient, on the basis that, as described before, tyrosine is involved in the rate-limiting step of melanogenesis. Therefore, these compounds containing tyrosine (or tyrosine derivatives, e.g., acetyl tyrosine) are thought to penetrate to the basal epidermis layer, where melanocytes are stimulated, producing melanin and therefore increasing the tan. Sometimes, riboflavin is added to these cosmetic preparations because it is known to activate the oxidation of tyrosine in melanogenesis, and thus it improves the efficacy of tyrosine-based preparations. Psoralens, which were incorporated several decades ago either as bergamot oil or as purified 5-methoxypsoralen, are no longer used in the cosmetics industry because of their prohibition motivated by their multiple side effects related to erythema and skin cancer (Herrmann et al., 2015).

These tanning agents are listed in Table 6.1.

Whitening Products

These products are designed to produce a whitening effect on the skin. They contain various ingredients that interfere in the melanogenesis process by means of different mechanisms, such as tyrosinase inhibition or melanosomal transfer disturbance, among others (Gillbro and Olsson, 2011; Burger et al., 2016; Desmedt et al., 2016).

Table 6.1 Tanning and whitening agents considered in this chapter

INCI name	Key[a]
Tanning agents	
Dihydroxyacetone	DHA
Erythrulose	
Tyrosine	
Psoralen	
Whitening agents	
Hydroquinone and its derivatives	
Hydroquinone	HQ
Hydroquinone monomethyl ether	HQMM
Hydroquinone monoethyl ether	HQME
Hydroquinone monobenzyl ether	HQMB
Resorcinol	RS
Arbutin	ARB
Mercury derivatives	
Mercury chloride	
Mercury iodide	
Retinoids	
Retinoic acid (tretinoin)	RA
Retinyl palmitate	RP
Ascorbic acid and its derivatives	
Ascorbic acid	AA
Ascorbyl palmitate	AP
Ascorbyl dipalmitate	ADP
Ascorbyl stearate	AS
Magnesium or sodium ascorbyl phosphate	MAP, SAP
Ascorbyl glucoside	AG
Others	
Kojic acid	KA
Kojic dipalmitate	KDP
Azelaic acid	AZA
Niacinamide	NC

INCI, International Nomenclature of Cosmetic Ingredients.
[a]Key system adopted by the authors.

The most popular active ingredient traditionally used in this type of cosmetic product is hydroquinone (HQ). However, different dermatological (dermatitis, ochronosis, permanent depigmentation) and carcinogenic side effects have been associated with these compounds (Couteau and Coiffard, 2016; Naidoo et al., 2016). In this sense, HQ is no longer used in the cosmetic industry (but it is still largely used in the pharmaceutical industry), not even its derivatives traditionally used in the past, such as its monomethyl (HQMM), monoethyl (HQME) and monobenzyl ether (HQMB). Resorcinol

(RS), a positional isomer of HQ, was also used in the past, but it is currently not used. However, arbutin (ARB), which is a natural HQ derivative found in different plants, has been most popularly accepted by the cosmetics industry rather than its parent compound, as it is less harmful (Gillbro et al., 2011; Naidoo et al., 2016). In fact, it is added to cosmetics as a pure compound or through extracts from *Arctostaphylos uva-ursi* and *Arbutus unedo* (Chisvert et al., 2007).

Some mercury derivatives, such as mercury chloride and mercury iodide, were used in the past to bleach the skin but are no longer used in skin whitening cosmetic products because mercury is broadly not allowed in cosmetic products (Naidoo et al., 2016), with few exceptions like thimerosal or phenyl mercury that are used as preservatives in some countries.

Other controversial whitening agents are those derived from vitamin A (retinoids), especially retinoic acid (RA), also known as tretinoin, and its derivative retinyl palmitate. However erythema, desquamation and teratogenic effects have been associated with RA, and thus it is not currently used in cosmetic preparations (Couteau and Coiffard, 2016; Naidoo et al., 2016; Desmedt et al., 2016).

Some corticosteroids, such as fluocinonide, betamethasone dipropionate and clobetasol propionate, among others, have been also used in the past as whitening agents in some countries.

Nowadays, other than ARB, the most popular whitening agents found in cosmetic products are ascorbic acid (AA), kojic acid (KA), azelaic acid and niacinamide. It should be emphasized that owing to its labile oxidative properties, KA is usually added to cosmetics by means of its dipalmitic ester, i.e., as kojic dipalmitate. The same happens with AA, which is usually found as ascorbyl palmitate, ascorbyl dipalmitate, ascorbyl stearate, magnesium (or sodium) ascorbyl phosphate and ascorbyl glucoside (Chisvert et al., 2007). Moreover, these derivatives have different solubility properties compared to the parent compound, and thus, they are employed in the formulation of new preparations having different properties (Chisvert et al., 2007).

All these whitening agents are listed in Table 6.1.

Usually, whitening cosmetic products will also contain peeling chemicals, such as α-hydroxy acids (glycolic, lactic or malic acids) or β-hydroxy acids (salicylic acid), to improve their effectiveness, because these chemicals remove the dead skin cells, making the task of the whitening agents easier. Also sunscreen agents are added to protect users' skin from sunlight to avoid tanning (Chisvert et al., 2007).

For all the aforementioned reasons, analytical methods are needed to control the concentration of tanning and whitening agents and thus to ensure the efficacy and the safety of cosmetic products that contain them as ingredients.

The aim of this chapter is to familiarize the reader with the different compounds used as tanning and whitening agents and the legislation regulating these compounds and especially to review the analytical methods for tanning and whitening agents determination in cosmetic products since 2006. Those articles published earlier (i.e., before 2006) were reviewed in the first edition of this book (Salvador and Chisvert, 2007).

REGULATORY ASPECTS

In contrast to UV filters, described in Chapter 5, there are no positive lists for tanning and whitening agents. Some of these active ingredients have been banned because of their side effects, whereas others have been subjected to restrictions. In any case, their prohibition/restriction or not depends on the legislation in force in each country, which differs from one part of the world to the other, as was mentioned in Chapters 1 and 2.

Tanning and Whitening Agents in the European Union

Products containing these ingredients can be considered pharmaceuticals or cosmetics depending on how they are declared, i.e., following guidelines established in Regulation EC 726/2004 (related to pharmaceutical products) or in Regulation EC 1223/2009 (known as the EU Cosmetics Regulation). In the first case, products are intended to correct skin disorders in small areas and are prescribed by a doctor, whereas in the last case products are used for aesthetic reasons on large skin areas and are freely available. In this regard, it should be emphasized that the active agents allowed in both types of products could be completely different, cosmetic products being subjected to more restrictions than pharmaceutical products. As example, it should be said that a user could find a pharmaceutical product containing HQ, whereas this compound is banned as a whitening agent in cosmetic products. Its use in cosmetics, and that of its monomethyl ether (HQMM), is restricted to artificial nail systems according to Annex III of the EU Cosmetics Regulation. Its positional isomer RS is also prohibited as a skin whitening agent and its use is restricted to oxidative hair dyes and other hair care products. In the case of mercury derivatives, they are banned (except for thimerosal or phenyl mercury used as preservatives), since the use of mercury-based ingredients is not allowed in cosmetic products according to Annex II of the EU Cosmetics Regulation. Similarly, RA, corticosteroids (e.g., fluocinonide, betamethasone dipropionate and clobetasol propionate) or psoralens (e.g., 5-methoxypsoralen) are completely banned in cosmetic products, as is stated in Annex II of the EU Cosmetics Regulation. However, market surveillance has shown an illegal inclusion and product mislabelling in some countries (Naidoo et al., 2016; Desmedt et al., 2016). No prohibition or restriction has been found for the other tanning or whitening agents described earlier.

Tanning and Whitening Agents Outside the European Union

In other countries, such as the countries of the Association of Southeast Asian Nations and the Southern Common Market in South America, regulation on cosmetics converges to the EU, and thus all the aforementioned could be applicable to these.

However, in the United States, according to the Food and Drug Administration (FDA), Title 21 of the Code of Federal Regulations (CFR), the unique reference to tanning agents is related to DHA, which is considered a colour additive exempt from certification, being allowed in cosmetics (21 CFR 73.2150). Regarding whitening agents, as expected, the use of mercury derivatives is not allowed based on the known hazards of mercury and its questionable efficacy as a skin bleaching agent (21 CFR 700.13). Any other reference regarding whitening agents has not been found. However, the FDA issued in 2006 a notice through the Federal Register (2006) in which they proposed to withdraw the rule in force on skin bleaching drug products for over-the-counter human use, based on new data and information on the safety of the only active ingredient that had been proposed for inclusion in these products, i.e., HQ. This proposal intended to consider all skin bleaching drug products to be new drugs that require approval as a new drug application to continue on the market. In Japan, HQMB is not allowed in cosmetic products (MHWN, 2000).

ANALYTICAL METHODS FOR TANNING AND WHITENING AGENTS DETERMINATION IN COSMETIC PRODUCTS

All the aforementioned restrictions established for these active ingredients need to be accompanied by efficient and reliable analytical methods to control them. This is even more so, if it is taken into account that market inspections have shown an illegal inclusion of certain compounds and product mislabelling in some countries (Naidoo et al., 2016; Desmedt et al., 2016).

Determination of Tanning Agents

There are no official analytical methods for the determination of any of the scarce aforementioned tanning agents. Moreover, as said before, DHA is the most popular one, and no serious side effects have been associated with its use. For this same reason the analytical community has paid very little attention to developing analytical methods for its control. A detailed review of those papers published until 2006 was carried out in the first edition of this book, revealing very few and old publications on this matter (Salvador and Chisvert, 2007). Because of its polar character, it was usual to derivatize DHA to a less polar or more volatile derivative so that it could be determined by reversed-phase liquid chromatography (LC) or gas chromatography (GC), respectively (Chisvert et al., 2007). Since 2006, only one publication regarding the determination of DHA by LC has been published (Biondi et al., 2007). In this paper, the cream samples are cleaned up by liquid–liquid extraction using aqueous sodium chloride and dichloromethane, and then the DHA is derivatized with pentafluorobenzylhydroxylamine after 5 min at room temperature, which in the authors' words is more rapid and selective than the methods published before.

Determination of Whitening Agents

Regarding whitening agents, there is an official analytical method published in the EU framework under Commission Directive 95/32/EC. This method focuses on the determination of HQ, HQMM, HQME and HQMB, and it is based on their identification by means of thin-layer chromatography (TLC), followed by their quantitative determination using LC with ultraviolet/visible (UV/VIS) detection, for which the sample is extracted with a water/methanol mixture under heating. As of this writing, in 2017, to update and extend this official method for the determination of HQ and its ethers, but also for the most common corticosteroids used illegally in cosmetic products, a new proposal is being approved as an EU standard (FprEN 16956, WI 00392023, 2017) (see Chapter 4). This standard, based on LC–UV/VIS, also describes a screening method for the identification of HQ, its ethers, and 38 corticosteroids, which relies on the paper published by Gimeno et al. (2016).

As can be found in the analytical databases, there are different published papers dealing with the determination of whitening agents in cosmetic products. A detailed and exhaustive review of those methods published up to 2006 can be found in the first edition of this book (Salvador and Chisvert, 2007). On that occasion, a detailed study of those published papers revealed that most of them were focused on HQ and its ether, whereas other whitening agents were scarcely considered. Since 2006, the demand for reliable and broad-spectrum analytical methods to improve and facilitate quality control in the cosmetics industry seems to have been heard. A detailed review of those papers published since the last edition has been carried out here. Table 6.2 summarizes the experimental details and some interesting remarks of these published papers dealing with the determination of whitening agents.

Skin whitening cosmetic products usually contain mixtures of whitening agents, so their direct measurement by, for example, UV/VIS spectrometry without a previous separation step is difficult. In this regard chromatographic techniques are usually employed to cover the determination of a high number of these active agents. Nevertheless, some specific colorimetric reactions or chemometric strategies have been proposed to directly determine individual whitening agents or mixtures of these compounds. In this sense, Uddin et al. (2011) developed a selective method to determine HQ based on its oxidation to benzoquinone catalysed by metavanadate. Calaca et al. (2011) determined mixtures of KA and HQ by using multivariate calibration based on the reaction with Fe^{3+}, so that Fe^{3+} is complexed by KA, whereas HQ reduces Fe^{3+} to Fe^{2+} and it is further complexed by phenanthroline. Also applying chemometrics, Elzanfaly et al. (2012) determined mixtures of RA and HQ by using the ratio difference method. Esteki et al. (2016) used another chemometric strategy to determine HQ in the presence of the preservative methylparaben, based on successive projections algorithm–multiple linear regression. Attenuated total reflectance–infrared spectroscopy and subsequent chemometric data

Table 6.2 Published papers from January 2006 to December 2016 concerning whitening agents determination in cosmetic products (chronological order)

Authors	Target whitening agents[a]	Type of matrix[b]	Sample preparation[c]	Analytical technique[c]	Remarks[d]
Varvaresou et al. (2006)	MAP	EM	Sample is dispersed in acetonitrile:water by sonication (15 min), diluted with water and filtered	**LC–UV/VIS**, C_{18} column, isocratic acetonitrile: phosphate buffer containing tetrabutylammonium hydroxide as mobile phase	LOD: 0.08% R: 99%–103%
Wang and Wu (2006)	AA, AP, MAP	LO	Sample is dispersed in water:chloroform:methanol by sonication (10 min) and centrifuged	**MEKC–UV/VIS**, fused-silica capillary, borate buffer containing sodium dodecyl sulphate as running buffer	LOD: no data R: >95%
Thongchai et al. (2007)	ARB	EM	Sample is dispersed in methanol by sonication (30 min), centrifuged and filtered	**LC–UV/VIS**, C_{18} column, isocratic methanol: water:hydrochloric acid (aq) as mobile phase	LOD: 0.0001% R: 100%
Lin et al. (2007a)	AG, KA, NC	EM, LO, OI	Sample is dispersed in water, vortexed, then submitted to microdialysis and injected online	**LC–UV/VIS**, Pf column, isocratic methanol:phosphate buffer as mobile phase	LOD: no data R: 92%–106%
Lin et al. (2007b)	ARB, HQ, KA	GE, OI	Sample is dispersed in water by sonication (30 min) and centrifuged	**CE–UV/VIS**, uncoated fused-silica capillary, phosphate buffer as running buffer	LOD: 0.07%–0.14% R: >99%
Wei et al. (2007)	AA, ARB	EM, GE, LO	Sample is dispersed in methanol, centrifuged and diluted with methanol: phosphate buffer	**LC–CL**, C_{18} column, isocratic methanol:phosphate buffer as mobile phase	LOD: 0.0002%–0.0003% R: 98%–106%

Shih et al. (2007)	KA	EM	Sample is dispersed in water and filtered	**FI–AD**, carbon-nanotube modified screen-printed electrode	LOD: no data R: 94%–107%
Balaguer et al. (2008a)	KDP	EM	Sample is dispersed in tetrahydrofuran by sonication (10 min) and filtered	**LC–UV/VIS**, PLGel Mixed-D column, tetrahydrofuran as mobile phase	LOD: 0.04%–0.15% R: 99%–102%
Balaguer et al. (2008b)	AA, AG, AP, MAP	EM	Sample is dispersed in ethanol:phosphate buffer by sonication (10 min) and filtered	**LC–UV/VIS**, C$_8$ column, gradient ethanol:phosphate buffer as mobile phase	LOD: 0.003%–0.12% R: 97%–99%
Hubinger (2009)	RA, RP	EM, LO	Sample is mixed with diatomaceous earth, then eluted with hexane:isopropanol:ethyl acetate	**LC–UV/VIS**, C$_{18}$ column, gradient methanol:ammonium acetate buffer:dichloromethane as mobile phase	LOD: 0.00003%–0.0001% R: 94%–103%
Chisvert et al. (2010)	ARB, AZA, HQ, KA, RS	EM	Sample is dissolved in dimethyl formamide, diluted and derivatized with BSTFA	**GC–MS**, 95% dimethyl–5% diphenyl polysiloxane column, helium as carrier gas	LOD: 0.0005%–0.24% R: 98%–103%
Cheng et al. (2010)	ARB, NC		Sample is extracted with chloroform:water mixture	**LC–UV/VIS**, C$_{18}$ column, isocratic methanol:water as mobile phase	LOD: no data R: 92%–110%
Calaca et al. (2011)	HQ, KA		Sample is solved in Clark–Lubs buffer, then mixed with Fe^{3+} and phenanthroline	**UV/VIS** (multivariate calibration method)	LOD: no data R: 98%–107%
Uddin et al. (2011)	HQ	EM	Sample is dispersed in isopropanol and filtered, then mixed with ammonium metavanadate	**UV/VIS**	LOD: no data R: no data

Continued

Table 6.2 Published papers from January 2006 to December 2016 concerning whitening agents determination in cosmetic products (chronological order)—cont'd

Authors	Target whitening agents[a]	Type of matrix[b]	Sample preparation[c]	Analytical technique[c]	Remarks[d]
Yang et al. (2011)	NC	EM	Sample is dispersed in methanol by vortex and filtered	LC–UV/VIS, C_{18} column, water as mobile phase	LOD: no data R: 100%
Gao and Legido-Quigley (2011)	HQ	EM, LO	Sample is dispersed with methanol:water by sonication (30 min), centrifuged and filtered	LC–UV/VIS, C_{18} column, isocratic methanol: ammonium formate buffer as mobile phase	LOD: 0.0005% R: 103%–116%
Elzanfaly et al. (2012)	HQ, RA	EM	Sample is dispersed in methanol by sonication (10 min), then filtered	UV/VIS (ratio difference method)	LOD: no data R: 100%
Desmedt et al. (2013, 2014)	ARB, HQ, KA, NC, RA (and some corticosteroids)	EM, LO, SO	Soaps are dissolved in water and neutralized with hydrochloric acid; neutralized soaps, EMs and LOs are dispersed in acetonitrile by sonication (30 min) and filtered	LC–UV/VIS, C_{18} column, gradient acetonitrile: ammonium borate buffer as mobile phase	LOD: 0.00005%–0.00075% R: 90%–103% LC–MS was also used for screening purposes
Jin et al. (2013)	ARB, HQ, KA, RS	EM, LO	Sample is dispersed in ethanol or ethanol:water by sonication (20 min), centrifuged and filtered	MEKC–AD, fused-silica capillary, borate buffer containing sodium dodecyl sulphate as running buffer	LOD: no data R: 100%
Thongchai and Liawruangrath (2013)	ARB, HQ	EM	Sample is dispersed in acetonitrile:phosphate buffer by sonication (30 min), centrifuged and filtered	LC–UV/VIS, C_{18} column, isocratic acetonitrile: phosphate buffer containing Brij 35 as mobile phase	LOD: 0.002%–0.003% R: 96%–99%

Sun and Wu (2013)	NC	EM	Sample is dispersed in trichloromethane:water by sonication (30 min) and the water phase was collected	**MEKC–UV/VIS,** fused-silica capillary, borate buffer containing sodium dodecyl sulphate as running buffer	LOD: 0.0004% R: 87%–108%
Jeon et al. (2014)	ARB, NC	EM, LO	Sample is dispersed in methanol by sonication (10 min), diluted with water and centrifuged	**LC–UV/VIS,** C_{18} column, gradient methanol:water as mobile phase	LOD: 0.006%–0.04% R: 101%–107%
Tidjarat et al. (2014)	HQ, RA	EM	Sample is mixed with ethanol and sonicated (10 min)	**TLC,** cellulose acetate nanofibers as stationary phase, methanol:water:acetic acid as mobile phase	Only screening purposes
Deconinck et al. (2014)	ARB, HQ, KA, NC, RA	EM, GE, LO, OI, SO	Sample is directly brought on the crystal and pressed	**ATR–IR** (k–nearest neighbours method)	Only screening purposes
Tsai et al. (2014)	ARB, KA, RS	EM, LO	Sample is dissolved in ethanol by sonication (10 min), then centrifuged and diluted with water	**MEKC–UV/VIS,** fused-silica capillary, borate buffer containing sodium dodecyl sulphate as running buffer	LOD: 0.005%–0.04% R: 85%–115%
Chen et al. (2015a)	AA, AG, MAP	EM, GE, LO	Low-fat samples are extracted with phosphate buffer, whereas high-fat samples are dispersed in dichloromethane and extracted with phosphate buffer, then centrifuged and filtered	**LC–UV/VIS,** C_{18} column, gradient methanol:phosphate buffer as mobile phase	LOD: 0.004%–0.008% R: 96%–101%
Chen et al. (2015b)	AG, ARB, HQ, KA, MAP	EM, LO	Sample is dispersed in water by stirring, then submitted to microdialysis and injected online	**LC–AD,** C_{18} column, isocratic citrate buffer containing tetrabutylammonium chloride as mobile phase	LOD: no data R: 91%–108%

Continued

Table 6.2 Published papers from January 2006 to December 2016 concerning whitening agents determination in cosmetic products (chronological order)—cont'd

Authors	Target whitening agents[a]	Type of matrix[b]	Sample preparation[c]	Analytical technique[c]	Remarks[d]
Jeon et al. (2015)	ARB, HQ	EM	Sample is dispersed in methanol by sonication (30 min), diluted with water and sonicated (30 min), then centrifuged and filtered	**LC–UV/VIS**, C_{18} column, gradient methanol:water as mobile phase	LOD: 0.0004%–0.0005% R: 90%–103%
Esteki et al. (2016)	HQ	EM, GE, LO	Sample is dispersed in methanol:water by sonication (30 min), centrifuged and filtered	**UV/VIS** (successive projections algorithm method)	LOD: no data R: 84%–114%
Galimany-Rovira et al. (2016)	HQ, KA	EM	Sample is dispersed in acetonitrile:water by sonication (15 min), then centrifuged	**LC–UV/VIS**, Phenyl column, gradient acetonitrile:acetic acid (aq) as mobile phase	LOD: no data R: 98%–101%
Shams et al. (2016)	ARB (and some corticosteroids)	EM	Sample is dispersed in acetonitrile:methanol:water, by stirring (30 min), then filtered	**LC–UV/VIS**, C_{18} column, isocratic acetonitrile:methanol:water as mobile phase	LOD: 0.0006% R: 96%–99%
Gimeno et al. (2016)	HQ, HQMM, HQME, HQMB (and some corticosteroids)	EM, OI	Sample is dispersed in methanol:water by shaking (5 min), diluted, stirred (30–60 min) and filtered	**LC–UV/VIS**, C_{18} column, gradient acetonitrile:acetic acid (aq) as mobile phase	LOD: <0.005% R: no data

[a]See Table 6.1 for key to abbreviations.
[b]*EM*, emulsion; *GE*, gel; *LO*, lotion; *OI*, oil; *SO*, soap.
[c]*AD*, amperometric detection; *ATR–IR*, attenuated total reflectance–infrared spectrometry; *BSTFA*, N,O-bis(trimethylsilyl)trifluoroacetamide; C_{18}, octadecylsilica; *CE*, capillary electrophoresis; *CL*, chemiluminescence; *FI*, flow injection; *GC*, gas chromatography; *LC*, liquid chromatography; *MEKC*, micellar electrokinetic chromatography; *MS*, mass spectrometry; *Pf*, perfluorinated phenyl; *TLC*, thin-layer chromatography; *UV/VIS*, ultraviolet/visible spectrometry.
[d]*LOD*, limit of detection (referred to sample, not to sample solution); *R*, recovery.

treatment based on k-nearest neighbours has also been used for screening whitening agents in cosmetic samples (Deconinck et al., 2014).

On just one occasion, electroanalytical detection, such as amperometric detection (AD), was used to determine an individual whitening agent (i.e., KA) without a previous separation (Shih et al., 2007). In this case, solutions were propelled by using a flow-injection system.

With regard to chromatographic techniques, LC is the most commonly employed because whitening agents present relatively high boiling points for employing GC. Usually, a UV/VIS spectrometric detector is preferred because most of them present chromophore moieties, although chemiluminescence (Wei et al., 2007) and AD (Chen et al., 2015b) have also been used. In 2014, mass spectrometry (MS) detection was used for identification purposes (Desmedt et al., 2014).

GC has been used on just one occasion (Chisvert et al., 2010), on which a previous derivatization of the whitening agents with N,O-bis(trimethylsilyl)trifluoroacetamide was carried out to convert them to the more volatile trimethylsilyl derivatives. MS was used as the detector to unequivocally determine HQ and RS.

Capillary electrophoresis and micellar electrokinetic chromatography have also been employed with very good analytical features for the quantitative determination of different whitening agents (Wang and Wu, 2006; Lin et al., 2007b; Jin et al., 2013; Sun and Wu, 2013; Tsai et al., 2014).

Finally, it should be said that TLC was used as a screening method for HQ and RA (Tidjarat et al., 2014).

Regarding sample preparation, no complex operations are required. Most of the cases consisted just in adding an appropriate solvent to the cosmetic sample and then sonicating it for several minutes to dissolve the sample totally or at least to leach the target compounds. Centrifugation or filtration is sometimes required to obtain clear solutions prior to measurement. The use of sample preconcentration techniques is not really necessary in this case, since whitening agents are major ingredients, even when used fraudulently.

REFERENCES

Balaguer, A., Salvador, A., Chisvert, A., 2008a. Talanta 75, 407.
Balaguer, A., Chisvert, A., Salvador, A., 2008b. J. Sep. Sci. 31, 229.
Balogh, T.S., Pedriali, C.A., Gama, R.M., de Oliveira, C.A.S., Bedin, V., Villa, R.T., Kaneko, T.M., Consiglieri, V.O., Velasco, M.V.R., Baby, A.R., 2011. Int. J. Cosmet. Sci. 33, 359.
Biondi, P.A., Passeró, E., Soncin, S., Bernardi, C., Chiesa, L.M., 2007. Chromatographia 65, 65.
Brown, D.A., 2001. J. Photochem. Photobiol. B 63, 148.
Burger, P., Landreau, A., Azoulay, S., Michel, T., Fernandez, X., 2016. Cosmetics 3, 36.
Calaca, G.N., Stets, S., Nagata, N., 2011. Quim. Nova 34, 630.
CFR (Code of Federal Regulations). https://www.ecfr.gov; https://www.accessdata.fda.gov/scripts/cdrh/cfdocs/cfcfr/cfrsearch.cfm.
Chang, T.-S., 2009. Int. J. Mol. Sci. 10, 2440.
Chen, P., Yan, Z., Tu, X., Xiao, F., Liang, H., 2015a. Chin. J. Chromatogr. 33, 771.

Chen, R.X., Wang, L., Wang, J., Xu, F.Q., 2015b. Anal. Lett. 48, 2159.
Cheng, P., Chen, M., Zhu, Y., 2010. Chin. J. Chromatogr. 28, 89.
Chisvert, A., Balaguer, A., Salvador, A., 2007. Tanning and whitening agents in cosmetics. Regulatory aspects and analytical methods. In: Salvador, A., Chisvert, A. (Eds.). Elsevier, Amsterdam.
Chisvert, A., Sisternes, J., Balaguer, A., Salvador, A., 2010. Talanta 81, 530.
Commission Directive 95/32/EC of 7 July 1995, Relating to Methods of Analysis Necessary for Checking the Composition of Cosmetic Products.
Couteau, C., Coiffard, L., 2016. Cosmetics 3, 27.
Deconinck, E., Bothy, J.L., Desmedt, B., Courselle, P., De Beer, J.O., 2014. J. Pharm. Biomed. Anal. 98, 178.
Desmedt, B., Rogiers, V., Courselle, P., De Beer, J.O., De Paepe, K., Deconinck, E., 2013. J. Pharm. Biomed. Anal. 83, 82.
Desmedt, B., Van Hoeck, E., Rogiers, V., Courselle, P., De Beer, J.O., De Paepe, K., Deconinck, E., 2014. J. Pharm. Biomed. Anal. 90, 85.
Desmedt, B., courselle, P., De Beer, J.O., Rogiers, V., Grosber, M., Deconinck, E., De Paepe, D., 2016. Eur. Acad. Dermatol. Venereol. 30, 943.
Elzanfaly, E.S., Saad, A.S., Abd-Elaleem, A.E.B., 2012. Saudi Pharm. J. 20, 249.
Esteki, M., Nouroozi, S., Shahsavari, Z., 2016. Int. J. Cosmet. Sci. 38, 25.
Federal Register, 2006. Skin Bleaching Drug Products for Over-the-Counter Human Use; Proposed Rule. 71, p. 51147.
FprEN 16956 (WI 00392023), CEN/TC 392-Cosmetics, 2017. Analytical Methods – HPLC/UV Method for the Identification and Assay of Hydroquinone, Ethers of Hydroquinone and Corticosteroids in Skin Whitening Cosmetic Products. https://standards.cen.eu.
Galimany-Rovira, F., Pérez-Lozano, P., Suñé-Negre, J.M., García-Montoya, E., Miñarro, M., Ticó, J.R., 2016. Anal. Methods 8, 1170.
Gao, W., Legido-Quigley, C., 2011. J. Chromatogr. A 1218, 4307.
Gillbro, J.M., Olsson, M.J., 2011. Int. J. Cosmet. Sci. 33, 210.
Gimeno, P., Maggio, A.-F., Bancilhon, M., Lassu, N., Gornes, H., Brenier, C., Lempereur, L., 2016. J. Chromatogr. Sci. 54, 343.
Herrmann, J.L., Cunningham, R., Cantor, A., Elewski, B.E., Elmets, C.A., 2015. J. Am. Acad. Dermatol. 72, 99.
Hubinger, J.C., 2009. J. Cosmet. Sci. 60, 485.
Jeon, J.S., Lee, M.J., Yoon, M.H., Park, J.A., Yi, H., Cho, H.J., Shin, H.C., 2014. Anal. Lett. 47, 1650.
Jeon, J.S., Kim, B.H., Lee, S.H., Kwon, H.J., Bae, H.J., Kim, S.K., Park, J.A., Shim, J.H., Abd El-Aty, A.M., Shin, H.C., 2015. Int. J. Cosmet. Sci. 37, 567.
Jin, W., Wang, W.Y., Zhang, Y.L., Yang, Y.J., Chu, Q.C., Ye, J.N., 2013. Chin. Chem. Lett. 24, 636.
Lenz, D., Vollhardt, J., Imfeld, D., Moser, H., 2010. SOFW J. 136, 16.
Lin, C.H., Wu, H.L., Huang, Y.L., 2007a. Anal. Chim. Acta 581, 102.
Lin, Y.H., Yang, Y.H., Wu, S.M., 2007b. J. Pharm. Biomed. Anal. 44, 279.
Ministry of Health and Welfare Notification (MHWN), 2000. Standard for Cosmetics, No.331 of 2000. Japan.
Naidoo, L., Khoza, N., Dlova, N.C., 2016. Cosmetics 3, 33.
Pantini, G., Ingoglia, R., Brunetta, F., Brunetta, A., 2007. Int. J. Cosmet. Sci. 29, 201.
Regulation (EC) No 1223/2009 of the European Parliament and of the Council of 30 November 2009 on Cosmetic Products, and its Successive Amendments. http://eur-lex.europa.eu/legal-content/EN/ALL/?uri=CELEX%3A32009R1223.
Regulation (EC) No 726/2004 of the European Parliament and of the Council of 31 March 2004 Laying Down Community Procedures for the Authorisation and Supervision of Medicinal Products for Human and Veterinary Use and Establishing a European Medicines Agency, and its Successive Amendments. https://ec.europa.eu/health/sites/health/files/files/eudralex/vol-1/reg_2004_726/reg_2004_726_en.pdf.
Salvador, A., Chisvert, A. (Eds.), 2007. Analysis of Cosmetic Products. Elsevier.
Shams, A., Khan, I.U., Iqbal, H., 2016. Int. J. Cosmet. Sci. 38, 421.
Shih, Y., Zen, J.M., Ke, J.H., Hsu, J.C., Chen, P.C., 2007. J. Food Drug Anal. 15, 151.

Sun, H., Wu, Y., 2013. Anal. Methods 5, 5615.
Thongchai, W., Liawruangrath, B., 2013. Int. J. Cosmet. Sci. 35, 257.
Thongchai, W., Liawruangrath, B., Liawruangrath, S., 2007. J. Cosmet. Sci. 58, 35.
Tidjarat, S., Winotapun, W., Opanasopit, P., Ngawhirunpat, T., Rojanarata, T., 2014. J. Chromatogr. A 1367, 141.
Tsai, I.C., Su, C.Y., Hu, C.C., Chiu, T.C., 2014. Anal. Methods 6, 7615.
Uddin, S., Rauf, A., Kazi, T.G., Afridi, H.I., Lutfullah, G., 2011. Int. J. Cosmet. Sci. 33, 132.
Varvaresou, A., Tsirivas, E., Iakovou, K., Gikas, E., Papathomas, Z., Vonaparti, A., Panderi, I., 2006. Anal. Chim. Acta 573, 284.
Wang, C.C., Wu, S.M., 2006. Anal. Chim. Acta 18, 124.
Wei, Y., Zhang, Z., Zhang, Y., Sun, Y., 2007. Chromatographia 65, 443.
Yang, Y., Strickland, Z., Kapalavavi, B., Marple, R., Gamsky, C., 2011. Talanta 84, 169.

CHAPTER 7

Colouring Agents in Cosmetics: Regulatory Aspects and Analytical Methods

Adrian Weisz, Stanley R. Milstein, Alan L. Scher[†], Nancy M. Hepp
U.S. Food and Drug Administration, College Park, MD, United States

INTRODUCTION

Application of colour is the main purpose of many cosmetic products such as lipsticks, blushers, eye shadows, eyeliners and nail polishes. All of these products contain one or more colouring agents—dyes, pigments, or other substances—for providing the desired colours.

The aim of this chapter is to review regulatory information concerning colouring agents in cosmetic products, as well as the methodologies involved in their analysis.

REGULATORY ASPECTS OF COLOURING AGENTS IN COSMETIC PRODUCTS

Colouring agents are subject to a wide range of regulatory restrictions across countries. As mentioned in Chapter 1, positive lists of colouring agents that may be used in cosmetic products have been published by three main regulatory authorities—the US Food and Drug Administration (FDA) in the United States of America, the European Commission (EC) in the European Union (EU), and the Ministry of Health, Labour and Welfare (MHLW) in Japan. Other countries permit colouring agents approved in the United States and/or the EU with certain variations.

The number and identity of colouring agents permitted for cosmetic use vary among countries. Table 7.1 shows the number of these ingredients listed for use in cosmetic products by the three aforementioned regulatory authorities.

United States Regulatory Requirements for Colouring Agents

In the United States, colouring agents are known as colour additives, which must comply with the requirements of the federal Food, Drugs, and Cosmetics Act (FD&C Act) and its implementing regulations. The term colour additive is defined in Section 201(t) of the FD&C Act as:

[†]Deceased.

Table 7.1 Number of colouring agents permitted for use in cosmetics by the three main regulatory authorities

Country	No. of colouring agents permitted
United States	65[a]
European Union	153[b]
Japan	83[c]

[a]FDA 21 CFR Parts 73 and 74.
[b]Annex IV, Part 1 of the EU Directive (Council Directive 76/768/EEC and its amendments).
[c]Ordinance No. 30/1966 from the Ministry of Health and Welfare (as amended by the Ministry of Health, Labour and Welfare Nos. 55/1972 and 126/2003); Rosholt, 2007.

… a material which (A) is a dye, pigment, or other substance made by a process of synthesis or similar artifice, or extracted, isolated, or otherwise derived, with or without intermediate or final change of identity, from a vegetable, animal, mineral, or other source, and (B) when added or applied to a food, drug, or cosmetic, or to the human body or any part thereof, is capable (alone or through reaction with other substance) of imparting a color thereto.… The term 'color' includes black, white, and intermediate grays.

Colour additives permitted in the United States are classified from a regulatory perspective as those subject to batch certification by the FDA and those exempt from certification. The majority of the colour additives required to be certified (see Tables 7.2–7.9) are synthetic aromatic organic compounds (also known as coal-tar colours) and their lakes, which are water-soluble dyes (called straight colours) extended onto various substrata by adsorption, co-precipitation, or chemical combination. The certified colour additives are listed in Title 21 of the Code of Federal Regulations (CFR) in 21 CFR Part 74 (dyes, pigments and a few lakes) and 21 CFR Part 82 (most lakes). They are certified by the FDA to ensure that their composition complies with requirements given in their listing regulations, to protect the public health. The FDA assigns a unique certification lot number to each certified batch. Certification-exempt colour additives (see Table 7.10) include a wide variety of substances that are derived from inorganic, plant or animal sources, and they are listed in 21 CFR Part 73. A Summary of Color Additives for Use in the United States in Food, Drugs, Cosmetics, and Medical Devices has been published on the FDA website (http://www.fda.gov/ForIndustry/ColorAdditives/ColorAdditiveInventories/ucm115641.htm).

Certifiable and certification-exempt colour additives must undergo the FDA's pre-market approval process and be listed in the CFR to be used in FDA-regulated cosmetics and other products. A proposal to list a new colour additive or a new use for a listed colour additive is made by petitioning the FDA as described in 21 CFR Part 71. Descriptions of the petition process can be found on the FDA website (http://www.fda.gov/ForIndustry/ColorAdditives/ColorAdditivePetitions/default.htm). The colour additive listing regulations describe the identity of each colour additive, specifications, uses and restrictions, labelling requirements and the requirement for or exemption from

Table 7.2 US certifiable monoazo colour additives for cosmetic use

Azo-enol structure: $R_1-N=N-R_2$

US listed name[a]	Common names[b]	CI No.[c]	CAS No.[d]	R₁	R₂
D&C Orange No. 4	CI Acid Orange 7 Orange II	15510	633-96-5	NaO₃S–C₆H₄–	HO-naphthyl
FD&C Red No. 4	CI Food Red 1 Ponceau SX	14700	4548-53-2	NaO₃S, H₃C, CH₃ substituted phenyl	HO-naphthyl-SO₃Na
D&C Red No. 6	CI Pigment Red 57 Lithol Rubine B	15850	5858-81-1	SO₃Na, H₃C substituted phenyl	HO, CO₂Na naphthyl
D&C Red No. 7	CI Pigment Red 57:1 Lithol Rubine B Ca	15850:1	5281-04-9	SO₃Ca₁/₂, H₃C substituted phenyl	HO, CO₂Ca₁/₂ naphthyl
D&C Red No. 31	CI Pigment Red 64:1 Brilliant Lake Red R	15800:1	6371-76-2	phenyl	HO, CO₂Ca₁/₂ naphthyl
D&C Red No. 33	CI Acid Red 33 Acid Fuchsin D	17200	3567-66-6	phenyl	H₂N, HO, SO₃Na, NaO₃S naphthyl
D&C Red No. 34	CI Pigment Red 63:1 Deep Maroon	15880:1	6417-83-0	SO₃Ca₁/₂ naphthyl	HO, CO₂Ca₁/₂ naphthyl
D&C Red No. 36	CI Pigment Red 4 Flaming Red	12085	2814-77-9	O₂N, Cl substituted phenyl	HO naphthyl
FD&C Red No. 40	CI Food Red 17 Allura Red AC	16035	25956-17-6	H₃C, NaO₃S, H₃CO substituted phenyl	HO, SO₃Na naphthyl
FD&C Yellow No. 5	CI Acid Yellow 23 CI Food Yellow 4 Tartrazine	19140	1934-21-0	NaO₃S–C₆H₄–	SO₃Na phenyl, HO, NaO₂C pyrazole
FD&C Yellow No. 6	CI Food Yellow 3 Sunset Yellow FCF	15985	2783-94-0	NaO₃S–C₆H₄–	HO, SO₃Na naphthyl

[a]Names assigned by the FDA after certification and listed in 21 CFR Part 74.
[b]Not used in the United States for the names of certified colour additives.
[c]CI, Colour Index. Not used in the United States for certified colour additives.
[d]CAS, Chemical Abstracts Service.

Table 7.3 US certifiable bisazo colour additives for cosmetic use

Bisazo-enol structure: $R_1-N=N-R_2-N=N-R_3$

US listed name[a]	Common names[b]	CI No.[c]	CAS No.[d]	R₁	R₂	R₃
D&C Brown No. 1	CI Acid Orange 24, Resorcin Brown	20170	1320-07-6	NaO₃S–C₆H₄–	dimethyl-dihydroxyphenyl	–C₆H₄–N(CH₃)₂
D&C Red No. 17	CI Solvent Red 23, Sudan III, Toney Red	26100	85-86-9	C₆H₅–	–C₆H₄–	HO-naphthyl

[a]Names assigned by the FDA after certification and listed in 21 CFR Part 74.
[b]Not used in the United States for the names of certified colour additives.
[c]CI, Colour Index. Not used in the United States for certified colour additives.
[d]CAS, Chemical Abstracts Service.

Table 7.4 US certifiable triphenylmethane colour additives for cosmetic use

Triphenylmethanium resonance structures

US listed name[a]	Common names[b]	CI No.[c]	CAS No.[d]	R₁	R₂
FD&C Blue No. 1	CI Acid Blue 9 (sodium salt), CI Food Blue 2, Brilliant Blue FCF	42090	3844-45-9	–SO₃Na	–H
D&C Blue No. 4	CI Acid Blue 9 (ammonium salt), Alphazurine FG, Erioglaucine	42090	6371-85-3	–SO₃NH₄	–H
FD&C Green No. 3	CI Food Green 3, Fast Green FCF	42053	2353-45-9	–SO₃Na	–OH

[a]Names assigned by the FDA after certification and listed in 21 CFR Part 74.
[b]Not used in the United States for the names of certified colour additives.
[c]CI, Colour Index. Not used in the United States for certified colour additives.
[d]CAS, Chemical Abstracts Service.

Table 7.5 US certifiable fluoran colour additives for cosmetic use

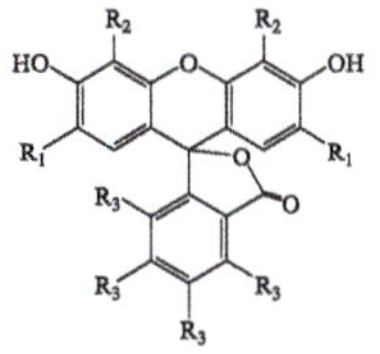

Fluoran structure

U.S. listed name[a]	Common names[b]	CI No.[c]	CAS No.[d]	R₁	R₂	R₃
D&C Orange No. 5	CI Solvent Red 72 Dibromofluorescein	45370:1	596-03-2	–H	–Br	–H
D&C Orange No. 10	CI Solvent Red 73 Diiodofluorescein	45425:1	38577-97-8	–H	–I	–H
D&C Red No. 21	CI Solvent Red 43 Tetrabromofluorescein	45380:2	15086-94-8	–Br	–Br	–H
D&C Red No. 27	CI Solvent Red 48 Tetrabromotetrachlorofluorescein	45410:1	13473-26-2	–Br	–Br	–Cl
D&C Yellow No. 7	CI Solvent Yellow 94 Fluorescein	45350:1	2321-07-5	–H	–H	–H

[a] Names assigned by the FDA after certification and listed in 21 CFR Part 74.
[b] Not used in the United States for the names of certified colour additives.
[c] CI, Colour Index. Not used in the United States for certified colour additives.
[d] CAS, Chemical Abstracts Service.

Table 7.6 US certifiable xanthene colour additives for cosmetic use

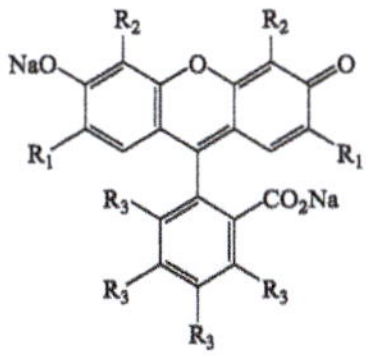

Xanthene structure

US listed name[a]	Common names[b]	CI No.[c]	CAS No.[d]	R₁	R₂	R₃
D&C Orange No. 11	CI Acid Red 95 Erythrosine Yellowish Na	45425	33239-19-9	–H	–I	–H
D&C Red No. 22	CI Acid Red 87 Eosin Y	45380	17372-87-1	–Br	–Br	–H
D&C Red No. 28	CI Acid Red 92 Phloxine B Cyanosine	45410	18472-87-2	–Br	–Br	–Cl
D&C Yellow No. 8	CI Acid Yellow 73 Uranine	45350	518-47-8	–H	–H	–H

[a] Names assigned by the FDA after certification and listed in 21 CFR Part 74.
[b] Not used in the United States for the names of certified colour additives.
[c] CI, Colour Index. Not used in the United States for certified colour additives.
[d] CAS, Chemical Abstracts Service.

Table 7.7 US certifiable quinoline colour additives for cosmetic use

Quinoline tautomeric structures

US listed name[a]	Common names[b]	CI No.[c]	CAS No.[d]	R
D&C Yellow No. 10	CI Acid Yellow 3 Quinoline Yellow WS	47005	8004-92-0	Mixture of 6'- and 8'- $-SO_3Na$
D&C Yellow No. 11	CI Solvent Yellow 33 Quinoline Yellow SS	47000	8003-22-3	$-H$

[a]Names assigned by the FDA after certification and listed in 21 CFR Part 74.
[b]Not used in the United States for the names of certified colour additives.
[c]CI, Colour Index. Not used in the United States for certified colour additives.
[d]CAS, Chemical Abstracts Service.

Table 7.8 US certifiable anthraquinone colour additives for cosmetic use

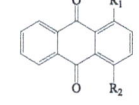

Anthraquinone structure

US listed name[a]	Common names[b]	CI No.[c]	CAS No.[d]	R_1	R_2
D&C Green No. 5	CI Acid Green 25 Alizarine Cyanine Green F	61570	4403-90-1	NaO_3S—⌬—CH_3, HN—	HN—⌬—CH_3, NaO_3S
D&C Green No. 6	CI Solvent Green 3 Quinizarin Green SS	61565	128-80-3	—⌬—CH_3, HN—	HN—⌬—CH_3
D&C Violet No. 2	CI Solvent Violet 13 Alizurol Purple SS	60725	81-48-1	—⌬—CH_3, HN—	$-OH$
Ext. D&C Violet No. 2	CI Acid Violet 43 Alizarine Violet	60730	4430-18-6	NaO_3S—⌬—CH_3, HN—	$-OH$

[a]Names assigned by the FDA after certification and listed in 21 CFR Part 74.
[b]Not used in the United States for the names of certified colour additives.
[c]CI, Colour Index. Not used in the United States for certified colour additives.
[d]CAS, Chemical Abstracts Service.

Colouring Agents in Cosmetics: Regulatory Aspects and Analytical Methods   129

Table 7.9 US certifiable nitro, pyrene, thioindigoid, and carbon colour additives for cosmetic use

US listed name[a]	Common names[b]	CI No.[c]	CAS No.[d]	Dye classification	Chemical structure
D&C Black No. 2	Carbon black (high-purity furnace black)	77266	1333-86-4	Inorganic pigment	C (carbon)
D&C Black No. 3	Bone black		8021-99-6	Inorganic pigment	Calcium hydroxyapatite
D&C Green No. 8	CI Solvent Green 7 Pyranine	59040	6358-69-6	Pyrene	
D&C Red No. 30	CI Vat Red 1 Helidone Pink CN	73360	2379-74-0	Thioindigoid	
Ext. D&C Yellow No. 7	CI Acid Yellow 1 Naphthol Yellow S	10316	846-70-8	Nitro	

[a]Names assigned by the FDA after certification and listed in 21 CFR Part 74.
[b]Not used in the United States for the names of certified colour additives.
[c]CI, Colour Index. Not used in the United States for certified colour additives.
[d]CAS, Chemical Abstracts Service.

Table 7.10 US certification-exempt colour additives for cosmetic use

US listed name[a]	CI No.[b]	CAS No.[c]	Chemical structure
Aluminum powder[d]	77000	4729-90-5	Al
Annatto (extract from *Bixa orellana* L.)[d]	75120	1393-63-1	
Bismuth citrate[e,f]	–	813-93-4	
Bismuth oxychloride[d]	77163	7787-59-9	BiOCl
Bronze powder[d]	77400	7440-50-8	CuZn alloy

Continued

Table 7.10 US certification-exempt colour additives for cosmetic use—cont'd

US listed name[a]	CI No.[b]	CAS No.[c]	Chemical structure
Caramel[d]	–	8028-89-5	(Heat treated glucose or sucrose saturated solutions)
Carmine (Al or Ca–Al lake on Al(OH)$_3$ of aqueous extract of cochineal)[d]	75470	1390-65-4	
β-Carotene (synthetic and natural)[d]	40800 75130	7235-40-7	
Chromium hydroxide green[d]	77289	12001-99-9	$Cr_2O_3 \cdot xH_2O$
Chromium oxide greens[d]	77288	1308-38-9	Cr_2O_3
Copper powder[d]	77400	7440-50-8	Cu
Dihydroxyacetone[d,g]	–	96-26-4	
Disodium EDTA-copper[d,g]	–	14025-15-1	
Ferric ammonium ferrocyanide[d]	77510	25869-00-5	$NH_4Fe[Fe(CN)_6] \cdot xH_2O$
Ferric ferrocyanide[d]		14038-43-8	$Fe_4[Fe(CN)_6]_3 \cdot xH_2O$
Guaiazulene[d,g]	–	489-84-9	
Guanine[d] (from fish scales)	75170	73-40-5	
Henna[d,e]	75480	83-72-7	

Table 7.10 US certification-exempt colour additives for cosmetic use—cont'd

US listed name[a]	CI No.[b]	CAS No.[c]	Chemical structure
Iron oxides[d] (synthetic)	77489[f] 77491 77492 77499	See Rosholt, 2007	$Fe_xO_y \cdot zH_2O$
Lead acetate[f,g]	–	6080-56-4	$Pb(O_2CCH_3)_2 \cdot 3H_2O$
Luminescent zinc sulphide	–	1314-98-3	$ZnS + 100 \pm 5$ mg/kg copper.
Manganese violet[d]	77742	10101-66-3	$MnNH_4P_2O_7$
Mica[d,g] (from muscovite mica)	77019	12001-26-2	$K_2Al_4(Al_2Si_6O_{20})(OH)_4$ or $H_2KAl_3(SiO_4)_3$
Potassium sodium copper chlorophyllin[d] (chlorophyllin-copper complex)	75810	11006-34-1	[chlorophyllin-copper complex structure]
Pyrophyllite[d,g] (mineral)	–	12269-78-2	$Al_2O_3 \cdot 4SiO_2 \cdot H_2O$
Silver[d] (crystalline powder)	77820	7440-22-4	Ag
Titanium dioxide[d] (synthetic)	77891	13463-67-7	TiO_2
Ultramarines[d]	77007		$Na_v(Al_wSi_xO_y)S_z$
Ultramarines[d,e]	77013		
Zinc oxide[d] (French process)	77947	1314-13-2	ZnO

[a] 21 CFR Part 73.
[b] Colour Index (CI) number. Not used in the US for certification-exempt colour additives.
[c] Chemical Abstracts Service (CAS) number.
[d] Treated as a cosmetic ingredient, not as a colour additive in Japan.
[e] Not permitted in the EU.
[f] Not permitted in Japan.
[g] Treated as a cosmetic ingredient, not as a colour additive in the EU.

batch certification. In addition, a listing regulation must specifically authorize the use of the colour additive in the area of the eye [21 CFR Section 70.5(a)], in injections [21 CFR Section 70.5(b); as of this writing, no colour additive is listed for use in injections] or in surgical sutures [21 CFR Section 70.5(c)].

Cosmetic products are required to comply with the requirements of the FD&C Act and its implementing regulations, as well as other applicable laws and regulations. A cosmetic product (with the exception of coal-tar hair dyes, discussed in Chapter 8) containing an unlisted colour additive or a listed colour additive that does not conform to the requirements of its listing regulation is considered adulterated under Sections 601(e) and 721(a) of the FD&C Act.

In the United States, cosmetic products that are offered for retail sale are subject to the provisions of the federal Fair Packaging and Labeling (FP&L) Act. Under the authority of the FP&L Act, 21 CFR 701.3 requires the label of a cosmetic product to bear a declaration of the ingredients, usually in descending order of predominance, as mentioned in Chapter 1. However, 21 CFR Section 701.3(f) states that colour additives are permitted to be declared as a group at the end of the ingredient statement, without respect to order of predominance. This requirement for colour additive declaration does not apply to professional-use-only (or salon) products unless specifically required by regulation. In addition, colour additives present in only some shaded products in a line of products with similar composition and intended for the same use may be declared on cosmetic labels by preceding the colour additive names with 'may contain' [21 CFR Section 701.3(g)].

Colour additives may be declared either by their listed names or, for certifiable colour additives, by abbreviated names formed by omitting 'FD&C' or 'D&C' and 'No.' but including 'Ext.' (external) and 'Lake'. The FDA has stated that the agency 'does not intend to object to the immediate use of abbreviated labeling for declaring the presence of certified color additives in cosmetics' (FDA, 1999).

Both the FD&C Act and the FP&L Act provide authority to the FDA to regulate cosmetic products. Failure to comply with the requirements for cosmetic safety and labelling may render a cosmetic adulterated under Section 601 of the FD&C Act and/or misbranded under Section 602 of the FD&C Act or Section 1454(c)(3)(B) of the FP&L Act.

Regulatory requirements for the marketing of cosmetics in the United States have been presented previously (Milstein et al., 2006). Further details about colour additives permitted in the United States may be found on the FDA website (http://www.fda.gov/ForIndustry/ColorAdditives/ColorAdditiveInventories/ucm115641.htm).

European Union Regulatory Requirements for Colouring Agents in Cosmetic Products

Within the EU Cosmetics Directive (EU Directive) (i.e., Council Directive 76/768/EEC), recast as a regulation in 2009 (1223/2009/EC) (EU Regulation), all colouring agents, except those intended to colour hair, and their field of application and other requirements are listed under Annex IV. It should be emphasized that in the EU Inventory of Cosmetic Ingredients (Commission Decision, 2006/257/EC), colouring agents are divided into cosmetic colourants, which 'colour cosmetics and/or impart colour to the skin and/or its appendages', and hair dyes, which 'colour hair', As mentioned previously, hair dyes are described in Chapter 8, and are not the subject of this chapter.

According to the EC Regulation recast, as previously mentioned in Chapter 1, labels of cosmetic products marketed in the EU are required to declare their ingredients in descending order of predominance using the International Nomenclature of Cosmetic Ingredients (INCI) names. However, colouring agents may be listed in any order after other components. They are listed in the EU Regulation's Annex IV by their Colour Index (CI) number or denomination, which is the INCI name for that cosmetic ingredient. In the special case of decorative cosmetic products marketed in several colour shades, all colouring agents used in the collection may be listed, provided that the words 'may contain' or the symbol '+/−' is added.

Annex IV is divided into two parts: Part 1 lists colouring agents that are currently allowed for use in cosmetics, whereas Part 2, for provisionally allowed colouring agents, is empty as of this writing. Footnote 1 to Annex IV permits the lakes or salts of the permanently listed straight colouring agents also to be used as cosmetic ingredients provided they are prepared from substances not prohibited under Annex II or excluded under Annex V of the EU Directive. The lakes or salts have the same CI numbers as the corresponding straight colouring agents.

Japan Regulatory Requirements for Colouring Agents in Cosmetic Products

In Japan, colouring agents are called colourants. As mentioned in Chapter 1, a positive list of synthetic organic colourants was created for the first time in 1966 by the Ministry of Health and Welfare (MHW, 1966), and amended by the MHLW in 1972 and 2003. Japan uses its own names, not INCI names, for colouring agents in cosmetics that are marketed domestically.

It should be emphasized that only synthetic organic (or coal-tar) compounds are listed as colourants by the MHLW. These colourants do not need to be certified, but they must conform to specifications. Inorganic, plant and animal substances are regulated as cosmetic ingredients, but may be used as colourants. According to Rosholt, inorganic colourants and organic substances of 'natural' origins that are used to impart colour to cosmetic products are not treated specifically as 'colourants' for purposes of general safety assessment but as cosmetic ingredients; they may be used without prior approval in cosmetic products as long as their safety can be substantiated; they must be granted prior approval by the MHLW, however, if they are to be used as colourants in 'quasi-drug' products (Rosholt, 2007).

Other International Regulatory Requirements for Colouring Agents

Many countries have enacted legislation and issued regulations for approving and listing colouring agents and declaring colouring agents on the labels of cosmetics. Rosholt (2007) and Faulkner (2012) present detailed discussions of specific requirements, by country. Kumar and Gupta (2012) also published an excellent review in which the respective approaches to regulating colour additives in cosmetics in the United States, the EU, Canada, India, and the Association of Southeast Asian Nations are discussed and contrasted. Other countries have chosen approaches to control the use of colouring agents in cosmetics that reflect, to a greater or lesser extent, either the US or the EU regulatory models. Otterstatter (1999) notes that this is done in some cases by incorporating into national regulations references to lists of approved colouring agents that are virtually identical to the lists of the United States or those of the EU. Alternatively, some countries have adopted parallel regulatory approaches for the approval of such colouring agents, whereby equivalence can be established.

Some countries require colouring agents to be declared on the labels of cosmetic products in their primary national language(s). Cosmetic products also may have ingredient labelling in several languages if they are marketed in more than one country.

United States and European Union International Harmonization Efforts for Cosmetic Labelling

The use of CI numbers has been an approach for harmonizing the declaration of colour additives in cosmetic products marketed in the Member States of the EU. Since 1995, the FDA has permitted the use of dual declaration of colouring agents on the labels of cosmetic products marketed in the United States. Dual declaration of a colouring agent would consist of the US listed name followed by the CI number in parentheses.

The FDA also stated that manufacturers of finished cosmetic products (other than hair dyes) intended for sale in the United States should be alerted that, although a dual declaration for the colour additive name may be used for cosmetic labelling purposes, US law requires that they use in their products only colour additives that are in full compliance with the applicable regulations. The use in a cosmetic product of an uncertified and, therefore, unapproved colour additive subject to batch certification renders such product adulterated within the meaning of the FD&C Act. Certification-exempt colour additives also must be in full compliance with the identity, uses and restrictions, specifications and labelling requirements of their listing regulations. When such certification-exempt colour additives are used, it is the responsibility of the manufacturer of the finished cosmetic product to ensure that these requirements are met. See http://www.fda.gov/ForIndustry/ColorAdditives/ColorAdditivesinSpecificProducts/InCosmetics/ucm110032.htm.

DETERMINATION OF COLOURING AGENTS

Throughout this section of the chapter, US-certifiable colour additives (dyes and pigments, Scheme 7.1) are designated by their names as listed in the CFR. Their respective CI number, common name, and chemical structure are presented in Tables 7.2–7.9 to facilitate cross-referencing. Colouring agents that are not permitted in the United States for use in cosmetics are identified by their common names and CI numbers.

The Importance of Determining Colouring Agents in Cosmetic Products

The reasons for determining colouring agents in cosmetics have regulatory, forensic or manufacturing significance. These purposes are presented below.

1. *To ensure that only permitted colouring agents are added to a cosmetic product*

    As was discussed earlier, colouring agents permitted in one country are sometimes not approved in others. For example, erythrosine is permitted as a colouring agent in cosmetics in the EU (CI 45430) and as a colourant in Japan (Aka3), but although it is certifiable as FD&C Red No. 3 for use in foods and drugs, it is not permitted for use in cosmetics in the United States [21 CFR 81.30(u)].

    A different case is the colouring agent Quinoline Yellow. This dye consists of a mixture of mono-, di-, and trisulphonated quinophthalone positional isomers, the relative proportions of which depend on the degree of sulphonation during its manufacture. A mixture of the monosodium salts of the 6'- and 8'-monosulphonic acids

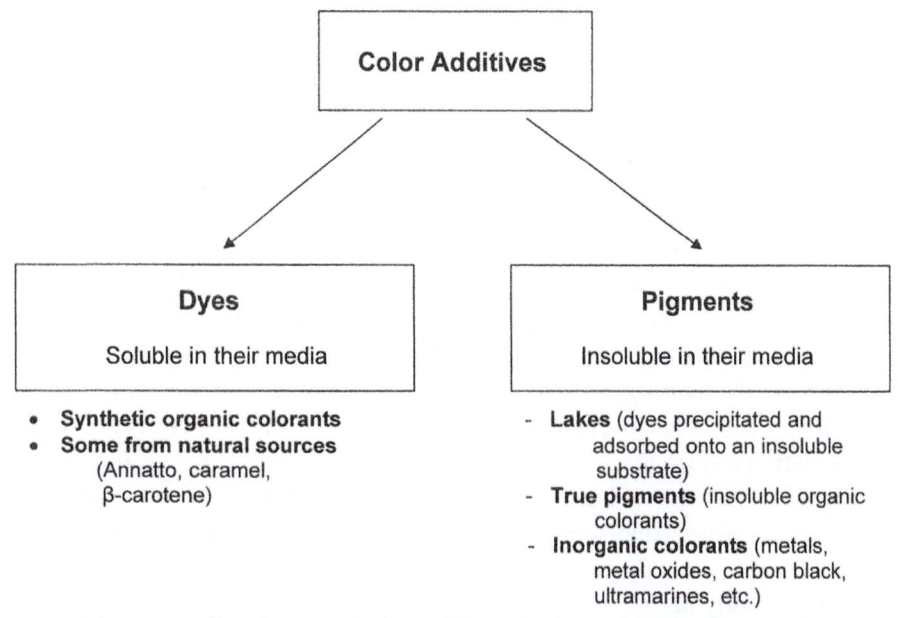

Scheme 7.1 Classification of colour additives by their solubility in their media.

Color Additives

Dyes — Soluble in their media
- Synthetic organic colorants
- Some from natural sources (Annatto, caramel, β-carotene)

Pigments — Insoluble in their media
- **Lakes** (dyes precipitated and adsorbed onto an insoluble substrate)
- **True pigments** (insoluble organic colorants)
- **Inorganic colorants** (metals, metal oxides, carbon black, ultramarines, etc.)

with up to 15% of the disodium salts of the disulphonated isomers is certifiable in the United States as the colour additive D&C Yellow No. 10. A mixture that contains mostly di- and trisulphonated components is permitted as a cosmetic colouring agent in the EU (CI 47005) and as a colourant in Japan (Ki203) (Rosholt, 2007). Even though both of these variant forms are indexed as CI 47005, the latter mixture (i.e., consisting mainly of di- and trisulphonated components) is not certifiable as D&C Yellow No. 10 and is therefore not permitted in cosmetics that are imported and sold in the United States.

2. *To ensure that the information on a cosmetic label is complete and correct.*
In the United States, the colouring agents that are part of a cosmetic product offered for retail sale must be declared by name on the product label. Analysis of the cosmetic will verify the colour additive content and may find undeclared colour additives. If the label is not complete or not correct, the product may be considered adulterated or misbranded under the FD&C Act or the FP&L Act.

3. *To investigate the cause of allergic and dermatologic reactions*
Contact of the human body with certain colouring agents, their impurities, or their decomposition products (that may appear during processing or storage of the cosmetic product) can produce allergic reactions, sensitization or photosensitization in susceptible people. Such agents include the US certification-exempt colour additive carmine (Table 7.10) and certain xanthene-type dyes and tattoo components (Rosenthal et al., 1988; Wei et al., 1994, 1995; Mselle, 2004; Antonovich and Callen, 2005; Klontz et al., 2005; Suzuki et al., 2011; Quint et al., 2012). Determination of the colouring agent(s) present in the cosmetic product used may provide a clue to the source of the unexpected reaction.

4. *To help in forensic investigations*
Lipstick smears left on drinking glasses, cups and cigarette butts can link a suspect to a crime scene. When found on a suspect's clothing, they can prove a link between the suspect and the victim. Results obtained from the analysis of lipstick smears in a forensic science laboratory are often found to be important evidence in criminal cases (Barker and Clarke, 1972; Andrasco, 1981; Russel and Welch, 1984; Gennaro et al., 1994; Griffin et al., 1994; Ehara and Marumo, 1998; Rodger et al., 1998; Gardner et al., 2013; Salahioglu et al., 2013).

5. *To determine the stability of a colouring agent added to various matrices*
The stability of a colouring agent can be affected by many factors during storage of the cosmetic product. Such factors are light, heat, pH, the nature of the packaging, the nature of the product base, etc. (Rush, 1989; Otterstatter, 1999; Faulkner, 2002). Therefore, cosmetic manufacturers may conduct stability studies in various matrices before marketing a product.

6. *To conduct quality control*
Cosmetic manufacturers must determine the colouring agents present in their products to ensure that quality standards are consistently maintained (Rodger, 1998). This

quality control may be conducted at various stages in the production and pre-marketing process. The sample tested is then compared with a standard using various analytical techniques such as colorimetry and spectroscopy (Dragocolor, 2004).

Composition of Colouring Agents

Generally, colouring agents used in cosmetics are prepared by either of two means. One process, which applies to the mono- and bisazo dyes (Tables 7.2 and 7.3), involves azo coupling reactions. The other, for dyes such as those in the fluorescein, quinoline and triphenylmethane classes, is a manufacturing process that involves several steps: condensation of the starting materials, partial purification of the resulting condensation product and introduction of functional groups to increase the colouring agent's solubility in water (e.g., by sulphonation or carboxylation) or its solubility in organic solvents (e.g., by halogenation, nitration or alkylation). As an example, Fig. 7.1 schematically shows the preparation of the halogenated fluorescein dyes D&C Red No. 27, D&C Red No. 28 and Rose Bengal (CI 45440, approved for cosmetic use in Japan, South Korea and Taiwan).

During manufacture, in addition to the main component of a colouring agent, various impurities may be produced, depending on the purity of the starting materials used and the conditions under which the technological process was performed. These impurities can consist of intermediates (compounds from which a colouring agent is directly or indirectly synthesized), side-reaction products and subsidiary colours (Leatherman et al., 1977; Abrahart, 1968; Marmion, 1991). A subsidiary colour is a structural variant

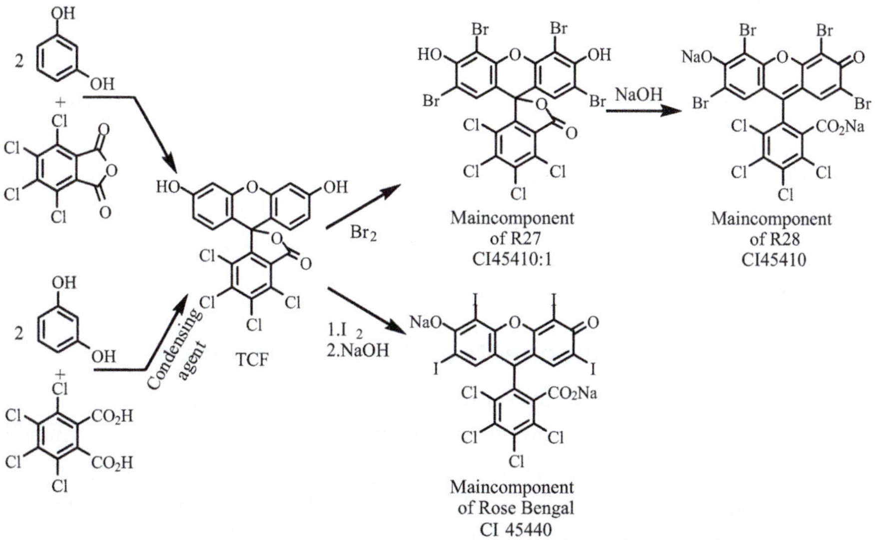

Figure 7.1 Preparation of D&C Red Nos. 27 and 28 and Rose Bengal. *Adapted from Weisz, A., Andrzejewski, D., Shinomiya, K., Ito, Y., 1995. Modem countercurrent chromatography. In: Conway, W.D., Petroski, R.J. (Eds.), Preparative Separation of Components of Commercial Tetrachlorofluorescein by pH-Zone-Refining Countercurrent Chromatography, ACS Symposium Series No. 593, American Chemical Society, Washington, D.C. (Chapter 16).*

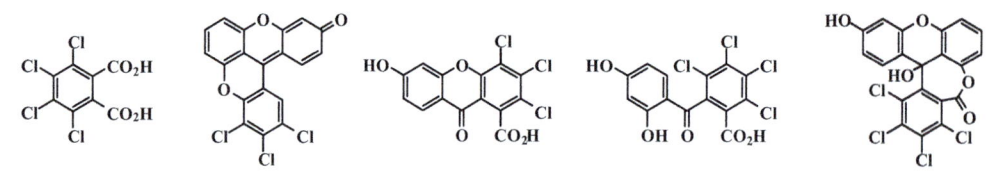

Figure 7.2 Impurities isolated from commercial 4,5,6,7-tetrachlorofluorescein. *Adapted from Weisz, A., Andrzejewski, D., Shinomiya, K., Ito, Y., 1995. Modem countercurrent chromatography. In: Conway, W.D., Petroski, R.J. (Eds.), Preparative Separation of Components of Commercial Tetrachlorofluorescein by pH-Zone-Refining Countercurrent Chromatography, ACS Symposium Series No. 593, American Chemical Society, Washington, D.C. (Chapter 16).*

of the main colour component that varies in the position, number or length of the substituent group(s). D&C Red Nos. 27 and 28, for example, may contain CFR-limited amounts of lower-halogenated subsidiary colours (Weisz et al., 1992, 1994a, 1996). An intermediate of those colours, 4,5,6,7-tetrachlorofluorescein (TCF), was found to contain impurities, as shown in Fig. 7.2 (Weisz et al., 1995, 1998). These TCF impurities, including the unreacted starting material tetrachlorophthalic acid, can be brominated or iodinated during the manufacturing process and carried into the final product. Often analyses reveal the presence of contaminants that vary in nature across batches of the same dye obtained from different suppliers (Van Liedekerke and De Leenheer, 1990; Gagliardi et al., 1995; Weisz et al., 1995). Such variability is evident from the chromatograms shown in Fig. 7.3 that were obtained by liquid chromatography (LC) of commercial batches of D&C Red No. 28 from three different sources.

The application of improved technologies has enabled identification and quantification of impurities that are not specified in the CFR (Yamada et al., 1996; Ishimitsu et al., 1997; Weisz, 1997; Andrzejewski and Weisz, 1999; Ngang et al., 2001; Matsufuji et al., 2002; Weisz and Andrzejewski, 2003; Weisz et al., 2004, 2006, 2015). If such a batch were tested in FDA certification laboratories, it would then be denied certification based on considerations of good manufacturing practices. In some cases, the toxicity of impurities may be assessed by the FDA and specifications limiting their presence in colour additives may be added to the CFR.

Preparation of Colour Components as Reference Materials

To develop methods of analysing dyes, purified components and purified contaminants of the dyes are needed for use as reference materials. Such compounds are typically not available commercially. In a review on the purity of dyes used for biological staining, Lyon (2002) summarized the situation: '*Pure dyes are still difficult or impossible to purchase.*' That statement remains valid today (Keck-Wilhelm et al., 2015).

Owing to this lack of pure dyes on the market, it is necessary to purify them in the laboratory. Throughout the 1980s, the few reports in the literature that addressed dye purification involved only small amounts of subsidiary colours that were separated from

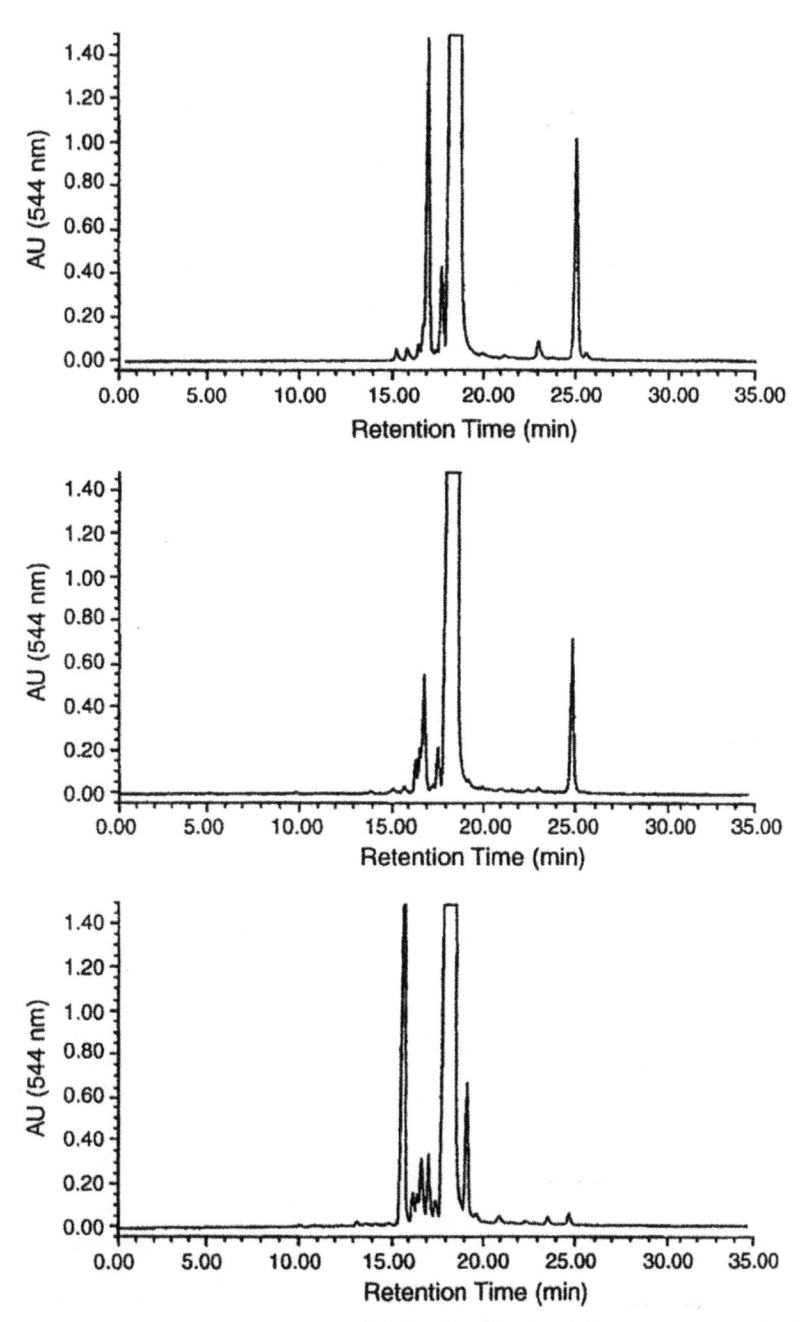

Figure 7.3 Reversed-phase liquid chromatography chromatograms of portions from batches of D&C Red No. 28 obtained from three different sources. *Adapted from Weisz, A., Wright, P.R., Andrzejewski, D., Meyers, M.B., Glaze, K., Mazzola, E.J., 2006. J. Chromatogr. A 1113, 186.*

dye mixtures by chromatographic methods (Fales et al., 1985; Freeman and Williard, 1986; Freeman et al., 1988; Zollinger, 2003). Among those techniques was high-speed countercurrent chromatography (HSCCC), which since then has been further developed. In both its conventional and its modified modes, HSCCC has become one of the most effective methods for the separation and purification of preparative amounts of dyes (Conway, 1990; Ito, 1996, 2013). A general approach to the separation of dyes by HSCCC is presented in Fig. 7.4. The experimental conditions for this approach were previously described (Weisz and Ito, 2000).

A modified form of conventional HSCCC, named pH-zone-refining CCC, was developed in 1993 and applied from the outset to separate components of multigram quantities of

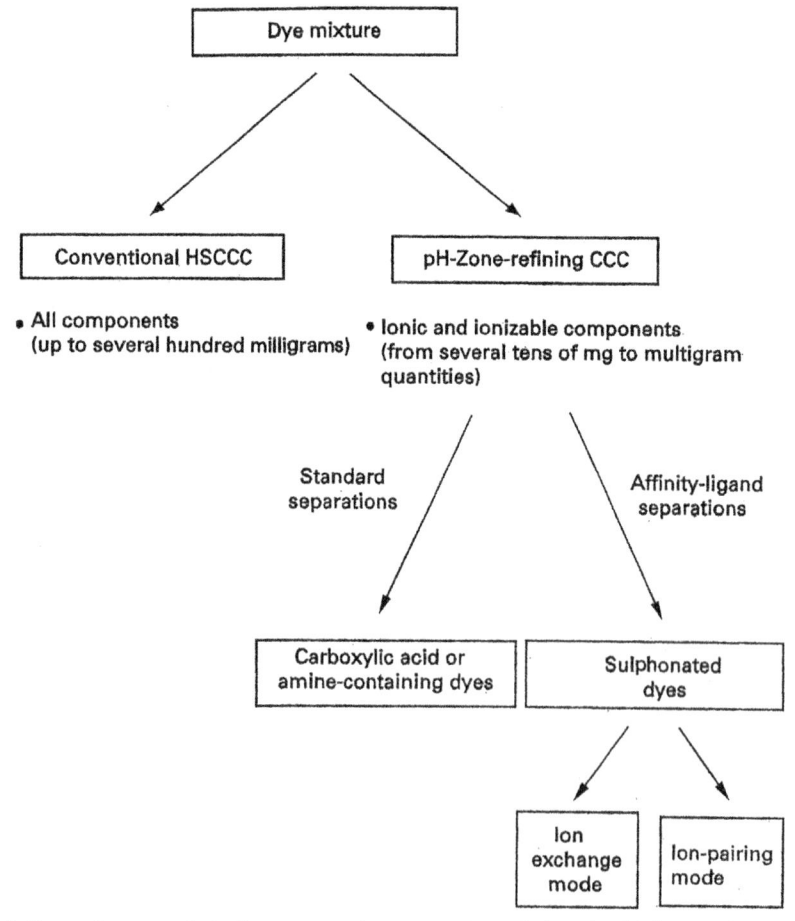

Figure 7.4 General approach to the preparative separation of dyes by high-speed countercurrent chromatography (HSCCC). *Adapted from Weisz, A., Ito, Y., 2000. Encyclopedia of separation science. In: Wilson, I.D., Adlard, E.R., Cooke, M., Poole, C.F. (Eds.), Dyes—High-Speed Countercurrent Chromatography, vol. 6. Academic Press, London.*

dye mixtures (Weisz et al., 1994b; Ito et al., 1995; Weisz, 1996; Ito, 2013). As an example, Fig. 7.5 shows the pH-zone-refining CCC separation of 5 g of D&C Orange No. 5 and the LC analysis of the separated components. This separation resulted in 4′,5′-dibromofluorescein (2.83 g), 2′,4′,5′-tribromofluorescein (1.52 g) and 2′,4′,5′,7′- tetrabromofluorescein (0.26 g) of adequate purity (>99.5%) to be used as reference materials.

Affinity-ligand pH-zone-refining CCC is a variant procedure that is effective for separating gram quantities of highly polar sulphonated dye components (Weisz and Ito, 2000). The monosulphonated components of D&C Yellow No. 10 and some di- and

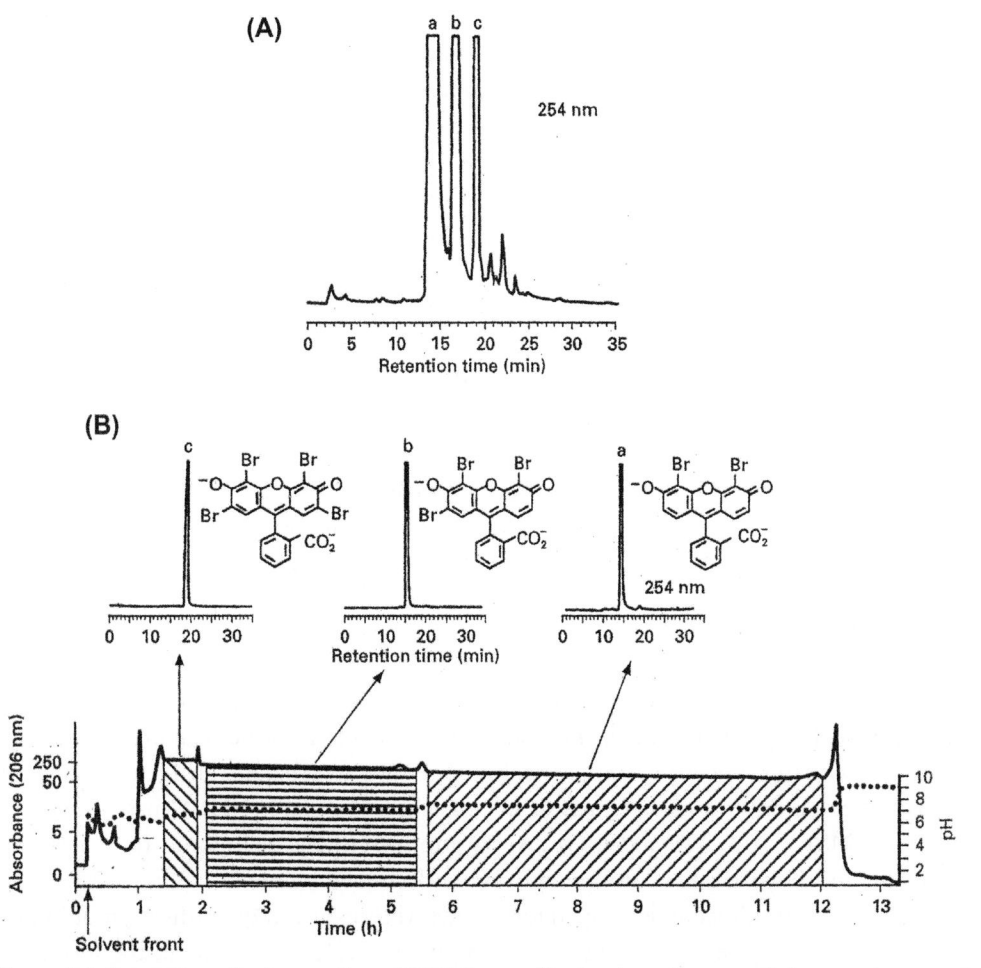

Figure 7.5 Separation of a 5-g portion of D&C Orange No. 5 using pH-zone-refining countercurrent chromatography (CCC). (A) Reversed-phase liquid chromatography (LC) analysis of the portion. (B) pH-zone-refining CCC chromatogram of the separation, and reversed-phase LC chromatograms of the components in the combined fractions of each *hatched region*. *Adapted from Weisz, A., Scher, A.L., Shinomiya, K., Fales, H.M., Ito, Y., 1994b. J. Am. Chem. Soc. 116, 704.*

trisulphonated components of the Japanese colourant Ki203 (Cl 47005) were separated by this technique to obtain reference materials (Weisz et al., 2001a, 2009). Fig. 7.6 shows the separation of 1.8 g of D&C Yellow No. 10, which resulted in 0.6 g of the 6′-monosulphonated isomer and 0.18 g of the 8′-monosulphonated isomer, both >99% pure. The usefulness of this technique is further demonstrated by the separation of 1,3,6-pyrenetrisulphonic acid (P3S) from D&C Green No. 8 (Fig. 7.7). P3S, the component needed as a reference material, was present in the dye at a level of only 3.5%. A typical separation involving 20.3 g of dye yielded ~0.58 g P3S of greater than 99% purity (Weisz et al., 2011).

Dye contaminants needed as reference materials also can be isolated from the dyes by HSCCC (Weisz et al., 1998) or by other chromatographic methods. Alternatively, some can be synthesized (Weisz et al., 2006; Weisz and Andrzejewski, 2003; Weisz, 1997), and others can be obtained by purifying purchased material of technical grade (Andrzejewski and Weisz, 1999; Weisz et al., 2004; Weisz and Ito, 2008).

Analytical Methods for Determining Colouring Agent Components

Analytical methods are continuously being developed for the FDA's colour additive batch certification program. These methods are used to enforce the limiting specifications listed in 21 CFR Parts 74 and 82 for both organic and inorganic impurities and to identify contaminants not specified in the CFR. Some methods have been presented in detail by Leatherman et al. (1977), Marmion (1991) and Latimer (2012). Since those publications appeared, new technologies have been developed, analytical instrumentation has been improved, and as a result, some of the described methods have been replaced. This section focuses primarily on reviewing the methods for analysing impurities in colour additives that have been published since the appearance of Marmion's book. The determination of colouring agents in cosmetic products is described further on.

Organic Impurities

The organic impurities in colour additives have varied sources, including unreacted starting materials, components present in low-grade starting materials, subsidiary colours and synthetic by-products, all of which can be carried over into the final product during the manufacturing process. Common techniques used to separate, characterize and/or determine these impurities make use of LC, gas chromatography (GC)/mass spectrometry (GC/MS) or LC/MS and CCC. Applications of these techniques to analyses of impurities in US-certifiable colour additives are described below.

Early LC methods were developed for the identification and quantification of impurities in azo dye-type colouring agents. An impurity was found by LC in the monoazo colour additive FD&C Red No. 40 and identified by GC/MS as 4-nitro-*p*-cresidine (2-methoxy-5-methyl-4-nitrobenzenamine). The impurity was detected in all 28 certified batches of dye analysed and, along with other aromatic amines (*p*-cresidine and aniline), was quantified at parts per billion (µg/kg) levels using a previously developed

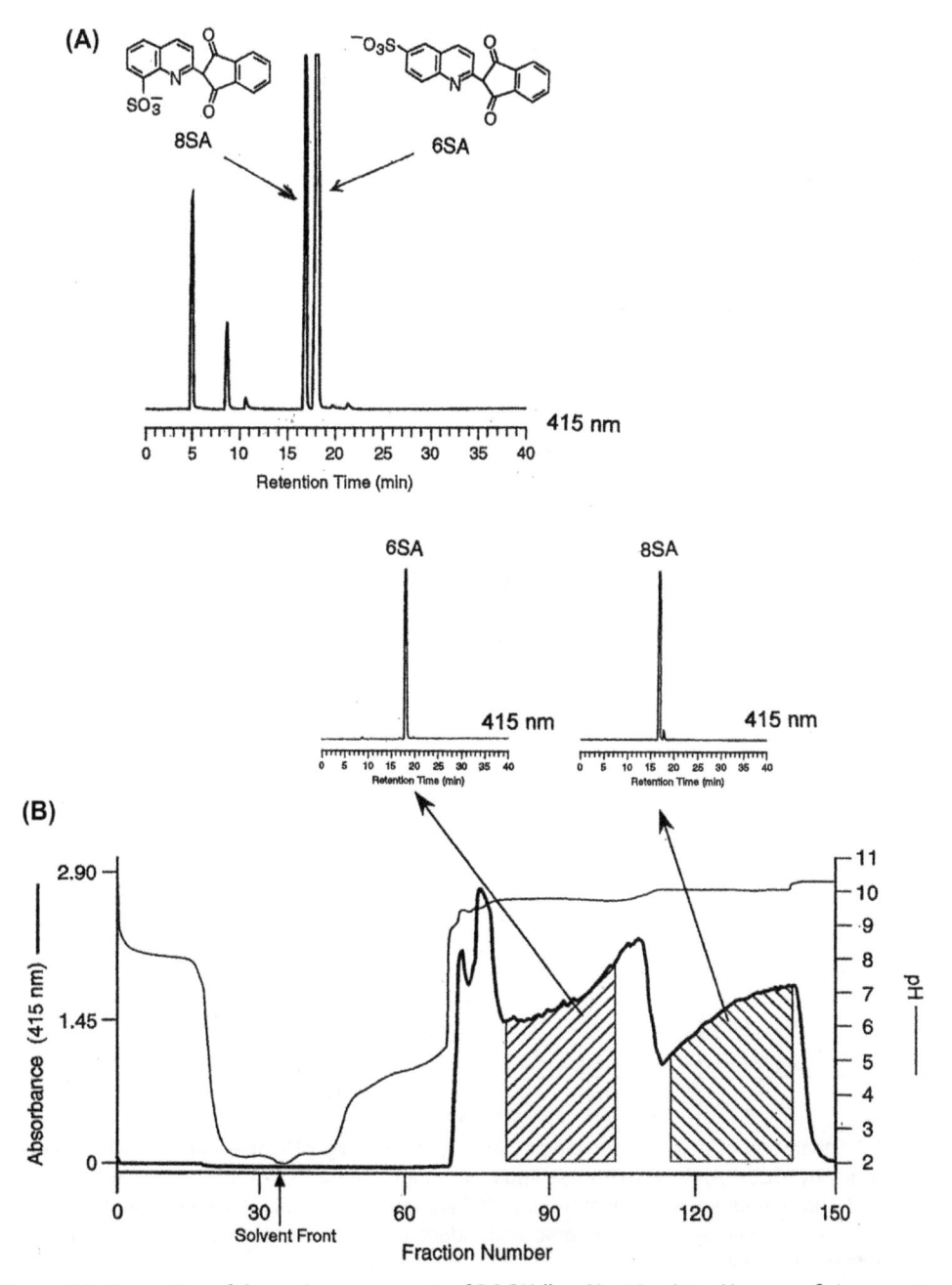

Figure 7.6 Separation of the main components of D&C Yellow No. 10 using pH-zone-refining countercurrent chromatography (CCC). (A) Reversed-phase liquid chromatography (LC) analysis of the colour additive. (B) pH-zone-refining CCC chromatogram of the separation of a 1.8-g portion of the colour additive, and LC analyses of the separated components. *Adapted from Weisz, A., Mazzola, E.P., Matusik, J.E., Ito, Y., 2001a. J. Chromatogr. A 923, 87.*

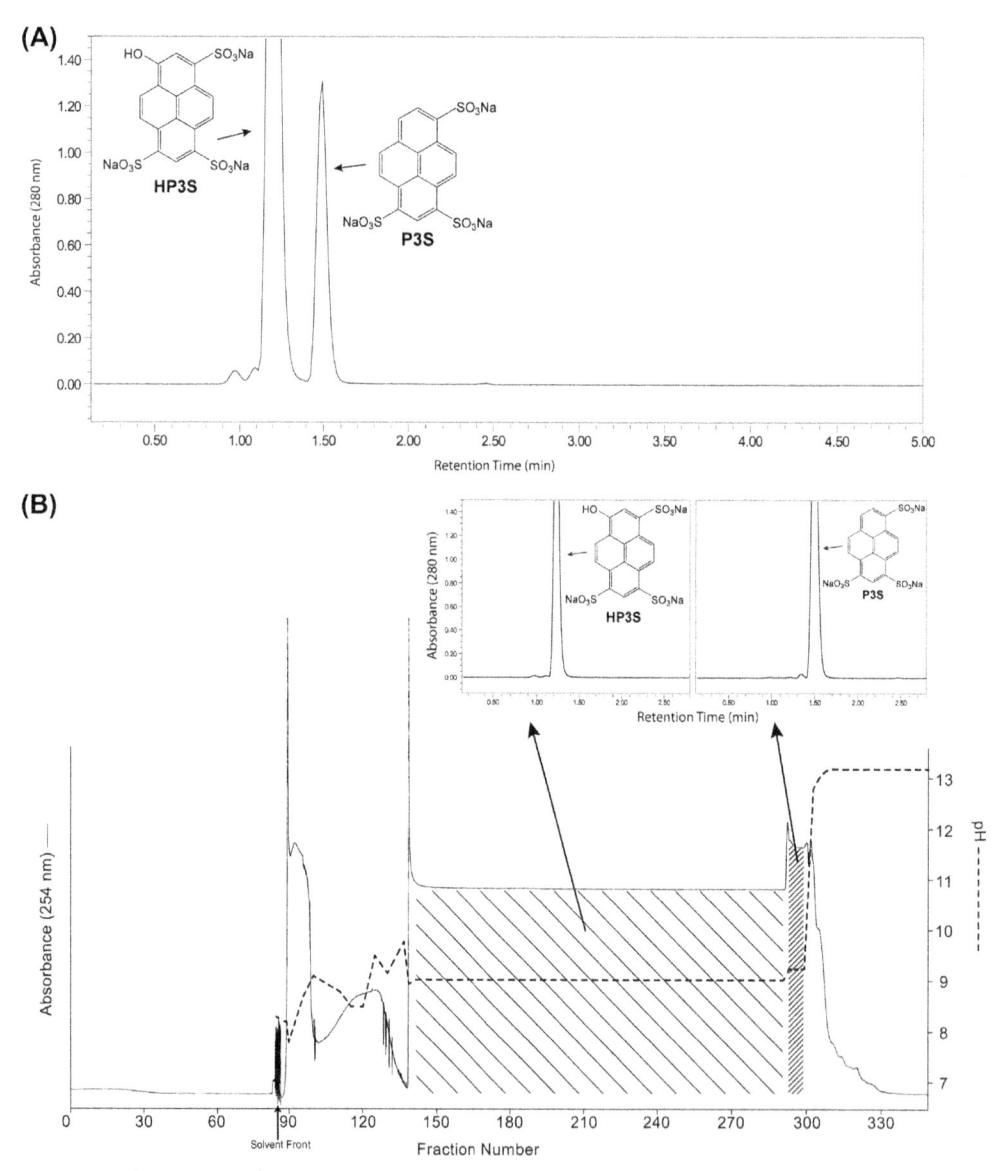

Figure 7.7 Separation of a 20.3-g portion of D&C Green No. 8 by pH-zone-refining countercurrent chromatography (CCC). (A) Reversed-phase liquid chromatography (LC) of the portion. (B) pH-zone-refining CCC chromatogram of the separation, and LC analyses of the separated components for use as reference materials. *P3S*, pyrenetrisulphonic acid. *Adapted from Weisz, A., Mazzola, E.P., Ito, Y., 2011. J. Chromatogr. A 1218, 8249.*

LC method (Richfield-Fratz et al., 1989). In another monoazo colour additive, D&C Red No. 36, the intermediates (2-chloro-4-nitroaniline and 2-naphthol) and an impurity (2,4-dinitroaniline) were determined by an LC method (Scher and Adamo, 1993).

Two LC methods were developed to determine, at trace levels (μg/kg), the total quantity of benzidine (free aromatic amine and combined forms) in the colour additives FD&C Yellow No. 5 and FD&C Yellow No. 6. These methods have several steps in common: reduction (with sodium dithionite) of any combined benzidine, present in the colour additive as azo and/or bisazo dyes, to free benzidine; an extraction step; diazotization and coupling with pyrazolone-T (for FD&C Yellow No. 5) or with 2-naphthol-3,6-disulphonate (for FD&C Yellow No. 6) and LC analysis of the coupled product (Davis and Bailey, 1993; Prival et al., 1993; Peiperl et al., 1995).

A method was developed that uses thin-layer chromatography (TLC) to separate organic components in D&C Red Nos. 27 and 28 and videodensitometry to quantify them. This method permits in situ quantification of lower-halogenated subsidiary colours (e.g., 2′,4′,5′-tribromo-4,5,6,7-TCF) or the ethyl ester of the main component 2′,4′,5′,7′-tetrabromo-4,5,6,7-TCF in multiple dye samples on the same analytical TLC plate. The maximum time necessary for the analysis of five standards and four samples applied to each plate is 45 min (Wright et al., 1997).

MS was shown to be a useful technique in the structural assignment of isomeric mono- and disulphonic acid components of the colouring agent Quinoline Yellow, for which, as discussed above, the predominantly monosulphonated form is certifiable as D&C Yellow No. 10 (Weisz et al., 2001b, 2002).

LC methods were also developed for the identification and quantification of subsidiary colours in triphenylmethane colouring agents. Specifically, Matsufuji et al. (1998) determined five subsidiary colours in Brilliant Blue FCF, certifiable as FD&C Blue No. 1, and Tsuji et al. (2006) determined subsidiary colours in Fast Green FCF, certifiable as FD&C Green No. 3.

An LC method was developed for the determination in one analysis of intermediates (5-amino-2,4-dimethyl-1-benzenesulphonic acid and 4-hydroxy-1-naphthalenesulphonic acid) and subsidiary colours of the monoazo colour additive FD&C Red No. 4. For use as reference materials, one of the intermediates and a subsidiary colour were purified using HSCCC (Vu et al., 2011).

A relatively new type of LC technology known as ultra-high performance LC was implemented in developing two methods for determining intermediates (2-amino-1-naphthalene sulphonic acid and 3-hydroxy-2-naphthoic acid) and subsidiary colours in the monoazo colour additive D&C Red No. 34 and its lakes (Harp et al., 2011; Belai et al., 2012).

Conventional LC was used for determining synthetic by-products (halogenated benzoylbenzoic acids and several variously brominated resorcinols) and intermediates (phthalic acid and 3,4,5,6-tetrachlorophthalic acid) in the xanthene and fluoran colour additives D&C Orange No. 5, D&C Red Nos. 27 and 28 and their lakes (Yang and Weisz, 2012; Weisz and Krantz, 2014).

Separation and determination of two closely related, very polar subsidiary colours of D&C Green No. 8 (1,3,6-P3S acid trisodium salt and 1,3,6,8-pyrenetetrasulphonic acid tetrasodium salt) were achieved with the development of a new method using a conventional LC instrument. The compounds were separated in less than 4 min using the new method (Jitian et al., 2014).

Impurities not specified in the CFR were determined in the fluorescein-type colour additives D&C Red Nos. 21 and 22 and D&C Red Nos. 27 and 28 using various techniques. Solid-phase microextraction combined with GC/MS was used to determine 2,4,6-tribromoaniline in D&C Red Nos. 21 and 22 (Weisz et al., 2004) and hexachlorobenzene and 2-bromo-3,4,5,6-tetrachloroaniline in D&C Red Nos. 27 and 28 (Andrzejewski and Weisz, 1999; Weisz and Andrzejewski, 2003). LC methods were used to quantify l-carboxy-5,7-dibromo-6-hydroxy-2,3,4-trichloroxanthone and the decarboxylated 2′,4′,5′,7′-tetrabromo-4,5,6,7-TCF in D&C Red Nos. 27 and 28 (Weisz, 1997; Weisz et al., 2006). More recently, three impurities in D&C Red No. 33 that are not specified in the CFR but often observed in submitted batches were separated by spiral HSCCC and characterized by MS and nuclear magnetic resonance spectroscopy as subsidiary colours (Weisz et al., 2015).

Inorganic Impurities

All US-certifiable colour additives have CFR-specified limits for arsenic and lead, most have specified limits for mercury and a few also have specified limits for other elements such as chromium and manganese. Sources of such impurities include the starting materials and catalysts used in the technological processes. For example, the triphenylmethane dyes used as colour additives are generally prepared in two steps: a condensation reaction that results in a colourless intermediate, which is a leuco base, and an oxidation reaction of the leuco base that results in the coloured material (Fierz-David and Blangey, 1949). The oxidizing agents used for the second step are typically manganese dioxide or a dichromate salt, and traces of manganese and chromium may remain in the final product. Presented here are methods for analysing these elemental impurities in US-certifiable colour additives.

Since the 1960s, methods for determining whether colour additives met the limiting specifications for lead and arsenic used wavelength-dispersive X-ray fluorescence (WDXRF) spectrometry, while chromium, manganese and mercury were determined by atomic absorption (AA) methods. As instrumentation became more advanced, it became feasible to develop WDXRF methods for the determination of chromium (Hepp, 1996) and for manganese (Hepp, 1998) in FD&C Blue No. 1. The analyses were completely automated and could be performed in conjunction with lead and arsenic determinations in the same sample portion, requiring about 5 min per element.

Because XRF was not sufficiently sensitive to determine mercury at the typical CFR limit of 1 ppm (calculated as elemental mercury), an improved AA method was developed that uses microwave digestion of the sample prior to determining mercury by cold-vapour AA spectrometry (Hepp et al., 2001). That method was later modified and extended to the determination of mercury in the newly certifiable colour additive D&C Black No. 2 (Hepp, 2006). It should be noted that this method of mercury determination cannot be applied to the colour additives that contain iodine, D&C Orange No. 10 and D&C Orange No. 11, because digestion produces iodine, which penetrates Teflon tubing and subsequently binds to mercury (Hepp et al., 2001).

An alternative method was developed to determine arsenic (calculated as elemental arsenic) at levels well below the specified limit of 3 ppm. It uses dry ashing followed by hydride-generation AA (Hepp, 1999).

Since 2006, advances in inductively coupled plasma (ICP)–MS, microwave digestion, and XRF spectrometry have allowed the development of improved methods for the determination of inorganic components of colour additives. Toward that effort, a WDXRF method was developed for determining all specified elements, including mercury, in most colour additives. Also, a method for the simultaneous determination of arsenic, chromium, lead, manganese and mercury in all certifiable colour additives by ICP–MS was developed and evaluated by comparison to the WDXRF method (Hepp, 2015). Advances in polarized energy-dispersive (PED) XRF technology enabled the development of a simple, quantitative XRF technique for determining arsenic, lead and mercury in certifiable colour additives (Hepp and James, 2016). In this study, the authors found that the PEDXRF technique yielded results similar to those from the WDXRF and ICP–MS methods, with the advantage that PEDXRF permits use of a smaller sample portion size than does WDXRF and is more precise than ICP–MS.

Determination of Colouring Agents in Cosmetic Products

The previous section presented the analysis of colouring agents themselves. This section describes the isolation/separation of colouring agents from cosmetic products (sample preparation) and the analysis of the isolated dyes and pigments.

Cosmetic products vary widely in their colouring agent contents. Decorative cosmetics contain the highest percentages of colouring agents, frequently present as mixtures of multiple colouring agents. As a result, such products—lipsticks, blushes, face powders, mascaras, eye shadows, eyeliners and nail polishes—have been the subjects of most analytical studies. Their colouring agent content ranges between 1% and 25% (Gagliardi et al., 1995; Schlossman, 2000; Wilkinson and Moore, 1982). In contrast, cosmetics such as shampoos, bubble baths, creams and oil-based lotions generally contain between 0.01% and 0.3% colour additives (Dragocolor, 2004).

The colouring agents present in most cosmetic products must be isolated from their matrices prior to their identification and quantification. The colouring agents in

Table 7.11 Types of colouring agents that might be used in selected cosmetic products

Cosmetic product	Dye Water soluble	Dye Oil soluble	Pigment Lake	Pigment Organic	Pigment Inorganic
Shampoo	✓				
Bubble bath	✓				
Make-up powder			✓	✓	✓
Lipstick		✓	✓	✓	✓
Nail polish and enamel			✓		✓
Cream	✓	✓			✓
Blusher		✓			✓
Eye shadow			✓		✓
Fragrance preparation	✓				

clear-liquid cosmetics or in products that can be easily dissolved generally do not require such isolation as long as the ultraviolet/visible (UV/VIS) spectrophotometric analysis of each additive can be achieved without interference from that of the others. Pawliszyn (2002) and Pawliszyn and Lord (2010) offer a comprehensive treatment of modern sample preparation techniques. Although these books pertain mainly to biological, food and environmental matrices, the described techniques are adaptable to the analysis of the colouring agents in cosmetic product matrices.

Generally, isolation of a colouring agent depends on the cosmetic matrix and on the solubility and other physical and chemical properties of the colouring agent. The solubility of cosmetic dyes in various solvents has been tabulated (Holtzman, 1962; Zuckerman, 1974; Zuckerman and Senackerib, 1979; Marmion, 1993). Table 7.11 shows the types of colouring agents that may be used in selected cosmetic products, with dyes grouped by their solubility.

The methods used most often for the separation of colouring components from a product are liquid–liquid extraction (with two-phase solvent systems) and various adsorption techniques such as solid-phase extraction (SPE). Pigments cannot be separated by adsorption from the matrix; rather they are collected as residues after fat and other components, such as soluble dyes, are eliminated. Details and examples of how to use these techniques for sample preparation have been previously presented (Bell, 1977; Leatherman et al., 1977; Lehmann, 1986; Marmion, 1991; Otterstatter, 1999). Fig. 7.8 (adapted from Etournaud and Aubort, 1983) shows a general method for the extraction of various colouring agents present in lipsticks, and it is also applicable to the extraction of colouring agents from fatty and non-fat-based make-up and mascaras. Official methods for the extraction of some colouring agents from cosmetic products are also available in several countries (e.g., Pharmaceutical Society of Japan, 2000; European Commission, 1999; Latimer, 2012).

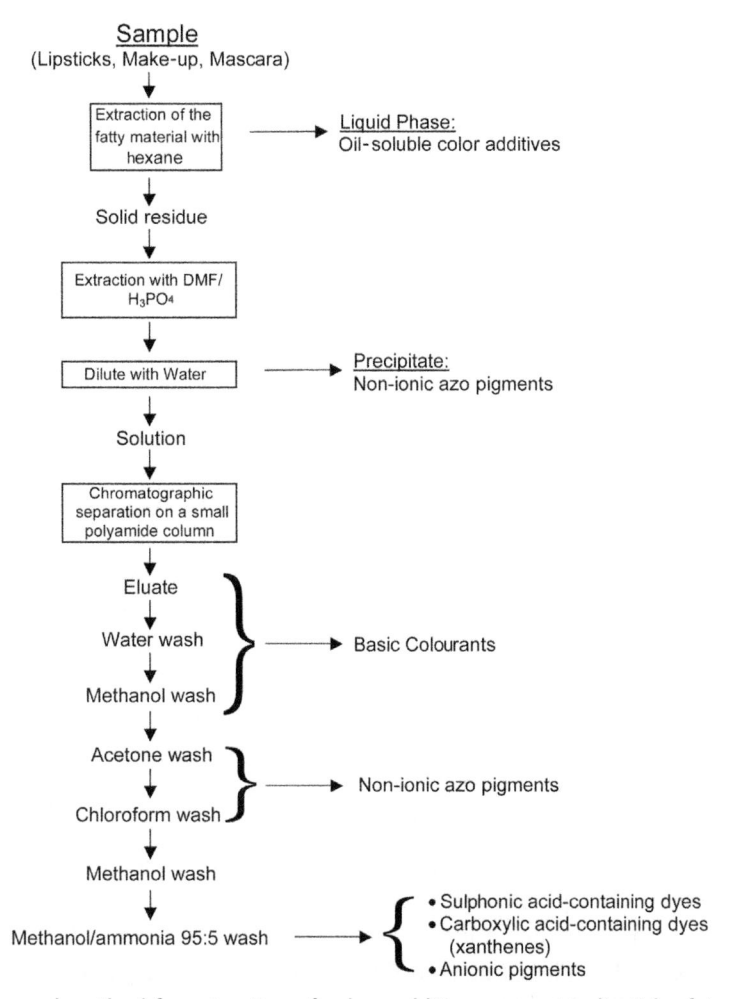

```
                    Sample
          (Lipsticks, Make-up, Mascara)
                       ↓
          ┌──────────────────────┐
          │ Extraction of the    │       Liquid Phase:
          │ fatty material with  │ ────→ Oil-soluble color additives
          │ hexane               │
          └──────────────────────┘
                       ↓
                Solid residue
                       ↓
          ┌──────────────────────┐
          │ Extraction with DMF/ │
          │ H₃PO₄                │
          └──────────────────────┘
                       ↓
          ┌──────────────────────┐       Precipitate:
          │ Dilute with Water    │ ────→ Non-ionic azo pigments
          └──────────────────────┘
                       ↓
                  Solution
                       ↓
          ┌──────────────────────┐
          │ Chromatographic      │
          │ separation on a small│
          │ polyamide column     │
          └──────────────────────┘
                       ↓
                   Eluate
                       ↓
               Water wash    ⎫
                       ↓     ⎬ ────→ Basic Colourants
               Methanol wash ⎭
                       ↓
               Acetone wash  ⎫
                       ↓     ⎬ ────→ Non-ionic azo pigments
              Chloroform wash⎭
                       ↓
               Methanol wash
                       ↓                    • Sulphonic acid-containing dyes
         Methanol/ammonia 95:5 wash ────→   • Carboxylic acid-containing dyes
                                              (xanthenes)
                                            • Anionic pigments
```

Figure 7.8 General method for extraction of colour additives present in lipsticks, fat- and non-fat-based make-up and mascaras. *DMF*, dimethylformamide. *Adapted from Etournaud, A., Aubort, J.D., 1983. Trav. Chim. Aliment. Hyg. 74, 372.*

Once the colouring agents are extracted from a cosmetic product, they can be further separated from one another, identified and quantified using analytical methods developed for those purposes. A wealth of published work is available for the analysis of colouring agents in cosmetics, especially lipsticks. Marmion (1991) reviewed the literature that was published from the mid-1960s through the 1980s. The remainder of this section highlights selected early studies that are now considered classics and reviews methods published since Marmion's work, grouped by the analytical technique employed.

Thin-Layer Chromatography

Widespread use of TLC as a means of separating the components of mixtures began in the mid-1960s, and its basic procedure and variations have been well documented by many, including Touchstone (1992). TLC remains a common technique today for separating multiple dyes from one another. Silica gel is the most typical adsorbent, followed by alumina and microcrystalline cellulose. In the TLC technique, a sample solution is spotted or streaked onto the lower part of a TLC plate. The plate is developed with a suitable solvent system and dried. The separated component bands are individually removed by scraping, and the dyes are extracted from the adsorbent and identified and quantified by UV/VIS spectrophotometry. The use of TLC has been described for the separation of synthetic dyes (Wall, 2000; Cserhati and Forgacs, 2001; Gupta, 2003; Vlajkovic et al., 2013).

When applied to cosmetic products, TLC has been used primarily for the separation of colouring agents present in lipsticks. Silk (1965) developed a method whereby 15 colouring agents were analysed without the need for a preliminary clean-up step because the lipsticks were directly applied to a warm silica gel TLC plate. The colouring agents were separated in two steps: elution with methylene chloride separated the oil-soluble dyes from one another and left the water-soluble dyes and the pigments at the origin; then the water-soluble dyes were separated from one another by elution with ethyl acetate:methanol:8.7% ammonium hydroxide (15:3:3). The separated bands were removed by scraping, and the dyes were extracted from the silica gel and then identified and quantified by UV/VIS spectrophotometry. This method was also applied to the analysis of dyes in nail polishes and blushes (Leatherman et al., 1977). By modifying the solvent systems used to develop the TLC plate, the method was extended to the separation of other oil-soluble, fluorescein-type, sulphonated and basic colouring agents from lipsticks, blushes, make-up and nail polish (Bell, 1977). Sjoberg and Olkkonen (1985) also analysed colouring agents in lipsticks after separating them by direct application of the sample onto a TLC plate, but then determined the extracted colouring agents by using LC. Direct application of lipsticks onto a TLC plate combined with developing the plate with a series of selective eluants of increasing polarity or, alternatively, the use of solvent extraction with dimethylformamide (DMF) followed by TLC enabled Perdih (1972) to separate more than 150 dyes, 37 of which were found in lipsticks.

Gagliardi et al. (1995) presented the ratio-to-front (also called retention factor) values obtained for 20 colouring agents most frequently encountered in cosmetic products following their development on silica gel TLC plates with eight different solvent systems. Those authors also reported the volume of solvent needed for the elution of the colouring agents with three LC solvent systems. The described methods were applied to the analysis of the colouring agents present in 25 cosmetic products, including lipsticks, mouthwashes, toothpastes, eye shadows and blushes.

Ohno et al. (1996) developed a reversed-phase TLC method using octadecylsilica (C_{18}) gel that complementarily employed four solvent systems to separate 45 water-soluble dyes, most of which are used for colouring cosmetics or food in Japan. The method was combined with scanning densitometry to separate and identify the dyes in a cosmetic lotion, a bath preparation and imported candies. Another reversed-phase TLC–scanning densitometry method, which involved two developing solvent systems, was used by Ohno et al. (2003) to separate and identify 11 oil-soluble cosmetic dyes. That method was applied to the separation and identification of the colouring agents in two kinds of nail polish and other cosmetic products.

Liquid Chromatography

LC combined with UV/VIS spectrophotometric detection is the most widely used analytical technique for the determination of dyes and pigments. Ion-exchange LC requires the use of anion-exchange columns (strong ones for separating sulphonated dyes, weak ones for azo dyes) and gradient elution with buffered eluants of increasing ionic strength. Reversed-phase LC requires the use of columns packed with short-chain alkyl-bonded silica gel [e.g., octyl (C_8), octadecyl (C_{18}), amino-bonded] and either isocratic or gradient elution. Depending on the composition of the eluant (usually buffered to obtain an appropriate pH level and modified with an organic solvent such as methanol or acetonitrile), the affinity of the analyte for the column packing material can be influenced by ion suppression or an ion-pairing mechanism. The preferred method of detection is with a UV/VIS photodiode array (PDA) detector, which is capable of simultaneously recording absorbance data from 190 to 800 nm. Another advantage of a PDA detector is to enable the identification of eluted peaks by matching their spectra with spectral libraries of previously analysed standard compounds.

Wegener et al. (1987) characterized 126 colouring agents by their retention times and UV/VIS spectra obtained with an ion-pair reversed-phase LC system connected to a diode array detector (DAD, predecessor of the PDA detector). A C_{18}-bonded silica-packed column and gradient elution were used with an eluant made of distilled water and a dilute solution of tetrabutylammonium hydroxide (an ion-pairing reagent) in aqueous methanol (pH 7.0 adjusted with phosphoric acid). The method was applied to the determination of colouring agents in 45 cosmetic products including lipstick, nail polish, shampoo, bath foam, face powder, eye shadow, after-sun cream and bar soap. Sample preparation included heating the sample with DMF that contained 5% phosphoric acid followed by filtration. The filtrate was diluted with aqueous 0.1 M tetrabutylammonium hydroxide and extracted twice with chloroform. The combined extracts were concentrated and analysed by LC. This general extraction procedure was slightly modified according to the type of cosmetic product processed (e.g., lipstick was defatted by extracting the acidic DMF solution with *n*-hexane before filtration).

Rastogi et al. (1997) used an ion-pair reversed-phase LC–DAD method to build a spectral library consisting of retention times and UV/VIS spectra of 130 organic cosmetic colouring agents. An analytical column was packed with a polymeric material and gradient elution was used with a mobile phase that consisted of three solvents: citrate buffer containing tetrabutylammonium hydroxide as the ion-pairing reagent (pH 9.0

adjusted with concentrated ammonia), acetonitrile and tetrahydrofuran. The method was applied to the analysis of colouring agents in 139 cosmetic products that were collected from 52 manufacturers representing 12 European countries and the United States. Among the products analysed were different types of lipstick, nail polish, eye make-up, shampoo, body lotion and aftershave. Detailed sample preparation procedures were presented for the various cosmetic products, including an SPE method for the extraction of colouring agents from cosmetics with complex matrices.

Several LC methods have been developed to identify xanthene dyes in cosmetics. Gagliardi et al. (1988) analysed 99 lipsticks for the presence of xanthene dyes by a reversed-phase LC method. A C_{18}-bonded silica-packed column and gradient elution were used with an eluant made of water (pH 3 adjusted with glacial acetic acid) and acetonitrile. Detection was performed with a variable-wavelength UV/VIS spectrophotometric detector. The dyes were extracted from the lipsticks following the sample-preparation methods described by Etournaud and Aubort (1983) and Lehmann (1986). Later, Gagliardi et al. (1996) focused on the aminoxanthene dye Rhodamine B (CI 45170), which is prohibited as a colouring agent in cosmetic products in both the United States and the EU. Various cosmetic products (e.g., shampoos, lipsticks, foam bath products) were spiked with the dye, which was extracted according to the type of cosmetic (i.e., anhydrous or aqueous formulations). To determine the dye, a reversed-phase LC–DAD method was developed that used a C_{18} silica column and gradient elution with a mobile phase composed of acetonitrile and 0.1 M aqueous sodium perchlorate (pH 3 adjusted with perchloric acid). The intended use of the method was to verify that cosmetics do not contain the prohibited Rhodamine B as a colouring agent.

Scalia and Simeoni (2001) developed an assay of six xanthene dyes in lipsticks using an inverse supercritical fluid extraction (SFE) method for sample preparation. The SFE extraction produced recoveries that were comparable to those obtained with a conventional liquid–liquid extraction method. Separation of the extracted dyes was performed by LC with a cyano-propyl packed column, which was eluted isocratically with aqueous sodium acetate (0.02 M, pH 4.5):acetonitrile:methanol (55:35:10). The spectra were recorded with a variable-wavelength UV/VIS spectrophotometric detector.

Elslande et al. (2008) identified several anthraquinone-type dyes in archaeological microsamples of cosmetic powders from the Greco-Roman period. Two complementary techniques were used to identify the dyes: LC coupled to electrospray ionization with high-resolution MS and direct analysis of the microsample using laser desorption ionization–MS.

Miranda-Bermudez et al. (2014) developed a simple and efficient LC method that enables qualitative analysis of 29 colouring agents in lip products, nail polishes, eye products, blushes, body glitter, face paints, bath gels, shampoos, bar soaps, creams and toothpastes. The colouring agents were extracted with small amounts of methylene chloride, methanol, acetic acid and water and were identified by LC with PDA detection.

The addition of tandem MS (MS/MS) to LC techniques for analysing colouring agents in cosmetics has increased the selectivity and sensitivity of developed methods. Xian et al. (2013) used LC–MS/MS to determine in one analysis 11 dyes in cosmetic products such as lipstick, lip gloss and eye shadow. The determinations were fast, requiring 4 min per analysis, and reached detection limits of sub–parts-per-billion levels for the dyes tested. Guerra et al. (2015) introduced micro-matrix solid-phase dispersion as a procedure to extract colouring agents from different cosmetic matrices. Using this sample preparation followed by LC–MS/MS, the authors developed a method for the simultaneous determination of nine water-soluble colouring agents in many cosmetic products such as lipstick, nail polish, perfume and regenerating cream. Keck-Wilhelm et al. (2015) investigated colouring agents in face paints using preparative TLC to extract the dyes and pigments and LC–MS/MS for confirmatory analyses. Among their findings was the EU-prohibited red colouring agent CI 15585 (Pigment Red 53:1), which was identified in 40% of the analysed samples.

Spectrophotometry

Simultaneous determination of up to four colouring agents—a quinoline, a triphenylmethane and two azo dyes—in cosmetic products was demonstrated in the mid-1990s by Capitan-Vallvey and colleagues using various UV/VIS spectrophotometric techniques. The colouring agents were isolated from the cosmetic products by liquid–liquid extraction with an ethanol:water:methylene chloride two-phase solvent system. The aqueous phase contained the dyes of interest, and the interfering compounds remained in the organic phase. The absorbance of the colouring agents was measured directly in the aqueous phase or after isolation by SPE in Sephadex diethylaminoethyl A-25 gel. In one study, solid-phase spectrophotometry was used to simultaneously determine the quinoline and triphenylmethane dyes in colognes, aftershave lotions and shampoo gels (Capitan-Vallvey et al., 1996). In other studies, first-derivative spectrophotometry enabled simultaneous determination of triphenylmethane and one of the azo dyes in colognes (Capitan-Vallvey et al., 1995) and of both azo dyes in shampoos, bath gels and colognes (Capitan-Vallvey et al., 1997a). A third variation, called partial least-squares multivariate-calibration UV/VIS spectrophotometry, was implemented for the simultaneous determination of all four dyes in colognes, bath salts, aftershaves, deodorants, facial tonics, bath gels and shampoos (Capitan-Vallvey et al., 1997b).

The colouring agents in the mixtures extracted from the cosmetic products in the aforementioned spectrophotometry studies were of comparable levels of concentration. By contrast, Jitian et al. (2014) found that if one component within a colour-mixture solution was at a disproportionately high concentration level (~60%) relative to each of the other dye components (<1% and <6%, respectively), then derivative spectrophotometry did not yield reliable results. The spectrum of the highly concentrated component overwhelmed the spectra of the others, preventing accurate quantification of the dye components.

Other Methods

Desiderio et al. (1998) reported a quantitative method for analysing dyes in lipstick using micellar electrokinetic capillary chromatography with diode array UV/VIS detection. This electrophoretic method was optimized for the rapid (approximately 3 min) separation of four xanthene and three azo dyes. The dyes in the lipstick samples were extracted using a shortened version of the Etournaud and Aubort (1983) general sample preparation method. A related technique called micellar-enhanced spectrofluorimetry was developed by Wang et al. (2008) for the determination of a xanthene dye in lipsticks.

Rodger et al. (1998) demonstrated the use of surface-enhanced resonance Raman scattering (SERRS) spectroscopy, without any separation procedure, to analyse dyes and pigments in lipsticks. Lipsticks smeared on glass and cotton surfaces required treatment with a surfactant [e.g., poly(L-lysine)] and silver colloid prior to the analysis. This in situ SERRS method was applied to six commercial lipstick samples. Discrimination between the various samples and identification of some pigments present in individual samples were achieved. The method is qualitative in nature and was suggested to have potential for forensic and quality control applications.

ACKNOWLEDGEMENTS

The authors would like to express their appreciation to Julie N. Barrows for her review of the manuscript and her valuable editorial assistance.

REFERENCES

Abrahart, E.N., 1968. Dyes and Their Intermediates. Pergamon Press, Oxford.
Andrasco, J., 1981. Forensic Sci. Int. 17, 235.
Andrzejewski, D., Weisz, A., 1999. J. Chromatogr. A 863, 37.
Antonovich, D.D., Callen, J.P., 2005. Arch. Dermatol. 141, 869.
Barker, A.M.L., Clarke, P.D.B., 1972. J. Forensic Sci. Soc. 12, 449.
Belai, N., Harp, B.P., Mazzola, E.P., Lam, Y.-F., Abdeldayem, E., Aziz, A., Mossoba, M.M., Barrows, J.N., 2012. Dyes Pigments 95, 304.
Bell, S.J., 1977. Determination of Colors in Cosmetics, Newburger's Manual of Cosmetic Analysis. In: Senzel, A.J. (Ed.), second ed. AOAC, Washington D.C.
Capitan-Vallvey, L.F., Navas, N., de Orbe, I., Avidad, R., 1995. Analusis 23, 448.
Capitan-Vallvey, L.F., Navas Iglesias, N., de Orbe Paya, I., Avidad Castaneda, R., 1996. Talanta 43, 1457.
Capitan-Vallvey, L.F., Navas Iglesias, N., de Orbe Paya, I., Avidad Castaneda, R., 1997a. Mikrochim. Acta 126, 153.
Capitan-Vallvey, L.F., Navas, N., Avidad, R., de Orbe, I., Berzas-Nevado, J.J., 1997b. Anal. Sci. 13, 493.
Commission Decision 2006/257/EEC Dated 09.02.2006 Amending Decision 96/335/EEC Establishing an Inventory and a Common Nomenclature of Ingredients Employed in Cosmetic Products. http://europa.eu.int/comm/enterprise/cosmetics/html/cosm_inci_index.htm.
Conway, W.D., 1990. Countercurrent Chromatography. VHS, New York.
Council Directive 76/768/EC Dated 27.07.1976 on the Approximation of the Laws of the Member States Relating to Cosmetic Products, and its Successive Amendments and Adaptations. http://europa.eu.int/comm/enterprise/cosmetics/html/consolidated_dir.htm.
Cserhati, T., Forgacs, E., 2001. Encyclopedia of Chromatography, Thin-layer Chromatography of Synthetic Dyes. In: Cazes, J. (Ed.). Marcel Dekker, New York-Basel.

Davis, V.M., Bailey Jr., J.E., 1993. J. Chromatogr. 635, 160.
Desiderio, C., Marra, C., Fanali, S., 1998. Electrophoresis 19, 1478.
Dragocolor, 2004. Dictionary of Colors. Symrise GmbH&Co. KG, Holzminden.
Ehara, Y., Marumo, Y., 1998. Forensic Sci. Int. 96, 1.
Elslande, E., Guerineau, V., Thirioux, V., Richard, G., Richardin, P., Laprevote, O., Hussler, G., Walter, P., 2008. Anal. Bioanal. Chem. 390, 1873.
Etournaud, A., Aubort, J.D., 1983. Trav. Chim. Aliment. Hyg. 74, 372.
European Commission, 1999. The Rules Governing Cosmetic Products in the European Union, Vol. 2: Methods of Analysis. European Commission, Bruxelles. http://www.leffingwell.com/cosmetics/vol_2en.pdf.
Fales, H.M., Pannell, L.K., Sokoloski, E.A., Carmeci, P., 1985. Anal. Chem. 57, 376.
Faulkner, E.B., 2002. Chemistry and manufacture of cosmetics. In: Schlossman, M.L. (Ed.), third ed. .
Faulkner, E.B., 2012. Coloring the Cosmetic World. Using Pigments in Decorative Cosmetic Formulations (Chapter 2). Allured Books, Carol Stream, IL.
and Drug Administration, Code of Federal Regulations, Title 21, Parts 70–82 for Color Additives. http://www.accessdata.fda.gov/scripts/cdrh/cfdocs/cfcfr/cfrsearch.cfm.
FDA-Food and Drug Administration, 1995. http://www.cfsan.fda.gov/~acrobat/cosltr03.pdf.
FDA-Food and Drug Administration, 1999. FDA Response to CTFA Request Regarding the Use of Abbreviated Labeling for Declaring Certified Color Additives in Cosmetics. http://www.cfsan.fda.gov/~dms/col-ltr.html.
FDA-Food and Drug Administration Website Concerning Color Additives. http://www.cfsan.fda.gov/~dms/col-toc.html.
Fierz-David, H.E., Blangey, L., 1949. Fundamental Processes of Dye Chemistry. Interscience Publishers, New York.
Freeman, H.S., Williard, C.S., 1986. Dyes Pigments 7, 407.
Freeman, H.S., Hao, Z., McIntosh, S.A., Mills, K.P., 1988. J. Liq. Chromatogr. 11, 251.
Gagliardi, L., Amato, A., Cavazzutti, G., Tonelli, D., Montanarella, L., 1988. J. Chromatogr. 448, 296.
Gagliardi, L., De Orsi, D., Cozzoli, O., 1995. Cosmet. Toilet. Ed. Ital. 16, 31.
Gagliardi, L., De Orsi, D., Cavazzutti, G., Multari, G., Tonelli, D., 1996. Chromatographia 43, 76.
Gardner, P., Bertino, M.F., Weimer, R., Hazelrigg, E., 2013. Forensic Sci. Int. 232, 67.
Gennaro, M.C., Abrigo, C., Cipolla, G., 1994. J. Chromatogr. A 674, 281.
Griffin, R.M.E., Speers, S.J., Elliott, L., Todd, N., Sogomo, W., Kee, T.G., 1994. J. Chromatogr. A 674, 271.
Guerra, E., Celeiro, M., Lamas, J.P., Llompart, M., Garcia-Jares, C., 2015. J. Chromatogr. A 1415, 27.
Gupta, V.K., 2003. Handbook of thin-layer chromatography, part II (Chapter 31). In: Sherma, J., Fried, B. (Eds.), Synthetic Dyes. Marcel Dekker, New York.
Harp, B.P., Belai, N., Barrows, J.N., 2011. J. AOAC Int. 94, 1548.
Hepp, N.M., 1996. J. AOAC Int. 79, 1189.
Hepp, N.M., 1998. J. AOAC Int. 81, 89.
Hepp, N.M., 1999. J. AOAC Int. 82, 327.
Hepp, N.M., 2006. J. AOAC Int. 89, 192.
Hepp, N.M., 2015. J. AOAC Int. 98, 160.
Hepp, N.M., James, I.C., 2016. X-ray Spectrom 45, 330.
Hepp, N.M., Cargill, A.M., Shields, W.B., 2001. J. AOAC Int. 84, 117.
Holtzman, H., 1962. The chemistry and manufacture of cosmetics. In: DeNavarre, M.G. (Ed.), Cosmetic Colors, vol. 1. second ed. Van Nostrand, Princeton.
Ishimitsu, S., Mishima, I., Tsuji, S., Shibata, T., 1997. Kokuritsu Iyakuhin Shokuhin Eisei Kenkyusho Hokoku 115, 175.
Ito, Y., 1996. High-speed countercurrent chromatography, principle, apparatus, and methodology of high-speed countercurrent chromatography. In: Ito, Y., Conway, W.D. (Eds.). John Wiley, New York.
Ito, Y., 2013. J. Chromatogr. A 1271, 71.
Ito, Y., Shinomiya, K., Fales, H.M., Weisz, A., Scher, A.L., 1995. Modern countercurrent chromatography, pH-zone-refining countercurrent chromatography: a new technique for preparative separation. In: Conway, W.D., Petroski, R.J. (Eds.), ACS Symposium Series 593. American Chemical Society, Washington D.C.
Jitian, S., White, S.R., Yang, H.-H.W., Weisz, A., 2014. J. Chromatogr. A 1324, 238.

Keck-Wilhelm, A., Kratz, E., Mildau, G., Ilse, M., Schlee, C., Lachenmeier, D.W., 2015. Int. J. Cosmet. Sci. 37, 187.
Klontz, K.C., Lambert, L.A., Jewell, R.E., Katz, L.M., 2005. Arch. Dermatol. 141, 918.
Kumar, S., Gupta, R.N., 2012. Der Pharm. Lett. 4, 181.
Official methods of analysis of AOAC international (Chapter 46). In: Latimer Jr., G.W. (Ed.), 2012. Colour Additives, 2012. nineteenth ed. AOAC International, Gaithersburg.
Leatherman, A.B., Bailey, J.E., Bell, S.J., Watlington, P.M., Cox, E.A., Graichen, C., Singh, M., 1977. The analytical chemistry of synthetic dyes (Chapter 17). In: Venkataraman, K. (Ed.), Analysis of Food, Drug, and Cosmetic Colors. Wiley, New York.
Lehmann, G. (Ed.), 1986. Identifizierung von Farbstoffen in Kosmetika. VCH, Weinheim.
Lyon, H.O., 2002. Biotech. Histochem. 77, 57.
Marmion, D.M., 1991. Handbook of U.S. Colorants, Foods, Drugs, Cosmetics and Medical Devices, third ed. Wiley, New York.
Marmion, D., 1993. Kirk-othmer encyclopedia of chemical technology. In: Kroschwitz, J.I., Howe-Grant, M. (Eds.), Colorants for Foods, Drugs, Cosmetics, and Medical Devices, vol. 6. fourth ed. Wiley-Interscience, Hoboken.
Matsufuji, H., Kusaka, T., Tsukuda, M., Chino, M., Kato, Y., Nakamura, M., Goda, Y., Toyoda, M., Takeda, M., 1998. J. Food Hyg. Soc. Jpn. 39, 7.
Matsufuji, H., Ngang, E., Chino, M., Goda, Y., Toyoda, M., Takeda, M., 2002. Jpn. J. Food Chem. 9, 107.
MHW-Ministry of Health and Welfare, 1966. Ordinance No. 30/1966: Ordinance to Regulate Coal-Tar Colors Permitted for Use in Drugs, Quasi-drugs and Cosmetics (As Amended by MHLW-Ministry of Health, Labor, and Welfare Ordinances Nos. 55/1972 and 126/2003).
Milstein, S.R., Halper, A.R., Katz, L.M., 2006. Handbook of cosmetic science and technology (Chapter 65). In: Paye, M., Barel, A.O., Maibach, H.I. (Eds.), Regulatory Requirements for the Marketing of Cosmetics in the United States, second ed. Taylor & Francis, Boca Raton.
Miranda-Bermudez, E., Harp, B.P., Barrows, J.N., 2014. J. AOAC Int. 97, 1039.
Mselle, J., 2004. Trop. Doct. 34, 235.
Ngang, E., Matsufuji, H., Chino, M., Goda, Y., Toyoda, M., Takeda, M., 2001. J. Food Hyg. Soc. Jpn. 42, 298.
Ohno, T., Ito, Y., Mikami, E., Ikai, Y., Oka, H., Hayakawa, J., Nakagawa, T., 1996. Jpn. J. Toxicol. Environ. Health 42, 53.
Ohno, T., Mikami, E., Matsumoto, H., 2003. J. Health Sci. 49, 401.
Otterstatter, G., 1999. Coloring of Food, Drugs, and Cosmetics (A. Mixa, Trans.). Marcel Dekker, New York.
Pawliszyn, J. (Ed.), 2002. Wilson and Wilson's Comprehensive Analytical Chemistry, Vol. XXXVII: Sampling and Sample Preparation for Field and Laboratory. Elsevier, Amsterdam.
Pawliszyn, J., Lord, H.L. (Eds.), 2010. Handbook of Sample Preparation. Wiley-Blackwell, Hoboken.
Peiperl, M.D., Prival, M.J., Bell, S.J., 1995. Food Chem. Toxicol. 33, 829.
Perdih, A., 1972. Z. Anal. Chem. 260, 278.
Pharmaceutical Society of Japan, 2000. Methods of Analysis in Health Science. Kanehara Press, Tokyo.
Prival, M.J., Peiperl, M.D., Bell, S.J., 1993. Food Chem. Toxicol. 31, 751.
Quint, K.D., Genders, R.E., Vermeer, M.H., 2012. Dermatol. Surg. 38, 951.
Rastogi, S.C., Barwick, V.J., Carter, S.V., 1997. Chromatographia 45, 215.
Richfield-Fratz, N., Baczynskyj, W.M., Miller, G.C., Bailey Jr., J.E., 1989. J. Chromatogr. 467, 167.
Rodger, C., Rutherford, V., Broughton, D., White, P.C., Smith, W.E., 1998. Analyst 123, 1823.
Rosenthal, I., Yang, G.C., Bell, S.J., Scher, A.L., 1988. Food Addit. Contam. 5, 563.
Rosholt, A.P. (Ed.), 2007. CTFA International Color Handbook, fourth ed. The Cosmetic, Toiletry, and Fragrance Association, Washington D.C.
Rush, S., 1989. Cosmet. Toilet. 104, 47.
Russel, L.W., Welch, A.E., 1984. Forensic Sci. Int. 25, 105.
Salahioglu, F., Went, M., Gibson, S.J., 2013. Anal. Methods 5, 5392.
Scalia, S., Simeoni, S., 2001. Chromatographia 53, 490.
Scher, A.L., Adamo, N.C., 1993. J. AOAC Int. 76, 287.
Schlossman, M.L., 2000. Cosmeceuticals. Drugs vs. Cosmetics (Chapter 19). In: Eisner, P., Maibach, H.I. (Eds.), Decorative Products. Marcel Dekker, New York.
Silk, S.R., 1965. (Chapter 7) AOAC 48, 838.

Sjoberg, A.M., Olkkonen, C., 1985. J. Chromatogr. 318, 149.
Suzuki, K., Hirokawa, K., Yagami, A., Matsunaga, K., 2011. Dermat. Contact Atop. Occup. Drug 22, 348.
Touchstone, J.C., 1992. Practice of Thin Layer Chromatography, third ed. Wiley, New York.
Tsuji, S., Yoshii, K., Tonogai, Y., 2006. J. Chromatogr. A 1101, 214.
Van Liedekerke, B.M., De Leenheer, A.P., 1990. J. Chromatogr. 528, 155.
Vlajkovic, J., Andric, F., Ristivojevic, P., Radoicic, A., Tesic, Z., Milojkovic-Opsenica, D., 2013. J. Liq. Chromatogr. Rel. Technol. 36, 2476.
Vu, N.T., Rickard, J.D., Sullivan, M.P., Richfield-Fratz, N., Weisz, A., 2011. J. Liq. Chromatogr. Rel. Technol. 34, 106.
Wall, P.E., 2000. Encyclopedia of separation science. In: Wilson, I.D., Adlard, E.R., Cooke, M., Poole, C.F. (Eds.), Thin-Layer (Planar) Chromatography, vol. 6. Academic Press, London.
Wang, C.C., Masi, A.N., Fernandez, L., 2008. Talanta 75, 135.
Wegener, J.W., Klamer, J.C., Govers, H., Brinkman, U.A.Th., 1987. Chromatographia 24, 865.
Wei, R.R., Warner, W.G., Bell, S.J., Kornhauser, A., 1994. Photochem. Photobiol. 59, 31S.
Wei, R.R., Warner, W., Bell, S., Kornhauser, A., 1995. Photochem. Photobiol. 61, F35.
Weisz, A., 1996. In: Ito, Y., Conway, W.D. (Eds.), High-Speed Countercurrent Chromatography, Separation and Purification of Dyes by Conventional High-speed Countercurrent Chromatography and PH-Zone-Refining Countercurrent Chromatography, Chemical Analysis Series, vol. 132. John Wiley, New York.
Weisz, A., 1997. Dyes Pigments 35, 101.
Weisz, A., Andrzejewski, D., 2003. J. Chromatogr. A 1005, 143.
Weisz, A., Ito, Y., 2000. Encyclopedia of separation science. In: Wilson, I.D., Adlard, E.R., Cooke, M., Poole, C.F. (Eds.), Dyes – High-Speed Countercurrent Chromatography, vol. 6. Academic Press, London.
Weisz, A., Ito, Y., 2008. J. Chromatogr. A 1198–1199, 232.
Weisz, A., Krantz, Z.B., 2014. Food Addit. Contam. Part A 31, 979.
Weisz, A., Scher, A.L., Andrzejewski, D., Shibusawa, Y., Ito, Y., 1992. J. Chromatogr. 607, 47.
Weisz, A., Andrzejewski, D., Ito, Y., 1994a. J. Chromatogr. a. 678, 77.
Weisz, A., Scher, A.L., Shinomiya, K., Fales, H.M., Ito, Y., 1994b. J. Am. Chem. Soc. 116, 704.
Weisz, A., Andrzejewski, D., Shinomiya, K., Ito, Y., 1995. Modem countercurrent chromatography (Chapter 16). In: Conway, W.D., Petroski, R.J. (Eds.), Preparative Separation of Components of Commercial Tetrachlorofluorescein by PH-Zone-Refining Countercurrent Chromatography, ACS Symposium Series No. 593. American Chemical Society, Washington D.C.
Weisz, A., Scher, A.L., Ito, Y., 1996. J. Chromatogr. A 732, 283.
Weisz, A., Andrzejewski, D., Highet, R.J., Ito, Y., 1998. J. Liq. Chromatogr. Rel. Technol. 21, 183.
Weisz, A., Mazzola, E.P., Matusik, J.E., Ito, Y., 2001a. J. Chromatogr. A 923, 87.
Weisz, A., Andrzejewski, D., Fales, H.M., Mandelbaum, A., 2001b. J. Mass Spectrom. 36, 1024.
Weisz, A., Andrzejewski, D., Fales, H.M., Mandelbaum, A., 2002. J. Mass Spectrom. 37, 1025.
Weisz, A., Andrzejewski, D., Rasooly, I.R., 2004. J. Chromatogr. A 1057, 185.
Weisz, A., Wright, P.R., Andrzejewski, D., Meyers, M.B., Glaze, K., Mazzola, E.J., 2006. J. Chromatogr. A 1113, 186.
Weisz, A., Mazzola, E.P., Ito, Y., 2009. J. Chromatogr. A 1216, 4161.
Weisz, A., Mazzola, E.P., Ito, Y., 2011. J. Chromatogr. A 1218, 8249.
Weisz, A., Ridge, C.D., Mazzola, E.P., Ito, Y., 2015. J. Chromatogr. A 1380, 120.
Harry's cosmeticology (Chapter 19). In: Wilkinson, J.B., Moore, R.J. (Eds.), 1982. Coloured Make-up Preparations, seventh ed. Chemical Publishing, New York.
Wright, P.R., Richfield-Fratz, N., Rasooly, A., Weisz, A., 1997. J. Planar Chromatogr. 10, 157.
Xian, Y., Wu, Y., Guo, X., Lu, Y., Luo, H., Luo, D., Chen, Y., 2013. Anal. Methods 5, 1965.
Yamada, M., Nakamura, M., Yamada, T., Maitani, T., Goda, Y., 1996. Chem. Pharm. Bull. (Tokyo) 44, 1624.
Yang, H.-H.W., Weisz, A., 2012. Food Addit. Contam. Part A 29, 1386.
Zollinger, H., 2003. Color Chemistry, third ed. Verlag Helvetica Chimica Acta, Zurich, and Wiley-VCH, Weinheim.
Zuckerman, S., 1974. Cosmetics science and technology. In: Balsam, M.S., Sagarin, E. (Eds.), Color in Cosmetics, vol. 3. second ed. Wiley, New York.
Zuckerman, S., Senackerib, J., 1979. Kirk-othmer encyclopedia of chemical technology. In: Colorants for Foods, Drugs, and Cosmetics, vol. 6. third ed. Wiley, New York, pp. 570–573.

CHAPTER 8

Hair Dyes in Cosmetics: Regulatory Aspects and Analytical Methods

Alberto Chisvert, Pablo Miralles, Amparo Salvador
University of Valencia, Valencia, Spain

INTRODUCTION

Hair dye cosmetic products are those cosmetic products used for colouring hair. These products were initially designed for women to hide grey hair, but they are increasingly being used by men and, in fact, nowadays various hair dye products exclusively designed for men can be found on the market. The reasons for using this type of product may be different now, not just to hide grey hair, but also to potentiate the natural hair colour or to change it completely. In any case, the main reasons behind all of these are fashion and feeling more attractive.

Hair dye products are mainly classified into two main categories according to their duration in the hair, i.e., temporary and permanent, which in turn are subdivided into temporary and semi-permanent, and permanent and demi-permanent hair dye products, respectively. This classification is in line with the type of active ingredients involved in the dyeing process and with the dyeing process itself, known commonly as non-oxidative and oxidative hair dye products. The colouring ingredients used in each type of hair dye product and the dyeing process involved are described next.

Non-Oxidative Hair Dye Products

Temporary hair dye products colour hair for a relatively short period of time, in such a way that the artificial colour disappears after one shampooing (i.e., temporary hair dye products themselves) or several shampooings (i.e., semi-permanent hair dye products). Both of them fall within the non-oxidative hair dye products group.

Specifically, temporary hair dye products are marketed as lotions and gels, but foam preparations are undoubtedly the easiest way to apply them on wet hair. They can also be found in spray formulations to be applied on dry hair. These products are designed to provide a very fleeting colour change, in such a way that the provided change disappears after the first shampooing (or after brushing, also in the case of spray formulations). They are especially employed for a social (e.g., a party) or professional (e.g., a theatre actor in a performance) occasion when the user expects to recover his or her natural hair colour after that event. Spray formulations are also designed for people using permanent hair dye

Acid Orange 7

(Sodium 4-[(2-hydroxy-1-naphthyl)azo]benzenesulphonate)

Basic Blue 99

(3-[(4-amino-6-bromo-5,8-dihydro-1-hyydroxy-8-imino-5-oxo-2-naphthyl)amino]-N,N,N-trimethylaanilinium chloride)

N,N'-bis(2-hydroxyethyl)-2-nitro-p-phenylenediamine

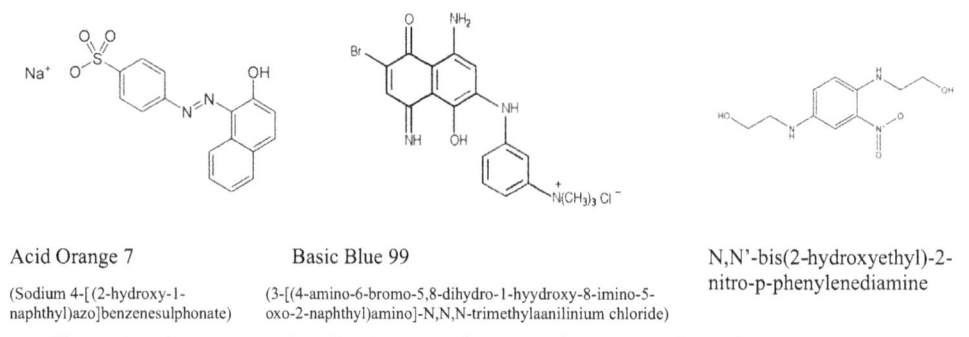

Figure 8.1 Some examples of active ingredients used in non-oxidative hair dye products.

products to conceal roots easily and instantly, by applying them directly on the non-coloured roots. Temporary hair dye products are formulated with combinations of the so-called non-oxidative hair dyes, also known as direct hair dyes, which can be classified into three groups: acid dyes (e.g., Acid Orange 7) that contain acid moieties like sulphonic (—SO$_3$H) or carboxylate (—COOH), basic dyes (e.g., Basic Blue 99) that contain quaternary amine moieties, and nitro dyes [e.g., N,N'-bis(2-hydroxyethyl)-2-nitro-p-phenylenediamine] that contain nitro moieties (see Fig. 8.1). The first two groups belong to various chemical families, such as azo dyes, triphenylmethane-derived dyes, fluoran-derived dyes and anthraquinone-derived dyes, amongst others (see Chapter 7), whereas the last ones are mainly substituted phenylenediamines and aminophenols containing nitro moieties in such a way that the substituents largely determine the shade and intensity of the colour provided by the dye (Ghosh and Sinha, 2008). The combination of several of these colouring agents provides a wide range of shades when deposited on the cuticle, i.e., the outer part of the hair, and they are not able to cross it and thus penetrate into the cortex. Therefore, their deposition on the outside of the hair and their relatively high water solubility make these products easily removable during the first shampooing (da França et al., 2015). In the case of spray formulations, they can also contain solid dyes, such as iron oxides, titanium dioxide and aluminium lakes, which are easily removed by shampooing or brushing.

Regarding semi-permanent hair dye products, different presentations, such as lotions, emulsions, foams, etc., can be found on the market. They are easily applied onto wet hair and left during a defined period of time (c. 10–30 min). After that, the excess is carefully rinsed with water. These products are designed to provide a colour change that fades progressively after successive washes (between three and eight). There exists a greater variety of shades compared to the temporary ones, and the dyeing process is less aggressive than that of the permanent ones (described below), so they constitute the preferred option for those people reluctant to use permanent hair dye products. These products are also formulated with non-oxidative hair dyes as temporary hair dye products, but in this case, the active ingredients are required to hold onto hair for a longer period of time.

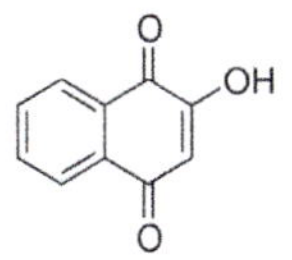

2-hydroxy-1,4-naphthoquinone (lawsone) 1,3,4-trihydroxyflavone (apigenin)
(active component of henna)   (active component of chamomile)

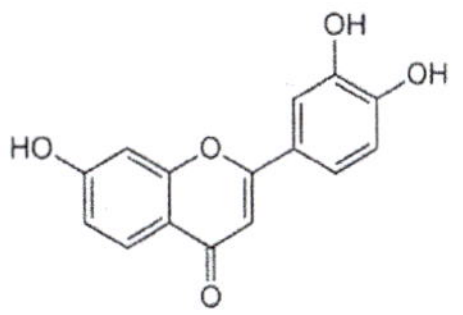

Figure 8.2 Some examples of vegetable-based dyes used in non-oxidative hair dye products.

This effect is achieved by formulating the product with a higher pH value (c.9–10) by using ammonia. Sodium hydroxide and ethanolamine have also been employed and marketed by claiming 'ammonia free'. The alkaline pH promotes the cuticle opening, thus allowing these compounds to penetrate into the cortex. Afterwards, a neutral or slightly acid rinse is performed to close the cuticle and thus to entrap the compounds. Thus, these colouring agents are not removed in the first shampooing, unlike temporary hair dye products, but they are progressively released in the subsequent washes (between three and eight).

Vegetable-based dyes, such as henna and chamomile, could also be used to confer a semi-permanent effect. The first is mainly obtained from *Lawsonia* genus plants, and provides reddish to brown shades due to 2-hydroxy-1,4-naphthoquinone (known as lawsone) (see Fig. 8.2), which is a colouring agent with a high affinity for keratin. Regarding chamomile, it is mainly obtained from *Anthemis nobilis* and *Matricaria chamomilla* plants. It provides a slight clearing and an unusual brightness to the hair due to its active ingredient known as apigenin (1,3,4-trihydroxyflavone) (see Fig. 8.2) (Ghosh and Sinha, 2008). These two are the most commonly known but there are other vegetable-based dyes that provide different semi-permanent shades on the hair. Those readers interested in this topic are invited to read the interesting review by Ghosh and Sinha (2008).

Oxidative Hair Dye Products

Permanent hair dye products and demi-permanent hair dye products constitute this second group.

Regarding permanent hair dye products, they colour hair permanently, so that if people want to recover their natural colour they need to let their hair grow. They are based on non-coloured active ingredients, which are further oxidized by a chemical reaction to develop the colour (da França et al., 2015). These active ingredients are known as oxidative hair dyes, although it is better to refer them as intermediate dyes, because they are not true colouring agents, but they participate in a complex oxidative dyeing process to develop the colour. Actually, the oxidative hair dyes should be considered as a binary mixture of the so-called dye precursor (also known as the base or primary intermediate) and the modifier (also known as the coupler). Precursors are typically *para-* and *ortho*-phenylenediamines

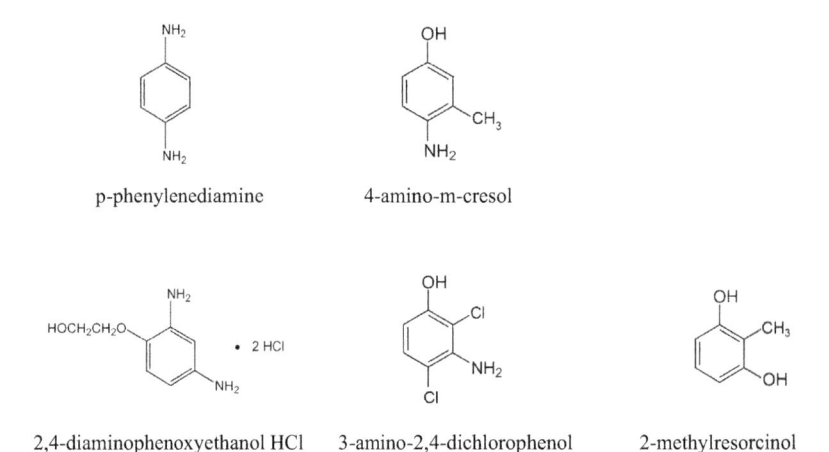

Figure 8.3 Some examples of precursors (top) and couplers (bottom) used in oxidative hair dye products.

(e.g., *p*-phenylenediamine) and aminophenols (e.g., 4-amino-*m*-cresol), substituted in such a way that the two donor moieties in the *para* and *ortho* positions confer the property of easy oxidation, whereas modifiers are substituted *meta*-phenylenediamines (e.g., 2,4-diaminophenoxyethanol HCl), aminophenols (e.g., 3-amino-2,4-dichlorophenol) and hydroxyphenols (e.g., 2-methylresorcinol), which, unlike the precursors, do not present an easy oxidation (da França et al., 2015) (see Fig. 8.3).

In both cases, the substituents play a crucial role in the shade and intensity of the developed colour (Ghosh and Sinha, 2008). Therefore, it is evident that the dyeing process requires an oxidizing agent (also known as a developer), and hydrogen peroxide is the preferred option. An alkaline medium, typically achieved with ammonia, is also needed because it is responsible for opening the cuticle to promote the entrance of the other three constituents. Once in the cortex, the oxidizing agent has a double function, i.e., to oxidize the precursor molecules, thus forming imines (see Fig. 8.4), but also to decolour melanin (responsible for the natural colour), thus lightening the natural colour, which allows it to radically change the hair colour. The imines formed react with the modifier molecules, thus forming dinuclear compounds called indo dyes (see Fig. 8.4). These indo dyes, depending on the substituents of the *meta* difunctional coupler, could undergo further coupling reactions to yield trinuclear and even polynuclear compounds (Ghosh and Sinha, 2008).

So, oxidative hair dye products are marketed in two separated bottles. One of them contains an emulsion with the precursors, the couplers and the base, whilst the other contains an emulsion with the developer. Often, non-oxidative hair dyes are also included to create nuances or to make some shades stand out, although obviously they do not participate in the oxidative process. Both bottles are mixed just before use.

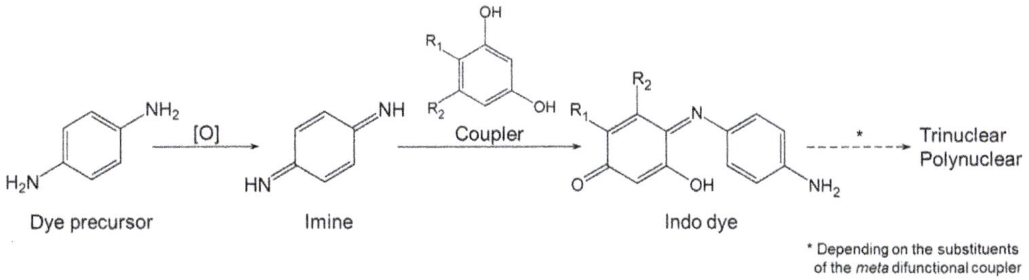

Figure 8.4 A simplified scheme of the reactions involved in the formation of coloured compounds in oxidative hair dye products.

Regarding demi-permanent hair dye products, they contain the same active ingredients as permanent hair dye products. The difference is that they contain the oxidizing agent in a minor concentration. At this low concentration the melanin is not decoloured, so it is not possible to colour hair to a lighter shade than the natural colour.

Despite their increasingly widespread use, many concerns about hair dye safety have been raised throughout the years. Contact dermatitis is the most often reported side effect (Søsted et al., 2013; SCCS, 2013; Schuttelaar and Vogel, 2016), but there are other studies showing more harmful effects, such as risk of developing cancer; especially there is evidence showing an increased risk of bladder cancer associated with the use of oxidative hair dye products (Gago-Domínguez et al., 2001, 2003; Koutros et al., 2011; NCI, 2017). However, there are contradictory results, since other reports show that the carcinogenic risk in humans under realistic use and exposure conditions appears negligible (Nohynek et al., 2004; Turati et al., 2014). Nevertheless, those compounds that can be used as hair dyes are regulated by the legislation in force in each country, in such a way that maximum allowed concentrations have been established for each of them.

For all the aforementioned reasons, analytical methods are needed to control the concentration of hair dyes and thus to ensure the efficacy and safety of hair dye cosmetic products.

The aim of this chapter is to familiarize the reader with the different compounds used as hair dyes and the legislation regulating these compounds, and especially to review the analytical methods for hair dye determination in cosmetic products since 2006. Those articles published before 2006 were reviewed in the first edition of this book (Salvador and Chisvert, 2007).

REGULATORY ASPECTS

As was already mentioned, to guarantee the safety of hair dye products, the compounds that can be used as hair dyes, and their maximum allowed concentrations, are regulated by the legislation in force in each country. As mentioned in Chapter 1, the regulatory systems for cosmetic products differ from one part of the world to the other, and thus the hair dyes allowed in various countries could be different.

Hair Dyes in the European Union

In contrast to other cosmetic ingredients, such as UV filters, described in Chapter 5, there are no positive lists for hair dyes. In the European Union (EU), hair dye products are considered cosmetics and are regulated under the EU Cosmetics Regulation (Regulation EC 1223/2009). According to this regulation, both non-oxidative and oxidative hair dyes are considered 'colorants', which are defined as '*substances which are exclusively or mainly intended to colour the cosmetic product, the body as a whole or certain parts thereof, by absorption or reflection of visible light; in addition, precursors of oxidative hair colorants shall be deemed colorants*'. This regulation contains a positive list of colorants in Annex IV, but according to the authors' interpretation, this list ignores hair dyes, and moreover it could lead to confusion. For example, Acid Blue 9, a typical non-oxidative hair dye, appears in Annex IV as entry No. 63 without any restriction. Thus, it could be assumed that it can be used in hair dye products without any restriction. However, it appears again in Annex III, devoted to generally restricted substances, as entry No. 190, with a restriction for '*Hair dye substance in non-oxidative hair dye products*' with a maximum allowed concentration of 0.5%. Another example is Acid Orange 7, a very commonly used non-oxidative hair dye that appears in Annex IV, under entry No. 21, with the only restriction of '*Not to be used in eye products*'. Thus, it is assumed that it can be used in hair dye products without any restriction, despite the fact that there are conflicting opinions of the Scientific Committee on Consumer Safety (SCCS) questioning its use at 0.5% (SCCS, 2010, 2014).

Regarding safety concerns it should be mentioned that the European Commission agreed in 2003 on an overall strategy to regulate hair dye substances, mainly motivated by the results published by Gago-Domínguez et al. (2001) and the opinion of the Scientific Committee on Cosmetic Products and Non-food Products Intended for Consumers (SCCNFP, 2001, 2002), further replaced by the Scientific Committee on Consumer Products (SCCP) and nowadays by the SCCS. This strategy required producers to submit safety files on hair dyes to the SCCP, who evaluated the data, and on this basis, several substances were prohibited as hair dyes. The overall objective of this strategy was to establish a positive list of hair dyes within the framework of the legislation in force in that moment, which was the old EU Cosmetic Directive (i.e., Directive 76/768/EEC). Different substances were prohibited as hair dyes and incorporated into Annex II of Directive 76/768/EEC (see, for example, Directive, 2007/54/EC), but no positive list was created. With the implementation in 2013 of the Cosmetics Regulation, the positive list of hair dyes did not arrive, and nowadays, after various amendments it continues to be missing. Nevertheless, on the basis of the efforts to regulate these substances, there are 181 substances formerly used as hair dyes listed in Annex II, devoted to prohibited substances, whereas there are another 114 substances allowed for restricted use in hair dye products listed under Annex III, devoted to generally restricted substances (as in the aforementioned example of Acid Blue 9). To facilitate this information, separated lists of all these substances can be found on the European Commission's website devoted to hair dye cosmetic products (https://ec.europa.eu/growth/sectors/cosmetics/products/hair-dye_en). However, this last list should not be considered a positive list, because those hair dyes with no restrictions are not listed (as in the aforementioned example of Acid Orange 7). Based on this controversy, the authors consider that a positive list for hair dyes is necessary within the EU framework.

Hair Dyes Outside the European Union

In the United States, there is no positive list either. Colouring agents are commonly referred to as coal-tar if they are synthetic (i.e., not natural). According to Title 21 of the Code of Federal Regulations (CFR), Section 70.3 (21 CFR 70.3), any colouring agent, with the exception of coal-tar hair dyes, needs Food and Drug Administration (FDA) approval before it is permitted to be used in cosmetics, and moreover it needs to be batch certified by the FDA prior to addition to cosmetics (see Chapter 7). Hair dyes from plant or mineral sources are not included in this exception and they are regulated as the other colour additives. This exception is based on every hair dye product being labelled with the following special caution statement, related to the prevention of contact dermatitis: '*Caution—This product contains ingredients which may cause skin irritation on certain individuals and a preliminary test according to accompanying directions should first be made. This product must not be used for dyeing the eyelashes or eyebrows; to do so may cause blindness*'. Moreover, the product should contain adequate instructions for consumers to do a skin test before they dye their hair. With regard to the relationship between hair dyes and cancer commented on previously, the FDA published a regulation requiring a special warning statement for all hair dye products containing 4-methoxy-*m*-phenylenediamine (also known as 2,4-diaminoanisole) and 2,4-methoxy-*m*-phenylenediamine sulphate (also known as 2,4-diaminoanisole sulphate) (21 CFR 740.18). Since then, the cosmetics industry has reformulated its products in such a way that these two ingredients are no longer used in hair dye products. Nowadays, the FDA does not have reliable evidence showing a link between cancer and the hair dyes on the market (https://www.fda.gov/cosmetics/product singredients/products/ucm143066.htm).

In other countries, such as the countries of the Association of Southeast Asian Nations and the Southern Common Market (Mercosur) in South America, regulations on cosmetics converge with the EU. In Japan, hair dye products are considered quasi-drugs (MHWN, 2000), and those hair dyes allowed are listed in Article 3 of the ministerial ordinance No. 30 of the Ministry of Health and Welfare (1966). However, this list has not been accessible and thus it is not discussed here.

ANALYTICAL METHODS FOR HAIR DYE DETERMINATION IN COSMETIC PRODUCTS

All the aforementioned restrictions established for hair dyes need to be accompanied by efficient and reliable analytical methods to control them; thus, the development of analytical methods is encouraged, even more so bearing in mind the lack of official methods to control these active ingredients.

To our knowledge, in the EU framework, a method for the qualitative and semi-quantitative determination of certain oxidative hair dyes was published under Directive 82/434/EEC. The method proposes the extraction of the target analytes at pH 10 with ethanol by means of centrifugation, and afterwards the supernatant is run in either a one- or a two-dimensional thin-layer chromatography (TLC) plate. The identification and subsequent semi-quantitative determination are carried out by comparing the positions and intensities of the spots obtained with those of spots obtained with an appropriate concentration range of reference substances. It should be emphasized that this method is old-fashioned regarding both the analytes considered (some of them are no longer used) and the analytical technique employed (currently there are much more powerful analytical techniques).

In 2002, the Institute for Reference Materials and Measurements of the Joint Research Centre of the European Commission carried out an inter-laboratory validation of an analytical method developed years before (Pel et al., 1998; Vincent et al., 1999, 2002a,b) with the purpose of proposing it as a candidate reference method. The method was based on a three-step liquid–liquid extraction by using *n*-heptane as the extractant and subsequent measurement by liquid chromatography (LC) with a ultraviolet/visible spectrometry detector. Despite obtaining promising results (Vincent et al., 2004; Chisvert et al., 2007) it has not yet been published as an official method.

Fortunately, in the analytical chemistry databases there are published papers describing analytical methods for the determination of hair dyes in cosmetic products. A detailed and exhaustive review of those methods published up to 2006 can be found in the first edition of this book (Salvador and Chisvert, 2007). On that occasion, a detailed study of those published papers revealed that most of them were focused on oxidative hair dyes. Since 2006, the trend has continued, which is in line with people's concern about the relationship between oxidative hair dyes and cancer. A detailed review of the papers published since the last edition has been carried out here. Table 8.1 summarizes the experimental details and some interesting remarks on these published papers dealing with the determination of hair dyes.

As mentioned before, to obtain the various shades that people demand, especially those similar to natural ones, the hair dye cosmetic products contain mixtures of hair dyes. Sometimes, non-oxidative hair dyes are even added to oxidative hair dye products to create nuances or to make some shades stand out, although obviously they do not

Hair Dyes in Cosmetics: Regulatory Aspects and Analytical Methods 167

Table 8.1 Published papers from January 2006 to December 2016 concerning hair dye determination in cosmetic products (in chronological order)

Authors	Target hair dyes[a]	Type of matrix	Sample preparation[b]	Analytical technique[b]	Remarks[c]
Masukawa (2006)	Basic Red 76, Basic Brown 16, Basic Brown 17, Basic Yellow 57, Basic Blue 99	Hair dye product	Sample is dispersed in methanol:water by stirring and sonication (30 min) and filtered	CE–UV/VIS, uncoated fused-silica capillary, ammonium acetate buffer as running buffer	LOD: 0.002%–0.02% R: 93%–111%
Narita et al. (2007)	pAP, pMAP, mMAP, oAP, 2A4NP, 2A5NP, RES, pPD, 5A2MP	Hair dye product	Sample is dispersed in methanol, sonicated (1 min), highly diluted with methanol and filtered	LC–AD, C_{18} column and methanol:acetate buffer as mobile phase	LOD: 0.01%–0.04% R: 96%–100%
El-Shaer et al. (2007)	Lawsone	Henna powder	Sample is dispersed in sodium carbonate solution, left for 1 h and diluted with water; an aliquot is filtered, neutralized with HCl, subjected to LLE (with 3 × 10 mL of chloroform); the extract is evaporated and redissolved in methanol	TLC–DS, silica plates and chloroform: methanol as mobile phase	LOD: no data R: 94%–99%
Akyüz and Ata (2008)	T24D, T26D, T34D, 2NpPD, oPD, mPD, pPD, oAP, mAP, pAP and others	Hair dye product and henna powder	Sample is dispersed in HCl (aq) and sonicated (40°C, 15 min); pH is adjusted to 1 and dispersion cleaned up with 3 × 3 mL chloroform; pH is adjusted to 8 and subjected to LLE (with 3 × 3 mL chloroform containing bis-2-ethylhexylphosphate as ion-pair reagent); back-extracted with HCl (aq) by sonication (1 min); pH adjusted to 8 and LLE (with 3 × 3 mL chloroform containing bis-2-ethylhexylphosphate); dried with Na_2SO_4, evaporated to dryness and derivatized with isobutylchloroformate (10 min); evaporated and reconstituted with toluene, dried with Na_2SO_4, pH adjusted, centrifuged and evaporated to 1 mL	GC–MS, 5% phenyl/95% dimethyl polysiloxane capillary column	LOD: 0.000000002%–0.00000002% R: 92%–98%

Continued

Table 8.1 Published papers from January 2006 to December 2016 concerning hair dye determination in cosmetic products (in chronological order)—cont'd

Authors	Target hair dyes[a]	Type of matrix	Sample preparation[b]	Analytical technique[b]	Remarks[c]
Dong et al. (2008)	oPD, mPD, pPD, CAT, RES	Hair dye product	Sample is dispersed in ethanol, sonicated (25 min), filtered, diluted with borate–phosphate buffer and filtered	CE–AD, uncoated fused-silica capillary, borate–phosphate buffer as running buffer	LOD: no data R: 91%–108%
Bai et al. (2010)	pPD	Hair dye product	Sample is diluted with water	AD, β-MnO_2 nanowire-modified glassy carbon electrode as working electrode	LOD: no data R: no data
Li et al. (2010)	CAT, RES, HQ	Hair dye product	Sample is diluted with running buffer and filtered	CE–CL, uncoated fused-silica capillary, borate buffer containing luminol as running buffer; ferrate(VI) electrochemically generated	LOD: 0.007%–0.009% R: 91%–112%
Xiao et al. (2011)	2NA, 3NA, 4NA, 2,4DNA	Hair dye product	Sample is diluted with methanol, and centrifuged; an aliquot is diluted with phosphate buffer, sonicated and subjected to IT–SPME [poly(methacrylic acid–ethylene glycol dimethacrylate) monolith as extraction phase and methanol as elution solvent]	LC–UV/VIS, C_{18} column and isocratic methanol:water as mobile phase	LOD: 0.00005%–0.0002% R: 72%–114%
Wu et al. (2011)	oPD, mPD, pPD	Hair dye product	Sample is dispersed in ethanol, sonicated (10 min), diluted with water, filtered and subjected to SPE (coal cinders column and aqueous NaOH as elution solvent)	MEKC–UV/VIS, uncoated fused-silica capillary, phosphate buffer containing β-cyclodextrin and 1-butyl-3-methyl-imidazolium hexafluorophosphate as running buffer	LOD: no data R: 92%–106%

Dejmkova et al. (2011)	2A3NP, 2A4NP, 2A5NP, 4A2NP, 4A3NP	Hair dye product	Sample is dissolved in methanol, subjected to SPE (C$_{18}$ cartridge and methanol as elution solvent) and diluted with water	LC–AD, C$_{18}$ column with isocratic methanol:Britton–Robinson buffer as mobile phase	LOD: 0.002%–0.009% R: 97%–103%
He et al. (2012)	CAT, RES, HQ	Hair dye product	Sample is dispersed in ethanol, sonicated (60 min), filtered and diluted with running buffer	CEC–UV/VIS, gold nanoparticle-coated capillary with acetate buffer as running buffer	LOD: no data R: 93%–105%
Zhong et al. (2012)	2MR, 4NoPD, HQ, mPD, oPD, pPD, RES, 4ClR, DEpPD	Hair dye product	Sample is mixed with sulphite and diluted with methanesulphonic acid (aq); subjected to USA–MSPD [alumina and hexane added, heated (60°C), vigorously mixed, centrifuged and filtered]	IC–UV/VIS, carboxylate-functionalized cation-exchange column and gradient acetonitrile:methanesulphonic acid (aq) as mobile phase	LOD: 0.0002%–0.0007% R: 86%–107%
Hudari et al. (2014)	pPD, RES	Hair dye product	Sample is dispersed in water and stirred (30 min); an aliquot is added to the cell containing ammonium buffer	LSV, multi-walled carbon nanotube–chitosan-modified glassy carbon electrode as working electrode	LOD: 0.002% R: no data
Zhong et al. (2014)	4A3NP, oAP, mAP, pAP, pMAP, 4A3MP	Hair dye product	Sample is subjected to USA–DSPE [sulphite as antioxidant and disperser, cyclohexane:ethanol as solvent, heated (40°C, 1 min), vigorously mixed, diluted with sodium hydroxide (aq), sonicated again (4 min), centrifuged and filtered]	IC–UV/VIS, quaternary ammonium-functionalized anion-exchange column and isocratic sodium hydroxide (aq):methanol as mobile phase	LOD: 0.0001%–0.0005% R: 86%–101%

Continued

Table 8.1 Published papers from January 2006 to December 2016 concerning hair dye determination in cosmetic products (in chronological order)—cont'd

Authors	Target hair dyes[a]	Type of matrix	Sample preparation[b]	Analytical technique[b]	Remarks[c]
Asghari et al. (2014)	oAP, mAP, pAP	Hair dye product	Sample is diluted with water and subjected to USAEME [chloroform as extraction solvent, sonicated (5 min) and centrifuged]; sedimented phase is evaporated and reconstituted with methanol	LC-UV/VIS, C_{18} column and isocratic water:acetonitrile as mobile phase	LOD: 0.003%–0.011% R: 89%–101%
Franco et al. (2015)	Basic Blue 99, Acid Violet 43, Basic Brown 16, Basic Red 76, Basic Yellow 57	Hair dye product	Sample is diluted with water and subjected to SPE (cation-exchange cartridge and acetonitrile:water as elution solvent)	LC-UV/VIS, C_{18} column with isocratic acetonitrile:water containing 1-butyl-3-methylimidazolium bis(trifluoromethane-sulphonyl)imide as mobile phase	LOD: 0.00002%–0.0001% R: no data

[a] 2A3NP, 2-amino-3-nitrophenol; 2A4NP, 2-amino-4-nitrophenol; 2A5NP, 2-amino-5-nitrophenol; 2MR, 2-methylresorcinol; 2NA, 2-nitroaniline; 2NpPD, 2-nitro-p-phenylenediamine; 3NA, 3-nitroaniline; 4A2NP, 4-amino-2-nitrophenol; 4A3MP, 4-amino-3-methylphenol; 4A3NP, 4-amino-3-nitrophenol; 4ClR, 4-chlororesorcinol; 4N4, 4-nitroaniline; 4NoPD, 4-nitro-o-phenylenediamine; 5A2MP, 5-amino-2-methylphenol; 24DNA, 2,4-dinitroaniline; CAT, catechol; DEpPD, N,N-diethyl-p-phenylenediamine; HQ, hydroquinone; mAP, m-aminophenol; mPD, m-phenylenediamine; oAP, o-aminophenol; oPD, o-phenylenediamine; pAP, p-aminophenol; pMAP, p-methylaminophenol; pPD, p-phenylenediamine; RES, resorcinol; T24D, toluene-2,4-diamine; T26D, toluene-2,6-diamine; T34D, toluene-3,4-diamine.
[b] AD, amperometric detection; C_{18}, octadecylsilica; CE, capillary electrophoresis; CEC, capillary electrochromatography; CL, chemiluminescence; DS, densitometry; DSPE, dispersive solid-phase extraction; GC, gas chromatography; IC, ion chromatography; IT–SPME, in-tube solid-phase microextraction; LC, liquid chromatography; LLE, liquid–liquid extraction; LSV, linear sweep voltammetry; MEKC, micellar electrokinetic chromatography; MS, mass spectrometry; MSPD, matrix solid-phase dispersion; SPE, solid-phase extraction; TLC, thin-layer chromatography; USA, ultrasound-assisted; USAEME, ultrasound-assisted emulsification microextraction; UV/VIS, ultraviolet/visible spectrometry.
[c] LOD, limit of detection (referred to sample, not to sample solution); R, recovery.

participate in the oxidative process. Therefore, their direct measurement without a previous separation step is rather difficult. Thus, separation techniques, such as chromatography and electrophoresis, are usually employed to cover the determination of a high number of these active agents. Nevertheless, in some few cases, no separation technique is used, although the number of the measured compounds is rather limited. For example, Bai et al. (2010), taking advantage of the redox properties that the oxidative hair dyes exhibit, measured *para*-phenylenediamine (pPD) by amperometric detection (AD) using a β-MnO$_2$ nanowire-modified glassy carbon electrode as the working electrode, or Hudari et al. (2014) determined pPD and resorcinol by linear sweep voltammetry using a multi-walled carbon nanotube—chitosan-modified glassy carbon electrode as the working electrode.

Regarding chromatographic techniques, LC using conventional columns (Narita et al., 2007; Xiao et al., 2011; Dejmkova et al., 2011; Asghari et al., 2014; Franco et al., 2015) or ion-exchange columns, i.e., ion chromatography (Zhong et al., 2012, 2014), is the most commonly employed technique since both non-oxidative and oxidative hair dyes present relatively low volatility for employing gas chromatography (GC), used in just one case with mass spectrometry detection to determine oxidative hair dyes after derivatization with isobutylchloroformate to form the more volatile isobutyloxycarbonyl derivatives (Akyüz and Ata, 2008). The ionic properties of hair dyes have helped to determine them by capillary electrophoresis (Masukawa, 2006; Dong et al., 2008; Li et al., 2010), although capillary electrochromatography with a gold nanoparticle-coated capillary (He et al., 2012) and micellar electrokinetic chromatography with β-cyclodextrin and 1-butyl-3-methylimidazolium hexafluorophosphate as modifiers (Wu et al., 2011) have also been used with very satisfactory results. The ionic liquid 1-butyl-3-methylimidazolium hexafluorophosphate as modifier was also used in LC by Franco et al. (2015), who stated that this ionic liquid played a crucial role in the separation of their target analytes. Usually, a UV/VIS spectrometry detector is the most common choice owing to the high absorption properties of hair dyes, especially the acid and basic dyes (Masukawa, 2006; Franco et al., 2015). Nevertheless, AD has often been used in the determination of oxidative hair dyes owing to their redox properties (Narita et al., 2007; Dong et al., 2008; Dejmkova et al., 2011). It is worth mentioning the paper published by Li et al. (2010), who used a chemiluminescence (CL) detector to measure the quenched CL of luminol sensitized by ferrate(VI) caused by some hair dyes. Finally, it should be said that TLC with a densitometric detector was used in just one case to determine lawsone in henna powder (El-Shaer et al., 2007).

Regarding sample preparation, except for that described by Akyüz and Ata (2008), who performed various liquid–liquid extraction steps prior to derivatization for GC, no complex operations are usually required. In most of the cases it consists in adding an appropriate solvent to the cosmetic sample and then stirring or sonicating it for several minutes to totally dissolve the sample or at least to leach the target compounds.

Centrifugation or filtration is required to get a clear solution prior to its measurement. The use of sample pre-concentration techniques is not really necessary in this case, because hair dyes are used as major ingredients in the hair dye cosmetic products, and good sensitivity and thus limit of detection values that are low enough are obtained. Nevertheless, Xiao et al. (2011) used in-tube solid-phase microextraction using a poly(methacrylic acid–ethylene glycol dimethacrylate) monolith as sorbent; Dejmkova et al. (2011) and Franco et al. (2015) employed solid-phase extraction (SPE) with conventional C_{18} and cation-exchange cartridges, respectively, whereas Wu et al. (2011) used SPE with a column made of coal cinders; Zhong et al. (2012) proposed the use of ultrasound-assisted matrix solid-phase dispersion with alumina as sorbent, and later the same authors (Zhong et al., 2014) proposed ultrasound-assisted dispersive solid-phase extraction with sulphite as the disperser sorbent (in addition to its antioxidant properties); and Asghari et al. (2014) proposed the use of ultrasound-assisted emulsification microextraction with chloroform.

REFERENCES

Akyüz, M., Ata, S., 2008. J. Pharm. Biomed. Anal. 47, 68.
Asghari, A., Fazl-Karimi, H., Barfi, B., Rajabi, M., Daneshfar, A., 2014. Hum. Exp. Toxicol. 33, 863.
Bai, Y.H., Yi, J.L., Xu, J.J., Chen, H.Y., 2010. Electroanalysis 22, 1239.
Chisvert, A., Cháfer, A., Salvador, A., 2007. In: Salvador, A., Chisvert, A. (Eds.), Hair Dyes in Cosmetics. Regulatory Aspects and Analytical Methods. Elsevier, Amsterdam.
CFR (Code of Federal Regulations). https://www.ecfr.gov; https://www.accessdata.fda.gov/scripts/cdrh/cfdocs/cfcfr/cfrsearch.cfm.
da França, S.A., Dario, M.F., Esteves, V.B., Baby, A.R., Velasco, M.V.R., 2015. Cosmetics 2, 110.
Dejmkova, H., Barek, J., Zima, J., 2011. Int. J. Electrochem. Sci. 6, 3550.
Directive 2007/54/EC of 29 August, 2007 Amending Council Directive 76/768/EEC, Concerning Cosmetic Products, for the Purpose of Adapting Annexes II and III Thereto to Technical Progress.
Directive 76/768/EEC of the Council of 27 July, 1976 on the Approximation of the Laws of the Member States Relating to Cosmetic Products, and its Successive Amendments and Adaptations.
Directive 82/434/EEC of 14 May, 1982 on the Approximation of the Laws of the Member States Relating to Methods of Analysis Necessary for Checking the Composition of Cosmetic Products.
Dong, S., Chi, L., Zhang, S., He, P., Wang, Q., Fang, Y., 2008. Anal. Bioanal. Chem. 391, 653.
El-Shaer, N.S., Badr, J.M., Aboul-Ela, M.A., Gohar, Y.M., 2007. J. Sep. Sci. 30, 3311.
Franco, J.H., Silva, B.F., Zanoni, M.V.B., 2015. Anal. Methods 7, 1115.
Gago-Domínguez, M., Bell, D.A., Watson, M.A., Jian-Min, Y., Castelao, J.E., Hein, D.W., Chan, K.K., Coetzee, G.A., Ross, R.K., Yu, M.C., 2003. Carcinogenesis 24, 483.
Gago-Domínguez, M., Castelao, J.E., Yuan, J.M., Yu, M.C., Ross, K.K., 2001. Int. J. Cancer 91, 575.
Ghosh, P., Sinha, A.K., 2008. Anal. Lett. 41, 2291.
He, J.F., Yao, F.J., Cui, H., Li, X.J., Yuan, Z.B., 2012. J. Sep. Sci. 35, 1003.
Hudari, F.F., Almeida, L.C., Silva, B.F., Zanoni, M.V.B., 2014. Microchem. J. 116, 261.
Koutros, S., Silverman, D.T., Baris, D., Zahm, S.H., Morton, L.M., Colt, J.S., Hein, D.W., Moore, L.E., Johnson, A., Schwenn, M., Cherala, S., Schned, A., Doll, M.A., Rothman, N., Karagas, M.R., 2011. Int. J. Cancer 129, 2894.
Li, F., Hu, Y., Zhang, H., Zhang, J., 2010. J. Sep. Sci. 33, 631.
Masukawa, Y., 2006. J. Chromatogr. A 1108, 140.
Ministry of Health and Welfare Notification (MHWN), 2000. Standard for Cosmetics, No.331 of 2000, Japan.

Narita, M., Muramaki, K., Kauffmann, J.M., 2007. Anal. Chim. Acta 588, 316.
NCI, 2017. National Cancer Institute, Hair Dyes and Cancer Risks. https://www.cancer.gov/about-cancer/causes-prevention/risk/myths/hair-dyes-fact-sheet#r11.
Nohynek, G.J., Fautz, R., Benech-Kieffer, F., Toutain, H., 2004. Food Chem. Toxicol. 42, 517.
Pel, E., Bordin, G., Rodríguez, A.R., 1998. J. Liq. Chrom. Relat. Technol. 21, 883.
Regulation (EC) 1223/2009 of the European Parliament and of the Council of 30 November, 2009 on Cosmetic Products, and its Successive Amendments. http://eur-lex.europa.eu/legal-content/EN/ALL/?uri=CELEX%3A32009R1223.
Salvador, A., Chisvert, A. (Eds.), 2007. Analysis of Cosmetic Products. Elsevier.
SCCNFP, 2001. Cosmetic Products and Non-food Products Intended for Consumers, Opinion 484/01, the Use of Permanent Hair Dyes and Bladder Cancer Risk.
SCCNFP, 2002. Cosmetic Products and Non-food Products Intended for Consumers, Opinion 0553/02, Assessment Strategies for Hair Dyes.
SCCS, 2010. Scientific Committee on Consumer Safety, Opinion 1382/10, Opinion on Acid Orange 7.
SCCS, 2013. Scientific Committee on Consumer Safety, Memorandum on Hair Dye Chemical Sensitization. European Commission, Brussels, Belgium.
SCCS, 2014. Scientific Committee on Consumer Safety, Opinion 1536/14, Opinion on Acid Orange 7.
Schuttelaar, M.-L.A., Vogel, T.A., 2016. Cosmetics 3, 21.
Søsted, H., Rustemeyer, T., Gonçalo, M., Bruze, M., Goossens, A., Giménez-Arnau, A.M., Le Coz, C.J., White, I.R., Diepgen, T.L., Andersen, K.E., Agner, T., Maibach, H., Menné, T., Johansen, J.D., 2013. Contact Dermat. 69, 32.
Turati, F., Pelucchi, C., Galeone, C., Decarli, A., La Vecchia, C., 2014. Ann. Epidemiol. 24, 151.
Vincent, U., Bordin, G., Rodríguez, A.R., 1999. J. Cosmet. Sci. 50, 231.
Vincent, U., Bordin, G., Rodríguez, A.R., 2002a. J. Cosmet. Sci. 53, 43.
Vincent, U., Bordin, G., Rodríguez, A.R., 2002b. J. Cosmet. Sci. 53, 101.
Vincent, U., Bordin, G., Robouch, P., Rodríguez, A.R., 2004. A Reference Analytical Method for the Determination of Oxidative Hair Dye Intermediates in Commercial Cosmetic Formulations Final Report. IRMM-Institute for Reference Materials and Measurements, Geel.
Wu, Y., Jiang, F., Chen, L., Zheng, J., Deng, Z., Tao, Q., Zhang, J., Han, L., Wei, X., Yu, A., Zhang, H., 2011. Anal. Bioanal. Chem. 400, 2141.
Xiao, P., Bao, C., Jia, Q., Su, R., Zhou, W., Jia, J., 2011. J. Sep. Sci. 34, 675.
Zhong, Z., Li, G., Wu, R., Zhu, B., Luo, Z., 2014. J. Sep. Sci. 37, 2208.
Zhong, Z., Li, G., Wu, Y., Luo, Z., Zhu, B., 2012. Anal. Chim. Acta 752, 53.

CHAPTER 9

Preservatives in Cosmetics: Regulatory Aspects and Analytical Methods

Gerardo Alvarez-Rivera, Maria Llompart, Marta Lores, Carmen Garcia–Jares
Universidade de Santiago de Compostela, Santiago de Compostela, Spain

INTRODUCTION

Most cosmetic formulations, owing to their organic composition and high water content, are products that can be easily degraded by microorganisms. Microbial contamination in cosmetics represents an important risk for consumer health, because under these conditions, contaminated products can lead to irritation or infection, especially when they are applied on damaged skin, on the eyes or on babies (Lundov et al., 2009). Proof of this are the reported outbreaks of microbial infection, both in hospitalized individuals and in consumers' homes, caused by contaminated cosmetics (Alvarez-Lerma et al., 2008; Kutty et al., 2007; Molina-Cabrillana et al., 2006). A more detailed discussion can be seen in Chapter 5.

To prevent microbial growth, preservatives are substances frequently added to cosmetics for the primary purpose of inhibiting the development of microorganisms (antimicrobial function). However, they may also be added to preserve cosmetics against damage and degradation caused by exposure to oxygen (antioxidant function) or to UV light (photoprotective function). Although these uses are not strictly considered as preservatives in the European Union (EU) Regulation, commonly used antioxidants, such as butylated hydroxyanisole (BHA) and butylated hydroxytoluene (BHT), are included as target compounds in many of the methods developed for cosmetic preservatives determination, as discussed in this chapter. Regarding the UV light absorbers, they will be dealt with in detailed in Chapter 5.

As they are biologically active products, preservatives can act as toxic, irritating or sensitizing agents. Thus, the safety of these chemicals is always being called into question. Scientific evidence of the possible harmful effects that some preservatives could have on consumers' health has led to increasing restrictions by various international regulations. For instance, the most widespread family of preservatives, the parabens, has been reviewed and re-reviewed in the EU Regulation, and the use of some of them (e.g., isopropyl-, isobutyl-, phenyl-, benzyl- and pentylparaben) has been banned in cosmetic products (European Commission, 2014).

Although the positive lists (permitted) and the restrictions applied to cosmetic preservatives vary greatly between regulatory frameworks (see section on Regulatory Aspects), the marketing directives suggest the use of 'universally' acceptable ingredients, i.e., those permitted in restrictive markets such as Japan, China and the EU (Steinberg, 2012). However, nowadays, the negative public perception about traditional preservatives, even those included in the positive lists by different regulations, has substantially changed the way the cosmetic formulators are preserving their products. More and more manufacturers decide not to include traditional preservatives, which are in the spotlight, claiming that their products are 'free' from a potentially toxic chemical. In fact, the inclusion of the term 'paraben-free' or 'preservative-free' on the label of several cosmetic products has already been observed, as this is seen as a positive characteristic by consumers. Thus, the use of non-traditional preservatives as well as fragrances and natural oils, often acting as so-called 'natural' preservatives, is growing in use as an alternative preservation system in cosmetic products. In the same way, other common cosmetic ingredients traditionally used with functions other than as preservatives, and thus not listed in the positive lists, have shown an important antimicrobial activity, and are being used with preservation purposes in 'preservative-free' cosmetic products.

Types of Preservative Systems

Although there exist extensive lists of potential preservative chemicals, only a handful of them offer efficacy in common application (Geis, 2006). In fact, the US cosmetics industry employs about 60 preservatives; however, the US Food and Drug Administration (FDA) data on the frequency of preservative use show that less than one-third are commonly included in cosmetic formulations (Steinberg, 2012). The ingredients most frequently found as preservatives in cosmetic products are described next.

Antimicrobial Preservatives

The wide variety of these antimicrobial-type ingredients can be organized into different chemical categories, according to the molecular structure and the more characteristic functional groups. Table 9.1 shows some representative preservatives from each class, their structure, their International Nomenclature of Cosmetic Ingredients name and their Chemical Abstracts Service number.

Organic Acid Preservatives, and Their Salts

These are pH-dependent ingredients, as they show activity only in the acid form, although they are usually added to formulations as salts to facilitate incorporation. However, the salt itself does not show antimicrobial activity until the pH is lowered to release the free acid form (Steinberg, 2012).

Table 9.1 Preservatives in common use in cosmetics[a]

Organic acids

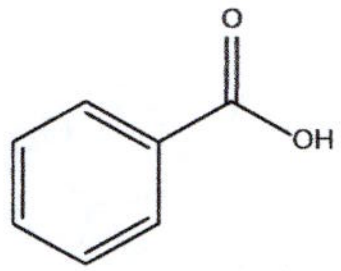 BA Benzoic acid [65-85-0]	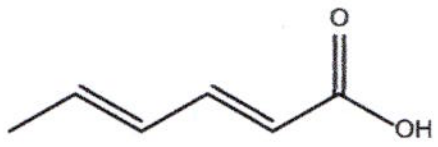 SO Sorbic acid [114-44-1]	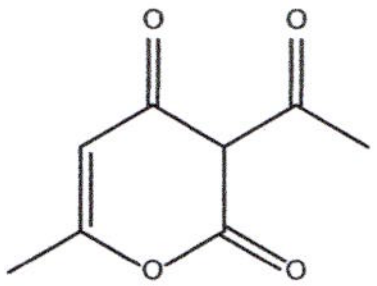 DHA Dehydroacetic acid [520-45-6]	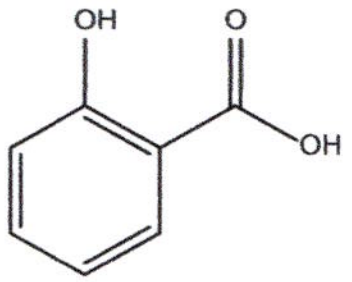 SA Salicylic acid [69-72-7]

Alcohols and derivatives

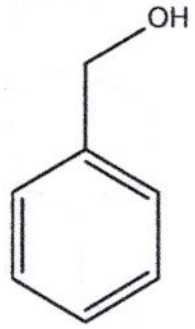

| BzOH
Benzyl alcohol
[100-51-6] | Cym
o-Cymen-5-ol
[3228-02-2]

R = CH$_3$

R = CH$_2$–CH$_3$

R = CH–(CH$_3$)$_2$

R = (–CH$_2$–)$_2$–CH$_3$
R = CH$_2$–CH–(CH$_3$)$_2$

R = (–CH$_2$–)$_3$–CH$_3$

R = CH$_2$–Ph | PhPhOH
o-Phenylphenol
[90-43-7]

MeP

EtP

i-PrP

PrP
i-BuP

BuP

BzP | PhEtOH
Phenoxyethanol
[122-99-6]
Methylparaben
[99-76-3]
Ethylparaben
[120-47-8]
i-Propylparaben
[4191-73-5]
Propylparaben
i-Butylparaben
[427-02-3]
Butylparaben
[944-26-8]
Benzylparaben
[94-18-8] |

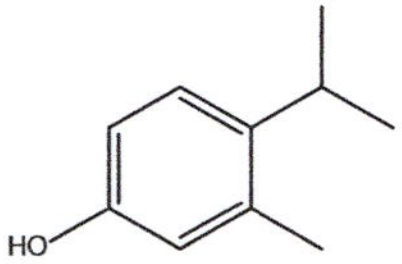

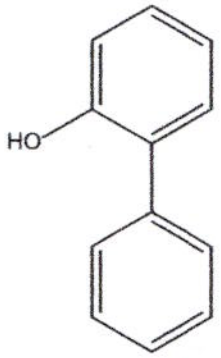

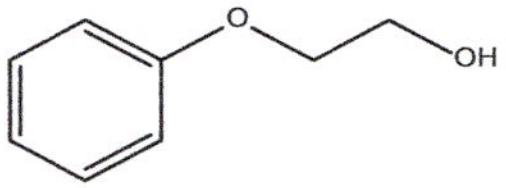

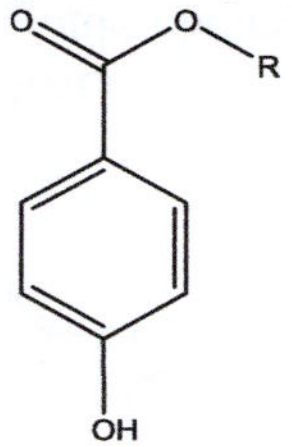

Continued

Table 9.1 Preservatives in common use in cosmetics[a]—cont'd

Formaldehyde releasers

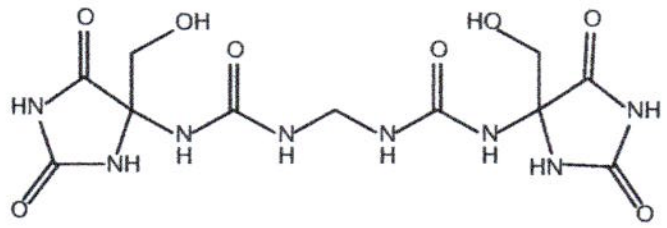 IMU Imidazolidinyl urea [39236-46-9]	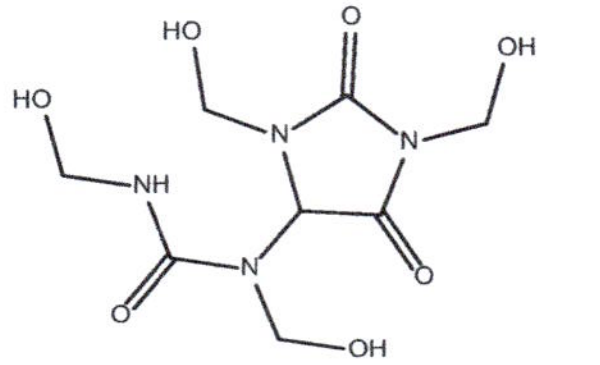 DIU Diazolidinyl urea [78491-02-8]	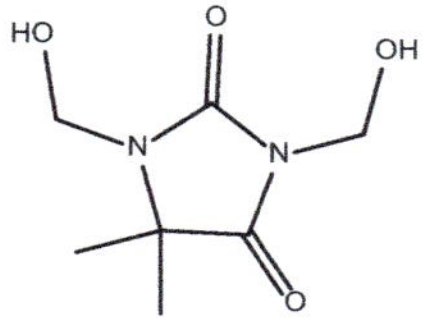 DMDM DMDM hydantoin [6440-58-0]
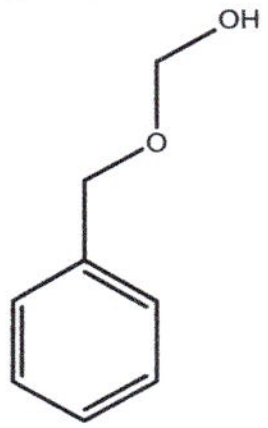 BzH Benzyl hemiformal [14548-60-8]	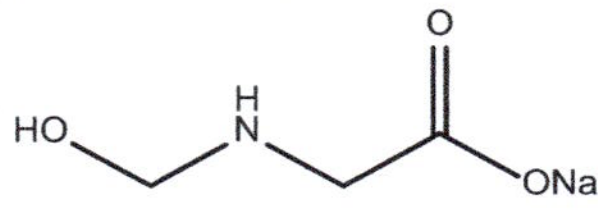 OHMeGlyNa Sodium hydroxymethylglycinate [6440-58-0]	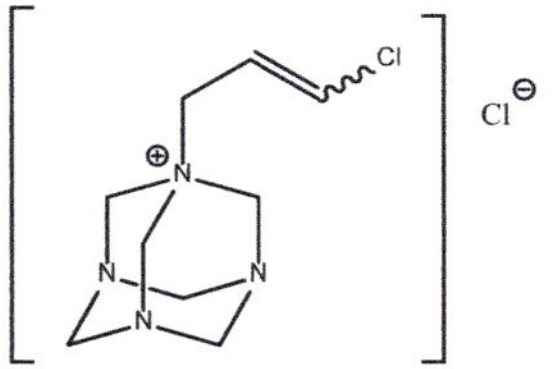 Q15 Quaternium-15 [51229-78-8]

Halogenated preservatives

| 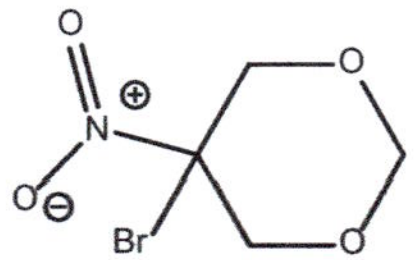 BRX
5-Bromo-5-nitro-1,3-dioxane
[30007-47-7] | 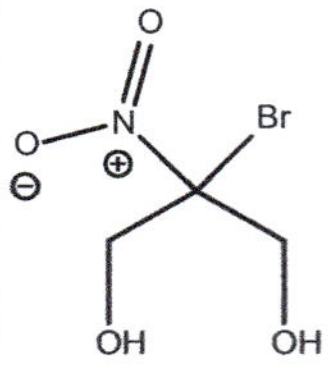 BRP
2-Bromo-2-nitro-propane-1,3-diol
[52-51-7] | 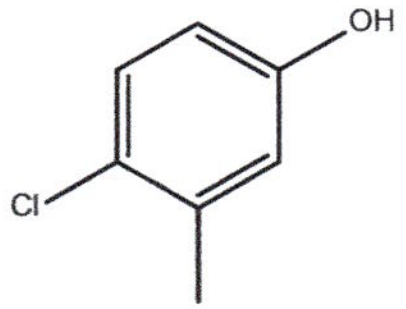 ClCr
p-Chloro-*m*-cresol
[59-50-7] | 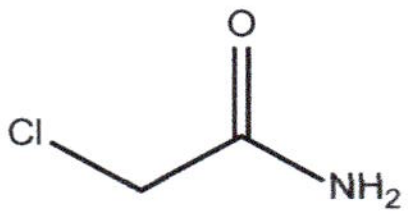 CAM
Chloroacetamide
[79-07-2] |

IPBC
Iodopropynyl butylcarbamate
[55406-53-6]

ClPh
Chlorphenesin
[104-29-0]

TCS
Triclosan
[3380-34-5]

CBZ
Climbazol
[38083-17-9]

Quaternary ammonium salts

BzKCl
Benzalkonium chloride
[8001-54-5]

BzTCl
Benzethonium chloride
[121-54-0]

CetCl
Cetrimonium chloride
[112-02-7]

Isothiazolinones

MI
Methylisothiazolinone
[2682-20-4]

CMI
Chloromethylisothiazolinone
[26172-55-4]

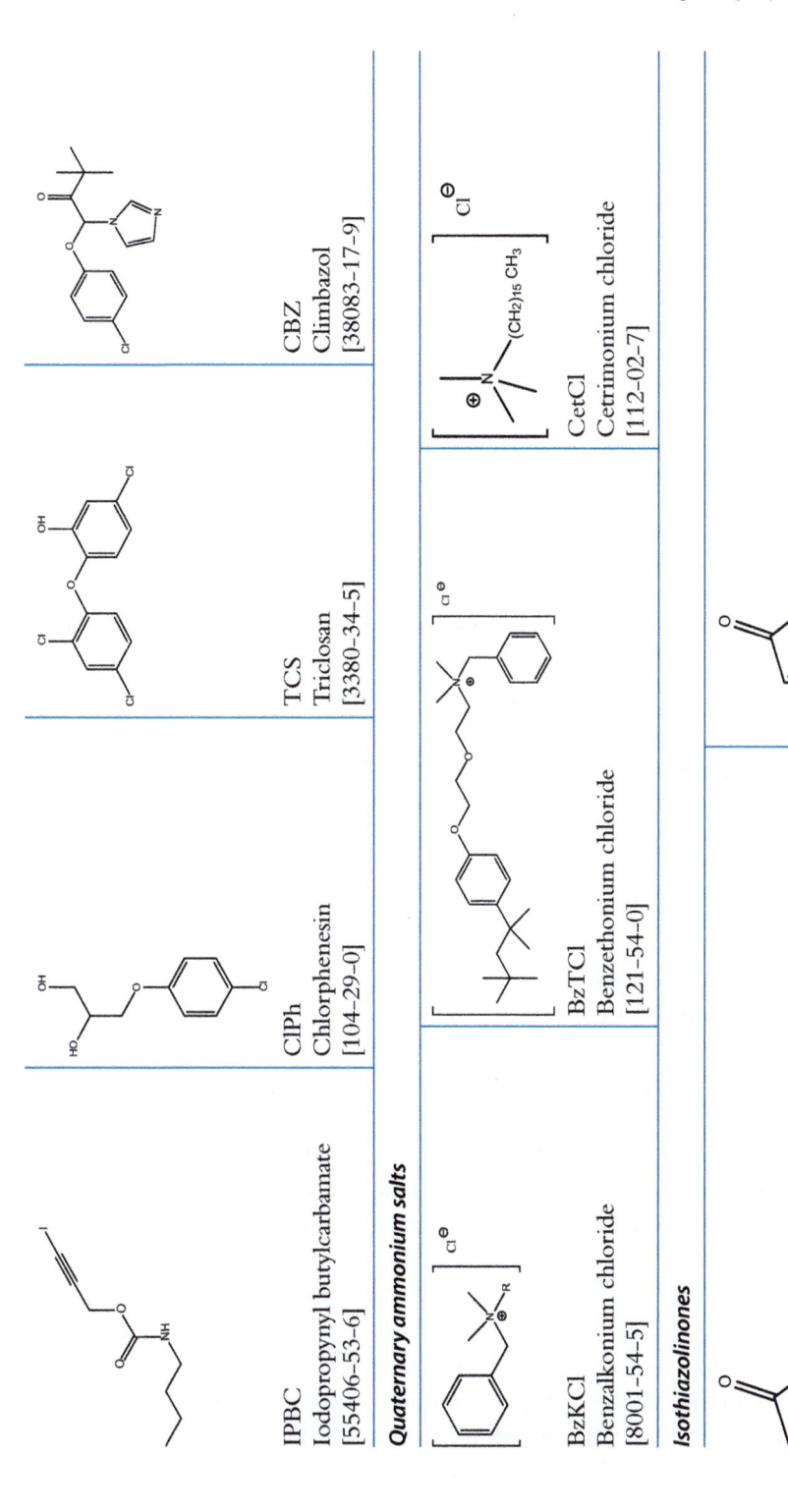

[a]Shown are the structure, abbreviation, INCI name and CAS number of each preservative.

Some typical organic acids used as preservatives are benzoic acid and its salt sodium benzoate, dehydroacetic acid and sodium dehydroacetate, formic acid, propionic acid and its salts, sorbic acid and potassium sorbate, salicylic acid and undecylenic acid.

Alcohols and Derivatives

Preservatives with hydroxyl groups (–OH) can have also acid properties although to a lesser extent than organic acid preservatives. This group of preservatives includes compounds such as benzyl alcohol, isopropyl methylphenol, phenoxyethanol, phenoxyisopropanol and *o*-phenylphenol and its sodium salt.

Particularly relevant are the alkyl esters of 4-hydroxybenzoic acid (4-HB, also known as parabens). The most common are the methyl, ethyl, propyl, isopropyl, butyl, isobutyl and benzyl esters, although the iso and benzyl forms are currently prohibited by the EU Cosmetics Regulation. These compounds exhibit antifungal properties and activity against gram-negative bacteria. The antimicrobial activity of these chemicals increases as the carbon number of the alkyl chain increases, while, by contrast, their water solubility decreases in parallel (Polati et al., 2007). Thus, the way they are incorporated into the formulation is particularly critical as they are active only in the water phase and are pH dependent (high pH values lead to paraben dissociation). For this reason, parabens are sometimes incorporated as their sodium and potassium salt forms.

Formaldehyde, Methylene Glycol and Formaldehyde Releasers

Formaldehyde is an anhydrous gas which can easily react with water yielding methylene glycol. The commercial solution of formaldehyde in water (i.e., methylene glycol) is commonly known as formalin (Steinberg, 2012). Thus, formaldehyde or its methylene glycol form is frequently added to water-based cosmetics products such as shampoo, conditioner, shower gel, liquid handwash and bubble bath. They show activity against bacteria but they are also active against fungi.

This group also includes many preservatives containing *n*-methylol groups in their chemical structure, which in polar solvents may act as formaldehyde donors or releasers. The most frequently found in cosmetic products are imidazolidinyl urea, diazolidinyl urea, sodium hydroxymethylglycinate, benzylhemiformal, DMDM hydantoin and Quaternium-15. According to the EU Regulation all finished products containing formaldehyde or substances which release formaldehyde must be labelled with the warning 'contains formaldehyde' when the concentration of formaldehyde in the finished product exceeds 0.05%.

Isothiazolinones

Isothiazolinone-type biocides are a group of effective preservatives used in a wide variety of aqueous-based industrial or domestic products and applications (Nakashima et al., 2000; Rafoth et al., 2007; Reinhard et al., 2001). These compounds are heterocyclic derivatives of 2H-isothiazolin-3-one with an active sulphur moiety capable of oxidizing thiol-containing residues, thereby offering a powerful preservation activity against a broad spectrum of fungi and bacteria (Collier et al., 1990). Isothiazolinones like 2-methyl-3-isothiazolinone (MI) and 5-chloro-2-methyl-3-isothiazolinone (CMI) are the active ingredients of a 3:1 CMI/MI mixture commercially sold under the name of Kathon CG (cosmetic grade). The cosmetics industry commonly includes Kathon CG in a wide range of both rinse-off and leave-on formulations such as shampoos, gels and hair and skin care products, because of their high effectiveness at very low concentrations (Fewings and Menné, 1999). The use of MI and CMI has dramatically risen, and even gained presence at much higher concentrations together with 1,2-benzisothiazolinone (BzI) and 2-octyl-3-isothiazolinone (OI) in cleaning agents and other household and industrial products (water-based coatings, paints, adhesives, wood preservatives, etc.) (Bester and Lamani, 2010; Thouvenin et al., 2002).

Halogenated Preservatives

Halogenated preservatives exhibit stronger activity, particularly against fungi. This group includes compounds such as 5-bromo-5-nitro-1,3 dioxane (bronidox) and 2-bromo-2-nitropropane-1,3-diol (bronopol). These can decompose, liberating nitrosating agents that can react with aliphatic amines like diethanolamine (DEA), triethanolamine (TEA) and monoethanolamine (MEA), which are commonly added to shampoos and other personal hygiene products (Polati et al., 2007).

This category also comprises other halogenated preservatives such as 2-chloroacetamide, *p*-chloro-*m*-cresol, chlorobutanol, chloroxylenol, chlorphenesin, dichlorobenzyl alcohol, iodopropynyl butylcarbamate and methyldibromo glutaronitrile. Despite their powerful antimicrobial activity, many of this compounds are poorly soluble in water, which makes their incorporation in the cosmetic matrix difficult.

On the other hand, triclosan, triclocarban and climbazol are preservatives containing chlorine which are employed as active ingredients to specifically kill microorganisms. Climbazol is an imidazole antifungal agent widely used in marketed antidandruff shampoos, whereas triclosan and triclocarban are two major antimicrobials, used in deodorants, soaps and toothpastes.

Quaternary Ammonium Salts

Quaternium compounds, such as benzalkonium chloride, benzethonium chloride and the series alkyl (C12–C22) trimethyl ammonium bromide and chloride, are frequently used at high levels in hair washing and conditioning products because of their anti-static and softening properties. These substances contain positively charged nitrogen, exhibiting strong antimicrobial activity at high pH.

Other Preservatives

Other compounds also used as cosmetic preservatives are hexamidine, dibromohexamidine, dibromopropamidine, chlorohexidine and cetylpyridinium chloride, among others. These ingredients are particularly used in mouthwash products.

Antioxidants and Other Preservation Systems

Cosmetic components such as fragrances and natural fats and oils are subject to autoxidation by exposure to air, causing undesirable odours and other instabilities. The presence of unsaturated or aldehyde groups intensifies this reaction, which begins with an attack by oxygen free radicals and propagates in a chain reaction (Steinberg, 2012).

Antioxidants are added to the raw material or the formulation to prevent degradation of cosmetic ingredients by oxidation. This type of preservative acts as a radical scavenger, avoiding the autoxidation and product rancidity. Thus, compounds such as citric acid and gallic acid and its esters, as well as vitamins such as tocopherol (vitamin E) and ascorbic acid (vitamin C), are well known by their general properties to inactivate free radicals. However, the most widely used antioxidant preservatives are BHT and BHA. They are widely employed for long-term preservation of cosmetics, food products and pharmaceuticals. The use of mixtures of them is very common, as this synergy enhances the antioxidant power (Polati et al., 2007).

Some companies are looking at alternative self-preserving formulations that reduce the need for preservatives. This so-called 'hurdle' technology has been used in the food industry since the 1970s (Leistner, 2000), and describes intelligent formulation using different preservation factors, including good manufacturing practices, appropriate packaging, careful choice of the form of emulsion, low water activity, and low or high pH values (Kabara and Orth, 1997). Some of these alternative strategies involve the use of multifunctional additives as antimicrobial stabilizers for preservative-free cosmetics. Multifunctional additives are molecules with more than one beneficial effect to the formulation or on the skin, e.g., glycols, glycerol ethers, fragrance ingredients and essential oils. Some of these widely used multifunctional additives also display a certain antimicrobial efficacy. The synergistic combination of multifunctional additives can be used to achieve the antimicrobial stabilization of cosmetic formulations instead of using traditional preservatives.

One of the proposed antimicrobial systems consists in the combination of enzymes (e.g., glucose oxidase and lactoperoxidase). Another way of preserving is lowering the water activity of a cosmetic by using solvents with strong water-binding capacity (e.g., glycerine, glyceryl polyacrylate, caprylyl glycol). Glycols are broadly used humectants in cosmetic products. Their antimicrobial efficacy increases with their chain length, whereas the water solubility decreases with longer chain lengths. The efficacy of glycols can be improved by combination with ethylhexylglycerin, which also gives them deodorizing, emollient and humectant properties, and/or with phenethyl alcohol, a fragrance ingredient that has light emollient and also antimicrobial activity.

Other preserving systems include vegetable oil–derived phospholipids (e.g., cocamidopropyl PG-dimonium chloride phosphate, sodium coco PG-dimonium chloride phosphate), which have been demonstrated to have a broad spectrum of antimicrobial activity against yeasts, moulds and both gram-positive and gram-negative bacteria (Steinberg, 2012).

Some fragrances and natural fragrance oils also show preservation activity. This is the case with aroma chemicals such as benzyl alcohol, citral, eugenol, farnesol, geraniol, as well as cinnamon oil, eucalyptus oil and lavender oil, among many others. Along this line, there is now a great demand for products containing natural extracts and blends of natural ingredients which are marketed with preservative function. Thus, essential oils from tea tree, thymus, rosemary or neem seed, as well as grapefruit seed extracts, have become natural preservative cocktails that can be found in cosmetic products. Although the preservation capacity of these systems is generally lower than that of the aforementioned synthetic compounds, cosmetic products labelled as containing 'natural' ingredients are much more appealing to the consumer, although sometimes, natural does not necessarily mean safer.

Safety of Cosmetic Preservatives

Despite the undeniable benefits that preservatives bring to cosmetic products, the use of cosmetics containing some of these compounds has been reported to cause several side effects on consumers' health. These undesirable effects may arise immediately after applying the product or years later, through continued use. Approximately 6% of the general population is allergic to preservatives and fragrance allergens. These substances may also lead to other negative effects, from soft skin irritation to strong oestrogenic activity, and some preservatives might even exhibit neurotoxic and genotoxic effects (Harvey, 2003; He et al., 2006a).

Parabens are suspected endocrine-disrupting compounds with oestrogenic- and anti-androgenic-like properties (Prusakiewicz et al., 2007; Terasaka et al., 2006). In addition, studies suggest a potential relationship between the application of paraben-containing products and some kinds of allergies (Savage et al., 2012), or even breast cancer (Darbre, 2006).

BHA and BHT may also cause serious long-term side effects. Particularly, BHA can act as a modulator and disruptor of the endocrine system (Hung et al., 2005), with the capacity to cause damage in lung tissues, and even deficiencies in reproductive system development (Jeong et al., 2005). As regards BHT, although the safety assessment report states that this preservative cannot be considered genotoxic, it can modify the genotoxicity of other agents (Lanigan and Yamarik, 2002).

The preservatives bronidox and bronopol are known to be effective nitrosating agents. Whenever these preservatives are co-formulated with constituents having secondary amine structures (e.g., DEA, TEA and MEA), the formation of substantial carcinogenic nitrosamine is to be expected (Matyska et al., 2000).

Some animal testing has demonstrated that triclosan can induce deregulation in the thyroid function, diminishing the thyroxine levels in plasma. Triclosan can also act as a sexual hormone disruptor, presenting anti-oestrogenic- and anti-androgenic-like properties (Crofton et al., 2007; Witorsch and Thomas, 2010). Triclosan has also been shown to react with chlorine in tap water, yielding chloroform gas (Rule et al., 2005). At this point, special attention must be paid to products like toothpastes and mouthwash, whose triclosan content is allowed up to a maximum concentration of 0.3% and 0.2% (w/w), respectively. On the other hand, the highest toxic potential of this compound is due to its decomposition or transformation processes, leading to byproducts such as chlorophenols, dioxins and polychlorinated compounds (Alvarez-Rivera et al., 2015; Lindstrom et al., 2002; Sanchez-Prado et al., 2006a,b).

3-Iodo-2-propynyl butylcarbamate (IPBC) does not present carcinogenic properties to the human being, but it is toxic to aquatic organisms (Juergensen et al., 2000). In addition, this compound can cause acute inhalation toxicity (Lanigan, 1998) and has been recognized as a contact allergen (Warshaw et al., 2013).

Preservatives included in the formaldehyde releasers group may cause eye irritation, nausea, difficulty breathing and allergies, and at high concentrations they can lead to headache and asthma attacks (Hauksson et al., 2015). Scientific studies indicate that formaldehyde is suspected of inducing cancer (Cole et al., 2010). Those products containing formaldehyde at concentrations higher than 0.2% are considered to be unsafe (Bergfeld et al., 2005).

Both MI and CMI have been shown to be strong skin sensitizers and allergens (Garcia-Gavin et al., 2010; Isaksson et al., 2004). The Scientific Committee on Consumer Safety reported that no adequate data have been given to support the safe use of the Kathon CG mixture in leave-on cosmetics (SCCS, 2009). Although the allowed maximum concentration for MI alone might be safe, being a weaker sensitizer compared to CMI, new cases of contact dermatitis have been reported (Garcia-Gavin et al., 2010). Scientific experiments also suggest that MI may be a neurotoxic biocide after prolonged exposures even at low levels (He et al., 2006a). Other evidence of contact allergy has been deduced from accumulated case reports associated with occupational or domestic exposure to products containing BzI and OI (Aalto-Korte et al., 2007; Magnano et al., 2009).

Finally, it is also important to highlight that there is also scientific evidence which indicates that the use or misuse of biocidal products may contribute to the increased occurrence of antibiotic-resistant bacteria, both in humans and in the environment (SCENIHR, 2009).

Regulatory Aspects

The use of preservatives as chemicals and as ingredients in finished products is subject to stringent regulatory oversight in different regions (Scott et al., 2006). In the European Community framework, a number of preservatives listed in Annex V of the EU

Regulation of Cosmetic Products (EC 1223/2009) constitute the group of allowed substances for cosmetic preservation from microbial spoilage. Table 9.2 summarizes the main regulatory issues of the preservatives included in this positive list.

Nevertheless, the regulatory issues concerning preservatives in cosmetics are different in other geographical areas, as for example in the United States, where they are regulated by the same governmental agency as are foods and pharmaceuticals, the FDA. Instead of proposing a positive list of preservatives (as Europe, Japan and most of the cosmetics legislations worldwide—although with different selections of compounds), the FDA does not preapprove preservatives (Steinberg, 2006). The FDA works to keep cosmetics safe with the cosmetics industry, consumers, and Cosmetic Ingredient Review (CIR) system. CIR studies individual chemical compounds as they are used in cosmetic products, reviewing and assessing their safety in an open, unbiased and expert manner and publishing the results in the peer-reviewed literature (CIR, 2016). CIR opinions can conclude that an ingredient (e.g., a preservative) is safe or unsafe with hints: ingredients found unsafe through this review process are not expected to be used in the United States, because no one would market or buy an unsafe product. Table 9.2 also includes the preservatives reviewed by the CIR and their regulatory status as of this writing. As shown in this table, the diversity of regulations can lead to situations somewhat incomprehensible to the consumer. Take the case of the controversial parabens: the isobutyl-, isopropyl-, phenyl- and benzylparabens have been forbidden in the EU and they are considered safe in the United States. Quaternium-15 is going to be banned in the EU under amendment No. 288/2015 (pending approval), but it is considered safe at concentrations lower than 0.2% in the United States. In Canada, without positive lists but with a cosmetic regulation proposing lists of banned and regulated ingredients, mercury-based preservatives are prohibited, while they are allowed with restrictions in the EU.

However, in other cases this trend to regulatory diversity is being corrected towards a desirable consistency in the legislations; such is the case of chloroacetamide. The prohibition of chloroacetamide in the EU (now permitted up to 0.3%) is nowadays under public consultation, because human data demonstrate that allergic reactions can be elicited at concentrations even lower than 0.3%; but it is already considered unsafe by the United States and it is, by the by, already prohibited in Canada.

ANALYTICAL METHODS

Since 2006, a relatively high number of articles have significantly contributed to improving the challenging task of preservative analysis in cosmetics. More and more authors are adopting efficient sample preparation and extraction procedures based on solid-phase extraction (SPE), pressurized liquid extraction (PLE), matrix solid-phase dispersion (MSPD) and advanced microextraction techniques such as solid-phase microextraction (SPME) and liquid–liquid microextraction (LLME), among others, to overcome the

Table 9.2 Preservatives allowed in the European Union for cosmetics use according to Cosmetics Regulation (EC) 1223/2009 and its amendments: comparison with US FDA regulation (CIR conclusions)

		EU Cosmetics Regulation (EC) 1223/2009 (Annex V)		CIR system conclusions (basis of preservative safe use in United States)	
Name of ingredient	CAS number	Current/upcoming limits on maximum concentration in ready-to-use preparation (%, w/w)	Conditions of use and warnings	Review conclusion[a]	Concentration (%, w/w) and other limitations
Benzoic acid	65-85-0	1. 2.5% (acid)	1. Rinse-off products, except oral products	S	0.000002%–5%
Sodium benzoate	532-32-1	2. 1.7% (acid)	2. Oral products		0.000001%–1%
		3. 0.5% (acid)	3. Leave-on products		
Other salts of benzoic acid and esters of benzoic acid					
Ammonium benzoate	1863-63-4	0.5% (acid)		—	—
Calcium benzoate	2090-05-3			S	0.002%–0.004%
Potassium benzoate	582-25-2			S	0.0002%–0.0003%
Magnesium benzoate	553-70-8			SQ[b,c,d]	Not in use at the time
MEA-benzoate	4337-66-0				
Methyl benzoate	93-58-3			S	0.0005%–0.3%
Ethyl benzoate	93-89-0			S	0.0008–0.01%
Propyl benzoate	2315-68-6			S	Not in current use
Butyl benzoate	136-60-7			S	Not in use at the time
Isobutyl benzoate	120-50-3			S	0.01%
Isopropyl benzoate	939-48-0			S	—
Phenyl benzoate	93-99-2			—	—
Propionic acid and its salts					
Propionic acid	79-09-4	2% (acid)		—	—
Ammonium propionate	17496-08-1				
Calcium propionate	4075-81-4				
Magnesium propionate	557-27-7				
Potassium propionate	327-62-8				
Sodium propionate	137-40-6				

Salicylic acid and its salts

Salicylic acid	69-72-7	0.5% (acid)	Not to be used in products for children under 3 years of age, except for shampoos Not to be used for children under 3 years of age	SQ
Calcium salicylate	824-35-1			SQ
Magnesium salicylate	18917-89-0			SQ
MEA-salicylate	59866-70-5			SQ[b,d]
Sodium salicylate	54-21-7			SQ
Potassium salicylate	578-36-9			SQ
TEA-salicylate	2174-16-5			SQ
				Safe when formulated to avoid irritation and to avoid increasing sun sensitivity, or when increased sun sensitivity would be expected, directions for use include the daily use of sun protection

Hexa-2,4-dienoic acid and its salts

Sorbic acid	110-44-1	0.6% (acid)		S
Calcium sorbate	7492-55-9			—
Sodium sorbate	7757-81-5			—
Potassium sorbate	24634-61-5		Not to be used in aerosol dispensers (sprays)	S
Formaldehyde[e]	50-00-0	0.1% (free formaldehyde) in oral products		SQ
				Up to 5%
				—
				—
				Up to 1%
				Safe for use in cosmetics when formulated to ensure use at the minimal effective concentration, but in no case should the formalin concentration exceed 0.2% (w/w), which would be 0.074% (w/w) calculated as formaldehyde or 0.118% (w/w) calculated as methylene glycol; safe in the present practices of use and concentration in nail hardening products
Paraformaldehyde	30525-89-4	0.2% (free formaldehyde) in other products		U
				Unsafe in the present practices of use and concentration in hair smoothing products

Continued

Preservatives in Cosmetics: Regulatory Aspects and Analytical Methods   187

Table 9.2 Preservatives allowed in the European Union for cosmetics use according to Cosmetics Regulation (EC) 1223/2009 and its amendments: comparison with US FDA regulation (CIR conclusions)—cont'd

Name of ingredient	CAS number	EU Cosmetics Regulation (EC) 1223/2009 (Annex V) Current/upcoming limits on maximum concentration in ready-to-use preparation (%, w/w)	Conditions of use and warnings	CIR system conclusions (basis of preservative safe use in United States) Review conclusion[a]	Concentration (%, w/w) and other limitations
Biphenyl-2-ol, and its salts					
o-Phenylphenol	90-43-7	0.2% (as phenol)		—	—
Sodium o-phenylphenate	132-27-4				
Potassium o-phenylphenate	13707-65-8				
MEA o-phenylphenate	84145-04-0				
Zinc pyrithione	13463-41-7	1.0% in hair products	Only in rinse-off products	—	—
		0.5% in other products	Not to be used in oral products	—	—
Inorganic sulphites and hydrogen sulphites					
Sodium sulphite	7757-83-7	0.2% (as free SO$_2$)		S	Up to 3%
Ammonium bisulphite	10192-30-0			S	Up to 32%
Ammonium sulphite	10196-04-0			S	Not in current use
Potassium sulphite	10117-38-1			S	Concentration not reported
Potassium hydrogen sulphite	7773-03-7			—	—
Sodium bisulphite	7631-90-5			S	Up to 0.7%
Sodium metabisulphite	7681-57-4			S	Concentration not reported
Potassium metabisulphite	16731-55-8			S	Concentration not reported
Chlorobutanol	57-15-8	0.5%	Not to be used in aerosol dispensers (sprays) 'Contains chlorobutanol'	—	—

4-Hydroxybenzoic acid and its salts and esters

4-Hydroxybenzoic acid	99-96-7	0.4% (as acid) for single ester, 0.8% (as acid) for mixtures	—	—
Potassium ethylparaben	36457-19-9			
Potassium paraben	16782-08-4			
Sodium methylparaben	5026-62-0			
Sodium ethylparaben	35285-68-8			
Sodium propylparaben	35285-69-9			
Sodium butylparaben	36457-20-2			
Sodium isobutylparaben	84930-15-4			
Sodium paraben	114-63-6			
Potassium methylparaben	2611-07-2			
Potassium butylparaben	38566-94-8			
Potassium propylparaben	84930-17-4			
Sodium propylparaben	35285-69-9			
Calcium paraben	69959-44-0			
Methylparaben	99-76-3	0.4% (as acid) for single ester	S	Up to 0.4% if used alone
Ethylparaben	120-47-8	0.8% (as acid) for mixtures[f]		Parabens mixture up to 0.8%
Butylparaben	94-26-8	0.14% (as acid) for the sum of individual concentrations	S	Up to 0.4% if used alone
Propylparaben	94-13-3	0.8% (as acid) for mixtures of substances mentioned in previous entry, where the sum of the individual concentrations of butyl- and propylparaben and their salts does not exceed 0.14%[f]		Parabens mixture up to 0.8% (BuP)

Continued

Table 9.2 Preservatives allowed in the European Union for cosmetics use according to Cosmetics Regulation (EC) 1223/2009 and its amendments: comparison with US FDA regulation (CIR conclusions)—cont'd

		EU Cosmetics Regulation (EC) 1223/2009 (Annex V)		CIR system conclusions (basis of preservative safe use in United States)	
Name of ingredient	CAS number	Current/upcoming limits on maximum concentration in ready-to-use preparation (%, w/w)	Conditions of use and warnings	Review conclusion[a]	Concentration (%, w/w) and other limitations
Isobutylparaben	4247-02-3	Prohibited[g]		S	Up to 0.4% if used alone Parabens mixture up to 0.8%
Isopropylparaben	4191-73-5			S	
Phenylparaben	17696-62-7			S	
Benzylparaben	94-18-8			S	
Dehydroacetic acid	520-45-6	0.6% (as acid)	Not to be used in aerosol dispensers (sprays)	S	Up to 0.7%
Sodium dehydroacetate	4418-26-2			S	Up to 0.6%
Formic acid	64-18-6	0.5% (as acid)		SQ	≤0.0064% of the free acid
Sodium formate	141-53-7			SQ	
Dibromohexamidine isethionate	93856-83-8	0.1%		—	—
Thimerosal	54-64-8	0.007% (of Hg) if mixed with other mercurial compounds authorized by this regulation, the maximum concentration of Hg remains fixed at 0.007%	Eye products ('Contains thimerosal')	—	—
Phenyl mercuric acetate	62-38-4	0.007% (of Hg) if mixed with other mercurial compounds authorized by this regulation, the maximum concentration of Hg remains fixed at 0.007%	Eye products ('Contains phenylmercuric compounds')	—	—
Phenyl mercuric benzoate	94-43-9				

Undec-10-enoic acid and its salts

Undecylenic acid Potassium undecylenate Sodium undecylenate Calcium undecylenate TEA-undecylenate MEA-undecylenate	112-38-9 6159-41-7 3398-33-2 1322-14-1 84471-25-0 56532-40-2	0.2% (as acid)	SQ[b,d] SQ[d]	Safe as used when formulated to be non-irritating and when the levels of free diethanolamine do not exceed the present practices of use and concentration of diethanolamine itself.
Hexetidine 5-Bromo-5-nitro-1,3-dioxane	141-94-6 30007-47-7	0.1% 0.1%	Avoid formation of nitrosamines	SQ	0.1%; but should not be used under circumstances where its actions with amines or amides can result in the formation of nitrosamines or nitrosamides
2-Bromo-2-nitropropane-1,3-diol	52-51-7	0.1%	Avoid formation of nitrosamines	SQ	0.1%; may contribute to endogenous nitrosamine formation; but should not be used under circumstances where its actions with amines or amides can result in the formation of nitrosamines or nitrosamides
Dichlorobenzyl alcohol Triclocarban	1777-82-8 101-20-2	0.15% 0.2%	Purity criteria: 3,3',4,4'-Tetrachloro-azobenzene <0.0001% 3,3',4,4'-Tetrachloroa-zoxybenzene <0.0001%	— —	

Continued

Table 9.2 Preservatives allowed in the European Union for cosmetics use according to Cosmetics Regulation (EC) 1223/2009 and its amendments: comparison with US FDA regulation (CIR conclusions)—cont'd

Name of ingredient	CAS number	EU Cosmetics Regulation (EC) 1223/2009 (Annex V) Current/upcoming limits on maximum concentration in ready-to-use preparation (%, w/w)	Conditions of use and warnings	CIR system conclusions (basis of preservative safe use in United States) Review conclusion[a]	Concentration (%, w/w) and other limitations
p-Chloro-m-cresol	59-50-7	0.2%	Not to be used in products applied on mucous membranes	SQ	Up to 5%
Triclosan	3380-34-5	0.3%[g]	Toothpastes, hand soaps, body soaps, shower gels, deodorants (non-spray), face powders and blemish concealers, nail products for cleaning the fingernails and toenails before the application of artificial nail systems Mouthwashes	S	0.04%–0.3%
Chloroxylenol	88-04-0	0.2%[g]		S	Up to 0.5%
Imidazolidinyl urea	39236-46-9	0.5%		S	Up to 1%
Polyaminopropyl biguanide	70170-61-5, 28757-47-3, 133029-32-0	0.6% 0.3%		—	—
Phenoxyethanol	122-99-6	1.0%		S	Less than or equal to 0.0002%–1%
Methenamine	122-99-6	0.15%		SQ	0.16%; but should not be used in products intended to be aerosolized

Quaternium-15	223-805-0	Prohibited[h]		SQ	Not to exceed 0.2%
Climbazole	38083-17-9	0.5%		–	–
DMDM Hydantoin	6440-58-0	0.6%		S	Up to 1%
Benzyl alcohol	100-51-6	1.0%		S	0.000006%–10%
1-Hydroxy-4-methyl-6-(2,4,4-trimethylpentyl) 2-pyridon, piroctone olamine	50650-76-5 68890-66-4	1.0% in rinse-off products 0.5% Other products		–	–
Bromochlorophene	15435-29-7	0.1%		–	–
o-Cymen-5-ol	3228-02-2	0.1%		SQ	Up to 0.5%
Methylchloroiso-thiazolinone (MCI) and methylisothiazolinone (MI)	26172-55-4, 2682-20-4, 55965-84-9	0.0015% (of a mixture 3:1, MCI:MI)[i] The use of the mixture of MCI:MI is incompatible with the use of MI alone in the same product	0.0015% (of a mixture 3:1, MCI:MI) Rinse-off products The use of the mixture of MCI:MI is incompatible with the use of MI alone in the same product	S	Rinse-off 0.0015% Leave-on 0.00075%
Methylisothiazolinone	2682-20-4	0.01% The use of the mixture of MCI:MI is incompatible with the use of MI alone in the same product	The use of the mixture of MCI:MI is incompatible with the use of MI alone in the same product	SQ	Safe at concentrations up to 0.01% and safe for use in leave-on cosmetic products when formulated to be non-sensitizing which may be determined based on a quantitative risk assessment
Chlorophene	120-32-1	0.2%		Z	
Chloroacetamide	79-07-2	0.3%	'Contains chloroacetamide'	U	

Continued

Table 9.2 Preservatives allowed in the European Union for cosmetics use according to Cosmetics Regulation (EC) 1223/2009 and its amendments: comparison with US FDA regulation (CIR conclusions)—cont'd

Name of ingredient	CAS number	EU Cosmetics Regulation (EC) 1223/2009 (Annex V) Current/upcoming limits on maximum concentration in ready-to-use preparation (%, w/w)	Conditions of use and warnings	CIR system conclusions (basis of preservative safe use in United States) Review conclusion[a]	Concentration (%, w/w) and other limitations
Chlorhexidine	55-56-1	0.3% (as chlorhexidine)		SQ	Up to 0.14%
Chlorhexidine diacetate	56-95-1			SQ	Up to 0.19%
Chlorhexidine digluconate	18472-51-0			SQ	Up to 0.20%
Chlorhexidine dihydrochloride	3697-42-5			SQ	Up to 0.16%
Phenoxyisopropanol	770-35-4	1.0%	Only for rinse-off products	—	—
Alkyl (C12–C22) trimethylammonium bromide and chloride					
Behentrimonium chloride	17301-53-0	0.1%[j]	While allowing the use of these substances for other uses than as preservatives at higher concentrations, the sum of these substances should be restricted to the maximum concentration indicated (individual concentrations or the sum of the individual concentrations)	SQ	
Cetrimonium bromide	57-09-0			SQ	
Cetrimonium chloride	112-02-7			SQ	
Laurtrimonium bromide	1119-94-4			SQ	
Laurtrimonium chloride	112-00-5			SQ	
Steartrimonium bromide	1120-02-1			SQ	
Steartrimonium chloride	112-03-8			SQ	Up to 0.25% for leave-on products (no rinse-off uses reported)

Preservatives in Cosmetics: Regulatory Aspects and Analytical Methods 195

Dimethyl oxazolidine	51200-87-4	0.1%	pH > 6	—
Diazolidinyl urea	78491-02-8	0.5%		SQ — Up to 0.5%
Hexamidine	3811-75-4	0.1%		SQ — Safe at concentrations less than or equal to 0.1%
Hexamidine diisethionate	659-40-5			SQ
Hexamidine paraben	93841-83-9			—
Glutaral	111-30-8	0.1%	Not to be used in aerosols (sprays) 'Contains glutaral'	SQ — <0.5% for rinse-off, but should not be used in products intended to be aerosolized (insufficient data to support safety in leave-on products)
7-Ethylbicyclo-oxazolidine	7747-35-5	0.3%	Not to be used in oral products and in products applied on mucous membranes	— —
Chlorphenesin	104-29-0	0.3%		S — 0.000008%–0.32%
Sodium hydroxymethylglycinate	70161-44-3	0.5%		— —
Silver chloride	7783-90-6	0.004% (as AgCl)	20% AgCl (w/w) on TiO$_2$; not to be used in products for children under 3 years of age, in oral products and in eye and lip products 1. Rinse-off products 2. Leave-on products other than oral products	—
Benzethonium chloride	121-54-0	0.1%		SQ — 0.5% skin; 0.02% eye area

Continued

Table 9.2 Preservatives allowed in the European Union for cosmetics use according to Cosmetics Regulation (EC) 1223/2009 and its amendments: comparison with US FDA regulation (CIR conclusions)—cont'd

Name of ingredient	CAS number	EU Cosmetics Regulation (EC) 1223/2009 (Annex V) Current/upcoming limits on maximum concentration in ready-to-use preparation (%, w/w)	Conditions of use and warnings	CIR system conclusions (basis of preservative safe use in United States) Review conclusion[a]	Concentration (%, w/w) and other limitations
Benzalkonium chloride	8001-54-5, 63449-41-2, 91080-29-4,	0.1% (as benzalkonium chloride)	Avoid contact with eyes	SQ	0.1% free active ingredient
Benzalkonium bromide				—	—
Benzalkonium saccharinate	68989-01-5, 68424-85-1, 68391-01-5, 61789-71-7, 85409-22-9			—	—
Benzylhemiformal	14548-60-8	0.15%	Rinse-off products Not to be used in oral and lip products	—	—
Iodopropynyl butylcarbamate	55406-53-6	1. 0.02% in rinse-off products 2. 0.01% in leave-on products 3. 0.0075% in deodorants/antiperspirants	1. Not to be used in products for children under 3 years of age, except in bath products/shower gels and shampoo 2. Not to be used in body lotion and body cream 3. Not to be used in products for children under 3 years of age	SQ	Safe for use at 0.1%; should not be used in products intended to be aerosolized

| Ethyl lauroyl arginate HCl | 60372-77-2 | 0.4%[k] | Not to be used in aerosol dispensers (sprays) Not to be used in oral and lip products | — | — |
| Citric acid (and) silver citrate | 460-890-5 | 0.2%, corresponding to 0.0024% of silver[j] | Not to be used in oral and eye products | — | — |

BuP, butylparaben; *MEA*, monoethanolamine; *TEA*, triethanolamine.

[a]*I*, the available data are insufficient to support safety; *S*, safe in the present practices of use and concentration; *SQ*, safe for use in cosmetics, with qualifications; *U*, the ingredient is unsafe for use in cosmetics; *UNS*, ingredients for which the data are insufficient and their use in cosmetics is not supported; *Z*, the available data are insufficient to support safety, but the ingredient is not in current use.

[b]Safe in the present practices of use and concentration (rinse–off products only) when formulated to be non-irritating.

[c]When used as a salicylate ingredient, safe when formulated to avoid irritation and to avoid increasing sun sensitivity, or when increased sun sensitivity would be expected, directions for use include the daily use of sun protection.

[d]Should not be used in cosmetic products in which *N*-nitroso compounds may be formed.

[e]'In Minnesota it must not be used in personal care products for children under 8 years of age, including shampoos, lotions and gels' (Minnesota Statutes, Chapter 325F: Consumer Protection; Products and Sales, Sections 325.176–178: Formaldehyde in children's products).

[f]Commission Regulation (EU) No. 1004/2014 of 18 September 2014.

[g]Commission Regulation (EU) No. 358/2014 of 9 April 2014.

[h]Commission Regulation (EU) No. 288/2015. Pending approval.

[i]Commission Regulation (EU) No. 1003/2014 of 18 September 2014. *Public consultation on MI*: (1) Banning MI in leave-on products. (2) Waiting for the final SCCS opinion on rinse-off and hair leave-on products, which is expected at the end of September 2015, before reducing the concentration from 0.01% to 0.0015%.

[j]Commission Regulation (EU) No. 866/2014 of 8 August 2014. Corrigendum, OJ L 254, 28.8.2014, p. 39 (866/2014).

[k]Commission Regulation (EU) No. 344/2013 of 4 April 2013.

drawbacks arising from the cosmetic matrix interference. Combinations of these procedures with chromatographic and electrophoretic techniques, coupled to different selective detectors, have been proposed as powerful tools to successfully tackle cosmetic preservative quality control. These new analytical developments are reviewed in the following subsections.

In addition to the methods reported in the literature, there also exist some official methods described in various EU directives for the analysis of preservatives in cosmetics, which were compiled in a book edited by the European Commission (1999). More recently, an official method has been reported for the determination of the preservative IPBC in cosmetics samples (EN 16343:2013). However, these methods still cover a relatively small number of ingredients, and they are not sufficient to carry out the necessary control of all the preservatives regulated by the EU Cosmetics Regulation at the time of this writing.

Table 9.3 summarizes the analytical methods reported in the literature for the determination of preservatives in cosmetics and personal care products since 2006. Those published before were reviewed in the first edition of this book (Salvador and Chisvert, 2007). For easy consultation the reported methods are organized by determination and extraction techniques, including the target preservatives, the cosmetic matrix, the extraction and determination approach and, when available, the analytical features for each method, in terms of recoveries (%), limit of detection (LOD) and relative standard deviation (RSD, %).

Analytical Techniques for Preservative Determination

The complexity of cosmetic matrices very often requires a sample preparation step to successfully accomplish the multi-component analysis of a broad variety of ingredients. Liquid chromatography (LC), thin-layer chromatography and electrophoretic techniques have been traditionally employed for separation, identification and quantification of cosmetic preservatives.

Nowadays, an increasing number of methods based on gas chromatography (GC) can be found in the literature, although the predominant methodologies for preservative determination are based on LC. Both techniques, along with capillary electrophoresis (CE) and micellar electrokinetic chromatography (MEKC), have been widely used for preservative determination in cosmetics, as evidenced in Table 9.3. Among the different classes of cosmetic preservatives, parabens are, by far, the compounds to which more analytical efforts have been devoted, whereas for the analysis of other preservatives the number of available methods is more limited.

Liquid Chromatography

Methods based on reversed-phase LC coupled to different detectors are the most used for the analysis of different classes of preservatives. UV detectors are very popular for LC

Table 9.3 Analytical methods for preservative determination in cosmetic products (from 2005 to 2016)

Analyte	Matrix	Remarks on sample preparation	Analytical technique	Analytical features	References
LC-based methods					
4-HB, MeP, EtP, PrP, i-PrP, BuP, BA	Shampoos, hand lotions, creams, bath foam	—	HPLC-UV	LOD: 25–250 ng/mL	Memon et al. (2005)
MI, CMI, BzOH, SB, MeP	Facial tonics, mouthwashes, toothpastes, creams, body lotions, shaving gels, shower gels, body oils, body sprays and face masks	UAE with MeOH	HPLC-UV	Recoveries: 69%–119% LOD: 0.06–4.38 µg/mL RSD: 1.30%–5.92%	Baranowska et al. (2014)
MeP, EtP, PrP, BuP, i-BuP	Toothpaste	SPE with C_{18} cartridges	HPLC-UV	Recoveries: 86%–113% LOD: 0.1–0.3 mg/L RSD <6.6%	Zotou et al. (2010)
TCS	Body wash cream, moisturizing face wash cream and handwash cream	IT–USA–SI–LLME with isopropanol and ammonium sulphate	HPLC-UV	Recoveries: 90.4%–98.5% LOD: 0.09 ng/mL RSD: 0.8%–5.3%	Chen et al. (2013)
MeP, EtP, PrP	Water-soluble cosmetics	MW-assisted IL-DLLME	HPLC-UV	Recoveries: 68.3%–124.5% LOD: 0.6–1.2 mg/L RSD: 4.9–5.1%	Cheng et al. (2011)
4-HB, MeP, EtP, PrP	Toothpaste, cream, shampoo (wastewater)	Two-phase DHF–LPME with octanol	HPLC-UV	Recoveries: 3.0%–16.2% LOD: 2–5 µg/L RSD: 4.1%–6.3%	Esrafili et al. (2014)

Continued

Table 9.3 Analytical methods for preservative determination in cosmetic products (from 2005 to 2016)—cont'd

Analyte	Matrix	Remarks on sample preparation	Analytical technique	Analytical features	References
4-HB, MeP, EtP, PrP	Toothpaste, cream, shampoo (wastewater)	Three-phase DHF–LPME with n-octanol and basic aqueous solution	HPLC–UV	Recoveries: 5.8%–28.5% LOD: 0.5–2 µg/L RSD: 4.3%–5.8%	Esrafili et al. (2014)
MeP, EtP, PrP	Sunblock, skin cream, aftershave	SFVCDME with decanoic acid and tetrabutylammonium	HPLC–UV	Recoveries: 92.2%–108.8% LOD: 0.2–0.5 µg/L RSD: 3.9%–11.9%	Moradi and Yamini, (2012)
MeP, EtP, PrP, 4-HB, BA	Toothpaste, shampoo, sun protection cream	USAEME with MeOH and buffer solution	HPLC–UV	Recoveries: 22.6%–102.1% LOD: 0.25–8.30 mg/L RSD ≤9.8%	Yamini et al. (2012)
MeP, EtP, PrP, BuP, i-PrP, i-BuP	—	—	UPLC–UV	Not provided RSD <1%	Pedjie (2010)
MeP, EtP, PrP, BuP	Body creams, antiperspirant creams, sunscreens	SBSE with PDMS stir bar	UPLC–UV	Recoveries: 17%–99% LOQ: 30–200 ng/mg RSD <5%	Melo and Queiroz (2010)
Phenethyl alcohol, methylpropanediol, phenylpropanol, caprylyl glycol, ethylhexylglycerin	Moisturizing creams, sunscreen creams, gels, bath gels	VALLsME	HPLC–UV	Recoveries: 84%–118% LOQ: 0.02–0.06 µg/mL RSD: 3.9%–9.5%	Miralles et al. (2016)

Analytes	Matrix	Sample preparation	Technique	Analytical performance	References
MI, CMI, BzOH, PS, SB, MeP	Creams, gels, lotions, masks, syrup, face tonic	UAE with MeOH	HPLC–UV	Recoveries: 69%–119% LOD: 0.15–5.3 µg/mL RSD <4.6%	Baranowska and Wojciechowska (2013)
BA, SA, MeP, EtP, PrP, BuP	Cosmetics (not specified)	UAE Sonication with MeOH:water (60:40 v/v), centrifugation and filtration	HPLC–UV	Recoveries: 83%–117% LOD: 15–200 ng/mL RSD <3.6%	Gao et al. (2012)
MeP, EtP, PrP, BzP, BA, SOA	Creams and lotions	UAE Sonication with 60% MeOH and 40% water (v/v), centrifugation and filtration	HPLC–UV	Recoveries: 86.5%–116.3% LOD: 0.05–1.0 µg/mL RSD <5.0%	Gao and Legido-Quigley (2011)
MeP, EtP, PrP, BzP	Sunblocks, lotions, creams	SPME PEG-DA fibre coating	HPLC–UV	Recoveries: 90.2%–97.7% LOD: 0.12–0.15 mg/mL RSD: 5.4%–6.9%	Fei et al. (2011)
MeP, EtP, PrP, BuP	Shampoo, body lotion, shower gel, facial cream, body lotion	UAE Vortex mixing with MeOH, sonication, centrifugation and dilution in mobile phase (SDS)	HPLC–micellar mobile phase (SDS)–UV	Recoveries: 92.4%–109.2% LOD: 0.04–0.1 mM RSD <3%	Youngvises et al. (2013)
MeP, EtP, PrP, i-PrP, i-BuP, BuP, BA, SA, SOA, DHA, PhEtOH	Cosmetic lotions, milky lotions, and cosmetic creams	UAE Sample dilution in MeOH, sonication, centrifugation and filtration	HPLC–UV	Recoveries: 92.8%–111.9% LOD: 0.20–1.0 µg/mL RSD <4.4%	Aoyama et al. (2014)

Continued

Table 9.3 Analytical methods for preservative determination in cosmetic products (from 2005 to 2016)—cont'd

Analyte	Matrix	Remarks on sample preparation	Analytical technique	Analytical features	References
MeP	Shampoo, shower gels, body lotions, balsams, body creams, sun creams, make-up removers	UAE Sample sonication with MeOH and filtration	UPLC–UV	Recoveries: 97.4%–102.6% LOD: 0.02 ng/µL RSD <2.3%	Mincea et al. (2009b)
MeP, EtP, PrP, BuP	Hair balsam, shower gel, cream	UAE Ultrasonication with methanol	UPLC–UV	Recoveries: 91.4%–105.8% LOD: 2.25–4.82 ng/mL	Mincea et al. (2009a)
MI, CMI, BzOH, BRP, PhEtOH, MeP, EtP, MeBz, i-PrP, PrP, 4-Cl-3-MePhOH, EtBz, 2-PhPhOH, i-BuP, BuP, 4-Cl-3,5-diMePhOH, PhBz, 2,4-diCl-3,5-diMePhOH, 2-Bz-4-ClPhOH, TCS, TCC	Creams, lotions, lipsticks	UAE with MeOH	UPLC–UV	Recoveries: 90.5%–97.8% LOD: 0.05–3.85 µg/mL RSD <3.2%	Wu et al. (2008)
MI, CMI	Shampoos, conditioners, skin care lotions, gels, moisturizers	Sample dilution (20% H$_2$O in MeOH), agitation and centrifugation, mixing with 50:50 H$_2$O:MeOH and filtration	HPLC–MS/MS	Recoveries: 93%–111% LOQ: 0.1 µg/g RSD <7%	Wittenberg et al. (2015)

Preservatives in Cosmetics: Regulatory Aspects and Analytical Methods 203

Analytes	Samples	Sample preparation	Technique	Analytical parameters	Reference
MeP, EtP, PrP, i-PrP, BzP, BuP, TCS, BHA, BHT	Foundations, lipstick, deodorant, hand lotion, hand sanitizer, toothpaste	UAE Sample sonication (1:1 (v/v) MeOH:ACN) for 10 min, centrifugation, supernatant filtration (0.2 µm)	HPLC–MS/MS	LOD: 0.91–4.19 µg/mL	Myers et al. (2015)
MI, CMI	Shampoo, face cleansing gel, dental cream, hair mask, baby liquid soaps, baby bath gels, baby soft shampoo, hand cream, fluid make-up, hair gel, baby body milks	MSPD	HPLC–MS/MS	Recoveries >80% (60% for MI) LOD: 0.0066–0.060 µg/g RSD <7%	Alvarez-Rivera et al. (2012)
MeP, EtP, PrP, BuP, BHA, BHT	Lanoline cream, skin milk, cream	SFE (MeOH as collection solvent)	HPLC–MS/MS	Not provided LOD: 4.7–142 ng g^{-1} RSD <18%	Lee et al. (2006)
MeP, EtP, PrP, BuP	Make-up, creams, shampoo, lotion, after sun	UAE Sample (1.0 g) dilution in ultrapure water (between 1:20 and 1:100, v/v), sonication and centrifugation. SPE using multi-walled carbon nanotubes	HPLC–CCAD	Recoveries: 82%–104% LOD: 0.5–2.1 mg/L RSD: 3.3%–7.6%	Márquez-Sillero et al. (2010)
MeP, EtP, PrP, BuP	Wash-off cosmetics	UAE Sample dilution in MeOH, sonication and filtration (0.22 µm)	HPLC–CL	Recoveries: 93.3%–105.9% LOD: 1.9–5.3 ng/mL RSD <4.5%	Zhang et al. (2005)

Continued

Table 9.3 Analytical methods for preservative determination in cosmetic products (from 2005 to 2016)—cont'd

Analyte	Matrix	Remarks on sample preparation	Analytical technique	Analytical features	References
MeP, EtP, PrP	Shampoo	SPE C$_{18}$ cartridges and acetonitrile for elution	HPLC–ECD	Recoveries: 93.1%–104.4% LOD: 0.01% (w/w) RSD: 2.3%–9.8%	Martins et al. (2011)
BRX, BRP	Shampoo, hand soap, body wash	UAE Sample dilution in mobile phase and vortex mixing, sonication, centrifugation and filtration	UPLC–ICP–MS	Not provided LOD: 3.3 μg Br/L RSD <2.2%	Bendahl et al. (2006)
Phenethyl alcohol, methylpropanediol, phenylpropanol, caprylyl glycol, ethylhexylglycerin	Moisturizing creams, sunscreen creams, gels, bath gels				

GC-based methods

Analyte	Matrix	Remarks on sample preparation	Analytical technique	Analytical features	References
MeP, EtP, PrP, BuP	Face wash, fairness cream, shaving cream, moisturizers, hair gel	UAE Vortex mixing with MeOH, sonication and centrifugation; isobutyl chloroformate derivatization and pre-concentration using DLLME	GC–FID Confirmation by GC–PCI–MS	Recoveries: 86.4%–101.4% LOD: 0.029–0.102 μg/mL RSD: 1.12%–6.86%	Jain et al. (2013)
MeP, EtP, PrP	Mouthwash solution, toothpaste, shampoo	AALLME Butylchloroformate as derivatization agent/ extraction solvent	GC–FID	Recoveries: 59%–116% LOD: 0.41–0.62 mg/L RSD <4.9%	Farajzadeh et al. (2013)

Analytes	Sample	Extraction	Determination	Analytical parameters	Reference
MeP, EtP, PrP,	Mouth rinse solution	DLLME (octanol)	GC-FID	Recoveries: 25%–72% LOD: 0.005–0.015 µg/mL RSD: 2%–3%	Farajzadeh et al. (2010)
BuP, PrP, EtP, MeP	Perfumes	Sample dilution in EtAc	GC-MS	Recoveries >88% LOD: 0.016–0.50 2 µg/g RSD: 1.6%–6.5%	Sanchez-Prado et al. (2011b)
MeP, EtP, PrP, BuP	Hair sprays, perfumes, deodorants, cream, lotion	UAE Sonication-assisted extraction with methanol and clean-up with C_{18} SPE	GC-MS	Recoveries: 85%–108% LOD: 10–200 µg/kg RSD: 4.2%–8.8%	Shen et al. (2007)
PhEtOH, BRX, MeP, EtP, PrP, BzP, BHA, BHT, TCS, i-PrP, i-BuP	Shampoo, shower gel, body milk, sunblock, among others, products intended for babies	In-vial MSPD	GC-MS	Recoveries: 80%–110% LOD: 0.0053–0.0595 µg/g (IPBC 0.150 µg/g) RSD <15%	Celeiro et al. (2014b)
MeP, EtP, PrP, BuP, i-PrP, i-BuP, BzP, BRX, BRP, BHA, BHT, IPBC, TCS	Body milk, moisturizing creams, anti-stretch mark creams, hand creams, make-up, sun milk, deodorant, shampoos and liquid soaps, hand soaps, etc.	MSPD Sorbent, Florisil; solvent, hexane:acetone (1:1); extracts acetylation	GC-MS	Recoveries >78% LOD: 0.15–11 ng/mL RSD <10%	Sanchez-Prado et al. (2011a)

Continued

Table 9.3 Analytical methods for preservative determination in cosmetic products (from 2005 to 2016)—cont'd

Analyte	Matrix	Remarks on sample preparation	Analytical technique	Analytical features	References
BRX, PhEtOH, MeP, BHA, BHT, EtP, i-PrP, PrP, IPBC, i-BuP, BuP, TCS, BzP	Baby wipes and wet toilet paper	PLE (MeOH, 110°C, 5 min)	GC-MS	Recoveries: 80%–115% LOD: 0.00077–0.051 µg/g (0.0030–0.200 µg/wipe) RSD <10%	Celeiro et al. (2015)
BRX, BRP, MeP, EtP, PrP, BzP, BHA, BHT, i-PrP, i-BuP, BHA, BHT, IPBC, TCS	Moisturizing and anti-wrinkle creams and lotions, hand creams, sunscreen and after-sun creams, baby lotions, products for hair care	PLE In-cell acetylation Dispersant, Florisil; solvent, ethyl acetate	GC-MS	Recoveries: 74%–110% LOD: 0.041–1.5 µg/g RSD <10%	Sanchez-Prado et al. (2010)
MeP, EtP, PrP, i-PrP, BuP, i-BuP, BHA, BHT	Emulsion, lotion, body cream	SFE–(HS-SPME) In situ derivatization (N,O-bis(trimethylsilyl)trifluoroacetamide with 0.1% trimethylchlorosilane)	GC-MS	Recoveries: 91.2%–161.9% LOD: 0.5–8.3 ng/g RSD <7.8%	Yang et al. (2010)
MeP, EtP, PrP, BuP, i-PrP, i-BuP, BHA, BHT, TBMP	Cosmetics (not specified)	SPME Silica fibre coated with polyacrylate	GC-MS	Recoveries: 83%–98% LOD: 0.4–8.5 ng/mL RSD <16%	Tsai and Lee (2008)
4-HB, MeP, EtP, PrP	Toothpaste, cream, shampoo (wastewater)	Three-phase DHF–LPME (n-dodecane and acetonitrile)	GC-MS	Recoveries: 8.4%–31.3% LOD: 0.01–0.2 µg/L RSD: 3.9%–6.0%	Esrafili et al. (2014)

Analytes	Sample	Extraction technique	Analytical technique	Analytical parameters	Reference
MeP, EtP, PrP, BuP, i-PrP	Make-up removing gel, mouthwash solution, hair gel	SDME In-syringe derivatization with N,O-bis(trimethylsilyl) acetamide	GC–MS	Recoveries: 92.4%–104.5% LOD: 1–15 ng/L RSD: 8.1%–13%	Saraji and Mirmahdieh (2009)
PhEtOH, BRX, MeP, EtP, PrP, BzP, BHA, BHT, TCS, i-PrP, i-BuP	Shampoos, toothpaste, shower gel, liquid soap, baby moisturizing lotion, body milks, sunblock, lipstick, gloss lipstick, deodorants, nail polish remover, regenerative cream	Micro-MSPD	GC–MS and GC–MS/MS	Recoveries: 83%–115% LOD: 0.006–0.100 µg⁻¹ (GC–MS analysis) and 0.00050–0.037 µg/g (GC–MS/MS analysis). RSD <10%	Celeiro et al. (2014a)
MeP, EtP, PrP, BuP	Cosmetics (not specified)	UAE Dilution in MeOH followed by vortexing, sonication, centrifugation, and filtration	GC–MS/MS	Recoveries: 97%–107% RSD: 0.45%–6.42%	Wang and Zhou (2013)
MeBz, EtBz, BuBz, PhBz, PhEtOH, BRX, MeP, EtP, PrP, BzP, BHA, BHT, TCS, i-PrP, i-BuP	Body milks, baby body milks, moisturizing cream, deodorants, sunscreen, baby after sun, moisturizing lotion, make-up, eye make-up remover, liquid soap, child bath gel, toothpaste, shampoo, hair conditioner, facial cleansing milk, aftershave	SPME 40°C, NaCl (20%, w/v), DVB/CAR/PDMS fibre coating	GC–MS/MS	Recoveries >85% LOD: 0.000092% (w/w) (0.00091% (w/w) for Bronidox) RSD <13%	Alvarez-Rivera et al. (2014))

Continued

Table 9.3 Analytical methods for preservative determination in cosmetic products (from 2005 to 2016)—cont'd

Analyte	Matrix	Remarks on sample preparation	Analytical technique	Analytical features	References
BRX	Shampoo, bath oil, facial exfoliant and gel	SPME PDMS/DVB fibre coating, 10 mL of diluted cosmetic, 20% NaCl, room temperature	GC–ECD	Recoveries ≥70% LOD: 2.2×10^{-5} (%, w/w) RSD <10%	Fernandez-Alvarez et al. (2010)
MeP, EtP, PrP, i-PrP, BuP	Mouthwash solution and hand cream	SPE Solvent-assisted dispersive micro-SPE; sonication with hexyl acetate (15 μL) as a solvent and aminopropyl-functionalized MNPs (5 μg) as a sorbent; ACN was added as a desorption solvent; derivatization (acetylation)	GC–PID	Recoveries: 87%–103% LOD: 50–300 ng/L RSD <8%	Abbasghorbani et al. (2013)

CE and MEKC-based methods

Analyte	Matrix	Remarks on sample preparation	Analytical technique	Analytical features	References
BZT and cetylpyridinium ions	Cosmetic powder sample and mouthwash	Cosmetic powder, dilution in deionized water and filtration; mouthwash, dilution in ACN–water	CE	Recoveries: 98%–102% LOD: 1.47–4.30 g/mL RSD <0.3%	Oztekin and Erim (2005)
MeP, EtP, PrP, BuP	Ointments, lotions	Water-based lotions, dilution with ACN; ointment and oil-based lotions, sonication with ACN and centrifugation	CZE	LOD: 0.06–0.15 mg/mL	Huang et al. (2006)

208   Analysis of Cosmetic Products

MeP, EtP, PrP, BuP	Cosmetics (not specified)	SPE Graphene cartridge and MeOH	CZE	LOD: 0.10–0.15 mg/L	Ye et al. (2013)
MeP, EtP, PrP, BuP	Shampoo, hair dyes, toothpaste, aftershave gel	SPE LC–18 solid-phase and MeOH	CZE	LOD: 1.42–2.86 mM	Uysal and Güray (2008)
MeP, EtP, PrP, BuP, SOA, BA, SA		DLLME (dispersive solution of trichloromethane and isopropyl alcohol)	HPCE–UV	Recoveries: 71.1%–112.6% LOD: 0.200–0.375 mg/kg RSD: 2.22%–4.89%	Xue et al. (2013)
Formaldehyde and glyoxal	Skin care products, baby lotion, toothpaste	UAE Sample dilution in water and ultrasonication; derivatization reaction (TBA) 40 min at 75°C; centrifugation and filtration	Mini–CE–AD	Recoveries: 94%–105% LOD: 1.64–2.80 ng/mL RSD: 3.5 and 2.1–1	Li et al. (2014)
Triclosan	Toothpaste, lotion, facial cleanser	UAE Sample dilution in MeOH and ultrasonication	NACE–UV	Recoveries: 94.2%–97.7% LOD: 0.075 µg/mL	Ma et al. (2014)
MeP, EtP, PrP, BuP	Cosmetics (not specified)	Sample dilution in MeOH:water (3:2 v/v)	LVSS–MEKC	LOD: 8.0×10^{-5} M RSD <3%	He et al. (2006b)
MeP, EtP, PrP, BzP, i-PrP, i-BuP	Lotions	Sample dilution in MeOH; successive dilution in water (1000-fold)	MEKC	Recoveries: 81.0%–113.6% LOD: 4.32–7.78 nM RSD <2.96%	Wu et al. (2014)
MeP, EtP, PrP, BuP PhEtOH	Creams, water-based lotions, moisturizers	Sample dilution with ethanol and centrifugation	MEKC	Recoveries: 84.1%–103.0% LOD: 0.31–1.52 mg/mL RSD <4.5%	Huang et al. (2013)

Continued

Table 9.3 Analytical methods for preservative determination in cosmetic products (from 2005 to 2016)—cont'd

Analyte	Matrix	Remarks on sample preparation	Analytical technique	Analytical features	References
MeP, EtP, BuPi-BuP, SOA, SA, BA, DHA	Creams	Sample vortex mixing with MeOH and ultracentrifugation; drying and reconstitution with deionized water	MEKC	LOD: 0.005–2 μg/mL	Cheng et al. (2012)
MeP, EtP, PrP, BuP, SOA, SA, BA	Shampoos, gels, soaps, perfumes, creams	UAE Sample dilution in 25% MeOH, ultrasonication and filtration (0.45-μm nylon filters)	MEKC	Recoveries: 89%–115% LOD: 1.10–11.04 μg/mL RSD: 2.4%–16.7%	Lopez-Gazpio et al. (2014)
MeP, EtP, PrP, BuP, IMU, BzOH, DHA, SOA, PhEtOH, BA, SA	Cosmetics (not specified)	UAE Dilution in MeOH, vortex mixing and ultrasonication	MEKC	Recoveries: 93.0%–102.7% LOD: 20–100 mg/g RSD <5%	Wang et al. (2012)
MeP, EtP, PrP, BuP	Creams, lotions, gels	FIA-SPE	MEKC	Recoveries: 92.2%–102% LOD: 0.07–0.1 mg/mL RSD <2.3%	Han et al. (2008)
MI, SOA, TCS, BHA, BHT, MeP, EtP, PrP, BuP, SA, BA	Shampoos, gels, soaps, perfumes, creams, toothpastes	UAE Sample dilution in MeOH, agitation, ultrasonication, filtration	MEKC–UV	Recoveries: 90%–115% LOD: 0.91–2.80 μg/mL RSD <9%	Lopez-Gazpio et al. (2015)
BHA, BHT, MeP, PrP, BuP	Facial mask, sunblock cream, facial cleanser, shower gel, two kinds of moisturizing lotions, two skin care oils	UAE Sample ultrasonication with anhydrous ethanol, centrifugation and filtration	MEKC–ED	Recoveries: 86.0%–101.6% LOD: $(1.1–12.0) \times 10^{-7}$ g/mL RSD: 0.9%–7.7%	Wang et al. (2010a)

Methods based on other determination techniques

MeP, phloroglucinol, 2,4-DHBA, SA, n-propyl gallate	Cleansing gel, callus corrector cream	Sample drying, ethyl acetate addition and ultrasonication; sample purification using SPE; solvent removal and reconstitution in ACN/water	FIA–CL	Recoveries: 92.9%–111.9% LOD: 0.052–4.1 μM RSD 2.8%–6.8%	Ballesta-Claver et al. (2011)
MeP, EtP, PrP, BuP	Cleansing mousse, cleansing towels	UAE Sonication with MeOH or ACN	FIA–CL	Recoveries: 92.0%–111.4% LOD: 0.02–0.04 μM RSD: 2.3%–9.3%	Ballesta Claver et al. (2009)
MeP, EtP, PrP, BuP	Hair foam, cleansing towels	Foam samples, LLE with diethyl ether; towel samples, sonication with ACN; gel samples, sonication with MeOH	FIA–UV	Recoveries: 89.0%–103.3% LOD: 11.2–33.7 μM RSD: 0.65%–3.55%	Garcia Jimenez et al. (2010)
MeP, EtP, PrP, BuP, BHA	Cleansing towels, hair mousse, toothpaste	UAE Towel sample cut into strips, ultrasonication in ACN and filtration; hair mousse, diethyl ether extraction; toothpaste, suspended in the carrier and filtered	FIA–UV	Recoveries: 90.9%–97.1% LOD: 0.02–1.60 μg/mL RSD: 1.14%–2.78%	García-Jiménez et al. (2007)
4-HB, MeP, EtP, PrP, BuP	Sun cream, sun milk, suntan oil, face cream, make-up, lipstick	—	DART–MS; confirmation by GC–MS	Not provided	Haunschmidt et al. (2011)
MeP, PrP	Creams	—	EC	Recoveries: 72%–122%	Behpour et al. (2015)

Continued

Table 9.3 Analytical methods for preservative determination in cosmetic products (from 2005 to 2016)—cont'd

Analyte	Matrix	Remarks on sample preparation	Analytical technique	Analytical features	References
BA, SA, SOA	Shampoos and a mouthwash	SPE C$_{18}$ cartridge and MeOH	IC–conductivity detector	Recoveries: 92.2%–98.7% LOD: 0.1–0.2 mg L RSD: 1.8%–4.9%	Sid et al. (2010)
PhPOH, benzyl-4-chlorophenol, hexetidine, MI, CBZ	Facial cream, sunscreen, moisturizer	—	MIPDI–MS	LOD: pg/mm^2	Zhao and Duan (2015)
MeP, EtP, PrP, BuP	In-lab prepared cosmetics	—	MIPs electrochemical sensor	Recoveries: 8.7%–101.8% LOD: 0.2–0.4 μM RSD <3.3% RSD 2.8%–6.8%	Wang et al. (2010b)

AALLME, air-assisted liquid–liquid microextraction; *ACN*, acetonitrile; *AD*, amperometric detection; *BA*, benzoic acid; *BHA*, butylated hydroxyanisole; *BHT*, butylated hydroxytoluene; *BRP*, 2-bromo-2-nitro-propane-1,3-diol; *BRX*, 5-bromo-5-nitro-1,3-dioxane; *BuBz*, butyl benzoate; *BuP*, butylparaben; *BzOH*, benzyl alcohol; *BzP*, benzylparaben; *BZT*, benzethonium; *CBZ*, climbazol; *CCAD*, corona-charged aerosol detector; *CE*, capillary electrophoresis; *CL*, chemiluminescence; *4-Cl-3-MePhOH*, 4-chloro-3-methylphenol; *4-Cl-3,5-diMePhOH*, 4-chloro-3,5-dimethylphenol; *CMI*, 5-chloro-2-methyl-3-isothiazolinone; *CZE*, capillary zone electrophoresis; *DHA*, dehydroacetic acid; *2,4-diCl-3,5-diMePhOH*, 2,4-dichloro-3,5-dimethylphenol; *2,4-DHBA*, 2,4-dihydroxybenzoic; *DHF–LPME*, dynamic hollow fiber-based liquid-phase microextraction; *DLLME*, dispersive liquid–liquid microextraction; *DVB/CAR/PDMS*, Divinylbenzene/Carboxen/Polydimethylsiloxane; *EC*, electrochemical detection; *ECD*, electron capture detector; *ED*, electrochemical detection; *EtAc*, ethyl acetate; *EtBz*, ethyl benzoate; *EtP*, ethylparaben; *FIA*, flow injection analysis; *FID*, flame ionization detector; *GC*, gas chromatography; *HS*, headspace; *4-HB*, 4-hydroxybenzoic acid; *HPCE*, high-performance capillary electrophoresis; *HPLC*, high-performance liquid chromatography; *i-BuP*, isobutylparaben; *IC*, ion chromatography; *ICP*, inductively coupled plasma; *IL*, Ionic liquid; *IMU*, imidazolidinyl urea; *IPBC*, iodopropynyl butylcarbamate; *i-PrP*, isopropylparaben; *IT–USA–SI–LLME*, in-tube-based ultrasound-assisted salt-induced liquid–liquid microextraction; *LC*, liquid chromatography; *LOD*, limit of detection; *LVSS*, large-volume sample stacking; *MeBz*, methyl benzoate; *MEKC*, micellar electrokinetic chromatography; *MeOH*, methanol; *MeP*, methylparaben; *MI*, 2-methyl-3-isothiazolinone; *MIPDI*, microwave-induced plasma desorption ionization; *MIP*, molecularly imprinted polymer; *MNP*, magnetite nanoparticle; *MS*, mass spectrometry; *MS/MS*, tandem mass spectrometry; *MSPD*, matrix solid-phase dispersion; *MW*, microwave; *NACE*, non-aqueous capillary electrophoresis; *PCI*, positive chemical ionization; *PEG-DA*, poly(ethylene glycol) diacrylate; *PhBz*, phenyl benzoate; *PhEtOH*, phenoxyethanol; *2-PhPhOH*, *o*-phenylphenol; *PhPOH*, 2-phenylphenol; *PID*, photoionization detector; *PLE*, pressurized liquid extraction; *PrP*, propylparaben; *PS*, potassium sorbate; *RSD*, relative standard deviation; *SA*, salicylic acid; *SB*, sodium benzoate; *SBSE*, stir bar sorptive extraction; *SDS*, sodium dodecyl sulphate; *SFE*, supercritical fluid extraction; *SFVCDME*, solidified floating vesicular co-acervative drop microextraction; *SOA*, sorbic acid; *SPE*, solid-phase extraction; *SPME*, solid-phase microextraction; *TBA*, 2-thiobarbituric acid; *TCC*, triclocarban; *TCS*, triclosan; *TBMP*, 2-tert-butyl methylphenol; *UAE*, Ultrasound assisted extraction; *USAEME*, ultrasound-assisted emulsification microextraction; *UPLC*, ultra-high-performance liquid chromatography; *UV*, ultraviolet; *VALLsME*, vortex-assisted liquid–liquid semi-microextraction; *2-Bz-4-ClPhOH*, 2-benzyl-4-chlorophenol.

coupling, although mass spectrometry (MS), chemiluminescence (CL), electrochemical (ED), corona-–charged aerosol (CCAD) and inductively coupled plasma MS (ICP–MS) detectors are also employed.

The identification and subsequent quantification of cosmetic preservatives can be a challenging task because of the possible presence of matrix components that might interfere in the analysis. Selective detectors based on MS allow solving the co-elution problems, leading to the unambiguous identification of target preservatives. However, the applications of LC–MS for preservative determination in cosmetic products are still quite limited (Ocaña-Gonzalez et al., 2015).

Lee et al. (2006) and Myers et al. (2015) developed determination methods coupling LC to a triple-quadrupole (QqQ) mass analyser. Operating in both positive and negative ionization modes, several preservatives, including parabens and triclosan (TCS), and antioxidants such as BHA and BHT, were determined in cosmetic samples of different natures.

Isothiazolinone-type preservatives are LC-amenable compounds that can be accurately analysed by LC–MS/MS (QqQ). Thus, operating in positive electrospray ionization mode, Wittenberg et al. (2015) determined MI and CMI in cosmetic samples after simple sample pre-treatment based on dilution, agitation and centrifugation, whereas Alvarez-Rivera et al. (2012) proposed MSPD as an extraction procedure before the analysis of MI, CMI, BzI and OI in cosmetic and household products. Lower detection limits were achieved using MSPD (0.0066–0.060 µg/g) compared to the sample dilution-based approach (0.1 µg/g).

Additionally, an official method has been reported for the determination of the preservative IPBC in cosmetics samples by LC–MS/MS (EN 16343:2013).

Based on the chemiluminescent enhancement of the cerium(IV)–rhodamine 6G system in a strong sulphuric acid medium, parabens can be determined by high-performance LC (HPLC)–CL (Zhang et al., 2005). Amperometric detection on a boron-doped diamond electrode by HPLC–ED was proposed for the same parabens by Martins et al. (2011). In this case, an efficient sample clean-up must be performed before the analysis to avoid passivation on the electrode surface. LC coupled to CCAD has demonstrated its capability for the determination of parabens (Márquez-Sillero et al., 2010), providing signals that are dependent on the analytes' particle size, rather than on individual spectroscopic analyte properties.

Bromine-containing preservatives such as bronopol, bronidox and methyldibromo glutaronitrile can be selectively analysed by LC coupled to ICP–MS, monitoring the ^{79}Br and ^{81}Br isotopes simultaneously (Bendahl et al., 2006).

Gas Chromatography

Nowadays, GC–MS and GC–MS/MS couplings represent affordable tools, which are becoming more and more popular alternatives to LC methods for the analysis of cosmetic preservatives. Other detection systems such as the flame ionization detector (FID), electron capture detector (ECD) and photoionization detector (PID) are also employed, although to a lesser extent.

For the GC analysis of some cosmetic preservatives, derivatization is highly recommended to improve chromatographic performance (e.g., peak separation and peak symmetry) (Saraji and Mirmahdieh, 2009; Yang et al., 2010). Acetylation is an efficient derivatization procedure, commonly applied for the determination of phenolic preservatives such as parabens, TCS and BHA (Abbasghorbani et al., 2013; Llompart et al., 1997; Regueiro et al., 2009; Sanchez-Prado et al., 2010, 2011a) in cosmetic samples. Acetylation is a low-cost alternative compared to other derivatization procedures based on silylation agents such as N,O-bis(trimethylsilyl) trifluoroacetamide (Saraji and Mirmahdieh, 2009; Yang et al., 2010). Alkylation using butylchloroformate (Farajzadeh et al., 2013) and isobutylchloroformate (Jain et al., 2013) has been reported as another alternative to paraben derivatization.

Mass analysers based on single quadrupole (Celeiro et al., 2014a,b, 2015; Sanchez-Prado et al., 2011a,b, 2010; Shen et al., 2007) and QqQ (Alvarez-Rivera et al., 2014; Celeiro et al., 2014a; Wang and Zhou, 2013) are the most frequently used, followed by ion trap (Yang et al., 2010), in GC–MS and GC–MS/MS applications.

Conventional GC detectors have also been proposed, although, in most cases, confirmation requires MS. GC–FID has been described for paraben determination in cosmetics (Farajzadeh et al., 2010, 2013), although the use of this technique has decreased since 2006 owing to the lack of unambiguous identification compared with GC–MS. In some cases, even when GC–FID was proposed as the determination method (Jain et al., 2013), the identification of the analytes was confirmed using GC–positive chemical ionization–MS. Alternatively, Fernandez-Alvarez et al. (2010) proposed a selective GC approach using µECD for the determination of bronidox in rinse-off cosmetic products. The improvement in the sensitivity was demonstrated compared to GC–MS in selected-ion monitoring mode.

Electrophoresis

Electrophoretic techniques were also shown to be valuable tools for the determination of both charged and hydrophobic compounds in cosmetic products. CE was proposed for the analysis of parabens (Huang et al., 2006; Li et al., 2014; Ma et al., 2014; Uysal and Güray, 2008; Xue et al., 2013; Ye et al., 2013), as well as ionic preservatives (e.g., benzethonium and cetylpyridinium) (Oztekin and Erim, 2005), using both fused-silica and monolithic capillary columns.

MEKC can be an alternative for both ionic and neutral preservative separation, using a micellar medium by addition of a surfactant (sodium dodecyl sulphate) to the carrier buffer. The analysis of preservatives such as parabens, imidazolidinyl urea, benzyl alcohol, dehydroacetic acid (DHA), phenoxyethanol, sorbic acid (SOA), benzoic acid (BA), salicylic acid (SA), TCS and MI was carried out using this technique.

Some authors developed online sample stacking methods for sensitivity enhancement in cosmetic preservative analysis. Thus, He et al. (2006b) obtained 300 times lower detection limits using a large-volume sample stacking–MEKC method, compared to normal MEKC techniques. On the other hand, Cheng et al. (2012) proposed a strategy based on a large volume of sample and analyte concentration through an electro-osmotic flow pump and sweeping for the analysis of cosmetic preservatives, including hydrophobic and hydrophilic compounds such as SOA, SA, BA and DHA.

Other Determination Techniques

Other techniques, such as electrochemical detection, flow injection analysis (FIA), direct analysis in real time–MS (DART–MS) and microwave-induced plasma desorption ionization (MIPDI)–MS, are worth mentioning as analytical alternatives, which have also been implemented in cosmetic preservative determination, as discussed below.

Two innovative proposals based on electrochemical detection for paraben analysis are reported in the literature. Wang et al. (2010b) developed a paraben sensor using molecularly imprinted polymers on film on a glassy carbon electrode for the determination of total paraben content, whereas Behpour et al. (2015) reported the use of multivariate voltammetric calibration for the simultaneous determination of methylparaben (MeP) and propylparaben (PrP) in cosmetics.

Despite the limited resolution capacity compared to HPLC, FIA has also been employed in several works to increase sample throughput. Monolithic columns are frequently used, as they provide efficient separations under low-pressure conditions. Thus, two methods were developed using an ultra-short monolithic column and an FIA configuration with UV detection for the determination of a mixture of four parabens (MeP, ethylparaben (EtP), PrP, butylparaben (BuP)), among other preservatives in commercial cosmetics samples (García-Jiménez et al., 2007, 2010). Similarly, Ballesta Claver et al. (2009) proposed an FIA method applying CL detection. Using Ce(IV) in the presence of Rhodamine 6G as reactants, an intense CL signal was obtained to enable a sensitive determination of the four parabens and phenolic preservatives such as MeP, 4-HB and SA, among others with lower detection limits. The automation, miniaturization and low mobile phase and sample consumption are the main advantages of these methods.

The semi-quantitative determination of parabens and 4-HB, along with UV filters, directly from a sample, was carried out by using DART–MS (Haunschmidt et al., 2011). This new analytical technique does not require sample pre-treatment, as it allows direct sample measurement. MIPDI–MS was proposed for the analysis of several preservatives including phenylphenol, benzyl-4-chlorophenol, hexetidine, MI and climbazole. The results obtained show that the semi-quantitation of the target preservatives in cosmetics is possible with detection limits at the picograms/square millimetre level (Zhao and Duan, 2015).

Sample Preparation

Cosmetic sample dilution in suitable solvents and subsequent homogenization by agitation or vortex mixing is a widely reported sample preparation procedure for the analysis of cosmetic preservatives. In most of these approaches, ultrasound is commonly used to facilitate the extraction of compounds from the sample. The most used solvent is methanol, followed by acetonitrile, diethyl ether, water, ethanol and mixtures of methanol/acetonitrile or methanol/water.

Sometimes, the so-obtained extracts are not sufficiently homogeneous and clean. The insoluble matrix components may cause contamination in the determination system after a few analyses, and the possible presence of interference with the target analytes cannot be ignored. Therefore, the implementation of efficient sample preparation techniques, involving a clean-up step beyond centrifugation and/or filtration, is very often required, to remove the insoluble fraction of the cosmetic matrix and reach the desired selectivity and sensitivity of the analysis.

Solid-Phase Extraction

Several studies reported the use of SPE to remove interference from dirty solutions obtained after cosmetic sample dilution or sonication. For cosmetic samples, a pre-treatment step (dilution, homogenization and centrifugation or filtration) is mandatory before SPE to avoid clogging the cartridges.

SPE procedures strongly depend on the nature of the sorbent used. Two reversed-phase cartridges from different manufacturers were compared for the analysis of five paraben preservatives in toothpaste samples (Zotou et al., 2010). The cartridge Abselut Nexus provided significantly lower recoveries for all analytes compared LiChrolut RP-18. The clear supernatant obtained after toothpaste sample dispersion in methanol and subsequent centrifugation was passed through the latter cartridges for clean-up before HPLC–UV. C_{18} cartridges were also used to isolate benzoic acid, salicylic acid and sorbic acid from shampoos and mouthwash samples (Sid et al., 2010). The analytes were then eluted with methanol and the separation was performed by ion-exclusion chromatography. The analysis of parabens in cosmetic samples using C_{18} cartridges (Martins et al., 2011; Uysal and Güray, 2008) was also accomplished. In both works, samples were diluted in water and sonicated before SPE. An automated approach for online sample pre-treatment and determination of three parabens (i.e., methyl-, ethyl- and propylparaben) in water-based lotions and oil-based cream, gel and lotion was also proposed (Han et al., 2008). An FIA was used to couple an SPE C_8-bonded silica column with MEKC separation. Recoveries ranging from 92.2% to 102% were obtained, avoiding running buffer contamination and reducing buffer consumption.

New materials with higher loading capacities and higher efficiency for retention and selectivity have been proposed for the extraction of cosmetic preservatives like parabens. After sample dilution in ultrapure water, followed by sonication and centrifugation, SPE using multi-walled carbon nanotubes was applied for purification before analysis by HPLC coupled to a CCAD, obtaining satisfactory recoveries (82%–104%) (Márquez-Sillero et al., 2010). The main advantages of this method rely on the low amount of stationary phase required and the reusability of the cartridge, which notably reduces the cost of the analysis. Graphene was also demonstrated to be a suitable sorbent for the SPE of parabens in cosmetic samples, followed by determination by CE (Ye et al., 2013). This material exhibited excellent adsorption capacity for sample preparation leading to mean recoveries between 63% and 100%.

Dispersive SPE has been applied to the analysis of parabens (MeP, EtP, PrP, i-PrP, BuP) (Abbasghorbani et al., 2013). The extraction is carried out by dispersing the sorbent in the sample solution. The sample is then sonicated in a conical tube containing hexyl acetate and aminopropyl-functionalized magnetite nanoparticles as sorbent. Dispersed magnetite was collected in the bottom of the tube using a strong magnet, and the analytes were subsequently desorbed with acetonitrile. After acetylation, the extracts were analysed by GC–PID with recoveries in the range of 87%–103%. The advantage of this procedure is the reduction of the amount of sorbent and solvent required, as well as the operation time.

Matrix Solid-Phase Dispersion

MSPD has also been shown to be a suitable extraction procedure for cosmetic preservatives. This technique involves blending a viscous, solid or semi-solid sample with a solid support, where sample components dissolve and disperse into the bound organic phase on the surface of the particle, leading to complete sample disruption and dispersion (Barker, 2000; Kristenson et al., 2006). The possibility of performing extraction and clean-up at the same time together with procedural simplicity is the main advantage of this technique.

MSPD has been applied for the determination of multi-class preservatives including two bromine-containing preservatives, seven parabens, IPBC, TCS and the antioxidant preservatives BHA and BHT, in a wide variety of both rinse-off and leave-on cosmetic samples (Sanchez-Prado et al., 2011a). Florisil was selected as the dispersive agent and the MSPD column was eluted with hexane/acetone (1:1). After derivatization (acetylation), the extract was directly analysed by GC–MS. Another MSPD method has been described for the analysis of isothiazolinones, including MI and CMI, in cosmetics (Alvarez-Rivera et al., 2012). The most suitable extraction conditions comprise the use of Florisil as the dispersive phase and methanol as the elution solvent. Subsequently, the extract was readily analysed by HPLC–MS/MS. Recoveries were in general higher than >80, whereas lower extraction yields were obtained for MI (60%).

To reduce the amount of generated residues and solvent consumption, two miniaturized MSPD approaches based on micro-MSPD (Celeiro et al., 2014a) and in-vial micro-MSPD (Celeiro et al., 2014b) were proposed. Using Pasteur glass pipettes as micro-MSPD columns, the procedure involved the use of only 0.1 g of sample and a small amount of dispersing sorbent (Florisil). Elution with 1 mL of ethyl acetate allowed a solvent volume reduction of five times, compared to the normal MSPD approach. GC–MS and GC–MS/MS analysis of the extracts showed mean recoveries of 90% and RSD values generally below 10%. The in-vial micro-MSPD allowed analyte extraction in less than 5 min, leading to a quick and low-cost extraction procedure without losses of very volatile compounds.

Pressurized Liquid Extraction and Supercritical Fluid Extraction
PLE is an efficient extraction procedure, which increases automation of the analysis, decreases the amount of organic solvents and allows controlling the selectivity of the extraction by loading different sorbents into the extraction cell.

The analysis of multi-class preservatives in leave-on cosmetics has been performed using a one-step sample preparation methodology applying PLE (Sanchez-Prado et al., 2010). Direct cosmetic preservative acetylation was carried out by simply adding the derivatization reagents (acetic anhydride and pyridine) into the PLE cell. Under the optimized conditions the sample was mixed with Florisil as the dispersing sorbent and extracted with ethyl acetate, operating at 120°C for 15 min. A multi-component method, based on PLE, was developed for the simultaneous determination of different cosmetic ingredients, including preservatives in wipes and wet toilet paper for children (Celeiro et al., 2015). The method exhibited a satisfactory performance with mean recoveries of 90% working with MeOH as the extraction solvent at 110°C for 5 min.

Supercritical fluid extraction (SFE) was proposed for the determination of several parabens, along with antioxidant preservatives in creams and skin milk samples (Lee et al., 2006). Operating with CO_2 supercritical fluid at high pressure (14,000 kPa) and 65°C, low detection limits were achieved (4.7–142 ng/g).

Microextraction-Based Methods
Solid-Phase Microextraction
Introduced by Pawliszyn's group in 1990, SPME is one the most miniaturized sample preparation techniques. The easy coupling of SPME with GC, and to a lesser extent with LC, allows complete automation of the extraction process.

Operating in direct sampling mode very often requires the dilution of the cosmetic sample to prevent fibre damage. Tsai and Lee (2008) evaluated the performance of different commercial fibre coatings (polyacrylate (PA), PDMS, PDMS/DVB, CAR/PDMS and CW/DVB) for the extraction of phenolic preservatives (parabens) and antioxidants in cosmetic samples. In this study, a simple pre-treatment that consisted in sample dilution with methanol was applied to improve matrix dispersion for subsequent dilution in water. PA fibres yielded the higher extraction efficiencies.

Coupling SFE and online headspace SPME to GC–MS, Yang et al. (2010) developed a hyphenated methodology for the determination of paraben preservatives and polyphenolic antioxidants in cosmetics. After SFE, preservatives were in situ silylated and the resulting derivatives were adsorbed 'online' on a PA SPME fibre exposed to the headspace, being subsequently analysed in the GC–MS system.

Direct SPME followed by GC–ECD was used for the selective determination of bronidox in rinse-off cosmetic samples (Fernandez-Alvarez et al., 2010). The samples diluted in water were extracted with a PDMS/DVB fibre in direct sampling mode. This brominated preservative was also determined, along with other regulated cosmetic preservatives, including benzoates, 2-phenoxyethanol, parabens and TCS, among others, by SPME in cosmetics (Alvarez-Rivera et al., 2014). In this method, in situ acetylation and subsequent organic modifier addition were implemented in the SPME process as an effective extractive strategy for matrix effect compensation and chromatographic performance improvement. This multi-component analytical method exhibited quantitative recoveries as well as high selectivity and sensitivity using GC–MS/MS as the determination technique.

An alternative to commercial SPME coatings was proposed for direct SPME of parabens in cosmetic samples (Fei et al., 2011). The new fibre coating was prepared through UV-induced polymerization of poly(ethylene glycol) diacrylate (PEG-DA). The PEG-DA polymer thin film, covalently attached to the fibre, was homogeneous and wrinkled, leading to a high surface area and high extraction efficiency.

Melo and Queiroz (2010) demonstrated the feasibility of stir bar solid extraction (SBSE) for the extraction and pre-concentration of parabens from cosmetic products. Liquid desorption was carried out under sonication at room temperature.

Dispersive Liquid–Liquid Microextraction

Compared to traditional liquid extraction, dispersive liquid–liquid microextraction (DLLME) allows a rapid mass transfer between the sample and the extraction solvent, leading to a higher pre-concentration and extraction efficiency (Rezaee et al., 2006). A DLLME method using a mixture of octanol and acetone as extraction and dispersive solvents, respectively, was developed (Farajzadeh et al., 2010). The method allowed the extraction and pre-concentration of some preservatives including methyl-, ethyl- and propylparaben from a mouth rinse solution, among other samples. After centrifugation, a portion of the collected solvent was removed from the water surface by a capillary tube and analysed by GC–FID.

Using an extraction solvent denser than water (trichloromethane) in combination with isopropyl alcohol as the dispersive solvent, a DLLME method was developed for the determination of four parabens, SOA, BA and SA, in cosmetic products (Xue et al., 2013). A suitable amount of lower layer was collected and evaporated to dryness. The reconstituted residue was analysed by high-performance CE.

The inclusion of an in situ derivatization step in DLLME procedures for the analysis of parabens in cosmetic samples was described. Using a mixture of chloroform, ethanol, isobutylchloroformate and pyridine as extraction, disperser, derivatization agent and catalyst, respectively, Jain et al. (2013) obtained satisfactory recoveries of 86.4%–101.4%. This derivatization procedure overcomes the disadvantages of widely used silylation, as it does not require heating and is completed within 1 min directly in the aqueous medium. In addition, the use of butylchloroformate as an extraction/derivatization solvent and picoline as a catalyst was also reported (Farajzadeh et al., 2013), in which an air-assisted approach (sucking and expelling the mixture with a syringe) was applied for the cloudy solution formation.

A microwave-assisted modification was introduced by M. Cheng et al. (2011) for the analysis of three parabens in water-soluble cosmetic samples, using ionic liquids as extraction solvents. This new proposal was shown to accelerate the dispersion of the extracting solvent, speeding up the mass transfer between phases.

Despite its advantages in terms of low solvent consumption and as an efficient extraction alternative, the several manual manipulations involved in DLLME make this technique difficult to automate. In addition, DLLME very frequently requires the use of internal standards and surrogates (Ramos, 2012).

Other Microextraction-Based Methods

Other approaches based on LLME, such as ultrasound-assisted emulsification microextraction (USAEME) and in-tube-based ultrasound-assisted salt-induced LLME (IT–USA–SI–LLME), have also been applied to isolate preservatives from cosmetic samples. These methods use ultrasound energy to increase the speed of mass transfer between the two immiscible liquid phases. Yamini et al. (2012) implemented USAEME for the determination of parabens, sodium benzoate and p-hydroxybenzoic acid, whereas TCS was individually determined using IT–USA–SI–LLME (Chen et al., 2013). This technique is based on the rapid phase separation of water-miscible organic solvent from the aqueous phase owing to the salting-out phenomenon.

Vortex-assisted liquid–liquid semi-microextraction was proposed by Miralles et al. (2016) for the analysis of alternative preservatives such as phenethyl alcohol, methylpropanediol, phenylpropanol, caprylyl glycol and ethylhexylglycerin. Because of the low UV/VIS absorbance of the target compounds, chromophoric derivatization with benzoyl chloride was applied, enhancing their analytical response during the LC–UV determination.

Three different methods based on dynamic hollow fiber–based liquid-phase microextraction (DHF–LPME) were described for the extraction and pre-concentration of parabens (Esrafili et al., 2014). The first two methods are based on two-phase and three-phase DHF–LPME, using *n*–octanol and a mixture of *n*–octanol with basic aqueous solution, respectively, whereas the third method was a three-phase DHF–LPME approach using two immiscible solvents (i.e., n–dodecane and acetonitrile). Higher extraction efficiency and lower LODs were obtained in this case.

Single-drop microextraction (SDME) was applied to the extraction of five parabens (Saraji and Mirmahdieh, 2009). A single micro-drop of a water-insoluble solvent (hexyl acetate) suspended at the tip of a GC microsyringe is exposed in a pre-treated cosmetic sample (sonication, water dilution and filtration) followed by in-syringe derivatization with N,O–bis(trimethylsilyl)acetamide.

Based on the principles of SDME, solidified floating vesicular co-acervative drop microextraction was proposed for the determination of MeP, EtP and PrP in body creams, antiperspirant creams and sunscreens (Moradi and Yamini, 2012). This technique uses a supramolecular solvent (e.g., tetrabutylammonium-induced vesicles of decanoic acid). The floating droplets obtained after solvent solidification in an ice bath can be easily collected and analysed by HPLC–UV.

REFERENCES

Aalto-Korte, K., Alanko, K., Henriks-Eckerman, M.L., Kuuliala, O., Jolanki, R., 2007. Contact Dermat. 56, 160.
Abbasghorbani, M., Attaran, A., Payehghadr, M., 2013. J. Sep. Sci. 36, 311.
Alvarez-Lerma, F., Maull, E., Terradas, R., Segura, C., Planells, I., Coll, P., Knobel, H., Vazquez, A., 2008. Crit. Care 12, R10.
Alvarez-Rivera, G., Dagnac, T., Lores, M., Garcia-Jares, C., Sanchez-Prado, L., Lamas, J.P., Llompart, M., 2012. J. Chromatogr. A 1270, 41.
Alvarez-Rivera, G., Llompart, M., Garcia-Jares, C., Lores, M., 2015. J. Chromatogr. A 1390, 1.
Alvarez-Rivera, G., Vila, M., Lores, M., Garcia-Jares, C., Llompart, M., 2014. J. Chromatogr. A 1339, 13.
Aoyama, A., Doi, T., Tagami, T., Kajimura, K., 2014. J. Chromatogr. Sci. 52, 1010.
Ballesta-Claver, J., Valencia, M.C., Capitan-Vallvey, L.F., 2011. Luminescence 26, 44.
Ballesta Claver, J., Valencia, M.C., Capitán-Vallvey, L.F., 2009. Talanta 79, 499.
Baranowska, I., Wojciechowska, I., 2013. Pol. J. Environ. Stud. 22.
Baranowska, I., Wojciechowska, I., Solarz, N., Krutysza, E., 2014. J. Chromatogr. Sci. 52, 88.
Barker, S.A., 2000. J. Chromatogr. A 885, 115.
Behpour, M., Masoum, S., Lalifar, A., Khoobi, A., 2015. Sens. Actuators B 214, 10.
Bendahl, L., Hansen, S.H., Gammelgaard, B., Sturup, S., Nielsen, C., 2006. J. Pharm. Biomed. 40, 648.
Bergfeld, W.F., Belsito, D.V., Marks Jr., J.G., Andersen, F.A., 2005. J. Am. Acad. Dermatol. 52, 125.
Bester, K., Lamani, X., 2010. J. Chromatogr. A 1217, 5204.
BS EN 16343, 2013. Cosmetics. Analysis of cosmetic products. Determination of 3-iodo-2-propynyl butylcarbamate (IPBC) in cosmetic preparations. LC-MS Methods.
Celeiro, M., Guerra, E., Lamas, J.P., Lores, M., Garcia-Jares, C., Llompart, M., 2014a. J. Chromatogr. A 1344, 1.
Celeiro, M., Lamas, J., Llompart, M., Garcia-Jares, C., 2014b. Cosmetics 1, 171.
Celeiro, M., Lamas, J.P., Garcia-Jares, C., Llompart, M., 2015. J. Chromatogr. A 1384, 9.
Chen, M.-J., Liu, Y.-T., Lin, C.-W., Ponnusamy, V.K., Jen, J.-F., 2013. Anal. Chim. Acta 767, 81.
Cheng, M.X.C., Jiang, X., Zhang, H., 2011. Jingxi Huagong 28, 568.
Cheng, Y.C., Wang, C.C., Chen, Y.L., Wu, S.M., 2012. Electrophoresis 33, 1443.
CIR, 2016. Cosmetics Ingredients Review. http://www.cir-safety.org.
Cole, P., Adami, H.O., Trichopoulos, D., Mandel, J., 2010. Regul. Toxicol. Parmacol. 58, 161.
Collier, P.J., Ramsey, A.J., Austin, P., Gilbert, P., 1990. J. Appl. Bacteriol. 69, 569.
Crofton, K.M., Paul, K.B., Devito, M.J., Hedge, J.M., 2007. Environ. Toxicol. Pharmacol. 24, 194.
Darbre, P.D., 2006. Best pract. Res. Clin. Endocrinol. Metab. 20, 121.
Esrafili, A., Yamini, Y., Ghambarian, M., Moradi, M., 2014. Chromatographia 77, 317.

European Commission, 2014. Commission Regulation (EU) No 1004/2014 of 18 September 2014 Amending Annex V to Regulation (EC) No 1223/2009 of the European Parliament and of the Council on Cosmetic Products (Parabens).
European Commission, 1999. Directive 82/434/EEC. Cosmetics legislation. Cosmetic products – Methods of analysis.
Farajzadeh, M.A., Djozan, D., Bakhtiyari, R.F., 2010. Talanta 81, 1360.
Farajzadeh, M.A., Khosrowshahi, E.M., Khorram, P., 2013. J. Sep. Sci. 36, 3571.
Fei, T., Li, H., Ding, M., Ito, M., Lin, J.M., 2011. J. Sep. Sci. 34, 1599.
Fernandez-Alvarez, M., Lamas, J.P., Sanchez-Prado, L., Llompart, M., Garcia-Jares, C., Lores, M., 2010. J. Chromatogr. A 1217, 6634.
Fewings, J., Menné, T., 1999. Contact Dermat. 41, 1.
Gao, W., Gray, N., Heaton, J., Smith, N.W., Jia, Y., Legido-Quigley, C., 2012. J. Chromatogr. A 1228, 324.
Gao, W., Legido-Quigley, C., 2011. J. Chromatogr. A 1218, 4307.
Garcia-Gavin, J., Vansina, S., Kerre, S., Naert, A., Goossens, A., 2010. Contact Dermat. 63, 96.
García-Jiménez, J.F., Valencia, M.C., Capitán-Vallvey, L.F., 2007. Anal. Chim. Acta 594, 226.
Garcia Jimenez, J., Carmen Valencia, M., Capitan-Vallvey, L., 2010. J. Anal. Chem. 65, 188.
Geis, P.A., 2006. Cosmetics microbiology. A practical approach (Chapter 7). In: Geis, P.A. (Ed.), Preservation Strategies. Taylor & Francis, New York.
Han, F., He, Y.-Z., Yu, C.-Z., 2008. Talanta 74, 1371.
Harvey, P.W., 2003. J. Appl. Toxicol. 23, 285.
Hauksson, I., Ponten, A., Isaksson, M., Hamada, H., Engfeldt, M., Bruze, M., 2015. Contact Dermat. 2015, 12493.
Haunschmidt, M., Buchberger, W., Klampfl, C.W., Hertsens, R., 2011. Anal. Methods 3, 99.
He, K., Huang, J., Lagenaur, C.F., Aizenman, E., 2006a. J. Pharmacol. Exp. Ther. 317, 1320.
He, S., Zhao, Y., Zhu, Z., Liu, H., Li, M., Shao, Y., Zhuang, Q., 2006b. Talanta 69, 166.
Huang, H.Y., Huang, I.Y., Lin, H.Y., 2006. J. Sep. Sci. 29, 2038.
Huang, J.Q., Hu, C.C., Chiu, T.C., 2013. Int. J. Cosmet. Sci. 35, 346.
Hung, H., Blanchard, P., Halsall, C.J., Bidleman, T.F., Stern, G.A., Fellin, P., Muir, D.C., Barrie, L.A., Jantunen, L.M., Helm, P.A., Ma, J., Konoplev, A., 2005. Sci. Total Environ. 342, 119.
Isaksson, M., Gruvberger, B., Bruze, M., 2004. Dermatitis 15, 201.
Jain, R., Mudiam, M.K.R., Chauhan, A., Ch, R., Murthy, R.C., Khan, H.A., 2013. Food Chem. 141, 436.
Jeong, S.H., Kim, B.Y., Kang, H.G., Ku, H.O., Cho, J.H., 2005. Toxicology 208, 49.
Juergensen, L., Busnarda, J., Caux, P.-Y., Kent, R., 2000. Environ. Toxicol. 15, 201.
Kabara, J.J., Orth, D.S. (Eds.), 1997. Preservative-Free and Self-Preserving Cosmetics and Drugs. Marcel Dekker, New York.
Kristenson, E.M., Brinkman, U.A.T., Ramos, L., 2006. TrAC Trend Anal. Chem. 25, 96.
Kutty, P.K., Moody, B., Gullion, J.S., Zervos, M., Ajluni, M., Washburn, R., Sanderson, R., Kainer, M.A., Powell, T.A., Clarke, C.F., Powell, R.J., Pascoe, N., Shams, A., LiPuma, J.J., Jensen, B., Noble-Wang, J., Arduino, M.J., McDonald, L.C., 2007. Chest 132, 1825.
Lanigan, R.S., 1998. Int. J. Toxicol. 17 (S5), 1.
Lanigan, R.S., Yamarik, T.A., 2002. Int. J. Toxicol. 21 (S2), 19.
Lee, M.-R., Lin, C.-Y., Li, Z.-G., Tsai, T.-F., 2006. J. Chromatogr. A 1120, 244.
Leistner, L., 2000. Int. J. Food Microbiol. 55, 181.
Li, Y., Chen, F., Ge, J., Tong, F., Deng, Z., Shen, F., Gu, Q., Ye, J., Chu, Q., 2014. Electrophoresis 35, 419.
Lindstrom, A., Buerge, I.J., Poiger, T., Bergqvist, P.-A., Muller, M.D., Buser, H.-R., 2002. Environ. Sci. Technol. 36, 2322.
Llompart, M.P., Lorenzo, R.A., Cela, R., Paré, J.R.J., Bélanger, J.M.R., Li, K., 1997. J. Chromatogr. A 757, 153.
Lopez-Gazpio, J., Garcia-Arrona, R., Millan, E., 2014. Anal. Bioanal. Chem. 406, 819.
Lopez-Gazpio, J., Garcia-Arrona, R., Millan, E., 2015. Electrophoresis 36, 1064.
Lundov, M.D., Moesby, L., Zachariae, C., Johansen, J.D., 2009. Contact Dermat. 60, 70.
Ma, H., Wang, L., Liu, H., Luan, F., Gao, Y., 2014. Anal. Methods 6, 4723.
Magnano, M., Silvani, S., Vincenzi, C., Nino, M., Tosti, A., 2009. Contact Dermat. 61, 337.
Márquez-Sillero, I., Aguilera-Herrador, E., Cárdenas, S., Valcárcel, M., 2010. J. Chromatogr. A 1217, 1.

Martins, I., Carreira, F.C., Canaes, L.S., de Souza Campos Junior, F.A., da Silva Cruz, L.M., Rath, S., 2011. Talanta 85, 1.
Matyska, M.T., Pesek, J.J., Yang, L., 2000. J. Chromatogr. A 887, 497.
Melo, L.P., Queiroz, M.E.C., 2010. J. Sep. Sci. 33, 1849.
Memon, N., Bhanger, M.I., Khuhawer, M.Y., 2005. J. Sep. Sci. 28, 635.
Mincea, M., Lupşa, I., Talpoş, I., Ostafe, V., 2009a. Acta Chromatogr. 21, 591.
Mincea, M.M., Lupşa, I.R., Cinghita, D.F., Radovan, C.V., Talpos, I., Ostafe, V., 2009b. J. Serb. Chem. Soc. 74, 669.
Miralles, P., Vrouvaki, I., Chisvert, A., Salvador, A., 2016. Talanta 154, 1.
Molina-Cabrillana, J., Bolanos-Rivero, M., Varez-Leon, E.E., Martin Sanchez, A.M., Sanchez-Palacios, M., Alvarez, D., Saez-Nieto, J.A., 2006. Infect. Control Hosp. Epidemiol. 27, 1281.
Moradi, M., Yamini, Y., 2012. J. Chromatogr. A 1229, 30.
Myers, E.A., Pritchett, T.H., Brettell, T.A., 2015. LC-GC Eur.
Nakashima, H., Matsunaga, I., Miyano, N., Kitagawa, M., 2000. J. Health Sci. 46, 447.
Ocaña-Gonzalez, J.A., Villar-Navarro, M., Ramos-Payan, M., Fernandez-Torres, R., Bello-Lopez, M.A., 2015. Anal. Chim. Acta 858, 1.
Oztekin, N., Erim, F.B., 2005. J. Pharm. Biomed. Anal. 37, 1121.
Pedjie, N., 2010. LC-GC Eur.
Polati, S., Gosetti, F., Gennaro, M.C., 2007. Analysis of cosmetic products (Chapter 5). In: Salvador, A., Chisvert, A. (Eds.), Preservatives in Cosmetics. Regulatory Aspects and Analytical Methods. Elsevier, Amsterdam.
Prusakiewicz, J.J., Harville, H.M., Zhang, Y., Ackermann, C., Voorman, R.L., 2007. Toxicology 232, 248.
Rafoth, A., Gabriel, S., Sacher, F., Brauch, H.J., 2007. J. Chromatogr. A 1164, 74.
Ramos, L., 2012. J. Chromatogr. A 1221, 84.
Regueiro, J., Becerril, E., Garcia-Jares, C., Llompart, M., 2009. J. Chromatogr. A 1216, 4693.
Reinhard, E., Waeber, R., Niederer, M., Maurer, T., Maly, P., Scherer, S., 2001. Contact Dermat. 45, 257.
Rezaee, M., Assadi, Y., Milani Hosseini, M.-R., Aghaee, E., Ahmadi, F., Berijani, S., 2006. J. Chromatogr. A 1116, 1.
Rule, K.L., Ebbett, V.R., Vikesland, P.J., 2005. Environ. Sci. Technol. 39, 3176.
Salvador, A., Chisvert, A. (Eds.), 2007. Analysis of Cosmetic Products. Elsevier, Amsterdam.
Sanchez-Prado, L., Alvarez-Rivera, G., Lamas, J.P., Lores, M., Garcia-Jares, C., Llompart, M., 2011a. Anal. Bioanal. Chem. 401, 3293.
Sanchez-Prado, L., Lamas, J.P., Lores, M., Garcia-Jares, C., Llompart, M., 2010. Anal. Chem. 82, 9384.
Sanchez-Prado, L., Llompart, M., Lamas, J.P., Garcia-Jares, C., Lores, M., 2011b. Talanta 85, 370.
Sanchez-Prado, L., Llompart, M., Lores, M., Fernandez-Alvarez, M., Garcia-Jares, C., Cela, R., 2006a. Anal. Bioanal. Chem. 384, 1548.
Sanchez-Prado, L., Llompart, M., Lores, M., Garcia-Jares, C., Bayona, J.M., Cela, R., 2006b. Chemosphere 65, 1338.
Saraji, M., Mirmahdieh, S., 2009. J. Sep. Sci. 32, 988.
Savage, J.H., Matsui, E.C., Wood, R.A., Keet, C.A., 2012. J. Allergy Clin. Immunol. 130, 453.
SCCS, 2009. Opinion on the Mixture of 5-chloro-2-methylisothiazolin-3(2H)-one and 2-methylisothiazolin-3(2H)-one. European Commission, Brussels.
SCENIHR, 2009. Assessment of the Antibiotic Resistance Effects of Biocides.
Scott, E.H., Kanti, A., McNamee, M.P., William, A.A., 2006. Cosmetics microbiology. A practical approach (Chapter 9). In: Geis, P.A. (Ed.), Consumer Safety Considerations of Cosmetic Preservation. Taylor & Francis Group, New York.
Shen, H.Y., Jiang, H.L., Mao, H.L., Pan, G., Zhou, L., Cao, Y.F., 2007. J. Sep. Sci. 30, 48.
Sid, K.H., Rafiei, J., Bani, F., Khanchi, A.R., Hoveidi, H., 2010. Int. J. Environ. Res. 4, 289.
Steinberg, D., 2006. Cosmetics microbiology. A practical approach (Chapter 10). In: Geis, P.A. (Ed.), Global Regulation of Preservatives and Cosmetic Preservatives. Taylor & Francis Group, New York.
Steinberg, D.C. (Ed.), 2012. Preservatives for Cosmetics, third ed. Allured Books.
Terasaka, S., Inoue, A., Tanji, M., Kiyama, R., 2006. Toxicol. Lett. 163, 130.
Thouvenin, M., Peron, J.-J., Charreteur, C., Guerin, P., Langlois, J.-Y., Vallee-Rehel, K., 2002. Prog. Org. Coat. 44, 75.

Tsai, T.-F., Lee, M.-R., 2008. Chromatographia 67, 425.
Uysal, U.D., Güray, T., 2008. J. Anal. Chem. 63, 982.
Wang, J., Zhang, D., Chu, Q., Ye, J., 2010a. Chin. J. Chem. 28, 313.
Wang, P., Ding, X., Li, Y., Yang, Y., 2012. J. AOAC Int. 95, 1069.
Wang, P.G., Zhou, W., 2013. J. Sep. Sci. 36, 1781.
Wang, Y., Cao, Y., Fang, C., Gong, Q., 2010b. Anal. Chim. Acta 673, 145.
Warshaw, E.M., Belsito, D.V., Taylor, J.S., Sasseville, D., DeKoven, J.G., Zirwas, M.J., Fransway, A.F., Mathias, C.G., Zug, K.A., DeLeo, V.A., Fowler, J.F., Marks, J.G., Pratt, M.D., Storrs, F.J., Maibach, H.I., 2013. Dermatitis 24, 50.
Witorsch, R.J., Thomas, J.A., 2010. Crit. Rev. Toxicol. 40 (Suppl. 3), 1.
Wittenberg, J.B., Canas, B.J., Zhou, W., Wang, P.G., Rua, D., Krynitsky, A.J., 2015. J. Sep. Sci. 38, 2983.
Wu, C.-W., Lee, J.-Y., Hu, C.-C., Chiu, T.-C., 2014. J. Chin. Chem. Soc. 61, 453.
Wu, T., Wang, C., Wang, X., Ma, Q., 2008. Int. J. Cosmet. Sci. 30, 367.
Xue, Y., Chen, N., Luo, C., Wang, X., Sun, C., 2013. Anal. Methods 5, 2391.
Yamini, Y., Saleh, A., Rezaee, M., Ranjbar, L., Moradi, M., 2012. J. Liq. Chromatogr. Relat. 35, 2623.
Yang, T.-J., Tsai, F.-J., Chen, C.-Y., Yang, T.C.-C., Lee, M.-R., 2010. Anal. Chim. Acta 668, 188.
Ye, N., Shi, P., Li, J., Wang, Q., 2013. Anal. Lett. 46, 1991.
Youngvises, N., Chaida, T., Khonyoung, S., Kuppithayanant, N., Tiyapongpattana, W., Itharat, A., Jakmunee, J., 2013. Talanta 106, 350.
Zhang, Q., Lian, M., Liu, L., Cui, H., 2005. Anal. Chim. Acta 537, 31.
Zhao, Z., Duan, Y., 2015. RSC Adv. 5, 40636.
Zotou, A., Sakla, I., Tzanavaras, P.D., 2010. J. Pharm. Biomed. 53, 785.

CHAPTER 10

Perfumes in Cosmetics: Regulatory Aspects and Analytical Methods

Alberto Chisvert[1], Marina López-Nogueroles[1,2], Pablo Miralles[1], Amparo Salvador[1]
[1]University of Valencia, Valencia, Spain; [2]Instituto de Investigación Sanitaria La Fe (IIS La Fe), Valencia, Spain

INTRODUCTION

Odours play a crucial role in human behaviour. Whilst a pleasant scent can have a calming effect or make one feel better, unpleasant odours can alter our mood negatively and produce anxiety and discomfort. The importance odours have in human behaviour has been scientifically reported. An interesting review on this subject (Angelucci et al., 2014) reported that inhalation of some odorants can modulate different physiological pathways.

The first evidence of humans using perfumes is from thousands of years ago, when Egyptians used plants, gums and resins in religious rites (Pybus and Sell, 1999; Salvador and Chisvert, 2005). Nowadays, perfumes are part of our lives and our welfare. The word 'perfume' is derived from the Latin words 'per' (through) + 'fumare' (to smoke), based on its incense origins. However, when we use this word in our time we usually refer to a fragrant liquid that diffuses a pleasant smell. This definition accords with that given by some prestigious dictionaries, such as the Thesaurus or Oxford dictionary. Perfumes are responsible for providing us with a pleasant redolence that we smell every day. Each perfume is made up of hundreds of aromatic chemicals (also referred to as fragrance chemicals), each of which contributes its characteristic odour, together giving the characteristic aroma of the perfume.

The aim of this chapter is to give an overview of the determination of fragrance ingredients in cosmetic products. The fragrance types, their legislation and the analytical methods focussed on fragrance ingredients in perfumes are discussed. This chapter contains information about the methods published in the scientific literature concerning the determination of fragrance chemicals since 2006. Previous publications were compiled in the first edition of this book (Chisvert and Salvador, 2007; Chaintreau, 2007).

Nature of the Perfume: Natural Versus Synthetic Perfumes

Depending on their origin, perfumes can be classified as natural or synthetic. When the perfume raw material is of natural origin, therefore obtained from plants or animals, it is called an extract or essential oil. If not, concentrate is a more general term. All of these words are used to designate the concentrated liquid containing the volatile fragrance.

Natural Perfumes: Preparation Procedures

Essential oils or extracts are obtained from different parts of plants, such as flowers, fruits, roots, leaves, wood, bark, resin and seeds, or from whole plants. Also they may be obtained from animal glands and organs. All of these natural raw materials are obtained from their sources using different extraction procedures, such as hydro-distillation, steam distillation, solvent extraction, enfleurage, maceration, percolation, expression, supercritical fluid extraction, etc. The selected process depends on the natural product and on the chemicals responsible for the odour. Moreover, other operations may be needed to obtain good quality extracts (Chisvert and Salvador, 2007; Chisvert et al., 2013). The quality of the resultant essential oil is greatly affected by the method applied, and in this sense, analytical techniques can contribute to selecting the best experimental conditions (López-Nogueroles et al., 2010).

Synthetic Perfumes

Synthetic perfumes are a mixture of fragrance chemicals synthesized in a laboratory either with the objective of creating an odour similar to a natural fragrance or in search of something new and original. These synthetic ingredients appeared as a consequence of the high demand for perfumes in the 20th century, which made their cost increase, whilst the availability of some of them decreased for ethical and/or safety reasons. For example, natural musk, very valuable for its unique odour, is very difficult to obtain as a result of the Washington Treaty, which bans international trade in protected species of wild animals and plants (Mitsui, 1998). Therefore, synthetic musks, with different chemical structure but possessing musk-like odour properties, were developed.

The main advantage of using synthetic ingredients is that it decreases perfume costs compared to natural perfumes. Moreover, they do not entail problems related to poor crop quality or lack of supplies. Thus, no market variations affecting quality are expected. Another aforementioned advantage is the possibility of synthesizing new chemicals and, thus, developing new scents not found naturally. Unfortunately, some drawbacks are also worth mentioning. Because a natural perfume could contain hundreds, sometimes thousands, of fragrance chemicals, it is difficult to reproduce the desired scent exactly. A possibility would be to create the synthetic perfume by mixing just the main ingredients (i.e., those in higher concentration), but all the components, including those present at trace levels, have a synergistic effect on the final odour that is difficult to imitate synthetically. Thus, the resulting mixture would have a slightly different smell (Scott, 2005). An additional disadvantage could be present if the chemical compound responsible for a characteristic odour were only one of the two possible isomeric forms or, even worse, the other isomeric form is responsible for a different, and sometimes unpleasant, odour. For example, D-linalool has a floral scent with a woody note, whilst L-linalool has a sweet floral scent. Therefore, it is of interest to develop chiral synthesis methods using either optically active catalysts or enantiomeric separation strategies.

Odour of the Perfume: The Fragrance Wheel

Perfumes, and their essential oils or concentrates, can be classified according to the note type they provide, i.e., according to the fragrance type. So, different notes are found, such as floral, citrus, fruity, green, woody, oriental, spice, animal, leather, etc. In this sense, the fragrance expert Michael Edwards designed his own scheme of fragrance classification in 1983 taking into account the type of odour that a fragrance provided, and it is still a reference (Fragrances of the World website; Edwards, 2015). The fragrance wheel is a fragrance classification chart created to simplify fragrance classification and naming schemes. The wheel charts differences amongst four main groups (floral, fresh, woody and oriental) that lead to 14 different families and establishes the relationships amongst them.

Another important aspect to bear in mind is how long the scent persists on the body. There are three main notes in a perfume, responsible for the major or minor persistence. The so-called top notes are responsible for the first impression, which lasts from a couple of minutes to approximately an hour, and are generated by high-volatility fragrance chemicals. The middle notes, composed by medium-volatility fragrance chemicals, play the most important role in a perfume, because they are responsible of the identity of the perfume. Finally, the base notes last for more than 6 h and are composed by low-volatility chemicals. The balance amongst all these notes enables perfumes to be more or less persistent. Adding fragrance modifiers and fixatives helps to increase perfume persistence.

Fragrance Ingredients

Amongst the complex mixtures that comprise a perfume, the fragrance chemicals they contain can be classified into different families according to their chemical structure. It is usual to find the five-carbon isoprene unit in most of them, giving them the name of terpenes. So, one can find monoterpene hydrocarbons (e.g., limonene), sesquiterpene hydrocarbons (e.g., α-farnesene), alcohols (e.g., *cis*-3-hexenol), monoterpene alcohols (e.g., linalool), sesquiterpene alcohols (e.g., farnesol), phenols (e.g., eugenol), aldehydes (e.g., 2,6-nonadienal), terpene aldehydes (e.g., citral), ketones (e.g., cyclohexanone), terpene ketones (e.g., β-ionone), lactones (e.g., γ-undecalactone), esters (e.g., methyl salicylate), terpene esters (e.g., linalyl acetate), oxides (e.g., eucalyptol), etc.

The International Fragrance Association (IFRA) (IFRA website) listed the usual fragrance ingredients used by customers worldwide. Also, the Scientific Committee on Cosmetic Products and Non-Food Products [SCCNFP; now the Scientific Committee on Consumer Safety (SCCS)], intended for consumers of the European Commission (EC), established an inventory and the common nomenclature of perfume and aromatic raw materials usually employed in cosmetic products (SCCNFP/0389/00). In addition, general information on all cosmetic ingredients can be found in the Cosmetic Ingredient Database (CosIng) of the EC (on the EC website), with information on cosmetic substances and ingredients. Nevertheless, it is important to take into account that any substance may be used as an ingredient as long as it is not banned.

Cosmetics Containing Perfumes: Fragrance Content

The perfume raw materials are usually pure fragrance chemicals or crude extracts obtained by different procedures above mentioned. Perfumers mix these raw materials to create a scent that fulfils the established fragrance requisites, which usually fall in line with a market study, the type of cosmetic to which they are to be added, the cosmetic's image, the target consumer (age, sex, etc.), originality, fashion, etc.

Perfumes (also called fine fragrances), i.e., hydro-ethanolic solutions of essential oils or concentrates, can be subdivided according to the content of the concentrate in the finished product. Table 10.1 shows this classification. This table also shows the typical concentrate content found in other types of cosmetic products. Almost all other cosmetic products contain fragrance chemicals, but obviously in a much lower content than in the fine fragrances, the main function of which is to transmit a pleasant redolence to the user.

Perfume raw materials are commonly added to other cosmetic products so that their nice smell makes cosmetic users feel clean, comfortable and attractive. In addition, they are sometimes added to cosmetics to mask undesirable odours caused by other cosmetic ingredients. The type of cosmetic product conditions its aroma. In the case of perfumes, the scent employed should be, depending on the specific case, strong and seductive or, in contrast, sweet and refreshing. In the case of a skin or hair care product, sweet or tenuous notes are usually used, whereas for toothpastes highly refreshing power notes are preferred. In addition, depending on the formulation, there are perfumes that could be incompatible with the cosmetic itself, owing to solubilization problems (which affect the manufacturing process), chemical and/or physical interactions with other ingredients, etc.

Table 10.1 Perfume contents usually found in cosmetic products

Cosmetic product	Approximate content (%)
Perfumes (also called fine fragrances)	
Baby cologne	1–2
Cologne	2–3
Eau de cologne	3–4
Eau fraiche	4–5
Eau de toilette	5–15
Eau de parfum	15–20
Parfum	20–40
Other cosmetic products	
Skin care products	0.01–0.5
Hair care products	0.01–1
Bath preparations	0.1–3
Toothpastes	0.5–1

REGULATORY ASPECTS AND ANALYTICAL METHODS FOR FRAGRANCE INGREDIENTS IN COSMETICS

General Considerations

Legislations in force in the three principal markets regarding cosmetic products, i.e., the European Union, the United States and Japan, establish that all of the ingredients for cosmetics should be indicated on the label (see Part I of this book). This could be an extremely difficult task, because, as mentioned previously, a perfume could be composed of hundreds of chemicals, which should be mentioned on the label together with the other cosmetic ingredients.

According to the EU Regulation on Cosmetic Products (Regulation EC 1223/2009) (it replaced the former EU Directive on Cosmetic Products), 'perfume and aromatic compositions and their raw materials shall be referred to by the terms *parfum* or *aroma*'. However, substances classified as potentially allergenic must be declared on the label (see next section). In the same sense, the US Food and Drug Administration (FDA) establishes that fragrances or flavours may be listed under the word 'fragrance' or 'flavour' (FDA, Code of Federal Regulations, Title 21).

Potentially Allergenic Substances Used as Fragrances

Skin irritations or allergic reactions due to fragrance chemicals are relatively common. Various effects have been reported, such as skin sensitivity, rashes, dermatitis, coughing or asthma attacks. In 1999, the SCCNFP identified 26 fragrance allergens (SCCNFP/0017/98) and consequently measures were introduced into the legislation. These measures aimed to improve the diagnosis of contact allergies amongst consumers and facilitate these users in protecting themselves from these compounds. The 26 fragrance substances classified as **potentially allergenic substances** (PASs) were introduced into Annex III of the EU Directive on Cosmetic Products (Directive 76/768/EEC) by the seventh amendment (Directive, 2003/15/EC). Since then, any cosmetic product containing any of these 26 substances has to declare its presence on the label when present at a higher concentration than 0.001% in those products that remain on the skin or 0.01% in those intended to be rinsed off. The 26 substances listed in Table 10.2 are currently declared as PASs by the EU Regulation on Cosmetic Products (Regulation EC 1223/2009).

In 2012, the EU SCCS (formerly SCCNFP) adopted its opinion on fragrance allergens in cosmetics (SCCS/1459/11). The document SCCS/1459/11 provides interesting information in five tables devoted to:
- established contact allergens in humans (54 individual chemicals and 28 natural extracts) (the 26 PAS are included here);
- fragrance substances categorized as established contact allergens in animals (18 individual chemicals and 1 natural extract);

Table 10.2 Potentially allergenic substances according to the EU Regulation on Cosmetic Products

Common name	Abbreviation used in this book	CAS number
Amyl cinnamal	AC	122-40-7
Amylcinnamyl alcohol	ACA	101-85-9
Anisyl alcohol	AA	105-13-5
Benzyl alcohol	BA	100-51-6
Benzyl benzoate	BB	120-51-4
Benzyl cinnamate	BC	103-41-3
Benzyl salicylate	BS	118-58-1
Cinnamal	CN	104-55-2
Cinnamyl alcohol	CA	104-54-1
Citral	CT	5392-40-5
Citronellol	CTN	106-22-9
Coumarin	CM	91-64-5
Eugenol	EG	97-53-0
Farnesol	FA	4602-84-0
Geraniol	GE	106-24-1
Hexyl cinnamaldehyde	HCN	101-86-0
Hydroxycitronellal	HC	107-75-5
Hydroxymethylpentylcyclohexenecarboxaldehyde	HMPC	31906-04-4
Isoeugenol	IG	97-54-1
Lilial	LI	80-54-6
D-Limonene	LM	5989-27-5
Linalool	LN	78-70-6
Methyl heptine carbonate	MHC	111-12-6
3-Methyl-4-(2,6,6-trimethyl-2-cyclohexen-1-yl)-3-buten-2-one	MTC	127-51-5
Oak moss[a]		90028-68-5
Tree moss[a]		90028-67-4

[a]Natural extracts.

- fragrance substances categorized as likely contact allergens by combination of evidence (26 substances);
- fragrance substances categorized as possible contact allergens (35 individual chemicals and 13 natural extracts);
- established fragrance contact allergens of special concern (12 single chemicals) (all them included in the list of the 26 PASs).

In 2014, the EC proposed a change to a tighter regulation (European Commission, 2014). The proposal is to add an extra list of allergens to the previous list of 26 allergens, the presence of which has to be mentioned on the label, and to ban some others. Public consultation was carried out to which any interested parties, including authorities of the Member States, manufacturers of cosmetic products, producers of the substances concerned, relevant industry and consumer associations, could send any comment. In this new proposal, there are three compounds planned to be banned in cosmetic products, hydroxyisohexyl-3-cyclohexenecarboxaldehyde, atranol and chloroatranol. The proposal, with some changes if necessary, should become a measure in force in the future.

In addition to compliance with the legislation in force in each country, the fragrance industry is in some way self-regulated by some independent organizations. The Research Institute for Fragrance Materials (RIFM; RIFM website) evaluates and distributes scientific data on the safety assessment of fragrance substances found in cosmetics and other products. In fact, the IFRA (IFRA website) establishes usage guidelines for fragrance ingredients based on RIFM evaluation results and recommends avoiding many ingredients. The information on allergen substances given on the website of the International Dialogue for the Evaluation of Allergens (IDEA project website) may also be interesting for readers. It was designed to provide a broadly agreed-upon framework for assessing fragrance sensitizers globally.

Other Potentially Toxic Substances

The EU Regulation prohibits the use of more than 50 fragrance substances (including some extracts) in cosmetic products, under its Annex II. Special attention is paid to synthetic musks, especially musk ambrette, which are shown to be linked to different types of dermatitis, carcinogenic effects and endocrine dysfunction (Parker et al., 1986; Lovell and Sanders, 1988; Dietrich, 1999; Eisenhardt et al., 2001). So, musk ambrette, musk tibetene and musk moskene are prohibited in cosmetics, whereas musk xylene and musk ketone are permitted with the restrictions laid down in Annex III, on the basis of the SCCNFP recommendations (SCCNFP/0817/04).

The FDA list of prohibited ingredients has only a few compounds (FDA, Code of Federal Regulations, Title 21) (see Part I of this book) and none of them are used as fragrance ingredients. However, in the FDA guide to inspections of cosmetics (FDA, Guide To Inspections) there are some references to the need to investigate and document any use of fragrance allergenic ingredients.

Other entities concerned with the safety of perfumes are the aforementioned IFRA and RIFM. The RIFM maintains the largest database on toxicology for flavour and fragrance materials available worldwide, classifying more than 4500 materials. The database can be accessed online (RIFM website).

DETERMINATION OF FRAGRANCE INGREDIENTS IN COSMETICS

General Analytical Considerations

Analytical aspects related to perfumes involve, overall, characterizing the extracts obtained by perfume manufacturers to check whether they fulfil the desired quality requirements (ratio of fragrance ingredients, presence/absence of undesired compounds or contaminants, etc.), i.e., for quality control and also to characterize new extracts obtained from different sources or obtained by different methods. Also, quality control should be performed when the extracts and/or pure fragrance ingredients (synthetic or natural) are blended to create the perfume raw material (fragrance compound), which will be later sold to cosmetics manufacturers.

In this last case, additional quality control of perfume raw materials by cosmetics manufacturers is not needed, because perfume manufacturers issue a certificate that guarantees the quality of the perfume raw material. Nevertheless, cosmetics manufacturers can obviously perform a quality control of the raw material that they are buying, to avoid quality variations in their final product. Different assays can be carried out. Measurements of physical properties such as refractive index, optical rotation, density, colour and solubility in different solvents are commonly applied to perfumes and perfume raw materials. Also, the 'noses' of trained personal are useful for checking the notes of the perfume.

From a chemistry standpoint, the determination of acidity, the measurement of saponification and the carbonyl indices provide general information on the global quality control of perfumes. The use of spectroscopic analytical techniques, like ultraviolet/visible spectrometry, infrared spectrometry and nuclear magnetic resonance, also provides valuable information about quality. However, these measures do not provide feasible qualitative and quantitative information on the fragrance chemicals the perfumes contain. They are limited to providing general qualitative information on the perfume as a whole. So, more sophisticated analytical techniques are necessary, especially taking into account that a perfume could be made up of hundreds of fragrance ingredients.

Separation techniques, like chromatography, are the most suitable analytical techniques for these purposes. Bearing in mind that the fragrance chemicals usually have a low boiling point, gas chromatography, both by injection or in headspace mode, is the most widely used technique in the perfume industry. After appropriate sample preparation, optimization of experimental conditions and use of flame ionization or thermal conductivity detectors the individual Kovats index for each compound can be established. This represents a relative measurement of the retention time with respect to a group of known hydrocarbons. The identification is carried out by comparing the experimentally determined Kovats index with database values. However, sometimes this can be a hard task, because there could be several peaks with very close Kovats indices. In addition, although databases are frequently updated, new ingredients might be present, and thus the database will not be able to identify these new compounds. Nevertheless, a mass spectrometry detector coupled with gas chromatography can sometimes help to solve this problem, because the chemical structure can be elucidated by studying the mass spectra of the compound, with the use of databases that identify predefined compounds. Moreover, the mass spectrometry detector provides greater sensitivity and higher selectivity than the other aforementioned detectors. The use of gas chromatography coupled with electronic noses can also be useful in some cases.

Gas chromatography is also used for quantitative purposes. Chiral columns can be used to separate and quantify the different optical isomers from the same chemical, which is very important, because, as mentioned previously, there are compounds that exhibit different properties depending on their isomeric form. Liquid chromatography and thin-layer chromatography have also been applied for quantitative and qualitative purposes in perfume analysis more specifically, to determine both low volatile or thermolabile fragrance chemicals.

An interesting article on experimental precautions for flavour and fragrance determination by chromatography techniques with mass spectrometry detection has been published by Begnaud and Chaintreau (2016).

For this chapter, we reviewed selected published papers (since 2006) on fragrance ingredients determination in cosmetics or raw materials. Table 10.3 summarizes these works (in chronological order). Comments can be found in the following sections of the chapter.

Papers devoted to perfume characterization have not been reviewed here, because this chapter is focussed on the determination of fragrance chemicals in cosmetic products. Readers interested in this topic can find various review papers in the scientific literature (Marriot et al., 2001; Van Asten, 2002; Schulz and Baranska, 2005; Jalali-Heravi and Parastar, 2011). Also some articles such as those of Haddad et al. (2008), Tobolkina et al. (2013), Gebicki et al. (2015), Gomes et al. (2015, 2016) and Godinho et al. (2016) are worth mentioning.

Determination of Potentially Allergenic Substances Used as Fragrances

Many efforts have been focussed on developing methods to determine fragrance allergen ingredients restricted under Annex III of the EU Regulation (Regulation EC 1223/2009). The 26 substances listed in Table 10.1 are declared as PASs by the EU Regulation. Twenty-four of these 26 substances are chemically defined volatile compounds (hereinafter, the 24 EU volatile PASs), whereas the other two substances are natural moss extracts (hereinafter, the 2 EU PA moss extracts) and thus, not defined chemicals, but a natural mixture of many of them. It is important to note that within the 24 EU volatile PASs different classes of compounds from a chemical standpoint are found, such as alcohols, carbonyl compounds, esters or phenols.

Volatile Potentially Allergenic Substances

Most of the published works focus on the determination of the 24 EU volatile PASs, because the natural conditions of the two natural moss extracts, without standardized industrial processing, make the composition of these extracts variable and their accurate determination complicated. Their volatile characteristics make gas chromatography the

Table 10.3 Published articles dealing with the determination of fragrance ingredients in cosmetic products

	Analytes[a]						Analytical method[b]	
	EU volatile PASs	EU moss extract PASs	Other PASs	Musks	Other ingredients	Analysed samples	Sample preparation	Analytical technique
Bassereau et al. (2007)	24					Perfume raw materials		GC–MS
Chen et al. (2007)	GE		BAT CE ES MEG M2N PA			Shampoo	SPME	GC×GC–MS GC–FID
Cordero et al. (2007)	24					Perfume raw materials	Dilution with cyclohexane	GC×GC–FID GC×GC–MS
Mondello et al. (2007)	24					Perfumes	Direct injection	GC–MS
Rastogi et al. (2007)					Atranol Chloratranol			
Roosens et al. (2007)				AHTN HHCB MK MX		Perfumes, deodorants, hair care products, shower and bath products, body lotions	LLE with hexane	GC–MS
Villa et al. (2007)	24					Massage oil, moisturizing creams, hair conditioner	Dilution with acetonitrile	LC–UV
Tranchida et al. (2008)	24					Perfumes		GC×GC–MS
Ma et al. (2009)				MX		Cream, lotion, powder, shampoo, lipstick	SPE–isotope dilution	GC–MS/MS
Augusto et al. (2010)					Rosemary oil	Perfumes		GC×GC–FID

Reference	N			Analytes	Sample	Sample preparation	Analytical technique
Culea et al. (2010)				Volatile allergens, characterization	Perfumes, essential oils, soaps	Dilution in EtOH	HS–GC–MS
Del Nogal et al. (2010)	24				Perfumes, anti-hair loss products, post-depilation mousse, deodorant, creams	Dilution with propyl acetate	HS–GC–FID HS–GC–MS
Furlanetto et al. (2010)	18	MEG PN			Shampoo and bath gel	Dilution with water	CE–UV
Lamas et al. (2010)	24				Moisturizing creams and lotions, hand creams, sunscreen, after-sun creams	MSPD–PLE	GC–MS
Martínez-Girón et al. (2010)			HHCB AHTN AITI AHDI		Perfumes	LLE with hexane	CE–MEKC
Fonseca de Godoy et al. (2011)				Essential oils	Perfumes		GC × GC–FID
Chaintreau et al. (2011)	24				Perfumes		GC–MS
Lu et al. (2011)			AHTN HHCBMK MX HHCB-lactone		Toothpastes, shampoos, hair conditioners, body washes, hand sanitizers, facial cleansers, toilet soaps, face cream, body lotions, liquid foundation, nail polish	LLE and SPE	GC–MS

Continued

Table 10.3 Published articles dealing with the determination of fragrance ingredients in cosmetic products—cont'd

	Analytes[a]						Analytical method[b]	
	EU volatile PASs	EU moss extract PASs	Other PASs	Musks	Other ingredients	Analysed samples	Sample preparation	Analytical technique
Sanchez-Prado et al. (2011a)	24			ADBI AHDI AITI HHCB AHTN AMBT MA MX MM MT MK		Perfumes	Diluted with EtAc	GC–MS
Sanchez-Prado et al. (2011b)	24		MEG			Moisturizing creams and lotions, anti-cellulite creams, hand creams, shampoos and gels, hair conditioners, hand soaps	MSPD	GC–MS
Devos et al. (2012)	24		PA ES NE M2N MEG			Perfumes, body creams	Dilution with acetone	HS–GC–MS
Lopez-Gazpio et al. (2012)				AHTN HHCB AITI		Perfumes	Dilution with running electrolyte	MEKC
Ng et al. (2012)					Camphor and other	Talcum powder	HS–SPME	GC–MS
Wang (2012)				MA MX MM MT MK		Perfume	SPE	GC–MS/MS

Wang et al. (2012)			ADBI AHTN MA MX MK	Perfumes	HS-SPME	GC–MS
Wang and Liu (2012)	CA EG IG	MEG and other		Essential oils		LC–FLD
Zhou et al. (2012)						
Corbi et al. (2013)						
Homem et al. (2013)			DMPI ADBI AHDI EXAL HHCB AHTN MA MX MM MT MK	Body and hand cleansers, toilet soaps, skin moisturizers, roll-on deodorants, toothpaste	QuEChERS extraction	GC–MS
Llompart et al. (2013)			7 polycyclic musks and 5 nitro musks		MSPD	GC–MS
Minematsu et al. (2013)		VA		Perfumes	Sample is diluted with running electrolyte	CE–UV
Rudbäck et al. (2013)	LN LM	LNA		Essential oils	Dilution with methanol: water	LC–MS/MS
Sanchez-Prado et al. (2013)	24	MEG		Child care products	PLE and/or MSPD	GC–MS

Continued

Table 10.3 Published articles dealing with the determination of fragrance ingredients in cosmetic products—cont'd

	Analytes[a]					Analytical method[b]		
	EU volatile PASs	EU moss extract PASs	Other PASs	Musks	Other ingredients	Analysed samples	Sample preparation	Analytical technique
Celeiro et al. (2014)	24		MEG			Body milk, moisturizing milk, nail polish remover, toothpaste, hand cream, lipstick, sunblock, deodorants, shampoos, liquid soaps	MSPD	GC–MS
Debonneville and Chaintreau (2014)	24					Cream and shampoo	Online clean-up by a liner packed with polydimethylsiloxane and a programmed temperature vaporization injector	GC–MS
Dong et al. (2014)				MA MT MM MX MK AHDI AHTN		Cream	SLE–SPE	GC–MS/MS
Famiglini et al. (2014)	CT LM LN					Perfume, hand cream, shower gel, orange oil	Dilution with methanol:water	LC–MS
Kern et al. (2014)	LN					Antiperspirant, hydro-alcoholic products	Dilution with hexane or pentane	GC–FID GC–MS
Li et al. (2014)								LC–MS

Lopez-Gazpio et al. (2014)	AA BA CM EG IG CT LN GE			Perfumes, shampoos, gels, soaps, creams	Dilution with methanol:water and filtration	MEKC–UV
López-Nogueroles et al. (2014)		Atranol, chloroatranol		Perfumes	Sample clean-up by LLE with hexane and water, then simultaneous derivatization and DLLME process prior to GC–MS analysis	GC–MS
Rudbäck et al. (2014)	LM LN GE			Essential oils	Standard addition using toluene as solvent and derivatization	GC–MS
Wang et al. (2014)	AC CA CT GE EG		ES IEG MEG	Essential oils		ATR–IR
Celeiro et al. (2015)	24		MEG PN	Baby wipes and wet toilet paper Creams, lotions, gels	PLE	GC–MS
Desmedt et al. (2015)	24			Baby oil, lip balm, olive cream, deodorant stick, antiperspirant, face mask, face lotion, shampoo, toothpaste	Several extraction steps with hexane	HS–GC–MS
Divisova et al. (2015)	24				HS–SPME	GC–FID

Continued

Table 10.3 Published articles dealing with the determination of fragrance ingredients in cosmetic products—cont'd

	Analytes[a]					Analytical method[b]	
	EU volatile PASs	EU moss extract PASs	Other PASs	Musks	Other ingredients	Analysed samples	Sample preparation	Analytical technique
Gavris et al. (2015)			VA, EVA cis-Methyl dihydro-jasmonate	Muskolactone Muscone		Perfumes		GC–MS
Hou et al. (2015)	EG CTN GE		TP Dodecanol Hexadecanol			Shampoo, skin care water	HS–SPME	GC–FID
Ma et al. (2015)	CM					Creams, lotions, shampoos	Several steps (dilution, centrifugation, evaporation and reconstitution) and SPE	UPLC–MS/MS
Nakata et al. (2015)				Musk T Musk MC-4 Habanolide Cervolide AMBT EXAL Exaltone Civetone Muscone HHCB AHTN OTNE HHCB-lactone MK MX		Perfumes, body fragrances, shampoos, body lotions, body soap, antiperspirants, toilet deodorants, hair liquid, sunscreens, tooth powder	Extraction with hexane and SPE	GC–MS

Reference			Sample	Sample preparation	Technique
Perez-Outeiral et al. (2015)	18		Perfumes, body milk		LC–UV
Rey et al. (2015)	24	32	Perfumes	Direct injection	GC–GC–MS
Ruzik et al. (2015)			Perfumed waters	Sample is diluted with methanol	LC–UV/VIS CA
			Assessment of repeatability of composition of perfumed waters		
Amenduni et al. (2016)			Perfumes VOCs		GC–MS combined with olfactometry
Shibuta et al. (2016)	21	5	Perfumes	Direct dilution with acetone	GC–MS
Tranchida et al. (2016)	24	30	Perfumes		GC × GC–MS
Boeye et al. (2017)			Perfumes, aftershave lotion	Direct injection	LC–EL–UV
			Atranol Chloroatranol		

EtAc, ethyl acetate; EtOH, ethanol; PAS, potentially allergenic substance.
[a]AA, anisyl alcohol; AC, amyl cinnamal; ADBI, celestolide; AHDI, phantolide; AHTN, tonalide; AITI, traseolide; AMBT, ambrettolide; BA, benzyl alcohol; BAT, benzyl acetate; CA, cinnamyl alcohol; CE, cetalox; CM, coumarin; CT, citral; CTN, citronellol; DMPI, cashmeran; EG, eugenol; ES, estragol; EVA, ethylvanillin; EXAL, exaltolide; GE, geraniol; HHCB, galaxolide; IG, isoeugenol; LM, D-limonene; LN, linalool; LNA, linalyl acetate; M2N, methyl 2-nonynoate; MA, musk ambrette; MEG, methyleugenol; MK, musk ketone; MM, musk moskene; MT, musk tibetene; MX, musk xylene; NE, neral; OTNE, Iso E Super; PA, phenylacetaldehyde; PN, pinene; TP, terpineol; VA, vanillin; VOCs, volatile organic compounds.
[b]ATR, attenuated total reflectance; CE, capillary electrophoresis; DLLME, dispersive liquid–liquid microextraction; EL, electrochemical detection; FID, flame ionization detector; FLD, fluorimetric detector; GC, gas chromatography; HS, headspace; IR, infrared spectroscopy; LC, liquid chromatography; LLE, liquid–liquid extraction; MEKC, micellar electrokinetic chromatography; MS, mass spectrometry; MS/MS, tandem mass spectrometry; MSPD, matrix solid-phase dispersion; PLE, pressurized liquid extraction; QuEChERS, quick, easy, cheap, effective, rugged and safe; SLE, supported liquid extraction; SPE, solid-phase extraction; SPME, solid-phase microextraction; UPLC, ultra-high performance liquid chromatography; USAEME, ultrasound-assisted emulsification microextraction; UV, ultraviolet spectroscopy.

better technique of analysis, especially using flame ionization or mass spectrometry detectors. Published papers dealing with the determination of the 24 EU volatile PASs can be easily found in the first column of Table 10.2. Works dealing with other PASs are in the second column (the 2 EU PA moss extracts) and in the third column (other substances not included in the EU list).

As shown in Table 10.2, there are few studies using gas chromatography with flame ionization detector (Chen et al., 2007; Divisova et al., 2015; Hou et al., 2015). Some authors used it in addition to gas chromatography with mass detectors (Cordero et al., 2007; Culea et al., 2010; Kern et al., 2014). In fact, most of the published papers use gas chromatography coupled to mass spectrometry. Mass spectrometry is a selective and very useful detector, as one of the biggest difficulties to face in this determination is the co-elution of the target compounds with others of the many components of the perfume. Co-elution leads to false positives and negatives, which published works try to solve by operating in either selected-ion monitoring or extracted-ion chromatogram modes. The use of comprehensive/bidimensional gas chromatography coupled to mass spectrometry (Bassereau et al., 2007; Cordero et al., 2007; Tranchida et al., 2008, 2016; Rey et al., 2015) or flame ionization detectors (Cordero et al., 2007) is also a good solution in some cases. In addition, the use of clean-up steps before the instrumental step solves a good part of these problems.

Other chromatographic techniques, such as liquid chromatography, are less frequently used. This technique has been used with an ultraviolet/visible detector (Villa et al., 2007; Perez-Outeiral et al., 2015), a fluorimetric detector (Wang and Liu, 2012; Kern et al., 2014) or tandem mass spectrometry detectors (Rudbäck et al., 2013). In addition, some other techniques have been used in a few cases, such as capillary electrophoresis with ultraviolet detection (Furlanetto et al., 2010), electrokinetic micellar chromatography (Lopez-Gazpio et al., 2014) and attenuated total reflectance infrared spectroscopy (Wang et al., 2014).

In a high number of papers the authors have developed methods to determine the 24 volatile PASs declared as allergens in the EU Regulation (Bassereau et al., 2007; Cordero et al., 2007; Mondello et al., 2007; Villa et al., 2007; Tranchida et al., 2008; Del Nogal et al., 2010; Chaintreau et al., 2011; Sanchez-Prado et al., 2011a,b, 2013; Devos et al., 2012; Celeiro et al., 2014, 2015; Debonneville and Chaintreau et al., 2014; Desmedt et al., 2015; Divisova et al., 2015; Perez-Outeiral et al., 2015; Rey et al., 2015; Shibuta et al., 2016; Tranchida et al., 2016). The work from Chaintreau et al. (2011) is worth mentioning, in which the validation results obtained on 'ready to inject' samples under reproducibility conditions following inter-laboratory ring-testing are presented. This work derives from the IFRA method based on gas chromatography–mass spectrometry working in selected-ion monitoring (Chaintreau et al., 2003). In this work, the investigation of an automated data treatment procedure to aid the analyst during the interpretation of the analytical results was proposed.

In other cases, only some of the 24 EU volatile allergens are determined in addition to others, not included in the list, but also considered potentially allergenic. Some of these substances are covered in the latest opinion of the SCCS (SCCS/1459/11), such as benzyl acetate (established fragrance contact allergen of special concern), phenylacetaldehyde (fragrance substance categorized as established contact allergen in animals), pinene and vanillin (established contact allergens in humans) and many others, such as methyleugenol and methyl 2-nonynoate, that are restricted substances according to Annex III of the EU Regulation.

Regarding the nature of the samples, as shown in Table 10.2, PASs have been determined mainly in perfumes and perfume raw materials, including essential oils, but have also been determined in other types of products such as bath products, oral hygiene products, skin care products, sun products, hair products, products related to hair removal, deodorants, etc.

Different options for sample preparation are used, from direct analysis to complex methods of extraction and pre-concentration. In this sense, Mondello et al. (2007) and Rey et al. (2015) proposed methods based on the direct injection of the samples for the analysis of perfumes by gas chromatography and two-dimensional gas chromatography, respectively, with mass spectrometry detection in both cases. Other works, based on the use of capillary electrophoresis with ultraviolet detection or micellar electrokinetic chromatography, such as Furlanetto et al. (2010), Minematsu et al. (2013) and Famiglini et al. (2014), did not require the use of organic solvents either. In a high number of publications, sample treatment was based on direct dilution in organic solvents such as acetone for the analysis of perfumes and body creams (Devos et al., 2012; Shibuta et al., 2016); acetonitrile for the analysis of massage oil, moisturizing creams and hair conditioner (Villa et al., 2007); cyclohexane for the analysis of perfume raw materials (Cordero et al., 2007); ethanol for the analysis of perfumes, essential oils and soaps (Culea et al., 2010); ethyl acetate for the analysis of various cosmetics (Sanchez-Prado et al., 2011b); hexane and pentane for the analysis of antiperspirants and hydro-alcoholic products (Kern et al., 2014); methanol:water mixtures for the analysis of perfumes, shampoos, gels and creams (Rudbäck et al., 2013; Famiglini et al., 2014; Lopez-Gazpio et al., 2014); propyl acetate for the analysis of perfumes, anti-hair loss products, post-depilation mousses, deodorants and creams (Del Nogal et al., 2010); toluene for the analysis of essential oils (Rudbäck et al., 2014); etc.

Regarding extraction techniques, liquid–liquid extraction has been used in very few works as the only pre-treatment technique. In this sense, Roosens et al. (2007) used it for the analysis of perfumes and Desmedt et al. (2015) proposed a method based on several extraction steps with hexane for the analysis of creams, lotions and gels. Solid-phase extraction was used for the analysis of different types of cosmetics generally in combination with liquid–liquid extraction or with other extraction techniques (Lu et al., 2011; Wang et al., 2012; Dong et al., 2014; Ma et al., 2015; Nakata et al., 2015). Solid-phase microextraction has been used in various works for the analysis of shampoo, creams, deodorant, toothpaste and other products (Chen et al., 2007; Ng et al., 2012; Wang et al., 2012; Hou et al., 2015; Divisova et al., 2015). Headspace is used in different publications (Culea et al., 2010; Del Nogal et al., 2010; Devos et al., 2012; Ng et al., 2012; Wang, 2012; Desmedt et al., 2015; Divisova et al., 2015; Hou et al., 2015). Online clean-up using a liner packed with polydimethylsiloxane foam and thermal desorption was also used (Debonneville and Chaintreau, 2014). Matrix solid-phase dispersion has been used for the analysis of very different types of cosmetics (Lamas et al., 2010; Sanchez-Prado et al., 2011a,b, 2013; Celeiro et al., 2014; Divisova et al., 2015). Pressurized liquid extraction has been used alone (Celeiro et al., 2015) or in combination with matrix solid-phase dispersion in some cases (Lamas et al., 2010; Sanchez-Prado et al., 2013). Ultrasound-assisted emulsification microextraction has been used for the analysis of perfumes and body milk (Perez-Outeiral et al., 2015).

Natural Moss Extracts

Concerning the determination of the two natural moss extracts restricted as potentially allergenic (treemoss and oakmoss), the number of publications is lower. The industrial processing varies considerably but basically harvested lichens, from oak trees in the case of oakmoss and pine and cedar trees in the case of treemoss, are extracted, usually with hexane or more polar solvents. Then they are further diluted in ethanol and usually submitted to physical treatments intended to remove their original colour. Taking into account their natural conditions and that their industrial processing is not standardized, it is not surprising to find variability in the chemical composition of these extracts. Therefore, the quantification of some of their individual components is not indicative of their total amount. Joulain and Tabacchi (2009a,b) published two comprehensive reviews on the composition of these extracts. One hundred seventy constituents were identified in oakmoss extracts and 90 in the case of treemoss.

From all the different compounds present in oakmoss and treemoss extracts, atranol and chloroatranol are very potent allergens (Menné Bonefeld et al., 2014). They are degradation products, formed after transesterification and decarboxylation of the lichen depsides atranorin and chloroatranorin, and are formed in the moss absolute production (Hiserodt et al., 2000). As already specified, although these compounds are not listed in the EU Cosmetic Regulation, the EC proposed in February of 2014 a change to a tighter regulation regarding fragrance allergens (European Commission, 2014). The proposal includes the prohibition of atranol and chloroatranol in cosmetic products, because of their highly allergenic features. Therefore, these compounds will, most assuredly, be banned or at least restricted in the near future. Therefore, works dealing with their determination will be useful. Rastogi et al. (2007) quantified the presence of these two allergenic compounds in different commercial perfumes by liquid chromatography–tandem mass spectrometry. López-Nogueroles et al. (2014) developed and validated a method to determine both compounds in perfumes using simultaneous derivatization and dispersive liquid–liquid microextraction and gas chromatography with mass spectrometry detection. Boeye et al. (2017) also determined both compounds in perfumes and aftershave lotion by liquid chromatography with electrochemical and ultraviolet detection.

Determination of Other Potentially Toxic Substances

The determination of musks arouses special interest because, as mentioned previously, several side effects have been described. Moreover, the presence of some of them is banned or restricted in cosmetic products. Some early publications on this topic can be found in the scientific literature. Roosens et al. (2007) determined two polycyclic and two nitro musks in a high number of cosmetic products, including perfumes, using gas chromatography with mass spectrometry detection, and combined these results with the average usage to estimate exposure profiles through dermal application. Ma et al. (2009) determined musk xylene in cosmetics with a solid-phase extraction–isotope dilution with gas chromatography–tandem mass spectrometry detection method. Martinez-Girón et al. (2010) achieved the enantiomeric separation of four chiral polycyclic musks using capillary electrophoresis and determined them in perfumes. Lu et al. (2011) investigated the concentrations of four musks amongst other compounds in a high number of cosmetic products from China using gas chromatography with mass spectrometry detection, concluding that polycyclic musks were more commonly used than nitro musks. Sanchez-Prado et al. (2011a) developed a method that allows the simultaneous determination of many compounds, including nitro musks and polycyclic musks, in perfumes. The method consists in diluting the sample in ethyl acetate and injecting it directly into a gas chromatography with mass spectrometry detection system. Wang et al. (2012) determined five musks in perfumes by headspace solid-phase microextraction and gas chromatography with mass spectrometry detection. Later, Wang (2012) proposed a method to determine the five nitro musks in different cosmetics by gas chromatography–tandem mass spectrometry; samples were cleaned up using a solid-phase extraction cartridge. Also, Lopez-Gazpio et al. (2012) developed a nonaqueous micellar electrokinetic chromatography method for the determination of polycyclic musks in perfumes. Homem et al. (2013) developed a method based on 'quick, easy, cheap, effective, rugged and safe' extraction followed by gas chromatography with mass spectrometry detection for the analysis of 12 musks in personal care products. Llompart et al. (2013) proposed a matrix solid-phase dispersion and gas chromatography with mass spectrometry detection method for the determination of seven polycyclic musks and five nitro musks amongst other compounds in many different cosmetic formulations. Dong et al. (2014) determined seven synthetic musks in cream by means of supporting liquid extraction coupled with solid-phase extraction and then followed by gas chromatography–tandem mass spectrometry. Finally, Nakata et al. (2015) determined 15 musks, including macrocyclic, polycyclic and nitro musks, in cosmetic products from Japan using gas chromatography with mass spectrometry detection.

Camphor and other perfume-related compounds have been determined in talcum powder by using headspace solid-phase microextraction and gas chromatography with mass spectrometry detection (Ng et al., 2012).

In addition to the fragrance ingredients, other substances (preservatives, colouring agents, contaminants, etc.) are present in perfumes as in other cosmetic products. Published papers focussing on the determination of these other compounds are discussed in other chapters of this book.

REFERENCES

Amenduni, A., Brattoli, M., de Gennaro, G., Massari, F., Palmisani, J., Tutino, M., 2016. Environ. Eng. Manag. J. 15, 1963.
Angelucci, F., Silva, V.V., Dal Pizzol, C., Spir, L.G., Praes, C.E.O., Maibach, H., 2014. Int. J. Cosmet. Sci. 36, 117.
Augusto, F., Poppi, R.J., Pedroso, M.P., de Godoy, L.A.F., Hantao, L.W., 2010. LCGC Eur. 23, 430.
Bassereau, M., Chaintreau, A., Duperrex, S., Joulain, D., Leijs, H., Loesing, G., Owen, N., Sherlock, A., Schippa, C., Thorel, P., Vey, M., 2007. J. Agric. Food Chem. 55, 25.
Begnaud, F.A., Chaintreau, A., 2016. Philos. Trans. R. Soc. A 374 20150365.
Boeye, G., Gismera, M.J., Sevilla, M.T., Procopio, J.R., 2017. Electroanalysis 29, 116.
Celeiro, M., Guerra, E., Lamas, J.P., Lores, M., Garcia-Jares, C., Llompart, M., 2014. J. Chromatogr. A 1344, 1.
Celeiro, M., Lamas, J.P., Garcia-Jares, C., Llompart, M., 2015. J. Chromatogr. A 1384, 9.
Chaintreau, A., Joulain, D., Marin, C., Schmidt, C.O., Vey, M., 2003. J. Agric. Food Chem. 51, 6398.
Chaintreau, A., Cicchetti, E., David, N., Earls, A., Gimeno, P., Grimaud, B., Joulain, D., Kupfermann, N., Kuropka, G., Saltron, F., Schippa, C., 2011. J. Chromatogr. A 1218, 7869.
Chaintreau, A., 2007. Analytical methods to determine potentially allergenic fragrance-related substances in cosmetics. In: Salvador, A., Chisvert, A. (Eds.), Analysis of Cosmetic Products. Elsevier, Amsterdam, p. 257.
Chen, Y., Begnaud, F., Chaintreau, A., Pawliszyn, J., 2007. J. Sep. Sci. 30, 1037.
Chisvert, A., Salvador, A., 2007. Perfumes in cosmetics. Regulatory aspects and analytical methods for fragrance ingredients and other related chemical in cosmetics. In: Salvador, A., Chisvert, A. (Eds.), Analysis of Cosmetic Products. Elsevier, Amsterdam, p. 243.
Chisvert, A., López-Nogueroles, M., Salvador, A., 2013. Essential oils: analytical methods to control the quality of perfumes. In: Ramawat, K.G., Mérillon, J.M. (Eds.), Handbook of Natural Products. Springer-Verlag Berlin Heidelberg, p. 3287.
Corbi, E., Peres, C., David, N., 2013. Flavour Fragr. J. 29, 173.
Cordero, C., Bicchi, C., Joulain, D., Rubiolo, P., 2007. J. Chromatogr. A 1150, 37.
CosIng (Cosmetic Ingredient Database). https://ec.europa.eu/growth/sectors/cosmetics/cosing_es.
Culea, M., Iordache, A., Cozar, O., 2010. J. Environ. Prot. Ecol. 11, 523.
Debonneville, C., Chaintreau, A., 2014. Flavour Fragr. J. 29, 267.
Del Nogal, M., Perez-Pavon, J.L., Cordero, B.M., 2010. Anal. Bioanal. Chem. 397, 2579.
Desmedt, B., Canfyn, M., Pype, M., Baudewyns, S., Hanot, V., Courselle, P., De Beer, J.O., Rogiers, V., De Paepe, K., Deconinck, E., 2015. Talanta 131, 444.
Devos, C., Ochiai, N., Sasamoto, K., Sandra, P., David, F., 2012. J. Chromatogr. A 1255, 207.
Dietrich, D.R., 1999. Toxicol. Lett. 111, 1.
Directive 2003/15/EC of the European Parliament and of the Council of 27 February 2003, Amending Council Directive 76/768/EEC on the Approximation of the Laws of the Member States Relating to Cosmetic Products. http://eur-lex.europa.eu/LexUriServ/LexUriServ.do?uri=OJ:L:2003:066:0026:0035:en:PDF.
Directive 76/768/CEE of the Council of 27 July 1976 on the Approximation of the Laws of the Member States Relating to Cosmetic Products. http://eur-lex.europa.eu/eli/dir/1976/768/oj.

Divisova, R., Vitova, E., Divis, P., Zemanova, J., Omelkova, J., 2015. Acta Chromatogr. 27, 509.
Dong, H., Tang, H., Chen, D., Xu, T., Li, L., 2014. Talanta 120, 248.
Edwards, M., 2015. Fragrances of the World, thirty first ed. (Fragrances of the World Ed, Sidney).
Eisenhardt, S., Runnebaum, B., Bauer, K., Gerhard, I., 2001. Environ. Res. 87, 123.
European Commission, 2014. MEMO/14/108, Questions and Answers: Commission Launches Consultation on Fragrance Allergens. http://europa.eu/rapid/press-release_MEMO-14-108_en.htm.
Famiglini, G., Termopoli, V., Palma, P., Capriotti, F., Cappielo, A., 2014. Electrophoresis 35, 1339.
FDA (U.S. Food and Drug Administration), Code of Federal Regulations, Title 21, Parts 70–82 for Colorants, Parts 330-360 for OTC Drugs, Parts 700–740 for Cosmetics. http://www.accessdata.fda.gov/scripts/cdrh/cfdocs/cfcfr/cfrsearch.cfm.
FDA (U.S. Food and Drug Administration), Guide to Inspections of Cosmetic Product Manufacturers. http://www.fda.gov/ICECI/Inspections/InspectionGuides/ucm074952.htm.
Fonseca de Godoy, L.A., Hantao, L.W., Pedroso, M.P., Poppi, R.J., Augusto, F., 2011. Anal. Chim. Acta 699, 120.
Fragrances of the World Website. http://www.fragrancesoftheworld.com/fragrancewheel.aspx.
Furlanetto, S., Orlandini, S., Giannini, I., Pasquini, B., Pinzauti, S., 2010. Talanta 83, 72.
Gavris, I., Bodoki, E., Verite, P., Oprean, R., 2015. Farmacia 63, 760.
Gebicki, J., Szulczynski, B., Kaminski, M., 2015. Meas. Sci. Technol. 26, 125103.
Godinho, R.B., Santos, M.C., Poppi, R.J., 2016. Spectrochim. Acta A 157, 158.
Gomes, C.L., Lima, A.C.A., Candido, C.L., Silva, A.B.R., Loiola, A.R., Nascimento, R.F., 2015. J. Braz. Chem. Soc. 26, 1730.
Gomes, C.L., Lima, A.C.A., Loiola, A.R., Silva, A.B.R., Candido, M.C.L., Nascimento, R.F., 2016. J. Forensic Sci. 61, 1074.
Haddad, R., Catharino, R.R., Marques, L.A., Eberlin, M.N., 2008. Rapid Commun. Mass Spectrom. 22, 3662.
Hiserodt, R.D., Swijter, D.F.H., Mussian, C.J., 2000. J. Chromatogr. A 888, 103.
Homem, V., Silva, J.A., Cunha, C., Alves, A., Santos, L., 2013. J. Sep. Sci. 36, 2176.
Hou, X., Wang, L., Tang, X., Xiong, C., Guo, Y., Liu, X., 2015. Analyst 140, 6727.
IDEA (International Dialogue for the Evaluation of Allergens). http://www.ideaproject.info/news.
IFRA (International Fragance Association) website. http://www.ifraorg.org/en-us/ingredients.
Jalali-Heravi, M., Parastar, H., 2011. Talanta 85, 835.
Joulain, D., Tabacchi, R., 2009a. Flavour Fragr. J. 24, 49.
Joulain, D., Tabacchi, R., 2009b. Flavour Fragr. J. 24, 105.
Kern, S., Dkhil, H., Hendarsa, P., Ellis, G., Natsch, A., 2014. Anal. Bioanal. Chem. 406, 6165.
Lamas, J.P., Sanchez-Prado, L., Garcia-Jares, C., Lores, M., Llompart, M., 2010. J. Chromatogr. A 1217, 8087.
Li, X.L., Meng, D.L., Zhao, J., Yang, Y.L., 2014. Chin. Chem. Lett. 25, 1198.
Llompart, M., Celeiro, M., Lamas, J.P., Sanchez-Prado, L., Lores, M., Garcia-Jares, C., 2013. J. Chromatogr. A 1293, 10.
Lopez-Gazpio, J., Garcia-Arrona, R., Ostra, M., Millan, E., 2012. J. Sep. Sci. 35, 1344.
Lopez-Gazpio, J., Garcia-Arrona, R., Millan, E., 2014. Anal. Bioanal. Chem. 406, 819.
López-Nogueroles, M., Chisvert, A., Salvador, A., 2010. J. Chromatogr. A 1217, 3150.
López-Nogueroles, M., Chisvert, A., Salvador, A., 2014. Anal. Chim. Acta 826, 28.
Lovell, W.W., Sanders, D.J., 1988. Int. J. Cosmet. Sci. 10, 271.
Lu, Y., Yuan, T., Wang, W., Kannan, K., 2011. Environ. Pollut. 159, 3522.
Ma, Q., Bai, H., Wang, C., Ma, W., Zhang, Q., Xiao, H.Q., Zhou, X., Dong, Y.Y., Wang, B.L., 2009. Chin. J. Anal. Chem. 37, 1776.
Ma, Q., Xi, H., Ma, H., Meng, X., Wang, Z., Bai, H., Li, W., Wang, C., 2015. Chromatographia 78, 241.
Marriot, P.J., Shellie, R., Cornwell, C., 2001. J. Chromatogr. A 936, 1.
Martinez-Girón, A.B., Crego, A.L., Gonzalez, M.J., Marina, M.L., 2010. J. Chromatogr. A 1217, 1157.
Menné Bonefeld, C., Nielsen, M.M., Gimenéz-Arnau, E., Lang, M., Vennegaard, M.T., Geisler, C., Johansen, J.D., Lepoittevin, J.-P., 2014. Contact Dermat. 70, 282.
Minematsu, S., Xuan, G.S., Wu, X.Z., 2013. J. Environ. Sci. 25 (Suppl.), S8.
New cosmetic science (Chapter 4). In: Mitsui, T. (Ed.), 1998. Cosmetics and Fragrances, Elsevier, Amsterdam.
Mondello, L., Sciarrone, D., Casilli, A., Tranchida, P.Q., Dugo, P., Dugo, G., 2007. J. Sep. Sci. 30, 1905.

Nakata, H., Hinosaka, M., Yanagimoto, H., 2015. Ecotoxicol. Environ. Saf. 111, 248.
Ng, K.H., Heng, A., Osborne, M., 2012. J. Sep. Sci. 35, 758.
Parker, R.D., Buehler, E.V., Newmann, E.A., 1986. Contact Dermat. 14, 103.
Perez-Outeiral, J., Millan, E., Garcia-Arrona, R., 2015. J. Sep. Sci. 38, 1561.
Pybus, D.H., Sell, C.S., 1999. The Chemistry of Fragrances. Royal Society of Chemistry, Cambridge.
Rastogi, S.C., Johansen, J.D., Bossi, R., 2007. Contact Dermat. 56, 201–204.
Regulation (EC) 1223/2009 of the European Parliament and the Council of 30 November 2009 on Cosmetic Products. http://eur-lex.europa.eu/eli/reg/2009/1223/2016-08-12.
Rey, A., Corbi, E., Peres, C., David, N., 2015. J. Chromatogr. A 1404, 95.
RIFM (Research Institute for Fragrance Materials), Website. http://www.rifm.org.
Roosens, L., Covaci, A., Neels, H., 2007. Chemosphere 69, 1540.
Rudbäck, J., Islam, N., Nilsson, U., Karlberg, A.T., 2013. J. Sep. Sci. 36, 1370.
Rudbäck, J., Ramzy, A., Karlberg, A.-T., Nilsson, U., 2014. J. Sep. Sci. 37, 982.
Ruzik, L., Obarski, N., Papierz, A., Mojski, M., 2015. Int. J. Cosmet. Sci. 37, 348.
Salvador, A., Chisvert, A., 2005. Encyclopedia of Analytical Science, Perfumes. In: Worsfold, P., Townshend, A., Poole, C. (Eds.). Elsevier, Amsterdam.
Sanchez-Prado, L., Llompart, M., Lamas, J.P., Garcia-Jares, C., Lores, M., 2011a. Talanta 85, 370.
Sanchez-Prado, L., Lamas, J.P., Alvarez-Rivera, G., Lores, M., Garcia-Jares, C., Llompart, M., 2011b. J. Chromatogr. A 1218, 5055.
Sanchez-Prado, L., Alvarez-Rivera, G., Lamas, J.P., Llompart, M., Lores, M., Garcia-Jares, C., 2013. Anal. Methods 5, 416.
SCCNFP/0017/98, Draft Pre-opinion Concerning Fragrance Allergy in Consumers. http://ec.europa.eu/health/ph_risk/committees/sccp/documents/out93_en.pdf.
SCCNFP/0389/00, Opinion Concerning the 1st Update of the Inventory of Ingredients Employed in Cosmetic Products. Section II: Perfume and Aromatic Raw Materials. http://ec.europa.eu/health/archive/ph_risk/committees/sccp/documents/out131_en.pdf.
SCCNFP/0817/04, Musk Xylene and Musk Ketone. http://ec.europa.eu/health/archive/ph_risk/committees/sccp/documents/out280_en.pdf.
SCCS/1459/11, Opinion on Fragrance Allergens in Cosmetic Products. http://ec.europa.eu/health/scientific_committees/consumer_safety/docs/sccs_o_102.pdf.
Schulz, H., Baranska, M., 2005. Perf. Flavor. 30, 28.
Scott, R.P.W., 2005. In: Worsfold, P., Townshend, A., Poole, C. (Eds.), Encyclopedia of Analytical Science, Essential Oils. Elsevier, Amsterdam.
Shibuta, S., Imasaka, T., Imasaka, T., 2016. Anal. Chem. 88, 10693.
Tobolkina, E., Qiao, L., Xu, G., Girault, H.H., 2013. Rapid Commun. Mass Spectrom. 27, 2310.
Tranchida, P.Q., Dugo, P., Sciarrone, D., Dugo, G., Casilli, A., Mondello, L., 2008. LCGC Eur. 21, 130.
Tranchida, P.Q., Maimone, M., Franchina, F.A., Bjerk, T.R., Zini, C.A., Purcaro, G., Mondello, L., 2016. J. Chromatogr. A 1439, 144.
Van Asten, A., 2002. TrAC Trends Anal. Chem. 21, 698.
Villa, C., Gambaro, R., Mariani, E., Dorato, S., 2007. J. Pharm. Biomed. 44, 755.
Wang, L.H., Liu, H.J., 2012. J. Anal. Chem. 67, 64.
Wang, G., Tang, H., Chen, D., Feng, J., Li, L., 2012. Chin. J. Chromatogr. 30, 135.
Wang, L.H., Chen, J.X., Wang, C.C., 2014. J. Essent. Oil Res. 26, 185.
Wang, Z., 2012. Chin. J. Chromatogr. 30, 1178.
Zhou, C., Tong, S., Chang, Y., Jia, Q., Zhou, W., 2012. Electrophoresis 33, 1331.

CHAPTER 11

Surfactants in Cosmetics: Regulatory Aspects and Analytical Methods

M. Carmen Prieto-Blanco, María Fernández-Amado, Purificación López-Mahía, Soledad Muniategui-Lorenzo, Darío Prada-Rodríguez
Universidade da Coruña, A Coruña, Spain

INTRODUCTION

Surfactants are synthetic compounds whose specific properties of diminishing surface tension and adsorbing onto surfaces turn them into objects of study in various scientific and technical disciplines. The close relationship between surfactants and cosmetic products considered in terms of scientific research can be understood by the relative research production of the largest companies. Manufacturers (Dow, Merck, Abbot) and consumers (Procter, Unilever) of surfactants or related products (as cosmetic products) (Henkel, Kao and Firmenich), along with universities and state-owned research centres, are important agents in the production of scientific literature on surfactants.

A relevant part of this investigation is related to analytical techniques (Bailón-Moreno et al., 2006).

This chapter reviews the literature concerning analytical methodologies for the determination of surfactants in cosmetic products, especially since 2006, as the previous literature was reviewed in the first edition of this book. Although the thread running through this chapter deals with analytical techniques, the main classes of surfactants (anionic, cationic, amphoteric and non-ionic surfactants) and different types of cosmetic products are studied. The regulatory aspects are also discussed. Of particular interest is the determination of surfactant mixtures and residual products which may be incorporated into cosmetic products.

REGULATORY ASPECTS OF SURFACTANTS IN COSMETIC PRODUCTS

Regulations on cosmetic products concerning surfactants and related compounds published since 2009 are shown in Table 11.1.

Within the European Union (EU) framework, Commission Decision 2006/257/EC specified the ingredients employed in cosmetic products with a surfactant function. Moreover, many of the listed surfactants may have other functions (preservative, foaming, cleansing, etc.) and more than one of them may be used in a cosmetic

Table 11.1 Regulations on cosmetic products which affect surfactants and related compounds

Compound	Reference number	Regulation	Annex
Dimethyl sulphate	750	Annex II to Regulation (EC) No. 1223/2009 of the European Parliament and of the Council	Annex II: List of substances prohibited in cosmetic products
Benzyl chloride	650		
1,4-Dioxane	343		
α,α-Dichlorotoluene	1126		
Ethylene oxide	182		
Dimethylamine	142		
Nitrosamines	410: Nitrosamines, e.g., NDMEA, NDPLA, NDELA		
Alkylamines, alkylamides, alkanolamines and alkanolamides	411: Secondary alkyl- and alkanolamines and their salts	Annex II to EU Regulation (EC) No. 1223/2009 of the European Parliament and of the Council	
	60: Fatty acid dialkylamides and dialkanolamides	Annex III to EU Regulation (EC) No. 1223/2009 of the European Parliament and of the Council	Annex III: List of substances which cosmetic products must not contain except subject to the restrictions laid down/Part 1
	61: Monoalkylamines, monoalkanolamines and their salts		
	62: Trialkylamines, trialkanolamines and their salts		
Hexadecyl ammonium fluoride	36	Commission Regulation (EU) No. 344/2013 amending Annexes II, III, V and VI to Regulation (EC) No. 1223/2009 of the European Parliament and of the Council	
3-(N-hexadecyl-N-2-hydroxyethylammonio) propyl-bis(2-hydroxyethyl) ammonium difluoride	37		
N,N',N'-tris(polyoxyethylene)-N-hexadecylpropylenediamine dihydrofluoride	38		
9-Octadecen-1-amine hydrofluoride	39		
C16,18-Alkyltrimethylammonium chloride	286	Corrigendum to Commission Regulation (EU) No. 866/2014 amending Annexes III, V and VI to Regulation (EC) No. 1223/2009 of the European Parliament and of the Council	
C22-Alkyltrimethylammonium chloride	287		

Benzethonium chloride	53	Annex IV to Regulation (EC) No. 1223/2009 of the European Parliament and of the Council	Annex V: List of preservatives allowed in cosmetic products
Benzalkonium chloride	54		
Alkyl (C12–22) trimethylammonium bromide and chloride	44	Corrigendum to Commission Regulation (EU) No. 866/2014	
Alkylamine	—	Ministry of Health of the People's Republic of China hygienic standard for cosmetics	Other regulations and technical documents
Diethanolamides	—	2011 Amended Final Safety Assessment, Diethanolamides as Used in Cosmetics	Cosmetic Ingredient Review Expert Panel members
Ethanolamides	—	2012 Safety Assessment of Ethanolamides as Used in Cosmetics	
Alkyl betaines	—	2014 Safety Assessment of Alkyl Betaines as Used in Cosmetics	

NDELA, *N*-nitrosodiethanolamine; *NDMEA*, *N*-nitrosodimethylamine; *NDPLA*, *N*-nitrosodi-*n*-propylamine.

product. One of the objectives of the current European Cosmetics Regulation (EU Regulation, 2009) is '*to ensure ... a high level of protection of human health.*' In Chapter 4, restrictions for certain substances are established. A group of prohibited substances is listed in Annex II. The cosmetic products must not contain these substances, but according to Article 17, their unintended presence would be permitted if it is technically unavoidable and the cosmetic product is safe under reasonable conditions of use. Table 11.1 shows those substances deriving from the surfactants which may be present in cosmetic products. Another group contains the substances which can be used with restrictions. Annex III provides information on these substances, including the maximum concentration in ready-for-use preparations according to the cosmetic product type. Alkylamines, alkylamides and any cationic surfactants are the surfactant types considered restricted substances (see Table 11.1). New substances have been included in further amendments to Annexes II, III and IV of the EU Regulation, such as Commission Regulations No. 344/2013 and No. 866/2014. Moreover, some surfactants that act as preservatives are listed in Annex IV with their maximum concentration for use.

The establishment of the frame formulation is another important aspect of the EU Regulation. There it is defined as '*a formulation which lists the category or function of ingredients and their maximum concentration in the cosmetic product*'. Various frame formulations in which surfactants are one of the major ingredients are classified in 17 sections (skin care, mouthwash, etc.) in the Cosmetic Products Notification Portal (CPNP, 2014). Other technical documents elaborated by EU or American organizations discuss the role of some surfactants, such as the document concerning cetylpyridinium (CP) elaborated by Cosmetics Europe—The Personal Care Association, formerly named Cosmetics Europe Personal Care Association [Scientific Committee on Consumer Safety (SCCS), 2015], or those elaborated by Cosmetic Ingredient Review (CIR) on alkyl betaines (CIR, 2014), diethanolamides (CIR, 2011) or ethanolamides (CIR, 2012).

ANALYTICAL METHODS FOR THE DETERMINATION OF SURFACTANTS IN COSMETICS AND RAW MATERIALS

International Standardized Analytical Methods

As was mentioned in Chapters 3 and 4 of this book, some analytical methods for cosmetic products have been established by different international organizations such as the International Organization for Standardization (ISO) or the European Committee for Standardization (CEN). The CEN National Members implement European standards as national standards, distribute and sell them; an example of such National Members is the Spanish Association for Standardization and Certification (AENOR). In the United States, some analytical methods from the United States Pharmacopeia (USP) are used according to the US Food and Drug Administration (FDA) regulations.

Owing to the lack of international methods for determining surfactants in cosmetic products, analytical methods proposed for other samples are considered in this section when the target analytes are surfactants usually employed in cosmetic products. They are summarized in Table 11.2.

Note that ISO (code) is the ISO nomenclature for methods, while EN (code) is used for CEN methods. When a method has been adopted by both organizations, it is named by both acronyms.

Some of these methods are minor revisions of the previous versions, such as the determination of active matter in cationic surfactants (EN-ISO 2871-1 and 2871-2) or anionic surfactants (EN-ISO 2870). In other methods (i.e., EN 14480, EN 14669, see Table 11.2), the potentiometric titration, which allows automation (removing errors due to the operator), was introduced, and chloroform was replaced by other less toxic solvents. In general, these are methods for the determination of the active matter expressed

Table 11.2 International official methods which affect surfactants used as raw materials in cosmetic products

Type	Code	Determination	Technique
Anionic surfactant			
EN ISO	2870:2009	Active matter hydrolysable and non-hydrolysable	Sample treatment (hydrolysis) and potentiometric or direct two-phase titration
EN	14669:2005	Content of anionic surfactants and soap	Potentiometric two-phase titration
EN	14480:2004/ AC:2006	Content of anionic surfactants	Potentiometric two-phase titration
EN	14670:2005	Active content and molar mass of sodium dodecyl sulphate	Gas chromatography and extraction
EN ISO	8799:2009	Unsulphated active matter	Ion-exchange chromatography
Cationic surfactant			
EN ISO	2871–1:2010	High-molecular-mass active matter	Two-phase titration
EN ISO	2871–2:2010	Low-molecular-mass active matter	Two-phase titration
EN	14668:2005/ AC:2006	Content of quaternary ammonium surfactants	Potentiometric two-phase titration
Amphoteric surfactant			
EN	15109:2006	Active matter	Non-aqueous titration

as millimoles per 100 g (i.e., EN 14668), or as a percentage by mass (i.e., EN ISO 8799, EN 14670). Sometimes the molecular mass is known (i.e., EN ISO 15109).

Analytical Methods Published in the Scientific Literature

This section contains information about the methods published in the scientific literature from 2005 to 2015, concerning the determination of surfactants in cosmetic products. Previous publications were compiled in the first edition of this book. To facilitate reading, a classification based on the analytical technique used is shown in the following.

Electroanalytical Techniques

The potentiometry of surfactant-sensitive electrodes was devoted to simplifying, automating and miniaturizing the standard procedures used for surfactant titration in cosmetic formulations. Therefore, faster procedures and more adequate routine analysis can be achieved. Other characteristics such as lower consumption of toxic organic solvents, or a more reliable determination of the titration endpoint (mainly in turbid and coloured solutions), make the potentiometric methods more environmentally friendly and reliable than some first versions of standard methods with which they are compared. Some disadvantages such as limited life span of the electrodes and reduced selectivity and sensitivity (especially for direct determination) are also reported (Sánchez and del Valle, 2005; Bazel et al., 2014). In this section, the technical information on commercial electrodes and the development of new electrodes are discussed (see Table 11.3).

Commercial Electrodes for the Determination of Ionic Surfactants

Most commercial electrodes can be used as potentiometric indicators in titrations of ionic surfactants in cosmetic matrices, but not in direct measurements. Some of the companies which market these electrodes provide technical notes with the parameters and the conditions most suitable for the analysis of different formulations (Metrohm, 233/3, 269/4, T-10, T-12, T-46, T-48). Ion-selective electrodes with a polyvinyl chloride (PVC) membrane are employed for surfactant titration in aqueous media and aqueous mixtures of methanol or ethanol at controlled concentration of alcohol (Metrohm, 233/3). This type of electrode contains an ionophore in the membrane which establishes an equilibrium reaction with the analyte surfactant in solution. As a result, a difference in electrical potential in the membrane against the reference electrode is measured. The titrant type and concentration are chosen according to the analysed surfactant and the cosmetic matrix. The anionic surfactants are titrated with 1,3-didecyl-2-methylimidazolium chloride (DDMICl) in shower gel and shampoo (see Table 11.3). Sodium dodecyl sulphate (SDS) is used for the analysis of cationic surfactants in most formulations except for those containing amine fluoride (gargle solution and mouthwash), in which bis(2-ethylhexyl) sodium sulphosuccinate is the selected titrant (see Table 11.3). In any case,

Table 11.3 References on the determination of surfactants in cosmetic products by electroanalytical techniques according to the class of surfactant and type of cosmetic product

Class	References	Compound	Matrix	Type of electrode	Electroactive material	Plasticizer
Cationic	Madunić-Čačić et al. (2008)	CPC, CTAB, benzethonium chloride	Mouthwash	PVC membrane	DDMICl–TPB	o-NPOE
	Devi and Chattopadhyaya (2012)	CPC			CP–SnP	DOP
	Mostafa (2006)	CP			s-BT-TPB	DOP
	Sanan and Mahajan (2013)	CPC	Mouthwash, nasal drops	PVC membrane ion pair (ionic liquid)	C12MeIm-DDS	DOP
	Khedr et al. (2014)	TTAB (C12–C14) Dimethyl ethanol ammonium chloride	Shampoo	Modified carbon paste PVC membrane	ADHA-PM	DBP
	Ibrahim and Khorshid (2007)[a]	CTAB	Mouthwash		CTA-TCIP	TCP
	Khaled et al. (2008)	CPC HTMABr		Screen-printed carbon paste		o-NPOE
	Mohamed et al. (2010)	CPC Hexadecyltrimethylammonium		Screen-printed (in situ, modified and unmodified)	CP-TPB	o-NPOE
	Masadome et al. (2006)	CPC	Dental rinses	Cationic surfactant-ISFET with microfluidic polymer chip	C14-BAK-derivatized TPB	
	Samardzic et al. (2011)	CPC	Commercial disinfectants		TPB	
	Metrohm, AB-233	CPC	Mouth/dental wash	PVC membrane	SDS	
	Metrohm, AB-233, T-12	Total cationic surfactants	Hair conditioner		SLS	
	Metrohm, AB-233		Toothpastes (amine fluoride) Gargle solution (benzalkonium) Mouth/dental wash (amine fluoride)		DOSS	
	Metrohm, AB-269		Hair conditioner	Surfactrode		

Continued

Table 11.3 References on the determination of surfactants in cosmetic products by electroanalytical techniques according to the class of surfactant and type of cosmetic product—cont'd

Class	References	Compound	Matrix	Type of electrode	Electroactive material	Plasticizer
Anionic	Madunić-Čačić et al. (2012)	SLES	Shampoo, foam bath, shower gel	PVC membrane	HTA-TPB	o-NPOE
	Madunić-Čačić et al. (2011)	SLES	Shower gel, liquid soap		HTA-TPB	o-NPOE
	Devi and Chattopadhyaya (2013)	SDS	Toothpastes		CTA + -TPB-	DOP
	Mostafa (2008)	SDS	Liquid soap		CTA + -DDS-	o-NPOE
	Makarova and Kulapina (2015a)	Homologous sodium alkylsulphates	Shampoo, shower gel, raw materials		[CuSalen]DDS2, [Cu(Dipyr)2]DDS2, [Cu(Phen)2]DDS2	DBP
	Makarova and Kulapina (2015b)	Homologous sodium alkylsulphates	Shower gel, raw materials	Screen-printed	CP-DDS	o-NPOE
	Metrohm, AB-233, AN T-10	Total anionic surfactants	Shower gel, shampoo	PVC membrane	[CuSalen]DDS2, [Cu(Dipyr)2]DDS2, [Cu(Phen)2]DDS2 TEGOtrant	DBP
	Metrohm, AB-233	Total anionic surfactants	Toothpastes, mouthwash (lauryl sulphate)		TEGOtrant	
	Metrohm, AN T-46	Soap content	Soap noodles			
	Metrohm, AN T-48	SDS	Shower oil			
	Metrohm, AB-269	Total anionic surfactants	Cosmetic oil baths and shower oils	Surfactrode	TEGOtrant A100	

ADHA, alkyl dimethyl hydroxyethyl ammonium chloride; *BAK*, benzalkonium chloride; *C12MeIm*, dodecylmethylimidazolium; *CP*, cetylpyridinium; *CPC*, cetylpyridinium chloride; *CTA*, cetyltrimethylammonium; *CTAB*, cetyltrimethylammonium bromide; *DBP*, dibutylphthalate; *DDMICl*, 1,3-didecyl-2-methylimidazolium chloride; *DDS*, dodecyl sulphate; *DOP*, dioctylphthalate; *DOSS*, bis(2-ethylhexyl)-sodium sulfosuccinate; *HTA*, hexadecyltrioctadecylammonium; *HTMABr*, hexadecyltrimethylammonium bromide; *ISFET*, ion-selective field-effect transistor; *o-NPOE*, o-nitrophenyl octyl ether; *PM*, phosphomolybdic acid; *PVC*, polyvinyl chloride; *s-BT*, s-benzylthiuronium; *SDS*, sodium dodecyl sulphate; *SLES*, sodium lauryl ether sulphate; *SLS*, sodium lauryl sulphate; *SnP*, Sn(IV) phosphate; *SPCPE*, screen-printed carbon paste electrode; *SPE*, screen-printed electrode; *TClP*, tetrachloropalladate(II); *TCP*, tricresyl phosphate; *TTAB*, tetradecyltrimethylammonium bromide; *TPB*, tetraphenyl borate; *[CuSalen]DDS₂*, compound of dodecyl sulphate with a cationic complexe of copper (II) with N,N′-bis(salicylidene)ethylenediamine; *[Cu(Dipyr)₂]DDS₂*, compound of dodecyl sulphate with a cationic complexe of copper (II) with α,α′-dipyridyl; *[Cu(Phen)₂]DDS₂*, compound of dodecyl sulphate with a cationic complexe of copper (II) with 1,10-phenanthroline.

[a] Flow injection analysis.

the pH range in which the titration can be performed depends on the type of analysed surfactant and the other surfactants present in the formulation. However, this PVC membrane electrode is incompatible with fatty or oily formulations and neither can it be used in two-phase titration (Epton's titration) since the organic solvents swell or dissolve the PVC membrane. Commercial surfactant electrodes with the ionophore in chloroform-resistant graphite are applied in two-phase titration for the determination of anionic and cationic surfactants in shower oil and hair conditioner (see Table 11.3). The replacement of chloroform (a toxic solvent) in the titration with a mixture of ethanol and methyl isobutyl ketone is recommended. Using these solvents, surfactant electrodes with the ionophore in refill paste can be employed. In potentiometric two-phase titration, the type of titrant and the pH range are also important parameters which have to be selected to achieve a good extraction and steeper potential jump of the titration curve. The equivalence point is more precise using a commercial titrator than if it is determined visually. A special rod stirrer which provides an emulsion without trapped air bubbles allows automating the entire process. The result of titration can be expressed as the total surfactant concentration or single surfactant concentration if the composition of the matrix and the potential interferers are known. For industrial monitoring, the concentration of total surfactants is useful because it provides information on detergent efficiency of the formulation (Mostafa, 2008).

Development of Surfactant-Sensitive Electrodes for the Determination of Anionic Surfactants

New PVC membrane electrodes were developed for the determination of alkyl sulphates and alkyl ether sulphates in very different cosmetic products. This type of electrode has three components: PVC, ionophore and plasticizer. The ionophore must be stable, soluble in the membrane and lipophilic to avoid leaching to the sample solution. It is essential that it has rapid exchange kinetics (Khedr et al., 2014). One of the functions of the plasticizer is to solubilize the ionophore in the membrane, and it also must be compatible with the PVC. The combination of plasticizer and ionophore is a parameter of study because it has a strong influence on the analytical response of the membrane (Sánchez and del Valle, 2005). As shown in Table 11.3, the electrodes developed have as ionophores complexes from among cationic surfactants and anionic surfactants or metal complexes.

Madunić-Čačić et al. (2011) quantified sodium lauryl ether sulphate (SLES) in shower gel and liquid soap with results comparable to those obtained with both a commercial sensor and the standard two-phase titration. A sensor with a highly lipophilic ionophore (see Table 11.3) allows the determination of alkyl sulphates, alkyl ether sulphates and the homologous series of C_7–C_{12} sulphonates. Several cationic titrants were tested, but DDMICl provides the high jump curves at the equivalence point. Moreover, this sensor was employed to quantify SLES in formulations containing cocamidopropyl betaine (CAPB; an amphoteric surfactant used in conjunction to SLES because of its synergic effects). To avoid betaine interference and therefore to achieve good recoveries, the titration must be performed at pH 4. In addition, because of this interfering effect, the method can be applied only to formulations with sodium lauryl sulphate (SLS)/CAPB ratios in the range of 1:1 to 2:1 (Madunić-Čačić et al., 2012).

Another sensor with fewer lipophilic electroactive components [cetyltrimethylammonium (CTA)–tetraphenyl borate (TPB)] and a different plasticizer (see Table 11.3) was proposed for the determination of SDS in toothpaste by Devi and Chattopadhyaya (2013). CP chloride (CPC) was used as titrant, obtaining a near-Nernstian monovalent response, in the range of the studied SDS concentration and pH (4–8.5). Only interference by dodecyl benzene sulphonate was found. The results of titration are in agreement with those obtained by the spectrometric method using methylene blue. Mostafa (2008) proposed a PVC membrane electrode for the determination of SDS in liquid soap (and other environmental waters and pharmaceuticals). The determination was carried out by direct and standard addition techniques with results comparable with those obtained by the two-phase titration method (Li and Rosen, 1981). Moreover, the electrode can serve as an endpoint indicator in titrations using CPC or CTAB (cetyltrimethylammonium bromide). Of all the tested plasticizers, o-nitrophenyl octyl ether was the one that improved selectivity owing to better dissolution of the ionophore in the membrane. The performance of the sensor in terms of sensitivity, long life span, short response, reproducibility and selectivity was demonstrated. The limit of detection was 3×10^{-6} M (Mostafa, 2008).

The total anionic surfactant concentration was determined in several raw materials and shampoo and shower gel samples using new sensors designed by Makarova and Kulapina (2015a). They are based on PVC membranes with various ionophores [dodecyl sulphate (DDS)–cationic complex of copper with some organic reagents] (see Table 11.3). The standard addition method and potentiometric titration with CPC were used for the quantification of homologous alkyl sulphates, alkyl sulphonates and alkylbenzene sulphonates. Wide pH (3–12) and concentration ranges for homologous sodium alkyl sulphates and a very low limit of detection for SDS of 1×10^{-7} M were obtained. Other features were a fast and stable response, a nearly Nernstian slope and selectivity with the compounds found in formulations (phosphates, citrates, sulphate, etc.), and the analysis time was between 30 and 40 min. These authors (Makarova and Kulapina, 2015b), using the same ionophores, as well as CP–DDS, designed miniaturized planar surfactant sensors which were applied to determine alkyl sulphates in cosmetic matrices and raw materials. They are screen-printed sensors (SPSs) with polymeric polyurethane substrate and home-made ink containing materials such as graphite or carbon nanotubes. The carbon ink composition, the effects of the type and content of the plasticizer and the effects of the type and content of the ion associates were examined for the construction of the electrode. The determination was carried out by titration in a 10-min analysis time and the method was applied to shower gel and raw materials containing alkyl sulphonates. Smaller volumes can be used with SPSs, but the limit of detection and work concentration range were similar to those of PVC membrane electrodes.

Development of Electrodes for the Determination of Cationic Surfactants

CP and benzalkonium chloride (BAK) are the most commonly analysed cationic surfactants in mouthwash using PVC membrane electrodes and chemically modified carbon paste electrodes (CMCPEs), although electrodes based on screen-printed technology have also been proposed. CP has been determined in mouthwash with three electrodes containing different electroactive materials (some of them highly lipophilic; Mostafa, 2006) and plasticizer (Madunić-Čačić et al., 2008; Devi and Chattopadhyaya, 2012) (see Table 11.3). The CP determination was performed by direct determination (Mostafa, 2006; Devi and Chattopadhyaya, 2012) and using the standard addition method (Mostafa, 2006). All proposed electrodes can be employed as indicators for potentiometric titration using SDS or TPB as titrant. The results obtained were in a satisfactory agreement with those obtained by the British Pharmacopoeia method (Mostafa, 2006), by the two-phase titration method (Devi and Chattopadhyaya, 2012) and by using a commercial sensor (Madunić-Čačić et al., 2008). A wide working pH range (2–11), a short response time (4–6 s) and a Nernstian response for all cationic surfactants are the distinguishing characteristics of the sensor proposed by Madunić-Čačić et al. (2008). An ionic liquid, dodecylmethylimidazolium, in conjunction with DDS was used as the ionophore for the construction of the PVC membrane sensor, which showed potentiometric response to three surfactants (CPC, tetradecyltrimethylammonium bromide and BAK) with a super-Nernstian slope. The three cationic surfactants were determined by titration and direct measurement. The results were concordant with those obtained by the two-phase method and with the declared content in formulations of mouthwash, antiseptic and nasal drops (Sanan and Mahajan, 2013).

Another kind of electrode, the CMCPE, in which the PVC membrane is substituted by carbon paste, has been developed. In general, it provides a greater stability, reproducibility and lower limit of detection than PVC membrane electrodes in the analysis of surfactants. Ibrahim and Khorshid (2007) compared the potentiometric features of CMCPEs and PVC membrane electrodes. Several counter-anions of ionophores (using CTA as counter-cation), plasticizers and additives to increase the lipophilicity were tested. The lower limit of detection (about 10^{-7} M) and a more stable response were obtained when using CMCPEs rather PVC membrane electrodes. The additives improve the electrode selectivity for CTA when other cationic surfactants of a similar structure are present in the formulation. CTA in mouthwash and gargle solution was quantified using this optimized electrode in an online flow injection analysis (FIA) system by means of the standard addition method. Khedr et al. (2014) determined alkyl dimethyl hydroxyethyl ammonium chloride in shampoo with a CMCPE by standard addition and curve calibration, although it can be used in potentiometric titration with TPB. For the construction of this electrode, different heteropoly acids used as counter-anions in the ionophore and plasticizers were examined. Moreover, the amounts of ionophore and plasticizers, as well as the graphite/plasticizer relationship, were optimized.

Khaled et al. (2008) developed disposable screen-printed carbon paste electrodes (SPCPEs) for the potentiometric titration of CPC and CTA with sodium TPB in mouthwash. The disposable strips containing work and reference electrodes were constructed by the optimization of several parameters (plasticizer content and type, binding material type and carbon content). Their performance was compared with that of other types of electrodes [screen-printed electrodes, carbon paste electrodes (CPEs) and PVC]. Their advantages include rapid response time and stability of potential reading, allowing the titration in 5 min versus 10 min for PVC membrane or 15 min for the two-phase titration. Furthermore, they can be used in field measurements, as they measure on small volumes and are expected to last approximately 3 months without any performance change. The results are comparable with those obtained when using a commercial electrode and the British Pharmacopoeia method. A similar study concerning the analysed surfactants and cosmetic products was carried out comparing SPCPEs, CPEs and PVC membrane electrodes which were prepared with different methods (in situ, modified and soaking) for the incorporation of the ion pair (ionophore). Modified and soaked SPCPEs presented the best performance and in situ PVC a better response (Mohamed et al., 2010).

Other approaches, such as microchips integrated with electrochemical sensors, specifically with ion-selective field-effect transistors, were applied to the determination of CPC in mouthwash. The electrode produced and described by Masadome et al. (2006) also responded effectively to other cationic surfactants, such as stearyltrimethylammoniun and benzylcetyldimethylammonium.

Spectroscopic Techniques

In recent years, UV/VIS spectrophotometric titration of ionic surfactants has involved the reaction of UV/VIS-undetectable surfactants (analyte or titrant) with chromoionophores and subsequent measurement using sensors. Dürüst et al. (2014) determined anionic surfactants in several cosmetic products (see Table 11.4) using polymer film-based optodes containing 2′,7′-dichlorofluorescein octadecyl ester.

The polymer film showed a colour shift (from yellow to pink) in the presence of excess cationic titrant (Fig. 11.1).

The optodes were used in the format of a microtitre plate with an absorbance reader so that the method could be suitable for routine control. In this way, 96 samples can be analysed in less than 5 min with good recovery and a detection limit of 1 ppm. Kuong et al. (2007) proposed chemically modified (with an anionic protective group) gold nanoparticles as probes for the semi-quantitative determination of alkytrimethylammonium surfactants in hair conditioner within 10 min. A colour shift is produced from red

Table 11.4 References on the determination of surfactants in cosmetic products by spectroscopic techniques according to the class of surfactant and type of cosmetic product

Class	References	Compound	Matrix	Technique
Cationic	Cumming et al. (2010)	Polyquaternium	Raw materials	Spectrophotometric titration
	Kuong et al. (2007)	C10,12,14,16,18-Trimethylammonium	Hair conditioner	Sensor/nanoparticles
	Ensafi et al. (2009)	CTAB, TBAC, CPC	Conditioner shampoo	FIA
	Afkhami et al. (2011)	CTAB, DTAB, CPB	Conditioner shampoo	UV spectroscopy
	Karimi et al. (2008)	DTAB, CTAB, CPC	Shampoo, mouthwash	UV spectroscopy
Anionic	Dürüst et al. (2014)	NaDBS	Shampoo, shower gel, toothpaste, liquid soap, face cleaning gel	Potentiometric sensor/titration
Non-ionic	Pasieczna-Patkowska and Olejnik (2013)	Cetyl alcohol, PEG	Depigmentation cream, black eye pencil	Infrared spectrometry

CPB, cetylpyridinium bromide; *CPC*, cetylpyridinium chloride; *CTAB*, cetyltrimethylammonium bromide; *DTAB*, dodecyltrimethylammonium bromide; *FIA*, flow injection analysis; *NaDBS*, sodium dodecylbenzenesulfonate; *PEG*, polyethylene glycol; *TBAC*, tetra-n-butyl ammonium chloride.

Figure 11.1 Microplate wells after the addition of cationic titrant showing some colour shift on the polymer film-based optodes (from initial yellow to pink) (Dürüst et al., 2014).

to indigo/purple in a wide concentration range, and beyond a critical micelle concentration the colour shifts back to red. The method is very simple and no instrumentation is required because the estimation is performed by comparing a series of diluted hair conditioners with those containing surfactant standards. No interference of anionic surfactant was found.

Another kind of cationic surfactant, polymeric quaternary ammonium salts (polyquaterniums), of relatively low charge density, are also used in cosmetic products because of their properties of adsorption to skin or hair, forming a film with anti-static properties. Titration of polyquaterniums for the determination of charge density and concentration was carried out using the potassium salt of polyvinyl sulphate (PVSK) as titrant and o-toluidine blue as an indicator. The latter is a metachromatic dye which changes from blue to pink when it reacts with the PVSK after all polyquaternium has been consumed in the titration. Normalities as low 3×10^{-5} N can be determined with the visual endpoint and 3×10^{-6} N with the spectrometric endpoint. Anionic surfactants such as SDS do not interfere, but the titration interferes with other cations and it may be not possible to differentiate among distinct polyquaterniums (Cumming et al., 2010).

Very common cationic surfactants in cosmetic preparations, such as CTAB, can be determined in shampoo by means of the spectrophotometric method, based on the formation of a complex between a cationic surfactant and a dye. The Orange II ion pair with CTAB after extraction in chloroform allowed a limit of CTAB detection of 0.012 µg/mL (Afshar and Parham, 2010). Another method for the flow injection spectrophotometric determination of cationic surfactants using Eriochrome Black T was developed by Ensafi et al. (2009). In this case, the complex can be measured directly in aqueous medium avoiding the step of the liquid–liquid extraction with organic solvents. The chemical parameters (pH, reagent concentration and ionic strength) and FIA variables (reaction coil length, pump flow rate and sample volume) were optimized. The method, which has a throughput of 35 ± 5 samples per hour, was applied to two types of conditioner shampoo for CTAB determination with good recovery. Another method, developed by Afkhami et al. (2011), allows the determination of CTAB in conditioner shampoo based on the effects that cationic surfactants cause on the absorption spectra of the complexes between Al^{3+} or Be^{3+} and Chrome Azurol S (Chrome AS). Better limits of detection (LODs) are obtained with complexes of Al^{3+} or Be^{3+} and the method has been applied to other cationic surfactants (see Table 11.4). Simultaneous binary mixtures can be analysed based on this principle (complexes of beryllium with Chrome AS) using continuous wavelet transforms (Afkhami et al., 2009). CTAB in shampoo and CPC in mouthwash are determined using a method with previous treatment before the spectrometric determination. Cationic surfactants are separated from SDS by adsorption of the latter on γ-alumina and after being concentrated on C_{18} cartridges. In the final step, the competition between methylene blue and cationic surfactants for the formation of a complex with SDS is used by the spectrometric measurement, because the substitution by the cationic surfactants within the complex results in a quantitative increase in absorption (Karimi et al., 2008).

Pasieczna-Patkowska and Olejnik (2013) proposed infrared (IR) spectrometry to assess the composition of cosmetic products with regard mainly to their major components. Two non-ionic surfactants were identified in cosmetic preparations, polyethylene glycol in black eye pencil by its characteristic bands from 3347 to 887 cm^{-1} and cetyl alcohol in depigmentation cream by bands from 2955 to 719 cm^{-1}.

Moreover, a Japanese patent determined the composition of ethoxylated non-ionic surfactant using matrix-assisted laser desorption/ionization–time-of-flight–mass spectrometry (Yoshida, 2006).

Chromatography and Related Techniques

The studies collected in this section pay particular attention to the determination of individual surfactants in mixtures similar to cosmetic preparations which contain various classes of surfactants. The identification of surfactants of special structural complexity as non-ionic surfactants is another focus of interest. Moreover, the use of the liquid chromatography (LC) as the technique of choice is remarkable in most cases.

Ionic and Amphoteric Surfactants

Only a few works were reported during the target period on the analysis of ionic and amphoteric surfactants in cosmetic samples (Table 11.5). They are mainly focused on the determination of a large number of cationic or amphoteric surfactants or on a more complex analysis of cationic polymers using capillary electrophoresis (CE) and LC.

Six types of cationic surfactants—benzylalkyldimethylammonium, imidazolinium, alkyltrimethylammonium, dialkyldimethylammonium, trialkylmethylammonium and tetraalkylammonium (14 individual compounds)—are separated using a C_{18} column and methanol/water (0.15% trifluoroacetic acid, TFA) gradient elution within 52 min. Evaporative light-scattering detection (ELSD) provides a suitable sensitivity for quality control of the surfactants without a chromophore group. The method was applied to hair rinse, quantifying C16- and C18-trimethylammonium at 1% and 0.2%, respectively (Ryu et al., 2007). To evaluate the efficacy of conditioning agents in shampoo, two cationic polysaccharides were determined on hair after treatment with these conditioning compounds by LC–tandem mass spectrometry (MS/MS) (Ungewiβ et al., 2005). Polyquaternium-10 and guar hydroxypropyltrimonium chloride were extracted in acid medium with TFA to break the ionic bonds on the hair surface and were hydrolysed under optimized conditions (100°C, 20 h) to obtain the monomers. The separation was performed using diol stationary phase and gradient elution with *tert*-butylmethyl ether, methanol and water. The fragmentation mechanism was studied with electrospray ionization (ESI) and the transition of the cationic monomer to the trimethylammonium ion was used. The method was applied for the determination of both polymers on

Table 11.5 References on the determination of ionic and amphoteric surfactants in cosmetic products by chromatographic techniques according to the class of surfactant and type of cosmetic product

Class	References	Compound	Matrix	Technique
Cationic	Ungewiß et al. (2005)	Polyquaternium-10, Guar hydroxypropyltrimonium chloride	Hair	LC–MS, LC–MS/MS
	Ryu et al. (2007)	C12,14-Benzyldimethylammonium C34-, C36-, C38-imidazolinium C12,14,16,18-trimethylammonium Di(C12,14,16,18)-dimethylammonium Tri(C8,7)-methylammonium Tetra(C8)-ammonium	Hair rinse	LC–ELSD
	Mohammad and Mobin (2015)	C16-trimethylammonium chloride	Shampoo	TLC
Amphoteric	Koike et al. (2007)	C8–C18-AO C8–C18-APAO C8–C18-Bt C8–C18-APB	Shampoo	CE

AO, alkyldimethylamine N-oxide; *APAO*, alkylamidopropylamine N-oxide; *APB*, alkylamidopropylbetaine; *Bt*, alkylbetaine; *CE*, capillary electrophoresis; *ELSD*, evaporative light-scattering detection; *LC*, liquid chromatography; *MS*, mass spectrometry; *MS/MS*, tandem mass spectrometry; *TLC*, thin-layer chromatography.

undamaged and bleached hair, obtaining values in the range 150–300 µg/g. Moreover, thin-layer chromatography was used to separate a mixture of three cationic surfactants (cetrimonium chloride, benzyl trimethylammonium chloride and tetramethylammonium hydroxide). Cetrimonium chloride was identified in a shampoo sample by applying this method developed by Mohammad and Mobin (2015). CTAB was also analysed in hair conditioner using capillary isotachophoresis with conductivity detection within 15 min. The concentration of ethanol added to the electrolyte has to be optimized to avoid the formation of micelles (Praus, 2005).

Concerning the analysis of amphoteric surfactants, a method for the determination of six homologues of four types of amphoteric surfactants (24 individual compounds) with an analysis time of 17 min by CE using indirect UV detection was proposed by Koike et al. (2007). One of the analysed types was alkylamine oxides, a subclass which does not present a formal charge and acts as a cationic or non-ionic surfactant depending on pH. Under neutral and alkaline conditions they behave as non-ionic surfactants, but under pH <3 they are protonated and act as cationic surfactants. However, they are classified by some authors (Koike et al., 2007) as amphoteric surfactants considering the dipolar moment between their nitrogen and oxygen atoms. Thus, two of the betaines most commonly used in cosmetic products (alkyl and alkylamidopropyl betaines) and amine oxides were separated taking into account the effect of pH, the organic modifier on the background solution and the type of chromophore. Acetonitrile 50%, pH 2.0, and benzyltrimethylammonium chloride as chromophore were the optimized conditions which allowed satisfactory parameters of validation. A single homologue (C12) of amidopropyl betaine was identified and quantified at 5% in a shampoo sample by applying the proposed method.

Finally, a study is reported on the analysis of anionic surfactants in cosmetic products. Tian and Qin (2014) proved that a complex of dimethylformamide and acetic acid can be used as a background electrolyte for CE-coupled contactless conductivity detection. The new electrolyte was applied to the identification of C12- and C16-alkyl sulphates in a shampoo sample.

Non-Ionic Surfactants

Non-ionic surfactants are ingredients of cosmetic products carrying out diverse functions (as foam stabilizers, thickeners, emulsifiers, dispersing agents, etc.). From the point of view of their analysis this may be the most complex class of surfactants. Many of them are polymers synthesized from natural acids or oil, fatty alcohols and acids, carbohydrates and ethylene oxide (EO) and/or polypropylene oxide. The end product is a mixture of hundreds of individual compounds including homologues of alkyl chain size, oligomers with different EO or glucose units, stereoisomers and ring isomers (Nasioudis et al., 2011; Beneito-Cambra et al., 2013). In addition to this drawback, most of them lack a chromophoric or fluorophoric group for detection and lack a standard for quantification. LC with ELSD or charged aerosol detection (CAD) is a good solution when only some compounds (oligomers, isomers) are determined. Both ELSD and CAD are universal detectors applied to non-UV-absorbing compounds by nebulization and evaporation of the mobile phase. These detectors give no linear response between signal and concentration or mass, but by changing to log–log curves, a linear response is obtained. However, for a more complete characterization (but usually it is partial), LC coupled with mass spectrometry (MS) is the chosen technique. In the target period, methods for different purposes (characterization, quality control, etc.) for alkylpolyglucosides, fatty acid esters of sugar or polyethylene glycol and castor oil ethoxylates were developed (see Table 11.6).

Alkylpolyglucosides (APGs), synthesized from glucose and fatty alcohols, are carbohydrate-derived surfactants. They are commonly used in shampoo preparations and other products of personal care because of their skin compatibility and synergic effect with other surfactants (Beneito-Cambra et al., 2007; Hübner et al., 2006). From the mixtures of individual compounds which constitute the industrial APGs, it is necessary to distinguish the alkylmonoglycosides with different chain lengths attached to glycosyl

Table 11.6 References on the determination of non-ionic surfactants in cosmetic products by chromatographic techniques according to the class of surfactant and type of cosmetic product

Subclass	References	Compound	Matrix	Technique
APGs	Hübner et al. (2006)	C8,10,12-α,β-APG	Shampoo, technical mixtures	CE–PAD
	Beneito-Cambra et al. (2007)	C1–C16-G1	Baby shampoo, hand cream, raw material	LC–ESI–MS
PEG fatty acid esters	Lee et al. (2008)	PEG-40,55 stearates	Lotion, cream	LC–ELSD
Castor oil ethoxylates	Nasioudis et al. (2011)	Castor oil ethoxylates	Raw material	LC–ESI–MS
Sugar fatty acid ester	Lie et al. (2013)	Sucrose caprate		LC–CAD

APG, alkylpolyglucoside; *CAD*, charged aerosol detection; *CE*, capillary electrophoresis; *ELSD*, evaporative light-scattering detection; *ESI*, electrospray ionization; *G1*, one glucose unit in the alkylpolyglycoside; *LC*, liquid chromatography; *MS*, mass spectrometry; *PAD*, pulsed amperometric detection; *PEG*, polyethylene glycol.

units, their stereoisomers (α- and β-epimers), the ring isomers (pyranoside and furanoside forms) and positional isomers. Moreover, di- and polyglycosides can be formed. Using an LC–ESI–MS system, two chromatographic columns (cyano and alkylamide columns) with acetonitrile/water were tested by Beneito-Cambra et al. (2007). A good resolution between α- and β-epimers and ring isomers of alkylmonoglycosides was achieved using isocratic conditions and an alkylamide column. Using a cyano column and gradient elution, the ring isomers were resolved in a shorter time and less equilibration time was required. The detection was performed in positive mode with adducts of sodium ions. Differences in factor response were found according to the chain length (increasing from C1 to C10 and decreasing for C12 and C14). APGs were analysed in technical mixtures, a baby shampoo and a hand cream. In baby shampoo, monoalkylglycosides with C18, C10 and C12 alkyl chains were quantified at a concentration range of 0.2%–0.6%; moreover, the isomers of dialkylglycosides were partially resolved. Differences between the furanoside/pyranoside relationship were observed in two cosmetic products (shampoo and hand cream).

Micellar electrokinetic chromatography coupled to pulsed amperometric detection is another technique that allows separating APGs according to alkyl chain length and stereoisomers. For the coupling, a special configuration with micro-cylindrical gold electrodes was constructed by Hübner et al. (2006). Two electrolytes were optimized, one distinguishing the short-chain monoalkylglycosides (up to C10) along with the stereoisomers, and the other separating only homologues up to C12. The method applied to a technical mixture and a shampoo sample showed the possibility of identifying the producer of APGs for comparing them to the electropherograms. Other carbohydrate-derived surfactants are sugar fatty acid esters used as additives in cosmetic products for their antimicrobial and emulsifying properties. They are synthesized by esterification of sucrose with fatty acids, forming a high number of isomers. The separation of the eight regioisomers of the sucrose caprate was performed in a commercial sample by reversed-phase LC with CAD. Considering the resolution and time analysis, the best conditions were optimized using the design of experiments approach and a multivariate regression analysis. The analysis time using gradient elution was 24 min and the sensitivity provided was in the range 10–100 ng (Lie et al., 2013).

Another type of ester non-ionic surfactant, polyethylene glycol (PEG) stearates, is also used as emulsifiers in cosmetic products. Lee et al. (2008) developed two methods for the separation of highly ethoxylated stearates: PEG-25 (17–38 EO units), PEG-40 (27–52 EO units), and PEG-55 (37–68 EO units) by LC–ELSD. The first method used a C_8 column and water/acetonitrile gradient elution to separate the oligomers in order of increasing size. A single peak for each surfactant (in any case two or three from homologues) was obtained with a C_{18} stationary phase and water/methanol gradient elution for the optimized second method. The application to cosmetic samples gave values of 1% PEG-40 in lotion and 1.5% PEG-55 in cream. No interference with other ingredients was found and the detection limit was around 50 μg/mL.

The ethoxylate derivatives of castor oil (rich in ricinoleic acid) are used as solubilizers in cosmetic products. The characterization of castor oil ethoxylates from technical samples was investigated by Nasioudis et al. (2011) using LC coupled to several MS instruments (ion traps, time-of-flight mass analysers and Orbitrap). Of the ionization techniques tested, ESI gave higher response rates, and atmospheric-pressure chemical ionization completed the structural information. Moreover, the analysis of the positive and negative modes provided more complete information, detecting in the negative mode the presence of di-, tri-, and tetraricinoleate. In addition, different adduct ions were tested for optimization of MS detection and multistage MS fragmentation. A unique fragmentation pathway of ricinoleic acid, which could be used to evaluate the polymerization, was also proposed.

Mixtures Containing Different Classes of Surfactants

The separation of surfactant mixtures can arise with different levels of complexity. The simplest approach would be the separation of several classes of surfactants which are part of the specific cosmetic formulations. The most common formulations are mixtures of anionic surfactants (alkyl sulphate and/or alkyl ether sulphates), amphoteric surfactants (CAPB) and non-ionic surfactants (i.e., APGs), used for shower gel preparations, or mixtures of cationic surfactants (i.e., alkyltrimethylammonium) and non-ionic surfactants (i.e., fatty alcohols) for the preparation of hair conditioners (Terol et al., 2010). On a second level, the simultaneous separation of alkyl homologues of different surfactant classes (Park and Rhee, 2004) should be pointed out, in which various mechanisms of chromatographic retention have to be controlled. The simultaneous separation of several types of surfactants according to alkyl homologues and degree of ethoxylation (oligomers) is a process of great complexity. In this case, it may be necessary to separate more than 100 compounds (Elsner et al., 2012) (see Table 11.7).

LC with UV, refractive index (RI) or ELSD, as well as LC–MS, is employed in most analyses with a previous simple sample treatment (dilution and filtration). A more expensive and complex instrumentation is needed for comprehensive two-dimensional chromatography [LC × LC and gas chromatography (GC) × GC] applied to the separation of homologues and oligomers in mixtures of several surfactant classes. Differences in analytical strategy are found between LC × LC and GC × GC. For the former, direct sample analysis (sometimes with dilution) can be performed. However, a previous online or offline treatment is required for GC × GC, because of the low volatility of the surfactants. Offline derivatization, depending on the surfactant class (silylation for nonionic and silylation and hydrolysis for anionic and thermal decomposition in the injector for amphoteric and cationic surfactants), is one of the options employed (Hübner et al., 2007; Wulf et al., 2010). Ripoll-Seguer et al. (2013) addressed a previous separation of surfactant classes in the sample treatment by solid-phase extraction (SPE), carrying out the separation of homologues and oligomers in a simple and inexpensive chromatographic system. In this section, the separation of raw material mixtures is also included when they are likely to be applied to cosmetic products (Elsner et al., 2012; Wulf et al., 2010; Lobachev and Kolotvin, 2006).

Analytical methods using reversed-phase LC combined with ELSD were proposed for the determination of various surfactant classes (most of them not detectable by UV) with acceptable separation of homologues (see Table 11.7). A method for the simultaneous determination of eight surfactants of the non-ionic class (APG, fatty alcohol ethoxylates, coconut diethanolamide) and ionic class (alkyl sulphate and alkyl sulphonate) for quality control of the formulations was proposed by Park and Rhee (2004). Using a methanol/water gradient, the mixture was separated within 30 min, although two anionic surfactants could not be quantified and the alkyl chain of three ethoxylated surfactants was not separated. When the optimized method was applied, SLS in toothpaste (at around 3%) and coconut diethanolamide in body cleanser (around 2%) were quantified. Cationic, 17 non-ionic and amphoteric surfactants (any ethoxylated ones) were simultaneously separated within 60 min according to the alkyl chain using a methanol/water mobile phase containing 0.2% TFA (ion-pairing reagent) in a gradient elution (Park et al., 2006). The method was proposed for routine analysis and its applicability in cosmetic products was demonstrated with the quantification of CAPB and coconut diethanolamide in liquid body cleanser. Im et al. (2008) optimized the separation of four surfactant classes (13 compounds, including the homologues) using different columns, mobile phase and percentage of TFA. The best results were obtained using a C_{18} column with high carbon content and under a gradient elution with a ternary mobile phase

Table 11.7 References on the determination of surfactant mixtures in cosmetic products by chromatographic techniques according to the class of surfactant and type of cosmetic product

Class	References	Compound	Matrix	Technique
Cationic	Im et al. (2008)	CTAC, STAC, BTAC	Shampoo, hair conditioner	LC–ELSD, LC–MS
Anionic		SLS		
Amphoteric		CAPB		
Non-ionic		CME		
Cationic	Park et al. (2006)	DMDS, DMDP, DMDM, DMDL, BAK, IMD	Body cleanser	LC–ELSD
Amphoteric		CAPB		
Non-ionic		A/O, CDE		
Cationic	Afkhami et al. (2009)	CTAB, DTAB, CPB	Hair conditioner, mouthwash	UV spectroscopy
Non-ionic		Triton X-100		
Anionic	Ripoll-Seguer et al. (2013)	AES ($12 \leq n \leq 18$) and average EO number $m=3$, LAS (mixture of $10 \leq n \leq 13$ oligomers)	Shampoo, shower gel	LC–UV
Non-ionic		FAEs with $12 \leq n \leq 18$, $m=7$		
Anionic	Park and Rhee (2004)	SLS, AOS	Body cleanser, tooth paste	LC–ELSD
Non-ionic		CDE, LAE9-cap, LAE7, NPE7, C8,10-APG		
Cationic	Terol et al. (2010)	HTMABr	Hair conditioner	LC–RI
Non-ionic		1-Octadecanol, 1-hexadecanol, cetearyl alcohol		
Cationic	Wulf et al. (2010)	C12-, C14-BAK	Raw materials	GC × GC
Non-ionic		C10–C16 FAAO		
Non-ionic	Hübner et al. (2007)	n-alkyl-β-maltosides (C10-M, C12-M, C14-M, C16-M) n-alkyl-α(β)-glucopyranosides (C8–P, C10–P, C12–P)	Shampoo, shower gel	GC × GC–TOF–MS
Anionic		Amidoamines (C8–C18-ADA) and ethylhexylsulphate		
Amphoteric		CAPB		

A/O, amine oxide; *ADA*, amidoamine; *AES*, alkyl ethoxylated sulphates; *AOS*, alpha olefin sulfonate; *APG*, alkylpolyglucosides; *BAK*, benzalkonium chloride; *BTAC*, behenyltrimethylammonium chloride; *CAPB*, cocamidopropyl betaine; *CDE*, coconut diethanolamide; *CME*, cocomonoethanolamide; *CTAB*, cetyltrimethylammonium bromide; *CTAC*, cetyltrimethyl ammonium chloride; *DMDL*, dimethyldilaurylammonium chloride; *DMDM*, dimethyldimyristylammonium chloride; *DMDP*, dimethyldipalmitylammonium chloride; *DMDS*, dimethyldistearylammonium chloride; *DTAB*, dodecyltrimethylammonium bromide; *ELSD*, evaporative light-scattering detection; *EO*, ethylene oxide; *FAAO*, fatty alcohol alkoxylates; *FAE*, fatty alcohol ethoxylate; *GC*, gas chromatography; *HTMABr*, hexadecyltrimethylammonium bromide; *IMD*, immidazolinium salt; *LAE*, lauryl alcohol ethoxylate; *LAS*, linear alkylbenzene sulphonate; *LC*, liquid chromatography; *MS*, mass spectrometry; *NPE7*, nonylphenyl 7 mol ethoxylate; *RI*, refractive index; *SLS*, sodium lauryl sulphate; *STAC*, stearyltrimethylammonium chloride; *TOF*, time of flight.

(tetrahydrofuran, acetonitrile, water and 0.5% TFA). Raw materials used for the method optimization were characterized by fast-atom bombardment–MS and elution peaks were identified by LC–ESI–MS. For the common surfactants present in hair conditioner, all homologues were separated. For those present in shampoo, although the dominant homologues were separated, a complete resolution for all homologues was not possible. From the analysis of three different commercial shampoos and four hair conditioners, different compositions can be observed in regard to the surfactants found and to their concentration. For the three methods described, the calibration curves were linear in the log–log plot of the peak area versus the injected amount (micrograms) taking the peak area from the dominant homologues. Furthermore, LODs were at the micrograms/microliters level and no matrix effect was observed.

Another method suitable for quality control using reversed-phase LC with RI detector is that proposed by Terol et al. (2010). A cationic surfactant (cetrimonium chloride) and four anionic surfactants (1-tetradecanol, 1-hexadecanol, 1-octadecanol and 1-eicosanol) were determined in nine commercial samples of hair conditioner of different types and brands (see Table 11.7). Chromatographic conditions involved the use of the C_{18} column, methanol/water as the mobile phase, acid pH and $NaClO_4$ as the ion-pairing reagent. In three types of hair conditioner—daily rinse-off conditioner, usual treatment or mask and leave-on treatment—two to four surfactants were found, with values for each surfactant of less than 5%.

Among the methods that used SPE for the sample treatment, those developed by Lobachev and Kolotvin (2006) for five non-ionic surfactants and an amphoteric surfactant (CAPB) should be mentioned. A preliminary fractionation with ion-exchange SPE was performed, and a subsequent separation with reversed-phase LC using two detectors in series (UV and RI). The effect of the mobile phase was examined, and the ternary mobile phase (water/acetonitrile/methanol) in the presence of acid pH and $NaClO_4$ allowed the separation of the predominant homologues of individual surfactants. The method is recommended by the quality control of the finished cosmetic products.

Ripoll-Seguer et al. (2013) carried out a several-step sample treatment: solid-phase on weak ion exchanger, evaporation to dryness, dissolution in 1,4-dioxane and derivatization. In the first step, two fractions are collected, the anionic surfactants in a basic medium and non-ionic surfactants in an acid medium. Each fraction is derivatized (further evaporation) by esterification of the alcohols and transesterification of the anionic surfactants except for linear alkylbenzene sulphonate (LAS). However, two runs are necessary for each class; the chromatographic conditions are optimized for saving time using the same mobile phase for both analyses. The mobile phase consisted of a water/acetonitrile mixture containing 0.1% acetic acid and 0.1 M $NaClO_4$. In the anionic fraction, the selectivity between LAS and alkyl ethoxylated sulphates (AES) was achieved by the addition of sodium perchlorate in the mobile phase. Moreover, an excellent separation of derivatives of the fatty alcohol ethoxylates (FAEs) and AES oligomers in the non-ionic fraction was obtained under the optimized conditions. Three samples of shampoo with different compositions and one of shower gel were analysed. Other components such as CAPB were eluted with a lower retention time than FAEs.

Comprehensive two-dimensional LC coupled to quadrupole time-of-flight (TOF)–MS was used for the separation developed by Elsner et al. (2012) and applied to mixtures of anionic, non-ionic and amphoteric surfactants, a total of 110 compounds, including the homologues and oligomers of each subclass (APGs, FAEs, fatty alcohol sulphates, fatty alcohol ether sulphates and CAPB). Hydrophilic-interaction chromatography, in the first dimension, was employed for separation by degree of ethoxylation and reversed-phase LC, in the second dimension, for separation by alkyl chain length. For detection, TOF–MS with ESI in positive mode was used. The optimization was performed in a one-dimensional system testing different columns. A gradient elution of acetonitrile and water was established for the first dimension and methanol and water for the second dimension, using ammonium acetate buffer in both gradients. The method can be transferred to another system (ultra-high-performance LC–LC × LC–MS) with modifications which allow a decrease in analysis time (from 170 to 120 min) and improvement in the sensitivity and peak shape. All the analysed surfactants may be ingredients in cosmetic products such as shampoo or shower gel. Direct application to cosmetic products was developed with a similar mixture of surfactant classes and subclasses using comprehensible GC × GC–TOF–MS (Hübner et al., 2007) (see Fig. 11.2).

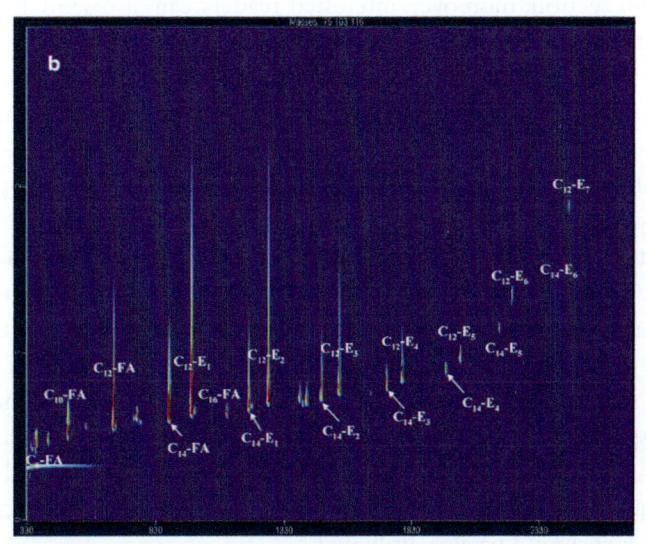

Figure 11.2 Determination of fatty alcohol sulphates and fatty alcohol ether sulphates in shower gel by two-dimensional gas chromatography–time-of-flight–mass spectrometry after hydrolysis and silylation (Hübner et al., 2007).

To detect the three surfactant classes, an aliquot sample was silylated with N,O-bis-tri methylsilyltrifluoracetamide/*N*-methyl-*N*-trimethylsilyltrifluoracetamide and subsequently a second sample was hydrolysed and silylated in this way. In the case of the first aliquot, the non-ionic and amphoteric surfactants were analysed by suitable thermal decomposition in an injector. In the case of the second aliquot, fatty alcohol sulphates and fatty ether sulphates were determined as derivatives of their corresponding alcohols. The temperature for decomposition of betaines to amidoamines was optimized. Two columns, a non-polar one in the first dimension and a moderately polar column in the second dimension, separated the surfactants by alkyl chain length and by degree of ethoxylation. It is worth noting that other components of shampoo, in addition to surfactants, were identified: scents, anti-degradants, additives and solvents. There were even components not listed on the label. The characterization of another mixture type as cationic and non-ionic surfactants was proposed by Wulf et al. (2010) using GC × GC–TOF–MS. A cleaner containing fatty ether alcohols and BAK was analysed. BAK was decomposed to degradation products by Hoffmann elimination in the injector and alcohols were converted to their silylated derivatives. The method determined FAEs that contain an ethoxy and a propoxy group.

DETERMINATION OF RESIDUAL PRODUCTS FROM SURFACTANTS
Internationally Standardized Analytical Methods

Standardized methods for contaminant determination in surfactant raw materials are considered in this section; moreover, interested readers can also see Chapter 16 of this book for furthermore information about contaminants in cosmetics.

Two ISO methods describe the detection and determination of *N*-nitrosodiethanolamine (NDELA) in cosmetic products or raw materials using high-performance LC (see Table 11.8).

The ISO 10130 method employs coupling with post-column photolysis and derivatization to obtain a specific detection. Nitrosamine is converted to nitrite ion, which is derivatized by the Griess reaction and determined at 540 nm. The ISO 15819 method uses LC coupled with MS/MS. The identification is performed with one molecular ion and two daughter ions, and the quantification is done by the ratio of the major fragment ions of NDELA and the deuterated standard (d8-NDELA), which acts like an internal standard. For both methods, the sample treatment depends on whether the sample is dispersible in water. In this case, an aqueous extraction and a clean-up with SPE (C_{18}) are carried out. If the sample is not dispersible in water, a liquid–liquid extraction (LLE) using dichloromethane should be employed. The matrix effect for each type of cosmetic product should be also evaluated.

Various ISO and CEN methods deal with the analysis of residual products from betaine synthesis (see Table 11.8). ISO 17293–1 and ISO 17293–2 determine

Table 11.8 International official methods which affect residual products from surfactants

Type	Code	Matrix	Residual product	Technique
EN	12974:2000	Alkyl ethoxylated sulphates	1,4-Dioxane	HS–CG–FID
ISO	17280:2015	Alkyl ether sulphates and alcohol ethoxylates	1,4-Dioxane	HS–CG–FID
EN	12139:1999	Non-ionic ethoxylated	Polyethylene glycol	LC/GPC
EN	12582:1999	Non-ionic ethoxylated	Polyethylene glycol	LC–ELSD
EN	13268:2001/ AC:2002	Adducts of ethylene oxide and propylene oxide	Ethylene oxide, propylene oxide groups	CG–FID
EN	13320:2002	Ethoxylates (i.e., alcohol ethoxylates)	Ethylene oxide	GC
ISO	16560:2015	Non-ionic ethoxylated	Polyethylene glycol	LC
EN	14881:2005	Alkylamidopropyl betaines	N-(3-dimethylaminopropyl) alkylamide	CG–FID
EN	15608:2009	Alkylamidopropyl betaines	Fatty acid	CG–FID
ISO	17293–1:2014	Anionic and amphoteric surfactants	Chloroacetic acid, dichloroacetic acid	LC
ISO	17293–2:2014	Anionic and amphoteric surfactants	Chloroacetic acid	IC
ISO	10130:2009	Cosmetic products	NDELA	LC derivatization
ISO	15819:2014	Cosmetic products	NDELA	LC–MS/MS

ELSD, evaporative light-scattering detection; *FID*, flame ionization detector; *GC*, gas chromatography; *HS*, headspace; *IC*, ion chromatography; *LC*, liquid chromatography; *MS/MS*, tandem mass spectrometry; *NDELA*, N-nitrosodiethanolamine.

monochloroacetic acid and dichloroacetic acid (DCA) in amphoteric surfactants (alkyl imidazoline carboxylate, alkyldimethyl betaine, and fatty acetyl propyl dimethyl betaine) and anionic surfactants (alkyl(phenyl) ethoxylated carboxylate). Ion chromatography (IC) and LC are the proposed analytical techniques. The LOD and limit of quantification (LOQ) are better using the IC method. Other residuals such as alkylamides and fatty acids are determined by GC with a flame ionization detector (FID) (see Table 11.8) using methods EN 14881 and EN 15608, respectively. The contents of residual products which may be determined with each method were established within the ranges 0.02–3.0 g of (C6 to C20) fatty acids per 100 g of betaines and 0.02–1.0 g of amidoamines per 100 g.

The analysis of residual products (1,4-dioxane, EO and PEG) from ethoxylated surfactant synthesis is described in ISO methods over the period under study and in CEN methods from the previous years (see Table 11.8). ISO 17280 and EN 12974 methods allow for the determination of 1,4-dioxane in a similar range of concentrations (5–100 mg/kg), but with respect to the matrices, EN 12974 is applicable only for ethoxylated anionic surfactants (see Table 11.8). Two methods of the USP, USP 467 and 228, are used by the FDA for the determination of 1,4-dioxane. The former has two sample treatments depending on whether the analysed product is soluble or not in water. The extract is analysed by static headspace GC with an FID. Using the USP 228 method, 1,4-dioxane and EO are determined in products synthesized from EO by static headspace GC. PEG contents greater than or equal to 0.1% can be determined in non-ionic ethoxylated surfactants by both ISO (ISO 16560) and CEN methods (EN 12139, EN 12582, see Table 11.8). Finally, EN 13268 describes the analysis of EO and propylene oxide groups in polyethers and polyglycol esters and EN 13320 the content of free EO (1–100 mg/kg) for several ethoxylated surfactants (alcohols, alkylphenol and fatty acid polyglycol esters).

Analytical Methods Published in the Scientific Literature

This section contains information about analytical methods published in the scientific literature from 2005 to 2015, concerning the determination of contaminants in surfactant raw materials or cosmetics manufactured from them. 1,4-Dioxane, *N*-nitrosamines, alkylamines and alkanolamines, and other residual products which can be formed from raw materials of surfactants during their synthesis, are examined (see Table 11.9).

In addition to the analytical methods, other relevant aspects such as the monitoring of several products from world markets or alternative production methods aimed at minimizing residual products are briefly presented. Nevertheless, readers can find more information on analytical methods for some of these contaminants (1,4-dioxane, nitrosamines) and others (metals and metalloids, hydroquinone, formaldehyde, acrylamide) in cosmetic samples in Chapter 16.

1,4-Dioxane

It is well known that the content of 1,4-dioxane can be minimized during the industrial synthesis of ethoxylated surfactants. Matheson et al. (2009) suggested that in ethoxylated alcohol production under conditions of alkaline catalysis, its formation is not favourable and its final concentration is below 5 ppm. However, the synthesis of AES from alcohols, under highly acidic conditions, must be strictly controlled to reduce the 1,4-dioxane formation. Moreover, the well-known stripping process, suitable sulphation equipment and optimized process conditions (rapid neutralization after sulphation, reactor loading, etc.) make it possible for the final values of 1,4-dioxane in AES to be below 100 ppm, which involves values below 10 ppm in cosmetic formulation. Production methods for AES which minimize 1,4-dioxane are also patented (Tahara et al., 2013).

Another relevant aspect in the evaluation of this impurity is the monitoring of cosmetic products on different national markets. Although on the American market in the

Table 11.9 References on the determination of residual products in cosmetic products according to the residual compound and type of cosmetic product

Author	Compound	Matrix	Technique	LOD (µg·g⁻¹)	LOQ (µg·g⁻¹)
Tanabe and Kawata (2008)	1,4-Dioxane	Body wash, hair conditioner, hand wash, shampoo	GC–MS	n.p.	0.05
Huang et al. (2012)		Shampoo	GC–MS	n.p.	n.p.
Davarani et al. (2012)		Shampoo, raw materials	HS-SPME–GC–FID	1.5[a]	5[a]
Tahara et al. (2013)		Shampoo, hand soap, body cleanser, raw materials	HS–GC–MS	0.5	1.0
Bunte and Born (2005)		Raw materials	IR	n.p.	n.p.
Mohammadbeigi and Kahe (2011)		Raw materials	GC	n.p.	n.p.
Li et al. (2013a)		Lotion, cream, shampoo	HS–GC–MS	n.p.	2.5
Li et al. (2013b)		Cosmetic products	HS–GC–MS	1.3[a]	4.3[a]
Zhao et al. (2010)		Shampoo, bath lotion	HS–GC–MS	1.0	n.p.
Lin et al. (2010)		Toiletries	HS–GC–MS	0.5	n.p.
Wang et al. (2005)		Cosmetic products	HS–GC–FID	2.5	n.p.
Cui et al. (2012)		Cosmetic products	HS–GC–MS/MS	0.010	0.035
Tay et al. (2014)		Raw materials	GC–FID	10	30
Flower et al. (2006)	NDELA	Make-up remover, moisture lotion, cucumber cleansing lotion, shampoo, hair dye, raw material, wheat germ hair conditioner	HPLC–UV	n.p.	10–40
Ghassempour et al. (2008)		Hair and body shampoo, hand and face cream, hand liquid soap	LC–MS	2.5[b]	10[b]
Joo et al. (2015)		Skin care, make-up, shampoo, hair products, raw materials	UPLC–MS/MS	10[a]	20[a]
Davarani et al. (2013)		Shampoo, body shampoo, hand washing liquid	HS-SPME–GC–MS	1[a]	3[a]
Wang et al. (2006)	NDELA, NBHPA, NDMEA, NDPLA, NDPHLA	Hair shampoo, skin cream, foam bath	HPLC–DAD	0.02–0.03[b]	n.p.
Ma et al. (2011)	NDELA, NDMA, NDPLA, NMOR, NPYR, NPIP, NDBA, NDPHLA, N-NO-DCHA, NDBzA	Cream, lotion, powder, shampoo, lipstick	GC–MS/MS	2.5	10

Continued

Table 11.9 References on the determination of residual products in cosmetic products according to the residual compound and type of cosmetic product—cont'd

Author	Compound	Matrix	Technique	LOD (µg·g⁻¹)	LOQ (µg·g⁻¹)
Zhong et al. (2012b)	MA, EA, DMA, trimethylamine, propylamine, butylamine	Cleansing products, essence, sun block products, skin care cream and whitening lotion, hair dye and hair growing agents	IC	45–75[c]	150–250[c]
Zhong et al. (2012a)			IC	23–38[c]	100–120[c]
Zhong et al. (2013)	MA, diethylamine, DMA, TEA, DEA	Sunscreen cream, cleansing lotion, cosmetic products	IC	72–120[c]	240–420[c]
Ghassempour et al. (2006)	MCA, DCA	Raw material	HPLC–DAD	0.1–0.15[b]	1–3[b]
Lei et al. (2010)			IC	11–32[c]	n.p.
He et al. (2008)			HS–GC	6.3–7.0	n.p.
Shi et al. (2011)	GL, MCA, DCA		IC	17–90[d]	n.p.
Soleimani et al. (2013)			HPLC–UV	3.7–13[d]	n.p.
Duan et al. (2010)			IEC	1–5	n.p.
Xia et al. (2013)	DMPA		GC–FID	0.2	n.p.
Prieto Blanco et al. (2009)	Benzaldehyde, toluene, α,α-dichlorotoluene, 2-chlorotoluene		HPLC–DAD	11–34[c]	36–110[c]
Schäfer and Zöllner (2013)	Dimethyl sulphate		GC–MS/MS	0.24	0.48
Bengoechea and Fernández (2008)	Sulphones		HT–GC	n.p.	n.p.
Zhang et al. (2013)	Ethylene oxide	Shampoo, bath foam	HS–GC–MS	0.02[e]	0.07[e]

DAD, diode array detector; DCA, dichloroacetic acid; DEA, diethanolamine; DMA, dimethylamine; DMPA, dimethylaminopropylamine; EA, ethylamine; FID, flame ionization detector; GC, gas chromatography; GL, glycolic acid; HPLC, high-performance liquid chromatography; HS, headspace; IC, ion chromatography; IR, infrared spectrometry; LC, liquid chromatography; LOD, limit of detection; LOQ, limit of quantification; MA, methylamine; MCA, monochloroacetate; MS, mass spectrometry; MS/MS, tandem mass spectrometry; NBHPA, N-nitroso-bis(2-hydroxypropyl)amine; NDELA, N-nitrosodiethanolamine; NDPHLA, N-nitrosodiphenylamine; NDPLA, N-nitrosodi-n-propylamine; n.p, not provided; SPME, solid-phase micro-extraction; TEA, triethanolamine; UPLC, ultra-high-performance liquid chromatography; UV, ultraviolet.

[a] ×10⁻³.
[b] µg/mL.
[c] ng/mL.
[d] ng/L.
[e] µg.

late 1990s, the FDA survey found values ranging from 14 to 79 µg/g in some products, a more recent survey carried out by the Campaign for Safe Cosmetics reported values that did not exceed 12 ppm in children products and 23 ppm in adult products (Agency for Toxic Substances and Disease Registry (ATSDR), 2012). On the Japanese market in 2008, Tanabe and Kawata (2008) quantified values from 0.05 to 33 µg/g (for 40 samples) and Tahara et al., in 2013, values below 5 µg/g (for 12 samples). Concerning the European market, the German Cosmetic, Toiletry, Perfumery and Detergent Association (IKW) recommended a maximum of 10 ppm in derivatives of PEG for use in cosmetics (Fruijtier-Pölloth, 2005). Chung et al. (2014) assessed the incidence of several contaminants in cosmetic products in Taiwan for longer periods of time. In a monitoring process conducted during 2011, the 1,4-dioxane content was in compliance with the national regulations, which stipulated it should not exceed 100 ppm.

In the period considered, the technique chosen for the analysis of 1,4-dioxane was GC with FID or coupled with MS. Using GC–FID, Davarani et al. (2012) developed a method in which 1,4-dioxane was extracted from SLES and shampoo samples by headspace solid-phase micro-extraction (SPME) using an aluminium hydroxide coating on fused silica fibre. The coating fibre acts as a Lewis acid and oxophile, interacting with oxygen atoms of 1,4-dioxane. The parameters of extraction, temperature and time of extraction (34°C and 4 min), equilibrium time (13 min) and salt content (25%), were optimized. The fibre is reusable for more than 80 extractions and it has a good thermal stability. Excellent LOD and LOQ at levels of nanograms/gram are obtained (see Table 11.9). As a result of the application of this method to cosmetic products and raw materials, the obtained 1,4-dioxane values were from 12 to 6 ppm using for quantification the standard addition method. A GC–FID system with direct injection was used by Tay et al. (2014) for the analysis of palm-based FAEs. The heavier ethoxylated compound remained in the glass wool of the liner and the more volatile ones, which passed to the chromatographic column, had to be separated from 1,4-dioxane. Commercial FAEs with different degrees of ethoxylation (1–20) were analysed and no 1,4-dioxane was detected. Other methods which analyse non-ionic surfactants do not use headspace, but a treatment with distillation (Mohammadbeigi and Kahe, 2011). Tanabe and Kawata (2008), employed a multi-step sample treatment with SPME, freezing and concentration by nitrogen gas and redissolution. The chromatographic analysis was performed by GC–MS using an internal standard (1,4-dioxane-d8) for quantification. Of 51 samples (including household products) of products from the Japanese market analysed, 40 samples containing 1,4-dioxane were found. The raw materials (AES) presented a higher mean (7.2 µg/g) than those (2.7 µg/g) quantified in finished products. In 2013, Tahara et al. also applied a new method to monitoring products from the Japanese market. A GC–MS methodology without previous treatment performed the extraction in headspace after the optimization of the temperature and the time of this step. Moreover, fluorobenzene and 1,4-dioxane-d8 can be used under optimized conditions as internal standards for quantification by standard addition. Although in all the raw materials (lauryl ether and sodium polyoxyethylene lauryl ether sulphate) and cosmetic products analysed 1,4-dioxane was detected, its concentration was below 10 µg/g.

Several papers written in Chinese (Li et al., 2013a,b; Cui et al., 2012; Zhao et al., 2010; Lin et al., 2010) describe the method of headspace (HS)–GC–MS, except for one, which used GC–FID (Wang et al., 2005). LODs were at levels around 0.5–2.5 µg/g, but Cui et al. (2012) obtained 10 ng/g by HS–GC–MS/MS using the multiple-reaction monitoring mode for detection. Li et al. (2013b) found a group of samples containing between 2.5 and 15 µg/g and another between 30 and 50 µg/g. Another option, different to chromatographic analysis, is the determination in non-ionic and ionic ethoxylated surfactants using IR spectroscopy after a treatment which separates 1,4-dioxane and other components in the gas phase (Bunte and Born, 2005). Moreover, Huang et al. (2012) studied the exposure to dioxane in shampoo via the skin route, establishing an exposure evaluation model and calculating the exposure dose based on the 1,4-dioxane concentration in shampoo.

N-*nitrosamines*

Most works referred to one of the nitrosamines, polar and non-volatile NDELA present in cosmetic products owing to the reaction under specific conditions of diethanolamine (DEA) and triethanolamine (TEA) (extensively used) with preservatives such as bronopol. NDELA can be also introduced into cosmetic products from contaminated raw materials (SCCS, 2012). Moreover, some studies addressed the determination of mixtures of N-nitrosamines. Although the presence of nitrosamines must be controlled in finished products, they also could be formed during storage or use. Other than the case of 1,4-dioxane, no monitoring studies of cosmetic products from different markets have been found.

The determination of NDELA by chromatographic techniques involves some difficulties which must be considered. To improve its retention and sensitivity a derivatization step can be included. In many cosmetic matrices, a clean-up step, to reduce the concentration of other components which are extracted along with NDELA, simplifies the chromatographic analysis. Moreover, false positive and negative results can be avoided by reinforcing the identification step and controlling the sample preparation.

Different sample treatments according to whether the sample is dispersible or not in water, are applied by the method developed by Flower et al. (2006). In the first case, after dispersion and centrifugation, sample cleaning is performed by means of SPE. In the second type the sample was dispersed in dichloromethane and extracted with water. The separation was carried out by reversed-phase LC, and post-column derivatization by Griess reaction allowed the detection by UV at 540 nm (see also ISO 10130). During the Griess reaction, NDELA undergoes photolysis for denitrosation, diazotization and coupling reaction to form a purple azo compound. The validation included inter-laboratory exercises. The method is applicable to a broad range of matrices except for some, such as hair dyes and lipstick. Another approach to NDELA separation and detection is the formation of an ion pair with a complexation agent (Ghassempour et al., 2008). The sample treatment is similar to that proposed by other authors: SPE for a product soluble in water and LLE for a less soluble one. The molar ratio of sodium 1-octanesulphonate and NDELA was examined and the formation of the ion pair complex was confirmed by LC–MS and MS/MS. The highest retention, sensitivity and repeatability were obtained for NDELA when it was complexed. The NDELA values quantified by LC–UV in cosmetic products are within the range 0.05–16 μg/g. The highest values are found in expired cosmetic products.

Owing to the great variety of cosmetic products (with diverse components and physical states) which can be susceptible to containing NDELA, one of the analytical challenges is to achieve a selective and sensitive method applicable to more complex matrices. A systematic sample treatment according to the type of product was proposed by Joo et al. (2015). In accordance with the kind of product (water-soluble product, water-insoluble product containing polymers and wax, soap type and water-dispersible emulsion) being analysed, a proper solvent extraction and/or LLE was recommended. In the second step, a clean-up process using a mixed-mode SPE with a sorbent-containing C_8, strong anion and cation exchange, reducing the interferences, was recommended. Ultra-high-performance LC separation was optimized to achieve a good retention of NDELA and separation of monoethanolamine, TEA and the internal standard of d8-NDELA, as well as good MS sensitivity. A porous graphite carbon column and elution gradient gave the best results within 7 min. The detection was carried out by multiple-reaction monitoring in positive ESI. LOQ and LOD at the level of micrograms/kilogram were obtained. The optimized method was applied to 103 cosmetic products and 12 raw materials, detecting NDELA in 13 and 4 of them, respectively. The quantified levels were from 20 μg/kg (LOQ) to 214.2 μg/kg. An HS–SPME–GC–MS method applicable to shampoo samples, with a good analytical performance (wide linear range, low LOD and good recoveries) and which was organic solvent free, was developed by Davarani et al. (2013). For the extraction, they used a fibre with characteristics similar to those used for the determination of 1,4-dioxane (Davarani et al., 2012). Several parameters (salting-out effect, stirring rate, sample volume, extraction temperature and time, desorption temperature and time) were optimized. The NDELA concentrations found in cosmetic products were from 50 to 160 μg/kg.

Concerning the determination of mixtures, five volatile and non-volatile nitrosamines (see Table 11.9) were simultaneously determined by LC–diode array detector (Wang et al., 2006). A cyano column and gradient elution with methanol and 1 mM buffer at pH 4 were used for a separation of 15 min. Extraction with ethyl acetate, clean-up by SPE, evaporation and redissolution were the steps of the sample treatment. Four nitrosamines (NDELA, N-nitroso-bis(2-hydroxypropyl) amine, N-nitrosodi-n-propylamine and N-nitrosodiphenylamine) were quantified in several cosmetic products.

Ma et al. (2011) determined 10 volatile nitrosamines by GC–MS/MS (see Table 11.9). The ultrasonic extraction was optimized with different solvent mixtures according to the type of cosmetic product. SPE was performed to reduce interference and the concentration to increase the sensitivity before the chromatographic determination. The application of this method to 20 samples detected none of the studied nitrosamines.

Alkylamines and Alkanolamines

These two groups of amines may be present in cosmetic preparations from different paths. Alkylamines can be introduced from impure raw materials or can be formed by decarboxylation of amino acids (Zhong et al., 2013). Although alkanolamides can also be present from raw materials, in most cases they are part of preparations with specific functions such as pH adjuster or hair fixative. In addition, their salts can act as preservatives or surfactants (Fiume et al., 2015). Their concentrations are regulated (see *Regulatory Aspects of Surfactants in Cosmetic Products*) mainly because of the risk of forming nitrosamines with nitrosation agents under specific conditions.

Three papers analysed mixtures of amines covering very different matrices by IC with suppressed conductivity detection. The main efforts were addressed to achieving a rapid, economic and efficient sample treatment. The extract obtained must not contain inorganic salts or organic compounds which interfere in chromatographic analysis. Zhong et al. (2012a) determined six alkylamines (see Table 11.9) using a specially designed distillation system, which allowed, after optimization, distillation within 3 min versus 65 min for conventional distillation. After the application of the method to 76 cosmetic products, it turned out that methylamine was the most abundant, followed by ethylamine and dimethylamine, and a different profile was found according to the type of cosmetic product. For the same compounds, Zhong et al. (2012b) used a laboratory-made multi-channel purge and trap system for the treatment process of a large number of samples with good reproducibility and efficiency. This miniaturized system reduces the amount of absorbent agents and enhances the concentration, and the distilled amines are obtained within 10 min. The design and the conditions of distillation were optimized, as well as the chromatographic conditions. Results similar to those reported by Zhong et al. (2012a), regarding profile and abundance, were obtained in the analysis of 40 commercial products. A third type of treatment, ultrasound-assisted low-density solvent dispersive LLE, was applied to the determination of three alkanolamines and two alkylamines by Zhong et al. (2013) (Fig. 11.3).

The treatment combines in one step the extraction of amines (in acid solution) and the clean-up of the other components of the matrix (in disperser solvent) employing simple equipment. Using ultrasound radiation a more efficient extraction is accomplished in a shorter time (10 min). The extraction parameters were optimized by orthogonal array design. The analysis of 15 commercial products provided in five of the samples higher values for DEA, ranging from 20.3 to 109 mg/kg, and TEA, from 822 to 1880 mg/kg.

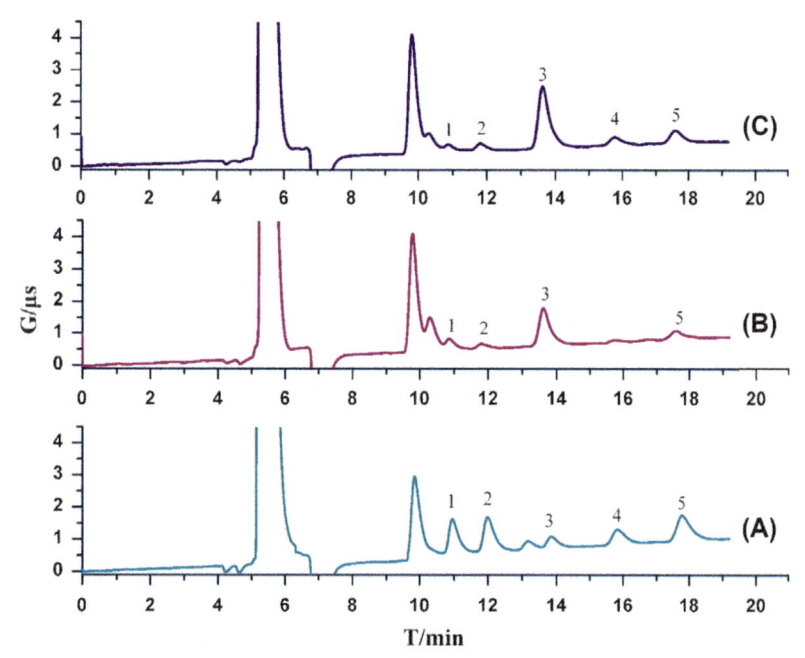

Figure 11.3 Ion chromatography chromatograms of three alkanolamines (monoethanolamine, diethanolamine, triethanolamine, diethylamine) and two alkylamines (dimethylamine, diethylamine) (A) in standard solution, (B) in a sunscreen sample, and (C) in a spiked sunscreen sample. Peaks 1, monoethanolamine; 2, dimethylamine; 3, triethanolamine; 4, diethanolamine; 5, diethylamine. (Zhong et al., 2013).

Residual Products of Betaines

CAPB is possibly the most commonly used betaine in cosmetic products, in terms of both its production volume and the number of products which contain it as an ingredient. Its industrial synthesis is carried out in a two-batch process. The first is for the formation of cocaminopropylamide (CAPA) from dimethylaminopropylamine (DMPA) and coconut oil and the second involves the reaction of CAPA with sodium monochloroacetate (MCA) under alkaline conditions to synthesize CAPB at around 30% in water (Mostafalu et al., 2015). Residual products in CAPB can be derived from the aminic compounds and from MCA (Human and Environmental Risk Assessment (HERA), 2005). Residual CAPA and DMPA in betaines have been associated with skin sensitization reactions (Jacob and Amini, 2008; Schnuch et al., 2011). Xia et al. (2013) proposed a method for the determination of DMPA in CAPB by GC–FID. Other authors suggest that residual DMPA could react with MCA to give DMPA–betaine. This by-product acts as interference in the determination of CAPB using potentiometry titration, but not if the two-phase titration is used (Mostafalu et al., 2015).

Concerning the residual products from MCA, three products with different degrees of toxicity were studied, unreacted MCA, DCA (usually an impurity of MCA) and glycolic acid (GL) (Soleimani et al., 2013). The last can be formed by partial hydrolysis of

MCA (HERA, 2005). Various methods were developed for the determination of three compounds in CAPB in the studied period. Ghassempour et al. (2006) combined extraction and derivatization of MCA and DCA using naphthylamine as the derivatizing agent. The separation was performed in reversed phase using a methanol/water mixture in isocratic mode and detection of the derivatives at 222 nm. For the application of the method it needs 1 mL of sample, 1 h for derivatization and 20 min of chromatographic analysis per sample. Shi et al. (2011) used SPE and anion-exchange chromatography with suppression conductivity detection for the determination of MCA, GL and DCA. To protect the analytical column, a strong cation-exchange SPE to reduce betaine concentration and the removal of a high concentration of chloride by clean-up cartridge were conducted. From the application to three betaine samples, GL (1.1–1.3 mg/g) and DCA (28–9.3 µg/g) were determined but no MCA was detected.

Soleimani et al. (2013) synthesized a new sorbent (amino-modified nanoporous silica) for the extraction of MCA, GL and DCA, with good recovery (88%–93%). Parameters such as pH, flow rate of the sample and eluent and type and volume of the eluent were optimized, and a concentration factor of 129 was achieved. In addition, a previous sample treatment similar to those conducted by Shi et al. (2011) was carried out. After the treatment, the separation was performed by IC with UV detection at 205 nm. The concentration values found for three CAPB samples for GL, MCA and DCA were 2.8–4.8 mg/g, 2.9–8.6 µg/g and 7.3–14.9 µg/g, respectively. Other authors proposed methods for the determination of GL, MCA and DCA using different instrumental techniques: ion-exclusion chromatography (Duan et al., 2010), IC (Lei et al., 2010) and HS–GC (He et al., 2008).

Other Toxic Residual Products

Benzyl chloride and dimethyl sulphate (DMS) are compounds included in the list of substances banned for use in cosmetic products. They can be used as quaternizing agents in the synthesis of cationic surfactants. Schäfer and Zöllner (2013) developed a method of GC for the determination of DMS in the presence of monomethyl sulphate. Prieto-Blanco et al. (2009) determined compounds such as α,α-dichlorotoluene and benzaldehyde, which are present in technical benzyl chloride in wide range of concentrations, according to its commercial sources. The aforementioned impurities may remain as residual products in cationic surfactants as a result of the synthesis using benzyl chloride. These methods could serve as a basis for further application to cosmetic products. Other residual products such as EO and propylene oxide, which are derived from ethoxylated surfactants, can also be found in cosmetic products. Zhang et al. (2013) applied a method for their analysis in shampoo and bath foam by GC–MS. Finally, Bengoechea and Fernández (2008) proposed a new method using high-temperature GC for the determination of sulphones and anhydrides in the sulphonation process of LAS.

REFERENCES

Afkhami, A., Nematollahi, D., Madrakian, T., Abbasi-Tarighat, M., Hajihadi, M., 2009. J.Hazard. Mater. 166, 770.
Afkhami, A., Nematollahi, D., Madrakian, T., Hajihadi, M., 2011. Clean Soil Air Water 39, 171.
Afshar, Z.G., Parham, H., 2010. Asian J. Chem. 22, 3009.
Agency for Toxic Substances, Disease Registry (ATSDR), 2012. Toxicological Profile for 1,4-dioxane. http://www.atsdr.cdc.gov/toxprofiles/tp187.pdf.
Bailón-Moreno, R., Jurado-Alameda, E., Ruiz-Baños, R., 2006. J. Am. Soc. Inf. Sci. Technol. 57, 949.
Bazel, Y.R., Antal, I.P., Lavra, V.M., Kormosh, Z.A., 2014. J. Anal. Chem. 69, 211.
Beneito-Cambra, M., Bernabé-Zafón, V., Herrero-Martínez, J.M., Ramis-Ramos, G., 2007. Talanta 74, 65.
Beneito-Cambra, M., Herrero-Martínez, J.M., Ramis-Ramos, G., 2013. Anal. Methods 5, 341.
Bengoechea, C., Fernández, A., 2008. J. Surfactants Deterg. 11, 103.
Bunte, R., Born, D., 2005. Application: DE DE Patent No. 2004–102004009306 102004009306.
Chung, M.H., Huang, W.S., Chang, Y.C., Chen, Y.H., Lee, M.S., Huang, S.C., Chen, Y.P., Shih, D.Y.C., Cheng, H.F., 2014. J. Food Drug Anal. 22, 399.
Commission Decision 2006/257/EC of 9 February 2006 Amending Decision 96/335/EC Establishing an Inventory and a Common Nomenclature of Ingredients Employed in Cosmetic Products. Official Journal of the European Union, L 97/1.
Commission Regulation (EU) No 1197/2013 of 25 November 2013 Amending Annex III to Regulation (EC) No 1223/2009 of the European Parliament and of the Council on Cosmetic Products. Official Journal of the European Union, L 315/34.
Commission Regulation (EU) No 344/2013 of 4 April 2013 Amending Annexes II, III, V and VI to Regulation (EC) No 1223/2009 of the European Parliament and of the Council on Cosmetic Products. Official Journal of the European Union, L 114/1.
Commission Regulation (EU) No 866/2014 of 8 August 2014 Amending Annexes III, V and VI to Regulation (EC) No 1223/2009 of the European Parliament and the Council on Cosmetic Products. Official Journal of the European Union, L 238/3.
Corrigendum to Commission Regulation (EU) No 866/2014 of 8 August 2014 Amending Annexes III, V and VI to Regulation (EC) No 1223/2009 of the European Parliament and the Council on Cosmetic Products. Official Journal of the European Union, L 254/39.
Cosmetic Ingredient Review (CIR) Expert Panel members, 2011. Diethanolamides as Used in Cosmetics. http://www.cir-safety.org/sites/default/files/amides092011FAR.pdf.
Cosmetic Ingredient Review (CIR) Expert Panel members, 2012. On the Safety Assessment of Ethanolamides as Used in Cosmetic. http://www.cir-safety.org/sites/default/files/amides032012FAR.pdf.
Cosmetic Ingredient Review (CIR) Expert Panel members, 2014. Safety Assessment of Alkyl Betaines as Used in Cosmetics. http://www.cir-safety.org/sites/default/files/alkbet032014final_0.pdf.
Cosmetic Products Notification Portal (CPNP) Article 13 User Manual, 2014. http://ec.europa.eu/growth/sectors/cosmetics/cpnp/index_en.htm.
Cui, J., Yang, J-l., Liu, X., Hong, Y.-C., Tong, L-l., 2012. (in Chinese) Fenxi Ceshi Xuebao 31, 1446.
Cumming, J.L., Hawker, D.W., Matthews, C., Chapman, H.F., Nugent, K., 2010. Toxicol. Environ. Chem. 92, 1595.
Davarani, S.S.H., Masoomi, L., Banitaba, M.H., Zhad, H.R.L.Z., Sadeghi, O., Samiei, A., 2012. Chromatographia 75, 371.
Davarani, S.S.H., Masoomi, L., Banitaba, M.H., Zhad, H.R.L.Z., Sadeghi, O., Samiei, A., 2013. Talanta 105, 347.
Devi, S., Chattopadhyaya, M.C., 2012. J. Surfactants Deterg. 15, 387.
Devi, S., Chattopadhyaya, M.C., 2013. J. Surfactants Deterg. 16, 391.
Duan, G., Huang, Z.-M., Wu, H-l., Wang, X.-Y., Sun, F.-E., 2010. (in Chinese) Riyong Huaxue Gongye 40, 465.
Dürüst, N., Naç, S., Ünal, N., 2014. Sensors Actuators B Chem. 203, 181.
Elsner, V., Laun, S., Melchior, D., Köhler, M., Schmitz, O.J., 2012. J. Chromatogr. A 1268, 22.
EN 12139:1999. Surface Active Agents – Determination of the Total Polyethylene Glycol Content of Nonionic Surface Active Agents (EO Adducts) by HPLC/GPC. https://standards.cen.eu/dyn/www/f?p=204:110:0::::FSP_PROJECT,FSP_ORG_ID:13913, 6257&cs=1D111964351E124928B6037A9193286FA.

EN 12582:1999. Surface Active Agents – Determination of the Polyethylene Glycol Content According to Molar Mass in Non-ionic Surface Agents (Ethoxylated) by HPLC/ELSD. https://standards.cen.eu/dyn/www/f?p=204:110:0::::FSP_PROJECT,FSP_ORG_ID:13928,6257&cs=19879E8953E1433F97A476D36760A4D18.

EN 12974:2000. Surface Active Agents – Determination of 1,4-dioxane in Alkyl-ethoxy-sulfate Products by GLC/HEAD Space Procedure. https://standards.cen.eu/dyn/www/f?p=204:110:0::::FSP_PROJECT,FSP_ORG_ID:13915, 6257&cs=15FBF4AE1652317B0A6612EA9E0B920E5.

EN 13268:2001/AC:2002. Surface Active Agents – Determination of Ethylene Oxide and Propylene Oxide Groups in Ethylene Oxide and Propylene Oxide Adducts. https://standards.cen.eu/dyn/www/f?p=204:110:0::::FSP_PROJECT,FSP_ORG_ID:22034, 6257&cs=13431A44BFD5DFFEEFA8C928173E1A636.

EN 13320:2002. Surface Active Agents – Gas Chromatographic Trace Determination of Free Ethylene Oxide in Ethoxylates. https://standards.cen.eu/dyn/www/f?p=204:110:0::::FSP_PROJECT,FSP_ORG_ID:13929, 6257&cs=168323F1CA36516131B52161436F022A0.

EN 14480:2004/AC:2006. Surface Active Agents – Determination of Anionic Surface Active Agents – Potentiometric Two-phase Titration Method. https://standards.cen.eu/dyn/www/f?p=204:110:0::::FSP_PROJECT,FSP_ORG_ID:26389, 6257&cs=19F9F1DB593E8C825655DBB238C90597D.

EN 14668:2005/AC:2006. Surface Active Agents – Determination of Quaternary Ammonium Surface Active Agents in Raw Materials and Formulated Products – Potentiometric Two-phase Titration Method. https://standards.cen.eu/dyn/www/f?p=204:110:0::::FSP_PROJECT,FSP_ORG_ID:26390, 6257&cs=12B4D7FC8964FBA14424E7E7053F35CC9.

EN 14669:2005. Surface Active Agents – Determination of Anionic Surface Active Agents and Soaps in Detergents and Cleansers – Potentiometric Two-Phase Titration Method. https://standards.cen.eu/dyn/www/f?p=204:110:0::::FSP_PROJECT,FSP_ORG_ID:22232, 6257&cs=16A3DDB9833A4DA8AC3042C2B06BAA01F.

EN 14670:2005. Surface Active Agents – Sodium Dodecyl Sulfate – Analytical Method. https://standards.cen.eu/dyn/www/f?p=204:110:0::::FSP_PROJECT,FSP_ORG_ID:22233, 6257&cs=1D287408F489ECA305F9D985DEE4FC4A8.

EN 14881:2005. Surface Active Agents – Determination of N-(3-dimethylaminopropyl)-alkylamide Content in Alkylamidopropylbetaines – Gas Chromatographic Method. https://standards.cen.eu/dyn/www/f?p=204:110:0::::FSP_PROJECT,FSP_ORG_ID:23486, 6257&cs=1165F4C7DE2CF55131A62268E92D54162.

EN 15109:2006. Surface Active Agents – Determination of the Active Matter Content of Alkylamidopropylbetaines. https://standards.cen.eu/dyn/www/f?p=204:110:0::::FSP_PROJECT,FSP_ORG_ID:24605, 6257&cs=1715329AF9C5900BC2412C0B15620012B.

EN 15608:2009. Surface Active Agents – Quantitative Determination of Free Fatty Acid in Alkylamidopropylbetaines – Gas-chromatographic Method. https://standards.cen.eu/dyn/www/f?p=204:110:0::::FSP_PROJECT,FSP_ORG_ID:27818, 6257&cs=1C2F2F949BEBF84DAC28E6DFD1D274FD9.

EN ISO 2870:2009. Surface Active Agents – Detergents – Determination of Anionic-Active Matter Hydrolysable and Non-Hydrolysable Under Acid Conditions. http://www.iso.org/iso/iso_catalogue/catalogue_tc/catalogue_detail.htm?csnumber=45657.

EN ISO 2871–1:2010. Surface Active Agents – Detergents – Determination of Cationic-Active Matter Content – Part 1: High Molecular-Mass Cationic Active Matter. http://www.iso.org/iso/iso_catalogue/catalogue_tc/catalogue_detail.htm?csnumber=53739.

EN ISO 2871–2872:2010. Surface Active Agents – Detergents – Determination of Cationic-Active Matter Content – Part 2: Cationic-Active Matter of Low Molecular Mass (Between 200 and 500). http://www.iso.org/iso/iso_catalogue/catalogue_tc/catalogue_detail.htm?csnumber=53740.

EN ISO 8799:2009. Surface Active Agents – Sulfated Ethoxylated Alcohols and Alkylphenols – Determination of Content of Unsulfated Matter. http://www.iso.org/iso/iso_catalogue/catalogue_tc/catalogue_detail.htm?csnumber=52126.

Ensafi, A.A., Hemmateenejad, B., Barzegar, S., 2009. Spectrochim. Acta Part A Mol. Biomol. Spectrosc. 73, 794.

EU Regulation, 2009. Regulation (EC) No 1223/2009 of the European Parliament and of the Council of 30 November 2009 on Cosmetic Products. Official Journal of the European Union, L 342/359.

Fiume, M.M., Heldreth, B.A., Bergfeld, W.F., Belsito, D.V., Hill, R.A., Klaassen, C.D., Liebler, D.C., Marks Jr., J.G., Shank, R.C., Slaga, T.J., Snyder, P.W., Andersen, F.A., 2015. Int. J. Toxicol. 34, 84S.

Flower, C., Carter, S., Earls, A., Fowler, R., Hewlins, S., Lalljie, S., Lefebvre, M., Mavro, J., Small, D., Volpe, N., 2006. Int. J. Cosmet. Sci. 28, 21.

Fruijtier-Pölloth, C., 2005. Toxicology 214, 1.

Ghassempour, A., Chalavi, S., Abdollahpour, A., Mirkhani, S.A., 2006. Talanta 68, 1396.

Ghassempour, A., Abbaci, M., Talebpour, Z., Spengler, B., Römpp, A., 2008. J. Chromatogr. A 1185, 43.

He, Q.-G., Yao, C.-Z., Yan, F., 2008. (in Chinese) Riyong Huaxue Gongye 38, 409.

Huang, G., Bu, H., Sun, S., Chen, A., Zhou, Y., 2012. Procedia Eng. 43, 407.

Hübner, J., Nguyen, A., Turcu, F., Melchior, D., Kling, H.W., Gäb, S., Schmitz, O.J., 2006. Anal. Bioanal. Chem. 384, 259.

Hübner, J., Taheri, R., Melchior, D., Kling, H.W., Gäb, S., Schmitz, O.J., 2007. Anal. Bioanal. Chem. 388, 1755.

Human and Environmental Risk Assessment (HERA) on ingredients of household cleaning products, 2005. Cocamidopropyl Betaine (CAPB). Edition 1.0 http://www.heraproject.com/files/45-hh-e101023f-d12f-6a30-deb0770e9bf8e4d0.pdf.

Ibrahim, H., Khorshid, A., 2007. Anal. Sci. 23, 573.

Im, S.H., Jeong, Y.H., Ryoo, J.J., 2008. Anal. Chim. Acta 619, 129.

ISO 10130:2009. Cosmetics – Analytical Methods – Nitrosamines: Detection and Determination of N-nitrosodiethanolamine (NDELA) in Cosmetics by HPLC, Post-Column Photolysis and Derivatization. http://www.iso.org/iso/catalogue_detail.htm?csnumber=45840.

ISO 15819:2014. Cosmetics – Analytical Methods – Nitrosamines: Detection and Determination of N-nitrosodiethanolamine (NDELA) in Cosmetics by HPLC-MS-MS. http://www.iso.org/iso/catalogue_detail.htm?csnumber=62042.

ISO 16560:2015. Surface Active Agents – Determination of Polyethylene Glycol Content in Nonionic Ethoxylated Surfactants – HPLC Method. http://www.iso.org/iso/iso_catalogue/catalogue_tc/catalogue_detail.htm?csnumber=57112.

ISO 17280:2015. Surface Active Agents – Determination of 1,4-dioxan Residues in Surfactants Obtained from Epoxyethane by Gas Chromatography. http://www.iso.org/iso/iso_catalogue/catalogue_tc/catalogue_detail.htm?csnumber=59514.

ISO 17293–1:2014. Surface Active Agents – Determination of Chloroacetic Acid (Chloroacetate) in Surfactants – Part 1: HPLC Method. http://www.iso.org/iso/iso_catalogue/catalogue_tc/catalogue_detail.htm?csnumber=59516.

ISO 17293–2:2014. Surface Active Agents – Determination of Chloroacetic Acid (Chloroacetate) in Surfactants – Part 2: Ionic Chromatographic Method. http://www.iso.org/iso/iso_catalogue/catalogue_tc/catalogue_detail.htm?csnumber=60484.

Jacob, S.E., Amini, S., 2008. Dermatitis 19, 157.

Joo, K.M., Shin, M.S., Jung, J.H., Kim, B.M., Lee, J.W., Jeong, H.J., Lim, K.M., 2015. Talanta 137, 109.

Karimi, M.A., Behjatmanesh-Ardakani, R., Goudib, A.A., Zalib, S., 2008. J. Braz. Chem. Soc. 19, 1523.

Khaled, E., Mohamed, G.G., Awad, T., 2008. Sensors Actuators B Chem. 135, 74.

Khedr, A.M., Shawish, H.M.A., Gaber, M., Almonem, K.I.A., 2014. J. Surfactants Deterg. 17, 183.

Koike, R., Kitagawa, F., Otsuka, K., 2007. J. Chromatogr. A 1139, 136.

Kuong, C.L., Chen, W.Y., Chen, Y.C., 2007. Anal. Bioanal. Chem. 387, 2091.

Lee, Y.H., Jeong, E.S., Cho, H.E., Moon, D.C., 2008. Talanta 74, 1615.

Lei, X.-Y., Xia, X.-Y., Fang, L.-D., 2010. (in Chinese) Riyong Huaxue Gongye 40, 141.

Li, Z.-P., Rosen, M.J., 1981. Anal. Chem. 53, 1516.

Li, J., Ma, Q., Li, W., Wang, C., Bai, H., Ma, H., Cai, T., Jiao, Y., Zhang, X., 2013a. (in Chinese) Chin. J. Chromatogr. (Se Pu) 31, 481.

Li, Y., Tang, Y., Liang, J., Luo, L.-J., Zhu, Z.-J., 2013b. (in Chinese) Guangzhou Huaxue 38, 27.

Lie, A., Wimmer, R., Pedersen, L.H., 2013. J. Chromatogr. A 1281, 67.

Lin, X.-S., Wu, H.-Q., Huang, X.-L., Ma, Y.-F., Huang, F., Zhu, Z.-X., Den, X., Luo, H.-T., 2010. (in Chinese) Lihua Jianyan, Huaxue Fence 46, 938.

Lobachev, A.L., Kolotvin, A.A., 2006. J. Anal. Chem. 61, 622.
Ma, Q., Xi, H.W., Wang, C., Bai, H., Xi, G.C., Su, N., Xu, L.Y., Wang, J.B., 2011. Fenxi Huaxue/Chin. J. Anal. Chem. 39, 1201.
Madunić-Čačić, D., Sak-Bosnar, M., Galović, O., Sakač, N., Matešić-Puač, R., 2008. Talanta 76, 259.
Madunić-Čačić, D., Sak-Bosnar, M., Matešić-Puač, R., 2011. Int. J. Electrochem. Sci. 6, 240.
Madunić-Čačić, D., Sak-Bosnar, M., Matešić-Puač, R., Samardžić, M., 2012. Int. J. Electrochem. Sci. 7, 875.
Makarova, N.M., Kulapina, E.G., 2015a. Electroanalysis 27, 621.
Makarova, N.M., Kulapina, E.G., 2015b. Sensors Actuators B Chem. 210, 817.
Masadome, T., Yada, K., Wakida, S.I., 2006. Anal. Sci. 22, 1065.
Matheson, L., Russell, G., MacArthur, B., Sheats, W.B., 2009. 1,4-Dioxane—A Current Topic for Household Detergent and Personal Care Formulators. 100th AOCS Annual Meeting.
Metrohm. Application Bulletin No. 233/3 e. Titrimetric/Potentiometric Determination of Anionic and Cationic Surfactants. http://partners.metrohm.com/GetDocument?action=get_dms_document&docid=693052.
Metrohm. Application Bulletin No. 269/4 e. Potentiometric Determination of Ionic Surfactants by Two-Phase Titration Using Surfactrodes. http://partners.metrohm.com/GetDocument?action=get_dms_document&docid=2075026.
Metrohm. Ti Application Note No. T-10. Anionic Surfactants in Shower Lotions and Shampoos. http://partners.metrohm.com/GetDocument?action=get_dms_document&docid=696394.
Metrohm. Ti Application Note No. T-12. Cationic Surfactants in Hair Conditioner. http://partners.metrohm.com/GetDocument?action=get_dms_document&docid=696166.
Metrohm. Ti Application Note No. T-46. Soap Content of Soap Noodles. http://partners.metrohm.com/GetDocument?action=get_dms_document&docid=696404.
Metrohm. Ti Application Note No. T-48. Anionic Surfactants in Shower Oil by Potentiometric Two-Phase Titration. http://partners.metrohm.com/GetDocument?action=get_dms_document&docid=696057.
Mohamed, G.G., Ali, T.A., El-Shahat, M.F., Al-Sabagh, A.M., Migahed, M.A., Khaled, E., 2010. Anal. Chim. Acta 673, 79.
Mohammad, A., Mobin, R., 2015. J. Planar Chromatogr. Mod. TLC 28, 17.
Mohammadbeigi, S.M.B.T., Kahe, B.S., 2011. SÖFW-J. Int. J. für Angew. Wiss. 137, 30.
Mostafa, G.A.E., 2006. J. Pharm. Biomed. Anal. 41, 1110.
Mostafa, G.A.E., 2008. Int. J. Environ. Anal. Chem. 88, 435.
Mostafalu, R., Banaei, A., Ghorbani, F., 2015. J. Surfactants Deterg. 18, 919.
Nasioudis, A., Van Velde, J.W., Heeren, R.M.A., Van den Brink, O.F., 2011. J. Chromatogr. A 1218, 7166.
Park, H.S., Rhee, C.K., 2004. J. Chromatogr. A 1046, 289.
Park, H.S., Ryu, H.R., Rhee, C.K., 2006. Talanta 70, 481.
Pasieczna-Patkowska, S., Olejnik, T., 2013. Annales Universitatis Mariae Curie-Skłodowska Lublin – Polonia. Sectio AA, LXVIII, 95.
Praus, P., 2005. Talanta 65, 281.
Prieto-Blanco, M.C., López-Mahía, P., Prada-Rodríguez, D., 2009. J. Chromatogr. Sci. 47, 121.
Ripoll-Seguer, L., Beneito-Cambra, M., Herrero-Martínez, J.M., Simó-Alfonso, E.F., Ramis-Ramos, G., 2013. J. Chromatogr. A 1320, 66.
Ryu, H.R., Park, H.S., Rhee, C.K., 2007. Bull. Korean Chem. Soc. 28, 85.
Samardžić, M., Sak-Bosnar, M., Madunić-Čačić, D., 2011. Simultaneous potentiometric determination of cationic and ethoxylated nonionic surfactants in liquid cleaners and disinfectants, Talanta, 83(3), 789–794.
Sanan, R., Mahajan, R.K., 2013. J. Colloid Interf. Sci. 394, 346.
Sánchez, J., del Valle, M., 2005. Crit. Rev. Anal. Chem. 35, 15.
Scientific Committee on Consumer Safety (SCCS), 2012. Opinion on Nitrosamines and Secondary Amines in Cosmetic Products (SCCS/1458/11). http://ec.europa.eu/health/scientific_committees/consumer_safety/docs/sccs_o_090.pdf.
Scientific Committee on Consumer Safety (SCCS), 2015. Opinion on Cetylpyridinium Chloride – Submission II (SCCS/1548/15), COLIPA No P97. http://ec.europa.eu/health/scientific_committees/consumer_safety/docs/sccs_o_171.pdf.
Schäfer, C., Zöllner, P., 2013. J. Chromatogr. A 1289, 139.
Schnuch, A., Lessmann, H., Geier, J., Uter, W., 2011. Contact Dermat. 64, 203.

Shi, H., Wang, Z., Wang, H., Zhao, R., Ding, M., 2011. Chin. J. Chem. 29, 778.
Soleimani, M., Khani, A., Moazzen, E., Ebrahimzadeh, H., Samiei, A., Masoomi, L., 2013. Chromatographia 76, 33.
Tahara, M., Obama, T., Ikarashi, Y., 2013. Int. J. Cosmet. Sci. 35, 575.
Tanabe, A., Kawata, K., 2008. J. AOAC Int. 91, 439.
Tay, B.Y.P., Maurad, Z., Muhammad, H., 2014. J. Am. Oil Chem. Soc. 91, 1103.
Terol, A., Gómez-Mingot, M., Maestre, S.E., Prats, S., Luis Todolí, J., Paredes, E., 2010. Int. J. Cosmet. Sci. 32, 65.
Tian, Z., Qin, W., 2014. Anal. Methods 6, 5353.
Ungewiß, J., Vietzke, J.P., Rapp, C., Schmidt-Lewerkühne, H., Wittern, K.P., Salzer, R., 2005. Anal. Bioanal. Chem. 381, 1401.
United States Pharmacopeial Convention (USP), 2012a. General Information Chapter, Ethylene Oxide and Dioxane <228>. USP 35 – NF 30. USP, Rockville, MD, p. 143.
United States Pharmacopeial Convention (USP), 2012b. General Information Chapter, Residual Solvents <467>. USP 35 – NF 30. USP, Rockville, MD, p. 185.
Wang, C., Wang, X., Ji, M-q., Chen, W., Cai, T-p., Liu, J., 2005. (in Chinese) Zhipu Xuebao 26, 254.
Wang, L., Hsia, H., Wang, C., 2006. J. Liq. Chromatogr. Relat. Tech. 29, 1737.
Wulf, V., Wienand, N., Wirtz, M., Kling, H.W., Gäb, S., Schmitz, O.J., 2010. J. Chromatogr. A 1217, 749.
Xia, X., Gu, X., Lei, X., 2013. (in Chinese) Guangdong Huagong 40, 152.
Yoshida, S., 2006. Application: JP JP Patent No. 2005–37944 2006226717 (in Japanese).
Zhang, J.-X., Li, H., Cai, L.-P., Fan, B., Zhang, Y., 2013. (in Chinese) Fenxi Huaxue 41, 1293.
Zhao, T.T., Xi, S.F., Li, H.Y., Wang, J.C., Tan, J.H., Jia, F., Song, W., 2010. (in Chinese) Modern Chem. Ind. 30, 91.
Zhong, Z., Li, G., Zhu, B., Luo, Z., 2012a. J. Liq. Chromatogr. Relat. Tech. 35, 1719.
Zhong, Z., Li, G., Luo, Z., Zhu, B., 2012b. Anal. Chim. Acta 715, 49.
Zhong, Z., Li, G., Zhong, X., Luo, Z., Zhu, B., 2013. Talanta 115, 518.

FURTHER READING

Commission Implementing Decision of 25 November 2013 on Guidelines on Annex I to Regulation (EC) No 1223/2009 of the European Parliament and of the Council on Cosmetic Products (2013/674/EU). Official Journal of the European Union, L 315/82.

CHAPTER 12

Nanomaterials in Cosmetics: Regulatory Aspects

Diana M. Bowman[1], Nathaniel D. May[1], Andrew D. Maynard[2]
[1]Arizona State University, Phoenix, AZ, United States; [2]Arizona State University, Tempe, AZ, United States

INTRODUCTION

The aim of this chapter is to discuss the existing regulatory structures for cosmetic products containing nanomaterials. It does this by first providing a snapshot of the types of nanomaterials that are being incorporated into products today, and the reasons for doing so. Second, by drawing on the current and projected product landscape, it presents what is new, and what are the implications for regulation. These questions underpin our considerations about regulatory approaches and developments in the United States (US) and the European Union (EU)—the two largest cosmetics markets in the world (by dollar value).

As the chapter shows, there is no one-size-fits-all approach to regulating a class of product, and factors beyond just science and scientific uncertainty will drive the design and implementation of a regulatory regime. This is arguably more so, though, with nanotechnology in the US and the EU, given the historical context in which nano-based products are being placed.

NANOMATERIALS IN COSMETICS

Cosmetics are big business. Global sales in 2014 were estimated at US$460 billion, with market analysts predicting that this figure will grow to US$675 billion by 2020 (Business Wire, 2015). Pressure to hold and/or increase market share has been a key driver for innovation within the sector. Nanotechnology, and the use of nanomaterials more specifically, can be viewed as one of the latest evolutions within the field.

What seems to make this shift to nano-based cosmetics different from earlier innovations, however, is twofold: one, the potential human health risks associated with certain families of nanomaterials being used in cosmetic products are uncertain or unknown, and two, some nano-based cosmetic products have the capacity to blur the boundary between what is considered—for regulatory purposes—a cosmetic and what is considered a therapeutic good (or drug) ('cosmeceutical'). These functional cosmetics claim to enhance the physiological condition of skin (Kligman, 2000; Brandt et al., 2011) by, for example, improving skin texture and decreasing the appearance of wrinkles.

Both advances give rise to questions for regulators, industry and consumers alike.

As of this writing, nanomaterials are being utilized by the cosmetics industry in a number of different ways. The EU's Scientific Committee on Consumer Products (SCCP) in 2007 divided them into two distinct categories of nanomaterials. The Scientific Committee on Consumer Safety (SCCS), which superseded the SCCP, updated these categories slightly in 2012, to more closely align with the EU definition of materials for regulatory purposes. Under this revision, the words 'intentionally made' were added to the first category, so as to read as 'intentionally made, insoluble/partially soluble and/or biopersistent nanomaterials' (SCCS, 2012). Fig. 12.1 shows this categorization and it will be employed in this chapter.

Pursuant to the SCCS (2012), the first category of nanoparticles includes intentionally manufactured nanoscale silver, gold, titanium dioxide, zinc oxide and fullerenes. Nano-silver and nano-gold exhibit superior antimicrobial properties (compared to their macro-scale equivalents), and have been used in deodorants (silver) (Raj et al., 2012), toothpaste (gold) (Patel et al., 2011), and anti-ageing creams (gold) (Lohani et al., 2014). Nano-titanium dioxide (nano-TiO_2) and nano-zinc oxide (nano-ZnO) are routinely used in sunscreen products, having replaced their macro-scale/bulk equivalents. According to Newman et al. (2009):

this new formulation has resolved the problem of the unsightly white film of traditional sunscreens and created a vehicle that is more transparent, less viscous, and blends into the skin more easily.

Nano-TiO_2 and nano-ZnO also exhibit higher efficacy in reflecting UVA and UVB rays than their non-nano equivalents (Nohynek et al., 2007; Tyner et al., 2011).

Fullerenes, C-60, are spherical, cage-like molecules composed completely of carbon atoms. They have been employed in a number of anti-ageing products because of their antioxidant properties and their capacity to carry active ingredients for deeper skin penetration (Bakry et al., 2007; Lohani et al., 2014). Their use is not, however, without

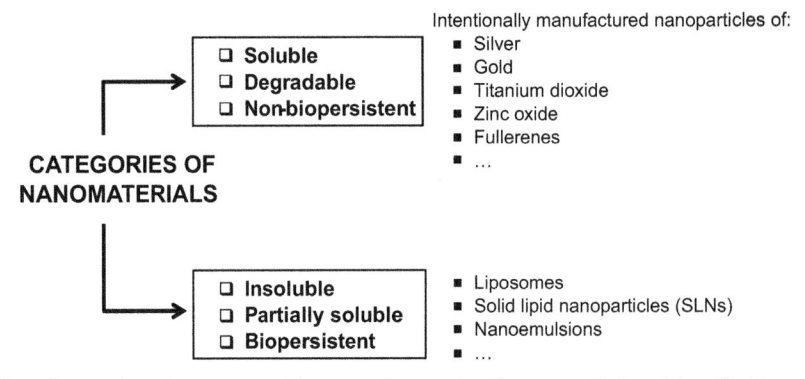

Figure 12.1 Categories of nanomaterials according to the European Union Scientific Committee on Consumer Safety.

controversy owing to concerns over the potential health risks that they may pose (Mu and Sprando, 2010).

The second category of nanomaterials—soluble/biodegradable nanoparticles—includes liposomes, solid lipid nanoparticles (SLNs) and nanoemulsions. Liposomes, closed phospholipid vesicles, have been incorporated into numerous cosmetic products. Their popularity is largely due to their ability to be loaded with active ingredients, including vitamins A and E, for easier transport and increased stability across cell membranes (Bombardelli, 1991; Raj et al., 2012; Nohynek et al., 2007). They can be found in, for example, many types of anti-ageing and moisturizing products (Lohani et al., 2014).

SLNs are similarly being employed by cosmetic and drug companies as novel carrier systems (Müller et al., 2002). SLNs are 'submicron colloidal carriers whose size ranges from 50 to 1000 nm and are composed of physiological lipid, dispersed in water or in aqueous solution of surfactant' (Lohani et al., 2014). SLNs are commonly found in moisturizing and anti-ageing products as they increase the volume of active ingredients delivered to a site, and can do so in a more controlled manner. SLNs also exhibit UV-resistant properties (Raj et al., 2012).

Nanoemulsions are 'dispersions of nanoscale droplets of one liquid within another' that permit increased stability and result in longer shelf life for cosmetic products (Sonneville-Aubrun et al., 2004; Raj et al., 2012). It is for this reason that they are commonly found in moisturizing products (Lohani et al., 2014).

As this cursory overview illustrates, cosmetic chemists have been able to incorporate different families of nanomaterials into a wide range of cosmetic and personal care applications. As the technology matures, it is likely that the types of nanomaterials used in cosmetic products, and the ways in which they are used, will evolve. This has the potential to challenge regulators and regulatory instruments in ways that cannot be imagined at this time.

With this in mind, the next part will now turn to examining how the US and EU regulatory frameworks are approaching nano-based cosmetics, and the efficacy of these two very different approaches.

WHAT'S NEW TO REGULATE, AND IMPLICATIONS FOR REGULATION

Cosmetics were some of the first nano-enabled products available to consumers, regardless of whether they knew it (or not). As such, there has now been over a decade of debate over the use of different types of nanomaterials in cosmetic products, the potential risks posed by certain types of nanomaterials, and the adequacy of existing regulatory frameworks to deal effectively with nano-enhanced cosmetics and new cosmetic products that are nano-based.

While certain non-governmental organizations have been the most vocal in expressing their concerns [see, for example, the Erosion, Technology and Concentration (ETC)

Group, 2004; Friends of the Earth, 2006; Miller, 2006], they have not been the only ones. A significant body of literature now exists that illustrates the varying concerns of members of the scientific community, academics and policy makers (see, for example, SCCP, 2007; Gulson et al., 2010; Prow et al., 2012; Monteiro-Riviere, 2013).

These concerns have evolved over time, reflecting the evolving state of the scientific art. Documenting each of their concerns is, however, beyond the scope of this chapter. Our focus here, instead, shall be on identifying the most important of these that relate to regulation. In our view there are five such concerns:

1. Are nanoparticles, which are topically applied to the skin as part of a cosmetic formulation, able to permeate the skin and enter the vascular system of the body or otherwise interfere metabolic pathways?
2. Are conventional risk assessment protocols (that underpin regulatory frameworks) adequate for determining potential risks associated with those nanomaterials used in cosmetic and personal care products?
3. The regulatory framework for cosmetics in most jurisdictions is often considered to be weak, with limited pre- and post-market powers. Are light-touch regulatory approaches generally suitable for cosmetics and, in particular, for current and new nano-based cosmetics?
4. Should conventional cosmetics, which have been subsequently enhanced through the inclusion of nanomaterials, be considered new for regulatory purposes?
5. Should industry be required to identify the presence of nanomaterials in their products through mandatory labelling requirements?

The first concern—the ability of nanoparticles to pass through the skin, enter the vascular system and travel to organs within the body and potentially interfere with metabolic pathways—is relevant to the regulatory debate as it raises questions of potential human health risks. The question itself, however, sits firmly within the scientific domain. Of particular concern, and therefore controversy, has been the question of whether insoluble and/or biopersistent particles, such as nano-TiO_2 and nano-ZnO, are able to permeate the dermal barrier, given their widespread use (see, for example, Nohynek et al., 2007; Osmond-McLeod et al., 2014; Gulson et al., 2015; Grande and Tucci, 2016), and what health impacts may they have, if any, should this occur. There have been numerous studies that have sought to answer the question definitively (see, for example, Nanoderm, 2007; Pinheiro et al., 2007; Wu et al., 2009; Monteiro-Riviere et al., 2011); while the weight of evidence suggests permeation does not present a substantial health risk, there remain scientific uncertainties that need to be resolved (Gulson et al., 2015).

Should conclusive evidence be presented that certain-sized insoluble/partially soluble and/or biopersistent nanoparticles, when topically applied in a cosmetic, are able to permeate the skin and may pose a risk to human health, the onus shall be placed on regulators to take appropriate and proportional steps. What these steps may look like, however, will need to be informed by science.

The second concern—adequacy of conventional risk assessment paradigms that are based on mass metrics—is similarly a scientific question that has implications for regulatory science. While there are still a number of questions that remain to be answered in relation to appropriateness of current risk assessment protocols, the SCCS has stated that for soluble and/or biodegradable nanoparticles, conventional methods 'may be adequate'. The committee went on to say that, 'for the insoluble particles other metrics, such as the number of particles, and their surface area as well as their distribution are also required' (SCCS, 2007).

There has been significant activity among the scientific community to further evaluate conventional risk assessment protocols, with a focus on applicability to insoluble/partially soluble and/or biopersistent nanoparticles. Conclusions drawn from this body of work will need to inform the risk assessment protocols that are set out in regulatory instruments.

The third, fourth and fifth concerns previously noted are the most relevant to this chapter, and are considered next. As will be seen in the following section, which focuses on regulatory developments for nano-cosmetics, there is no one-size-fits-all approach to regulation. While the two approaches considered here have a scientific foundation, broader cultural, historical and political dimensions have all helped shape the regulatory regimes into what they are today. As such, divergence is likely between jurisdictions given their distinct histories, culture and experiences, thus giving rise to different regulatory approaches, despite the underlying science being the same. In the following section the focus is overwhelmingly on the US and EU approaches to nano-cosmetics.

REGULATORY DEVELOPMENTS FOR NANO-BASED COSMETICS

This section of the chapter examines the regulatory context and challenges posed by nano-cosmetics within the US and the EU. The regulatory philosophy of the two jurisdictions varies significantly in relation to cosmetics; the US have adopted what may be considered a light-touch approach to cosmetic products, while the EU has sought to ratchet up regulatory oversight. While the scope and powers of the regulatory regimes differ markedly, the effectiveness of each is dependent on sound scientific evidence.

United States

Food and Drug Administration Regulation of Nanotechnology in Cosmetics
The current FDA regulatory apparatus for nano-enabled cosmetics is in flux. In 2007, an FDA Task Force (2007) was assigned to discuss the future of nanotechnology products in the FDA. A guidance on nano-enabled cosmetics safety (FDA, 2014a) and a guidance on identifying nano-enabled products (FDA, 2014b) have since been released by the FDA to provide additional information to the industry on how the administration perceives

the use of nanomaterials in products that it regulates. As it stands, the FDA will continue to regulate products, such as cosmetics, under the traditional regulatory framework as provided for under the Federal Food, Drug, and Cosmetic Act of 1938 (FDA, 2014b). That being said, the FDA has made it known that it will monitor and adjust to changes in the knowledge of risks associated with nanotechnology products.

The FDA's approach to nano-enabled cosmetics is to continue to regulate products, rather than process. The FDA (2014a) considers 'the current framework for safety assessment sufficiently robust and flexible to be appropriate for a variety of materials, including nanomaterials'. Because the FDA does not recognize any specific definition of nanotechnology, the FDA will continue to assess the particular use of nanotechnology—and the safety of the resulting products—on a product-by-product basis.

The FDA provides two Points to Consider that manufacturers should use to determine whether their products involve nanotechnology. These include:

1. whether a material or end product is engineered to have at least one external dimension, or an internal or surface structure, in the nanoscale range (approximately 1–100 nm) and
2. whether a material or end product is engineered to exhibit properties or phenomena, including physical or chemical properties or biological effects, that are attributable to its dimensions, even if these dimensions fall outside the nanoscale range up to 1000 nm (FDA, 2014b).

The FDA typically undertakes this analysis at the pre-market review stage so as to inform manufacturers whether their product involves nanotechnology from a regulatory perspective. Because cosmetics are not subject to FDA pre-market review, these Points to Consider are to be used on a voluntary basis by manufacturers. And so, because manufacturers conduct their own safety substantiation tests, the FDA has relatively limited ability to ensure that the nano-related risks associated with nano-enabled cosmetics are quantified in these substantiation tests.

The FDA suggests that product-specific technical assessments should take into account the unique properties of nano-enabled products (FDA, 2014b). The size, structure and surface properties of nanomaterials introduce potential risks that are partly, but in many cases not fully, understood (Oberdorster et al., 2005; Scientific Committee on Emerging and Newly Identified Health Risks, 2009; Maynard et al., 2011; Luyts et al., 2013). The FDA's approach to such risks is to rely on the already available toxicological test data and conduct additional toxicological testing on new ingredients or formulations (FDA, 2014a). The FDA is aware that new testing methods may be required to assess nano-related risks associated with some families of nanomaterials, so the FDA is encouraging industry to come to the agency to discuss how best to substantiate their cosmetic safety claims.

The FDA (2014a) advises that the safety of nano-enabled cosmetics:

should be assessed through fully describing the nanomaterial and evaluating a wide range of physical and chemical properties, as well as through the assessment of impurities, if present...

The safety factors to consider include physico-chemical characteristics, agglomeration and size distribution of particles, impurities, routes of exposure and in vitro and in vivo toxicological data on nanomaterial ingredients (FDA, 2014a).

Because the FDA will continue to rely on the traditional regulatory regime—a common regulatory default scenario for emerging technologies that has been coined by Stokes (2012) as 'regulatory inheritance'—FDA oversight of nano-enabled cosmetics will be just as limited (or light) as for traditional cosmetics. The FDA Task Force (2007) suggested that the best steps forward for the FDA would be to evaluate current testing approaches that assess safety, effectiveness and quality, as well as promoting new assessment tools. This involves consulting and relying on manufacturers. The main challenge for the FDA will be to adequately assess risks. It would appear that data collection is a primary goal for the FDA moving forward, so that it can better regulate the safety of nanomaterials in cosmetics, as well as other nano-enabled products (Holdren et al., 2011).

Regulatory Challenges

To supplement the advisory options discussed in the preceding section, it has been suggested that the FDA should focus on developing its in-house nano-expertise, as well as researching analytical tools for nano-enabled products (Taylor, 2006; Hodge et al., 2010). Without congress vesting the FDA with serious enforcement powers over cosmetics, the next best option would be to focus on building up its information on nanotechnology analysis and risk assessment. Manufacturers are certainly interested in developing safe products, so having an agency available to discuss best practices and tools for nano-enabled products appears pragmatic and useful. Coordination between industry and the FDA may allow for the development of standard risk analysis frameworks for nanotechnology applications in consumer products.

Because federal legislation to expand the FDA's oversight of cosmetics under the Federal Food, Drug, and Cosmetic Act is unlikely, the FDA could seek to expand its advisory role so as to combat the challenges of assessing the safety of nanomaterials in cosmetic products. That being said, serious concerns exist as to the safety of nano-enabled cosmetics. Without real enforcement powers, and having to shoulder the burden of proof, the FDA will be hard pressed to mitigate health and safety risks from nano-enabled cosmetics.

European Union

European Union Regulation of Nanomaterials in Cosmetics

As part of the recast of the EU Regulation (2009), the European Parliament and Council voted to differentiate cosmetic products containing nanomaterials from their conventional cosmetic counterparts. In doing so, the Parliament and Council opted not to create a *sui generis* regulatory framework for nanotechnology and/or nanomaterials—as they had with genetically modified organisms—but rather to continue to regulate nanocosmetics as a cosmetic product, applying additional regulatory hurdles to their entry onto the internal market. This decision, as illustrated by Clause 30 of the Preamble, which talks about 'inadequate information on the risks associated with nanomaterials', appears to be one driven by broad concerns over the potential health consequences of such products to their citizens. Clause 30 of the Preamble stresses the need for regular review by the commission 'on nanomaterials in light of scientific progress', while Clause 35 requires the SCCS to 'give opinions where appropriate on the safety of use of nanomaterials in cosmetic products'. These requirements stand in stark contrast to the approach that has been the position in the US.

Pursuant to Article 2 (1)(k), for the purposes of the EU Regulation, a 'nanomaterial' is defined as 'an insoluble or biopersistent and intentionally manufactured material with one or more external dimensions, or an internal structure, on the scale from 1 to 100 nm;...' This includes, for example, nano-TiO_2, nano-ZnO and fullerenes. It does not, however, include SLNs and liposomes; manufacturers of finished cosmetics that contain SLNs and/or liposomes are therefore not required to conform to the nano-specific requirements set out in Articles 13, 16 and 19.

This narrow definition of what constitutes a 'nanomaterial' for the purposes of the EU Regulation has raised concerns with some commentators (Bowman et al., 2010); it is, however, consistent with the earlier opinions published by the SCCS, which had clearly stated that insoluble/partially soluble and/or biopersistent nanomaterials are more likely to raise concerns than families of soluble and/or biodegradable nanoparticles when used in cosmetic products. Arguably of greater concern is the very sharp cut-off point for size: 1–100 nm. This means that TiO_2 that has a size range of 120 nm, and exhibits the same functionality and/or novelty as TiO_2 with a size range of 90 nm, will not be considered a 'nanomaterial' for the purpose of the regulation (Maynard, 2011; Maynard et al., 2011). The EU Regulation does, however, require the commission to modify this definition in light of 'technical and scientific progress and to definitions subsequently agreed at international level' [Article 2 (3)].

As noted earlier, for finished cosmetic products that do contain nanomaterials, the responsible party will be required to meet the regulatory requirements set out in Articles 13 (Chapter III, Safety Assessment, Product Information File, Notification), 16 (Chapter IV, Restrictions for Certain Substances) and 19 (Chapter VI, Consumer Information). Each shall be discussed in turn.

Article 13 sets out notification requirements that apply to all cosmetic products prior to their placement on the market; this information is required to be submitted to the commission by the responsible party through electronic means. This includes, among other things, general product information such as product category, country of origin, contact details of the responsible party, name and Chemical Abstract Service number for certain types of chemicals and photographic images of the packaging.

Article 13 (1)(f) addresses nanomaterials specifically. Pursuant to the provision, the electronic dossier must contain information regarding the 'presence of substances in the form of nanomaterials…'. Information supplied to the commission under Article 13, including Article 13(1) (f), is then provided to all competent national authorities [Article 13(5)], for certain limited purposes, including 'market surveillance, market analysis, evaluation and consumer information'. The commission is also required to share this information with 'poison centres or similar bodies' [Article 13(6)] for the purposes of rendering medical assistance.

Article 16 deals exclusively with nanomaterials. Article 16(3) builds on the notification requirements set out in Article 13, and can be said to reflect the Parliament and Council's push for greater transparency with cosmetic products generally, and those that contain nanomaterials specifically. Pursuant to Article 16(3):

> cosmetic products containing nanomaterials shall be notified to the Commission by the responsible person by electronic means six months prior to being placed on the market, except where they have already been placed on the market by the same responsible person before 11 January 2013.

Article 16(3) then goes on to specify the types of information that the responsible party must provide to the commission including, for example, estimated quantity of the nanomaterial, physical and chemical properties, toxicological profile, safety data and 'reasonably foreseeable exposure conditions' [Article 16(3)(a)–(f)].

Article 16(4) provides the mechanism by which the commission can, when they have concerns over the safety of a nanomaterial, request the SCCS to provide 'its opinion on the safety of such nanomaterial for use in the relevant categories of cosmetic products and on the reasonably foreseeable exposure conditions'. The opinion must be rendered within 6 months of the request and made available to the public. The provision provides the commission with the power to require additional data from the responsible party should the SCCS find the initial submission lacking.

The creation of a publically available, and searchable, catalogue of nanomaterials used in cosmetic production that are available on the EU market, and regularly updated, is set out in Article 16(10)(a). The catalogue was to be available by 11 January 2014 (which it was not). Article 16(10) also requires annual status reports regarding the use of nanomaterials to be submitted to the Parliament and Council; the provision

> shall summarise, in particular, the new nanomaterials in new categories of cosmetic products, the number of notifications, the progress made in developing nano-specific assessment methods and safety assessment guides, and information on international cooperation programmes.

Together, these requirements are mechanisms to increase the transparency between the relevant stakeholders in the EU, including the commission, Parliament and Council, competent national authorities, consumers and industry itself. Data generated through these processes shall build a useful information base on which subsequent decisions may then be made.

Labelling requirements for all cosmetic products are set out in Article 19 of the EU Regulation. Pursuant to Article 19(1), the responsible party is required to, among other things, include name and address details of the responsible person, information regarding the nominal content of the product at the time of packaging by mass, date of minimum durability, appropriate precautions and function for a cosmetic product. These requirements apply to all finished cosmetic products and not just those that contain nanomaterials.

Article 19(1)(g) does, however, include an extra requirement for nano-based cosmetics:

... All ingredients present in the form of nanomaterials shall be clearly indicated in the list of ingredients. The names of such ingredients shall be followed by the word "nano" in brackets ....

For example, where a moisturizer contains nano-TiO_2, the responsible party would need to ensure that the label looks as follows: 'Titanium dioxide (nano); ...'.

In doing so, it would appear to be another mechanism by which the commission and council are promoting transparency regarding the use of nanomaterials in cosmetics. It is unclear, however, how this tool may influence the buying behaviour of consumers at this time.

Regulatory Challenges

The regulatory approach adopted by the European Parliament and Council in relation to cosmetics is one of precaution. The EU Regulation is, itself, underpinned by the precautionary principle (Clause 36 of the Preamble). This approach by the Parliament and Council is not unique to cosmetics, with the other key instruments recast since the early 2000s also underpinned by the precautionary principle [see, for example, Article 3(1) of Regulation (EC) No. 1907/2006, which regulates chemical substances and states that 'Its provisions are underpinned by the precautionary principle']. From a practical standpoint, this has resulted in a ratcheting up of the various regulatory regimes, with a greater emphasis placed on the timely provision of data to the regulatory agencies, the creation of feedback loops between industry and regulators, and greater transparency among stakeholders within the EU market.

Despite some early calls for a moratorium on nanotechnology and/or the introduction of nano-specific legislative instruments (see, for example, ETC Group, 2005; Friends of the Earth, 2007), the European Parliament and Council have chosen to instead include nano-specific provisions within product-focused legislation. In addition to the EU Regulation—which was the first such instrument in the world to do so—such nano-specific provisions may also be found in Regulation (EU) No. 1169/2011 on the provision of food information to consumers and Regulation (EU) No. 2015/2283 of the European Parliament and of the Council of 25 November 2015 on novel foods. This approach, arguably, recognizes the absence of scientific evidence of harm to humans as a result of nano-enabled products (including cosmetics), and the scientific uncertainties that still remain in relation to the use of nanoparticles in these very products. This is a very different approach compared to that found in the US, for example.

A core tenet of the EU Regulation is to provide a higher level of protection to consumers within the EU and enhance transparency in relation to cosmetic products. The recast of the EU Regulation provides the commission and competent national authorities with additional powers to do this. It remains to be seen, though, whether the narrow definition of 'nanomaterial' adopted in the EU Regulation may undermine this to some degree.

As noted earlier, the narrow definition adopted in the EU Regulation purposely excluded a number of families of nanomaterials. While the weight of scientific evidence would suggest that this shall not be problematic from a risk perspective, it does undermine the push for greater transparency around the use of nanomaterials in the Community market.

While responsible parties will have to comply with the general notification requirements set out in Article 13, the inclusion of, for example, nanosomes or a nanoemulsion in a product will not trigger the additional nano-specific notification requirements. This means that the commission, national competent authorities and other relevant parties will have incomplete information regarding the market. The commission shall similarly be working in an environment of incomplete information when it comes to the creation of the public catalogue for nanomaterials and the annual reports that it is required to provide the Parliament and Council (as required under Article 16).

Article 16(3) requires the responsible party to provide the commission with technical and scientific data relating to each and every nanomaterial used in their product; this information is used to determine if an independent safety assessment is needed. The onus is placed on the responsible party to supply high-quality data to the commission. The intent of the provision may, however, be undermined in some instances because of the burden that this places on the responsible party. For example, it presumes that the responsible party has the instrumentation to accurately measure particle size, that they have employed appropriate analytical methods for the safety data and are aware of the potential risks—if indeed there are any—of the nanoparticles in the formulation.

Bowman and Hodge (2009) suggest that the narrow definition of 'nanomaterial' in the regulation may result in some manufacturers reformulating existing nano-based cosmetics so as to avoid triggering the nano-specific requirements. They suggest that where, for example, a concealer containing nano-TiO_2 at 90 nm could be reengineered so as to incorporate TiO_2 at 120 nm, while still exhibiting the same functionality at a similar cost, some manufacturers may choose to do so. This would enable them to avoid the additional regulatory requirements, thereby avoiding additional scrutiny and compliance costs thereof. This hypothetical scenario is further complicated by all nanomaterials comprising nanoparticles having a range of particle sizes, meaning that manufacturers may adjust the size distribution so that the median size (for example) lies above 100 nm, but the majority of the particles lie below this. While it is unclear whether this has occurred, any such action would go against the spirit of the EU Regulation.

Finally, the effectiveness of these nano-specific provisions will be dependent on the commission and other bodies, such as the SCCS, having the necessary resources to perform their work. In particular, we would argue that it is incumbent on the commission to undertake randomized blind testing on finished cosmetic products to determine if they contain nanomaterials. Where nanomaterials are identified as an ingredient in a product, and the responsible party has not complied with Articles 13, 16 and 19, appropriate enforcement action—as provided by the regulation—should be pursed. Teeth and claws will both be important for ensuring the effectiveness of the EU Regulation.

CONCLUSION

Cosmetics should not pose any risks to the individual when used as intended. Nor should a consumer be required to do a risk–benefit analysis prior to using a cosmetic product, as they may do in relation to, for example, a drug or medical device.

A number of different nanomaterials are now routinely incorporated into hundreds of cosmetic and personal care products. They offer the industry an opportunity to further enhance current products and develop revolutionary new products, as well as increasing profit levels and/or market share. For consumers, nano-cosmetics promise superior products that enhance beauty, augment fashion and provide a range of (potential) skin benefits. These benefits can be fully realized only if, however, consumers accept the technology and the products themselves.

As this chapter has sought to illustrate, there are a range of regulatory approaches that can be, and have been, adopted to oversee cosmetics generally, and those that contain nanomaterials more specifically. The scope of these regimes varies from jurisdiction to jurisdiction: from light-touch approaches—under which regulatory agencies have limited pre- and post-market powers—through to more firm approaches. Each approach has its supporters and its critics; there has yet to be a 'perfect' regulatory regime.

With this in mind, the divergence between the approaches adopted by the US and the EU for overseeing the entry of nano-cosmetics onto the market provides a real-time comparative experiment as to the strengths and weaknesses of the regimes. The lessons learnt shall be important for the jurisdictions themselves, as well as other jurisdictions that are looking to the US and the EU for guidance on how they could and should regulate products containing nanomaterials.

REFERENCES

Bakry, R., Vallant, R.M., Najam-ul-Haq, M., Rainer, M., Szabo, Z., Huck, C.W., Bonn, G.K., 2007. Medicinal applications of fullerenes. Int. J. Nanomed. 2 (4), 639.
Bombardelli, E., 1991. Boll. Chimico Farm. 130, 431.
Bowman, D.M., Hodge, G.A., 2009. Regul. Gov. 3, 145.
Bowman, D.M., Van Calster, G., Friedrichs, S., 2010. Nanomaterials and regulation of cosmetics. Nature Nanotechnol. 5 (2), 92.

Brandt, F.S., Cazzaniga, A., Hann, M., 2011. Sem. Cutan. Med. Surg. 30, 141–143.

Business Wire, 2015. Global Cosmetics Market – By Product Type, Ingredient, Geography, and Vendors – Market Size, Demand Forecasts, Industry Trends and Updates, Supplier Market Shares 2014–2020. Research and Markets, New York.

ETC Group, 2004. Nanotech: Unpredictable and Un-Regulated: New Report from ETC Group. ETC Group, Ottawa. http://www.etcgroup.org/content/nanotech-unpredictable-and-un-regulated.

ETC Group, 2005. NanoGeoPolitics. ETC Group Communique No.89. Canada: July/August ETC Group, Ottawa. http://www.etcgroup.org/sites/www.etcgroup.org/files/publication/51/01/com89specialnanopoliticsjul05eng.pdf.

Food and Drug Administration, 2014a. Guidance for Industry: Safety of Nanomaterials in Cosmetic Products. Center for Food Safety and Applied Nutrition. http://www.fda.gov/Cosmetics/GuidanceRegulation/GuidanceDocuments/ucm300886.htm.

Food and Drug Administration, 2014b. Office of the Commissioner, FDA, Guidance for Industry: Considering whether an FDA-Regulated Product Involves the Application of Nanotechnology. http://www.fda.gov/RegulatoryInformation/Guidances/ucm257698.htm.

Food and Drug Administration Task Force, 2007. Nanotechnology Task Force, FDA, Nanotechnology.

Friends of the Earth, 2006. Nanomaterials, Sunscreens and Cosmetics: Small Ingredients, Big Risks. FOE, Melbourne. http://libcloud.s3.amazonaws.com/93/ce/0/633/Nanomaterials_sunscreens_and_cosmetics.pdf.

Friends of the Earth, 2007. Nanotechnology Policy Statement. FOE, Melbourne. http://emergingtech.foe.org.au/wp-content/uploads/2014/08/FoEA-Nanotechnology-Policy-May-2007.pdf.

Grande, F., Tucci, P., 2016. Mini Rev. Med. Chem. 16, 762.

Gulson, B., McCall, M., Korsch, M., Gomez, L., Casey, P., Oytam, Y., Taylor, A., McCulloch, M., Trotter, J., Kinsley, L., Greenoak, G., 2010. Toxicol. Sci. 118, 140.

Gulson, B., McCall, M.J., Bowman, D.M., Pinheiro, T., 2015. Arch. Toxicol. 89, 1909.

Hodge, G.A., Bowman, D., Maynard, A., 2010. International Handbook on Regulating Nanotechnologies. Edward Elgar Publishing Ltd.

Holdren, J.P., Sunstein, C.R., Siddiqui, I.A., 2011. Memorandum for the Heads of Executive Departments and Agencies.

Kligman, D., 2000. Dermatol. Clin. 18, 609.

Lohani, A., Verma, A., Joshi, H., Yadav, N., Karki, N., 2014. Nanotechnology-based cosmeceuticals. ISRN Dermatol. 2014.

Luyts, K., Napierska, D., Nemery, B., Hoet, P.H., 2013. Environ. Sci. Process. Impacts 15, 23.

Maynard, A.D., 2011. Nature 475, 31.

Maynard, A.D., Warheit, D., Philbert, M.A., 2011. Tox. Sci. 120, S109.

Miller, G., 2006. Chain React. 97, 21.

Monteiro-Riviere, N.A., Wiench, K., Landsiedel, R., Schulte, S., Inman, A.O., Riviere, J.E., 2011. Safety evaluation of sunscreen formulations containing titanium dioxide and zinc oxide nanoparticles in UVB sunburned skin: an in vitro and in vivo study. Toxicol. Sci. 123 (1), 264–280.

Monteiro-Riviere, N.A., 2013. In: Nasir, A., Friedman, A., Wang, S. (Eds.), Skin Penetration of Engineered Nanomaterials, Nanotechnology in Dermatology. Springer, New York.

Mu, L., Sprando, R.L., 2010. Pharm. Res. 27, 1746.

Müller, R.H., Radtke, M., Wissing, S.A., 2002. Adv. Drug Deliv. Rev. 54, S131.

Nanoderm, 2007. Quality of Skin as a Barrier to Ultra-fine Particles: Final Report. QLK4-CT-2002–02678 European Commission, Leipzig. http://www.uni-leipzig.de/~nanoderm/Downloads/Nanoderm_Final_Report.pdf.

Newman, M.D., Stotland, M., Ellis, J.I., 2009. J. Am. Acad. Dermatol. 61, 685.

Nohynek, G.J., Lademann, J., Ribaud, C., Roberts, M.S., 2007. Crit. Rev. Toxicol. 37, 251.

Oberdorster, G., Oberdorster, E., Oberdorster, J., 2005. Environ. Health Perspect. 113, 823.

Osmond-McLeod, M.J., Oytam, Y., Kirby, J.K., Gomez-Fernandez, L., Baxter, B., McCall, M.J., 2014. Nanotoxicology 8, 72.

Patel, A., Prajapati, P., Boghra, R., 2011. Overview on application of nanoparticles in cosmetics. Asian J. Pharm. Sci. Clin. Res. 1, 40.

Pinheiro, T., Allon, J., Alves, L.C., Filipe, P., Silva, J.N., 2007. Nucl. Instrum. Methods Phys. Res. B 260, 119.

Prow, T.W., Monteiro-Riviere, N.A., Inman, A.O., Grice, J.E., Chen, X., Zhao, X., Sanchez, W.H., Gierden, A., Kendall, M.A., Zvyaqin, A.V., Erdmann, D., Riviere, J.E., Roberts, M.S., 2012. Nanotoxicology 6, 173.

Raj, S., Jose, S., Sumod, U.S., Sabitha, M., 2012. J. Pharm. BioAllied Sci. 4, 186.

EU Regulation: Regulation (EC) No.1223/2009 of the European Parliament and of the Council of 30 November 2009 on Cosmetic Products Regulation (EC) No.1223/2009 of the European Parliament and of the Council of 30 November 2009 on Cosmetic Products. http://eur-lex.europa.eu/legal-content/EN/TXT/HTML/?uri=CELEX:02009R1223-20150416&from=EN.

Scientific Committee on Consumer Products, 2007. Opinion on Safety Assessment of Nanomaterials in Cosmetics. SCCP/1484/12. European Commission, Brussels.

Scientific Committee on Consumer Safety, 2012. Guidance on the Safety Assessment of Nanomaterials in Cosmetics. SCCS/1147/07. European Commission, Brussels.

Scientific Committee on Emerging and Newly Identified Health Risks, 2009. Risk Assessment of Products of Nanotechnologies. European Commission, Brussels.

Sonneville-Aubrun, O., Simonnet, J.T., L'Alloret, F., 2004. Adv. Colloid Interf. Sci. 108, 145.

Stokes, E., 2012. J. Law Soc. 39, 93.

Taylor, M.R., 2006. Woodrow Wilson International Center for Scholars, Regulating the Products of Nanotechnology: Does FDA Have the Tools It Needs? Project on Emerging Technologies, Washington, DC. http://www.nanotechproject.org/file_download/files/PEN5_FDA.pdf.

Tyner, K.M., Wokovich, A.M., Godar, D.E., Doub, W.H., Sadrieh, N., 2011. Int. J. Cosmet. Sci. 33, 234.

Wu, J., Liu, W., Xue, C., Zhou, S., Lan, F., Bi, L., Xu, H., Yang, X., Zeng, F.D., 2009. Toxicol. Lett. 191, 1.

CHAPTER 13

Green Cosmetic Ingredients and Processes

Carla Villa
University of Genova, Genova, Italy

INTRODUCTION

The investigation and application of the green chemistry principles have led to a high level of change in the manufacturing and design of chemical products, with the development of cleaner and more benign processes as human and environmental safety demand.

The cosmetics industry is subject to these changing attitudes, and personal care chemists need to take measures to improve the green credentials of their products.

This chapter will cover our studies in the cosmetic field concerning the use of solvent-free procedures under microwave irradiation as very useful and innovative methodologies for a green approach in various concerns. Eco-friendly procedures under dielectric heating have been successfully applied to conventional organic reactions such as esterification, Knoevenagel condensation, alkylation and ketalization for the synthesis of new potential cosmetic ingredients, as well as useful commercial compounds, as a simplification and improvement of the conventional procedures. More recently green microwave extractions were applied to several organic matrices as alternative and sustainable procedures for the exploitation of green renewable sources.

A related aspect is the interest in using green methods (environmentally friendly and safe for operator) in cosmetics analysis as was said in Chapter 4 of this book.

GREEN COSMETIC INGREDIENTS AND PROCESSES
Eco-friendly Organic Synthesis

Economic, legislative (EC Directive, 2006; EC Regulation, 2006; COM, 2001a,b) and consumer pressures are forcing an unprecedented level of attention on human and environmental safety of chemical products. Therefore the sustainable development of the chemical industry has become imperative, and new safer compounds and technologies are the subject of intense studies in both the academic and the industrial world.

Green chemistry is a highly effective approach to ecological problems because it applies innovative scientific solutions to real-world environmental situations. The 12 Principles of Green Chemistry, originally published by Anastas and Warner (1998), provide a road map for chemists to implement green chemistry. The investigation and application of these principles have led to high level of change in the manufacturing and design of chemical products, with the development of cleaner and more benign chemical processes.

The cosmetics industry is subject to these changing attitudes, and personal care chemists need to take measures to improve the green credentials of their products in all life cycle stages: sustainability of raw materials, product safety when being used and once released into the environment, energy efficiency of chemical processes and reduction of manufacturing cost. In this context, scientific research on cosmetic products has been ongoing.

This section will cover studies in the field of organic synthesis concerning the development of cleaner synthesis methods and of greener compounds based on the 'benign by design' ideology. To attain these goals such strategies are followed as drastic reduction of organic solvents, application of microwaves as an alternative energy source, use of safe reagents and application of in vitro cytotoxicity tests to assess the safety of the synthesized compounds.

The potentiality of solvent-free conditions and microwave (MW) activation is investigated, comparing the results obtained with those achieved by conventional procedures.

The use of MW activation to directly heat chemical reactions has become an increasingly popular technique in the scientific community (Kingston and Haswell, 1997; Loupy, 2002; Hayes, 2002; Lidström and Tierney, 2005; Larhed and Olofsson, 2001).

MW dielectric heating, as an efficient alternative to conventional conductive heating, is based on the properties of some solids and liquids to transform electromagnetic energy into heat. This in situ mode of energy conversion is very attractive in organic synthesis.

MW-assisted organic synthesis can be applied to a variety of chemical transformations, and the original kind of interaction of MWs and material (selective absorption of MW energy by polar molecules) causes thermal effects (connected to dipolar and charge space polarization) and sometimes specific, not purely thermal ones, which lead to a general speeding up (acceleration) of several chemical reactions and selective transformations. Although there is still debate on the existence and nature of so-called 'non-thermal' effects of MWs that could provide a rationalization of the rate enhancement in organic reactions (Perreux and Loupy, 2001), there is no doubt that a significant energy savings can be expected from MW technology on a laboratory scale compared to the energy efficiency of conventional conductive heating.

The possibility of performing reactions in a very short time by the direct interaction of MWs with the reaction mixture, in comparison with the indirect transfer of energy from a conductive thermal source, leads to the consideration of MW chemistry as a part of green chemistry: reduction of energy consumption, time savings, increased efficiency.

The use of solvents is often an integral part of chemical processes and can have important social, economical and environmental implications.

One of the key areas of green chemistry is the elimination of solvents in chemical processes or the replacement of hazardous solvents with environmentally benign chemicals. This need for developing unconventional media for chemical reactions has led to an increasing interest in the use of solvent-free procedures which are particularly adapted to MW activation. When MW applications were first being used, the solvent-free approach was popular because of safety problems in using domestic household MW ovens and open-vessel technology. More recently it is manifest that there are several advantages to using MW activation in the absence of organic solvents: increased safety, cost reduction and respect for the environment. Solvent-free procedures under MW irradiation seem to be very useful and innovative methodologies for greener organic synthesis.

In our studies, solvent-free procedures under dielectric heating have been successfully applied to conventional reactions such as esterification, Knoevenagel condensation, alkylation, etc., for the synthesis of potential new cosmetic ingredients, as well as useful commercial compounds, as a simplification and improvement of conventional procedures.

The reactions in dry media considered here are accomplished by three solvent-free methods (Loupy et al., 1998):

- reagents impregnated on mineral oxides such as alumina, silica and clays, as neutral, basic or acid supports, also in the presence of strong bases (KF/Al_2O_3, KOH/Al_2O_3) or acidic catalysts (p-toluenesulphonic acid/Al_2O_3);
- solid–liquid solvent-free phase-transfer catalysis (PTC), a specific method for anionic reactions using catalytic amounts of tetraalkylammonium salts (Aliquat 336, octyltrimethylammonium chloride; TBAB, tetra-n-butylammonium bromide) as the transfer agents under basic conditions (KOH, K_2CO_3);
- simple mixture of neat reactants in the presence of a catalyst.

To check the possible intervention of non-thermal MW effects, experiments in dry media were generally carried out under conductive heating (Δ) in a thermostated oil bath, under the same conditions as with MWs (time, temperature, vessel), with an accurate control of the profile temperature increase under both conditions.

The equipment used in our studies to supply MW energy to the reaction bulks are scientific monomode ovens, specific for organic synthesis. These MW reactors allow for a homogeneous distribution of the electric field and they can be used with low emitted power and high energetic yield. For these reasons the monomode reactors offer increased efficiency and reliability compared to the multimode commercial systems (domestic ovens). Moreover, these scientific tools present a number of advantages: temperature measurement by infrared detection or by optical fibre, temperature control by power modulation, monitoring of the reaction by a specific software, use of open or closed vessels, magnetic or mechanical stirring with adjustable rotation speeds.

Synthesis of Benzylidene Derivatives: UV Filters

It is well established that solar radiation is one of the most important environmental stress agents for human skin; today photoprotection has become an essential topic. These findings promoted research towards the development of efficient and safe UV cosmetic sunscreens, leading to the synthesis of some benzylidene derivatives of (+)1,3,3-trimethyl-2-oxabicyclo[2.2.2]octan-6-one (**1**) by conventional procedures (Mariani et al., 1993, 1994, 1996, 1998). These compounds, UVB and UVA filters, are related to 3-benzylidene camphor derivatives, a well-known class of UV cosmetic filters (Bouillon et al., 1982), taking into account the structural correlation between the bicyclic rings of bornane and cineole. The new compounds were analysed for their in vitro phototoxicity and percutaneous absorption (Mariani et al., 1993, 1998).

The synthesis of these safe compounds was revised (Mariani et al., 2000) and new benzylidene derivatives were synthesized and also assessed as regards their photostability (Villa et al., 2001).

MW activation coupled with solvent-free PTC proved to be a very efficient method to obtain compounds **3a–g** by Knoevenagel condensation of **1** with four-substitute benzaldehydes (**2a–c**) and 3,4-dialkoxybenzaldehydes (**2d–g**) (Scheme 13.1).

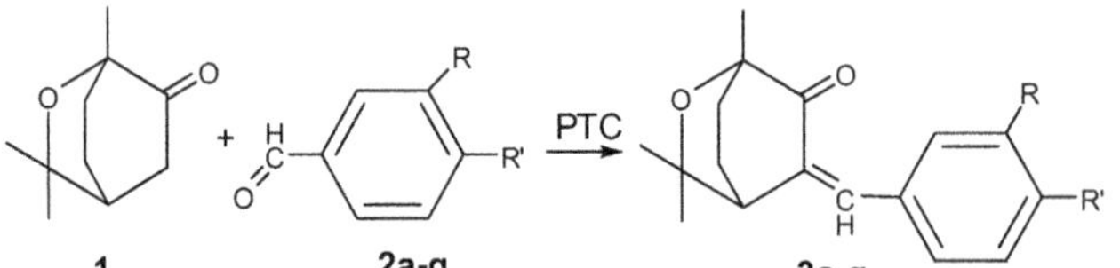

1 **2a-g** **3a-g**

	R	R'
a	H	CH_3
b	H	$N(CH_3)_2$
c	H	$OC_{16}H_{33}$
d	$OC_{12}H_{25}$	$OC_{12}H_{25}$
e	$OC_{14}H_{29}$	$OC_{14}H_{29}$
f	$OC_{16}H_{33}$	$OC_{16}H_{33}$
g	$OC_{18}H_{37}$	$OC_{18}H_{37}$

Scheme 13.1 Schematic solvent-free synthesis of benzylidene derivatives **3a–g**.

Considerable improvements over classical procedures, in terms of reaction times and yields, have been noted when using, under MW activation (Table 13.1): Aliquat 336 as the phase-transfer agent and KOH as the base, for compounds **3a–c**, and TBAB as the phase-transfer agent and KOH + K$_2$CO$_3$ as the base for compounds **3d–g**.

Comparing yields obtained under classical heating (Δ), all conditions being equal to those under MW activation (reaction time, temperature, vessel), a very strong specific MW effect is evident (compound **3d**, yield % MW/Δ = 80/13). Such a conclusion is supported by examination of the identical temperature rise profiles studied and here reported as an example for compound **3d** (Fig. 13.1).

The lipophilic aromatic aldehydes **2c–g** used for the synthesis of the corresponding benzylidene derivatives were obtained by alkylation reaction of 4-hydroxybenzaldehyde (In Scheme 13.2, as an example, the synthetic reaction for compound **2c** is reported) (Mariani et al., 2000) and 3,4-dihydroxybenzaldehyde (Scheme 13.3) (Villa et al., 2001) with the appropriate alkyl halides by PTC/MW.

Using PTC/MW, higher yields are obtained compared to PTC/Δ; moreover, a dramatic acceleration of the reaction is achieved in comparison to classical methods (compound **2c**, MeONa/dimethylformamide, 6 h, 60%; compounds **2d–g**, acetone/K$_2$CO$_3$, 24 h, 49%–66%).

Table 13.1 Syntheses of benzylidene derivatives **3a–g**

	Method	Time (min)	T (°C)	Yield (%)
3a	Classical: KOH/ethanol	720	RT	30
	MW: KOH/Aliquat 336	2	180	90
	Δ: KOH/Aliquat 336	2	180	20
3b	Classical: CH$_3$ONa/methanol	60	Reflux	51
	MW: KOH/Aliquat 336	2	180	94
	Δ: KOH/Aliquat 336	2	180	40
3c	Classical: CH$_3$ONa/methanol	720	Reflux	79
	MW: KOH/Aliquat 336	6	200	94
	Δ: KOH/Aliquat 336	6	200	12
3d	Classical: MeONa/THF	1440	Reflux	51
	MW: K$_2$CO$_3$–KOH/TBAB	60	140	80
	Δ: K$_2$CO$_3$–KOH/TBAB	60	140	13
3e	Classical: MeONa/THF	1440	Reflux	60
	MW: K$_2$CO$_3$–KOH/TBAB	60	140	78
3f	Classical: MeONa/THF	1440	Reflux	65
	MW: K$_2$CO$_3$–KOH/TBAB	60	140	81
3g	Classical: MeONa/THF	1440	Reflux	68
	MW: K$_2$CO$_3$–KOH/TBAB	60	140	87
	Δ: K$_2$CO$_3$–KOH/TBAB	60	140	19

Aliquat 336, octyltrimethylammonium chloride; *MeONa*, sodium methoxide; *MW*, microwave; *RT*, room temperature; *TBAB*, tetra-*n*-butylammonium bromide; *THF*, tetrahydrofuran; Δ, classical heating.

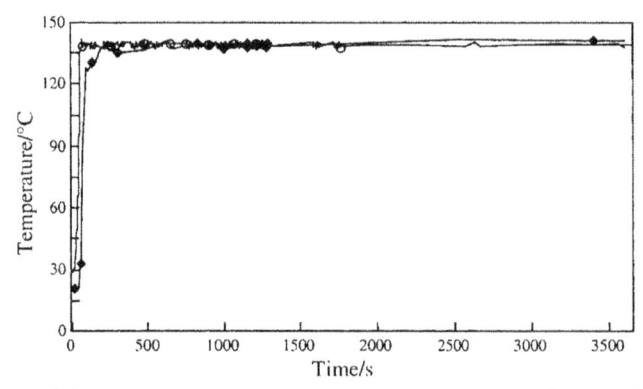

Figure 13.1 Profiles of the rise in temperature by microwave irradiation (○) and classical heating (♦) for the synthesis of compound **3d**.

Scheme 13.2 Solvent-free synthesis of compound **2c**.

Scheme 13.3 Solvent-free synthesis of compounds **2d–g**.

C$_n$H$_{2n+1}$	C$_{12}$H$_{25}$	C$_{14}$H$_{29}$	C$_{16}$H$_{33}$	C$_{18}$H$_{37}$
Yield%	86 (Δ 18)	80	81	80 (Δ 10)

The photochemical behaviour of the new compounds was investigated by irradiating with increasing UVA and UVB light doses, in a range from 0.3 to 32 J/cm^2 (1 J = 1 W/s).

The system was aerated to avoid thermal influence. Samples were placed 25 cm from UVA and UVB lamps, the intensities of which were 4.0 and 2.5 mW/cm^2, respectively.

The isomerization was monitored by thin-layer chromatography [chloroform/cyclohexane (65/35)] on precoated aluminium sheet silica gel 60F254 and the samples were then processed by reversed-phase liquid chromatography (RP-LC) using a mobile phase of acetonitrile/tetrahydrofuran (75/25) (flow 0.5 mL/min) to quantify the percentage of the two isomers. With regard to the UVB radiation, an increasing photoinduced *cis/trans* isomerization of compound **3a** was observed (Fig. 13.2). The *cis/trans* percentage ratio was detected by RP-LC. A 12% *cis* conversion was observed after an irradiation dose of 0.3 J/cm^2. This conversion increased until a photostationary equilibrium was reached at 18 J/cm^2 (*trans/cis*: 45/55) (Fig. 13.2). No degradation products were observed for the entire range of irradiation doses.

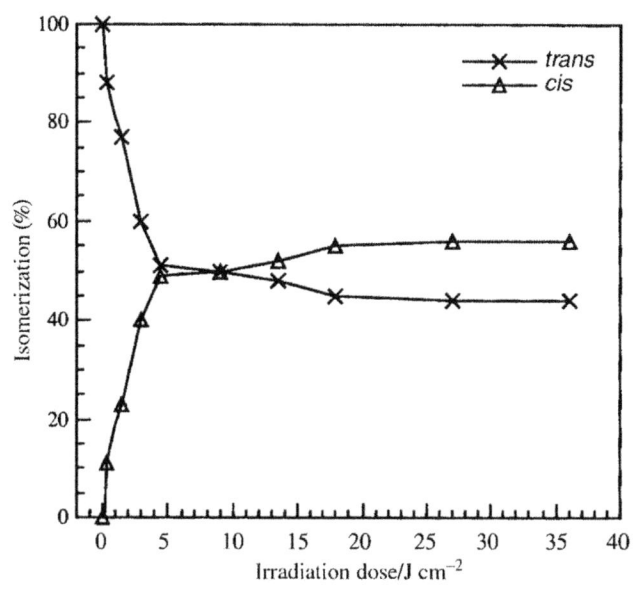

Figure 13.2 Isomerization percentage of compound **3a** as a function of UVB dose.

UVA radiation leads to a slower isomerization, which can be detected only at 32 J/cm² (*cis* conversion: 5%). No degradation products were observed for the entire range of irradiation doses. The isomerization was studied also by ¹H NMR: the condensation of the cineole ketone **1** with the appropriate dialkoxybenzaldehyde leads exclusively to the *trans* isomer, which is thermodynamically more stable. The *trans* structure is attributed by ¹H NMR spectral data considering the chemical shift of the vinylic proton. As an example, the ¹H NMR spectrum of compound **3a** in the *trans* and *cis* forms is reported (Fig. 13.3).

Synthesis of Cyclic Ketals: Fragrances

Alicyclic acetals and ketals are a class of compounds interesting to the cosmetics field for their olfactive features (Buchbaue and Lux, 1991) and are classically prepared by reaction of the carbonyl compounds with diols, in the presence of an acidic catalyst, in benzene or toluene as solvent and azeotropic removal of water.

The best performing solvent-free conditions for the synthesis of some cyclic ketals are here reported as clean methodologies for obtaining purer odorous compounds.

Cyclic Ketals From a Cineole Ketone

New cyclic ketals, 1,3-dioxolanes (**4a–e**) and 1,3-dioxanes (**5a–c**) of **1**, were obtained in excellent yields by acid-catalysed condensation of the cineole ketone with several aliphatic diols, under MW activation (Scheme 13.4).

These new compounds can be considered interesting and safe cosmetic ingredients, as the odour evaluation and in vitro cytotoxicity assays showed (Genta et al., 2002a,b).

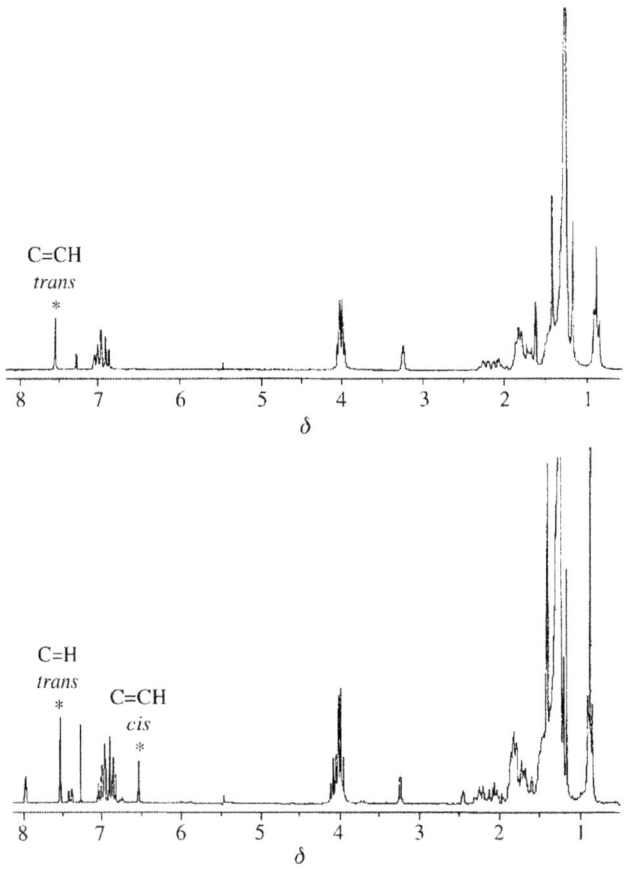

Figure 13.3 ¹H NMR spectra of pure *trans* isomer and *trans/cis* racemic mixture of compound **3a**.

	R	R¹		R²	R³	R⁴
4a	H	H	5a	CH₃	H	H
4b	H	CH₃	5b	CH₃	H	CH₃
4c	CH₃	CH₃	5c	H	CH₃	H
4d	H	C₂H₅				
4e	H	C₃H₇				

Scheme 13.4 Schematic solvent-free synthesis of cyclic compounds **4a–e** and **5a–c**.

Table 13.2 Most significant results related to the syntheses of compounds **4b** and **5b**

	Method	Time (min)	T (°C)	Yield (%)
4b	Classical: toluene/PTSA	360	Reflux	30
	MW: toluene/PTSA	15	110	90
	MW: acidic Al$_2$O$_3$/PTSA	30	110	78
	Δ: acidic Al$_2$O$_3$/PTSA	30	110	25
5b	Classical: toluene/PTSA	360	Reflux	12
	MW: toluene/PTSA	15	110	70
	MW: acidic Al$_2$O$_3$/PTSA	30	110	60
	Δ: acidic Al$_2$O$_3$/PTSA	30	110	19

MW, microwave; *PTSA*, *p*-toluenesulphonic acid; Δ, classical heating.

The results obtained using neat *p*-toluenesulphonic acid (PTSA) in toluene and impregnated on acid alumina in dry media (7% w/w; ketone/support = 1/4), at a temperature set point of 110°C, were compared. As examples the most significant results related to compounds **4b** and **5b** are reported in Table 13.2.

MW activation coupled with acidic catalysis in dry media (acidic Al$_2$O$_3$/PTSA) appeared to be an efficient and eco-friendly method to obtain terpenoidic cyclic ketals. The use of PTSA in toluene/MW led to quite the same results, but taking into account the interest in environmentally friendly processes, the solvent-free method should be considered the best.

Cyclic Ketals From 2-Adamantanone

MW irradiation coupled with acidic catalysis in dry media, without any solvent or support, was applied to the synthesis of some spiro[1,3-dioxolan]-2,2'-tricyclo[3.3.1.1^{3,7}] decane (compounds **7a–i**) promising cosmetic odorants (Genta et al., 2002a,b), by reacting 2-adamantanone (**6**) with different diols (Scheme 13.5).

The reactions were performed using heterogeneous mixtures of reactants and neat PTSA (10% by weight of ketone, at 140°C for 15 min) as an improvement over the use of mineral acidic supports reported in the previous paragraph.

The advantages of the MW activation-mediated method are highlighted compared with the conventional procedure (toluene/PTSA, azeotropic removal of water) or using classical heating (oil bath) under solvent-free conditions.

Example: compound **7b**

MW-PTSA 10%	15 min	94%
Δ–PTSA 10%	15 min	41%
Δ–PTSA 10%	360 min	60%
Classical	360 min	57%

Synthesis of Cosmetic Esters

In the cosmetics field, esters represent an important class of ingredients for different applications. Esters are classically prepared by esterification of carboxylic acids with

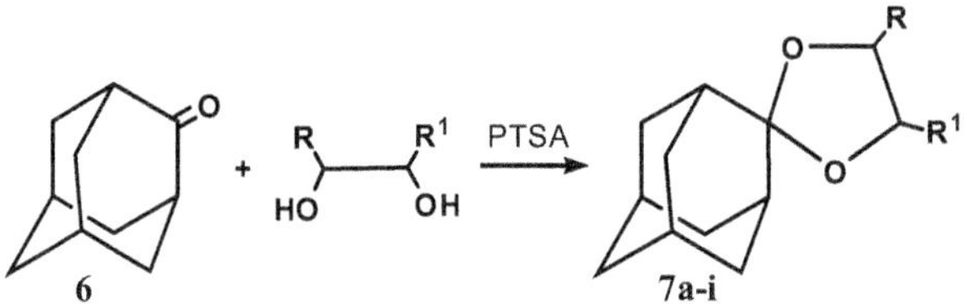

Scheme 13.5 Schematic solvent-free synthesis of cyclic compounds 7a–i.

	R	R¹
7a	H	H
7b	H	CH₃
7c	CH₃	CH₃
7d	H	C₂H₅
7e	H	C₃H₇
7f	H	CH₂OC₂H₅
7g	H	C₄H₉
7h	H	C₆H₁₃

alcohols using acidic catalysis, transesterification of methyl or ethyl esters or alkylation of carboxylate anions. Some reports refer to the use of PTSA, sulphonated resins, zeolites, zirconium oxide or supported acids and polymers, in non-conventional techniques for esterification reactions (Chemat et al., 1997; Okuyama et al., 2000; Ballini et al., 1998; Loupy et al., 1993). All these methodologies suffer from long reaction times, hard workup and the use of non-eco-friendly reagents and solvents.

In this review studies on the application of solvent-free MW-mediated methods as green chemistry synthesis of some esters of various chemical structures are reported.

Three kinds of reactions were applied:
- reaction of acyl chlorides with alcohols by basic catalysis with reagents supported on mineral oxides;
- esterification of carboxylic acids with alcohols in stoichiometric amounts by acidic catalysis;
- PTC alkylation of carboxylate anions (generated in situ) with alkyl halides.

Synthesis of 1,8-Cineole Derivatives: Fragrances

Some esters, 1,8-cineole derivatives, of interest as potential cosmetic compounds in perfumery, were synthesized (Mariani et al., 1995) under conventional procedures by reaction of the 2-cineolylols (**8**) (*endo/exo* mixture) in anhydrous pyridine with a number of acyl chlorides. As an improvement and simplification of the conventional procedure the synthesis of compound **9**, as an example, was reprocessed under MW activation in dry media (Mariani et al., 2000) using KF/Al₂O₃ (Scheme 13.6).

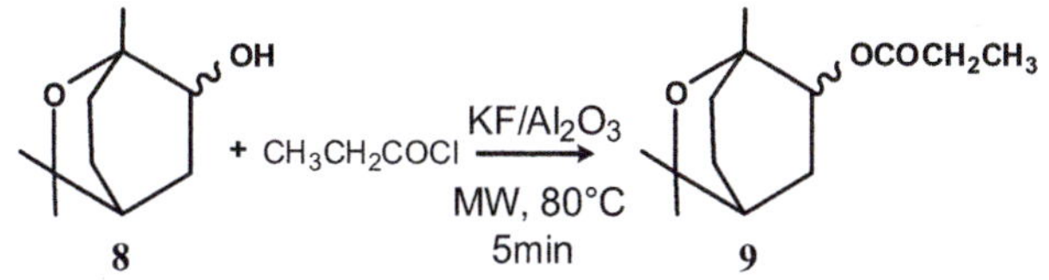

Scheme 13.6 Solvent-free synthesis of 1,8-cineole derivative, compound 9.

Table 13.3 Syntheses of compound **9**

Method		Time (min)	T (°C)	Yield (%)
KF/Al$_2$O$_3$	MW	5	80	98
KF/Al$_2$O$_3$	Δ	5	80	21
KF/Al$_2$O$_3$	Δ	300	80	75
Pyridine	Δ	60	100	71

MW, microwave; Δ, classical heating.

This method led to higher yields compared to the conventional procedure and to conductive heating (Table 13.3).

This improvement is connected to the intervention of highly polar reactants, proving consequently the development of strong MW interactions (Hamelin et al., 2002).

Fatty Esters: Lipophilic Ingredients

Long-chain aliphatic esters are important cosmetic and pharmaceutical ingredients applied as emollients, solubilizers, conditioning agents, spreading agents, etc. They have often been used to replace naturally occurring oils and waxes for ecological and above all economical reasons, based on the rather low cost of the raw materials.

A number of long-chain aliphatic esters, well-known lipophilic ingredients, derived from three monocarboxylic acids (pivalic, myristic and palmitic acids, compounds **10a–g**) and from a dicarboxylic acid (sebacic acid, compounds **10i–l**) have been synthesized under eco-friendly conditions (Villa et al., 2003) using two kinds of reactions under both MW activation and conventional heating (Δ):

- solvent-free PTC alkylation of carboxylate anions with alkyl bromides using Aliquat 336 as the transfer agent and K$_2$CO$_3$ as the base, at 140°C, 5 min MW, 15 min Δ;
- esterification of carboxylic acids with alcohols by simple mixture of reagents in the presence of neat PTSA without any support, at 160°C, 10 min MW, 15 min Δ.

The results are reported in Table 13.4.

For compounds **10c**, **10e** and **10i** the reactions were also carried out using two conventional procedures: H$_2$SO$_4$, the carboxylic acid and the appropriate alcohol in excess, under reflux, and anhydrous pyridine via acyl chloride and the corresponding alcohol at 100°C (Table 13.5).

Table 13.4 Solvent-free synthesis of compounds **10a–l**

	Compound		Yield (%) PTSA	Yield (%) PTC
10a	2-Ethylhexyl pivalate	Δ	90	91
		MW	91	93
10b	Isopropyl myristate	Δ	72	91
		MW	87	90
10c	Isopropyl palmitate	Δ	76	88
		MW	88	93
10d	Isobutyl palmitate	Δ	81	88
		MW	94	94
10e	Butyl palmitate	Δ	94	96
		MW	93	97
10f	2-Ethylhexyl palmitate	Δ	95	91
		MW	94	94
10g	Octyl palmitate	Δ	92	90
		MW	92	92
10h	Diisopropyl sebacate	Δ	65	61
		MW	73	63
10i	Dibutyl sebacate	Δ	87	85
		MW	91	91
10l	Bis(2-ethylhexyl) sebacate	Δ	91	86
		MW	94	94

MW, microwave; *PTC*, phase-transfer catalysis; *PTSA*, p-toluenesulphonic acid; Δ, classical heating.

Table 13.5 Conventional procedures for compounds **10c**, **10e** and **10i**

Compound	H_2SO_4 Yield (%) (time)	Pyridine Yield (%) (time)
10c	61 (2 h)	60 (6 h)
10e	73 (2 h)	87 (3 h)
10i	69 (2 h)	76 (18 h)

The solvent-free procedures under both MW activation and classical heating can be considered efficient, fast and eco-friendly methods to synthesize long-chain aliphatic esters.

Compared to the classical methods, the chemical transformations occur with higher efficiency, purity and easier workup, with further evident economic and ecological advantages also resulting from the stoichiometric use of reagents.

Aromatic Esters: UV Filters and Antimicrobial Agents

Solvent-free reactions were applied to the synthesis of some esters, compounds **11a–g**, derived from aromatic acids (Villa et al., 2003), with a noticeable improvement and simplification over classical procedures.

Table 13.6 Selected aromatic esters, compounds **11a–g**

	Name	Function
11a	3-Methylbutyl 4-methoxycinnamate	UV filter
11b	2-Ethylhexyl 4-methoxycinnamate	UV filter
11c	2-Ethylhexyl 4-(dimethylamino)benzoate	UV filter
11d	2-Ethylhexyl salicylate	UV filter
11e	4-Isopropylbenzyl salicylate	Cutaneous antilipoperoxidant
11f	Propyl 4-hydroxybenzoate	Antimicrobial agent
11g	Butyl 4-hydroxybenzoate	Antimicrobial agent

The selected compounds are widespread bioactive ingredients for cosmetic products: 3-methylbutyl 4-methoxycinnamate, 2-ethylhexyl 4-methoxycinnamate, 2-ethylhexyl 4-(dimethylamino)benzoate and 2-ethylhexyl salicylate, which are well-known UVB sunscreen filters; 4-isopropylbenzyl salicylate, a UV absorber and cutaneous antilipoperoxidant, and propyl 4-hydroxybenzoate and butyl 4-hydroxybenzoate (parabens), which are antimicrobial agents (Table 13.6).

Despite the importance of these compounds as raw materials for the cosmetics industry and the growth in demand, particularly for the UV filters, only a few examples of innovative synthetic methodologies are reported in the literature.

In this review the best performing solvent-free conditions for synthesis of the selected compounds are reported.

The synthesis of esters **11a–e** was achieved by carboxylate alkylation performed in situ with alkyl bromides using PTC with Aliquat 336 as the transfer agent and K_2CO_3 or K_2CO_3 + KOH as bases at a temperature set point of 160°C.

Compounds **11f–g** were obtained by acid-catalysed esterification of carboxylic acids and alcohols with neat PTSA, without any support, at the final temperature of 160°C.

The results obtained under MW activation and under conventional heating at different reaction times were compared (Table 13.7).

Generally the MW methods lead to higher yields within shorter reaction times as a consequence of the thermal effect consisting in a faster rise in temperature.

Esters of 3,3,5-Trimethylcyclohexanols

In the cosmetics field, 3,3,5-trimethylcyclohexanols (*cis, trans* epimers) (**12**) are alcohols of interest as cosmetic fragrances (European Commission, Commission Decision, 1996) and as starting compounds in the synthesis of active esters such as homosalate, an ester of salicylic acid and a well-known UVB sunscreen (Stockelbach, 1945).

In this review we report MW-mediated solvent-free procedures for the synthesis of some derived esters whose cosmetic interest was assessed (Gambaro et al., 2006a,b).

Seven esters, **13a–g**, widespread bioactive compounds and new potential actives, were selected: the esters from propanoic and butanoic acids (**13a** and **b**), already described as odour neutralizers; the esters from octanoic, 10-undecenoic and cyclopropanecarboxylic

Table 13.7 Syntheses of compounds **11a–g**: comparison of microwave activation and conventional heating data

	Method		Yield (%) 5 min	10 min	15 min
11a	Aliquat 336/K$_2$CO$_3$	MW	82	89	87
		Δ	58	85	87
11b	Aliquat 336/K$_2$CO$_3$	MW	88	92	90
		Δ	54	72	92
11c	Aliquat 336/K$_2$CO$_3$ + KOH	MW	85	90	90
		Δ	64	81	88
11d	Aliquat 336/K$_2$CO$_3$	MW	89	95	95
		Δ	84	85	88
11e	Aliquat 336/K$_2$CO$_3$	MW	70	89	84
		Δ	73	61	50
11f	PTSA	MW	72	71	70
		Δ	31	64	63
11g	PTSA	MW	77	76	76
		Δ	60	76	77

Aliquat 336, octyltrimethylammonium chloride; *MW*, microwave; *PTSA*, p-toluenesulphonic acid; Δ, classical heating.

acids (**13c–e**), potential new cosmetic ingredients; the ester from mandelic acid (**13f**), a well-known active, and the ester from salicylic acid (**13g**; homosalate), a UVB filter.

The esters were synthesized, under MW-mediated solventless conditions, by acidic catalysis (neat PTSA, neat graphite bisulphate, PTSA/acidic alumina, PTSA/mesoporous silica) and by basic catalysis (KOH/Al$_2$O$_3$, KF/Al$_2$O$_3$), making a comparative evaluation of the methods (Scheme 13.7).

In Table 13.8 the most significant results related to 3,3,5-trimethylcyclohexyl propanoate (**13a**) are reported.

These results can be considered representative for all compounds except for the very difficult case of homosalate (**13g**), obtained in good yield only using a basic catalyst on a solid mineral support (entry 6, yield = 72%; entry 8, yield = 65%).

The alcohols (**12**), not easily commercially available, were synthesized in the laboratory with a new green procedure, without energy consumption, by reduction of 3,3,5-trimethylcyclohexanone with sodium borohydride (NaBH$_4$) dispersed on neutral alumina in the solid state.

Reduction conditions were the following:

NaBH4 (10%)/neutral Al$_2$O$_3$, activity grade III (6% H$_2$O), reaction time 1 min at room temperature, yield 99% (Villa et al., 2008).

To evaluate the cosmetic interest of the studied compounds, the sweet-scented substances were submitted to an odour evaluation test; the most promising fragrances (**13a** and **13e**) and the ester from 10-undecenoic acid (**13d**), as an example of new lipophilic derivatives, were tested to assess their in vitro skin toxicity (Gambaro et al., 2006a,b).

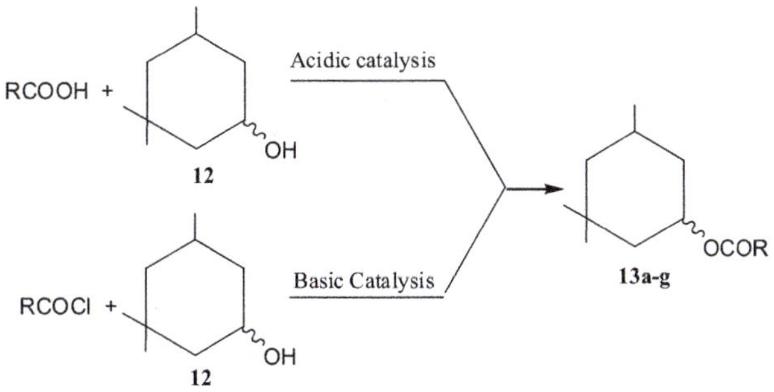

Scheme 13.7 Schematic solvent-free catalytic synthesis of esters **13a–g**.

	R
13a	CH$_3$CH$_2$
13b	CH$_3$(CH$_2$)$_2$
13c	CH$_3$(CH$_2$)$_6$
13d	CH$_2$=CH(CH$_2$)$_8$
13e	▷–
13f	C$_6$H$_5$CH(OH)
13g	2-(OH)C$_6$H$_4$-

Table 13.8 Microwave-mediated syntheses of compound **13a**

Entry	Method	T (°C)	Time (min)	Yield (%)
1	Neat PTSA	MW	130	30	98
2	Neat PTSA	Δ	130	30	87
3	PTSA/Al$_2$O$_3$	MW	130	30	40
4	PTSA/mesoporous silica	MW	130	30	45
5	Graphite bisulphate	MW	130	40	85
6	KF/Al$_2$O$_3$	MW	100	20	94
7	KF/Al$_2$O$_3$	Δ	100	20	77
8	KOH/Al$_2$O$_3$	MW	100	20	93

MW, microwave; *PTSA*, *p*-toluenesulphonic acid; Δ, classical heating.

Green Microwave Extractions

Nowadays the highest priorities for the chemical industry are process and product safety and respect for the environment. New technologies and methods for a sustainable cosmetics chemistry are the objectives of intense study and activity and in this context our research is addressed mainly to the study and development of new sustainable procedures and products involving the use of substrates from renewable sources, alternative energy (such as MW irradiation) and the application of mild conditions following the green chemistry and, more recently, the 'green extraction' targets (Villa et al., 2009; Chemat and Cravotto, 2013).

In an extractive context, studies on the exploitation and recycling of solid organic waste are gaining increasing interest. The agro–alimentary sector produces a large amount of organic residues that often are both highly polluting and quite expensive to treat.

Their conversion from worn-out plant matrices with high environmental impact to recycled sources with significant added value represents a great challenge to an eco-sustainable industrial context, owing to the general need for 'responsible care' of the environment together with high quality and safety of products and processes (Givel, 2007; Wijngaard et al., 2012).

The importance of the extraction step, in analytical procedures, and especially in the food, pharmaceutical and cosmetics industries, whose products have direct interactions with human health and consist in a diverse range of extracted plant components, has gained renewed attention in the past few decades. Existing conventional extraction methods are known for their economic impacts, due to high energy consumption (extraction step often uses more than 70% of the total process energy), large amounts of toxic solvents and time utilization in completion of the extraction step, and their environmental impacts due to CO_2 emissions and production of untreatable wastes.

The recently evolved concept of green extraction demands the development and utilization of techniques with a highly efficient approach to reducing energy consumption and generation and no or at least less utilization of solvents, with reduced generation of hazardous wastes.

The extraction of components by extensively used operations like solvent extraction or leaching is usually enhanced with the assistance of various processes like mechanical fragmentation, pressing, or heating, which, along with the generation of large amounts of heat and waste, also results in the degradation of sensitive components.

Several innovative extraction techniques have been introduced into green analytical chemistry (GAC) (Armenta et al., 2008; Smith, 2003; Ridgway et al., 2007), which typically use less solvent and energy, such as ultrasound-assisted extraction (Picò, 2013; Li et al., 2013a; Chemat and Zill-e-Huma Khan, 2011), supercritical fluid extraction (Reverchon, 1997), instant controlled pressure-drop process (Rezzoug et al., 2005; Allaf et al., 2013), accelerated solvent extraction (Brachet et al., 2001) and subcritical water extraction (Ozel et al., 2003).

Among these, MW-assisted extraction (MAE), supercritical fluid extraction, pressurized solvent extraction, pulsed electric field extraction and ultrasound-assisted extraction are considered the most promising in terms of better efficiency and reduced extraction times and solvent consumption (Chemat et al., 2012).

In the framework of these innovative extraction methodologies, MAE has received particular research effort attention. MWs are a non-contact heat source, which can not only make heating more effective and selective, but also help to accelerate energy transfer, start-up and response to heating control and to reduce thermal gradient, equipment size and operation units (Chemat and Lucchesi, 2006).

MAE is a key, sustainable technology in achieving the objectives of GAC. It has been rapidly developed as one of the hot-spot techniques for isolating interesting high-added-value compounds from solid samples. With the help of MWs, extraction or distillation can now be completed in minutes instead of hours, with various advantages (e.g., high reproducibility, less solvent and energy consumption, more compact procedures and greater purity of the final product). Several classes of cosmetic compounds (e.g., essential oils, antioxidants, pigments, aromas and other organic compounds) have been separated efficiently from raw materials, particularly natural plant resources.

Solvent-Free Microwave Extraction

Solvent-free MW extraction (SFME) was developed in 2004 by Chemat et al. (Lucchesi et al., 2004a,b). Based on a relatively simple principle, this process consists of the MW-assisted dry distillation of a fresh plant matrix without adding water or any organic solvent. SFME is neither a modified MAE, which uses organic solvents, nor a modified hydro-distillation (HD) method, which use a large quantity of water. The selective heating of the in situ water content of plant material causes tissues to swell and makes the glands and oleiferous receptacles burst. This process thus frees essential oil, which is evaporated by azeotropic distillation with the water present in the plant material (Li et al., 2013b). The water excess can be refluxed to the extraction vessel to restore the original water to the plant material. This process has been applied to several kinds of fresh and dry plants, such as spices (ajowan, cumin and star anise), aromatic herbs (basil, mint and thyme) and citrus fruits. The solvent-free method yields an essential oil with higher amounts of more valuable oxygenated compounds and allows substantial savings of costs, in terms of time, energy and plant material. Table 13.9 (Filly et al., 2014) summarizes the most important essential oils that have been extracted by SFME.

Solvent-Free Microwave Extraction of *Salvia somalensis* Vatke

The highest estimated worldwide production of valued essential oils is related to aromatic species belonging to the Lamiaceae family (Lawrence, 1992). The *Salvia* genus is its largest and includes about 900 species, spread throughout the world, some of which are economically important as they are used as spices and flavouring agents in foods as well as in cosmetics, perfumery and pharmaceuticals having biological properties (Hedge, 1992).

Taking into account the increasing interest in new and safe essential oils and considering that many species belonging to the *Salvia* genus have not been fully investigated yet, research has focused on *Salvia somalensis* Vatke, with the aim of assessing the potential cosmetic application of its essential oil. This sage is a wild plant native of Somalia, easy to grow and characterized by a pleasant perfume. It is well known as an ornamental plant and is well acclimatized to the temperate regions of the world such as the Mediterranean region, Italy in particular.

To evaluate the efficiency and reliability of the SFME procedure, the recovery of the essential oil was also processed by conventional HD and steam distillation (SD). The chemical composition of *S. somalensis* essential oil obtained by SFME was studied by gas chromatography/mass spectrometry (GC/MS) analysis and the results were compared with the data related to the oil obtained by HD. To our knowledge, only one report refers to the GC analysis of *S. somalensis* essential oil, obtained from dried plant material by SD (Chialva et al., 1992).

The potential cosmetic use of the essential oil obtained by SFME was assessed by an odour evaluation and by in vitro cytotoxicity assays. Its olfactive characteristics were evaluated in terms of note, intensity, evolution and persistence. Moreover, as regards essential

Table 13.9 Most important products extracted by solvent-free microwave extraction

Common name	Scientific name	SFME operating conditions	References
Orange	*Citrus sinensis* L.	T = 30 min, m = 200 g, P(atm) = 200 W	Ferhat et al. (2006)
Marjoram	*Origanum majorana* L.	T = 35 min, m = 150 g soaked in water for 1 h, P(atm) = 500 W	Bayramoglu et al. (2008)
Laurel	*Laurus nobilis* L.	T = 85 min, m = 150 g soaked in water for 1 h, P(atm) = 622 W	Bayramoglu et al. (2009)
Orange	*C. sinensis* L.	T = 10 min, m = 200 g, P(atm) = 200 W	Ferhat et al. (2008)
Lemon	*Citrus limon* L.	T = 30 min, m = 200 g, P(atm) = 200 W	Ferhat et al. (2007)
Basil	*Ocimum basilicum* L.	T = 30 min, m = 250 g, P(atm) = 500 W	Lucchesi et al. (2004a,b)
Mint	*Mentha crispa* L.	T = 30 min, m = 250 g, P(atm) = 500 W	
Thyme	*Thymus vulgaris* L.	T = 30 min, m = 250 g, P(atm) = 500 W	
Caraway	*Carum carvi* L.	T = 60 min, m = 250 g soaked in water for 1 h, P(atm) = 500 W	Lucchesi et al. (2004a,b)
Cumin	*Cuminum cyminum* L.	T = 60 min, m = 250 g soaked in water for 1 h, P(atm) = 500 W	
Anise or star anise	*Illicium verum*	T = 60 min, m = 250 g soaked in water for 1 h, P(atm) = 500 W	
Cardamom	*Elletaria cardamomum* L.	T = 75 min, m = 100 g soaked in water, P(atm) = 390 W	Lucchesi et al. (2007)
Rosemary	*Rosmarinus officinalis* L.	T = 40 min, m = 250 g, P(atm) = 500 W	Okoh et al. (2010)
Rosemary	*R. officinalis* L.	T = 30 min, m = 200 g, P(atm) = 200 W	Tigrine-Kordjani et al. (2006)
Laurel	*L. nobilis* L.	T = 50 min, m = 140 g soaked in water for 1 h, P(atm) = 85 W	Uysal et al. (2010)
Lemon balm	*Melissa officinalis* L.	T = 50 min, m = 280 g soaked in water, P(atm) = 85 W	

m, mass; *P(atm)*, atmospheric pressure; *SFME*, solvent-free microwave extraction; *T*, time.

oil safety, neutral red uptake (NRU) test and MTT [3-(4,5-dimethylthiazole-2-yl)-2,5-diphenyltetrazolium bromide] assays, performed on the human keratinocyte cell line NCTC 2544, were selected as these experimental in vitro models resulted in cost-effective and rapid methods in the prescreening phase of the development of new bioactive compounds (Genta et al., 2002a,b; Gambaro et al., 2006a,b; Burlando et al., 2008).

15 g of fresh botanical parts were placed in the MW reactor equipped with a modified Clevenger apparatus according to Chemat et al. (2004), without any solvent or water, and irradiated for an extraction time of 12 min (2 min up to 100°C and held at this temperature for 10 min). *S. somalensis* essential oils (SFME and HD oil) were analysed by GC/MS using a Varian CP-3800 gas chromatograph equipped with a DB-5 capillary column (30 m × 0.25 mm; coating thickness 0.25 μm) and a Varian Saturn 2000 ion trap mass detector. Analytical conditions were injector and transfer line temperatures of 220°C and 240°C, respectively; oven temperature programmed from 60°C to 240°C at 3°C/min; carrier gas helium at 1 mL/min; injection of 0.2 μL (10% hexane solution); split ratio 1:30 (Istituto Poligrafico Zecca dello Stato, 1991; Adams, 1995; Stenhagen et al., 1974; Connolly and Hill, 1991; Massada, 1976; Jennings and Shibamoto, 1980). The chemical compositions of the samples studied are reported in Table 13.10.

Both SFME and HD oil were characterized by higher amounts of monoterpenes than sesquiterpenes and they showed the same main volatile constituents (α-pinene, camphene, δ-3-carene, camphor and bornyl acetate) even if some differences in the relative percentages were observed.

Table 13.10 Gas chromatography/mass spectrometry analysis of *Salvia* solvent-free microwave-extracted essential oil

Compound	LRI	SFME oil	Hydro-distilled oil
Tricyclene	931	0.40	0.52
α-Pinene	942	5.94	8.96
Camphene	959	5.79	8.51
Sabinene	979	0.12	0.09
β-Pinene	984	2.12	2.85
Myrcene	992	1.02	1.73
α-Phellandrene	1010	0.32	0.23
δ-3-Carene	1013	5.19	6.72
α-Terpinene	1022	0.32	0.56
p-Cimene	1025	0.92	1.04
Sylvestrene	1031	0.87	0.78
Limonene	1035	3.77	4.33
β-Phellandrene	1037	0.77	1.13
(E)-Ocimene	1051	0.07	0.09
c-Terpinene	1064	0.82	1.20
cis-Sabinene hydrate	1075	0.26	0.32

Continued

Table 13.10 Gas chromatography/mass spectrometry analysis of *Salvia* solvent-free microwave-extracted essential oil—cont'd

Compound	LRI	SFME oil	Hydro-distilled oil
Isoterpinolene	1085	0.29	0.12
Terpinolene	1089	0.98	1.92
Linalool	1101	0.18	0.26
trans-Sabinene hydrate	1104	0.20	0.09
trans-p-Mentha-2,8-dien-1-ol★	1126	0.05	Tr
Camphor	1154	8.55	13.0
Borneol	1177	1.54	2.84
4-Terpineol	1184	0.41	0.76
α-Terpineol	1197	1.00	1.38
Bornyl acetate	1286	15.46	15.6
α-Cubebene	1349	0.63	0.36
Cyclosativene	1371	1.47	0.41
α-Copaene	1377	2.05	1.43
β-Bourbunene	1384	0.14	0.10
β-Cubebene	1388	0.26	0.14
Sativene	1392	0.09	0.43
cis-Jasmone	1394	0.52	0.07
α-Gurjunene	1407	1.05	0.75
Longifolene	1410	0.08	0.54
β-Caryophyllene	1420	5.69	4.44
β-Gurjunene	1431	0.45	0.08
Aromadendrene	1440	1.25	0.75
cis-Muurola-3,5-diene★	1447	0.34	Tr
α-Humulene	1457	0.60	0.37
Alloaromadendrene	1462	0.73	0.11
cis-Muurola-4,(14),5-diene★	1464	0.27	Tr
c-Muurolene	1476	0.86	0.42
trans-Cadina-1(6),4-diene★	1477	0.47	Tr
β-Selinene	1489	0.20	0.58
Curzerene	1493	1.06	1.10
α-Selinene	1496	0.99	1.02
α-Muurolene	1498	0.82	0.43
(E,E)-α-Farnesene	1503	0.94	0.32
cis-c-Cadinene	1511	2.69	1.71
trans-c-Cadinene	1513	1.62	3.68
D-Cadinene	1519	5.01	0.26
β-Sesquiphellandrene★	1525	0.10	0.14
trans-Cadina-1,4-diene	1535	0.47	0.21
α-Cadinene★	1539	0.58	Tr
Germacrene B	1563	1.01	1.01
trans-Nerolidol	1564	0.35	0.39
Caryophyllene alcohol	1584	0.13	0.16
Caryophyllene oxide	1587	0.24	0.16
s-Cadinol	1645	1.01	0.72
s-Muurolol	1646	2.71	1.48
β-Bisabolol oxide B	1659	0.96	0.54
β-Eudesmol	1687	2.13	2.10

SFME, solvent-free microwave extraction, *LRI*, linear retention indices relative to C_8–C_{23} n-alkanes on diphenylbenzene (DB 5%) MS capillary column; tr (traces), amount <0.01%.
★Tentative identification.

From a toxicological point of view, it is important to highlight that the essential oils of *S. somalensis* did not contain α- or β-thujone, two typical constituents of many sage essential oils such as that of *Salvia officinalis* (Perry et al., 1999; Bernotienė et al., 2007) that the European Commission requested to be monitored (EEC, Council Directive, 1988) in foodstuffs and flavours for their potential acute- and long-term neurotoxic effects.

Moreover, as regards the Seventh Amendment of the European Cosmetic Directive (Directive, 2003/15/EC), which identified 26 potential fragrance allergens (Villa et al., 2007), *S. somalensis* contains only limonene and linalool, a low level of allergenic fragrances compared with other *Salvia* species (Salameh and Dorđević, 2000). Considering the eco-sustainability of the SFME procedure, this SFME oil was selected to carry out studies on cosmetic functionality and safety. The essential oil obtained by SFME was submitted to an odour evaluation test. The sample diluted in dipropylene glycol 10% (w/w) was evaluated on specific blotting papers (*Tige*), after solvent evaporation, immediately and 10 min, 6 h and 24 h after impregnation, which identified a fresh highly aromatic top note with a very strong starting intensity not losing its characteristic note even after the initial evolution. The peculiar olfactive characteristics seem to indicate the potential cosmetic use of this oil not only in alcoholic male perfumery but also in hair and body detergents.

The results obtained from the evaluation of the cytotoxicity of the essential oil, as performed by means of the in vitro assays, are shown in Fig. 13.4.

The dose–response curves show very similar patterns for both tests. By studying the half-maximal inhibitory concentration (IC_{50}) values (the concentration that kills 50% of the cell population), it is evident that the volatile chemical mixture of the essential oil shows an intrinsic cytotoxicity 1000 times lower in comparison with the positive control, sodium dodecyl sulphate (MTT test IC_{50} mg/mL, 6294 vs. 0.004; NRU test IC_{50} mg/mL, 6315 vs. 0.006).

These promising results in terms of scent and safety seem to indicate this essential oil as an interesting potential green functional ingredient in the cosmetic field. Moreover the comparisons of the results obtained by SFME, HD and SD in terms of yields, extraction times, energy and solvent consumption show that the most efficient method is SFME: yields are clearly higher (1.21 against 0.58 and 0.28 for HD and SD, respectively; Fig. 13.5), within a shorter extraction time (15 min against 60 and 90 min; Fig. 13.6), with a consequent neat reduction in energy consumption (Fig. 13.7). Moreover, the Clevenger apparatus allows the elimination of solvents to collect the essential oil and, in the case of MW activation, no water is added to the extraction process (Fig. 13.8; Gambaro et al., 2006a,b).

Microwave Hydro-Diffusion and Gravity

According to the definition of green extraction (Chemat et al., 2012), MW hydro-diffusion and gravity (MHG) can be considered a green procedure possessing several

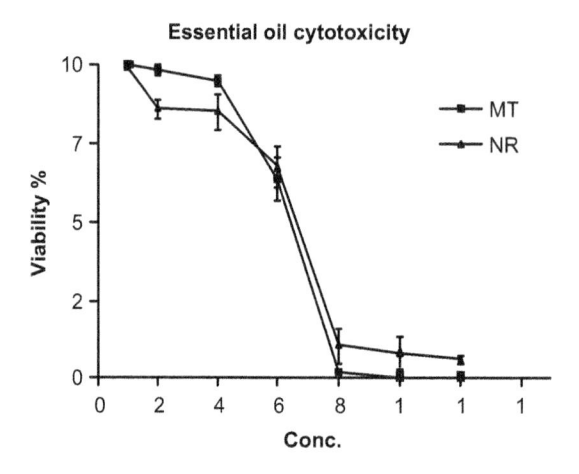

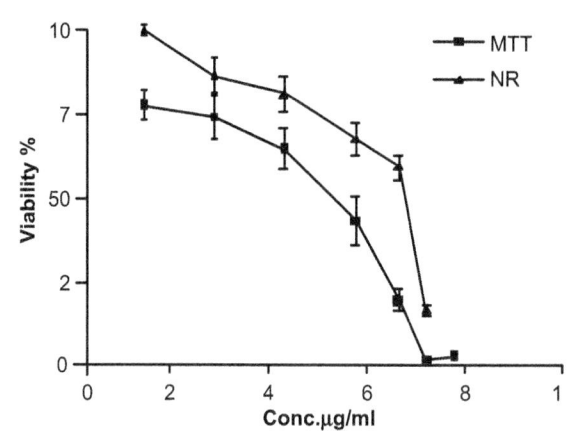

Figure 13.4 Cytotoxicity curve of *Salvia somalensis* solvent-free microwave-extracted oil against positive control (sodium dodecyl sulphate). *MTT*, 3-(4,5-dimethylthiazol-2-yl)-2,5-diphenyltetrazolium bromide; *NR*, neutral red.

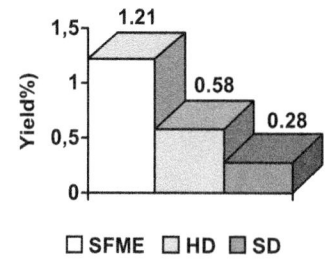

Figure 13.5 Yields obtained by solvent-free microwave extraction (*SFME*), hydro-distillation (*HD*) and steam distillation (*SD*).

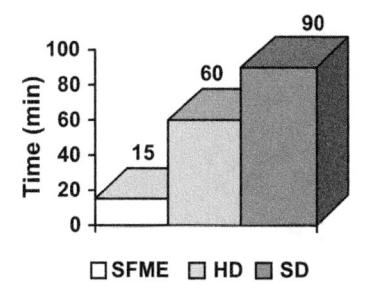

Figure 13.6 Extraction time for solvent-free microwave extraction (*SFME*), hydro-distillation (*HD*) and steam distillation (*SD*).

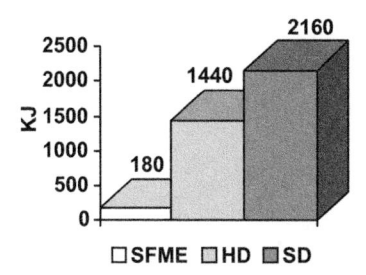

Figure 13.7 Energy consumption in solvent-free microwave extraction (*SFME*), hydro-distillation (*HD*) and steam distillation (*SD*).

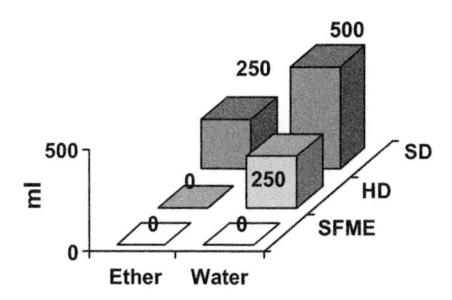

Figure 13.8 Amount of solvent and water used in solvent-free microwave extraction (*SFME*), hydro-distillation (*HD*) and steam distillation (*SD*).

advantages over more conventional extraction techniques, such as reduction of extraction times and energy consumption, solvent elimination, extraction efficiency and safety as well as high extract quality.

MHG is an original 'upside-down' MW alembic combining MW heating and earth gravity at atmospheric pressure. MHG was conceived for laboratory- and industrial-scale applications. It exploits warming of the water contained in the matrix, causing the expansion and consequent rupture of the matrix cells. This phenomenon, known as hydro-diffusion, allows the extract to diffuse outside the matrix. Moreover MWs increase tissue softness, cell permeability and cell disruption, thus enhancing the mass transfer

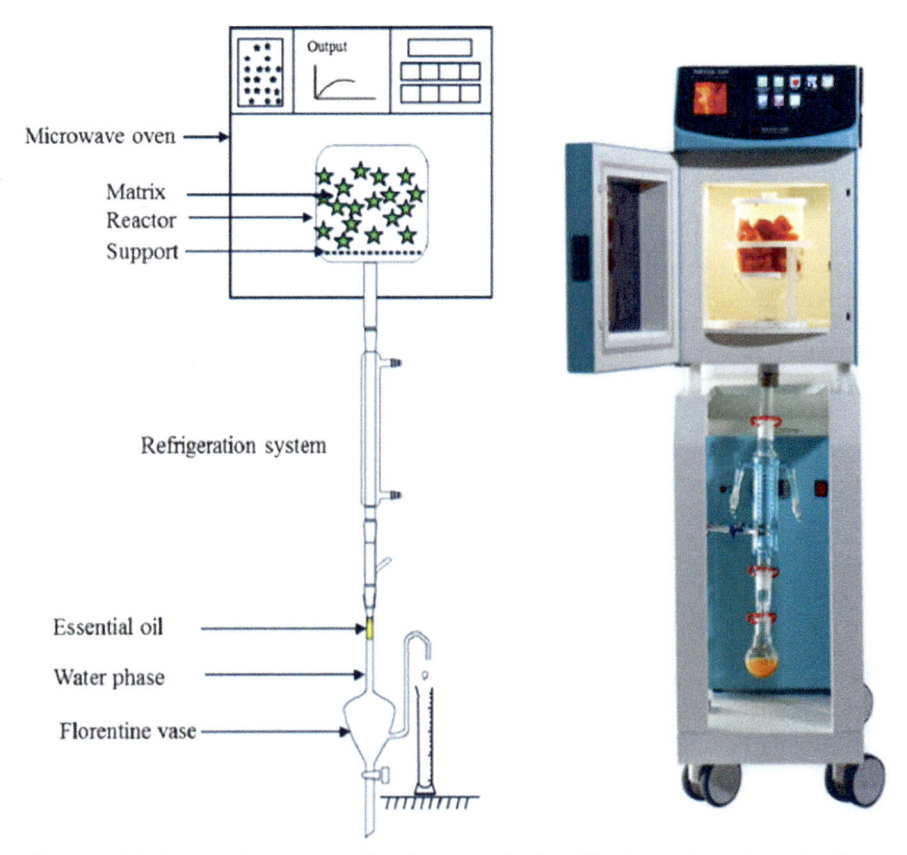

Figure 13.9 Schematic apparatus for microwave hydro-diffusion and gravity extraction.

within and outside the plant tissue. The extract is recovered by gravity, dropping out of the MW reactor via a hole positioned at the bottom of the applicator (Fig. 13.9; Chemat and Cravotto, 2013).

Microwave Hydro-Diffusion and Gravity Extraction of Green Rich Waters

Following this methodology, a multidisciplinary sustainable approach studied new green cosmetic extracts from exhaust organic matrices to obtain new bioactive compounds with potential nutraceutical and cosmeceutical properties (Chan et al., 2011; Ruberto et al., 2007). MHG without the addition of any water or solvent (Villa et al., 2012; Mandal et al., 2007; Leonelli and Villa, 2008) was exploited for the extraction of some worn-out plant matrices such as grape marc (from the winemaking process); apple, raspberry and pomegranate residues (from the fruit juice industry) and rosemary leaves (residues from foodstuff production), allowing the recovery of the inner water phases of the plant with different water-soluble principles (rich waters - RW) (Villa et al., 2014a, 2014b; Turrini et al., 2016).

Table 13.11 Main results related to microwave hydro-diffusion and gravity extraction of waste agro-food matrices

Waste matrix	Extraction time (min)	Rich waters recovery (%)	DPPH reduction (RSA%)	TPC (GAE mEq)
Rosemary	5	60	4.5	–
Apple	5	45	3.0	–
Raspberry	5	50	56.3	5.73
Blueberry	5	62	30.7	5.57
Grape	10	70	32.4	5.20
Pomegranate	10	50	83.8	30.4

DPPH, diphenylpicrylhydrazyl; *RSA*, radical scavenging activity; *TPC*, total phenolic content.

The peculiar green extracts quickly obtained (from 5 to 10 min) were characterized and tested for their potential cosmetic radical scavenging activity (RSA) by means of the free radical diphenylpicrylhydrazyl test (Brand-Williams et al., 1995), and the most promising ones (raspberry, blueberry, pomegranate and grape) were analysed with regard to the total phenolic content by means of the Folin–Ciocalteu method (Singleton et al., 1999).

To assess their safety and biological potential, several in vitro models that resulted in cost-effective and rapid methods for the prescreening phases of development of new bioactive compounds were applied. Normal human NCTC 2544 keratinocytes were exposed to selected RWs and we evaluated the cytotoxicity potential by means of NRU and MTT assays, proliferation by DNA content assay and antioxidant activity by commercial kits.

MW extraction offers several advantages such as short extraction times, no solvent consumption, high efficiency, high yield and reproducibility. The speediness of the process prevents the decomposition of principal constituents from thermal and hydrolytic reactions. Despite previous matrix exploitation, the green extracts obtained show a significant RSA and antioxidant activity (Table 13.11). Their polyphenol content together with their null cytotoxicity assessed by means of in vitro studies confirms their interest as new eco-sustainable cosmetic ingredients.

REFERENCES

Adams, R.P., 1995. Identification of Essential Oil Components by Gas Chromatography-Mass Spectroscopy. Allured Publ. Corp, Stream, Illinois.
Allaf, T., Tomao, V., Ruiz, K., Chemat, F., 2013. Ultrason. Sonochem. 20, 239.
Anastas, P.T., Warner, J.C., 1998. Green Chemistry: Theory and Practice. Oxford University Press, New York.
Armenta, S., Garrigues, S., de la Guardia, M., 2008. Trends Anal. Chem. 27, 497.
Ballini, R., Bosica, G., Carloni, S., Ciaralli, L., Maggi, R., Sartori, G., 1998. Tetrahedron Lett. 39, 6049.
Bernotienė, G., Nivinskienė, O., Butkienė, R., Mockutė, D., 2007. Chemija 18, 38.
Bayramoglu, B., Sahin, S., Sumnu, G., J, 2008. Food Eng 88, 535.
Bayramoglu, B., Sahin, S., Sumnu, G., 2009. Extraction of essential oil from Laurel leaves by using microwaves. Sep. Sci. Technol 44, 722–733. http://dx.doi.org/10.1080/01496390802437271.

Bouillon, C., Vayssie, C., Richard, F., 1982. Anti-solar Cosmetic Composition US Patent 4323549.
Brachet, A., Rudaz, S., Mateus, L., Christen, P., Veuthey, J.L., 2001. J. Sep. Sci. 24, 865.
Brand-Williams, W., Cuvelier, M.E., Berset, C., 1995. Lebensm. Wiss. Technol. 28, 25.
Buchbaue, G., Lux, C., 1991. Parfum. Kosmet. 72, 12.
Burlando, B., Parodi, A., Volante, A., Bassi, A.M., 2008. Toxicol. Lett. 15 (177), 144.
Chan, C.H., Yusoff, R., Ngoh, G.C., Kung, F.W.L., 2011. J. Chromatogr. A 1218, 6213.
Chemat, F., Cravotto, G. (Eds.), 2013. Microwave-Assisted Extraction of Bioactive Compounds: Theory and Practice. Springer, Food Engineering Series, New York.
Chemat, F., Lucchesi, M., 2006. Microwaves in Organic Synthesis. In: Loupy, A. (Ed.). Wiley-VCH, Weinheim, Germany, p. 959.
Chemat, F., Vian, M.A., Cravotto, G., 2012. Int. J. Mol. Sci 13, 8615.
Chemat, F., Zill-e-Huma Khan, M.K., 2011. Ultrason. Sonochem. 18, 813.
Chemat, F., Poux, M., Galema, S.A., 1997. J. Chem. Soc. Perkin Trans. 2, 2371.
Chemat, F., Lucchesi, M.E., Smadja, J., 2004. Solvent-Free Microwave Extraction of Volatile Natural Substances US patent 2004/0187340.
Chemat, F., Vian, M.A., Cravotto, G., 2012. Int. J. Mol. Sci. 13, 8615.
Chialva, F., Monguzzi, F., Manitto, P., 1992. J. Essent. Oil Res. 4, 447.
COM, February 2001a. 88 White Paper on the Strategy for a Future Chemicals Policy.
COM, July 2001b. 366 Green Paper Promoting a European Framework for Corporate Social Responsibility.
Connolly, J.D., Hill, R.A., 1991. Dictionary of Terpenoids. Chapman & Hall, London.
Official Journal of the European Communities, L132/1, Section II of the Annex. EC 1996, Commission Decision 96/335/EC of 8 May 1996 Establishing an Inventory and a Common Nomenclature of Ingredients Employed in Cosmetic Products, 1996. .
EC 2003, Directive 2003/15/EC of the European Parliament and of the Council of 27 February 2003 Amending Council Directive 76/768/EEC on the Approximation of the Laws of the Member States Relating to Cosmetic Products, L66/26, 2003. .
Official Journal of the European Union, L396/850. EC 2006, Directive 2006/121/EC of the European Parliament and of the Council of 18 December 2006 Amending Council Directive 67/548/EEC on the Approximation of Laws, Regulations and Administrative Provisions Relating to the Classification, Packaging and Labelling of Dangerous Substances in Order to Adapt it to Regulation (EC) No 1907/2006 Concerning the Registration, Evaluation, Authorisation and Restriction of Chemicals (REACH) and Establishing a European Chemicals Agency, 2006. .
EC 2006, Regulation (EC) No 1907/2006 of the European Parliament and of the Council of 18 December 2006 Concerning the Registration, Evaluation, Authorisation and Restriction of Chemicals (REACH), Establishing a European Chemicals Agency, Amending Directive 1999/45/EC and Repealing Council Regulation (EEC) No 793/93 and Commission Regulation (EC) No 1488/94 as Well as Council Directive 76/769/EEC and Commission Directives 91/155/EEC, 93/67/EEC, 93/105/EC and 2000/21/EC, 2006. .
Official Journal, L184, Annex II. EEC 1988, Council Directive 88/388/EEC on the Approximation of the Laws of the Member States Relating to Flavourings for Use in Foodstuffs and to Source Materials for Their Production, 1988.
Ferhat, M.A., Meklati, B.Y., Smadja, J., Chemat, F.J., 2006. Chromatogr. A 1112, 121.
Ferhat, M.A., Meklati, B.Y., Visinoni, F., Vian, M.A., Chemat, F., 2008. Chim. Oggi 8, 48.
Ferhat, A.M., Tigrine-Kordjani, N., Chemat, S., Chemat, F., 2007. Rapid extraction of volatile compounds using a new simultaneous microwave distillation: solvent extraction device. Chromatographia 65, 217–222. http://dx.doi.org/10.1365/s10337-006-0130-5.
Filly, A., Fernandez, X., Minuti, M., Visinoni, F., Cravotto, G., Chemat, F., 2014. Food Chem. 150, 193.
Gambaro, R., Villa, C., Baldassari, S., Mariani, E., Parodi, A., Bassi, A.M., 2006a. Int. J. Cosmet. Sci. 28 (6), 439.
Gambaro, R., Villa, C., Bisio, A., Mariani, E., Raggio, R., 24–26 Maggio 2006b. Solvent-free microwave extraction of essential oil from *Salvia somalensis* Vatke. In: Proceedings of Third National Meeting on Microwaves in the Engineering and in the Applied Sciences. Third National Meeting on Microwaves in the Engineering and in the Applied Sciences, p. 39Palermo.
Genta, M.T., Villa, C., Mariani, E., Longobardi, M., Loupy, A., 2002a. Int. J. Cosmet. Sci. 24, 257.

Genta, M.T., Villa, C., Mariani, E., Loupy, A., Petit, A., Rizzetto, R., Mascarotti, A., Morini, F., Ferro, M., 2002b. Int. J. Pharm. 231, 11.
Givel, M., 2007. Health Policy 81, 85.
Hamelin, J., Bazureau, J.P., Texier-Boullet, F., 2002. Microwaves in heterocyclic chemistry in: microwaves in organic synthesis. In: Loupy, A. (Ed.). Wiley-VCH, Weinheim.
Hayes, B.L. (Ed.), 2002. Microwave Synthesis: Chemistry at the Speed of Light. CEM Publishing, Matthews, NC.
Hedge, C., 1992. A global survey of the biogeography of labiatae in advances in labiatae science. In: Harley, R.M., Reynolds, T. (Eds.). Royal Botanic Gardens, Kew, UK.
Roma. Istituto Poligrafico Zecca Dello Stato, Farmacopea Ufficiale Della Repubblica Italiana, IX, 1991. .
Jennings, W., Shibamoto, T., 1980. Qualitative Analysis of Flavor and Fragrance Volatiles by Glass Capillary Gas Chromatography. Academic Press, New York.
Kingston, H.M., Haswell, S.J. (Eds.), 1997. Microwave-Enhanced Chemistry. Fundamentals, Sample Preparation and Applications. American Chemical Society, Washington, DC.
Larhed, M., Olofsson, K. (Eds.), 2001. Microwave Methods in Organic Synthesis. Springer, Berlin.
Lawrence, B.M., 1992. Chemical components of Labiatae oils and their exploitation. In: Harley, R.M., Reynolds, T. (Eds.), Advances in Labiatae Science. Royal Botanic Gardens, Kew, UK.
Leonelli, C., Villa, C., 2008. Applicazioni delle microonde in chimica analitica. In: Leoni, C. (Ed.), Il Riscaldamento a Microonde. Principi ed Applicazioni. Pitagora, Bologna, p. 205.
Li, Y., Fabiano-Tixier, A.S., Abert-Vian, M., Chemat, F., 2013a. Trends Anal. Chem. 47, 1.
Li, Y., Fabiano-Tixier, A.S., Tomao, V., Cravotto, G., Chemat, F., 2013b. Ultrason. Sonochem. 20, 12.
Lidström, P., Tierney, J.P. (Eds.), 2005. Microwave-Assisted Organic Synthesis. Blackwell Publishing, Oxford.
Loupy, A., Petit, A., Ramdani, M., Yvanaeff, C., Majdoub, M., Labiad, B., Villemin, D., 1993. Can. J. Chem. 71, 90.
Loupy, A., Petit, A., Hamelin, J., Texier-Boullet, F., Jacquault, P., Mathe, D., 1998. Synthesis 9, 1213.
Loupy, A. (Ed.), 2002. Microwaves in Organic Synthesis. Wiley-VCH, Weinheim.
Lucchesi, M.E., Chemat, F., Smadja, J., 2004a. Flavour Fragr. J. 19, 134.
Lucchesi, M.E., Chemat, F., Smadja, J., 2004b. J. Chromatogr. A 1043, 323.
Lucchesi, M.E., Smadja, J., Bradshaw, S., Louw, W., Chemat, F., 2007. J. Food Eng 79, 1058.
Mandal, V., Mohan, Y., Hemalatha, S., 2007. Pharmacogn. Rev. 1 (1), 7.
Mariani, E., Schenone, P., Guarrera, M., Dorato, S., 1993. Farmaco 48, 1687.
Mariani, E., Schenone, P., Bargagna, A., Longobardi, M., Dorato, S., 1994. Int. J. Cosmet. Sci. 16, 171.
Mariani, E., Neuhoff, C., Bargagna, A., Longobardi, M., Ferro, M., Gelardi, A., 1995. Int. J. Cosmet. Sci. 17, 187.
Mariani, E., Bargagna, A., Longobardi, M., Neuhoff, C., Rizzetto, R., Dorato, S., 1996. Boll. Chim. Farm. 135, 335.
Mariani, E., Neuhoff, C., Bargagna, A., Bonina, F., Giacchi, M., De Guidi, G., Velardita, A., 1998. Int. J. Pharm. 161, 65.
Mariani, E., Genta, M.T., Bargagna, A., Neuhoff, C., Loupy, A., Petit, A., 2000. Application of the microwave technology to synthesis and materials processing. In: Acierno, D., Leonelli, C., Pellacani, G.C., Mucchi (Eds.), ModenaISBN: 88-7000-346-9.
Massada, Y., 1976. Analysis of Essential Oils by Gas Chromatography and Mass Spectrometr. J. Wiley & Sons, New York.
Okuyama, K., Chen, X., Takata, K., Odawara, D., Suzuki, T., Nakata, S., Okuhara, T., 2000. Appl. Catal. A Gen. 190, 253.
Okoh, O.O., Sadimenko, A.P., Afolayan, A., 2010. Food Chem 120, 308.
Ozel, M.Z., Gogus, F., Lewis, A.C., 2003. Food Chem. 82, 381.
Perreux, L., Loupy, A., 2001. Tetrahedron 57, 9199.
Perry, N.B., Anderson, R.E., Brennan, N.J., Douglas, M.H., Heaney, A.J., Mc Gimpsey, J.A., Smallfield, B.M., 1999. J. Agric. Food Chem. 47, 2048.
Picò, Y., 2013. Trends Anal. Chem. 43, 84.
Reverchon, E., 1997. J. Supercrit. Fluid 10, 1.

Rezzoug, S.A., Boutekedjiret, C., Allaf, T., 2005. J. Food Eng. 71, 9.
Ridgway, K., Lalljie, S.P.D., Smith, R.M., 2007. J. Chromatogr. A 1153, 36.
Ruberto, G., Renda, A., Daquino, C., Amico, V., Spatafora, C., Tringali, C., De Tommasi, N., 2007. Food Chem. 100, 203.
Salameh, A., Dorđević, S., 2000. Facta Universitatis – Series: Working and Living Environmental Protection, vol. 1, p. 103.
Singleton, L., Orthofer, R., Lamuela-Raventós, M., 1999. J. Methods Enzym. 299, 152.
Smith, R.M., 2003. J. Chromatogr. A 1000, 3.
Stenhagen, E., Abrahamsson, S., McLafferty, F.W., 1974. Registry of Mass Spectral Data. Wiley & Sons, New York.
Stockelbach, F.E., 1945. Ultraviolet light filter compound. US Patent 2369084. Fries Bros., Inc., New York, NY.
Tigrine-Kordjani, N., Chemat, F., Meklati, B.Y., Tuduri, L., Giraudel, J.L., Montury, M., 2007. Anal. Bioanal. Chem 389, 631.
Turrini, F., Lacapra, C., Donno, D., Villa, C., Zunin, P., Signorello, M.G., Beccaro, G.L., Boggia, R., 31 May–1 June 2016. Green Extractions from Pomegranate C. Juice By-Products and Their Potential Use as Natural Food/Cosmetic Preservatives And/or Bioactive Ingredients, Proceedings of GENP 2016- Green Extraction of Natural Products II Edition. Turin. , pp. 80–81.
Uysal, B., Sozmen, F., Buyuktas, B.S., 2010. Nat. Prod. Commun 5, 111.
Villa, C., Genta, M.T., Bargagna, A., Mariani, E., Loupy, A., 2001. Green Chem. 3, 196.
Villa, C., Mariani, E., Loupy, A., Grippo, C., Grossi, G.C., Bargagna, A., 2003. Green Chem. 5, 623.
Villa, C., Gambaro, R., Mariani, E., Dorato, S., 2007. J. Pharm. Biomed. Anal. 44, 755.
Villa, C., Trucchi, B., Gambaro, R., Baldassari, S., 2008. Int. J. Cosmet. Sci. 30, 139.
Villa, C., Trucchi, B., Bertoli, A., Pistelli, L., Parodi, A., Bassi, A.M., Ruffoni, B., 2009. Int. J. Cosmet. Sci. 31, 55.
Villa, C., Boggia, R., Leardi, R., Leonelli, C., Rosa, R., Caponetti, E., Chillura Martino, D., September 24–26, 2012. Ecofriendly microwave-mediated approach for the extraction of bioactive compounds from waste matrices in the agro-alimentary sector. EXTECH 2012. In: 14th International Symposium on Advances in Extraction Technologies Messina, Italy.
Villa, C., Lacapra, C., Rum, S., Bassi, A., Danailova, J., Vernazza, S., Boggia, R., 2014a. Multidisciplinar sustainable approach for the study of green extracts from waste matrices: solvent-free microwave extraction and in vitro assays as potential cosmeceutical and nutraceutical ingredients. In: 28th Congress IFSCC Paris 2014 Full Proceedings. COURBEVOIE: Société Francaise de Cosmétologie P-191 P.2029-2038.
Villa, C., Lacapra, C., Rum, S., Boggia, R., Leonelli, C., Rosa, R., 25–26 settembre 2014b. Green microwave Extracts from Waste Matrices for Cosmeceutical and Nutraceutical Applications – Food Processing Innovation and Green Extraction Technologies: Recent Advances and Applications in Human Health. Università della Magna Grecia, Catanzaro.
Wijngaard, H., Hossain, M.B., Rai, D.K., Brunton, N., 2012. Food Res. Int. 46, 505.

FURTHER READING

Kappe, C.O., Stadler, A. (Eds.), 2005. Microwaves in Organic and Medicinal Chemistry. Wiley-VCH, Weinheim.
Villa, C., Baldassari, S., Gambaro, R., Mariani, E., Loupy, A., 2005. Int. J. Cosmet. Sci. 27, 11.

CHAPTER 14

Main Chemical Contaminants in Cosmetics: Regulatory Aspects and Analytical Methods

Isela Lavilla, Noelia Cabaleiro, Carlos Bendicho
University of Vigo, Vigo, Spain

INTRODUCTION

This chapter aims at providing an overview of the main contaminants that can be present in cosmetic products at trace levels. These may come from a large variety of sources that are considered in this work (e.g., impurities in raw materials, manufacturing process, storage conditions, potential migration from the final packaging to the product or chemical changes due to instability of the cosmetic). An overview of the regulations of various countries is included. The literature concerning analytical methods for determining contaminants such as some metals and metalloids, nitrosamines, 1,4-dioxane, hydroquinone, formaldehyde and acrylamide is reviewed. Perspectives on the need for a strict analytical control of these substances in the future to reduce risks to health and updated methods of analysis are also highlighted.

Impurities and Contaminants in Cosmetics

Cosmetics can become contaminated by a wide range of unintentional trace substances that do not appear in the labelling despite the fact that some of them may pose a certain health risk. Many of these compounds are forbidden or restricted in different legislations, but not all of them. Consequently, this issue is technically unavoidable.

Therefore, the potential toxicity of selected trace components with regard to their concentration must be considered. The International Cooperation on Cosmetics Regulation (ICCR), in the report 'Principle for the handling of traces of impurities and/or contaminants in cosmetic products', provides some general guidelines in this regard (ICCR, 2011). In these guidelines, it is made clear that manufacturers, distributors and/or importers that place the final product on the market are responsible for identifying those trace substances that can be a health and safety issue. Moreover, given that a search for all of them in cosmetic products is not possible, the guidelines provide some aspects that should be taken into account to establish these controls: the source of raw material, the manufacturing processes (especially the synthesis and extraction processes) and the interaction with packaging (primary packaging materials and formulations must be considered).

Strictly speaking, trace substances in cosmetics can be considered as impurities and/or contaminants according to their source of origin. In general, impurities are already present in raw materials and can also originate during the manufacturing process and/or normal storage conditions (usually by migration from the final packaging and/or chemical changes caused by contact of the cosmetics with the final packaging). In contrast, contaminants are derived from external sources and bad manufacturing practices (i.e., inadequate storage of raw materials, instability of primary packaging, etc.) (ICCR, 2011).

This chapter is focused on impurities or contaminants that could be unintentionally present in cosmetic products and are considered unsafe for human health according to scientific studies and current legislations.

Undoubtedly, the quality of raw materials is key to the final quality of cosmetic products. The number of raw materials used in the manufacture of cosmetics is very high, which makes it difficult to ensure the quality of all of them. In this regard, it should be mentioned that more than 16,000 compounds are included in the International Nomenclature of Cosmetic Ingredients list (INCI, 2015). Chemical contamination in cosmetics from the presence of impurities in raw materials could greatly be avoidable by adhering to high quality standards. However, there are no international official standards for ensuring ingredient purity and this question has become a decision of manufacturers (Seidel, 2013). In many cases, raw materials comply with published quality standards such as those of national pharmacopoeias and others proposed by private organizations for standardization.

According to the Environmental Working Group's study published in 2007, which is based on governmental (US Food and Drug Administration (FDA) and United Nations Economic Commission for Europe) and industrial (Cosmetic Ingredient Review (CIR)) sources, the majority of cosmetics have the potential to be contaminated with impurities from raw materials. It is estimated that 1 in 10 ingredients may contain impurities. Various compounds or groups of compounds were identified in this study as potential impurities associated with health concerns such as cancer, neurotoxicity and even reproductive problems (EGW's, 2007).

Table 14.1 shows some of these compounds, potential contamination sources and possible adverse effects (Department of the Environment Industry Profiles (DOE), 1995; EGW's, 2007).

The transportation and storage of raw materials are also crucial to prevent the contamination of cosmetics. The risk of contamination is higher in storage areas, pipework, pumps and handling areas. The ingredients (generally liquids or solids) are supplied in tankers or in drums (from 10 kg to 200 L) that are unloaded into warehouses or drum stores or are pumped into bulk storage tanks for liquids. These are transferred from the

Table 14.1 Some impurities and contaminants in cosmetics (DOE, 1995; EGW's, 2007)

Impurity/contaminant	Main ingredients that may contain it	Potential contamination causes	Legislation	Possible adverse effects
Metals and metalloids	Aluminium starch octenylsuccinate, D&C Red 6	Transportation, delivery and storage; manufacturing processes; cross-contamination by ingredients such as aluminium salts, iron oxides, titanium dioxide, zinc oxide, manganese dioxide, stannous fluoride, etc.	Prohibited[a,b,c,e] Restricted[a,b,c,d,e]	Some metals and metalloids such as Pb, As, Hg or Cd are among the most dangerous toxics; occupational hazards; persistent and bioaccumulative substances
Nitrosamines	Cocamidopropyl betaine, cocamide diethanolamine	Manufacturing and storage processes; formed by reactions between ingredients (a nitrosating agent and an amine, generally under acidic conditions)	Prohibited[a,c]	Most nitrosamines can be considered carcinogenic compounds
1,4-Dioxane	PEG-stearates, ceteareth-20, PEG-7 glyceryl cocoate, PEG-8	1,4-Dioxane can be a byproduct in ethoxylation reactions during the manufacturing of certain cosmetic ingredients (e.g., detergents, foaming agents, emulsifiers and certain solvents)	Prohibited[a,c]	Probable human carcinogen (carcinogen in animals) and irritant
Ethylene oxide	PEG-stearates, ceteareth-20, PEG-7 glyceryl cocoate, PEG-8	Used in ethoxylation, the finished products may contain traces of unreacted ethylene oxide; in addition, ethylene oxide can produce traces of 1,4-dioxane	Prohibited[a,c]	Probable carcinogen, mutagen and irritant
Hydroquinone	Tocopheryl acetate, tocopherol	It can be used in some cases as a skin lightener but may also be a contaminant in cosmetic ingredients	Prohibited[a,e] Restricted[c,d]	Nephrotoxicity and some evidence of carcinogenicity in rat have been reported
Formaldehyde	Diazolidinyl urea, DMDM hydantoin, imidazolidinyl urea, Quaternium-15	Released by some preservatives, FRPs, used to prevent bacteria growth in water-based products	Restricted[a,c,d] Prohibited[e]	Possible human carcinogen; human immune and respiratory toxicant and allergen
Acrylamide	Polyquaternium-7 and polyacrylamides	Polyacrylamides contain small amounts of unreacted acrylamide	Prohibited[a,c]	Possible human carcinogen; irritant; organ system toxicant (non-reproductive)

Continued

Table 14.1 Some impurities and contaminants in cosmetics (DOE, 1995; EGW's, 2007)—cont'd

Impurity/ contaminant	Main ingredients that may contain it	Potential contamination causes	Legislation	Possible adverse effects
Polycyclic aromatic hydrocarbons	Petrolatum, coal	From coal, crude oil, etc.; contamination from transportation, delivery and fuel storage areas, manufacturing processes and waste management processes can occur	Prohibited[a,c]	Some are known to be carcinogens
Polychlorinated biphenyls	D&C Red 6, hydrogenated oils (cottonseed, palm, etc.)	Though banned in many countries, they remain present in the environment; widely used in the past mainly in electrical equipment (electricity substations and transformer areas)	—	Persistent, bioaccumulative and toxic compounds
Pesticides	D&C Red 6, natural oils	Widely used in agricultural sector; some pesticides are persistent organic pollutants	Prohibited[a,c]	Human carcinogens; strong evidence of human neurotoxicity; toxicants or allergens
Dioxins	Triclosan	Byproduct of processes involving combustion	Prohibited[a,c]	Known human carcinogens; endocrine disruption; persistent and bioaccumulative
Chloroform	Triclosan	Contamination during transportation, delivery, storage and manufacturing processes; trace amounts from its use as a processing solvent during manufacture or as a byproduct from the synthesis of an ingredient	Prohibited[a,b,c,e]	Possible human carcinogen; irritant

D&C Red 6, barium(2+) (4Z)-4-[(4-methyl-2-sulphonatophenyl)hydrazinylidene]-3-oxonaphthalene-2-carboxylate; *DMDM*, 1,3-bis(hydroxymethyl)-5,5-dimethylimidazolidine-2,4-dione; *FRPs*, formaldehyde releasing preservatives; *PEG*, poly(ethylene glycol).
[a]EC 1223/2009.
[b]FDA.
[c]Canadian FDA.
[d]China (no access to prohibited).
[e]Japan.

warehouses to the production area in large enough quantities for 1 or 2 days. Overheating or other processes that might occur during storage of cosmetics or incoming supplies must be avoided (DOE, 1995; Seidel, 2013). In general, cleanliness and orderliness are essential to prevent mix-ups or cross-contamination between consumables, raw materials and intermediate and finished formulations. Buildings are also very important because facilities must be of suitable size, design and construction (FDA, 2013).

Despite the large number of compounds used in the cosmetics industry, manufacturing processes can be considered straightforward. For example, grinding, mixing, filtration, emulsion formation, distillation, extraction and, in some cases, different chemical reactions can be required (e.g., oxidation, condensation, hydrogenation, alkylation or nitration are usual in the production of synthetic or semi-synthetic fragrances). In general, contamination in production areas is less of an issue compared with the degree of contamination that can occur in transportation and storage areas (DOE, 1995). Notwithstanding, some contaminants can be formed during these processes when conditions become favourable. For example, nitrosamines can be formed when amines and nitrosating agents are included as ingredients in formulations or contaminants, and the ethoxylation processes may result in dioxane formation.

Given that some molecules like additives, monomers and other contaminants are able to migrate from the packaging to the cosmetics, different regulations such as the US Fair Packaging and Labeling Act (FP&L Act, 1967) or the European Cosmetics Regulation (EU Regulation, 2009) address specifically the classification, labelling and packaging material of ingredients and their mixtures. The industry must provide information on packaging (material components, purity and stability and possible interactions between the packaging and the content) to establish the product safety. Compatibility studies with defined types of formulations and control of finished products are required. Materials for cosmetic packaging are, in their majority, plastics such as high-density polyethylene and polyethylene and, to a lesser extent, glass and others as polyester, resin, steel or plastic mixtures. So, the main contaminants from packaging are different phthalates, anthracene, bisphenol A, musk xylene, 4,4′-diaminodiphenylmethane and formaldehyde. A 2014 study of content–container interactions for 79 cosmetic products showed that stick deodorants and perfumes were contaminated with traces of di-2-ethylhexyl phthalate from polypropylene packaging, although it does not pose a significant risk to the health (Thomas et al., 2014). Nevertheless, it would be appropriate to provide other chemical data concerning packaging such as potentially extractable compounds (Seidel, 2013).

To avoid all these problems in the cosmetics industry, manufacturers should comply with regulations and observe good manufacturing practices (GMP) guidelines. At present, these guidelines have been consolidated by the International Organization for Standardization (ISO) into a standard (ISO 22716:2007) (revised in 2011) that gives guidelines for the production, control, storage and shipment of cosmetic products. The FDA agrees with these GMP, but by incorporating, modifying or excluding some specific aspects (FDA, 2013). The follow-up of these practices can reduce the contamination risk of cosmetics. The FDA inspects cosmetics manufacturing to determine if the control and the practices are being followed properly. In particular, the FDA with the US Customs and Border Protection examines imported cosmetics. A number of import alerts regarding harmful contaminants can be found on the FDA website (FDA, 2015a). Some examples include lipsticks containing lead or bulk ingredients prone to contamination with organic solvents. On the other hand, the EU Regulation indicates that the safety of cosmetics must be ensured through GMP in accordance with those harmonized standards referred to in the *Official Journal of the European Union*. The monitoring and inspection of the cosmetics placed on the market (and therefore compliance with the EU Regulation) are performed by each Member State that appoints inspectors at the national level (Scientific and Technical Assessment, 2016). Any undesirable effect or presence in the cosmetic needs to be communicated, and for that purpose the Platform of European Market Surveillance Authorities for Cosmetics facilitates this exchange of information, including possible safety alerts.

OVERVIEW OF REGULATIONS CONCERNING CHEMICAL CONTAMINANTS IN COSMETICS

As has been said in previous chapters of this book, the importance of the cosmetics market is reflected by the €72.5 billion that were moved only in Europe in 2014 (Cosmetics Europe Personal Care Association (COLIPA), 2014). This sector is very dynamic and under continuous innovation. Part of this innovation lies in the continuous introduction of new ingredients and formulations, whose safety must be ensured. The presence of unwanted chemicals at trace levels is still possible even with strict controls performed through different physico-chemical tests. For most of these traces, there are no regulatory concentration limits available at the time of this writing.

In past years, efforts have been made to promote regulatory alignment through the creation of the ICCR (2015). The international group comprises regulatory bodies from the European Union (EU), the United States, Canada, Brazil and Japan, and serves as a link between regulations and cosmetics industry associations. The management of the presence of trace contaminants in cosmetics together with safety assessments is in the scope of the activities of the ICCR group (ICCR, 2011). Thus, the ICCR is also concerned with analytical methodologies for cosmetic quality control and safety assessments, including those for the detection of contaminants in the formulation. The standardization of analytical methods used is encouraged. Despite this, regulations concerning allowed ingredients and restricted/prohibited chemicals are still different depending on the country where the cosmetic is marketed. In the next paragraphs, an overview on how the main regulatory frameworks tackle the presence of chemical contaminants in cosmetics is given.

European Union

As of this writing, the EU Regulation is the most stringent in comparison with other regions. Despite this, although concentration restrictions exist for some ingredients, no regulatory concentration limits are available for many other substances and potential contaminants (EU Regulation, 2009; Commission Implementing Decision, 2013). This regulation allows 'the non-intended presence of small quantities of a prohibited substance, stemming from impurities of natural or synthetic ingredients, the manufacturing process, storage, migration from packaging, which is technically unavoidable in good manufacturing practice'. Traces occurring from the instability of ingredients, preservation problems or interactions of raw materials should be avoided through GMP or even reformulation of the final cosmetic.

Manufacturers have most of the responsibility for ensuring product safety in the EU. To this end, a safety assessment is required before each cosmetic is released to the market (pre-market approval). Information resulting from the assessment is included in a cosmetic safety report containing data regarding impurities, traces of prohibited substances and evidence of their technical unavoidability, among others. Production of such data requires the use of reliable analytical methods and in the EU the industry is free to select whatever analytical method it considers more adequate. To assist in this selection, a guidance document has been produced by the Scientific Committee on Consumer Safety (SCCS), the EU organism providing independent scientific support under the EC Directorate General for Health and Consumer Protection (European Commission, 2015). A compendium of 36 analytical methods for the identification/determination of different substances or groups of substances in cosmetics, compiling seven EU directives published since 1980 and 1996, is available (European Commission, 2000); however, these methods are not sufficient for the large amount of ingredients and possible contaminants and, in many cases, they are not adapted to current scientific–technical progress. In addition, as can seen in this document, the word 'official' does not appear, but only 'Methods of Analysis', as the EU scientific committees are revising these methods and studying new proposed methods to achieve the necessary 'harmonized' standards (interested readers can see more details on this subject in Chapter 3).

On the other side, COLIPA, now Cosmetics Europe, published in 2004 various analytical methods for cosmetics in an attempt to harmonize analytical methods for the identification/determination of substances listed in the annexes of the EU Cosmetics Directive. These, however, are not official methods. It is therefore obvious that these methods are insufficient, considering the continually growing numbers of ingredients (more than 1800 in the EU CosIng database), cosmetics types and potential impurities and contaminants (CosIng Database, 2015).

The United States of America

Cosmetics are regulated in United States by the Federal Food, Drug and Cosmetic Act (FD&C Act, Subchapter VI–Cosmetics, sections 361–364) under FDA authority (FD&C Act, 1960).

Under the FD&C Act a contaminant is defined as 'any ingredient of another ingredient or processing aid present at an insignificant level and having no technical or functional effect'. This regulation prohibits contamination or adulteration that may occur as a result of the presence of 'poisonous or deleterious substances, filthy, putrid, or decomposed substances, contamination from insanitary manufacturing or from the container' (FD&C Act, 1960).

Cosmetics are not subjected to pre-market approval and safety of the final product is assessed by manufacturers through available published data for ingredients and similar formulations. Any other additional testing required to ensure safety is the responsibility of the manufacturer, which involves the need for selecting the most adequate analytical methods (FDA, 2015c). However, no official methods for the analysis of cosmetics are available in the United States.

Methods described in the official compendia, e.g., Association of Official Analytical Chemists (AOAC), *United States Pharmacopeia*, *Pesticide Analytical Manual* and *Food Chemicals Codex*; specialized manuals (FDA manuals) or scientific journals (such as *Journal of the AOAC International*) can be used for the detection of contaminants or impurities (FDA, 2003). In particular, the FDA advocates utilizing the methods of the AOAC (FD&C Act, 2015a).

The FDA itself has supported cosmetic regulatory activities by conducting research on potential contaminants such as bovine spongiform encephalopathy, 1,4-dioxane or lead (FDA, 2015d).

Canada

In Canada, cosmetics are regulated by the Food and Drugs Act (Canadian FDA) (Food and Drugs Act, 2014) in the Cosmetic Regulations (Cosmetic Regulations Canada, 2007). Canadian regulations are somewhat similar to the EU legislation, and much more stringent than those of the United States. Information concerning contaminants should be included in the Cosmetic Notification form, which is required before marketing a cosmetic (Health Canada, 2014a). Under this regulation, 'incidental ingredients are defined as any processing aid added, removed, or converted to a declared ingredient, or any ingredient of another ingredient or processing aid present at an insignificant level and having no technical or functional effect'. Like in the EU and the United States, the responsibility of keeping impurities to a minimum level is held by the manufacturer, although it is not legally required to report impurities or byproducts.

Canadian legislation specifically bans the sale of cosmetics that contain any coal-tar dye (or coal-tar base or intermediate), mercury or its salts (unless used under restricted conditions), chloroform (as an ingredient) or oestrogenic substances (Cosmetic Regulations Canada, 2007). Although without specific mention of other types of contamination, the Canadian FDA has produced a guidance document on heavy metal impurities (Health Canada, 2012). No official methods of analysis have been reported in the legislation, neither for ingredients nor for impurities/contaminants.

Japan

Cosmetics are regulated in Japan through the Ministry of Health, Labour and Welfare (MHLW) by the Pharmaceutical Affairs Act. However, the regulation does not make reference to the definition of trace contaminants. Ensuring cosmetic safety lies on the manufacturer as well, it being their responsibility to ensure the use of adequate analytical methods, which should be conveniently reported to the pertinent body (PAL, 1960).

China

There are different regulations for the cosmetics marketed in China (i.e., Hygienic Standard for Cosmetics, 2007; Guideline for Risk Evaluation of Substances with Possibility of Safety Risk in Cosmetics, 2010; Cosmetics Technical Requirement Standard, 2011; etc.) mainly under the China Food and Drug Administration (CFDA) (CRS, 2013).

The approval (or hygiene licence) of each cosmetic product is subject to, among other issues, a safety assessment. A test report for the presence of prohibited substances and/or impurities is part of this assessment (Fellous and Zhongrui, 2013). Chinese authorities specifically mention chemicals such as phenol, dioxane, methanol, acrylamide and N-nitrosodiethanolamine (NDELA) and metals (Hg, Pb, As) as potential impurities, and cosmetics are consequently analysed for their possible presence in the final products (Cao, 2014). Official analytical methods for some of these compounds (e.g., mercury, lead) have been published in the Hygienic Standard for Cosmetics (2007) and should be used by CFDA-approved laboratories. Nevertheless, no official methods have been published for many other potential contaminants. In these cases, the manufacturer can use a suitable self-developed analytical methodology provided that it is validated (Cao, 2014).

ANALYTICAL METHODS FOR THE DETECTION OF MAIN CHEMICAL CONTAMINANTS IN COSMETICS AND RAW MATERIALS

As mentioned before, cosmetics can become contaminated by a wide range of unintentional trace substances. The major ones have been selected to be commented on in this chapter. Thus, literature concerning analytical methods for determining metals and metalloids, nitrosamines, 1,4-dioxane, hydroquinone, formaldehyde and acrylamide is considered here. Methods for other contaminants have been treated in other chapters of this book. Readers can search in the word index of the book for the compounds of interest to read the corresponding chapter in which they are considered.

Metals and Metalloids

Metals and metalloids can be found in cosmetics as ingredients, impurities from raw materials and/or contaminants considering their ubiquity in the environment and their participation in the composition of metallic devices used in the cosmetics industry. Table 14.2 shows the elements (and their compounds) used, according to the Inventory

Table 14.2 Some metal compounds that can be used as ingredients in cosmetics (Seidel, 2013)

Metal	Compound	Main uses
Al	Alumina	Absorbent, scrub agent, thickener/emulsifier
	Aluminium hydroxide	Absorbent, thickener/emulsifier
	Aluminium chlorohydrates and aluminium chloride	Antiperspirants, deodorants
	Magnesium aluminium silicate	Absorbent
	Aluminium silicate (kaolin)	Scrub agent, absorbent
	Aluminium stearate	Thickener/emulsifier
	Aluminium sulphate	Antibacterial/anti-acne, absorbent
	Aluminium starch octenylsuccinate	Thickener/emulsifier, absorbent
Ag	Colloidal silver	Antibacterial/anti-acne
	Silver nitrate	Pigment for eyelashes and eyebrows
Au	Colloidal gold and particulates	Anti-ageing
Ba	Barium sulphate	Colouring agent/pigment
Bi	Bismuth oxychloride	Thickener/emulsifier, absorbent, colouring agent/pigment
Ca	Calcium ascorbate	Antioxidant, vitamin
	Calcium carbonate (chalk)	Absorbent
	Calcium gluconate	Anti-irritant
	Calcium pantothenate	Vitamin
Cu	Copper gluconate	Antioxidant
	Copper peptides	Antioxidants
Cr	Chromium hydroxide green	Colouring agent/pigment
	Chromium oxide green	Colouring agent/pigment
Fe	Ferric ammonium ferrocyanide	Colouring agent/pigment
	Ferric ferrocyanide	Colouring agent/pigment
	Iron oxides	Colouring agent/pigment
Hg	Phenylmercuric salts (acetate, benzoate, borate)	Antimicrobial preservatives (eye area)
	Thimerosal	Antimicrobial preservatives (eye area)
Mg	Magnesium ascorbyl palmitate	Antioxidant, vitamin
	Magnesium ascorbyl phosphate	Antioxidant, vitamin
	Magnesium carbonate	Absorbent
	Magnesium gluconate	Antibacterial/anti-acne
	Magnesium hydroxide	Antibacterial/anti-acne, absorbent
	Magnesium laureth sulphate and oleth sulphate	Surfactant/detergent cleansing agent
	Magnesium stearate	Thickener/emulsifier
	Magnesium sulphate	Thickener/emulsifier
Mn	Manganese gluconate	Antioxidant
	Manganese violet	Colouring agent/pigment
Se	Selenium sulphide	Antidandruff agent
Sn	Tin(IV) oxide	Abrasive, bulking, and opacifying agent

Table 14.2 Some metal compounds that can be used as ingredients in cosmetics (Seidel, 2013)—cont'd

Metal	Compound	Main uses
Sr	Strontium chloride	Anti-irritant
	Strontium acetate	
	Strontium hydroxide	
Ti	Titanium dioxide	Thickener/emulsifier, sunscreen active ingredient, colouring agent/pigment
Zn	Zinc oxide	Antioxidant, thickener/emulsifier, anti-irritant, colouring agent/pigment
	Zinc stearate	Slip agent, thickener/emulsifier, colouring agent/pigment
	Zinc pyrithione	Preservative, antidandruff
	Zinc sulphate	Preservative
	Zinc chloride	Deodorant (mouth)
Zr	Aluminium zirconium chlorohydrates	Antiperspirants
	Zirconium silicate	Abrasive, opacifying agent

of Cosmetic Ingredients (European Commission, 2006), as ingredients in cosmetics as of this writing. In the past, even some metals considered highly toxic were used as cosmetic ingredients, e.g., lead acetate in hair dye. Today, some of them are still being used in some cases, e.g., phenylmercuric salts as antimicrobial preservatives.

Table 14.3 shows the elements (and their compounds) that are forbidden as cosmetic ingredients in the main legislations as of this writing. As mentioned, and as can be observed in Table 14.3, the EU has the strictest legislation on metals in cosmetics. Arsenic, cadmium, cobalt, chromium, mercury, nickel, lead and antimony are banned as ingredients, so they are limited as potential impurities at trace levels (EU Regulation, 2009). The Canadian FDA has banned As, Cd, Cr, Hg, Pb and Sb (Health Canada, 2014b). In contrast, the US FDA has banned only Hg and has established certain specifications for As, Hg and Pb in some colour additives (FD&C Act, 2015b). While Hg is banned in all legislations (United States and Canada establish limits for Hg as an impurity at 1 and 3 ppm, respectively), certain Hg compounds (Table 14.2) continue to be used as preservatives in certain eye-area cosmetics, with their concentration limited to 65–70 ppm.

Although the cosmetics industry is generally operating under GMP, metals can be found in all cosmetic products at trace levels. Numerous studies can be cited in this regard. For instance, in 2011 the Environmental Defence of Canada determined As, Be, Cd, Hg, Ni, Pb, Se and Tl in 49 face make-up products (foundations, concealers, powders, blushes or bronzers, mascaras, eyeliners, eye shadows, and lipsticks or glosses) manufactured in the United States, the EU, Canada and Korea. Ni was found in 100% of

Table 14.3 Elements and their compounds currently forbidden as cosmetic ingredients in the main legislations

Legislation	Prohibited ingredients	Restricted ingredients
EC 1223/2009	As, Cd, Pb, Sb and its compounds; Cr, chromic acid and its salts; Se (except Se$_2$S at 1% max.); Hg (except thimerosal, phenylmercuric salts, 0.007% in Hg max.); Co benzene sulphonate, dichloride and sulphate; Ni metal, mono-, di- and trioxide, disulphide, sulphide, dihydroxide, carbonate, sulphate and tetracarbonyl	Au, Ag, Cu, Al, Al hydroxide sulphate and Al stearate when are used as colourants; AlF and SnF$_2$ restricted to 0.15% max. as F; Al—Zr chloride hydroxide complexes in antiperspirants restricted to 20% (as anhydrous aluminium zirconium chloride hydroxide) and 5.4% (as Zr); Co–Al oxide when used as colourant; Fe oxides and ferric ammonium ferrocyanide when used as colourant; SrCl$_2$, Sr acetate in oral products and Sr(OH)$_2$ at max. 3.5%; SrCl$_2$ max. 2.1% in shampoos and face products; Sr peroxide max. 4.5%; Zn acetate, chloride, gluconate or glutamate, max. 1% as Zn; Zn phenol sulphonate. max 6%; Zn pyrithione max. 0.1% in leave-on hair products, 1% in hair products and 0.5% in others; ZnO$_2$ as UV filter; Zn stearate when used as colourant; TiO$_2$ as colourant or max. 25% as UV filter; Cr(III) oxide and Cr(III) hydroxide as colourants if they are free from chromate ion; ammonium Mn(III) diphosphate as colourant; max. of 4% AgNO$_3$ in colouring eyelashes and eyebrows; AgCl$_2$ max. 0.004%
US FDA	Zr complexes	Hg compounds max. 65 ppm; max. 1 ppm Hg and 3 ppm As and 0.6% free Pb in Pb acetate colour additive; max. 1 ppm Hg, 5 ppm As and 10 ppm Pb in Ag as colour additive in fingernail polish (max. 1% Ag); max. 1 ppm Hg, 3 ppm As and 20 ppm Pb in bismuth citrate colour additive; max. 100 ppm Cu in disodium–EDTA–Cu colour additive; max. 1 ppm Hg, 3 ppm As and 20 ppm Pb in guaiazulene colour additive; max. 3 ppm As and 20 ppm Pb in henna colour additive; max. 3 ppm Hg, 3 ppm As and 10 ppm Pb in iron oxides colour additive; max. 1 ppm Hg, 3 ppm As and 20 ppm Pb in ultramarines colour additive; max. 1 ppm Hg, 3 ppm As and 20 ppm Pb in manganese violet colour additive; max. 1 ppm Hg, 3 ppm As, 20 ppm Pb, 100.5 ppm Cu and 15 ppm Cd in luminescent zinc sulphide colour additive

Table 14.3 Elements and their compounds currently forbidden as cosmetic ingredients in the main legislations—cont'd

Legislation	Prohibited ingredients	Restricted ingredients
Canadian FDA	As, Cd, Pb (and its acetate), Sb, Hg and its compounds, chromic acid and its salts; Se (except Se$_2$S); Co benzene sulphonate; Au salts	Ag and its salts (≤0.04%), Al chloride, Al chlorohydrate and associated complexes, Al—Zr chloride hydroxide complexes; Zr and its compounds as colouring agents; Zn peroxide
China	No access	Max. 10 mg/kg Pb; max. 2 mg/kg As; max. 1 mg/kg Hg; max 5 mg/kg Cd; Au, Ag (and sulphate, chloride, oxide), Cu (sulphate, gluconate, aspartate, chlorophyll, acetylmethionate, PCA), Al and its salts (acetate, stearates, sulphate, chloride, chlorohydrate, hydroxide, etc.); Al—Zr chloride hydroxide complexes; Co-Al oxide; Fe gluconate and hydroxide, Fe—Ti oxide; iron oxides; ferric chloride and citrate, ferrocyanide; Zn acetate, chloride, gluconate, PCA, myristate, lactate, oxide, stearate, etc.; TiO$_2$, Ti citrate and hydroxide
Japan	Hg, Sr, Se and its compounds; Zr, Zn p-phenol sulphonate, Zn pyrithione; Zn hydroquinone monobenzyl ether	Al chlorhydroxyallantoinate; Ag—Cu zeolite

PCA, Pyrrolidone Carboxylic Acid.

cosmetics (up to 230 ppm), Pb in 96% (up to 110 ppm), Be in 90% (up to 8 ppm), Tl in 61% (up to 2 ppm), Cd in 51% (up to 3 ppm), As in 20% (up to 70 ppm) and Se in 14% (up to 40 ppm). Because in Canada, heavy metal impurity concentrations in cosmetic products are considered to be acceptable below 10 ppm for Pb, 5 ppm for Sb and 3 ppm for As, Cd and Hg, many of the analysed samples did not comply with this consideration. This study highlights that the highest levels of As, Cd, and Pb were found in lip glosses, and hence, metals could be ingested. A gloss was also the product containing seven of the eight determined metals (Environmental Defence, 2011). In general, impurities in colour additives may be the most important cause of this type of contamination. Although dyes and pigments used in cosmetics are strongly regulated, the permitted levels of some metals are still high in these ingredients, e.g., limits of 10–20 ppm of Pb (FD&C Act, 2015b; Health Canada, 2015). So, it is not surprising that Pb is found in most lipsticks although in general at levels that do not pose any risk for health. In a study carried out by the US FDA on 400 samples of lipsticks, lead concentrations were between <0.03 and 7.19 ppm, with a mean value of 1.11 ppm (FDA, 2015b). In contrast, Hg was not detected in any of the analysed samples. Although elevated mercury levels have been found in skin lightening soaps and creams, most of them from Africa, the Middle East, China and Latin America, this is considered a fraud. Numerous data on concentrations of different trace metals in cosmetic products can be found in a review by Bocca et al. (2014).

Consequently, the determination of some toxic metals in the manufacturing stages and in the final product becomes mandatory to ensure both quality and safety of cosmetics.

International Standardized Analytical Methods

The aforementioned methods of analysis compiled by the European Commission years ago covered a number of authorized cosmetic ingredients, either elements such as Ag, Se, Hg, Zr, Al, Zn, Ba, Sr and/or their compounds or anions such as ammonium, fluoride, sulphide, chlorate or iodate (European Commission, 2000); however, no methods for the determination of elements that could be unintentionally present were included in this publication.

At this writing, only an international official method has been published in this sense (ISO/TR 17276:2014) that includes most common and typical analytical approaches for screening and quantification of heavy metals of general interest in raw materials and finished products. This report covers from UV/VIS spectrophotometry to inductively coupled plasma (ICP)–mass spectrometry (MS), so that a suitable approach can be chosen.

Some methods are established to be used in individual countries, such as China, that cover As, Cd, Hg and Pb (CRS, 2015). Other methods are established for the determination of contaminant elements in raw materials, such those used in Spain for As (UNE 84055:1987), Cd (UNE 84055:1987), Pb (UNE 84061:2011, UNE 84059:2011, UNE 84058:2011, UNE 84057:2011, UNE 84056:2011), Sb (UNE 84741:2012) or Hg (UNE 84742:2010), published by the Spanish Association for Standardisation and Certification (AENOR).

Method validation in a control laboratory is crucial, and hence, the availability of certified reference materials (CRMs) and interlaboratory studies. Nowadays, there are very few cosmetic CRMs, e.g., the cosmetic cream CRM from the Health Sciences Authority of Singapore for As, Hg and Pb (Ng et al., 2015). Similarly, proficiency testing is carried out, e.g., in 2015, 217 laboratories in China participated in the determination of As and Pb in foundation cream cosmetics (Zhong et al., 2015).

Analytical Methods Published in the Scientific Literature

Table 14.4 provides an overview of published methods for contaminant trace elements determination in cosmetics, and also includes methods for the determination of elements used as ingredients, as most published papers include the simultaneous determination of both types of elements.

Table 14.4 Selected analytical methods for trace metal determination in cosmetics

Analyte	Matrix	Remarks on sample preparation	Technique	References
Conventional wet digestion				
Hg	Creams, washing milk, lotion	Sample digestion (1 g) with 3 mL HNO_3 and 0.5 mL $HClO_4$; heating in oven and in hot plate until evaporation of half the initial volume; after cooling, dilution with HCl	CV–AAS	De-Quiang et al. (1999)
As, Co, Cr, Ni, Pb	Eye shadows	Heat the sample (0.15 g) in a sand bath with 5 mL of HNO_3 and 5 mL of HCl nearly to dryness; add 2.5 mL of both acids after cooling; re-heat; dissolve the residue in an ultrasonic bath with 2 mL of HNO_3	ETAAS	Sainio et al. (2000)
Cd, Cr, Fe, Ni, Pb, Zn	Make-up	Dry samples (1 g) are digested with HNO_3; heat in a hot plate nearly to dryness; after cooling, digest with $HClO_4$	FAAS	Nnorom et al. (2005)
Pb	Lipstick	Sample (0.2 g) is digested with HNO_3 (4 h) and heated in oven overnight (85°C); after cooling, H_2O_2 is added and samples are heated for 1 h	ETAAS	Al-Saleh et al. (2009)
As, Cd, Co, Cr, Ni, Pb	Face powders	Pseudo-digestion with HNO_3 and H_2O_2 or HNO_3 and HCl; comparison with total digestion by HF	ICP–OES and ICP–MS	Capelli et al. (2014)
Cd, Co, Cr, Cu, Ni	Powdered eye shadows	Mineralization in a block digester with 5 mL of diluted HNO_3, 2 mL of H_2O_2 and 1 mL of Triton X-100 (5%, w/v) at 100°C for 180 min	ETAAS	Ferreira Batista et al. (2015)
Dry ashing				
Pb	Lipsticks	Sample (1 g) is ignited into ash at 500°C; addition of HCl in three steps for extraction	FAAS	Okamoto et al. (1971)

Continued

Table 14.4 Selected analytical methods for trace metal determination in cosmetics—cont'd

Analyte	Matrix	Remarks on sample preparation	Technique	References
Pb	Toothpaste	The sample (2–4 g) is heated in a hot plate with 1 mL of 1% (w/v) $NH_4H_2PO_4$, 10 mL of 10% (w/v) $Mg(NO_3)_2$ and ethanol; then, it is placed in a muffle at 200°C and kept at 450°C overnight; it is lixiviated with water and 5 M HNO_3	ETAAS	Khammas et al. (1989)

Microwave-assisted digestion

Analyte	Matrix	Remarks on sample preparation	Technique	References
Pb, As, Hg	Cosmetics	MW digestion with HNO_3 and HF (10:3) or with HNO_3 and H_2SO_4 (5:1)	ASV	Wang and Tien (1994)
27 elements	Lipstick Face powder Cream make-up	MW digestion (0.15 g sample) with 3 mL HNO_3; use of high-pressure closed-vessel digestion bomb; after cooling, dilution with water and filtration	ICP–OES	Besecker et al. (1998)
Hg	Eyeliner, shadow and pencil	Digestion of 0.25 g of sample with 1 mL HNO_3, 1 mL H_2O_2 and 3 mL H_2SO_4; heating for 45 min; after cooling, filtration and addition of 2.5 mL of APDC and water to 100 mL	CV–AFS	Gámiz-Gracia and Luque de Castro (1999a)
Se, Zn, Cd	Antidandruff shampoo	Digestion of 0.2 g sample with 1 mL HNO_3; dilution with water after cooling (Zn, Cd) or dilution with water and 2 g NaOH (Se)	ICP–OES and FAAS	Salvador et al. (2000a)
Cd, Co, Cr, Cu, Hg, Ir, Mn, Ni, Pb, Pd, Pt, Rh, V	Body creams	Digestion of sample (1 g) with 5 mL HNO_3 and 1 mL HF; MW irradiation about 20 min; dilution with water to 30 mL after cooling	ICP–MS	Bocca et al. (2007)
Pb	Lipsticks	MW digestion of 0.3 g of sample with 7 mL HNO_3 and 2 mL HF; after cooling and venting, addition of 30 mL of 4% (w/v) H_3BO_3; digestion at 180°C (10 min)	ICP–MS	Hepp et al. (2009)

Table 14.4 Selected analytical methods for trace metal determination in cosmetics—cont'd

Analyte	Matrix	Remarks on sample preparation	Technique	References
Cd, Cr, Co, Ni, Pb	Eye shadows	Addition of 5 mL of 67% (v/v) HNO_3 and 1 mL of 40% (v/v) HF to 1 g of sample; MW irradiation about 22 min	ICP–MS	Volpe et al. (2012)
Solid sampling				
Pb, Cu, Ag, Fe, Ca, Al, Si, Na, Sn, Zr, Sb	Kohl	Qualitative elemental analysis in kohl (stone) and quantitative determination of lead content by LIBS	LIBS	Haider et al. (2012)
Pb	Lipstick	Direct solid sampling high-resolution continuum source graphite furnace atomic absorption spectrometry; comparison with two digestion procedures	ETAAS	Lemaire et al. (2013)
Ge, As, Cd, Sb, Hg, Bi	Cosmetic lotions	A slurry sampling containing 2% (w/v) sample, 2% (w/v) thiourea, 0.05% (w/v) L-cysteine, 0.5 µg/mL Co(II), 0.1% (w/v) Triton X-100 and 1.2% (v/v) HCl; sample introduction by FI and VG	ICP–MS	Chen et al. (2015a)
Emulsification				
Cd	Cosmetic oils	Formation of a water emulsion	ETAAS	Vondruska (1995)
Fe and Zn	Sunscreens	0.04–0.2 g of sample, 2 mL of IBMK and 0.8 g of Nemol K-39 were shaken and diluted up to 50 mL with deionized water	FAAS	Salvador et al. (2000b)
As, Cd, Cr, Cu, Hg, Mg, Mn, Ni, Sr, Zn	Cosmetics	1 min of ultrasonic shaking and a dispersion medium containing 0.5% (w/v) SDS + 3% (v/v) HNO_3 or HCl for obtaining a stable emulsion	FAAS, ETAAS, ICP–OES, CV–AAS	Lavilla et al. (2009)
Ti, Al, Zn, Mg, Fe, Cu, Mn, Cr, Pb, B	Sunscreens	1% (w/v) sample is mixed with 0.5% (v/v) Triton X-100 and 0.5 M HNO_3; it is stirred, followed by sonication for 15 min	ICP–OES	Zachariadis and Sahanidou (2009)

Continued

Table 14.4 Selected analytical methods for trace metal determination in cosmetics—cont'd

Analyte	Matrix	Remarks on sample preparation	Technique	References
Extraction				
Hg^{2+}, $MeHg^+$	Cosmetics	Ultrasonic extraction (30 min in 4 mL aqueous medium); L-cysteine functionalized cellulose fibre used as adsorbent to pack a mini-column for online separation and pre-concentration in a sequential injection system	AFS	Chen et al. (2011a)
Hg^{2+}, $MeHg^+$, $EtHg^+$	Skin refreshers, hand lotion	DLLME; sample dilution with water and filtration; addition of APDC to 5 mL sample solution; addition of 0.5 mL methanol containing $[C_6MIM][PF_6]$; centrifugation and dissolution of sedimented phase in methanol	LC–ICP–MS	Jia et al. (2011)
As, Bi, Cd, Hg, Pb	Skin whitening cosmetics	SPE on multi-walled carbon nanotubes packed into a micro-column after MW acid digestion	ICP–OES	ALqadami et al. (2013)
Pb, Cd	Lipsticks, eyeliner and eye shadow	SPE using pyridine-2,6-diamine-functionalized Fe_3O_4 nanoparticles as sorbent for pre-concentration ions in digested cosmetic samples	FAAS	Ebrahimzadeh et al. (2013)
Pb	Lipsticks and hair dyes	MA–DLLME with floating organic drop; after digestion with 7 mL HNO_3 and 2 mL HF in a first step and 30 mL of 4% (w/v) boric acid in a second step, microextraction with 500 μL of acetone, containing 30 μL of 1-undecanol and diethyl dithiophosphoric acid	ETAAS	Sharafi et al. (2015)

AFS, atomic fluorescence spectrometry; *APDC*, ammonium pyrrolidine dithiocarbamate; *ASV*, anodic stripping voltammetry; *CV–AAS*, cold vapour atomic absorption spectrometry; *CV–AFS*, cold vapour atomic fluorescence spectrometry; *DLLME*, dispersive liquid–liquid microextraction; *ETAAS*, electrothermal atomic absorption spectrometry; *FAAS*, flame atomic absorption spectrometry; *FI*, flow injection; *IBMK*, isobutyl methyl ketone; *ICP–MS*, inductively coupled plasma–mass spectrometry; *ICP–OES*, inductively coupled plasma–optical emission spectrometry; *LC–ICP–MS*, high-performance liquid chromatography–inductively coupled plasma–mass spectrometry; *LIBS*, laser-induced breakdown spectroscopy; *MA–DLLME*, microwave-assisted dispersive liquid–liquid microextraction; *MW*, microwave; *SDS*, sodium dodecyl sulphate; *SPE*, solid-phase extraction; *VG*, vapour generation.

In general, acid digestion is commonly used prior to elemental determination by atomic techniques as FAAS (flame atomic absorption spectrometry), cold vapour–atomic absorption spectrometry, ETAAS (electrothermal atomic absorption spectrometry), ICP–optical emission spectrometry (OES) or ICP–MS. Nevertheless, because the matrix of cosmetics is not simple, different sample treatments are used for their degradation or elimination. In particular, drastic procedures for difficult-to-dissolve samples such as foundation creams and lipsticks are necessary.

Conventional wet digestion procedures with traditional heating systems such as oven, hot plate or sand bath are still used. Different oxidizing acids or mixtures such as HNO_3, HNO_3 with $HClO_4$, and H_2SO_4 with H_2O_2 are used. In these cases, digestion is carried out in open vessels, having important disadvantages such as the use of large amounts of acid and its continuous replacement, the risk of analyte volatilization and contamination, and the limited oxidizing power of acids used at atmospheric pressure. Thus, $HClO_4$ can be useful for the digestion of poorly oxidizable cosmetic samples such as creams, make-ups, etc. However, the risk from the use of this acid, i.e., explosion, should be considered in routine analysis (Nnorom et al., 2005). In addition, HF can be necessary when refractory minerals such as silica, alumina, titanium dioxide and mica are present in cosmetic formulations (Capelli et al., 2014). In general, these procedures are too long and tedious.

Despite their simplicity and suitability for samples containing large amounts of organic matter, dry ashing procedures are less used in cosmetic analysis given that typical temperatures (about 450–500°C) may result in loss of elements by volatilization. Dry ashing must be limited to those elements stable at high temperatures. Risk of ignition can also appear when cosmetics with a high content of fats and oils are analysed, thus provoking analyte losses. In addition, having in mind that silicated compounds (e.g., dimethicone) are frequently used in cosmetics, a silica residue is likely to remain after destruction of the matrix, metals being occluded in this insoluble residue (Okamoto et al., 1971; Khammas et al., 1989).

Microwave-assisted digestion (MAD) provides faster and more efficient cosmetic matrix degradation in comparison with conventional dissolution procedures, even for very fatty cosmetics. Digestion is significantly improved by the use of closed systems, which increases the reactivity of the acids used and at the same time avoids analyte losses. In addition, microwaves heat samples and acids without previous transfer of heat to the vessel walls, and consequently, high temperatures can be reached in the liquid phase, thus enhancing the acid reactivity and avoiding the use of large volumes of acids with high oxidizing power. HNO_3 or mixtures of this acid with H_2O_2, HCl or, in some cases, H_2SO_4 or HF are used for dissolving most cosmetics. For instance, very fatty matrices such as lipstick and foundations can be dissolved with only HNO_3 in high-pressure digestion vessels (Besecker et al., 1998). In some cases, the implementation of an extra step is necessary for complete solubilization of the sample (Hepp et al., 2009).

Despite these advantages, complete solubilization of the sample is not always possible after acid digestion. Thus, some alternatives to MAD without sample degradation have been proposed, such as solid sampling or emulsification. Nevertheless, these strategies have been scarcely used for cosmetics. Solid sample analysis without or with minimal sample pre-treatment includes direct solid sampling and slurry sampling. In this sense, laser ablation is a powerful tool for direct solid sampling owing to its great versatility. Haider et al. (2012) have proposed qualitative elemental analysis and Pb determination by laser-induced breakdown spectroscopy (LIBS) in kohl samples. Another interesting option for cosmetic analysis using techniques such as atomic absorption spectrometry, ICP–OES and ICP–MS is the direct vaporization and atomization of metals from a solid when this is introduced into an electrothermal atomizer/vaporizer. Direct solid sampling by ETAAS has been applied to determining Pb in lipsticks (Lemaire et al., 2013). In the same way, slurry sampling has been used together with flow injection (FI) and vapour generation for sample introduction in ICP–MS to determine Ge, As, Cd, Sb, Hg and Bi in cosmetics (Chen et al., 2015a).

Emulsification procedures can be used mainly for the elemental analysis of liposoluble and viscous samples (some cosmetics meet these requirements). It allows a direct and rapid sample preparation avoiding the use of concentrated acids. Direct emulsification is a possible strategy with techniques such as FAAS, ETAAS or even ICP–OES, because sample decomposition occurs in the cell. The important work of optimization is always necessary to eliminate interference. Emulsion formation is not a spontaneous process, agitation being mandatory to achieve a complete homogenization of the system. It can be done manually, mechanically or by means of ultrasonic irradiation (ultrasound-assisted emulsification; USAE). When mechanical or manual stirring is used, continuous agitation is required during the analysis (Vondruska, 1995; Salvador et al., 2000b). In contrast, a stable and homogeneous emulsion can be quickly obtained when ultrasound energy is used with this purpose (Lavilla et al., 2009; Zachariadis and Sahanidou, 2009). In addition, USAE can reduce the amount of surfactant needed for the formation of a homogeneous system.

Despite the evident usefulness of extraction techniques for trace metal determination, only in some articles have different extraction procedures been proposed for cosmetic analysis. Solid-phase extraction (SPE) with nanoparticles has been used for pre-concentration of different trace metals in digested cosmetic samples for subsequent determination by ICP–OES (ALqadami et al., 2013) or FAAS (Ebrahimzadeh et al., 2013). In the same way, dispersive liquid–liquid microextraction (DLLME) has been used for enhancing sensitivity to determine Pb in lipsticks by ETAAS (Sharafi et al., 2015). Speciation of mercury in cosmetics has been carried out using L-cysteine-functionalized cellulose fibres as adsorbents for online separation and pre-concentration of mercury species in a sequential injection system along with atomic fluorescence spectrometry determination (Chen et al., 2011a). A DLLME procedure has been also developed for speciation of this element by liquid chromatography (LC)–ICP–MS after dilution of skin refreshener and hand lotions (Jia et al., 2011).

When there are manufacturing problems in the cosmetics industry, contamination by metal particles can occur as a result of metal wear products or corrosion. In these cases, it is important to know the contamination origin. After their isolation, metal particles are analysed by energy dispersive X-ray spectrometry to identify the type of alloy (Martin, 2012).

Nitrosamines

The presence of nitrosamines in cosmetics was first documented in 1977. NDELA was found in 27 cosmetic samples from a total of 29 analysed samples (concentration ranged from 10 ppb to 50 ppm) (Fan et al., 1977). Up to 150 ppm of NDELA was found in cosmetics between 1978 and 1980. In 1979, the FDA expressed its concern about the presence of nitrosamines in cosmetics in a notice published in the Federal Register. Since then, numerous programs have been developed by industry and regulators, which have resulted in a reduction of nitrosamine levels in cosmetics. Thus, for example, the FDA found less than 3 ppm NDELA in 65% of total cosmetics analysed (FDA, 1998). In addition, in 1998 the UK Department of Trade and Industry discovered that nitrosamines could increase their levels after cosmetics are opened (more than two times in 4 months and more than four times in 17 months) (DTI, 1998). More recently, the European SCCS evaluated possible health risks associated with the presence of nitrosamines and secondary amines in cosmetics. The committee concluded that the current European limit of 50 µg/kg for nitrosamines should apply to raw materials as well as all nitrosamines potentially formed in the finished cosmetic (SCCS, 2012).

Despite the important reduction in nitrosamine levels that has occurred, there are still some concerns about these compounds as impurities in cosmetics because there are multiple indications of their relationship with the cancer and their absorption through the skin (US FDA, 1998; EGW's, 2007). In particular, two nitrosamines can be found fairly frequently in cosmetics at trace levels, NDELA and N-nitroso-bis(2-hydroxypropyl) amine (NBHPA). Other nitrosamines such as N-nitrosodimethylamine (NDMA), N-nitrosodiethylamine (NDEA), N-nitrosodiisopropylamine (NDPrA), N-nitrosomethyldodecylamine or 2-ethylhexyl 4-(N-nitroso-N-methylamine) benzoate (NMPABAO) have been also found in cosmetics and/or raw materials, yet less often. Primary, secondary and tertiary amines and nitrosating agents are considered precursors for the generation of nitrosamines. The extent of nitrosamine formation depends mainly on the type of amine (in general, secondary amines are the most reactive), their structure, the concentration of nitrosating agent and the pH (SCCS, 2012). The cosmetics industry uses amines and amino derivatives such as di- and triethanolamine that in the presence of a nitrosating agent, either as a contaminant (e.g., nitrogen oxides, nitrites) or as an ingredient [e.g., 2-bromo-2-nitropropane-1,3-diol (bronopol, Onyxide 500),

5-bromo-5-nitro-1,3-dioxane (Bronidox C) or tris(hydroxymethyl)nitromethane (Tris Nitro)], can result in the formation of nitrosamines. In the same way, secondary amines can also be found as impurities (especially in raw materials such as trialkylamines and trialkanolamines), hence contributing to the formation of nitrosamines (US FDA, 1998; BSI, 2013).

The industry should control and restrict any incompatibility of ingredients, but also minimize impurities and contaminants that can produce nitrosation. In general, no levels of nitrosamines have been established in legislations for cosmetic products. The EU limits the N-nitrosodialkanolamine content of fatty acid dialkanolamides, monoalkanolamines and trialkanolamines used as raw materials and the N-nitrosodialkylamine content of fatty acid dialkylamides, monoalkylamines and trialkylamines and their salts because of their potential as nitrosamine precursors to 50 µg/kg (European Commission, 2003). Although the total elimination of nitrosamines is technically unfeasible, the ISO Technical Guidance Document for Minimizing and Determining N-nitrosamines in Cosmetics provides some strategies to minimize nitrosamine levels. They can be summarized as: (1) elimination of adventitious sources of nitrite and nitrogen oxides, (2) elimination and reduction of secondary amine contaminants, (3) use of raw materials that are free from nitrosamine contamination, (4) incorporation of an inhibitor of nitrosamine formation in the product formulation as an additional caution and (5) the use of very sensitive analytical methods for monitoring the effectiveness of these strategies and ensuring that nitrosamines are not formed during manufacture and/or storage of the product (ISO/TR 14735:2013).

So, the analytical control of nitrosamines in both cosmetics and raw materials during storage and manufacturing processes is crucial.

International Standardized Analytical Methods

No method was contemplated in the EU document (European Commission, 2000) in which old EU directives on cosmetics analysis were compiled. Nowadays, the ISO provides two different international standards for the determination of nitrosamines in cosmetics and raw materials (ISO 10130:2009; ISO 15819:2014). ISO 10130:2009 uses LC coupled with post-column photolysis and derivatization and ISO15819:2014 uses LC–MS/MS. Both standards specifically target NDELA in those products in which this compound is present as a contaminant from the ingredients or where it is formed as a result of interaction between ingredients.

Analytical Methods Published in the Scientific Literature

Different published analytical methods for total N-nitroso compounds and for specific nitrosamines are included in Table 14.5. During the analysis, in situ formation of nitrosamines has been reported in some cases. Two strategies can be used to eliminate or control this problem, i.e., the addition of a secondary amine as marker (the absence of

Main Chemical Contaminants in Cosmetics: Regulatory Aspects and Analytical Methods 353

Table 14.5 Selected analytical methods for nitrosamine determination in cosmetics and raw materials

Analyte	Matrix	Remarks on sample preparation	Technique	References
Screening methods				
Total N-nitroso compounds	Cosmetics	Sample was partitioned between methylene chloride and water to separate polar and non-polar N-nitroso compounds; each extract was analysed by adding the cleavage reagent	CL detection of nitric oxide	Chou et al. (1987)
Total N-nitroso compounds	Cosmetics	Samples were dissolved/suspended in water or THF; then, sulphamic acid was used to eliminate interference; the solution was denitrosated with HBr and acetic acid in refluxing with n-propyl acetate	CL detection of nitric oxide	Challis et al. (1995)
NDELA	Eye make-up remover, body lotion, hand cream, lipstick	1 g of sample was dispersed in 10 mL of buffer (pH 4.4) and extracted in a separatory funnel with three 20-mL portions of methylene chloride	OT–CEC–UV	Matyska et al. (2000)
Determination of specific nitrosamines				
NDELA	Cosmetics, lotions, shampoo	Sample was stirred with ammonium sulphamate and ethyl acetate, filtered, washed with ethyl acetate, cleaned up using a silica gel column, washed with acetone, evaporated and filtered and concentrated under a stream of nitrogen	LC–TEA	Fan et al. (1977)
NDELA	Raw materials and cosmetics	Different steps of extraction/dissolution as a function of sample composition; injection into a first column, collection of target area, evaporation, reconstitution and reinjection onto another column	LC–UV	Rosenberg et al. (1980)
NDELA	Creams and gels	Clean-up of sample by different extraction steps and using a silica gel column; chemical derivatization and photolysis were used to form a volatile derivate	GC–ECD	Rollmann et al. (1981)
NDELA	Emulsion cream and a hair grooming gel	Obtain NDELA fraction using a series of solvent extractions	LC–TEA	Ho et al. (1981)

Continued

Table 14.5 Selected analytical methods for nitrosamine determination in cosmetics and raw materials—cont'd

Analyte	Matrix	Remarks on sample preparation	Technique	References
NDELA and NBHPA	Raw materials (alkanolamines)	Dissolve sample in methanol/water and add an ion-exchange resin; transfer the suspension into a glass column, elute with methanol/water and evaporate; dissolve the residue with chloroform/acetone; apply to a clean-up column, wash and elute; evaporate the solvent with nitrogen followed by silylation	GC–TEA	Sommer et al. (1988a)
NDELA and NBHPA	Cosmetics	Dissolve cosmetic in water and add IS and ammonium sulphamate; saturate with NaCl and adjust pH; add chloroform; transfer to a Kieselguhr column, wash, elute with n-butanol and evaporate; apply the residue to a second column, wash and elute with acetone; evaporate with nitrogen followed by silylation	GC–TEA	Sommer et al. (1988b)
NDELA	Creams and lotions	Mix sample with NaCl, glacial acetic acid, 1,2-dichloroethane and ammonium sulphamate solution; multiple steps follow: mix, centrifuge, clean-up using column and SPE cartridges, dilution, evaporation and derivatization	GC–ELCD	Collier et al. (1988)
NDELA	Cosmetics	Disperse sample in water with ammonium sulphonate and saturate with NaCl and add chloroform; multiple steps follow: clean-up using two different glass packet columns, dilution, evaporation and dryness, and derivatization for GC–MS	LC–UV and GC–MS	Schwarzenbach and Schmid (1989)
NPABAO	Sunscreen products	Different clean-up steps to afford a non-aqueous extract from product emulsions (lotions, creams and gels); oils are dissolved in the mobile phase and can be analysed directly	LC–TEA	Meyer and Powell (1991)
NDELA and NMPABAO	Skin cream and skin lotion	Sample, nitrosation inhibitor and NaCl were vortexed, mixed with 5 mL hexane and centrifuged; the extract was evaporated and reconstituted with acetonitrile and centrifuged	LC–PB–TEA	Billedeau et al. (1994)
Nitrosation products	Raw materials: Padimate-O and Incronam-30	Study on nitrosation products: after nitrosation reaction, samples were extracted twice with hexane (3 + 2 mL), evaporated to dryness and taken up in acetonitrile	LC–MS, LC–MS/MS and LC–hv–MS	Volmer et al. (1996)

Analyte	Matrix	Method description	Technique	Reference
NDELA	Gels, shampoos, creams, milks, conditioners and foams	Samples were diluted with water, adsorbed onto a Kieselguhr column and eluted with *n*-butanol; extracts were transferred to a silica gel column and eluted with acetone; eluates were dried and redissolved in dichloromethane	GC–TEA and online CL detection	Schothorst and Stephany (2001)
NDELA	Cosmetics	Aqueous cosmetics, dilute in water and SPE; emulsions, oils and solids, dissolve in DCM and extract with water; UV photolysis for generating nitrite ion that is derivatized according to the Griess reaction	LC with visible detection	Flower et al. (2006)
NDELA	Cosmetics	Soluble cosmetics, mix with water and shake, clean up with a conditioned C$_{18}$ column; samples less soluble, use DCM; the mixture is shaken and centrifuged	LC–MS/MS	Schothorst and Somers (2005)
Five N-nitrosamines	Cosmetics	Use of MeOH and 1.0 mM K$_2$HPO$_4$ (pH 4.0) as solvent for sample extraction and mobile phase	LC–PAD	Wang et al. (2006)
NDELA	Cosmetics	HS–SPME with a new fibre (grafting aluminium tri-*tert*-butoxide on the surface of fused silica); extraction at 70°C, in 15 min, desorption for 5 min at 260°C	GC–MS	Davarani et al. (2013)
Thirteen N-nitrosamines	Skin care cream, skin care water and raw materials	Sample was dispersed with water; then, it was extracted and purified using salting-out/acetonitrile homogeneous extraction method; the obtained solution was concentrated by slow nitrogen gas blowing	GC–MS	Dong et al. (2015)
NDELA	Raw materials and cosmetics	Evaluation of different cartridges of SPE for sample clean-up and chromatographic columns	UPLC–MS/MS	Joo et al. (2015)
Seven volatile nitrosamines	Cream-type cosmetics	Sample is dissolved in DCM and methanol (1:1 v/v), sonicated for 30 min, heated in an oven, heated on a hot plate until evaporation of half initial volume followed by HS–SPME procedure with CAR/PDMS fibres	GC–MS	Choi et al. (2016)

CAR/PDMS, carboxen/polydimethylsiloxane; *CL*, chemiluminescence; *DCM*, dichloromethane; *GC–ECD*, gas chromatography with electron capture detector; *GC–ELCD*, gas chromatography with electrolytic conductivity detector; *GC–MS*, gas chromatography–mass spectrometry; *GC–TEA*, gas chromatography with thermal energy analyser; *HS–SPME*, headspace solid-phase microextraction; *IS*, internal standard; *LC–hv-MS*, liquid chromatography–mass spectrometry with online photolysis reactor; *LC–MS*, liquid chromatography–mass spectrometry; *LC–MS/MS*, high-performance liquid chromatography–tandem mass spectrometry; *LC–PAD*, high-performance liquid chromatography–photodiode array detector; *LC–PB–TEA*, high-performance liquid chromatography–particle beam–thermal energy analyser; *LC–TEA*, high-performance liquid chromatography with thermal energy analyser; *LC–UV*, high-performance liquid chromatography with ultraviolet Detection; *NBHPA*, N-nitrosodiisopropanolamine; *NDELA*, N-nitrosodiethanolamine; *NMPABAO*, N-nitrosomethyl-*p*-amino-2-ethylhexyl benzoate; *NPABAO*, 2-ethylhexyl-4-(N-methyl-N-nitrosoamino) benzoate; *OT–CEC-UV*, open tubular capillary electrochromatography with ultraviolet detection; *SPE*, solid-phase extraction; *THF*, tetrahydrofuran; *UPLC–MS/MS*, ultra-high-performance liquid chromatography–tandem mass spectrometry.

the *N*-nitroso derivative of the marker must be controlled) or the use of an inhibitor minimizing to undetectable levels the formation of nitrosamines.

Screening methods for total *N*-nitroso compounds were first used in the detection of nitrosamines in cosmetic samples. These are based on chemiluminescence (CL) measurements of nitric oxide liberated by the reductive cleavage of the *N*-nitroso group. Polar and non-polar *N*-nitroso compounds are partitioned using methylene chloride and water, after which a cleavage reagent is used in the different extracts. The method is not quantitative; recoveries for polar *N*-nitroso compounds range between 48% and 83% and for non-polar *N*-nitroso compounds between 58% and 70% at parts-per-billion levels. False-positive responses were observed for some cosmetic products (Chou et al., 1987). A modified procedure was proposed and evaluated in a collaborative study organized under the auspices of the UK Cosmetic Toiletry and Perfumery Association. No partitioning of the *N*-nitroso compounds was necessary. The sample was dissolved or suspended in water or in tetrahydrofuran (THF). Interference from nitrite or nitrite ester was removed using sulphamic acid. This solution was denitrosated with HBr/acetic acid in a refluxing of *n*-propyl acetate. It improved sensitivity, and quantitative recoveries could be obtained. However, false positives were obtained for some nitro compounds that can be present in cosmetics such as some hair dyes (Challis et al., 1995). So, this method is usually referred to as the 'apparent total nitrosamine content' (ATNC) method. Anyway, CL methods are advised for preliminary screening of cosmetics, but the positive confirmation must be made with alternative methods (BSI, 2013). More recently, a simple screening method for NDELA detection at parts-per-million levels based on open tubular capillary electrochromatography with UV detection has been proposed. The sample is dispersed in buffer and then it is subjected to a liquid–liquid extraction (LLE) procedure with methylene chloride. The prepared extract is directly injected onto a C_{18} modified etched capillary column. Recoveries obtained were in the range of 55%–90% (Matyska et al., 2000).

The determination of specific nitrosamines in cosmetics is performed using a chromatographic technique, i.e., gas chromatography (GC) or LC with different detectors. The use of any chromatographic technique depends on nitrosamine volatility. Nevertheless, derivatization of non-volatile amines has been implemented for GC. For example, NDELA has been determined using GC–ECD (GC with electron capture detector) after denitrosation with thionyl chloride in methylene chloride to obtain a volatile nitramine. The sensitivity of the method allows determination of 50 ppb of analyte (Rollmann et al., 1981).

A very popular detector for both LC and GC is the thermal energy analyser (TEA) because of its high selectivity and sensitivity for the *N*-nitroso group. After elution, nitrosamines pass into a pyrolysis system where they are decomposed to liberate NO. A cold trap helps to eliminate other products resulting from pyrolysis. NO reacts with ozone in a CL reaction to produce infrared radiation that is detected by a photomultiplier. In general, the high selectivity gives rise to a simplification of sample preparation. GC–TEA for NDELA and NBHPA determination in raw materials and cosmetics using different steps for sample clean-up was reported by Sommer et al. (1988) and Sommer and Eisenbrand (1988). The obtained residue was treated with N-methyl N-trimethylsilylheptafluorobutyramide to convert nitrosamines into volatile derivatives. Schothorst and Stephany (2001) made a modification and validation of the methodology proposal by Sommer et al. (1988) in order to use GC–TEA without derivatization for the determination of NDELA in cosmetics. After various clean-up processes, the limit of detection (LOD) was 3.5 ppb and the repeatability expressed as relative standard deviation (RSD) was 5%.

LC–TEA, less sensitive than GC–TEA, was used by Fan et al. (1977) to determine NDELA in cosmetics for the first time. Subsequently, the use of LC–TEA for routine analysis was proposed by Ho et al. (1981). A sample treatment using a series of solvent extractions to perform clean-up and analyte pre-concentration yielded LODs of about 20–30 ppb and recoveries between 71% and 103%. Nevertheless, the coupling of LC and TEA is not easy. In this sense, a based particle beam–LC interface was proposed for improving the LC–TEA coupling and applied to determining NDELA and NMPABAO in cosmetics. This interface incorporates a thermospray vaporizer, a desolvation chamber and a counter-flow gas diffusion cell to reduce the effluent to a dry aerosol, and a single-stage momentum separator to form a particle beam of the analyte. The solvent is efficiently removed and thus the analytical characteristics are improved (Billedeau et al., 1994). 2-Ethylhexyl-4-(N-methyl-N-nitrosoamino) benzoate has been also determined by LC–TEA in sunscreen products (Meyer and Powell, 1991). The main disadvantage of both GC–TEA and LC–TEA is the need for subsequent mass spectral confirmation to minimize false positives.

More conventional detectors such as the UV/VIS detector have been also used. The applicability of LC–UV for routine determination of NDELA in raw materials and cosmetics was demonstrated in an interlaboratory study (Rosenberg et al., 1980). Owing to the complexity of cosmetic matrices, the sample type is critical when UV detection is used (it is normally used for samples that are soluble in water or water/alcohol). Various extraction/dissolution procedures are necessary. In addition, after sample injection onto a first column, the target area is collected, evaporated, reconstituted and reinjected onto another column to determine NDELA with LODs on the order of parts per billion. As of this writing, LC with UV/VIS detection is more used for routine determination of NDELA after post-column photolysis to give nitrite and further Griess derivatization. Aqueous samples are prepared by dilution with water followed by SPE, and emulsions, oils or solid samples are dissolved in dichloromethane (DCM) and extracted with water. Reversed-phase LC is used and LODs at parts-per-billion levels can be obtained (Flower et al., 2006). The reliability of this method is such that it has served as the base for the aforementioned ISO 10130:2009. Other nitrosamines, both volatile and non-volatile, such as NDEA, NBHPA, NDMA, NDPrA and N-nitrosodiphenylamine, have been also determined by LC with photodiode array detector (PAD) (Wang et al., 2006).

MS detectors provide the identification and confirmation of nitrosamines with simpler sample treatments. Identification of nitrosamines in cosmetics and raw materials is important because it is estimated that only 10% of the ATNC in cosmetics corresponds to known nitrosamines (Gangolli et al., 1994). Techniques such as GC–MS and LC–MS allow one to separate, detect and confirm the presence of nitrosamine in one run. GC–MS is used for volatile and semi-volatile nitrosamines, whereas LC–MS is used without derivatization for nitrosamines with a wide volatility range. GC–MS has been proposed for determining a high number of nitrosamines simultaneously in cosmetics (Dong et al., 2015; Choi et al., 2016). In addition, miniaturized sample preparation techniques such as solid-phase microextraction (SPME) are being introduced for nitrosamines determination in cosmetics by GC–MS (Davarani et al., 2013; Choi et al., 2016). For instance, the identification and quantification of seven volatile N-nitrosamines in emulsion-type water-insoluble cosmetics using headspace (HS)–SPME mode with a carboxen/polydimethylsiloxane (PDMS) fibre for sample clean-up and analyte pre-concentration has been reported (Choi et al., 2016). In comparison with classical sorbent SPE cartridges, widely used for cosmetic clean-up, HS–SPME shows an efficient removal of both hydrophilic and lipophilic interfering substances from the matrix. The analytical characteristics are similar to or better than those obtained with standard methods.

LC–MS and LC–MS/MS are used for the characterization of nitrosation products in raw materials and finished products. As of this writing, LC–MS/MS is used in the standard ISO 15819:2014 for determining NDELA in cosmetics. The analyte is extracted with water in the presence of deuterated NDELA-d8 as the internal standard (IS). The sample clean-up is carried out with a C_{18} cartridge or by LLE with DCM, depending on the type of sample matrix. Then, the obtained extracts are directly analysed by LC–MS/MS. LODs are about 20–50 ppb depending on the instrument and cosmetic matrices. The use of LC coupled to MS to monitor fragmented ions provides a high degree of specificity for NDELA. Four identification points are required for confirmation purposes (Schothorst and Somers, 2005; ISO 15819:2014). Despite this, cosmetic complexity makes it difficult to work with a single sensitive and selective analytical method. In 2015, a systematic study on sample preparation to cover diverse types of cosmetic products was published by Joo et al. (2015). Various SPE cartridges and columns aimed at obtaining optimal chromatographic separation of NDELA were evaluated.

Given the difficulty for determining nitrosamines in cosmetics and raw materials, the ISO recommends an approach to the topic as shown in Fig. 14.1 (ISO/TR 14735:2013).

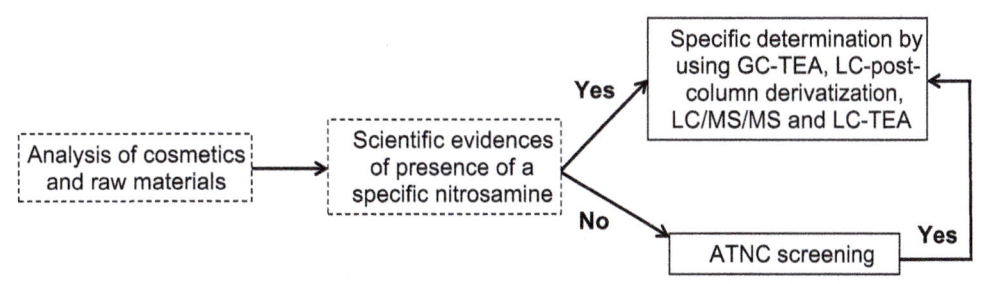

Figure 14.1 Approach to the determination of nitrosamines in cosmetics. *ATNC*, apparent total nitrosamine content; *GC*, gas chromatography; *HPLC*, high-performance liquid chromatography; *MS/MS*, tandem mass spectrometry; *TEA*, thermal energy analyser. *Adapted from ISO/TR 14735:2013. Cosmetics – Analytical methods – Nitrosamines: Technical Guidance Document for Minimizing and Determining N-nitrosamines in Cosmetics.*

1,4-Dioxane

The FDA considers 1,4-dioxane as a byproduct formed during the manufacture of cosmetics when ethoxylated raw materials are used as emulsifiers, foaming agents or solvents (usually identified by terms such as 'polyethylene', 'polyethylene glycol', 'polyoxyethylene', '-eth-' or '-oxynol-'). These raw materials are obtained by polymerization of ethylene oxide, and 1,4-dioxane arises upon dimerization of this oxide during ethoxylation processes, thus contaminating cosmetic raw materials. For example, sodium dodecyl sulphate is ethoxylated to form sodium lauryl sulphate (SLS), less abrasive and with enhanced foaming characteristics, that is widely used in shampoos and toothpastes as a foaming agent (FDA, 2014). In general, 1,4-dioxane is considered a possible carcinogen to humans that can penetrate the skin when it is applied in certain preparations, e.g., lotions. Despite this, the available amount of 1,4-dioxane for skin absorption in cosmetics is very small, even in cosmetics remaining on the skin for hours, because it evaporates (Bronaugh, 1982).

The contamination of ethoxylated surfactants with 1,4-dioxane was detected in 1979 for the first time (FDA, 1981). Since then, the FDA has encouraged manufacturers to remove 1,4-dioxane and monitor periodically the levels of this compound in cosmetics and raw materials. Changes have also been made in the manufacturing processes by using vacuum stripping to decrease the level of this contaminant. Black et al. (2001) reported data in regard to levels of 1,4 dioxane in the period 1979–97 in both raw materials and finished products. Up to 1410 ppm of 1,4-dioxane was found in raw materials (in particular in ammonium laureth sulphate and SLS) and up to 279 ppm in cosmetics (corresponding to a shampoo). These studies indicated that the industry had succeeded in reducing 1,4-dioxane in raw materials, but not in all cases. The FDA is concerned, in particular, with the control of these levels in children's products. Nevertheless, the FDA considers that the 1,4-dioxane levels usually found in cosmetics do not present a hazard to consumers (FDA, 2014).

In 2015 the European SCCS expressed its scientific opinion on the report of the ICCR working group, 'Considerations on Acceptable Trace Level of 1,4-Dioxane in Cosmetic Products', stating that a trace level lower than 10 ppm of 1,4-dioxane in the final cosmetic can be considered safe (SCCS, 2015c).

The level at which a compound can be considered harmful in cosmetics depends on the conditions of use. 1,4-Dioxane is found in cosmetic products that are rinsed off or are in contact with the skin for only a short period of time. Thus, there are not established limits for 1,4-dioxane in cosmetics, although it is true that the 10-ppm level of 1,4-dioxane is widespread today as a guidance level. On the other hand, a 2015 draft version of the Chinese Technical Safety Standard for cosmetics contemplates a limit of 30 mg/kg of this compound (CRS, 2015). In any case, continuous quality control of ethoxylated raw materials and finished products is necessary.

International Standardized Analytical Methods

There are different analytical methods approved by official agencies and organizations such as the US Environmental Protection Agency (EPA) or the US National Institute for Occupational Safety and Health for determining 1,4-dioxane in environmental and biological samples. However, there are no standard methods of analysis for cosmetics or ethoxylated raw materials.

Analytical Methods Published in the Scientific Literature

Some of the standard methods proposed for other matrices have been adapted to the analysis of cosmetics and raw materials. For example, EPA Method 5031 using azeotropic distillation was adapted by Black et al. (2001) for the determination of 1,4-dioxane in cosmetic raw materials and finished cosmetic products. The sample, with water, sodium carbonate, 2-octanol and Antifoam A, was heated to boiling and distilled at a moderate rate. The distillate was analysed by GC with flame ionization detector (GC–FID) using *n*-butanol as the IS. The LOD was 1 ppm and the obtained recoveries were between 77% and 98% (concentrations 20 and 100 ppm).

Table 14.6 shows some published analytical methods. 1,4-Dioxane is usually determined by GC–MS or GC–FID, although some analytical methods using LC have also been proposed. The determination of this compound at low concentrations is not easy because its miscibility with water and volatility can cause poor efficiency in pre-concentration and/or sample clean-up procedures, resulting in low recoveries. Different sample preparation methods have been developed to quantitatively extract this analyte from cosmetics and surfactants.

Generally, sample preparation procedures are more laborious for cosmetics than for raw materials, yet there are exceptions. For instance, Robinson and Ciurczak (1980) proposed the dilution of ethoxylated surfactants with water and/or ethanol, according to their viscosity, and direct injection into the chromatograph. A glass wool pre-column

Table 14.6 Selected analytical methods for 1,4-dioxane determination in cosmetics and raw materials

Matrix	Remarks on sample preparation	Technique	References
Ethoxylated surfactants	1,4-Dioxane is extracted from the ethoxylated surfactants using chlorobenzene, which is then diluted and injected directly into the chromatographic column	GC–FID	Stafford et al. (1980)
Ethoxylated surfactants	Most low-viscosity surfactants were injected as supplied onto the glass wool–filled pre-column; high-viscosity samples were diluted with water or methanol; KOH was added to prevent dioxane formation	GC–FID	Robinson and Ciurczak (1980)
Shampoos and bath preparations	HS methodology; standards are prepared in a product similar to the one being analysed but which does not contain ethoxylated material	GC–FID	Beernaert et al. (1987)
Sulphated polyoxyethylene fatty alcohols	Sample is dissolved with water; it is applied to a pre-conditioned C_{18} cartridge and eluted with acetonitrile/water (10:90, v/v); this fraction is directly analysed by RP–LC	LC–UV	Scalia (1990)
Day cream, aftershave emulsion, moisturizing lotion and sunscreen shampoo	Sample is extracted with 10% (v/v) DCM in hexane (twice) and applied to a cyano cartridge; it is eluted with 15% v/v acetonitrile in water; the eluate is passed through a C_{18} cartridge and analysed by RP–LC	LC–UV	Scalia et al. (1990)
Cosmetic products with polyethoxylated surfactants	HS methodology; isotope dilution method using 1,4-dioxane-d8); sample and standards are diluted in DCM introduced into an HS vial	GC–MS	Rastogi (1990)
Shampoo products	Sample is dispersed with isobutanol as IS and injected directly into the packed column	GC–FID	Italia and Nunes (1991)
Day cream, lotions, shampoo, bath foam	Extraction with 20% (v/v) DCM in hexane (twice); application to an SiOH cartridge; elution with acetonitrile; the eluate is passed through a C_{18} cartridge	GC–MS	Scalia et al. (1992)
Lotions, creams, shampoos, conditioners, cleansers	Extraction with hexane and methylene chloride (80:20 v/v); application to a C_8 cartridge and elution with acetonitrile; 1,4-dioxane-d8 used as IS	GC–MS	Song et al. (1996)

Continued

Table 14.6 Selected analytical methods for 1,4-dioxane determination in cosmetics and raw materials—cont'd

Matrix	Remarks on sample preparation	Technique	References
Cosmetic raw materials and finished products	Two sample preparation procedures were used: 1,4-dioxane was isolated by azeotropic atmospheric distillation or extracted by methylene chloride in hexane and then applied to a silica gel column and a C_{18} cartridge	GC–FID	Black et al. (2001)
Shampoos, liquid soaps, creams, sunscreens, deodorants	HS-SPME using 85-μm fibres with polyacrylate coating; vials were heated to 60°C for 10 min; microextraction time was 2 min; calibration was with a shampoo without ethoxylated ingredients	GC–FID	Silva et al. (2001)
Non-ionic surfactants as polyethylene oxide, poly(ethylene/propylene) oxide, polyhydric alcohol; cosmetics as shampoo, liquid soap, dishwashing detergent	HS–SPME with PDMS fibres; vials containing sample solution, deionized water, NaCl and THF as IS were sampled at room temperature for 2 min for surfactants and 10 min for cosmetics	GC–FID and GC–MS	Fuh et al. (2005)
Surfactants and detergents	HS–SPME with a new aluminium hydroxide coating on fused silica fibre; vials contained sample, water and NaCl; microextraction time was 13 min; standard addition was used for quantification in detergents	GC–FID	Davarani et al. (2012)
Cleansing products with polyoxyethylene-based surfactants, polyoxyethylene lauryl ether and sodium polyoxyethylene lauryl ether sulphate	HS methodology; fluorobenzene, p-bromofluorobenzene and 1,4-dioxane-d8 were used as ISs; HS vials were equilibrated for 20 min at 80°C before sampling	GC–MS	Tahara et al. (2013)

DCM, dichloromethane; *GC–FID*, gas chromatography with flame ionization detector; *GC–MS*, gas chromatography–mass spectrometry; *HS*, headspace; *HS–SPME*, headspace–solid-phase microextraction; *IS*, internal standard; *LC–UV*, liquid chromatography with ultraviolet detector; *PDMS*, polydimethylsiloxane; *RP–LC*, reversed-phase–high-performance liquid chromatography; *THF*, tetrahydrofuran.

allows the removal of interference. Nevertheless, the standard addition method is necessary for calibration. Good recoveries were obtained (92%–104%) for levels of 1,4-dioxane between 250 and 1000 ppm. Stafford et al. (1980) dissolved ethoxylated surfactants in chlorobenzene. The sample was directly injected into a chromatograph equipped with two columns in series, a first column packed with a non-polar stationary phase and a second with a highly polar stationary phase. The heavier ethoxylates and glycols were retained in the first column, and following 1,4-dioxane elution for the separation of lighter matrix compounds in the second column, the flow was reversed to flush these compounds. The LOD was 0.5 ppm and good recoveries were obtained. However, repeatability was on the order of 25% for levels of about 1–5 ppm. Italia and Nunes (1991) analysed shampoo products after dilution with water containing isobutanol as the IS and direct injection into a packed column. The reproducibility of the method was better than 7% and recoveries were in the range of 94%–105% (for concentrations from 1 to 250 ppm).

In addition to direct injection, HS sampling makes possible the development of very simple procedures. Samples are introduced into an HS vial with an IS followed by heating. For instance, Rastogi (1990) used 1,4-dioxane-d8 as the IS and heated the vials at 80°C for 16–18 h. The LOD was 0.3 ppm and recoveries were about 92% at a 50 ppm level with an RSD of 9.1%. Tahara et al. (2013) used fluorobenzene, p-bromofluorobenzene and 1,4-dioxane-d8 as ISs, and HS vials were equilibrated for only 20 min at 80°C before sampling. Similar LODs were obtained (0.5 ppm).

Despite these examples, the clean-up of samples for both cosmetics and raw materials is often needed. Within the conventional techniques, SPE is the most used with this purpose. Different solid phases can be used. Thus, Scalia (1990) used C_{18} cartridges for the analysis of ethoxylated fatty alcohol sulphates by LC–UV. Once the sample was solubilized in water, it was applied into the pre-conditioned cartridge and eluted with acetonitrile/water. Recoveries were about 96% for levels between 40 and 120 ppm with RSDs of 3.5% and a limit of quantification (LOQ) of 18 ppm. This procedure was adapted to cosmetic matrices, becoming larger and more tedious. A double extraction with 10% v/v DCM in hexane was necessary. Then, the combined supernatants were applied to a pre-conditioned cyano column. After centrifugation and elution with 15% v/v acetonitrile in water, the eluate was passed directly through a C_{18} cartridge. This final eluate was brought to volume and analysed by LC–UV (Scalia et al., 1990). Because the UV detection for 1,4-dioxane generally results in low selectivity, this procedure was adapted to GC–MS with some changes in sample preparation to improve specificity and sensitivity. In particular, cyano cartridges were replaced by SiOH cartridges because more effective clean-up is thus achieved, yielding more reproducible recoveries (91.1%–93.2%) with RSDs better than 4.3% and an LOQ of 3 ppm (Scalia et al., 1992). Sensitivity was slightly improved (LOQ 0.1 ppm) by Song et al. (1996), using an SPE procedure with a solid-phase C_8 after extraction with hexane/DCM (80:20 v/v) and 1,4-dioxane-d8 as IS. However, recoveries were only about 87%.

Most recently, HS–SPME has been introduced to analyse surfactant raw materials and cosmetics with high sensitivity. Silva et al. (2001) used a manual SPME holder with a polyacrylate coating for 10 min at 60°C to analyse different cosmetics by GC–FID (LOD 5 ppb). Fuh et al. (2005) analysed non-ionic surfactants and cosmetics by GC–FID and GC–MS using 100 μm PDMS fibres. The extraction time was only 2 min for surfactant samples and 10 min for cosmetics. Recoveries were higher than 96% with LODs between 0.06 and 0.51 ppm and RSDs smaller than 3%. Davarani et al. (2012) proposed the use of a new fibre for the analysis of some samples including ethoxylated fatty alcohol, sodium lauryl ether sulphate, dishwashing agents and shampoos. In this case, the surface of a fused silica capillary tubing was modified by means of aluminium tri-*tert*-butoxide, hence providing a Lewis acid–base interaction with analyte functional groups. The microextraction was carried out for 4 min at 34°C. The LOD and LOQ of the proposed method were estimated to be 0.0015 and 0.005 ppm, respectively. In general, the standard addition method is used when HS single-drop microextraction is applied.

Hydroquinone

Hydroquinone (HQ) is a cosmetic ingredient but also can be in cosmetic products as a contaminant.

HQ has been classically used in cosmetics as a skin lightener and also in hair dye products. However, as of this writing it is in Annex II of the EU Regulation as a prohibited substance with the only exception being its use in artificial nail systems (technical aid for polymerization) (max. 0.02%, after mixing, for professional use only, including also methyl ether HQ). This compound has been linked to cancer and organ system toxicity (IARC, 1999). However, the Scientific Committee on Cosmetic Products and Non-food Products intended for consumers concluded that HQ in artificial nail systems presents minimal risk to consumers (Commission Directive, 2003/83/EC). Canadian legislation restricts its use to hair dyes (0.3% maximum) and artificial nail systems (0.02%), and in addition it allows its use in cyanoacrylate-based adhesives at concentrations equal to or less than 0.1% (Health Canada, 2014b). China limits it to 0.3% for di-*t*-butylhydroquinone (CRS, 2014), whereas Japanese legislation bans the use of HQ monobenzyl ether in cosmetics (PAL, 1960; MHW, 1999).

No reference is made by FDA authorities, neither as a prohibited nor as a restricted ingredient.

HQ may also be considered as a contaminant. Main sources of it as an impurity are derived from the use of other ingredients such as tocopheryl acetate, tocopheral, arbutin, etc. For instance, the SCCS concluded in June 2015 that the use of deoxyarbutin up to 3% as an ingredient in face creams is unsafe, given the formation of HQ at levels that can compromise the safety of the cosmetic (SCCS, 2015a). Arbutin, known to hydrolyse and release HQ, has been considered safe in cosmetics up to 2% in face creams and 0.5% in body lotions because arbutin powder may contain about 0.15% HQ as an impurity (SCCS, 2015b).

International Standardized Analytical Methods

European methods of analysis were published for HQ identification by thin-layer chromatography and for its determination by LC–UV (European Commission, 2000). In the latter case, the method is valid for separation and determination of HQ, HQ monomethyl ether, HQ monoethyl ether and HQ monobenzyl ether. However, it is aimed at detecting these analytes in cosmetics for lightening the skin. Considering that the use of HQ for this purpose is no longer permitted, the method requires urgent revision, as low LODs are needed for forbidden substances, and therefore it does not fit for purpose. However, this method can be used to detect possible fraud in connection with the voluntary substitution of a labelled whitening ingredient with HQ, as the whitening efficacy of HQ is higher than that of other authorized ingredients. In these cases high contents of HQ could be found in the cosmetic product. The method consists in the extraction of the sample with 50% (v/v) methanol by shaking and warming at 60°C. After the lipid material is separated by filtration, the extract is analysed by LC with photodiode array detection. Separation is accomplished using THF/water 45:55 (v/v) in a Zorbax phenyl column and detection is performed at 295 nm. The method itself has several disadvantages such as the potential non-separation of the peaks, which could be circumvented by changing the column or adjusting the mobile phase. Possible peak asymmetry and paraben-caused interference make this method no longer suitable.

Analytical Methods Published in the Scientific Literature

On the other hand, several analytical methods have been developed for the determination for HQ in cosmetics. Table 14.7 shows some selected analytical methodologies. The majority of these methods are aimed at the analysis of semi-solid creams, emulsions, skin tonics and lotions. Sample dissolution/homogenization in buffer (Britton–Robinson, McIlvaine) (Xu et al., 2013; Calaça et al., 2015), alcoholic solutions (Firth and Rix, 1986; Cruz-Vieira and Fatibello-Filho, 2000; Qassim and Omaish, 2014), organic solvents such as dimethylformamide (Chisvert et al., 2010) or sulphuric acid solution (López-García et al., 2007) can be used as sample treatment. However, the vast majority of the analytical methods are based on extraction assisted by shaking, vortexing, sonication (or their combination) or microdialysis. In these cases, alcoholic extractants as methanol, ethanol, isopropanol (Sakodinskaya et al., 1992; Desiderio et al., 2000; Rueda et al., 2003; Guan et al., 2005; Sha et al., 2007; Mobin et al., 2010; Uddin et al., 2011; Jeon et al., 2015); buffers such as KH_2PO_4 (Huang et al., 2004; Guan et al., 2005); alcoholic–buffer mixtures (Wang et al., 2015); micellar phases (Thogchai and Liawruangrath, 2013) or water (Lin et al., 2005, 2007) are used. Long extraction times may be required even using ultrasound (Huang et al., 2004; Guan et al., 2005; Lin et al., 2007; Sha et al., 2007; Thogchai and Liawruangrath, 2013; Jeon et al., 2015; Wang et al., 2015). This may be related to the high content of lipids in the cosmetic. In most cases, additional filtration is required to avoid introduction of potential interferents into the detection system. For

Table 14.7 Selected analytical methods for hydroquinone determination in cosmetics

Matrix	Remarks on sample preparation	Technique	References
Skin toning creams	Sample is dissolved in methanol or methanol/light petroleum and directly injected into the chromatograph	LC	Firth and Rix (1986)
Skin toning cream	0.1 g of sample is mixed with 40 mL of solution containing 10% (v/v) methanol and 4×10^{-3} M caffeine (IS) under heating at 50°C, cooled and filtered; SDS is used as a background electrolyte for detection	MEKC	Sakodinskaya et al. (1992)
Skin toning cream	Skin tonic aliquots are mixed with an IS and extracted in 10% v/v methanol (2.5 mg cream/mL solution), vortexed at 50°C for 15 min, cooled and filtered	CEC–UV	Desiderio et al. (2000)
Creams	1.5–2 g of cream is dissolved in 40 mL methanol and stirred until dissolution; volume is made up to 50 mL with methanol; amperometric detection (paraffin/graphite modified electrode) is performed	ED	Cruz-Vieira and Fatibello-Filho (2000)
Suntan lotion	No sample pre-treatment indicated; FI; amperometric detection (hydroxybenzaldehyde/formaldehyde polymer-modified glassy carbon electrode)	ED	García and Ortiz (2000)
Bleaching cream	50 mg of cream is mixed with 10 mL of methanol, heated (40°C) under stirring, cooled and centrifuged; supernatant is made up to 10 mL with methanol; 100 µL of extract is made up to 25 mL with 2M acetic/acetate buffer (pH 4.8)/methanol 85:15 (v/v); 20 µL is injected onto an FI system	ED	Rueda et al. (2003)
Creams and lotions	1 g sample is mixed with IS and diluted 20-fold with 0.05 M KH_2PO_4 buffer (pH 2.5) and homogenized under sonication for 30 min; extract is filtered and diluted using the same buffer	LC–DAD	Huang et al. (2004)
Medicated skin cosmetic cream and lotion, non-medicated cosmetic cream and serum	Aliquots of sample are diluted with 100 mL water and transferred to 10-mL dialysis vials; microdialysis is done using a polycarbonate fibre probe and water as the perfusate	LC–UV	Lin et al. (2005)

Facial mask, bath oil, moisturizing sunblock lotion	Samples are sonicated with ethanol (30 min), diluted in buffer (20 mM borate buffer, pH 7.4, containing 25 mmol/L SDS) and filtered	MEKC–ED	Guan et al. (2005)
Cosmetics	30–47 mg sample is extracted with 10 mL water under sonication and centrifuged (10 min); lower aqueous layer is taken for analysis	MEKCE	Lin et al. (2007)
Cosmetics	Sample is dissolved in 0.1 mol/L phosphate buffer solution at pH 7.5	ED	Oliveira et al. (2007)
Moisturizing lotion, moisturizing whitening cream, moisturizing whitening mask, whitening sunblock lotion	1 mL ethanol is added to 0.3 g sample, sonicated (30 min), centrifuged and diluted to 5 mL with water	CE–ED	Sha et al. (2007)
Creams	20 mg is mixed with 50 mL of 0.05 M H_2SO_4, shaken (10 min) in a water bath at 40°C, made up to volume with same solvent and filtered; appropriate aliquots are diluted with 0.05 M H_2SO_4	UV/VIS spectrophotometry	López-García et al. (2007)
Skin whitening cosmetic creams	5 mL of DMF is added to up to 0.2 g sample, sonicated, diluted with DMF and filtered; DMF is added to an aliquot up to 200 μL; 50 μL of PCB solution and 100 μL of BSTFA are added to the vial and shaken	GC–MS	Chisvert et al. (2010)
Medicated cosmetics	50 mg of sample and 10 mL of methanol are warmed (40°C) and stirred until dissolution, cooled and centrifuged; the supernatant is transferred and made up to 10 mL with water; 100 μL of this solution is made up to 25 mL with phosphate buffer (pH 7); CPE modified with [Cu(μ2-hep-H)]$_2$·2PF$_6$ complex is used	SWV	Mobin et al. (2010)
Skin lightening creams	0.2 g sample is mixed with 2 mL iPr and filtered; the residue is washed three times with iPr, filtered and added to first filtrate, made up to 10 mL with iPr and mixed with ammonium *meta*-vanadate solution and iPr	UV/VIS spectrophotometry	Uddin et al. (2011)

Continued

Table 14.7 Selected analytical methods for hydroquinone determination in cosmetics—cont'd

Matrix	Remarks on sample preparation	Technique	References
Cosmetics	0.5 g of cream is dissolved in 25 mL of 1% (v/v) acetonitrile and 6 mM Brij 35 (pH 6.0); 10 mg/L resorcinol (IS) is added, sonicated (30 min), centrifuged (30 min) and filtered; supernatant is taken for analysis	MLC	Thogchai and Liawruangrath (2013)
Bleaching creams	10 mg of cosmetic sample is dispersed in 10 mL of B–R buffer with US (15 min); DPV-CNT-doped PEDOT-conducting polymer–modified CPE is used	ED	Xu et al. (2013)
Lightening cream	20 mL of methanol is added to 0.25 g of sample and mixed; 0.1 M NaOH is added; 1 mL of this solution is mixed with 1.5 mL of 0.02 M KMnO$_4$ and made up to 25 mL with water and held for 30 min	UV/VIS spectrophotometry	Qassim and Omaish (2014)
Creams	0.1 g sample is vortexed with 3 mL methanol and sonicated (30 min); 10 mL water is added, sonicated (30 min), centrifuged and filtered	LC-PAD	Jeon et al. (2015)
Creams, lotions, sera, foams, gels, mask sheets, soap	0.5 g sample is extracted with 20 mM NaH$_2$PO$_4$ with 10% (v/v) methanol and 3 g NaCl (soap and foam), sonicated (30 min) and centrifuged (30 min); 2 mL of extraction solvent is added to the residue, sonicated (30 min) and centrifuged; all extracts are combined and filtered	LC-UV	Wang et al. (2015)
Gels and creams	1 g sample is diluted to 100 mL with water; an aliquot is diluted to 10 mL with McIlvaine buffer (pH 8.0)	SWV	Calaça et al. (2015)
Bleaching cosmetics	Use of microdialysis for sample preparation; determination of different hydrophilic whitening agents	LC-ED	Chen et al. (2015b)

B–R buffer, 0.2 M Britton–Robinson buffer, pH 3.0; *BSTFA*, N,O-bis(trimethylsilyl)trifluoroacetamide containing 1% trimethylchlorosylane; *CEC*, capillary electrochromatography; *CE-ED*, capillary electrophoresis with electrochemical detection; *DAD*, diode array detector; *DMF*, dimethylformamide; *DPV-CNT-doped PEDOT-conducting polymer–modified CPE*, differential pulse voltammetry–carbon nanotube doped poly(3,4-ethylenedioxythiophene) conducting polymer–modified carbon paste electrode; *ED*, electrochemical detection; *FI*, flow injection; *GC-MS*, gas chromatography–mass spectrometry; *iPr*, isopropanol; *IS*, internal standard; *LC*, liquid chromatography; *LC-ED*, high-performance liquid chromatography with electrochemical detection; *LC-PAD*, high-performance liquid chromatography with photodiode array detection; *MEKC*, micellar electrokinetic chromatography; *MEKCE*, micellar electrokinetic capillary electrophoresis; *MEKC–ED*, micellar electrokinetic chromatography with electrochemical detection; *MLC*, micellar liquid chromatography; *PCB*, pentachlorobenzene; *SDS*, sodium dodecyl sulphate; *SWV*, square-wave voltammetry; *US*, ultrasound; *UV*, ultraviolet detector.

example, Huang et al. (2004) determined HQ decomposed from arbutin in skin whitening creams. Different creams and lotions (1 g) were extracted in 0.05 M KH_2PO_4 buffer (pH 2.5) using 30 min sonication. Samples were then filtered and diluted in the same buffer before detection by LC–PAD. Methanol was added to the mobile phase to avoid peak overlapping between arbutin and HQ. In many cases, temperatures between 40 and 50°C may be used for enhancing the sample preparation step for either homogenization or extraction of the sample.

Some semi-automated methods based on microdialysis sampling can be found in the literature (Lin et al., 2005; Chen et al., 2015b). Potential limitations regarding the type of cosmetic to be analysed need to be taken into account. Highly viscous cosmetics may clog the microdialysis semi-permeable membrane. Lin et al. (2005) reported an online method in which previous steps of sample pre-treatment such as 100-fold dilution with water and filtration were required. Recoveries in the range of 89%–112% may reflect some matrix effects (LOD 0.2 µM for HQ).

Instrumental separation techniques of choice have been mainly LC, micellar electrokinetic chromatography (MEKC) and capillary electrophoresis (CE) with UV or electrochemical detection (ED). LODs obtained were about 0.2 µM for LC–UV (Lin et al., 2007), 0.8 µM for MEKC–ED (Sha et al., 2007), 1.37 µg/mL for capillary electrochromatography (CEC)–UV (Desiderio et al., 2000) and 0.03 µg/mL for CEC–ED (Guan et al., 2005). GC–MS has been scarcely used, a LOD of 0.2 µg/mL being obtained (Chisvert et al., 2010).

Given the oxidation capabilities of HQ, electrochemical detection without separation has often been used (García and Ortiz, 2000; Cruz-Vieira and Fatibello-Filho, 2000; Mobin et al., 2010; Xu et al., 2013; Calaça et al., 2015). Major innovations have been introduced through the modification of carbon working electrodes with a conducting polymer formed by carbon nanotube-doped poly(3,4-ethylenedioxythiophene) (Xu et al., 2013), 4-hydroxybenzaldehyde/formaldehyde polymer-modified glassy carbon electrode (García and Ortiz, 2000), use of biomimetic electrochemical sensors employing Cu(II) complexes (Oliveira et al., 2007; Mobin et al., 2010) or enzyme-based biosensors (Cruz-Vieira and Fatibello-Filho, 2000). For instance, Mobin et al. (2010) developed a biomimetic sensor based on Cu(II) complexes immobilized on a carbon paste electrode that mimics the active site of the enzyme catechol oxidase. It allows the selective oxidation and further square-wave voltammetric detection of HQ. The sensor revealed a 16-fold enhancement in peak current in comparison with an unmodified carbon paste electrode. Interference from common cosmetic ingredients such as starch, methylparaben, Mg, K or ammonium were found not to affect the determination. The LOD was 0.015 µM and RSD 2.1% (Mobin et al., 2010).

To a lesser extent, UV/VIS spectrophotometric techniques can also be found in the literature for the detection of HQ in cosmetics. Usually, a colorimetric reaction based on the oxidation of HQ to *p*-benzoquinone using different oxidizers such as ammonium *meta*-vanadate (Uddin et al., 2011) or potassium permanganate (Qassim and Omaish, 2014) is used. Up to 30 min may be needed for colour development, and some interference due to the presence of glucose, maltose and lactose was found to occur (Qassim and Omaish, 2014). LODs were in the range of 0.012–7 μg/mL (Uddin et al., 2011; Qassim and Omaish, 2014).

Formaldehyde

Formaldehyde action when used in cosmetics is based on its preservative effects. However, it is scarcely added as such into cosmetic formulations (SCCS, 2002). The addition of so-called formaldehyde releasers or donors such as Quaternium-15, DMDM-hydantoin, imidazolidinyl urea, diazolidinyl urea and bronopol is preferred instead. In this way, formaldehyde present in water-based formulations is slowly released at low concentrations, but in large enough amounts to prevent microbial growth (SCCS, 2002). Typical products containing formaldehyde or formaldehyde releasers include shampoos, face and body creams, liquid soaps, hair cosmetics and nail hardeners.

This compound can also be present as a contaminant resulting from the degradation of other cosmetic ingredients or the cosmetic container. Degradation and autoxidation of poly(ethylene glycols) or surfactants as a result of temperature changes and storage may produce formaldehyde. Plastic cosmetic packages with melamine, melamine–formaldehyde or carbamide–formaldehyde coatings used for water-based cosmetics have also been linked to formaldehyde contamination (Malinauskiene et al., 2015). In addition, pH increases in the cosmetic, inappropriate long storage periods and high temperatures intensify the release of formaldehyde (SCCS, 2002).

Formaldehyde is considered a contact allergen that can induce dermatitis even below the regulated limits (Malinauskiene et al., 2015). It has also been classified in carcinogen category 1B under Classification, Labelling and Packaging Regulation (EC) No. 1272/2008. The CIR Expert Panel estimated in a report in 2011 that the safety of cosmetics is granted as long as the formaldehyde concentration does not exceed 0.074% (w/w). The same expert panel considered this compound unsafe for use under the actual allowed concentration in hair smoothing products (CIR, 2011). In addition, the former Scientific Committee on Cosmetic Products and Non-food Products established in 2014 that nail hardeners containing formaldehyde are safe under normal conditions of use (non-daily use) and maximum allowed concentration (SCCS, 2014).

So, the use of formaldehyde in cosmetics has been restricted following different legislations. The EU Regulation states that the maximum allowed concentration in nail hardeners is 5%, indicating that this ingredient should be stated on the cosmetic label. In oral products, the concentration of formaldehyde and *p*-formaldehyde is restricted to 0.1% as free formaldehyde, whereas in all other products this maximum is 0.2%. In addition, a label warning has to be included for those cosmetics containing formaldehyde or formaldehyde releasers if the concentration of formaldehyde in the final product is higher than 0.05% (EU Regulation, 2009). Restrictions from the Canadian FDA are similar. Formaldehyde is allowed at up to 0.01% in non-aerosol products and is banned in aerosol cosmetics. In the case of nail hardeners and oral cosmetics, the maximum concentrations are 0.5% and 0.1%, respectively. For non-oral cosmetics, the maximum concentration is restricted to 0.2% (Health Canada, 2014b). Chinese regulation states 0.5% as the maximum concentration, whereas Japan bans its presence in cosmetic products (CRS, 2014; MHW, 1999). Neither restriction nor ban exists on formaldehyde in cosmetics from the FDA.

International Standardized Analytical Methods
The Second Commission Directive 82/434/EEC provided methods for both identification and determination of free formaldehyde in all types of cosmetics (European Commission, 2000). The identification is based on a colorimetric reaction with the Schiff reagent, pararosaniline. Although valid for identification purposes, false-negative identification of formaldehyde may occur as a result of discoloration of the reagent solution caused by other ingredients present in the formulation (i.e., 2-propanol) (Malinauskiene et al., 2015). The quantitative method is based on sample solubilization with 100 mL of water and derivatization with pentane-2,4-dione in the presence of ammonium acetate. The derivative 3,5-diacetyl-1,4-dihydrolutidine is submitted to LLE with 10 mL of butan-1-ol. Measurement is also carried out by UV/VIS spectrophotometry (European Commission, 2000). The same directive provides an alternative for high concentrations of formaldehyde in the presence of formaldehyde releasers based on LC. In this case, cosmetic emulsions are extracted with DCM and HCl with the assistance of ultrasound. The aqueous phase is passed through a C_{18} cartridge. For gels and shampoos, samples are solubilized in the mobile phase (0.006 M disodium phosphate, pH 2.1) and then subjected to SPE with a C_{18} cartridge. Samples are finally analysed by LC with post-column derivatization (European Commission, 2000).

Analytical Methods Published in Scientific Literature
Additionally, several analytical methods for formaldehyde detection in different cosmetics such as shampoos, gels, creams, toothpaste, face cleansers, tonics or nail polish can be found in the literature. An overview of selected references is included in Table 14.8. In general, the direct determination of formaldehyde without or with minimum sample pre-treatment is difficult not only because of the complexity of the matrices, but also owing to the high volatility and reactivity of these analytes. Thus, direct pre-concentration of an untreated hair gel sample by HS–SPME in an *o*-(2,3,4,5,6-pentafluorobenzyl) hydroxylamine hydrochloride-coated fibre followed by direct GC–FID has been proposed. Concentrations less than 40 ppb can be determined, but with high RSDs (<12%) (Martos and Pawliszyn, 1998). Direct quantitative carbon-13 nuclear magnetic resonance (^{13}C NMR) has been used to assess 0.002% (w/w) formaldehyde levels (Emeis

Table 14.8 Selected analytical methods for formaldehyde determination in cosmetics

Matrix	Remarks on sample preparation	Technique	Reference
Hair gel	3 g of sample is placed in HS vial and allowed to equilibrate (1 min); PFBHA-coated fibre HS–SPME for 10 min	GC–FID	Martos and Pawliszyn (1998)
Oral solution, shampoo, bath gel, powder, toothpaste	Pervaporation; use of an FI manifold; evaporated formaldehyde diffuses through membrane and is derivatized with pararosaniline solution	UV/VIS spectrophotometry	Gámiz-Gracia and Luque de Castro (1999b)
Shampoo	1 g of sample is diluted to 10 mL with 45/55 (%, v/v) of acetonitrile and 0.025 M NaH_2PO_4; 0.4 mL of 0.1% DNPH is added to 1 mL of diluted sample solution, vortexed (1 min) and allowed to stand (2 min); 0.4 mL of 0.1 M phosphate buffer (pH 6.8) and 0.7 mL of 1 M NaOH are added	LC–UV/VIS	Vander Heyden et al. (2002)
Shampoo	1.0 g of shampoo is diluted to 10 mL with mobile phase; 0.4 mL of DNPH 0.1% (w/v) is added to 1.0 mL sample; two short columns are used	LC–UV/VIS	Nguyet et al. (2003)
Nail polish, shower gel, mascara, body cream	5 mL of water is added to 3 g of sample and sonicated (15 min); sample is made up to 10 mL with water, mixed and filtered; 0.5 mL of solution is transferred to an HS vial containing 0.6 g of NaCl; 1 mL of 1.5 mM PFPH solution and 0.5 mL of isotope-labelled formaldehyde (4 µg/mL) are added and sonicated (10 min); HS–SPME at 35°C for 15 min	GC–ID–MS	Rivero and Topiwala (2004)
Cream	No sample preparation	^{13}C NMR	Emeis et al. (2007)
Conditioner, baby and antidandruff shampoos, gels	USAEME simultaneous derivatization; 0.5 mL of 0.2% (w/v) acetyl acetone, 15% (w/v) ammonium acetate and 0.3% (v/v) glacial acetic acid and 0.5 mL of 30% (w/v) NaCl are added to 10–15 mg of sample; 80 µL of DCM is added, sonicated (5 min) and centrifuged (3 min)	UV/VIS micro-spectrophotometry	Lavilla et al. (2010)

Table 14.8 Selected analytical methods for formaldehyde determination in cosmetics—cont'd

Matrix	Remarks on sample preparation	Technique	Reference
Nail polish	0.1 mg of nail polish is dissolved in 15 mL DCM and 30 mL of 2 mM HCl, sonicated (30 min) and centrifuged; aqueous phase is diluted to 100 mL with 0.1 M NaOH as supporting electrolyte; FI	ED (Ni–BPE)	Chen et al. (2011b)
Hand and baby creams, shower gel, toothpaste, nail polish	Samples are mixed with 3 mL water and sonicated (30 min); 276 µL of 7.25×10^{-3} g/mL TBA, 100 µL HCl and water to 4 mL total volume are added, shaken (40 min, 75°C), centrifuged (10 min) and filtered	CE–ED	Li et al. (2014)
Hair creams	0.3 mg of sample is solubilized into derivatizing solution (5 mg of DNPH in 50 mL of acetonitrile, acidified with diluted $HClO_4$); 1 mL of this solution is made up to 10 mL with acetonitrile	GC–MS	Lobo et al. (2015)
Shampoos, shower gels, liquid soaps, face and peeling cleansers, hair conditioners, hair masks, make-up removers, tonics, hair conditioner and creams	1 g of sample is solubilized with 9 mL of THF/water (9:1); 1 mL of this solution is mixed with 0.4 mL of 0.1% (w/v) DNPH, vortex stirred (60 s), and stabilized with 0.4 mL of 0.1 M phosphate buffer (pH 6.8), 0.7 mL of 1 M NaOH and 2.5 mL of acetonitrile	LC–UV/VIS	Malinauskiene et al. (2015)

^{13}C NMR, carbon-13 nuclear magnetic resonance; *CE–ED*, capillary electrophoresis with electrochemical detection; *DCM*, dichloromethane; *DNPH*, 2,4-dinitrophenyl hydrazine; *ED (Ni–BPE)*, electrochemical detection using Ni-activated barrel plating electrode; *FI*, flow injection; *GC–FID*, gas chromatography with flame ionization detector; *GC–ID–MS*, gas chromatography–isotope dilution–mass spectrometry; *GC–MS*, gas chromatography–mass spectrometry; *HS*, headspace; *HS–SPME*, headspace–solid-phase microextraction; *LC–UV/VIS*, high-performance liquid chromatography with ultraviolet–visible detection; *PFBHA*, o-(2,3,4,5,6-pentafluorobenzyl)hydroxylamine hydrochloride; *PFPH*, pentafluorophenylhydrazine; *TBA*, 2-thiobarbituric acid; *THF*, tetrahydrofuran; *USAEME*, ultrasound-assisted emulsification microextraction.

et al., 2007). Similarly, Chen and Yang (2011b) developed an analytical method without derivatization, in which formaldehyde is extracted from nail polish by LLE for further detection of the aqueous extract by FI with electrochemical detection using an activated barrel plating nickel electrode. Recoveries in the range of 97.5%–104.7% with an LOD of 0.23 µg/mL were obtained (Chen et al., 2011b).

Commonly, colorimetric methods for formaldehyde have relied on derivatization with 2,4-dinitrophenylhydrazine (DNPH) (Vander Heyden et al., 2002; Nguyet et al., 2003; Malinauskiene et al., 2015; Lobo et al., 2015), pararosaniline (Gámiz-Gracia and Luque de Castro, 1999b), acetyl acetone (Lavilla et al., 2010) or chromotropic acid (Malinauskiene et al., 2015) to form UV-active compounds suitable for UV/VIS spectrophotometry or LC with UV detection. In general, LODs are between 0.001 (Malinauskiene et al., 2015) and 20 µg/mL (Gámiz-Gracia and Luque de Castro, 1999b). Only Lobo et al. (2015) analysed DNPH-derivatized samples by GC–MS. Even with the selectivity provided by the MS detector, matrix effects from skin whitening creams were observed (88%–115% recovery). Standard addition was required to correct for matrix effects. Other methods based on derivatization with thiobarbituric acid to yield electroactive formaldehyde adducts suitable for CE with electrochemical amperometric detection (Li et al., 2014) or on the formation of stable volatile derivatives with pentafluorophenylhydrazine suitable for GC–FID or GC–MS (Rivero and Topiwala, 2004) can also be found in the literature. Samples are commonly subjected to solubilization with the mobile phase used for chromatographic separation before derivatization (Vander Heyden et al., 2002; Nguyet et al., 2003; Malinauskiene et al., 2015), treated with water under sonication for 15–30 min (Rivero and Topiwala, 2004; Li et al., 2014) or added directly into the derivatizing solution (Lavilla et al., 2010; Lobo et al., 2015).

The low concentration of formaldehyde released from donor agents into a formulation also requires the application of highly sensitive methods. As a consequence, some of the aforementioned methodologies use analyte derivatization and pre-concentration steps based on LLE, liquid-phase microextraction (LPME) or SPME (Rivero and Topiwala, 2004; Lavilla et al., 2010). Additionally, the application of in situ pre-concentration steps allows an extra clean-up of the samples, hence avoiding potential interference from other ingredients of the cosmetic formulation and providing good LODs between 9 (Lavilla et al., 2010) and 40 µg/L (Rivero and Topiwala, 2004).

Acrylamide

Acrylamide is the basic monomer used for producing its water-soluble polymer, polyacrylamide. Its use in cosmetics is restricted to the polymeric form, with functions such as anti-static, binding or film-forming agent (CosIng Database, 2015). Polyacrylamides can be produced as a solid (microbeads or powder), an aqueous solution or an emulsion and are often used in leave-on cosmetics such as moisturizers, lotions, creams, hair and nail care products, and sunscreens (Anderson, 2005).

A number of other polymeric forms based on acrylamide are allowed to be used in cosmetics, such as ammonium acrylate copolymer, acrylamide/ethyltrimonium chloride, acrylate/ethalkonium chloride, acrylate copolymer, potassium acrylates/acrylamide copolymer, styrene/acrylamide copolymer or several polyacrylates and polyquaterniums, among others (CosIng Database, 2015). Therefore, acrylamide is likely to be present in cosmetics as a residue from polyacrylamides. In fact, acrylamide monomer residue has been found in polyacrylamide-based formulations in concentrations between 1 and 600 ppm, the highest concentrations being found in solid forms of the polymer (Anderson, 2005).

The possible relationship between acrylamide residues and risk of cancer has been a concern to the SCCS (SCCS, 1999b). The committee stated in 1999 the relationship between the long-term use of polyacrylamide-containing cosmetics and a high risk of cancer due to the presence of acrylamide residues in the formulation. Therefore, it is recommended that the concentration of acrylamide residues in cosmetics be less than 0.1 ppm in body care leave-on products and 0.5 ppm in other cosmetic products.

As a consequence, the EU directive bans the use of acrylamide as an ingredient in cosmetics and restricts the use of the polymeric form to maximum levels of acrylamide residue of 0.1 mg/kg for leave-on cosmetics and 0.5 mg/kg for the rest (European Regulation, 2009). A similar approach is used in the Canadian legislation, which bans the presence of the acrylamide, although without specifying restrictions on maximum residual levels (Health Canada, 2014b).

In 2013, the CIR Expert Panel (CIR, 2013) reviewed the safety of using polyquartenium-22 and polyquartenium-39 as cosmetic ingredients. They indicated that the maximum monomer content for these acrylamide-based polymers should be limited to 5 ppm in cosmetics. Despite the FDA acknowledging the human concern related to the presence of acrylamide in foods (FDA, 2015e), no restriction has been included so far regarding cosmetics. Chinese authorities set a maximum level for some polymers based on acrylamide such as methacrylamide/vinyl imidazole copolymer, acrylamide/ammonium acrylate copolymer, acrylamide copolymer, acrylate/diacetone acrylamide copolymer, acrylamide copolymer, among others, with maximum allowed concentrations ranging from 0.15% to 98.5% (CRS, 2014).

International Official Methods

Whereas no EU method of analysis exists for this analyte, the EPA has proposed an analytical method based on GC–ECD (EPA Method 8032A) with potential application to cosmetic analysis (EPA, 1996). Although initially aimed at the analysis of aqueous matrices such as waters, the agency states that it may be applicable to other matrices as well. Following this method, samples are derivatized for at least 1 h with potassium bromide, salted out and further extracted and made up to volume with ethyl acetate. Clean-up with a Florisil column is recommended when interference is observed. LODs of 0.032 µg/L in aqueous matrices are reported.

Analytical Methods Published in the Scientific Literature

Despite the lack of official methods, the scientific literature has hardly dealt with the detection of acrylamide in cosmetics. In all cases, given the water solubility of acrylamide, extraction from the cosmetic has been based on transferring the analyte to an aqueous phase suitable for analysis, e.g., using LC coupled to MS/MS (Ma et al., 2009; Jin et al., 2010) or UV (Wang et al., 2013).

Oily cosmetics are subjected to LLE with hexane/water mixtures, the aqueous extract being analysed by LC–MS/MS (Ma et al., 2009). A different strategy is followed for water-based cosmetics, which are first extracted in water and then the aqueous extract is subjected to SPE with strongly hydrophilic sorbents. In this way, possible matrix interference from lipophilic ingredients of the cosmetic is reduced (Ma et al., 2009; Wang et al., 2013). Another approach was used by Jin et al. (2010), for the extraction of acrylamide from water-based cosmetics. After extraction with a 0.1% (v/v) formic acid solution, cosmetics were defatted with petroleum ether before analysis by LC–MS/MS. However, even with the sample clean-up step, recoveries were in the range of 80.5%–102.5%, which may reveal some matrix effects. An LOD of 0.01 mg/kg for acrylamide was reported (Jin et al., 2010).

FUTURE PROSPECTS

As previously mentioned, the presence of contaminants in cosmetics is an unmanageable issue because of the great diversity and ubiquity of these compounds. Although it has been known for many years, it has not been treated at a global level until a few years ago. Contaminants in cosmetics are being addressed through the ICCR, though the concrete actions proposed so far are scarce. In addition to operating under GMP, the control of a pool of particular contaminants should be mandatory in all cosmetics products. These contaminants should be selected in accordance with their potential toxicity and frequency of occurrence in finished products, although this is far from easy. Legal limits should be clearly established for each contaminant selected. In this regard, a greater international unification would be desirable, given the global nature of the commerce in cosmetics and raw materials. In addition, it should be noted that many of these trace substances are impurities from the raw materials used, so the authorities should take measures aimed at establishing a standard of quality for these materials.

In any case, the degree of purity of raw materials and the identification/concentration of significant impurities/contaminants in cosmetics must be routinely controlled. With this purpose, simple, economic and rapid standard analytical methods must be implemented. However, the complexity of cosmetic matrices does not make it easy. Many of the analytical methods used today are slow and tedious and use expensive analytical tools. Advances in analytical chemistry, in particular in sample preparation, must be implemented to improve this situation. General trends such as automation, miniaturization and simplification have to be considered. Although some of these trends can be found in some published methods, they still have to be introduced into routine laboratory use.

Particular attention should be paid to the direct analysis of samples and the use of alternative sample treatments. Methods for direct analysis of metals without sample preparation by LIBS and ETAAS have been cited in this chapter. However, direct analysis has not been used for organic contaminants. No doubt this can be very difficult, since without removal of the matrix, interference is likely to occur. However, direct analysis could be advantageous to establish screening methods. Some analytical techniques such as NMR, infrared spectroscopy or Raman spectroscopy together with chemometrics would enable the direct analysis of solid and liquid cosmetic samples. Alternatives that entail a minimal sample preparation without total destruction of the matrix are also interesting for cosmetic analysis, e.g., emulsification.

In addition, multistep separation procedures frequently carried out in cosmetic analysis and based on analyte extraction and/or distillation can be refocused to improve these procedures. Thus, conventional procedures such as SPE and LLE could be replaced by miniaturized approaches, i.e., SPME and LPME, respectively. Only in certain cases have these new approaches been implemented. HS–SPME and DLLME have contributed to improvements in the cosmetic clean-up and analyte pre-concentration. So far, these developments have been few and other microextraction modes could be applied. In view of the this, the development of new analytical methods is recommended for determining contaminants in cosmetics. In particular, sample treatment must be improved to enhance rapidity and simplicity and to achieve better analytical characteristics. Appropriate method validation should also be considered so that procedures may be adopted by official organisms. Cosmetic CRMs should be provided for routine laboratory use.

REFERENCES

ALqadami, A.A., Abdalla, M.A., ALOthman, Z.A., Omer, K., 2013. Int. J. Environ. Res. Public Health 10, 361.
Al-Saleh, I., Al-Enazi, S., Shinwari, N., 2009. Regul. Toxicol. Pharmacol. 54, 105.
Anderson, F.A., 2005. Int. J. Toxicol. 24, 21.
Beernaert, H., Herpol-Borremans, M., De Cock, F., 1987. Belg. J. Food Chem. Biotechnol. 42, 131.
Besecker, K.D., Rhoades Jr., C.B., Jones, B.T., Barnes, K.W., 1998. At. Spectrosc. 19, 48.
Billedeau, S.M., Heinze, T.M., Wilkes, J.G., Thompson Jr., H.C., 1994. J. Chromatogr. A 688, 55.
Black, R.E., Hurley, F.J., Havery, D.C., 2001. J. AOAC Int. 84, 666.
Bocca, B., Forte, G., Petrucci, F., Cristaudo, A., 2007. J. Pharm. Biomed. Anal. 44, 1197.
Bocca, B., Pino, A., Alimonti, A., Giovanni, F., 2014. Regul. Toxicol. Pharmacol. 68, 447.
Bronaugh, R.L., 1982. Percutaneous absorption of cosmetic ingredients. In: Frost, P., Horwitz, S. (Eds.), Principles of Cosmetics for the Dermatologist. The C.V. Mosby Company, St. Louis.
Calaça, G.N., Machado, S., Wohnrath, K., Pessoa, C.A., Nagata, N., 2015. J. Electrochem. Soc. 162, H847.
Cao, E. (Ed.), 2014. China Cosmetics Legislation. https://chemlinked.com/chempedia/china-cosmetics-regulation#part3-cosmetic-ingredient.

Capelli, C., Foppiano, D., Venturelli, G., Carlini, E., Magi, E., Ianni, C., 2014. Anal. Lett. 47, 1201.
Challis, B.C., Colling, J., Cromie, D.D.O., Guthrie, W.G., Pollock, J.R.A., Taylor, P., 1995. Int. J. Cosmet. Sci. 17, 219.
Chen, M.-L., Ma, H.-J., Zhang, S.-Q., Wang, J.-H., 2011a. J. Anal. Spectrom. 26, 613.
Chen, P.-Y., Yang, H.-H., Zen, J.-M., Shih, Y., 2011b. J. AOAC Int. 94, 1585.
Chen, W.-N., Jiang, S.-J., Chen, Y.-L., Sahayam, A.C., 2015a. Anal. Chim. Acta 860, 8.
Chen, R.-X., Wang, L., Wang, J., Xu, F.-Q., 2015b. Anal. Lett. 48, 2159.
Chisvert, A., Sisternes, J., Balaguer, A., Salvador, A., 2010. Talanta 81, 530.
Choi, N.R., Kim, Y.P., Ji, W.H., Hwang, G.-S., Ahn, Y.G., 2016. Talanta 148, 69.
Chou, H.J., Yates, R.L., Wenninger, J.A., 1987. J. AOAC Int. 70, 960.
CIR-Cosmetic Ingredient Review, Expert Panel Meeting, 2011. Formaldehyde/Methylene Glycol. http://www.cir-safety.org/sites/default/files/119_final_formyl.pdf.
CIR-Cosmetic Ingredient Review, Expert Panel Meeting, 2013. Safety Assessment of Polyquaternium-22 and Polyquaternium-39 as Used in Cosmetics. http://www.cir-safety.org/sites/default/files/polyqu-092013final.pdf.
COLIPA, Cosmetics Europe, 2014. Cosmetics Europe Activity Report. https://www.cosmeticseurope.eu/publications-cosmetics-europe-association/annual-reports.html?view=item&id=101.
Collier, S.W., Milstein, S.R., Orth, D.S., Jayasimhulu, K., 1988. J. Soc. Cosmet. Chem. 39, 329.
Commission Directive 2003/83/EC, Commission Directive 2003/83/EC of 24 September 2003 Adapting to Technical Progress Annexes II, III and VI to Council Directive 76/768/EEC on the Approximation of the Laws of the Member States Relating to Cosmetic Products. http://eur-lex.europa.eu/legal-content/EN/TXT/?uri=celex:32003L0083R(01).
Commission Implementing Decision, 2013. European Commission Implementing Decision 25 November 2013, on Guidelines on Annex I to Regulation (EC) No 1223/2009 of the European Parliament and of the Council on Cosmetic Products, L 315/87. www.ust.is/library/Skrar/Atvinnulif/Efni/Snyrtivorur/Guideline%20on%20Safety%20report.pdf.
CosIng Database, 2015. European Commission Database for Information on Cosmetic Substances and Ingredients. http://ec.europa.eu/growth/tools-databases/cosing/.
Cosmetic Regulations Canada (C.R.C. c. 869), 2007. http://laws-lois.justice.gc.ca/eng/regulations/C.R.C.,_c._869/FullText.html.
CRS-Chemical Inspection and Regulation Service China, 2013. Cosmetics and Cosmetic Ingredient Law. http://www.cirs-reach.com/China_Chemical_Regulation/China_Cosmetics_Cosmetic_Ingredient_Law.html.
CRS-Chemical Inspection and Regulation Service China, 2014. Inventory of Existing Cosmetic Ingredients in China – IECIC (2014). http://www.cirs-reach.com/Cosmetic_Inventory/China_IECIC_Inventory_of_Existing_Cosmetic_Ingredients_in_China.html.
CRS-Chemical Inspection and Regulation Service China, 2015. Technical Safety Standard for Cosmetics (Draft) Issued for Public Comments for the Third Time. http://www.cirs-reach.com/news-and-articles/technical-safety-standard-for-cosmetics-%28draft%29-issued-for-public-comments-for-the-third-time.html.
Cruz-Vieira, I., Fatibello-Filho, O., 2000. Talanta 52, 681.
Davarani, S.S.H., Masoomi, L., Banitaba, M.H., Lotfi Zadeh Zhad, H.R., Sadeghi, O., Samiei, A., 2012. Chromatographia 75, 371.
Davarani, S.S.H., Masoomi, L., Banitaba, M.H., Zhad, H.R.L.Z., Sadeghi, O., Samiei, A., 2013. Talanta 105, 347.
De-Quiang, Z., Yang, L.-L., Sun, H.-W., 1999. Anal. Chim. Acta 395, 173.
Desiderio, C., Ossicini, L., Fanali, S., 2000. J. Chromatogr. A 887, 489.
DOE Industry Profiles – Department of the Environment Industry Profiles, 1995. Chemical Works. Cosmetics and Toiletries Manufacturing Works. Building Research Establishment, Ruislip, UK. https://www.gov.uk/government/uploads/system/uploads/attachment_data/file/290803/scho0195bjjz-e-e.pdf.
Dong, H., Guo, X., Xian, Y., Luo, H., Wang, B., Wu, Y., 2015. J. Chromatogr. A 1422, 82.
DTI-Department of Trade and Industry UK, 1998. In: Laboratory of the Government Chemist (Ed.), A Survey of Cosmetic and Certain Other Skin-Contact Products for N-nitrosamines. Consumer Safety Group, Lancashire.

Ebrahimzadeh, H., Moazzen, E., Amini, M.M., Sadeghi, O., 2013. Int. J. Cosmet. Sci. 35, 176.
EGW's, 2007. Skin Deep Cosmetics Database – Environmental Working Group's. http://www.ewg.org/skindeep/2007/02/04/impurities-of-concern-in-personal-care-products/.
Emeis, D., Anker, W., Wittern, K.-P., 2007. Anal. Chem. 79, 2096.
Environmental Defence, 2011. Heavy Metal Hazards. The Health Risks of Hidden Heavy Metals in Face. http://environmentaldefence.ca/reports/heavy-metal-hazard-health-risks-hidden-heavy-metals-in-face-makeup.
EPA, 1996. Method 5031, Volatile, Nonpurgeable, Water-soluble Compounds by Azeotropic Distillation. https://www.epa.gov/sites/production/files/2015-12/documents/5031.pdf.
EU Regulation, 2009. Regulation (EC) No 1223/2009 of the European Parliament and of the Council of 30 November 2009 on Cosmetic Products. http://eur-lex.europa.eu/LexUriServ/LexUriServ.do?uri=OJ: L:2009:342:0059:0209:en:PDF.
European Commission, 2000. The rules governing cosmetic products in the european union cosmetics legislation – cosmetic products. In: Methods of Analysis, vol. 2, pp. 1–187Luxembourg http://bookshop.europa.eu/en/methods-of-analysis-pbNB2699966/.
European Commission, 2006. 2006/257/EC: Commission Decision of 9 February 2006 Amending Decision 96/335/EC Establishing an Inventory and a Common Nomenclature of Ingredients Employed in Cosmetic Products.
European Commission, 2015. Scientific and Technical Assessment. http://ec.europa.eu/growth/sectors/cosmetics/assessment/index_en.htm.
Fan, T.Y., Goff, U., Song, L., Fine, D.H., Arsenault, G.P., Biemann, K., 1977. Food Cosmet. Toxicol. 15, 423.
FD&C Act-Federal Food, Drug and Cosmetic Act, 1960. Subchapter VI-Cosmetics (Sec. 361-Adulterated Cosmetics- Sec. 362-Misbranded Cosmetics). http://www.fda.gov/RegulatoryInformation/Legislation/FederalFoodDrugandCosmeticActFDCAct/FDCActChapterVICosmetics/ucm2016708.htm.
FD&C Act-Federal Food, Drug and Cosmetic Act, 2015b. Title 21: Food and Drugs, Chapter I, Subchapter A, Part 73, Subpart C-Cosmetics. http://www.accessdata.fda.gov/scripts/cdrh/cfdocs/cfCFR/CFRSearch.cfm?CFRPart=73&showFR=1.
FD&C Act-Federal Food, Drug and Cosmetics Act, 2015a. Title 21: Food and Drugs, Chapter I, Subchapter A, Part 2, Subpart A-General Provisions-2.19. Methods of Analysis. http://www.accessdata.fda.gov/scripts/cdrh/cfdocs/cfcfr/CFRSearch.cfm?CFRPart=2.
FDA Food and Drug Administration, 1979. Nitrosamine-contaminated cosmetics; call for industry action; request for data. Fed. Regist. 44 (70), 21365–21367.
FDA Food and Drug Administration, 1981. Progress Report on the Analysis of Cosmetics Raw Materials and Finished Cosmetics Products for 1 4 Dioxane.
FDA Food and Drug Administration, 1998. Guide to Inspections of Cosmetic Product Manufacturers. http://www.gmp-compliance.org/guidemgr/files/1-2-22.PDF.
FDA Food and Drug Administration, 2003. Orientation and Training, Volume IV – 1.3 Analytical Methods-Section 1-Laboratory Orientation. http://www.fda.gov/ScienceResearch/FieldScience/LaboratoryManual/ucm171912.ht.
FDA Food and Drug Administration, 2013. Draft Guidance for Industry: Cosmetic Good Manufacturing Practices. http://www.fda.gov/downloads/Cosmetics/GuidanceComplianceRegulatoryInformation/GuidanceDocuments/UCM358287.pdf.
FDA Food and Drug Administration, 2014. 1,4-Dioxane, a Manufacturing Byproduct. http://www.fda.gov/Cosmetics/ProductsIngredients/PotentialContaminants/ucm101566.htm.
FDA Food and Drug Administration, 2015a. Import Alert for Industry Cosmetics. http://www.accessdata.fda.gov/cms_ia/industry_53.html.
FDA Food and Drug Administration, 2015b. Lipstick & Lead: Questions & Answers. http://www.fda.gov/Cosmetics/ProductsIngredients/Products/ucm137224.htm.
FDA Food and Drug Administration, 2015c. Cosmetic Product Testing, Cosmetics-Science and Research. http://www.fda.gov/cosmetics/scienceresearch/producttesting/.
FDA Food and Drug Administration, 2015d. How FDA Evaluates Regulated Products: Cosmetics. http://www.fda.gov/AboutFDA/Transparency/Basics/ucm262353.htm.
FDA Food and Drug Administration, 2015e. Acrylamide Questions and Answers. http://www.fda.gov/Food/FoodborneIllnessContaminants/ChemicalContaminants/ucm053569.htm.

Fellous, R., Zhongrui, L., 2013. Speciality Chemicals Magazine-Cosmetic Legislation in China. http://www.specchemonline.com/articles/view/cosmetic-legislation-in-china#.VmMaSV76O-U.
Ferreira Batista, E., dos Santos Augusto, A., Rodrigues Pereira-Filho, E., 2015. Anal. Methods 7, 329.
Firth, J., Rix, I., 1986. Analyst 111, 129.
Flower, C., Carter, S., Earls, A., Fowler, R., Hewlins, S., Lalljie, S., 2006. Int. J. Cosmet. Sci. 28, 21.
Food and Drugs Act, 2014. R.S.C., 1985, C. F-27, Canada. http://laws-lois.justice.gc.ca/eng/acts/f-27/page-5.html#h-7.
FP&L Act, 1967. Fair Packaging and Labeling Act, 16 C.F.R. Parts 500, 501, 502, 503. https://www.ftc.gov/enforcement/rules/rulemaking-regulatory-reform-proceedings/fair-packaging-labeling-act.
Fuh, C.B., Lai, M., Tsai, H.Y., Chang, C.M., 2005. J. Chromatogr. A 1071, 141.
Gámiz-Gracia, L., Luque de Castro, M.D., 1999a. J. Anal. Spectrom. 14, 1615.
Gámiz Gracia, L., Luque de Castro, M.D., 1999b. Analyst 124, 1119.
Gangolli, S.D., van den Brandt, P.A., Feron, V.J., Janzowsky, C., Koeman, J.H., Speijers, G.J., 1994. Eur. J. Pharmacol. 292, 1.
García, C.D., Ortiz, P.I., 2000. Electroanalysis 12, 1074.
Guan, Y., Chu, Q., Fu, L., Ye, J., 2005. J. Chromatogr. A 1074, 201.
Haider, A.F.M.Y., Lubna, R.S., Abedin, K.M., 2012. Appl. Spectrosc. 66, 420.
Health Canada, 2012. Guidance on Heavy Metal Impurities in Cosmetics. http://www.hc-sc.gc.ca/cps-spc/pubs/indust/heavy_metals-metaux_lourds/index-eng.php.
Health Canada, 2014a. Guide to Completing Cosmetic Notification Forms. http://www.hc-sc.gc.ca/cps-spc/cosmet-person/notification-declaration/guide-eng.php.
Health Canada, 2014b. Cosmetic Ingredient Hotlist, Lists of Ingredients that are Prohibited and Restricted for Use in Cosmetic Products. http://www.hc-sc.gc.ca/cps-spc/cosmet-person/hot-list-critique/hot-list-liste-eng.php.
Health Canada, 2015. Natural and Non-prescription Health Products Directorate, Quality of Natural Health Products Guide. http://www.hc-sc.gc.ca/dhp-mps/prodnatur/legislation/docs/eq-paq-eng.php.
Hepp, N.M., Mindak, W.R., Cheng, J., 2009. J. Cosmet. Sci. 60, 405.
Ho, J.L., Wisneski, H.H., Yates, R.L., 1981. J. AOAC Int. 64, 800.
Huang, S.C., Lin, C.C., Huang, M.C., Wen, K.C., 2004. J. Food Drug Anal. 12, 13.
IARC, 1999. IARC Monograph on the Evaluation of Carcinogenic Risks to Humans, Re-evaluation of Some Organic Chemicals, Hydrazine and Hydrogen Peroxide, vol. 71, p. 691 (Part 2) Lyon http://monographs.iarc.fr/ENG/Monographs/vol71/mono71.pdf.
ICCR-International Cooperation on Cosmetic Regulation, Working Group on Traces, 2011. Report for International Cooperation on Cosmetic Regulation (ICCR). Principles for the Handling of Traces of Impurities And/or Contaminants in Cosmetic Products. http://ec.europa.eu/consumers/sectors/cosmetics/files/pdf/iccr5_contaminants_en.pdf.
ICCR-International Cooperation on Cosmetics Regulation, 2015. http://www.iccrnet.org/.
INCI-International Nomenclature of Cosmetic Ingredients, 2015. http://www.cirs-reach.com/Cosmetic_Inventory/International_Nomenclature_of_Cosmetic_Ingredients_INCI.html.
ISO 22716:2007-International Organization for Standardization 22716:2007, 2007. Cosmetics – Good Manufacturing Practices (GMP) – Guidelines on Good Manufacturing Practices. http://www.iso.org/iso/home/store/catalogue_tc/catalogue_detail.htm?csnumber=36437.
ISO 10130:2009-International Organization for Standardization 10130:2009, 2009. Cosmetics – Analytical Methods – Nitrosamines: Detection and Determination of N-nitrosodiethanolamine (NDELA) in Cosmetics by LC, Post-column Photolysis and Derivatization. http://www.iso.org/iso/catalogue_detail.htm?csnumber=45840.
ISO/TR 14735:2013-International Organization for Standardization/Technical Report 14735:2013, 2013. Cosmetics – Analytical Methods – Nitrosamines: Technical Guidance Document for Minimizing and Determining N-nitrosamines in Cosmetics. http://www.iso.org/iso/iso_catalogue/catalogue_tc/catalogue_detail.htm?csnumber=54399.
ISO/TR 17276:2014-International Organization for Standardization/Technical Report17276:2014, 2014a. Cosmetics – Analytical Approach for Screening and Quantification Methods for Heavy Metals in Cosmetics. http://www.iso.org/iso/catalogue_detail.htm?csnumber=59500.

ISO 15819:2014-International Organization for Standardization 15819:2014, 2014b. Cosmetics – Analytical Methods – Nitrosamines: Detection and Determination of N-nitrosodiethanolamine (NDELA) in Cosmetics by HPLC-MS-MS. http://www.iso.org/iso/catalogue_detail.htm?csnumber=62042.

Italia, M.P., Nunes, M.A., 1991. J. Soc. Cosmet. Chem. 42, 97.

Jeon, J.S., Kim, B.H., Lee, S.H., Kwon, H.J., Bae, H.J., Kim, S.K., 2015. Int. J. Cosmet. Sci. 37, 567.

Jia, X., Han, Y., Wei, C., Duan, T., Chen, H., 2011. J. Anal. Spectrom. 26, 1380.

Jin, W., Li, W., Jiang, W., Peng, X., Le, J., 2010. Chin. J. Pharm. Anal. 30, 1887.

Joo, K.-M., Shin, M.-S., Jung, J.-H., Kim, B.-M., Lee, J.-W., Jeong, H.-J., Lim, K.-M., 2015. Talanta 137, 109.

Khammas, Z.A.A., Farhan, M.H., Barbooti, M.M., 1989. Talanta 36, 1027.

Lavilla, I., Cabaleiro, N., Costas, M., de la Calle, I., Bendicho, C., 2009. Talanta 80, 109.

Lavilla, I., Cabaleiro, N., Pena-Pereira, F., De La Calle, I., Bendicho, C., 2010. Anal. Chim. Acta 674, 59.

Lemaire, R., del Bianco, D., Garnier, L., Beltramo, J.L., 2013. Anal. Lett. 46, 2265.

Li, Y., Chen, F., Ge, J., Tong, F., Deng, Z., Shen, F., 2014. Electrophoresis 35, 419.

Lin, C.H., Sheu, J.-Y., Wu, H.-L., Huang, Y.-L., 2005. J. Pharm. Biomed. Anal. 38, 414.

Lin, Y.-H., Yang, Y.-H., Wu, S.-M., 2007. J. Pharm. Biomed. Anal. 44, 279.

Lobo, F.A., Santos, T.M.O., Vieira, K.M., Osório, V.M., Taylor, J.G., 2015. Drug Test. Anal. 7, 848.

López-García, P., Miritello-Santoro, M.I.R., Singh, A.K., Kedor-Hackmann, E.R.M., 2007. Braz. J. Pharm. Sci. 43, 397.

Ma, Q., Wang, C., Bai, H., Wang, X., Zhang, Q., Xiao, H., 2009. Se Pu. 27, 856.

Malinauskiene, L., Blaziene, A., Chomiciene, A., Isaksson, M., 2015. Open Med. 10, 323.

Martin, K.A., 1 November, 2012. Cosmet. Toilet. http://www.cosmeticsandtoiletries.com/testing/invitro/premium-how-did-that-get-in-there-identifying-particulate-contamination-in-products-and-packaging-213004211.html.

Martos, P.A., Pawliszyn, J., 1998. Anal. Chem. 70, 2311.

Matyska, M.T., Pesek, J.J., Yang, L., 2000. J. Chromatogr. A 887, 497.

Meyer, T.A., Powell, J.B., 1991. J. AOAC Int. 74, 766.

MHLW, Ministry of Health, Labour and Welfare Law, 1999.

MHW-Japan Ministry of Health and Welfare, 1999. Standards for Cosmetics, Notification No. 331 of 2000. http://www.mhlw.go.jp/file/06-Seisakujouhou-11120000-Iyakushokuhinkyoku/0000032704.pdf.

Mobin, S.M., Sanghavi, B.J., Srivastava, A.K., Mathur, P., Lahiri, P.M.G.K., 2010. Anal. Chem. 82, 5983.

Ng, S.Y., Dewi, F., Wang, J., Sim, L.P., Shin, R.Y.C., Lee, T.K., 2015. Int. J. Mass Spectrom. 389, 59.

Nguyet, N.M., Tallieu, L., Plaizier-Vercammen, J., Massart, D.L., Heyden, Y.V., 2003. J. Pharm. Biomed. Anal. 32, 1.

Nnorom, I.C., Igwe, J.C., Oji-Nnorom, C.G., 2005. Afr. J. Biotechnol. 4, 1133.

Okamoto, M., Kanda, M., Matsumoto, I., Miya, Y., 1971. J. Soc. Cosmet. Chem. 22, 589.

Oliveira, I.R.W.Z., de Barros-Osório, R.E.M., Neves, A., Cruz-Vieira, I., 2007. Sens. Actuat. B-Chem. 122, 89.

PAL-Pharmaceutical Affairs Law, 1960. Law No. 145, Japan. www.jpma.or.jp/english/parj/pdf/2012_ch02.pdf.

Qassim, B.B., Omaish, H.S., 2014. J. Chem. Pharm. Res. 6, 1548.

Rastogi, S.C., 1990. Chromatographia 29, 441.

Rivero, R.T., Topiwala, V., 2004. J. Chromatogr. A 1029, 217.

Robinson, J.J., Ciurczak, E.W., 1980. J. Soc. Cosmet. Chem. 31, 329.

Rollmann, B., Lombart, P., Rondelet, J., Mercier, M., 1981. J. Chromatogr. A 206, 158.

Rosenberg, I.E., Gross, J., Spears, T., Rahn, P., 1980. J. Soc. Cosmet. Chem. 31, 237.

Rueda, M.E., Sarabia, L.A., Herrero, A., Ortiz, M.C., 2003. Anal. Chim. Acta 479, 173.

Sainio, E.L., Jolanki, R., Hakala, E., Kanerva, L., 2000. Contact Dermat. 42, 5.

Sakodinskaya, I.K., Desiderio, C., Nardi, A., Fanali, S., 1992. J. Chromatogr. A 596, 95.

Salvador, A., Pascual-Martí, M.C., Aragó, E., Chisvert, A., March, J.G., 2000a. Talanta 51, 1171.

Salvador, A., Pascual-Martí, M.C., Adell, J.R., Requeni, A., March, J.G., 2000b. J. Pharm. Biomed. Anal. 22, 301.

Scalia, S., Guarneri, M., Menegatti, E., 1990. Analyst 115, 929.

Scalia, S., Testoni, F., Frisina, G., Guarneri, M., 1992. J. Soc. Cosmet. Chem. 43, 207.

Scalia, S., 1990. J. Pharm. Biomed. Anal. 8, 867.

SCCS-Scientific Committee on Consumer Safety, 1999b. Opinion of the Scientific Committee on Cosmetic Products and Non-food Products Intended for Consumers Concerning Acrylamide Residues in Cosmetics Adopted by the Plenary Session of the SCCNFP of 30 September 1999. http://ec.europa.eu/health/scientific_committees/consumer_safety/opinions/sccnfp_opinions_97_04/sccp_out95_en.htm.

SCCS-Scientific Committee on Consumer Safety, 2002. Opinion on the Determination of Certain Formaldehyde Releasers in Cosmetic Products, SCCNFP/586/02. http://ec.europa.eu/food/fs/sc/sccp/out188_en.pdf.

SCCS-Scientific Committee on Consumer Safety, 2012. Opinion on Nitrosamines and Secondary Amines in Cosmetic Products. http://ec.europa.eu/health/scientific_committees/consumer_safety/docs/sccs_o_090.pdf.

SCCS-Scientific Committee on Consumer Safety, 2014. Opinion on the Safety of the Use of Formaldehyde in Nail Hardeners, SCCS/1538/14. http://ec.europa.eu/health/scientific_committees/consumer_safety/docs/sccs_o_164.pdf.

SCCS-Scientific Committee on Consumer Safety, 2015a. Opinion on Deoxyarbutin Tetrahydropyranyloxy Phenol, SCCS/1554/15. http://ec.europa.eu/health/scientific_committees/consumer_safety/docs/sccs_o_183.pdf.

SCCS-Scientific Committee on Consumer Safety, 2015b. Opinion on A-arbutin, SCCS/1552/15. http://ec.europa.eu/health/scientific_committees/consumer_safety/docs/sccs_o_176.pdf.

SCCS-Scientific Committee on Consumer Safety, 2015c. Scientific Opinion on the Report of the ICCR Working Group: Considerations on Acceptable Trace Level of 1,4-Dioxane in Cosmetic Products. The Report of the ICCR Working Group: Considerations on Acceptable Trace Level of 1,4-Dioxane in Cosmetic Products.

Schothorst, R.C., Somers, H.H.J., 2005. Anal. Bioanal. Chem. 381, 681.

Schothorst, R.C., Stephany, R.W., 2001. Int. J. Cosmet. Sci. 23, 109.

Schwarzenbach, R., Schmid, J.P., 1989. J. Chromatogr. 472, 231.

Scientific and Techncial Assessment. European Commission, 2016. http://ec.europa.eu/growth/sectors/cosmetics/assessment/index_en.htm.

Seidel, A. (Ed.), 2013. Kirk-Othmer Chemical Technology of Cosmetics. John Wiley & Sons, Inc., Hoboken, New Jersey.

Sha, B.-B., Yin, X.-B., Zhang, X.-H., He, X.-W., Yang, W.-L., 2007. J. Chromatogr. A 1167, 109.

Sharafi, K., Fattahi, N., Pirsaheb, M., Yarmohamadi, H., Fazlzadeh Davil, M., 2015. Int. J. Cosmet. Sci. 37, 489.

Silva, F.C., Faria, C.G., Gabriel, G.M., Cardeal, Z.L., 2001. Quim. Nova 24, 748.

Sommer, H., Eisenbrand, G., 1988a. Z. Lebensm. Unters. Forsch. 186, 235.

Sommer, H., Loeffler, H.-P., Eisenbrand, G., 1988b. J. Soc. Cosmet. Chem. 39, 133.

Song, D., Zhang, S., Zhang, W., Kohlhof, K., 1996. J. Soc. Cosmet. Chem. 47, 177.

Stafford, M.L., Guin, K.F., Johnson, G.A., Sanders, L.A., Rockey, S.L., 1980. J. Soc. Cosmet. Chem. 31, 281.

Tahara, M., Obama, T., Ikarashi, Y., 2013. Int. J. Cosmet. Sci. 35, 575.

Thogchai, W., Liawruangrath, B., 2013. Int. J. Cosmet. Sci. 35, 257.

Thomas, C., Siong, D., Pirnay, S., 2014. Int. J. Cosmet. Sci. 36, 327.

Uddin, S., Rauf, A., Kazi, T.G., Afridi, H.I., Lutfullah, G., 2011. Int. J. Cosmet. Sci. 33, 132.

UNE 84055:1987. Materias primas cosméticas. Determinación de arsénico en compuestos orgánicos.

UNE 84056:2011. Materias primas cosméticas. Determinación de trazas de plomo. Parte 1-A. Preparación de muestra. Plomo soluble en medio ácido presente en talcos, pigmentos y otros materiales insolubles en ácido.

UNE 84057:2011. Materias primas cosméticas. Determinación de trazas de plomo. Parte 1-B. Preparación de muestra. Plomo total en materiales inorgánicos solubles en ácido.

UNE 84058:2011. Materias primas cosméticas. Determinación de trazas de plomo. Parte 1-C. Preparación de muestra. Plomo total en materias primas orgánicas.

UNE 84059:2011. Materias primas cosméticas. Determinación de trazas de plomo. Parte 1-D. Preparación de muestra. Plomo total en líquidos orgánicos.

UNE 84061:2011. Materias primas cosméticas. Determinación de trazas de plomo. Parte 2: Desarrollo de color y medida.

UNE 84742:2010. Materias primas cosméticas. Filtros solares. Determinación de trazas de mercurio en dióxido de titanio por absorción atómica.

US EPA-U.S. Environmental Protection Agency, 1996. Method 8032A: Acrylamide by Gas Chromatography. http://www3.epa.gov/epawaste/hazard/testmethods/sw846/pdfs/8032a.pdf.
Vander Heyden, Y., Nguyen Minh Nguyet, A., Detaevernier, M.R., Massart, D.L., Plaizier-Vercammen, J., 2002. J. Chromatogr. A 958, 191.
Volmer, D.A., Lay, J.O., Billedeau, M., Vollmer, D.L., 1996. Rapid Commun. Mass Spectrom. 10, 715.
Volpe, M.G., Nazzaro, M., Coppola, R., Rapuano, F., Aquino, R.P., 2012. Microchem. J. 101, 65.
Vondruska, M., 1995. Chem. Listy 89, 383.
Wang, L.H., Tien, H.J., 1994. Int. J. Cosmet. Sci. 16, 29.
Wang, L.-H., Hsia, H.-C., Wang, C.-C., 2006. J. Liq. Chromatogr. Relate. Tech. 29, 1737.
Wang, Y., Du, R., Yu, T., 2013. Sci. Justice 53, 350.
Wang, Y.-H., Avonto, C., Avula, B., Wang, M., Rua, D., Khan, I.A., 2015. J. AOAC Int. 98, 5.
Xu, G., Li, B., Luo, X., 2013. Sens. Actuat. B-Chem. 176, 69.
Zachariadis, G.A., Sahanidou, E., 2009. J. Pharm. Biomed. Anal. 50, 342.
Zhong, Z., Li, G., Luo, J., Chen, W., Liu, L., He, P., Luo, Z., 2015. Anal. Methods 7, 3169.

FURTHER READING

Commission Directive 2000/6/EC, Twenty-fourth Commission Directive 2000/6/EC of 29 February 2000 Adapting to Technical Progress Annexes II, III, VI and VII to Council Directive 76/768/EEC on the Approximation of the Laws of the Member States Relating to Cosmetic Products. http://eur-lex.europa.eu/legal-content/GA/TXT/?uri=CELEX:32000L0006.
SCCS-Scientific Committee on Consumer Safety, 1999a. Opinion of Concerning 1,4-Dihydroxybenzene – Colipa n°A21-adopted by the Plenary Session of the SCCNFP of 17 February 1999. http://ec.europa.eu/health/scientific_committees/consumer_safety/opinions/sccnfp_opinions_97_04/sccp_out63_en.htm.
UNE 84060:2011. Materias primas cosméticas. Determinación de trazas de plomo. Parte 1-E. Preparación de muestra. Plomo total en estearatos metálicos.
UNE 84068:1987. Materias primas cosméticas. Determinación de cadmio en compuestos de zinc.

PART III

Analytical Methods for Monitoring of Cosmetic Ingredients in Biomedical and Environmental studies

CHAPTER 15

Human Biomonitoring of Select Ingredients in Cosmetics

Rajendiran Karthikraj[1], Kurunthachalam Kannan[1,2]
[1]Wadsworth Center, Albany, NY, United States; [2]University of New York at Albany, Albany, NY, United States

INTRODUCTION

As a part of our modern lifestyle, the production and usage of synthetic chemicals have increased globally, and humans are exposed to hundreds of chemicals in their daily lives. For instance, the products that we use to sanitize our hands, pack our foods, and keep our beds from going up in flames have seeped into our bodies in ways that were unconceivable a few decades ago (Herring, 2014). Although many of these chemicals are intended to perform certain functions that are beneficial for humans, they also have unintended consequences for human and environmental health.

Of particular concern is exposure of humans to chemicals present in cosmetics. Cosmetics are those products that are used in personal hygiene and for beautification. Cosmetic ingredients (the raw materials used in the production of cosmetics) are used primarily in the manufacture of skin care, hair care, oral care, fragrance and toiletry products (Diaz et al., 2014). Surveys have shown that 57% of cosmetics contain chemicals such as parabens, alkylphenols, and ultraviolet (UV) filters that act like oestrogens (http://www.ewg.org). Although ultra-trace levels of exposure (at a dose below the reference value) to a single chemical may not have an immediate effect on human health, exposure to multiple, potentially toxic chemicals present in several cosmetics, including deodorants, fragrances, lotions, shampoos, dyes, hair conditioners, make-up, and shaving creams, every day can have an eventual effect on health. In fact, the toxic/oestrogenic chemicals present in cosmetics are considered to be emerging environmental chemicals. These chemicals are reported to have adverse effects and potential accumulation in human tissues (Jares et al., 2009).

The use of cosmetics is rapidly increasing all over the world (see https://globenewswire.com/news-release/2016/04/29/834781/0/en/Global-Market-for-Personal-Care-Ingredients-to-Reach-US-11-76-bn-by-2023-Rising-Demand-for-Anti-aging-and-Personal-Care-Products-drives-industry-growth.html).

The cosmetics market is projected to expand by 5.2% between 2015 and 2023 and is expected to rise in value from US$7.4 billion to US$11.7 billion by 2023 (see http://www.transparencymarketresearch.com/article/personal-care-ingredients-market.htm).

A survey conducted by the Environmental Working Group in the United States indicated that women, on average, use 12 cosmetics daily, averaging 120 chemicals. Although cosmetics are used for such purposes as protection from sunlight, beauty care, cleanliness, and pleasant smell, some chemicals present in them pose undesirable health effects. The cosmetics industry is somewhat unregulated, and national and international regulatory agencies do not require all ingredients in cosmetics to be approved for safety (see http://www.ewg.org/skindeep/2011/04/12/why-this-matters/).

In this chapter, we have compiled biomonitoring studies that concern the exposure of humans to various known toxic/oestrogenic ingredients present in cosmetics. Attention is paid to preservatives, UV filters, fragrances, and plasticizers. A short description of these categories and other known classes of chemicals is provided in the following.

Antioxidants: These are natural or synthetic chemicals added to cosmetics to reduce free radical damage (and the reactions of several chemicals present in cosmetics) and environmental stress on skin. Antioxidants include vitamin C (ascorbic acid), vitamin E (tocopherol) and isoflavones, as well as synthetic phenolic derivatives (e.g., butylated hydroxy toluene).

Colouring and decorative agents: These are mainly dyes and oily substances added only to achieve visual perfection or attractiveness and are used in products such as lipsticks, mascara and nail polishes, e.g., carmine.

Plasticizers: These chemicals are added to plastics to impart malleability and flexibility. Phthalates are commonly used plasticizers and include dibutyl phthalate (DBP), used as a solubilizer and for chip resistance in nail polishes; dimethyl phthalate (DMP), used in hair spray to reduce stiffness; diethyl phthalate (DEP), used as a solvent and as an alcohol denaturant, and diethylhexyl phthalate (DEHP), used as a plasticizer in plastic products and toys (http://www.cosmeticsinfo.org/ingredient/dimethyl-phthalate-diethyl-phthalate-and-dibutyl-phthalate).

Preservatives: These chemicals are added to preserve products from decay or decomposition caused by microbial growth. The widely used preservatives in cosmetics are parabens (esters of *p*-hydroxybenzoic acid). Triclosan is also highly used because of its disinfectant properties.

Surfactants: These are chemicals that possess both a hydrophobic alkyl chain and a hydrophilic ionic or polar group. Specifically added to reduce surface tension and to maintain homogeneous formulations (mixing of water and oil/fat-soluble chemicals) of cosmetics, surfactants can act as detergents, emulsifiers, foaming agents and dispersants. Surfactants increase the contact between liquid and another substance, e.g., nonylphenol.

UV filters/blockers: These chemicals are added to protect the user from deleterious UV light (i.e., UV filter) or to protect any ingredient (e.g., active substances, colouring agents, scents) from photodegradation (i.e., UV blocker). An example of this group is benzophenones.

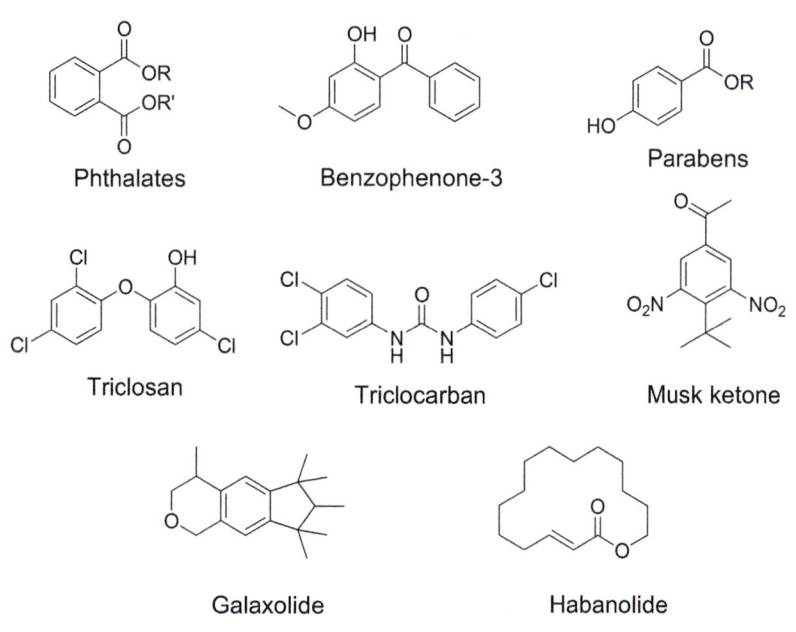

Figure 15.1 Chemical structures of some of the known oestrogenic chemicals present in personal care products (R and R' refer to alkyl/aryl moieties).

Although some of these chemicals are moderately toxic to humans, most are shown to be endocrine-disrupting chemicals (EDCs) (Fig. 15.1). EDCs are exogenous chemicals (natural or synthetic) that can mimic natural hormones by interacting with the endocrine system and eliciting adverse health effects. Studies have shown that exposure to EDCs can affect reproductive capacity (decrease in sperm count) and result in testicular or prostate cancer in men and an increased risk of breast cancer and reproductive abnormalities in women (Singh and Li, 2011). Exposure to EDCs has been linked to several diseases, such as diabetes, that are on the rise globally (http://www.psr.org/assets/pdfs/hazardous-chemicals-in-health-care.pdf).

BIOMONITORING AS AN APPROACH TO ASSESSING EXPOSURE

Humans are exposed to a myriad of undesired chemicals through various sources, such as foods, textiles, cosmetics, household articles, kitchen utensils, air and water. Assessment of the amounts and distribution of chemicals harmful to human populations is a critical component of chemical risk management. As long as disease prevention and health promotion are the principal tenets of public health, the assessment of doses of contaminant exposures is crucial. Further, assessment of human exposures to environmental chemicals is fundamental to the development of appropriate public health policies.

Direct measurements of suspect environmental chemicals in various sources of exposure, such as air, water and foodstuffs, are commonly used to establish whether and to what extent individuals are exposed to specific environmental chemicals. Nevertheless, this approach is neither affordable nor technically feasible for measuring exposures for everyone in all populations of interest (World Health Organization, 2000). With the advancements in highly sensitive analytical techniques since the late 1990s, biomonitoring has gained prominence in human exposure assessments. Biomonitoring is the direct measurement of environmental chemicals, or their products, in human biospecimens, such as blood and urine (Centers for Disease Control and Prevention, 2009). Biomonitoring involves the collection of bodily fluids (blood, saliva, urine, breast milk and seminal plasma) and/or tissues (muscle, fat, hair, bone and nail) from humans for the evaluation of the occurrence of environmental chemicals. Such assessments allow the measurement of internal doses of xenobiotic compounds that can be directly linked to possible adverse health effects (Hond et al., 2013). Biomonitoring takes into account the bioavailability of chemicals and doses found in target tissues. The application of biomonitoring approaches in public health has resulted in significant impacts on policy and intervention strategies for environmental contaminants, such as lead, tobacco smoke and pesticides (Sexton et al., 2004).

Biomonitoring also can permit the identification of novel chemicals and their biotransformation products in human bodies, trends in exposure levels and demographic factors that affect exposures. However, measurements of concentrations of chemicals in human biospecimens need to be converted into exposure doses to compare against reference doses or threshold/guideline values set by various health organizations. Such conversions require the use of pharmacokinetic information that allows for the calculation of concentrations in specific tissues/matrices to daily exposure doses. Such approaches can further enhance the usefulness of biomonitoring data in exposure assessments. Several developed countries have implemented national-scale biomonitoring programs to assess the exposure of populations to environmental chemicals as well as nutrients. Such programs provide valuable information for assessing chemical exposures and their link to environmental diseases and nutritional status in populations.

MATRIX OF CHOICE AND ANALYTICAL METHODS IN BIOMONITORING

The selection of appropriate tissues or body fluids in biomonitoring studies is based primarily on bioaccumulation properties of chemicals of interest and, in some cases, the time interval since the last exposure. Chemicals such as polychlorinated biphenyls have long biological half-lives in the body (months or years) because they are sequestered in lipid-rich adipose tissues. By contrast, chemicals such as bisphenol A (BPA), phthalates and parabens have relatively short biological half-lives (hours or days) and tend to be metabolized and excreted in urine (Bledzka et al., 2014; Johns et al., 2015). Although people are exposed to them on a daily basis, most EDCs that have been reported in cosmetics are short-lived. Such chemicals are excreted in urine, and, therefore, urine is the matrix of choice for biomonitoring studies. A few chemicals, such as polycyclic musks (e.g., Galaxolide), are lipophilic and tend to accumulate in lipid-rich tissues (Dietrich and Hitzfeld, 2004). Collection of fatty tissues can be invasive, while urine can be collected noninvasively.

As noted above, advancements in analytical technologies allow for sub-nanogram/millilitre concentrations of a wide range of environmental chemicals in a small volume of urine (often <1 mL) (Asimakopoulos et al., 2014b). Gas chromatography (GC) and liquid chromatography (LC), coupled with mass spectrometry (MS), have been the most frequently used analytical instruments in human biomonitoring programs. GC–MS with electron ionization (EI) and chemical ionization (CI) and LC–MS with electrospray ionization (ESI), atmospheric pressure CI and atmospheric pressure photoionization are commonly used techniques in human biomonitoring of cosmetics. In particular, LC coupled with negative/positive ESI–MS/MS (tandem MS) under reversed-phase chromatographic conditions has found major applications in the measurement of cosmetic chemicals in human specimens, as most of these chemicals bear an ester group (parabens, phthalates), electronegative atoms [triclosan (2,4,4'-trichloro-2'-hydroxy-diphenyl ether; TCS), triclocarban (TCC), nitro musks] or an acidic group (benzophenones). Multiple-reaction monitoring, coupled with isotope dilution–MS, using a triple-quadrupole mass spectrometer, is a powerful method for quantitative measurement of environmental chemicals. An isotope dilution method has been widely used for quantification by use of the relative response factors of appropriate, labelled internal standards. Similarly, the most widely used sample preparation methods in human biomonitoring studies are liquid–liquid extraction, solid–liquid extraction and solid-phase extraction (SPE).

Biomonitoring of Parabens

Parabens are esters of p-hydroxybenzoic acid, which are widely used as broad-spectrum antimicrobial preservatives (particularly against moulds and yeast) in cosmetics, beverages, foods and pharmaceuticals (Guo and Kannan, 2013, Guo et al., 2014; Liao et al., 2013; Moreta et al., 2015). Cosmetics that contain moisture and mineral nutrients require preservation from microbial growth and for enhancing stability, extending product lifetime, and avoiding undesired chemical reactions or degradation of main ingredients. Humans are largely exposed to parabens through multiple routes, such as ingestion (e.g., pharmaceuticals and foods), inhalation (e.g., fragrances) and dermal absorption (e.g., skin care products/cosmetics) (Soni et al., 2005). Parabens can act as oestrogen receptor agonists, androgen receptor antagonists and disruptors of lysosomal and mitochondrial functions and can induce DNA damage through the production of reactive oxygen species and nitric oxide (NO) (Darbre and Harvey, 2008; Handa et al., 2006; Tayama et al., 2008).

Methyl- (MeP), ethyl- (EtP), propyl- (PrP), isopropyl-, butyl- (BuP), isobutyl-, pentyl-, heptyl- and benzylparabens (BzP) are some of the alkyl/aryl esters of *para*-hydroxybenzoic acid or 4-hydroxybenzoic acid (4-HB) widely used alone or in combination in cosmetics such as fragrances, shampoos, soaps and shaving lotions (Dewalque et al., 2014; Guo and Kannan, 2013). Parabens are quite stable under acidic conditions but are hydrolysed to 4-HB and the corresponding alcohol under basic conditions. The increase in alkyl chain length of the ester group is directly proportional to the antibacterial activity and indirectly proportional to the water solubility of parabens (Bledzka et al., 2014). Approximately 80% of cosmetics in France were found to contain parabens (http://skinboutique.ca/wp-content/uploads/2012/09/Parabens-article-in-English.pdf), and 93% of the cosmetics from Denmark were found to contain parabens (Rastogi et al., 1995). Parabens were found in both rinse-off products (77%) and leave-on products (99%), and the total paraben contents in Danish cosmetic products were 0.01%–0.50% and 0.01%–0.59% by weight, respectively (except for a sunscreen product that contained 0.87% parabens) (Rastogi et al., 1995). The European Union (EU) and the US guidelines allow for the use of parabens by up to 0.4% for an individual paraben and 0.8% for paraben mixtures in cosmetics. The European Commission (in products sold after 16 April 2015) restricted the maximum permissible concentration of PrP and BuP to 0.14%, when used individually or in combination (http://europa.eu/rapid/press-release_IP-14-1051_en.htm). Soni et al. (2005) reported that the average daily exposure of parabens (a theoretical estimation) of an adult who weighed 60 kg was 76 mg (1.26 mg/kg bw/day); among several sources, cosmetics, pharmaceuticals and foodstuffs contributed 50 mg (0.833 mg/kg bw/day), 25 mg (0.417 mg/kg bw/day) and 1 mg (10–13 µg/kg bw/day) of parabens, respectively. Cosmetics are the major sources of paraben exposure in the general population.

Liao et al. (2013) measured parabens in foodstuffs from China and reported a daily dietary intake at ~1 µg/kg bw/day. Darbre et al. (2004) detected parabens in intact human breast tissue for the first time, with a mean concentration of 20.6 ± 4.2 ng/g tissue. Ye et al. (2008a) detected MeP, EtP and PrP in human blood with median concentrations of 10.9, 0.2 and 1.4 ng/mL, respectively. Frederiksen et al. (2011b) reported much lower concentrations (MeP, 0.89 ng/mL; EtP, below the limit of detection (LOD), and PrP, 0.23 ng/mL) in Danish men ($N=60$); the lower values found in men were attributed to gender-related differences in paraben exposures. Parabens also were found in breast milk (Ye et al., 2008b; Schlumpf et al., 2010) at a median concentration of between 1 and 1.5 ng/mL. The estimated daily intake (EDI) of parabens by infants through breast milk was calculated to be 193–301 ng/kg bw/day. Even at the highest EDI, the values were at least three orders of magnitude lower than the acceptable daily intake (ADI) of 10 mg/kg bw/day.

Parabens were frequently detected in human urine, with relatively higher concentrations in comparison to aforementioned matrices (Bledzka et al., 2014). Ye et al. (2006) determined urinary concentrations of parabens (conjugated and unconjugated) in a demographically diverse group of 100 US adults. Later, Calafat et al. (2010) measured urinary concentrations of four parabens in the US general population ($N=2548$), collected for the 2005–06 National Health and Nutrition Examination Survey (NHANES), and observed significant differences in concentrations between sexes and between races/ethnicities. Urine from non-Hispanic blacks had higher concentrations of MeP (5 times) and PrP (3.6 times) than that from non-Hispanic whites. Similarly, concentrations of MeP (3.2 times) and PrP (4.2 times) in females were higher than those in males. Greater urinary concentrations of parabens have been reported in women from several countries, which were attributed to higher amounts and frequency of use of cosmetics. Although biomonitoring studies measure parent parabens in urine, parabens are metabolized through several pathways, and it was suggested that metabolites also should be measured to accurately assess total paraben exposure (Wang and Kannan, 2013).

Parabens are rapidly hydrolysed to 4-HB and its glucuronide and sulphate conjugates in human bodies (Fig. 15.2). Several intermediates of paraben degradation, including protocatechuates, have been reported in human urine (Wang and Kannan, 2013). In fact, 4-HB and 3,4-dihydroxybenzoic acid accounted for >90% of the total paraben concentrations in urine (Fig. 15.3), which suggests the significance of measuring paraben metabolites to assess cumulative paraben exposure (Wang and Kannan, 2013). Nevertheless, 4-HB has been reported to be present in some foods naturally. Therefore, studies that distinguish natural and anthropogenic sources of 4-HB are warranted.

Wang and Kannan (2013) reported that the concentrations of methyl protocatechuate (OH-MeP) and ethyl protocatechuate (OH-EtP) were higher than those of their

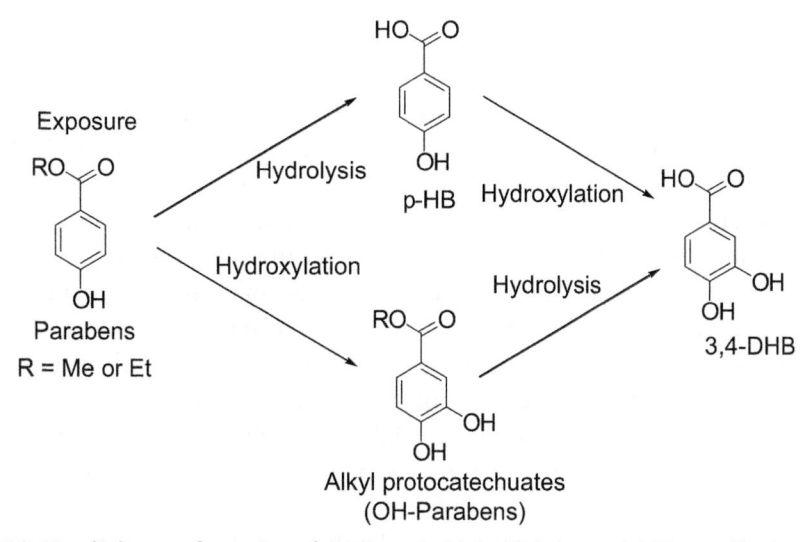

Figure 15.2 Metabolic transformation of parabens in biological systems. *DHB*, 3,4-dihydroxybenzoic acid; *p-HB*, p-hydroxybenzoic acid.

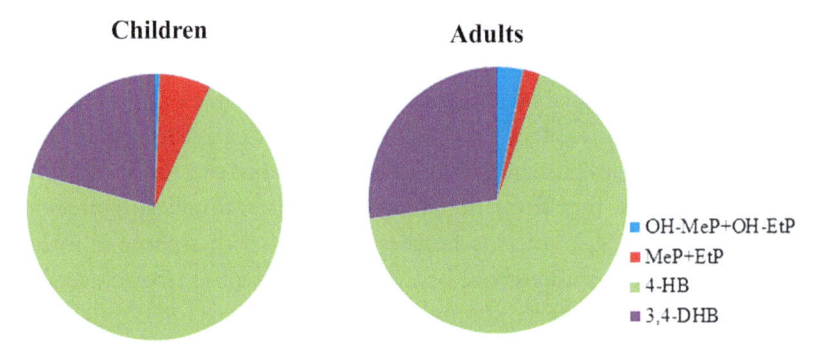

Figure 15.3 Content of parent parabens [methylparaben (*MeP*) and ethylparaben (*EtP*)], 4-hydroxybenzoic acid (*4-HB*), 3,4-dihydroxybenzoic acid (*3,4-DHB*) and alkyl protocatechuates (*OH-MeP* and *OH-EtP*) in urine from children and adults (Wang and Kannan, 2013).

corresponding parent compounds in urine and suggested the inclusion of these metabolites in exposure assessments. Alkyl protocatechuates are derived from 3,4-dihydroxybenzoic acid, another metabolite of parabens found in urine. Wang et al. (2015) also reported the occurrence of parabens in human adipose tissue collected from New York, USA, although at low nanogram/gram concentrations [1.38 ng/g, wet weight (ww)]. Nevertheless, the mean 4-HB concentration in adipose tissue was 4160 ng/g, ww (Wang et al., 2015). Sources and toxicity of 4-HB found in human adipose tissue warrant further studies. Further, a few studies have demonstrated the formation of halogenated (chlorinated and brominated) derivatives of parabens in water-treatment processes (Terasaki et al., 2009), and no studies on human exposure to halogenated parabens are available to date.

The reported concentrations of parabens and their metabolites measured in various human matrices collected from several countries are presented in Table 15.1.

Kang et al. (2016) revealed that the mean concentration of parabens measured in the Korean population (both children and adults) was higher than those reported for the United States, Denmark, China, Greece and India, but lower than the values reported for Spanish children. Frederiksen et al. (2011b) analysed parabens in urine, serum and seminal plasma from healthy Danish men and found that the concentrations in urine were much higher than in serum and seminal plasma. The median concentration of total parabens ($\sum$parabens = 5; MeP, EtP, PrP, BuP and BzP) measured in urine was 23.5 ng/mL, which was significantly higher than in serum (1.85 ng/mL) and seminal plasma (1.87 ng/mL). In a study of Swedish mother-and-children pairs, urinary paraben concentrations were higher in mothers than in children. Mothers' exposure to parabens was significantly associated with the use of cosmetics (especially make-up, lotion and mouthwash).

Wang et al. (2013) analysed parabens and 4-HB in urine collected from the United States (children, $N = 40$) and China [children ($N = 70$) and adults ($N = 26$)]. The median

Table 15.1 Reported paraben concentrations in human specimens collected from various countries

Matrix	Country	Study population	Method	Chemicals measured	LOD/LOQ	Geometric mean ∑paraben	References
Urine	USA	100 adults	Online SPE–LC–MS/MS	MeP, EtP, PrP, BuP, IBuP, BzP	LOD: 0.1–0.18 ng/mL	54.4 ng/mL (median)	Ye et al. (2006)
Serum	USA	11 women	Online SPE–LC–MS/MS	MeP, EtP, PrP	LOD: 0.1–0.2 ng/mL	12.5 ng/mL (median)	Ye et al. (2008a)
Milk	USA	4 men 4 women	Online SPE–LC–MS/MS	MeP, PrP	LOD: 0.1 ng/mL	1.3 ng/mL	Ye et al. (2008b)
Urine	USA	2548 individuals	Online SPE–LC–MS/MS	MeP, EtP, PrP, BuP	LOD: 0.2–1.0 ng/mL	72.2 ng/mL (median)	Calafat et al. (2010)
Milk	Switzerland	54 women	LLE followed by LC–MS/MS	MeP, EtP, PrP, BuP	LOD: 0.5–1 ng/mL	4.86 ng/mL (median)	Schlumpf et al. (2010)
Urine, serum, seminal plasma	Denmark	60 men	Automated SPE followed by LC–MS/MS	MeP, EtP, PrP, BuP, BzP	LOD: 0.02–0.41 ng/mL	Urine: 23.5 ng/mL Serum: 1.85 ng/mL Seminal plasma: 1.87 ng/mL (median)	Frederiksen et al. (2011b)
Urine	Denmark	145 women 143 children	Automated SPE followed by LC–MS/MS	MeP, EtP, PrP, IPrP, BuP, IBuP, BzP	LOD: 0.07–0.40 ng/mL	Women: 77.8 ng/mL Children: 21.3 ng/mL (mean)	Frederiksen et al. (2013b)
Urine	Flemish (Belgium)	210 adolescents; 204 adults	Direct analysis by LC–MS/MS	4-HB	LOD: 50 ng/mL	859 ng/mL (GM at 95% CI)	Hond et al. (2013)
Urine	Korea	1021 individuals	Online SPE–LC–MS/MS	MeP, EtP, PrP, BuP	LOQ: 0.2–1.0 ng/mL	Not reported	Lee et al. (2013)

Continued

Table 15.1 Reported paraben concentrations in human specimens collected from various countries—cont'd

Matrix	Country	Study population	Method	Chemicals measured	LOD/LOQ	Geometric mean Σparaben	References
Urine	Puerto Rico	105 pregnant women	Online SPE–LC–MS/MS	MeP, PrP, BuP	NA	171 ng/mL	Meeker et al. (2013)
Urine	USA	30 individuals	LLE followed by LC–MS/MS	MeP, EtP, OH-MeP, OH-EtP, 4-HB, 3,4-DHB	LOQ: 0.02–0.20 ng/mL	967.2 ng/mL	Wang and Kannan (2013)
Urine	USA	17 men 23 women	LLE followed by LC–MS/MS	MeP, EtP, PrP, BuP, HeP, BzP, 4-HB	LOQ: 0.02–0.20 ng/mL	6.45 μmol/L	Wang et al. (2013)
Urine	China	53 men 43 women	LLE followed by LC–MS/MS	MeP, EtP, PrP, BuP, HeP, BzP, 4-HB	LOQ: 0.02–0.20 ng/mL	15.7 μmol/L	Wang et al. (2013)
Urine	Greece	100 individuals	LLE followed by LC–MS/MS	MeP, EtP, PrP, BuP, HeP, BzP, OH–EtP	LOQ: 0.2–2.0 ng/mL	26.3 ng/mL	Asimakopoulos et al. (2014a)
Urine	Belgium	123 men 138 women	SPE followed by UPLC–MS/MS	MeP, EtP, PrP, BuP	NA	22.6 ng/mL	Dewalque et al. (2014)
Urine	Sweden	98 women 98 children	Automated SPE followed by LC–MS/MS	MeP, EtP, PrP, BuP, BzP	LOD: ≤0.1 ng/mL	55.1 ng/mL	Larsson et al. (2014)
Urine	Germany	59 children 59 women 39 men	Enzymatic deconjugation followed by LC/LC–MS/MS	MeP, EtP, PrP, IPrP, BuP, IBuP, PeP, HeP, BzP	LOQ: 0.5 ng/mL	27.1 ng/mL (median)	Moos et al. (2014)
Urine	Saudi Arabia	130 individuals	LLE followed by LC–MS/MS	MeP, EtP, PrP, BuP, HeP, BzP, 4-HB, 3,4-DHB, OH–MeP, OH–EtP	LOD: 0.003–0.59 ng/mL	161.8 ng/mL (median)	Asimakopoulos et al. (2016)
Urine	Australia	2400 individuals	Online SPE–LC–MS/MS	MeP, EtP, PrP, BuP	LOD: 0.1–1.0 ng/mL	330.4 ng/mL	Heffernan et al. (2015)

Human Biomonitoring of Select Ingredients in Cosmetics 397

Matrix	Country	Study population	Method	Chemicals measured	LOD/LOQ	Geometric mean ∑paraben	References
Urine	USA	34 women	Online SPE–LC–APCI/APPI–MS/MS	MeP, EtP, PrP, BuP	LOD: 0.2–1.0 ng/mL	178.3 ng/mL (median)	Hines et al. (2015)
Serum	USA	34 women	Online SPE–LC–APCI/APPI–MS/MS	MeP, PrP	LOD: 0.2–1.0 ng/mL	2.4 ng/mL (median)	Hines et al. (2015)
Urine	Brazil	30 postpartum women	SPE followed by UPLC–MS/MS	MeP, EtP, PrP, BuP, BzP	LOQ: 0.5 ng/mL	9.55 ng/mL (median)	Jardim et al. (2015)
Urine	Greece	239 women	SPE followed by LC–MS/MS	MeP, EtP, PrP, IPrP, BuP, IBuP	LOD: 0.04–0.13 ng/mL	114.3 ng/mL (median)	Myridakis et al. (2015)
Urine	Greece	239 children	SPE followed by LC–MS/MS	MeP, EtP, PrP, IPrP, BuP, IBuP	LOD: 0.04–0.13 ng/mL	19.5 ng/mL (median)	Myridakis et al. (2015)
Adipose tissue	USA	20 individuals	Solvent extractions followed by LC–MS/MS	MeP, EtP, PrP, BuP, HeP, BzP, 4-HB	LOQ: 0.24–0.80 ng/g ww	4161 ng/mL	Wang et al. (2015)
Urine	India	76 children	LLE followed by LC–MS/MS	MeP, EtP, PrP, BuP, HeP, BzP, 4-HB, 3,4-DHB	LOQ: 0.01–0.10 ng/mL	2264.2 ng/mL	Xue et al. (2015)
Urine	Korea	2541 individuals	SPE followed by LC–MS/MS	MeP, EtP, PrP, BuP	LOD: 0.04–0.30 ng/mL	152.8 ng/mL	Kang et al. (2016)

APCI, atmospheric pressure chemical ionization; *APPI*, atmospheric pressure photoionization; *BuP*, butylparaben; *BzP*, benzylparaben; *CI*, confidence interval; *3,4-DHB*, 3,4-dihydroxybenzoic acid; *EtP*, ethylparaben; *GM*, geometric mean; *4-HB*, 4-hydroxybenzoic acid; *HeP*, heptylparaben; *IBuP*, isobutylparaben; *IPrP*, isopropylparaben; *LC*, liquid chromatography; *LLE*, liquid–liquid extraction; *LOD*, limit of detection; *LOQ*, limit of quantification; *MeP*, methylparaben; *MS/MS*, tandem mass spectrometry; *NA*, not available; *PeP*, pentylparaben; *SPE*, solid-phase extraction; *UPLC*, ultra-high-performance liquid chromatography; *ww*, wet weight.

concentration of parabens in US children (54.6 ng/mL) was 5.5-fold higher than those in Chinese children (10.1 ng/mL). Nevertheless, remarkably elevated concentrations of parabens were found in a few Chinese female adults (1000–10,000 ng/mL). In the case of 4-HB, the concentrations found in Chinese children [geometric mean (GM) 1380 ng/mL] were twofold higher than in US children (GM 690 ng/mL). Hence, there was a significant difference in the ratio of total parabens to 4-HB in populations in the United States and China, which suggests various sources and pathways of exposure to parabens and 4-HB. Liao and Kannan (2014b) showed the presence of parabens in hygiene wipes used on babies in the United States. Based on the urinary concentrations of parabens and 4-HB, the daily intake (DI) was estimated for MeP and PrP. The GM DI_{MeP} was 0.5–0.7 mg/day for both US and Chinese children, which was much lower than the values estimated for Chinese adults (female 5.1 mg/day and male 1.6 mg/day). The estimated DI values, however, were still below the maximum DI_{MeP}. Similarly, the estimated GM DI_{PrP} value was 0.01 mg/day for US children, which was 20–30 times lower than that calculated for Chinese children (DI_{PrP} 0.2–0.3 mg/day). Nevertheless, extremely high exposure (DI_{PrP} 27 mg/day) was found in some Chinese female adults and children (Wang et al., 2013).

A summary of paraben concentrations reported for various populations is presented in Table 15.1. Overall, the reported urinary paraben concentrations were in the range of 2.4–2260 ng/mL (Table 15.1). The highest concentration (2260 ng/mL) was reported for Indian children (Xue et al., 2015) followed by populations in the United States (GM 967 ng/mL; Wang et al., 2013), Belgium (GM 859 ng/mL; Hond et al., 2013), and Australia (GM 330.4 ng/mL; Heffernan et al., 2015). The lowest concentration of parabens (9.55 ng/mL) was reported in urine from Brazilian postpartum women (Jardim et al., 2015). Wang et al. (2015) reported the occurrence of parabens (including 4-HB) in adipose tissue (4160 ng/g, ww) collected from the United States ($N=20$), which was the highest concentration ever reported for parabens in human specimens. On the whole, MeP was the predominant paraben found in human specimens from various countries, followed by PrP (Table 15.1). Despite widespread exposures, the measured values are below the ADI for parabens. Nevertheless, toxicological studies point to adverse outcomes at lower doses; therefore, continuous monitoring of parabens in human populations is necessary.

Biomonitoring of Benzophenone Ultraviolet Filters

The UV radiation from the sun can cause sunburn, skin cancer, photoageing and photodermatosis. To prevent or attenuate the harmful effects of the UV radiation on the skin and hair, manufacturers have used a variety of chemicals in sunscreen products and in other cosmetics (Kunisue et al., 2012). Benzophenone (BP)-type UV filters are extensively used in various cosmetics, such as lipsticks, facial creams and shampoos, and food packaging materials (Liao and Kannan, 2014a,b,c). Although restrictions have been placed on the use of some BPs, there are 12 different types of BP UV filters (BP-1 to BP-12) in addition to

2-hydroxybenzophenone, 4-hydroxybenzophenone (4-OH-BP) and 4-methylbenzophenone, which are widely used in cosmetics (Kawamura et al., 2003) (Fig. 15.4).

The most frequently used UV filter and photostabilizer in various cosmetics is BP-3 (2-hydroxy-4-methoxybenzophenone) because of its large molar absorptivity of both UVA (wavelength 320–400 nm) and UVB (290–320 nm) regions of sunlight. Although the use of BP-3 in cosmetics is allowed, the concentration in the final product is restricted by three major regulatory organizations (EU, US Food and Drug Administration and Japan's Pharmaceutical Affairs Law). The maximum allowable concentration of BP-3 in cosmetics ranges between 5% and 10% (ww) in final products. Dermal absorption of BP UV filters has been associated with many adverse effects, including allergies, endocrine disruption and carcinogenicity (Kim and Choi, 2014). Exposure to BP UV filters has been shown to cause androgenic and oestrogenic effects (Tarazona et al., 2013). Studies have shown that the oestrogenic potency of BP-3 is higher than that of BPA (Kawamura et al., 2003) and that exposure to BP-3-related UV filters was associated with endometriosis in women (Kunisue et al., 2012). It has been reported that BP-3 can be

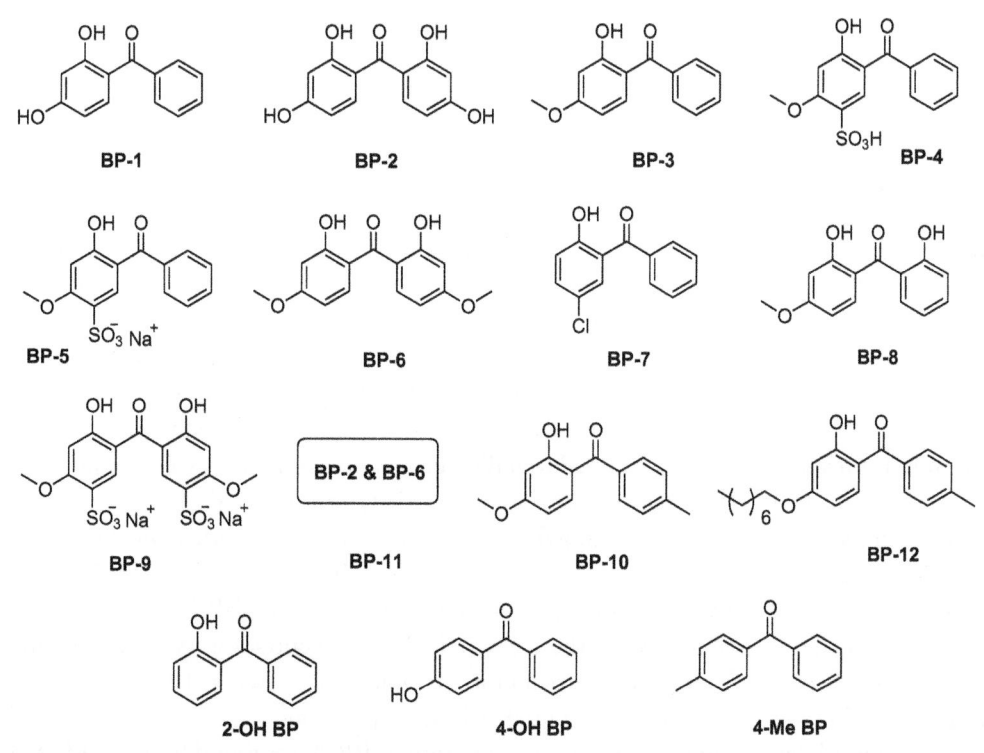

Figure 15.4 Chemical structures of benzophenone (*BP-1* to *BP-12*) UV filters and their metabolites/degradation products. *2-OH BP*, 2-hydroxybenzophenone; *4-OH BP*, 4-hydroxybenzophenone; *4-Me BP*, 4-methylbenzophenone.

transformed into three metabolites in vivo, i.e., BP-1, BP-8 and 2,3,4-trihydroxybenzophenone. Among them, BP-1 is the major metabolite, with longer half-life and greater oestrogenic potential than BP-3 (Chisvert et al., 2012).

The median urinary concentration of BP-3 found in Danish children was comparable to that reported for Spanish children but 7- to 10-fold lower than those reported for US children. BP-3 was found in >98% of urine samples collected in November in Denmark, even when the usage of sunscreen products was low, which suggested sources apart from sunscreen products that contribute to exposures (Frederiksen et al., 2013a). Zhang et al. (2013), for the first time, reported the occurrence of BP UV filters in paired urine and blood and in matched maternal and foetal cord blood samples and showed that the mean concentration of BP-3 was higher in women than in men. Further, a higher (2.5 times) concentration of BP-3 was found in urine than in blood (paired samples), which suggested rapid metabolism and urinary excretion of this chemical. The concentration ratios of BP-3 and 4-OH-BP between blood and urine were 0.21 and 0.36, respectively, which suggested that BP-3 is cleared faster than 4-OH-BP from human bodies. Both BP-3 and 4-OH-BP were found at higher levels in maternal blood than in cord blood.

The reported daily excretion dose of BPs by Chinese young adults was 0.12 mol/kg bw/day (Gao et al., 2015). The median urinary concentration of BP-3 found in Chinese young adults was 0.55 ng/mL, whereas much higher concentrations were reported in studies from Denmark (mother 3.7 ng/mL and child 1.8 ng/mL), Spain (pregnant women 3.4 ng/mL) and France (pregnant women 1.7 ng/mL) (Casas et al., 2011; Philippat et al., 2012). Xue et al. (2015) reported a GM concentration of BP-3 in Indian children at 0.91 ng/mL, which was 11 times lower than the concentration (9.97 ng/mL) reported for US children. The lower concentrations of BP-3 in India and China were attributed to lower usage of sunscreen products in those countries.

Calafat et al. (2008a) measured BP-3 in the US population as a part of the 2003–04 NHANES at concentrations that ranged from 0.4 to 21,700 ng/mL, with a GM of 22.9 ng/mL. Although BP-3 was detected in ~97% of the 2517 samples, the wide range in the measured concentrations was attributed to individuals' lifestyle differences. Dewalque et al. (2014) measured BP-3 in a Belgian population ($N=261$) and detected it in ~83% of the urine samples at concentrations ranging from <LOD to 663 ng/mL with a GM of 1.3 ng/mL. The measured BP-3 concentrations were four times higher in adolescents (12–19 years) than in adults. There was no significant difference, however, in BP-3 concentrations between males and females. Overall, the measured BP-3 concentrations were fairly comparable with those of other countries, except for the United States, where BP-3 concentrations were 10- to 20-fold higher than in the Belgian population.

An eastward decreasing trend in the concentration of BP-3 was found from the United States to Europe to Asia. The use of sunscreen products is greater among Caucasians than other ethnicities. Liao and Kannan (2014a) estimated the human exposure dose of BP-3 based on the concentration measured in urine. The calculated EDI$_{total}$ values for BP-3 in adult women from China and the United States were 1.47 and 46.1 µg/day, respectively. The contribution of cosmetics to total BP-3 exposure was 67% and 53% for adult women in China and the United States, respectively (Liao and Kannan 2014a).

A summary of the measured concentrations of BP-3 in human specimens from various countries is presented in Table 15.2. Overall, the highest concentration (143 ng/mL, GM at 95% CI) of BP-3 was reported in urine from France (Philippat et al., 2012), followed by Australia (mean 61.5 ng/mL; Heffernan et al., 2015). Klimova et al. (2015) reported the systemic exposure dosage (SED) based on a study that involved the mimicking of real-life consumer habits. Several cases were studied, including (1) a whole-body application with a sunscreen on a sunbathing day and (2) a facial treatment with a sunscreen on a non-specific day. The calculated SEDs for BP-3 for cases (1) and (2) were 4744 and 153 µg/kg bw/day, respectively. Based on the SED and no-adverse-effect level values (200 mg/kg bw/day) of BP-3, the margin of safety (MoS) values were calculated. The MoSs for BP-3 for cases (1) and (2) were 42 and 1307, respectively. If MoS was >100, then the compound was considered safe for use; however, for the case (1), it was below 100; hence, it was suggested to pose a health risk to humans.

Biomonitoring of Triclosan

TCS is commonly used as a disinfectant or antimicrobial/antibacterial agent in many cosmetics, such as toothpastes and soaps. TCS is a moderately lipophilic, phenolic compound (log K_{ow} = 4.76, pK_a = 7.9) and has been widely used since 1968 in Europe and from the mid-1990s in the United States. At high concentrations, TCS acts as a biocide with multiple cytoplasmic and membrane targets, and, at lower concentrations, found in cosmetics, TCS acts as a bacteriostatic by inhibiting fatty acid synthesis. TCS exposure has been shown to elicit androgenic and thyroid effects, skin irritation, contact dermatitis and endocrine disruption; thus, TCS is being regulated in many countries (Weiss et al., 2015). Cosmetics that contain TCS include soaps (0.1%–1%), detergents, toothpastes, mouthwashes, lotions, skin cleansers, sanitizers, body sprays (deodorants and perfumes) and hair sprays (Liao and Kannan 2014c). TCS also is used in kitchen utensils, toys, bedding, socks and trash bags. It has been reported that the major source of TCS exposure for Flemish adolescents was day/night skin cream and was liquid soaps for Puerto Ricans (Heffernan et al., 2015). Human exposure to TCS occurs mainly through dermal absorption (e.g., soaps) and oral administration (e.g., toothpastes).

TCS has been found in urine, serum, adipose tissue and breast milk. Among various matrices, urine is the most frequently used for TCS biomonitoring. Erici et al. (2002) demonstrated the occurrence of TCS in breast milk and fish from Sweden. High concentrations of TCS [ranging from 60 to 300 µg/kg lipid weight (lw)] were found in three of five tested breast milk samples. Allmyr et al. (2006) showed the presence of high concentrations (median) of TCS in breast milk (0.33–0.54 ng/g fresh weight) and plasma

Table 15.2 Reported concentrations of benzophenone UV filters in human specimens collected from various countries

Matrix	Country	Study population	Method	Chemicals measured	LOD/LOQ	Geometric mean	References
Urine	USA	2517 individuals	Online SPE–LC–MS/MS	BP-3	LOD: 0.5 ng/mL	22.9 ng/mL	Calafat et al. (2008a)
Milk	Switzerland	54 women	LLE followed by LC–MS/MS	BP-2 and BP-3	LOD: 1–2 ng/g lipid	BP-2, ND; BP-3, 26.7 ng/g lipid	Schlumpf et al. (2010)
Urine	Spain	1 man	SPE followed by LC–MS/MS	BP-1, BP-3, BP-8, THB	LOD: 0.027–0.103 ng/mL	Urine 1: BP-1, 3.8; BP-3, 1.8; BP-8, 0.3; THB, 0.17 μg (after 20h of application, total forms). Urine 2: BP-1, 1.3; BP-3, 2.0; BP-8, 0.3; THB, 0.06 μg (after 30h of application, total forms)	Leon et al. (2010)
Semen	Spain	1 man	SPE followed by LC–MS/MS	BP-1, BP-3, BP-8, THB	LOD: 1–3 ng/mL	Semen 1: BP-1, 0.01; BP-3, 0.08; BP-8, 0.03; THB, 0.15 μg (after first 24h of application, total forms). Semen 2: BP-1, 0.04; BP-3, 0.12; BP-8, 0.07; THB, 0.18 μg (after second 24h of application, total forms)	Leon et al. (2010)
Urine	Spain	120 women 30 children	LLE followed by LC–MS/MS	BP-3	LOD: 0.4 ng/mL	BP-3, 3.4 ng/mL (women) and 1.9 ng/mL (children)	Casas et al. (2011)
Urine	USA	625 individuals	LLE followed by LC–MS/MS	BP-1, BP-3, 4-HBP	LOD: 0.082–0.28 ng/mL	BP-1, 6.1; BP-3, 6.1; 4-HBP, 0.36 ng/mL (median)	Kunisue et al. (2012)
Urine	France	191 individuals	Online SPE–LC–MS/MS	BP-3	LOD: 0.4 ng/mL	143 ng/mL (at 95%)	Philippat et al. (2012)

Urine	Denmark	129 children	SPE followed by LC–MS/MS	BP-3	LOD: ≤0.13 ng/mL	Urinary excretion: 24 h, 6.52 ng/mL; 1st morning, 12.7; 2nd morning, 13.3 ng/mL (mean)	Frederiksen et al. (2013a)
Urine	Puerto Rico	105 pregnant women	Online SPE–LC–MS/MS	BP-3	NA	52.2 ng/mL (GM at 95% CI)	Meeker et al. (2013)
Serum	Spain	1 man 1 woman	DLLME followed by LC–MS/MS	BP-3	22–28 ng/mL	Between 6 and 9 h after applying cosmetics and after 24 h: 200 and 84 ng/mL for man; 304 and 206 ng/mL for woman	Tarazona et al. (2013)
Urine	USA	38 children 30 adults	LLE followed by LC–MS/MS	BP-1, BP-2, BP-3, BP-8, 4-HBP	0.07–0.20 ng/mL	BP-1, 4.21/4.37; BP-2, 0.24/0.29; BP-3, 9.97/15.7; BP-8, 0.26/0.16; 4-HBP, 0.91/1.04 ng/mL (children/adults)	Wang et al. (2013)
Urine	China	70 children 26 adults	LLE followed by LC–MS/MS	BP-1, BP-2, BP-3, BP-8, 4-HBP	0.07–0.20 ng/mL	BP-1, 0.16/0.93; BP-2, 0.22/0.79; BP-3, 0.62/0.98; BP-8, 0.09/0.18; 4-HBP, 0.08/0.08 ng/mL (children/adults)	Wang et al. (2013)
Urine	China	52 men 48 women	LLE followed by LC–MS/MS	BP-1, BP-2, BP-3, BP-8, 4-HBP	0.07–0.21 ng/mL	BP-1, 0.28; BP-2, ND; BP-3, 0.26; BP-8, ND; 4-HBP, 0.19 ng/mL	Zhang et al. (2013)
Blood	China	75 individuals	LLE followed by LC–MS/MS	BP-1, BP-2, BP-3, BP-8, 4-HBP	0.06–0.67 ng/mL	BP-1, 0.05; BP-2, ND; BP-3, 0.71; BP-8, ND; 4-HBP, 0.46 ng/mL	Zhang et al. (2013)
Urine	Belgium	123 men 138 women	SPE followed by UPLC–MS/MS	BP-3	LOD: 0.20 ng/mL	22.6 ng/mL (mean)	Dewalque et al. (2014)
Urine	Germany	59 children 59 women 39 men	LC/LC–MS/MS	BP-1, BP-3, BP-8	LOQ: 0.5–2.0 ng/mL	BP-1, <LOQ; BP-3, <LOQ; BP-8, <LOQ	Moos et al. (2014)

Continued

Table 15.2 Reported concentrations of benzophenone UV filters in human specimens collected from various countries—cont'd

Matrix	Country	Study population	Method	Chemicals measured	LOD/LOQ	Geometric mean	References
Serum	Spain	20 individuals	DLLME followed by LC–MS/MS	BP-1, BP-2, BP-3, BP-6, BP-8, 4-HBP	LOQ: 0.4–0.9 ng/mL	BP-1, 0.04 ng/mL; BP-3, 0.41 ng/mL (mean); others, ND	Soria et al. (2014)
Urine	Saudi Arabia	130 individuals	LLE followed by LC–MS/MS	BP-1, BP-2, BP-3, BP-8, 4-HBP	LOD: 0.004–0.28 ng/mL	BP-1, 4.85; BP-2, 1.03; BP-3, 20.4; BP-8, 0.11; 4-HBP, 0.45 ng/mL (mean)	Asimakopoulos et al. (2016)
Urine	China	68 men 41 women	SPE followed by LC–MS/MS	BP-1, BP-2, BP-3, BP-8, 4-HBP	LOD: 0.02–0.06 ng/mL	BP-1, 0.24; BP-2, 0.05; BP-3, 0.62; BP-8, <LOD; 4-HBP, 0.08 ng/mL	Gao et al. (2015)
Milk	Spain	10 individuals	DSPE followed by UPLC–MS/MS	BP-1, BP-3, BP-6, BP-8, 4-HBP	LOD: 0.3–0.6 ng/mL	BP-1, 0.06; BP-3, 5.65; BP-6, ND; BP-8, 0.07; 4-HBP, 0.66 ng/mL	Gomez et al. (2015)
Urine	Australia	2400 individuals	Online SPE–LC–MS/MS	BP-3	LOD: 0.2 ng/mL	61.5 ng/mL (mean)	Heffernan et al. (2015)
Urine, serum	USA	34 women	Online SPE–LC–MS/MS	BP-3	LOD, 0.34 ng/mL (urine); LOD, 0.5 ng/mL (serum)	Urine, 4.7 ng/mL; serum, 0.35 ng/mL (median)	Hines et al. (2015)
Milk	USA	9 women	Online SPE–LC–MS/MS	BP-3	LOD: 0.51 ng/mL	3.7 ng/mL (mean)	Hines et al. (2015)
Urine	USA	495 women	SPE followed by LC–MS/MS	BP-1, 4-HBP, BP-3	LOD: 0.08–0.28 ng/mL	BP1, 8.9 ng/mL; 4-HBP, 0.25 ng/mL; BP-3, 8.7 ng/mL	Pollack et al. (2015)
Urine	India	76 children	LLE followed by LC–MS/MS	BP-3	LOQ: 0.10 ng/mL	2.2 ng/mL (mean)	Xue et al. (2015)

BP, benzophenone; *CI*, confidence interval; *DLLME*, dispersive liquid–liquid micro-extraction; *DSPE*, dispersive solid-phase extraction; *GM*, geometric mean; *4-HBP*, 4-hydroxybenzoic acid; *LC*, liquid chromatography; *LC/LC*, two-dimensional liquid chromatography; *LLE*, liquid–liquid extraction; *LOD*, limit of detection; *LOQ*, limit of quantification; *MS/MS*, tandem mass spectrometry; *NA*, not available; *ND*, not detected; *SPE*, solid-phase extraction; *THB*, 2,3,4-trihydroxybenzophenone; *UPLC*, ultra-high-performance liquid chromatography.

(6.7–16.0 ng/g fresh weight). Median concentration ratios of TCS in paired milk and plasma samples (M/P) were between 0.01 and 2 (in only 2 of 51 cases was M/P >1). The larger amount of TCS in plasma (pH 7.5) than in milk (pH 6.5) was explained by its acidic nature (TCS).

The influence of age and gender on TCS concentrations was studied in Australia (Allmyr et al., 2008). The highest concentration of TCS was found in serum from 31- to 45-year-old males and females, and TCS levels were higher in males than in females. TCS concentrations in serum from Australians were twofold higher than those of the Swedes. Another study showed the highest TCS concentrations in the urine of Australian females (16–45 years) (Heffernan et al., 2015). A study from Korea showed that urinary TCS concentrations were significantly associated with cigarette smoking (Kim et al., 2011). A study from China showed the presence of TCS in the urine of children and students at a concentration range of 'not detected' to 558 µg/g creatinine (Li et al., 2013). This study also found higher TCS concentrations in females than in males.

Weiss et al. (2015) studied temporal variability and sources of TCS exposure during pregnancy among Canadian women. TCS was detected in 87% of the 1247 urine samples analysed, with a GM of 21.6 µg/L, which was 1.7 times higher than that reported previously (Arbuckle et al., 2015) for Canadian pregnant women. TCS concentrations were lower in samples collected between 1600 and 2359 hours in a given day and in women with a household income between US$80,000 and US$100,000 (Weiss et al., 2015), but high concentrations of TCS were found in samples collected in the fall and in women with first pregnancy. Xue et al. (2015) reported a TCS concentration in Indian children (GM 9.6 ng/mL) that was comparable to the concentrations reported for US (8.2 ng/mL) and Chinese children (7.5 ng/mL) (Calafat et al., 2008b; Li et al., 2013). The reported SED of TCS from the use of all cosmetics was 0.3795 mg/kg bw/day (SCCP, opinion on TCS). Table 15.3 shows the measured concentrations of TCS in various human specimens from various countries. In urine, the highest concentration of TCS was reported in Australia, with a GM of 87.7 ng/mL, and the lowest was reported in Korea (GM 1.68 ng/mL).

Biomonitoring of Triclocarban

Similar to TCS, TCC has been used as an antimicrobial and antifungal compound since the 1960s and is commonly found in soaps, lotions, deodorants, toothpastes and plastics. Nevertheless, the detection frequency and concentrations of TCC in human specimens were much lower than those for TCS (Asimakopoulos et al., 2016). For example, Asimakopoulos et al. (2014a) measured in human urine from Greece at a GM concentration of 8 ng/mL for TCS and 0.6 ng/mL for TCC. The frequency of occurrence of TCC in urine from the United States, Germany, Canada and Greece was 28%, 3.6%, 3.8% and 4%, respectively (Pycke et al., 2014). Pycke et al. reported metabolites of TCC, such as 2'-OH-TCC, 3'-OH-TCC and

Table 15.3 Reported concentrations of triclosan and triclocarban in human specimens collected from various countries

Matrix	Country	Study population	Method	Chemicals measured	LOD/LOQ	Geometric mean	References
Breast milk	Sweden	5 women	LLE followed by GC–MS	TCS	LOD:TCS, 20 µg/kg lw	TCS, 60–300 µg/kg lw	Erici et al. (2002)
Plasma, breast milk	Sweden	36 women	Derivatization followed by GC–ECNI/MS	TCS	LOD:TCS, 0.018 ng/g (milk), 0.009 ng/g (plasma)	TCS, <0.018–0.95 ng/g (breast milk), 0.010–38 ng/g (plasma)	Allmyr et al. (2006)
Breast milk	USA	62 women	LLE followed by GC–MS	TCS	LOQ:TCS, 150 ng/kg lw	TCS, 0–2100 ng/g lw	Dayan (2007)
Urine	USA	2517	Automated SPE followed by LC–MS/MS	TCS	LOQ:TCS, 2.3 ng/mL	TCS, 13.0 ng/mL	Calafat et al. (2008b)
Urine	Korea	1870 individuals	LLE followed by GC–MS	TCS	LOD:TCS, 0.05 ng/mL	TCS, 1.68 ng/mL	Kim et al. (2011)
Adipose tissue, liver, brain	Belgium	11 individuals	Derivatization followed by GC–ECNI/MS	TCS	TCS, 0.06 ng/g for adipose tissue and brain, 0.045 ng/g for liver	Mean:TCS, liver, 0.44 ng/g; adipose tissue, 0.28 ng/g; brain, <LOQ	Geens et al. (2012)
Urine	Denmark	129 children	SPE followed by LC–MS/MS	TCS TCC	LOD:TCS, 0.06 ng/mL TCC, 0.01 ng/mL	TCS, 1.45 ng/mL TCC, <LOD	Frederiksen et al. (2013a)
Urine	Denmark	145 women	SPE followed by LC–MS/MS	TCS	LOD:TCS, 0.06 ng/mL	TCS, 66 ng/mL (mothers), 43 ng/mL (children)	Frederiksen et al. (2013b)
		143 children		TCC	TCC, 0.01 ng/mL	TCC, 0.08 ng/mL (mothers), 0.07 ng/mL (children)	

Sample	Location	Subjects	Method	Analyte	LOD/LOQ	Concentration	Reference
Urine	Flanders (Belgium)	210 adolescents 204 adults	SPE, derivatization followed by GC–MS	TCS	LOQ:TCS, 0.1 ng/mL	TCS, 2.19 ng/mL	Hond et al. (2013)
Urine	USA	4037 individuals	Online SPE–LC–MS/MS	TCS	NA	TCS, 11.6 ng/mL	Lankester et al. (2013)
Urine	China	287 children	SPE, derivatization followed by GC–CI/MS	TCS	LOQ:TCS, 0.5 ng/mL	TCS, 3.77 ng/mL	Li et al. (2013)
Urine	Greece	50 women 50 men	LLE followed by LC–MS/MS	TCS TCC	LOQ:TCS, 0.5 ng/mL TCC, 0.5 ng/mL	TCS, 8.0 ng/mL TCC, 0.6 ng/mL	Asimakopoulos et al. (2014a)
Urine	Norway	105 women	Online SPE–LC–MS/MS	TCS	LOD:TCS, 2.3 ng/mL	TCS, <LOD	Bertelsen et al. (2014)
Urine	Sweden	98 women 98 children	Automated SPE followed by LC–MS/MS	TCS	LOD:TCS, 0.4 ng/mL	TCS, ND	Larsson et al. (2014)
Urine	Saudi Arabia	130 individuals	LLE followed by LC–MS/MS	TCS TCC	LOD:TCS, 0.019 ng/mL TCC, 0.004 ng/mL	TCS, 5.22 ng/mL TCC, 0.37 ng/mL (mean)	Asimakopoulos et al. (2016)
Urine	Australia	2400 individuals	Online SPE–LC–MS/MS	TCS	LOD:TCS, 1.0 ng/mL	TCS, 87.7 ng/mL	Heffernan et al. (2015)
Adipose tissue	USA	20 individuals	Solvent extraction followed by LC–MS/MS	TCS	LOQ:TCS, 0.80 ng/g	TCS, 7.21 ng/g	Wang et al. (2015)
Urine	Canada	1249 pregnant women	Solvent extraction followed by GC–MS/MS	TCS	LOQ:TCS, 3 ng/mL	TCS, 21.6 ng/mL	Weiss et al. (2015)
Urine	India	76 children	LLE followed by LC–MS/MS	TCS	LOQ:TCS, 0.10 ng/mL	TCS, 9.55 ng/mL	Xue et al. (2015)

CI, chemical ionization; *ECNI*, electron capture negative ion; *GC*, gas chromatography; *LLE*, liquid–liquid extraction; *lw*, lipid weight; *LOD*, limit of detection; *LOQ*, limit of quantification; *MS*, mass spectrometry; *MS/MS*, tandem mass spectrometry; *NA*, not available; *ND*, not detected; *SPE*, solid-phase extraction; *TCC*, triclocarban; *TCS*, triclosan.

3′-Cl-TCC, in urine from individuals from New York at mean concentrations of 0.24, 0.04 and 0.01 ng/mL, respectively. A study from China measured concentrations of TCS and TCC in matched human urine and nails (fingernails and toenails). The respective GM concentrations of TCS and TCC were 0.40 and 0.40 µg/g creatinine in urine, 5.67 and 41.5 µg/kg in toenails, and 13.6 and 84.7 µg/kg in fingernails. Both TCS and TCC were found at twofold higher concentrations in fingernails than in toenails. Further, the study found twofold higher concentrations of TCC in the urine and fingernails of females than in males (Yin et al., 2016).

Biomonitoring of Synthetic Musks

Synthetic musks (SMs) are artificial or human-made aroma chemicals, synthesized in large quantities as a replacement for expensive natural musks originally obtained from musk deer and musk ox. SMs are used in cosmetics such as detergents, perfumes, deodorants, scented consumer products, soaps, lotions and cleaning products (Tanabe, 2005). In 1981, for the first time, the presence of SMs in the environment (freshwater fish) was reported in Japan. Later, SMs were detected in almost all environmental matrices, such as water, sediment and aquatic organisms (Kannan et al., 2005; Taylor et al., 2014). Similarly, SMs have been found in various human matrices (milk, blood, urine and adipose fat) (Taylor et al., 2014). However, because of the lipophilic nature of SMs, they are expected to accumulate more in lipid-rich tissues (Dietrich and Hitzfeld, 2004).

SMs are classified into three groups as polycyclic musks, nitro musks, and macrocyclic musks. Galaxolide [1,3,4,6,7,8-hexahydro-4,6,6,7,8,8-hexamethylcyclopenta[g]2-benzopyrane (HHCB)] and Tonalide [7-acetyl-1,1,3,4,4,6-hexamethyltetrahydronaphthalene (AHTN)] are the most commonly used polycyclic musks (Tanabe, 2005). The less familiar polycyclic musks are Celestolide (4-acetyl-6-*tert*-butyl-1,1-dimethylindane), Traseolide (5-acetyl-1,1,2,6-tetramethyl-3-isopropylindane), and Phantolide (5-acetyl-1,1,2,3,3,6-hexamethylindane). In the case of nitro musks, musk xylene (MX) and musk ketone (MK) are used predominantly as fragrance chemicals. Although nitro musks include musk ambrette (MA), musk moskene and musk tibetene, these chemicals have been banned from use in products that contain fragrance because of their adverse reactions (e.g., allergic reactions) in humans. Following the restrictions on the use of nitro musks, the global usage of polycyclic musks increased by twofold from 1987 to 2000 (Christian et al., 1999). Although nitro musks are banned in several countries, production continues in China and India, and these musks are used in many non-cosmetic products. The use of polycyclic musks in cosmetics is banned in Europe but still not restricted in the United States.

Human exposure to SMs occurs through inhalation, ingestion and dermal absorption (Taylor et al., 2014). Although the use of fragrances that contain SMs is a major source of exposure, contamination of drinking water and aquatic animals, including fish, has been reported (Benotti et al., 2009; Kannan et al., 2005; Reiner and Kannan, 2011).

Among polycyclic musks, HHCB is the major compound, with an annual use of over 1 million pounds in the United States (US EPA High Production Volume Chemical List Database, 2007).

The occurrence of nitro musks has been reported in breast milk samples collected from southern Bavaria, Germany; MX, MK and MA were found at median concentrations of 70, 30 and 30 ng/g lw, respectively (Liebl and Ehrenstorfer, 1993). A study from Denmark showed a median HHCB concentration of 147 ng/g lw in breast milk (Olesen et al., 2005). HHCB and AHTN were found in human adipose tissues from New York at concentrations of 12–798 ng/g lw and <5–134 ng/g lw, respectively (Kannan et al., 2005). MX, MK, HHCB, AHTN and HHCB–lactone were reported in breast milk from the United States (Reiner et al., 2007), and the DI of MX, MK, HHCB, AHTN and HHCB by breast-fed infants was calculated to be 297 ± 229, 780 ± 805, 1830 ± 1170, 565 ± 614 and 649 ± 598 ng, respectively. The temporal variation in the concentrations of SMs in breast milk ($N = 101$) from Sweden and their association with the use of fragrances was studied (Lignell et al., 2008).

A positive correlation between HHCB and AHTN concentrations was found in most studies. Zhang et al. (2011) studied the occurrence of SMs in breast milk collected from three different cities in eastern China. HHCB was the major compound that accounted for 70% of the total SM concentrations, followed by MX (19%). HHCB concentrations in breast milk from China were similar to the concentrations found in Sweden but 2–10 times lower than those reported in the United States, South Korea and Denmark (Zhang et al., 2011). Zhang et al. also estimated the DI of SMs by infants via breast milk and found that the estimated doses were below the provisional tolerable DI (PTDI) values for adults. The calculated PTDI values for MX, MK, AHTN and HHCB were 7, 7, 50 and 500 µg/kg bw, respectively (Rimkus, 2004). Yin et al. (2012) studied the occurrence of SMs in breast milk collected from Sichuan, southwestern China, and found concentrations similar to those reported in aforementioned studies. The estimated maximum DI values (median) for HHCB and AHTN of infants through breast milk were 295 and 427 ng, respectively, whereas these intake values were 1650 and 250 ng for Swedish and 360 and 280 ng for Japanese infants.

Gatermann et al. (1998), for the first time, reported the amino metabolites of MX (*para*/4-amino MX and *ortho*/2-amino MX) and MK (4-amino MK) in sewage effluent and surface waters. The concentrations of amino metabolites in surface water were 4–40 times higher than those of their corresponding parent musks which showed nitro reduction during sewage treatment. Riedel et al. (1999) demonstrated haemoglobin binding activity of amino metabolites and their excretion in human urine. Nevertheless, studies on the metabolism of musks in humans are sparse. An acetyl derivative of the 4-amino metabolite of musks that was found in rat urine was not found in human urine (Rimkus, 2004). Hutter et al. (2005) examined the occurrence of polycyclic musks in the blood of healthy young adults ($N = 100$) from Austria. They found high concentrations of HHCB (median = 420 ng/L) with a detection rate of 91%, whereas AHTN was found at lower concentrations in only 17% of the samples. The concentrations of HHCB were significantly higher in women than in men.

Zhang et al. (2015) reported partitioning of SMs between blood (paired maternal blood and umbilical cord blood) and breast milk. The median concentrations in maternal blood, umbilical cord blood and breast milk were 71.2, 87.3 and 35.2 ng/glw, respectively. The distribution ratio between maternal blood and cord blood was 0.82 and between maternal blood and breast milk was 2.02. Lee et al. (2015) studied the occurrence, temporal trends and health risks of SMs in Korea. They found no correlation between the concentrations of SMs and maternal age or body mass index. Further, the mean concentrations of SMs were three orders of magnitude lower than the PTDI values suggested for adults.

Overall, perfumes, air fresheners, body creams and lotions are the major sources of exposure to HHCB, whereas detergents and soaps are the major route of exposure to MX (Hutter et al., 2010). It is important to note that dermal sorption from cosmetics is a major route of exposure to SMs for adults (Brunn et al., 2004; Slanina, 2004). For infants, dietary exposure via breast milk is the major route of exposure to SMs (Slanina, 2004).

Studies that report SMs in various human matrices are compiled and presented in Table 15.4. Although up to 11 SMs were covered in a few studies, most of the studies included the measurement of HHCB, AHTN, MX and MK. The highest median concentration (213 ng/g, lw) of HHCB in breast milk was reported in Korea (Lee et al., 2015), and the highest concentration in blood (420 ng/L) was reported in Austria (Hutter et al., 2005).

Biomonitoring of Phthalates

Phthalates are esters of phthalic acid (diesters of 1,2-benzenedicarboxylic acid), and synthetic organic chemicals used in a broad spectrum of industries, particularly as solvents, plasticizers and additives in cosmetics (Latini, 2005). Phthalates are nearly ubiquitous in the environment, which can be attributed to their very high production volumes. The annual production of phthalates was over 5 million tons globally in 2006 (Mackintosh et al., 2006). Another study indicated that global annual production and use of DEHP and total phthalates were over 2 million tons and 18 billion pounds, respectively (estimated in 2002) (Lorz et al., 2002). Phthalates have been used in various commercial applications, such as cosmetics, medical devices, food packaging, perfumes, toys, teethers, adhesives, paints, floorings, medical devices, lubricants, hair sprays, shampoos, soaps, nail polishes and detergents (Latini, 2005; Gimeno et al., 2012). In polyvinyl chloride plastics, DEHP, diisononyl phthalate and butyl benzyl phthalate (BBzP) are used as additive chemicals to increase flexibility, transparency, durability and longevity. Because phthalates are not covalently bound to the plastic polymer, they can easily be released into the environment by evaporation and abrasion (Gimeno et al., 2014). Humans are largely exposed to phthalates via ingestion, inhalation and dermal absorption (Guo and Kannan, 2013; Guo et al., 2014).

Table 15.4 Reported synthetic musk concentrations in human specimens collected from various countries

Matrix	Country	Study population	Method	Chemicals measured	LOD/LOQ	Reported concentration (geometric mean)	References
Breast milk	Germany	391 individuals	LLE, silica gel clean-up followed by GC–ECD	MX, MK, MA	NA	MX, 70 ng/g, fat; MK, 30 ng/g, fat; MA, 30 ng/g, fat (median)	Liebl and Ehrenstorfer (1993)
Blood	Austria	100 adults	LLE followed by GC–MS	HHCB, AHTN	LOQ: 31–62 ng/L	HHCB, 420 ng/L; AHTN, <LOD; MX, NM; MK, NM (median)	Hutter et al. (2005)
Adipose tissue	USA	12 men 37 women	Solvent extraction followed by GC–MS	HHCB, AHTN	LOQ: HHCB, 5.0 ng/g, ww AHTN, 5.0 ng/g, ww	HHCB, 149 ng/g, lw AHTN, 37.4 ng/g, lw (median)	Kannan et al. (2005)
Breast milk	Denmark	10 individuals	LLE, silica gel clean-up followed by GC–HRMS	HHCB, AHTN, MX, MK	LOD: 0.22–5.0 ng/g fat weight	HHCB, 147 ng/g, fat; AHTN, 17.5 ng/g, fat. MX, 9.4 ng/g, fat. MK, 14.9 ng/g, fat (median)	Olesen et al. (2005)
Breast milk	USA	39 women	LLE, GPC followed by GC–MS	HHCB, HHCB–lactone, AHTN, MX, MK	LOQ: 2–10 ng/g lw	HHCB, 136 ng/g lw; HHCB-lactone, 58.3 ng/g lw; AHTN, 53 ng/g lw; MX, 17 ng/g lw; MK, 58.2 ng/g lw (median)	Reiner et al. (2007)
Breast milk	Sweden	101 women	LLE, GPC followed by GC–MS	HHCB, AHTN, MX, MK	LOQ: 2–9 ng/g lw	HHCB, 63.9 ng/g lw; AHTN, 10.4 ng/g lw; MX, 9.5 ng/g lw; MK, <LOQ (median)	Lignell et al. (2008)
Breast milk	Germany	85 women	LLE, GPC followed by GC–MS	MX, MK	1–5 ng/g lw	MX, 8.0 ng/g lw; MK, ND (median)	Raab et al. (2008)

Continued

Table 15.4 Reported synthetic musk concentrations in human specimens collected from various countries—cont'd

Matrix	Country	Study population	Method	Chemicals measured	LOD/LOQ	Reported concentration (geometric mean)	References
Blood	Austria	114 adults	LLE, silica gel clean-up followed by GC–MS	HHCB, MX	5.6–124 ng/L	HHCB, 420.0 ng/L; MX, 11.0 ng/L (median)	Hutter et al. (2009)
Breast milk	Japan	20 women	GPC followed by GC–MS	HHCB, AHTN	LOD: 1 ng/g ww	HHCB, <50–440 ng/g, lipid; AHTN, <50–190 ng/g, lipid (median)	Ueno et al. (2009)
Adipose tissue	Japan	3 individuals	GPC followed by GC–MS	HHCB, AHTN	LOD: 0.1 ng/g ww	HHCB, <11–33 ng/g, lipid; AHTN, <9–13 ng/g, lipid (median)	Ueno et al. (2009)
Blood	China	204 individuals	LLE followed by GC–MS	HHCB, AHTN, MX, MK	0.13–0.15 ng/mL	HHCB, 0.85 ng/g; AHTN, 0.53 ng/g; MX, <LOQ; MK, <LOQ	Hu et al. (2010)
Blood	Austria	53 women (age >50)	LLE, silica gel clean-up followed by GC–MS	HHCB, AHTN, MX, MK	6–124 ng/L	ND—6900 ng/L (max. concentration) (11 musks)	Hutter et al. (2010)
Blood	Austria	55 women (age <50)	LLE, silica gel clean-up followed by GC–MS	HHCB, MK	6–124 ng/L	HHCB, 111 ng/L MK, 41.4 ng/L	Hutter et al. (2010)
Maternal serum	Korea	20 women	LLE, GPC followed by GC–MS	HHCB, HHCB–lactone, AHTN, MX, MK	1 ng/g lw	HHCB, 0.4 ng/g, lipid; HHCB-lactone, 0.2 ng/g, lipid; AHTN, 0.2 ng/g lipid; MX, 0.1 ng/g, lipid; MK, NM	Kang et al. (2010)

Cord serum 1 ng/g	Korea	20 women	LLE, GPC followed by GC–MS	HHCB, HHCB–lactone, AHTN, MX, MK	1 ng/g lw	HHCB, 0.7 ng/g, lipid; HHCB-lactone, 0.5 ng/g, lipid; AHTN, 0.4 ng/g lipid; MX, NM; MK, NM	Kang et al. (2010)
Breast milk	Korea	17 women	LLE, GPC followed by GC–MS	HHCB, HHCB–lactone, AHTN, MX, MK	0.5 ng/g lw	HHCB, 0.2 ng/g, lipid; HHCB-lactone, 0.01 ng/g, lipid; AHTN, 0.02 ng/g, lipid; MX, 0.02 ng/g, lipid; MK, 0.03 ng/g, lipid	Kang et al. (2010)
Breast milk	Switzerland	54 women	LLE, GPC followed by GC–MS	HHCB, AHTN, MX, MK	0.5–20 ng/g lw	HHCB, 36.1 ng/g lw; AHTN, 10.2 ng/g lw; MX, 1.3 ng/g lw; MK, 0.6 ng/g lw (median)	Schlumpf et al. (2010)
Breast milk	China	10 women	LLE followed by GC–MS/MS	HHCB, HHCB–lactone, AHTN, MK	0.6–3.5 ng/g lw	HHCB, 11.7–67.6 ng/g, lipid; HHCB-lactone, ND—70.6 ng/g, lipid; AHTN, 23–118 ng/g lipid; MX, NM; MK, ND (concentration ranges)	Wang et al. (2011)
Breast milk	China	100 women	LLE, GPC followed by GC–MS	HHCB, AHTN, MX, MK	4 ng/g lw	HHCB, 63 ng/g lw; AHTN, 5 ng/g lw; MX, 17 ng/g lw; MK, 4 ng/g lw (median)	Zhang et al. (2011)
Breast milk	China	110 women	LLE, GPC followed by GC–MS/MS	HHCB, HHCB–lactone, AHTN, MX, MK	0.6–5.4 ng/g lw	HHCB, 11.5 ng/g lw; HHCB-lactone, 7.9 ng/g lw; AHTN, 16.5 ng/g lw; MX, <LOQ; MK, <LOQ (median)	Yin et al. (2012)

Continued

Table 15.4 Reported synthetic musk concentrations in human specimens collected from various countries—cont'd

Matrix	Country	Study population	Method	Chemicals measured	LOD/LOQ	Reported concentration (geometric mean)	References
Breast milk	China	46 women	LLE, GPC followed by GC–MS	HHCB, HHCB–lactone, AHTN, MX, MK	LOD: 5–15 ng/mL	HHCB, 21.2 ng/g, lipid; HHCB-lactone, 10.8 ng/g, lipid; AHTN, 2.3 ng/g, lipid; MX, 11.6 ng/g, lipid; MK, 2.7 ng/g, lipid (median)	Zhou et al. (2012)
Blood	Belgium	210 adolescents 204 adults	SBSE followed by GC–MS	HHCB, AHTN	0.028–0.038 ng/mL	HHCB, 0.717 ng/mL; AHTN, 0.118 ng/mL (GM at 95% CI)	Hond et al. (2013)
Breast milk	Korea	208 women	LLE, GPC followed by GC–MS	HHCB, AHTN, MX, MK	2–5 ng/g lw	HHCB, 213 ng/g lw; AHTN, 15.8 ng/g lw; MX, ND, MK, ND (median)	Lee et al. (2015)
Maternal blood	China	42 women	LLE, silica gel clean-up followed by GC–MS	HHCB, HHCB–lactone, AHTN, MX, MK	3.0–8.0 ng/mL (LOD)	HHCB, 14.4 ng/g lw; HHCB-lactone, <LOD; AHTN, <LOD; MX, <LOD; MK, <LOD (median)	Zhang et al (2015)
Cord blood	China	42 women	LLE, silica gel clean-up followed by GC–MS	HHCB, HHCB–lactone, AHTN, MX, MK	3.0–8.0 ng/mL (LOD)	HHCB, 37.7 ng/g lw; HHCB-lactone, 15.3 ng/g lw; AHTN, <LOD; MX, <LOD; MK, <LOD (median)	Zhang et al (2015)
Breast milk	China	42 women	LLE, silica gel clean-up followed by GC–MS	HHCB, HHCB–lactone, AHTN, MX, MK	3.0–8.0 ng/mL (LOD)	HHCB, 9.9 ng/g lw; HHCB-lactone, 6.0 ng/g lw; AHTN, 0.9 ng/g lw; MX, 6.6 ng/g lw; MK, 4.8 ng/g lw (median)	Zhang et al (2015)

AHTN, Tonalide; *CI*, confidence interval; *ECD*, electron capture detector; *GC*, gas chromatography; *GM*, geometric mean; *GPC*, gel permeation chromatography; *HHCB*, Galaxolide; *HRMS*, high-resolution mass spectrometry; *LLE*, liquid–liquid extraction; *LOD*, limit of detection; *LOQ*, limit of quantification; *lw*, lipid weight; *MA*, musk ambrette; *MK*, musk ketone; *MS*, mass spectrometry; *MX*, musk xylene; *NA*, not available; *ND*, not detected; *NM*, not measured; *SBSE*, stir bar sorptive extraction; *uw*, wet weight.

Phthalates are reproductive and developmental toxicants (Kay et al., 2014). Studies have demonstrated that phthalate exposure is associated with oxidative stress in humans (Asimakopoulos et al., 2016; Ferguson et al., 2015; Kim et al., 2014). The low molecular weight phthalates such as DMP, DEP and DBP and diisobutyl phthalate (DIBP) are extensively used in cosmetics (Koniecki et al., 2011). Although DEHP is not used as an ingredient in cosmetics, the plastic containers that are used to store cosmetics can contribute to the occurrence of DEHP in some cosmetics. Although more than 25 different phthalates are registered for commercial use, our assessment here is restricted to DMP, DEP, DBP, DIBP and DEHP, which are the most commonly used phthalates in cosmetics (Gimeno et al., 2012). Various sources and routes of exposure to low-molecular-weight phthalates from cosmetics have been reported by Schettler (2006). Diet is the major source of exposure to DEHP, whereas the use of fragrances and other cosmetics accounts for a major source of exposure to DEP. Thus, apart from cosmetics, several other sources contribute to phthalate exposure in humans.

Reddy et al. (2006) measured phthalates in the plasma from 85 women with endometriosis in southern India and found a significant association between phthalate exposure and the risk of developing endometriosis. Specht et al. (2015) measured phthalates in the serum of couples from Greenland, Poland and Ukraine and found that high DEHP levels reduced time to achieve pregnancy. Hogberg et al. (2008) measured phthalate diesters and their metabolites in breast milk, serum and urine from Swedish women ($N=42$). In most milk and serum samples, the concentrations of phthalate diesters and their metabolites were below the LOD (0.12–3.0 ng/mL). However, detectable concentrations of phthalate metabolites were found in urine (0.5–761 ng/mL). Hogberg et al. concluded that measurements of phthalate diesters in breast milk or serum are prone to false positives due to background contamination and the sensitivity of the method. Similarly, challenges associated with low-level phthalate analysis in biological specimens have been described (Guo and Kannan, 2013).

Lin et al. (2011a) estimated the amount of parent phthalate to which humans are exposed from the concentrations of metabolites and excretion fractions in urine from Taiwan, as presented in the following equation:

$$\text{Parent compound} = \frac{\text{Metabolite concentration}}{\text{Excretion fraction}}$$

The EDI of parent phthalates was further calculated by considering the average weight of women and children with their average urinary excretion per day, as shown in next equation:

$$\text{Estimated daily intake (EDI)} = \frac{(\text{Estimated parent phthalate concentration}) \times (\text{Daily urine excretion})}{\text{Average body weight}}$$

The calculated EDI$_{max}$ values for DEHP and DBP were 8 and 0.08 µg/kg bw/day, respectively, which were much lower than the tolerable daily intake values suggested for DEHP (50 µg/kg bw) and DBP (10 µg/kg bw) by the European Food Safety Authority (Lin et al., 2011a).

Phthalate metabolites are one of the most-studied environmental chemicals in human biomonitoring studies because of their ubiquitous nature and exposure in humans. Phthalates are metabolized to monoesters, which are further conjugated with glucuronide or sulphate and excreted in urine and faeces (Silva et al., 2003). Enzymatic deconjugation followed by purification (mostly by SPE) was widely employed for the measurement of phthalate metabolites in urine. Human exposure assessment of phthalates is based mainly on the measurement of their urinary monoester metabolites (Dewalque et al., 2014). In general, DMP, DEP and DBP undergo degradation/hydrolysis in humans and form their corresponding monoesters, i.e., monomethyl phthalate (mMP), monoethyl phthalate (mEP) and monobutyl phthalate (mBP), respectively. Human metabolism of DEHP, however, is quite complex, and both hydrolysis and oxidation products are formed from this compound. The hydrolysis product of DEHP, mono(2-ethylhexyl) phthalate (mEHP), is not a major metabolite, which is different from that found for other phthalates. The oxidative metabolites of DEHP, such as mono(2-ethyl-5-oxohexyl) phthalate (mEOHP), mono(2-ethyl-5-hydroxyhexyl) phthalate (mEHHP), mono(2-ethyl-5-carboxypentyl) phthalate (mECPP) and mono(2-carboxymethylhexyl) phthalate, however, are the major contributors to DEHP exposure and are appropriate biomarkers of exposure (Silva et al., 2007). The general metabolic pathways of phthalate esters in human are depicted in Fig. 15.5.

Koch et al. (2003) studied phthalate exposures in the general population based on the measurement of primary and secondary metabolites in urine. Urine samples were collected from 32 men and 53 women (age group 7–64 years) from northern Bavaria (Germany). mBP (181 ng/mL), mEP (90.2 ng/mL) and major DEHP metabolites, such as mEHHP, 46.8 ng/mL, and mEOHP, 36.5 ng/mL, were found at high median concentrations. Barr et al. (2003) measured human exposure to phthalates in the United States from samples collected as a part of the NHANES. The median concentrations of DEHP metabolites, such as mEHP, mEOHP and mEHHP, were 4.5, 28.3 and 35.9 ng/mL, respectively, and all were highly intercorrelated (r^2 = 0.984). To obtain the most reliable biomarkers for DEHP exposure, Koch et al. (2003) calculated the concentration ratios mEHHP/mEOHP, mEHHP/mEHP and mEOHP/mEHP, which were 1.4, 8.2 and 5.9, respectively. These ratios suggested that mEHP was further oxidized to form mEHHP and mEOHP. Fromme et al. (2007) studied the occurrence of phthalate metabolites in Germans, and mBP was found at high concentrations in Germans (median = 49 ng/mL). There were no gender- or age-related differences in phthalate monoester concentrations in urine. Silva et al. (2007) demonstrated the possibility of measuring 22 phthalate metabolites in human urine at sub-parts-per-billion levels (ng/mL). Ye et al. (2008a,b,c)

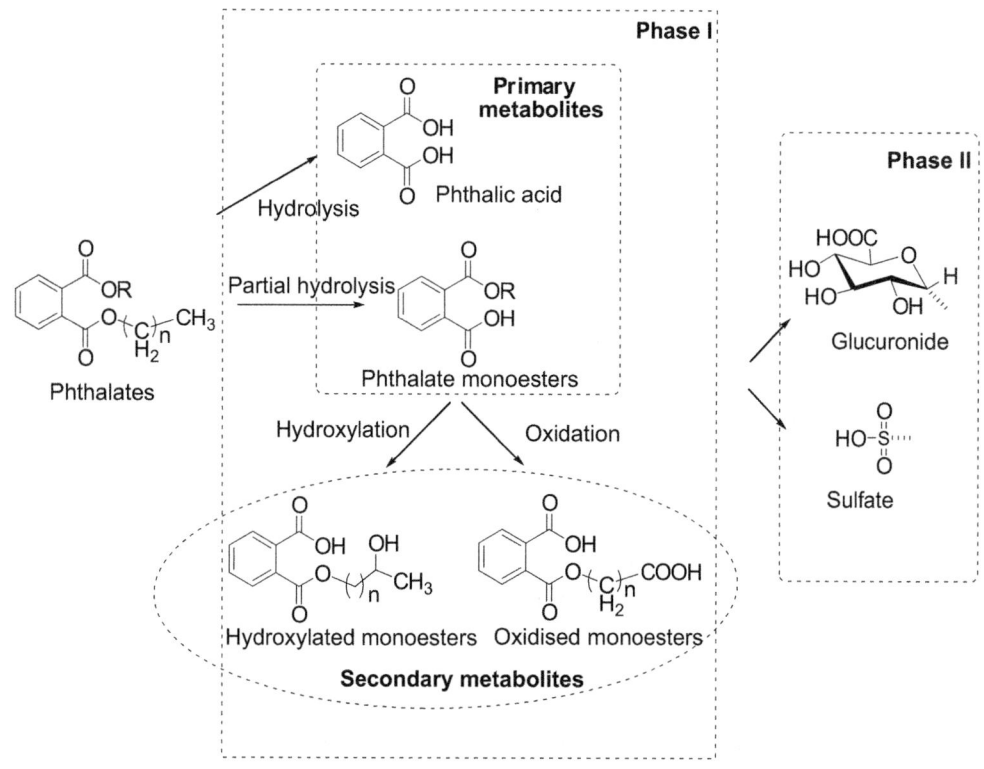

Figure 15.5 Metabolic pathways of phthalate esters in humans.

studied 14 phthalate metabolites in urine from 100 pregnant women from the Generation R study in the Netherlands. Twelve women had mEP concentrations above 1000 µg/g creatinine, with an overall median of 222 µg/g creatinine, which was the highest among all phthalate metabolites measured. Similarly, the median concentration of total DEHP metabolites (sum of five metabolites) was 88.4 µg/g creatinine.

Frederiksen et al. (2010) reported correlations between phthalate metabolites in urine, serum and seminal plasma in young Danish men ($N=60$). The mean concentrations of mEP, mBP and DEHP metabolites (sum of mEHP, MEHHP, mEOHP and mECPP) were considerably lower in serum (mEP 4.2, mBP 0.4 and DEHP 7.6 ng/mL) and seminal plasma (1.0, 0.8 and 0.6 ng/mL) than in urine (326, 42.5 and 115 ng/mL). Guo et al. (2011a) examined the occurrence of phthalate metabolites in seven Asian countries. The urine samples collected from Kuwait contained the highest median concentrations of mEP (391 ng/mL), mBP (94 ng/mL) and DEHP metabolites (202 ng/mL), and the total phthalate metabolite concentration (median) was 1050 ng/mL, which was the highest among all the countries studied. Overall, mEP was the predominant metabolite in the Indian and Kuwaiti urine samples (49% of the total); in China (52%), mBP was the major metabolite. In Korea (46%), Japan (31%) and Vietnam (52%), DEHP metabolites were the dominant ones. In contrast, mMP accounted for <8% of the total phthalate metabolite concentrations in all Asian countries, except for Japan, where it was 20%. Overall, mEP and DEHP metabolites were the major contributors to total phthalate metabolites in most Asian countries, which was similar to that found for the United States (Colacino et al., 2010).

The estimated mean daily exposures to DEP (100-fold below) and DBP (10-fold below) in the Asian countries and the United States were well below the reference doses (DEP 800, DBP 100 and DEHP 20 µg/kg bw/day) suggested by the US Environmental Protection Agency (EPA). The estimated daily exposure doses to DEHP in Kuwait and India, however, were close to the reference doses of the EPA. Similarly, high concentrations of DEHP metabolites (mean concentration of the sum of five metabolites = 338 ng/mL) were reported in the Saudi population (Asimakopoulos et al., 2016). A large number of studies have reported on the measurement of phthalate metabolites in human specimens collected from European (Germany, the Netherlands, Denmark, Norway, Sweden, Greece, the Czech Republic, Hungary, Slovakia and Spain) and Asian countries (Japan, China, South Korea, India, Taiwan, Vietnam, Saudi Arabia, Malaysia and Kuwait) (Table 15.5). Overall, a large number of studies are available from the United States, China and Germany. Since 2001, there has been clear evidence of a decline in DEP, DBP and DEHP exposure in the US and German populations, but, in the Chinese population, DEHP exposure has increased (Johns et al., 2015), and phthalate exposure levels are higher in children than in adults (Table 15.5).

CONCLUSIONS

Cosmetics are the major sources of exposure to several oestrogenic chemicals, and cosmetics contain a wide range of chemicals that are used for various reasons. In this chapter, we presented an overview of human exposure to various environmental oestrogenic chemicals that are present in cosmetics and noted that human biomonitoring is a valuable approach to assessing human exposures. We described the distribution of the chemicals, possible routes of exposure and their risks. It should be noted that the chemicals reviewed here are only a subset of those that are listed as ingredients in various cosmetics. A myriad of chemicals other than those that are reviewed in this chapter exist in cosmetics. The toxicity and exposure to all of the chemicals labelled as ingredients in cosmetics have not been thoroughly investigated. Therefore, a thorough investigation of the sources of and exposures to all of those ingredients in cosmetics, and their exposure levels and health risks, should be the focus of future research. Biomonitoring of exposure to other ingredients in cosmetics is also needed. Further, although a large number of human biomonitoring studies are available in developed countries, there is still a lack of data/literature on the exposure of populations in developing countries. Thus, epidemiological studies are needed to assess exposure to the environmental chemicals used in cosmetics and the health outcomes.

Table 15.5 Reported concentrations of phthalate metabolites in human specimens collected from various countries

Matrix	Country	Study population	Method	Chemicals measured	LOD/LOQ	Geometric mean	References
Urine	USA	2541 individuals	Enzymatic deconjugation, SPE followed by online LC–MS/MS	mEP, mBP, mDEHP	LOD: 0.6–1.2 ng/mL	mMP, NM; mEP, 305.0 ng/mL; mBP, 41.0 ng/mL; mIBP, NM; mDEHP, 68.7 ng/mL (median)	Barr et al. (2003)
Urine	Germany	53 women 32 men	Enzymatic deconjugation followed by online LC–MS/MS	mEP, mBP, mDEHP	LOD: 0.25–1.0 ng/mL	mMP, NM; mEP, 90.2 ng/mL; mBP, 181.0 ng/mL; mIBP, NM; mDEHP, 83.3 ng/mL (median)	Koch et al. (2003)
Urine	Germany	254 children	Enzymatic deconjugation followed by online LC–MS/MS	mDEHP	LOQ: 0.5 ng/mL	mMP, NM; mEP, NM; mBP, NM; mIBP, NM; mDEHP, 99.9 ng/mL	Becker et al. (2004)
Urine	USA	214 women	Enzymatic deconjugation, SPE followed by online LC–MS/MS	mEP, mBP, mIBP, mDEHP	LOD: 0.71 ng/mL	mMP, NM; mEP, 117.0 ng/mL; mBP, 16.2 ng/mL; mIBP, 2.5 ng/mL; mDEHP, 24.8 ng/mL	Swan et al. (2005)
Breast milk	Denmark	65 women	SPE followed by LC–MS/MS	mMP, mEP, mBP, mDEHP	LOD: 0.01–0.10 ng/mL	mMP, 0.1 ng/mL; mEP, 0.9 ng/mL; mBP, 4.3 ng/mL; mIBP, NM; mDEHP, 9.5 ng/mL	Main et al. (2006)
Breast milk	Finland	65 women	SPE followed by LC–MS/MS	mMP, mEP, mBP, mDEHP	LOD: 0.01–0.10 ng/mL	mMP, 0.1 ng/mL; mEP, 1.0 ng/mL; mBP, 12.0 ng/mL; mIBP, NM; mDEHP, 13.0 ng/mL	Main et al. (2006)

Continued

Table 15.5 Reported concentrations of phthalate metabolites in human specimens collected from various countries—cont'd

Matrix	Country	Study population	Method	Chemicals measured	LOD/LOQ	Geometric mean	References
Urine	Germany	399 individuals	Enzymatic deconjugation followed by online LC–MS/MS	mBP, mIBP, mDEHP	LOD: 0.25 ng/mL	mMP, NM; mEP, NM; mBP, 49.6 ng/mL; mIBP, 44.9 ng/mL; mDEHP, 38.8 ng/mL (median)	Fromme et al. (2007)
Urine	Japan	35 adults 1 children	Enzymatic deconjugation, SPE followed by online LC–MS/MS	mMP, mEP, mBP, mDEHP	NA	mMP, 33.0 ng/mL; mEP, 18.0 ng/mL; mBP, 36.0 ng/mL; mIBP, NM; mDEHP, 5.0 ng/mL (median)	Itoh et al. (2007)
Urine	Germany	634 individuals	Enzymatic deconjugation followed by online LC–MS/MS	mBP, mIBP, mDEHP	LOD: 0.25 ng/mL	mMP, NM; mEP, NM; mBP, 109.0 ng/mL; mIBP, 35.4 ng/mL; mDEHP, 45.3 ng/mL (median)	Wittassek et al. (2007)
Breast milk	Sweden	42 women	Enzymatic deconjugation, SPE followed by online LC–MS/MS	mEP, mBP, mIBP, mDEHP	LOD: 0.6–1.2 ng/mL	mMP, NM; mEP, ND; mBP, 0.5 ng/mL; mIBP, ND; mDEHP, 0.49 ng/mL (median)	Hogberg et al. (2008)
Serum	Sweden	36 women	Enzymatic deconjugation, SPE followed by online LC–MS/MS	mEP, mBP, mIBP, mDEHP	LOD: 0.6–1.2 ng/mL	mMP, NM; mEP, 0.5 ng/mL; mBP, 0.5 ng/mL; mIBP, 0.5 ng/mL; mDEHP, 0.5 ng/mL (median)	Hogberg et al. (2008)
Urine	Sweden	38 women	Enzymatic deconjugation, SPE followed by online LC–MS/MS	mMP, mEP, mBP, mIBP, mDEHP	LOD: 0.6–1.2 ng/mL	mMP, 1.2 ng/mL; mEP, 35.0 ng/mL; mBP, 46.0 ng/mL; mIBP, 16.0 ng/mL; mDEHP, 35.0 ng/mL (median)	Hogberg et al. (2008)

Sample	Country	Subjects	Method	Analytes	LOD/LOQ	Results	Reference
Urine	USA	35 children	Enzymatic deconjugation, SPE followed by online LC–MS/MS	mEP, mBP, mIBP, mDEHP	LOD: 0.25–0.9 ng/mL	mMP, NM; mEP, 177.7 ng/mL; mBP, 52.4 ng/mL; mIBP, 16.6 ng/mL; mDEHP, 1025.9 ng/mL	Teitelbaum et al. (2008)
Urine	Netherlands	100 women	Enzymatic deconjugation, online SPE–LC–MS/MS	mMP, mEP, mBP, mIBP, mDEHP	LOD: 0.5–2.0 ng/mL	mMP, ND; mEP, 112.0 ng/mL; mBP, 43.2 ng/mL; mIBP, 41.3 ng/mL; mDEHP, 61.8 ng/mL	Ye et al. (2008a,b,c)
Urine	Germany	599 children	Enzymatic deconjugation, online SPE–LC/LC–MS/MS	mBP, mIBP, mDEHP	LOQ: 0.25–1.0 ng/mL	mMP, NM; mEP, NM; mBP, 95.6 ng/mL; mIBP, 94.3 ng/mL; mDEHP, 174.6 ng/mL	Becker et al. (2009)
Urine	Japan	80 women (controls)	Enzymatic deconjugation, SPE followed by LC–MS/MS	mEP, mBP, mDEHP	NA	mMP, NM; mEP, 21.4 ng/mL; mBP, 84.3 ng/mL; mIBP, NM; mDEHP, 72.7 ng/mL (median)	Itoh et al. (2009)
Urine	Japan	57 women (cases)	Enzymatic deconjugation, SPE followed by LC–MS/MS	mEP, mBP, mDEHP	NA	mMP, NM; mEP, 39.6 ng/mL; mBP, 87.2 ng/mL; mIBP, NM; mDEHP, 89.3 ng/mL (median)	Itoh et al. (2009)
Urine	Norway	10 women	Enzymatic deconjugation, SPE followed by LC–MS/MS	mMP, mEP, mBP, mIBP, mDEHP	LOD: 0.5–2.0 ng/mL	mMP, 2.0 ng/mL; mEP, 310.0 ng/mL; mBP, 41.1 ng/mL; mIBP, 57.0 ng/mL; mDEHP, 112.3 ng/mL (mean)	Ye et al. (2009)

Continued

Table 15.5 Reported concentrations of phthalate metabolites in human specimens collected from various countries—cont'd

Matrix	Country	Study population	Method	Chemicals measured	LOD/LOQ	Geometric mean	References
Urine	Mexico	221 women	Enzymatic deconjugation, SPE followed by LC–MS/MS	mEP, mBP, mIBP, mDEHP	NA	mMP, NM; mEP, 107.0 ng/mg, C; mBP, 82.5 ng/mg, C; mIBP, 8.85 ng/mg, C; mDEHP, 166.3 ng/mg, C	Carrillo et al. (2010)
Urine	Mexico	233 women	Enzymatic deconjugation, SPE followed by LC–MS/MS	mEP, mBP, mIBP, mDEHP	NA	mMP, NM; mEP, 170.0 ng/mg, C; mBP, 63.0 ng/mg, C; mIBP, 7.8 ng/mg, C; mDEHP, 169.4 ng/mg, C	Carrillo et al. (2010)
Urine	USA	2350 individuals	Enzymatic deconjugation, SPE followed by LC–MS/MS	mMP, mEP, mBP, mIBP, mDEHP	LOD: 0.26–1.0 ng/mL	mMP, 1.8 ng/mL; mEP, 194.4 ng/mL; mBP, 20.7 ng/mL; mIBP, 3.7 ng/mL; mDEHP, 73.0 ng/mL	Colacino et al. (2010)
Urine	Denmark	60 men	Enzymatic deconjugation, SPE followed by LC–MS/MS	mEP, mBP, mIBP, mDEHP	LOD: 0.14–1.43 ng/mL	mMP, NM; mEP, 54.5 ng/mL; mBP, 36.8 ng/mL; mIBP, 47.3 ng/mL; mDEHP, 68.1 ng/mL (median)	Frederiksen et al. (2010)
Serum	Denmark	60 men	Enzymatic deconjugation, SPE followed by LC–MS/MS	mEP, mBP, mIBP, mDEHP	LOD: 0.28–0.82 ng/mL	mMP, NM; mEP, <LOD; mBP, ND; mIBP, <LOD; mDEHP, 8.4 ng/mL (median)	Frederiksen et al. (2010)
Seminal plasma	Denmark	60 men	Enzymatic deconjugation, SPE followed by LC–MS/MS	mEP, mBP, mIBP, mDEHP	LOD: 0.28–1.01 ng/mL	mMP, NM; mEP, <LOD; mBP, <LOD; mIBP, ND; mDEHP, <LOD (median)	Frederiksen et al. (2010)

Urine	Korea	25 adults	Enzymatic deconjugation, followed by online LC–MS/MS	mEP, mBP, mIBP, mDEHP	LOQ: 0.1–3.15 ng/mL	mMP, NM; mEP, 80.0 ng/mL; mBP, 134.0 ng/mL; mIBP, 40.4 ng/mL; mDEHP, 125.8 ng/mL (median)	Ji et al. (2010)
Milk	Switzerland	54 women	Enzymatic deconjugation, online SPE–LC/LC–MS/MS	mBP, mIBP, mDEHP	LOD: 0.5–1.0 ng/mL	mMP, NM; mEP, NM; mBP, 6.0 ng/mL; mIBP, 24.3 ng/mL; mDEHP, 26.2 ng/mL (median)	Schlumpf et al. (2010)
Urine	Spain	120 women	Enzymatic deconjugation, followed by online LC–MS/MS	mEP, mBP, mIBP, mDEHP	LOD: 0.2–1.2 ng/mL	mMP, NM; mEP, 324 ng/mL; mBP, 27.5 ng/mL; mIBP, 30.0 ng/mL; mDEHP, 69.6 ng/mL	Casas et al. (2011)
Urine	Spain	30 children	Enzymatic deconjugation, followed by online LC–MS/MS	mEP, mBP, mIBP, mDEHP	LOD: 0.2–1.2 ng/mL	mMP, NM; mEP, 755 ng/mL; mBP, 30.2 ng/mL; mIBP, 42.0 ng/mL; mDEHP, 223.2 ng/mL	Casas et al. (2011)
Urine	USA	7600–10,031 individuals	Enzymatic deconjugation, followed by online LC–MS/MS	mMP, mEP, mBP, mIBP, mDEHP	LOD: 0.1–1.0 ng/mL	mMP, 1.4 ng/mg, C; mEP, 167.0 ng/mg, C; mBP, 18.9 ng/mg, C; mIBP, 3.6 ng/mg, C; mDEHP, 73.1 ng/mg, C	Ferguson et al. (2011)
Urine	Mexico	108 women	Enzymatic deconjugation, followed by online LC–MS/MS	mEP, mBP, mIBP, mDEHP	NA	mMP, NM; mEP, 83.2 ng/mg, C; mBP, 72.4 ng/mg, C; mIBP, 8.4 ng/mg, C; mDEHP, 151.1 ng/mg, C	Franco et al. (2011)

Continued

Table 15.5 Reported concentrations of phthalate metabolites in human specimens collected from various countries—cont'd

Matrix	Country	Study population	Method	Chemicals measured	LOD/LOQ	Geometric mean	References
Urine	Denmark	129 children	Enzymatic deconjugation, followed by online LC–MS/MS	mEP, mBP, mIBP, mDEHP	LOD: 0.24–3.94 ng/mL	mMP, NM; mEP, 29.0 ng/mL; mBP and mIBP, 111.0 ng/mL; mDEHP, 107.0 ng/mL (median)	Frederiksen et al. (2011a)
Urine	Kuwait	22 women 24 men	Enzymatic deconjugation, SPE followed by LC–MS/MS	mMP, mEP, mBP, mIBP, mDEHP	LOQ: 0.1–0.5 ng/mL	mMP, 10.1 ng/mL; mEP, 411 ng/mL; mBP, 113 ng/mL; mIBP, 54.1 ng/mL; mDEHP, 180.4 ng/mL	Guo et al. (2011a)
Urine	South Korea	60 individuals	Enzymatic deconjugation, SPE followed by LC–MS/MS	mMP, mEP, mBP, mIBP, mDEHP	LOQ: 0.1–0.5 ng/mL	mMP, 10.0 ng/mL; mEP, 13.4 ng/mL; mBP, 16.7 ng/mL; mIBP, 4.5 ng/mL; mDEHP, 43.6 ng/mL	Guo et al. (2011a)
Urine	Vietnam	16 women 14 men	Enzymatic deconjugation, SPE followed by LC–MS/MS	mMP, mEP, mBP, mIBP, mDEHP	LOQ: 0.1–0.5 ng/mL	mMP, 8.4 ng/mL; mEP, 7.2 ng/mL; mBP, 19.1 ng/mL; mIBP, 13.6 ng/mL; mDEHP, 56.7 ng/mL	Guo et al. (2011a)
Urine	Malaysia	19 women 10 men	Enzymatic deconjugation, SPE followed by LC–MS/MS	mMP, mEP, mBP, mIBP, mDEHP	LOQ: 0.1–0.5 ng/mL	mMP, 6.3 ng/mL; mEP, 18.6 ng/mL; mBP, 10.5 ng/mL; mIBP, 10.8 ng/mL; mDEHP, 27.5 ng/mL	Guo et al. (2011a)
Urine	Japan	8 women 27 men	Enzymatic deconjugation, SPE followed by LC–MS/MS	mMP, mEP, mBP, mIBP, mDEHP	LOQ: 0.1–0.5 ng/mL	mMP, 18.2 ng/mL; mEP, 16.4 ng/mL; mBP, 17.7 ng/mL; mIBP, 7.5 ng/mL; mDEHP, 35.1 ng/mL	Guo et al. (2011a)

Urine	China	21 women 19 men	Enzymatic deconjugation, SPE followed by LC–MS/MS	mMP, mEP, mBP, mIBP, mDEHP	LOQ: 0.1–0.5 ng/mL	mMP, 16.5 ng/mL; mEP, 20.7 ng/mL; mBP, 49.6 ng/mL; mIBP, 44.0 ng/mL; mDEHP, 44.2 ng/mL	Guo et al. (2011a)
Urine	India	15 women 7 men	Enzymatic deconjugation, SPE followed by LC–MS/MS	mMP, mEP, mBP, mIBP, mDEHP	LOQ: 0.1–0.5 ng/mL	mMP, 8.6 ng/mL; mEP, 150 ng/mL; mBP, 13 ng/mL; mIBP, 18.3 ng/mL; mDEHP, 77.9 ng/mL	Guo et al. (2011a)
Urine	China	183 individuals	Enzymatic deconjugation, SPE followed by LC–MS/MS	mMP, mEP, mBP, mIBP, mDEHP	LOQ: 0.1–0.5 ng/mL	mMP, 14.6 ng/mL; mEP, 22.1 ng/mL; mBP, 63.5 ng/mL; mIBP, 57.1 ng/mL; mDEHP, 76.1 ng/mL	Guo et al. (2011b)
Urine	Germany	111 children (48 girls and 63 boys)	Enzymatic deconjugation, SPE followed by LC/LC–MS/MS	mBP, mIBP, mDEHP	LOQ: 0.25–0.5 ng/mL	mMP, NM; mEP, NM; mBP, 53.6 ng/mL; mIBP, 74.9 ng/mL; mDEHP, 130.1 ng/mL	Koch et al. (2011)
Urine	Taiwan	30 (children, age 2)	Enzymatic deconjugation, SPE followed by LC–MS/MS	mBP, mIBP, mDEHP	NA	mMP, NM; mEP, NM; mBP, 100.4 ng/mL; mIBP, 17.2 ng/mL; mDEHP, 195.8 ng/mL	Lin et al. (2011a)
Urine	Taiwan	59 (children, age 5)	Enzymatic deconjugation, SPE followed by LC–MS/MS	mBP, mIBP, mDEHP	NA	mMP, NM; mEP, NM; mBP, 75.2 ng/mL; mIBP, 25.2 ng/mL; mDEHP, 148.9 ng/mL	Lin et al. (2011a)

Continued

Table 15.5 Reported concentrations of phthalate metabolites in human specimens collected from various countries—cont'd

Matrix	Country	Study population	Method	Chemicals measured	LOD/LOQ	Geometric mean	References
Urine	Taiwan	100 women	Enzymatic deconjugation, SPE followed by LC–MS/MS	mBP, mIBP, mDEHP	NA	mMP, NM; mEP, NM; mBP, 72.3 ng/mL; mIBP, 12.5 ng/mL; mDEHP, 96.8 ng/mL	Lin et al. (2011a)
Urine	Taiwan	155 women	Enzymatic deconjugation, SPE followed by LC–MS/MS	mMP, mEP, mBP, mDEHP	LOD: 0.23–3.4 ng/mL	mMP, 5.7 ng/mL; mEP, 25.3 ng/mL; mBP, 80.0 ng/mL; mIBP, NM; mDEHP, 22.6 ng/mL	Lin et al. (2011b)
Urine	Denmark	143 children	Enzymatic deconjugation, SPE followed by LC–MS/MS	mEP, mBP, mIBP, mDEHP	LOD: 0.53–1.43 ng/mL	mMP, NM; mEP, 28.0 ng/mL; mBP, 39.0 ng/mL; mIBP, 74.0 ng/mL; mDEHP, 99.0 ng/mL (mean)	Frederiksen et al. (2013b)
Urine	Denmark	145 women	Enzymatic deconjugation, SPE followed by LC–MS/MS	mEP, mBP, mIBP, mDEHP	LOD: 0.53–1.43 ng/mL	mMP, NM; mEP, 74.0 ng/mL; mBP, 26.0 ng/mL; mIBP, 48.0 ng/mL; mDEHP, 50.7 ng/mL (mean)	Frederiksen et al. (2013b)
Urine	Belgium	123 men 138 women	Enzymatic deconjugation, SPE followed by UPLC–MS/MS	mEP, mBP, mIBP, mDEHP	LOD: 0.19–0.37 ng/mL	mMP, NM; mEP, 37.6 ng/mL; mBP, 31.3 ng/mL; mIBP, 26.2 ng/mL; mDEHP, 17.1 ng/mL	Dewalque et al. (2014)
Urine	Korea	39 children	Enzymatic deconjugation, followed by online LC–MS/MS	mEP, mBP, mIBP, mDEHP	LOQ: 0.5–1.9 ng/mL	mMP, NM; mEP, 19.2 ng/mL; mBP, 107.0 ng/mL; mIBP, 53.4 ng/mL; mDEHP, 145.6 ng/mL	Kim et al. (2014)

Urine	Germany	465 children	Enzymatic deconjugation, SPE followed by LC/LC–MS/MS	mMP, mEP, mBP, mIBP, mDEHP	LOQ: 0.2–1.0 ng/mL	mMP, 3.16 ng/mL; mEP, 25.8 ng/mL; mBP, 47.0 ng/mL; mIBP, 48.9 ng/mL; mDEHP, 82.8 ng/mL	Kasper-Sonnenberg et al. (2014)
Urine	Saudi Arabia	130 individuals	Enzymatic deconjugation, SPE followed by LC–MS/MS	mMP, mEP, mBP, mIBP, mDEHP	LOD: 0.009–0.21 ng/mL	mMP, 8.65 ng/mL; mEP, 47.5 ng/mL; mBP, 38.5 ng/mL; mIBP, 38.5 ng/mL; mDEHP, 117.1 ng/mL (median/GM)	Asimakopoulos et al. (2016)
Urine	Sweden	314 men	Enzymatic deconjugation, automated SPE followed by LC–MS/MS	mEP, mBP, mDEHP	LOD: 0.07–0.33 ng/mL	mMP, NM; mEP, 41.0 ng/mL; mBP, 47.0 ng/mL; mIBP, NM; mDEHP, 48.4 ng/mL (median)	Axelsson et al. (2015)
Urine	Czech Republic	117 women	Enzymatic deconjugation, SPE followed by LC–MS/MS	mMP, mEP, mDEHP	LOD: 0.09–0.5 ng/mL	mMP, ND; mEP, 56.7 ng/mL; mBP, NM; mIBP, NM; mDEHP, 32.2 ng/mL	Cerna et al. (2015)
		120 children	Enzymatic deconjugation, SPE followed by LC–MS/MS	mMP, mEP, mDEHP	LOD: 0.09–0.5 ng/mL	mMP, ND; mEP, 31.6 ng/mL; mBP, NM; mIBP, NM; mDEHP, 61.9 ng/mL	Cerna et al. (2015)

Continued

Table 15.5 Reported concentrations of phthalate metabolites in human specimens collected from various countries—cont'd

Matrix	Country	Study population	Method	Chemicals measured	LOD/LOQ	Geometric mean	References
Urine	Hungary	115 women	Enzymatic deconjugation, SPE followed by LC–MS/MS	mMP, mEP, mDEHP	LOD: 0.09–0.5 ng/mL	mMP, ND; mEP, 55.0 ng/mL; mBP, NM; mIBP, NM; mDEHP, 32.4 ng/mL	Cerna et al. (2015)
		117 children	Enzymatic deconjugation, SPE followed by LC–MS/MS	mMP, mEP, mDEHP	LOD: 0.09–0.5 ng/mL	mMP, ND; mEP, 47.0 ng/mL; mBP, NM; mIBP, NM; mDEHP, 56.7 ng/mL	Cerna et al. (2015)
Urine	Slovakia	125 women	Enzymatic deconjugation, SPE followed by LC–MS/MS	mMP, mEP, mDEHP	LOD: 0.09–0.5 ng/mL	mMP, ND; mEP, 54.8 ng/mL; mBP, NM; mIBP, NM; mDEHP, 36.7 ng/mL	Cerna et al. (2015)
		127 children	Enzymatic deconjugation, SPE followed by LC–MS/MS	mMP, mEP, mDEHP	LOD: 0.09–0.5 ng/mL	mMP, ND; mEP, 39.6 ng/mL; mBP, NM; mIBP, NM; mDEHP, 82.8 ng/mL	Cerna et al. (2015)
Urine	USA	482 individuals	Enzymatic deconjugation, SPE followed by LC–MS/MS	mEP, mBP, mIBP, mDEHP	LOD: 0.1–1.0 ng/mL	mMP, NM; mEP, 141 ng/mL; mBP, 17.8 ng/mL; mIBP, 7.6 ng/mL; mDEHP, 106.6 ng/mL	Ferguson et al. (2015)
Urine	Greece	239 women	Enzymatic deconjugation, SPE followed by LC–MS/MS	mEP, mBP, mIBP, mDEHP	LOD: 0.8–2.5 ng/mL	mMP, NM; mEP, 142.0 ng/mL; mBP, 32.1 ng/mL; mIBP, 36.7 ng/mL; mDEHP, 44.6 ng/mL	Myridakis et al. (2015)

Urine	Greece	239 children	Enzymatic deconjugation, SPE followed by LC–MS/MS	mEP, mBP, mIBP, mDEHP	LOD: 0.8–2.5 ng/mL	mMP, NM; mEP, 35.3 ng/mL; mBP, 23.3 ng/mL; mIBP, 36.0 ng/mL; mDEHP, 45.6 ng/mL	Myridakis et al. (2015)
Urine	China	430 children (208 girls and 222 boys)	Enzymatic deconjugation, SPE followed by LC–MS/MS	mMP, mEP, mBP, mDEHP	NA	mMP, 15.7 ng/mL; mEP, 4.14 ng/mL; mBP, 21.9 ng/mL; mIBP, NM; mDEHP, 14.3 ng/mL	Shen et al. (2015)
Urine	China	108 young adults	Enzymatic deconjugation, SPE followed by LC–MS/MS	mMP, mEP, mBP, mIBP, mDEHP	LOQ: 0.1–0.5 ng/mL	mMP, 31.8 ng/mL; mEP, 37.5 ng/mL; mBP, 67 ng/mL; mIBP, 57.2 ng/mL; mDEHP, 65.3 ng/mL	Gao et al. (2016)
Urine	South Korea	305 women	Enzymatic deconjugation, SPE followed by LC–MS/MS	mBP, mDEHP	LOQ: 0.7–1.0 ng/mL	mMP, NM; mEP, NM; mBP, 41.0 ng/mg, C; mIBP, NM; mDEHP, 23.7 ng/mg, C	Jo et al. (2016)

C, creatinine; GM, geometric mean; LC, liquid chromatography; LC/LC, two-dimensional liquid chromatography; LOD, limit of detection; LOQ, limit of quantification; mBP, monobutyl phthalate; mDEHP, sum of diethylhexyl phthalate metabolites; mEP, monoethyl phthalate; mIBP, monoisobutyl phthalate; mMP, monomethyl phthalate; MS/MS, tandem mass spectrometry; NA, not available; ND, not detected; NM, not measured; SPE, solid-phase extraction; UPLC, ultra-high-performance liquid chromatography.

REFERENCES

Allmyr, M., Erici, M.A., McLachlan, M.S., Englund, G.S., 2006. Sci. Total Environ. 372, 87.
Allmyr, M., Harden, F., Toms, L.M., Mueller, J.F., McLachlan, M.S., Erici, M.A., Englund, G.S., 2008. Sci. Total Environ. 393, 162.
Arbuckle, T.E., Marro, L., Davis, K., Fisher, M., Ayotte, P., Belanger, P., Dumas, P., LeBlanc, A., Berube, R., Gaudreau, E., Provencher, G., Faustman, E.M., Vigoren, E., Ettinger, A.S., Dellarco, M., MacPherson, S., Fraser, W.D., 2015. Environ. Health Perspect. 123, 277.
Asimakopoulos, A.G., Thomaidis, N.S., Kannan, K., 2014a. Sci. Total Environ. 470–471, 1243.
Asimakopoulos, A.G., Wang, L., Thomaidis, N.S., Kannan, K., 2014b. J. Chromatogr. A 1324, 141.
Asimakopoulos, A.G., Xue, J., Carvalho, B.P.D., Iyer, A., Abualnaja, K.O., Yaghmoor, S.S., Kumosani, T.A., Kannan, K., 2016. Environ. Res. 150, 573.
Axelsson, J., Rylander, L., Hydbom, A.R., Jonsson, B.A.G., Lindh, C.H., Giwercman, A., 2015. Environ. Res. 85, 54.
Barr, D.B., Silva, M.J., Kato, K., Reidy, J.A., Malek, D., Hurtz, M., Sadowski, L.L., Needham, L.L., Calafat, A.M., 2003. Environ. Health Perspect. 111, 1148.
Becker, K., Seiwert, M., Angerer, J., Heger, W., Koch, H.M., Nagorka, R., Robkamp, E., Schluter, C., Seifert, N., Ullrich, D., 2004. Int. J. Hyg. Environ. Health 207, 409.
Becker, K., Thomas, G., Seiwert, M., Conrad, A., Fub, H.P., Muller, J., Wittassek, M., Schulz, C., Gehring, M.K., 2009. Int. J. Hyg. Environ. Health 212, 685.
Bertelsen, R.J., Engel, S.M., Jusko, T.A., Calafat, A.M., Hoppin, J.A., London, S.J., Eggesbo, M., Aase, H., Zeiner, P., Ted, R.K., Knudsen, G.P., Guidry, V.T., Longnecker, M.P., 2014. J. Expo. Sci. Environ. Epidemiol. 24, 517.
Benotti, M., Trenholm, R., Vanderford, B., Halady, J., 2009. Environ. Sci. Technol. 43, 597.
Bledzka, D., Gromadzinska, J., Wasowicz, W., 2014. Environ. Int. 67, 27.
Brunn, H., Bitsch, N., Amberg, M.J., 2004. Toxicology of synthetic musk compounds in man and animals. In: The Handbook of Environmental Chemistry, vol. 3. Springer-Verlag, Berlin, Heidelberg, p. 259. Part X.
Calafat, A.M., Wong, L.Y., Ye, X., Reidy, J.A., Needham, L.L., 2008a. Environ. Health Perspect. 116, 893.
Calafat, A.M., Ye, X., Wong, L.Y., Reidy, J.A., Needham, L.L., 2008b. Environ. Health Perspect. 116, 303.
Calafat, A.M., Ye, X., Wong, L.Y., Bishop, A.M., Needam, L.L., 2010. Environ. Health Perspect. 118, 679.
Carrillo, L.L., Ramirez, R.U.H., Calafat, A.M., Sanchez, L.T., Portillo, M.G., Needham, L.L., Ramos, R.R., Cebrian, M.E., 2010. Environ. Health Perspect. 118, 539.
Casas, L., Fernandez, M.F., Llop, S., Guxens, M., Ballester, F., Olea, N., Irurzun, M.B., Rodriguez, L.S.M., Riano, I., Tardon, A., Vrijheid, M., Calafat, A.M., Sunyer, J., 2011. Environ. Int. 37, 858.
Cerna, M., Maly, M., Rudnai, P., Kozepesy, S., Naray, M., Halzlova, K., Jajcaj, M., Grafnetterova, A., Krskova, A., Antosova, D., Forysova, K., Hond, E.D., Schoeters, G., Joas, R., Casteleyn, L., Joas, A., Biot, P., Esteban, M., Koch, H.M., Kolossa-Gehring, M., Gutleb, A.C., Pavlouskova, J., Vrbik, K., 2015. Environ. Res. 141, 118.
Chisvert, A., Leon-Gonzalez, Z., Tarazona, I., Salvador, A., Giokas, D., 2012. Anal. Chim. Acta 752, 11.
Christian, M.S., Parker, R.M., Hoberman, A.M., Diener, R.M., Api, A.M., 1999. Toxicol. Lett. 111, 169.
Colacino, J.A., Harris, T.R., Schecter, A., 2010. Environ. Health Perspect. 118, 998.
Darbre, P.D., Harvey, P.W., 2008. J. Appl. Toxicol. 28, 561.
Darbre, P.D., Alijarrah, A., Miller, W.R., Coldham, N.G., Sauer, M.J., Pope, G.S., 2004. J. Appl. Toxicol. 24, 5.
Dayan, A.D., 2007. Food Chem. Toxicol. 45, 125.
Dewalque, L., Pirard, C., Charlier, C., 2014. Biomed. Res. Int. 2014, 1.
Diaz, I.J., Gomez, A.Z., Ballesteros, O., Navalon, A., 2014. Talanta 129, 448.
Dietrich, D.R., Hitzfeld, B.C., 2004. Bioaccumulation and ecotoxicity of synthetic musks in the aquatic environment. In: The Handbook of Environmental Chemistry, vol. 3. Springer-Verlag, Berlin, Heidelberg, p. 233. Part X.
Environmental health criteria 214, 2000. Human Exposure Assessment, International Program on Chemical Safety. UNEP/ILO/WHO, World Health Organization (WHO), Geneva.
Erici, M.A., Pettersson, M., Parkkonen, J., Sturve, J., 2002. Chemosphere 46, 1485.
Ferguson, K.K., Caruso, R.L., Meeker, J.D., 2011. Environ. Res. 111, 718.

Ferguson, K.K., McElrath, T.F., Chen, Y.H., Mukherjee, B., Meeker, J.D., 2015. Environ. Health Perspect. 123, 210.
Fourth National Report on Human Exposure to Environmental Chemicals, 2009. Department of health and human sevices, Centers for Disease Control and Prevention (CDC), Atlanta, GA. https//www.cdc.gov/exposurereport/pdf/fourthreport.pdf.
Franco, M.R., Ramirez, R.U.H., Calafat, A.M., Cebrian, M.E., Needham, L.L., Teitelbaum, S., Wolff, M.S., Carrillo, L.L., 2011. Environ. Int. 37, 867.
Frederiksen, H., Jorgensen, N., Andersson, A.M., 2010. J. Anal. Toxicol. 34, 400.
Frederiksen, H., Aksglaede, L., Sorensen, K., Skakkebaek, N.E., Juul, A., Andersson, A.M., 2011a. Environ. Res. 111, 656.
Frederiksen, H., Jorgensen, N., Andersson, A.M., 2011b. J. Expo. Sci. Environ. Epidemiol. 21, 262.
Frederiksen, H., Aksglaede, L., Sorensen, K., Nielsen, O., Main, K.M., Skakkebaek, N.E., Juul, A., Andersson, A.M., 2013a. Int. J. Hyg. Environ. Health 216, 710.
Frederiksen, H., Nielsen, J.K.S., Morck, T.A., Hansen, P.W., Jensen, J.F., Nielsen, O., Andersson, A.M., Knudsen, L.E., 2013b. Int. J. Hyg. Environ. Health 216, 772.
Fromme, H., Bolte, G., Koch, H.M., Angerer, J., Boehmer, S., Drexler, H., Mayer, R., Liebl, B., 2007. Int. J. Hyg. Environ. Health 210, 21.
Gao, C.J., Liu, L.Y., Ma, W.L., Zhu, N.Z., Jiang, L., Li, Y.F., Kannan, K., 2015. Environ. Pollut. 203, 1.
Gao, C.J., Liu, L.Y., Ma, W.L., Ren, N.Q., Guo, Y., Zhu, N.Z., Jiang, L., Li, Y.F., Kannan, K., 2016. Sci. Total Environ. 543, 19.
Gatermann, R., Huhnerfuss, H., Rimkus, G., Attar, A., Kettrup, A., 1998. Chemosphere 36, 2535.
Geens, T., Neels, H., Covaci, A., 2012. Chemosphere 87, 796.
Gimeno, P., Maggio, A.F., Bousquet, C., Quoirez, A., Civade, C., Bonnet, P.A., 2012. J. Chromatogr. A 1253, 144.
Gimeno, P., Thomas, S., Bousquet, C., Maggio, A.F., Civade, C., Brenier, C., Bonnet, P.A., 2014. J. Chromatogr. B 949–950, 99.
Gomez, R.R., Gomez, A.Z., Garcia, N.D., Ballesteros, O., Navalon, A., 2015. Talanta 134, 657.
Guo, Y., Kannan, K., 2013. Environ. Sci. Technol. 47, 14442.
Guo, Y., Alomirah, H., Cho, H.S., Minh, T.B., Mohd, M.A., Nakata, H., Kannan, K., 2011a. Environ. Sci. Technol. 45, 3138.
Guo, Y., Wu, Q., Kannan, K., 2011b. Environ. Int. 37, 893.
Guo, Y., Wang, L., Kannan, K., 2014. Arch. Environ. Contam. Toxicol. 66, 113.
Handa, O., Kokura, S., Adachi, S., Takagi, T., Naito, Y., Tanigawa, T., Yoshida, N., Yoshikawa, T., 2006. Toxicology 227, 62.
Heffernan, A.L., Baduel, C., Toms, L.M.L., Calafat, A.M., Ye, X., Hobson, P., Broomhall, S., Mueller, J.F., 2015. Environ. Int. 85, 77.
Herring, P., 2014. Oregon's Agricultural Progress Magazine.
Hines, E.P., Mendola, P., Ehrenstein, O.S.V., Ye, X., Calafat, A.M., Fenton, S.E., 2015. Reprod. Toxicol. 54, 120.
Hogberg, J., Hanberg, A., Berglund, M., Skerfving, S., Remberger, M., Calafat, A.M., Filipsson, A.F., Jansson, B., Johansson, N., Appelgren, M., Hakansson, H., 2008. Environ. Health Perspect. 116, 334.
Hond, E.D., Paulussen, M., Geens, T., Bruckers, L., Baeyens, W., David, F., Dumont, E., Loots, I., Morrens, B., de Bellevaux, B.N., Nelen, V., Schoeters, G., Larabeke, N.V., Covaci, A., 2013. Sci. Total Environ. 463–464, 102.
Hu, Z., Shi, Y., Niu, H., Cai, Y., Jiang, G., Wu, Y., 2010. Environ. Toxicol. Chem. 29, 1877.
Hutter, H.P., Wallner, P., Moshammer, H., Hartl, W., Sattelberger, R., Lorbeer, G., Kundi, M., 2005. Chemosphere 59, 487.
Hutter, H.P., Wallner, P., Moshammer, H., Hartl, W., Sattelberger, R., Lorbeer, G., Kundi, M., 2009. Sci. Total Environ. 407, 4821.
Hutter, H.P., Wallner, P., Hartl, W., Uhl, M., Lorbeer, G., Gminski, R., Sundermann, V.M., Kundi, M., 2010. Int. J. Hyg. Environ. Health 213, 124.
Itoh, H., Yoshida, K., Masunaga, S., 2007. Environ. Sci. Technol. 41, 4542.
Itoh, H., Iwasaki, M., Hanaoka, T., Sasaki, H., Tanaka, T., Tsugane, S., 2009. Sci. Total Environ. 4408, 37.
Jardim, V.C., Melo, L.D.P., Domingues, D.S., Queiroz, M.E.C., 2015. J. Chromatogr. B 974, 35.
Jares, C.G., Regueiro, J., Barro, R., Dagnac, T., Llompart, M., 2009. J. Chromatogr. A 1216, 567.

Ji, K., Kho, Y.L., Park, Y., Choi, K., 2010. Environ. Res. 110, 375.
Jo, A., Kim, H., Chung, H., Chang, N., 2016. Int. J. Environ. Res. Public Health 13, 680.
Johns, L.E., Cooper, G.S., Galizia, A., Meeker, J.D., 2015. Environ. Int. 85, 27.
Kang, C.S., Lee, J.H., Kim, S.K., Lee, K.T., Lee, J.S., Park, P.S., Yun, S.H., Kannan, K., Yoo, Y.W., Ha, J.Y., Lee, S.W., 2010. Chemosphere 80, 116.
Kang, H.S., Kyung, M.S., Ko, A., Park, J.H., Hwang, M.S., Kwon, J.E., Suh, J.H., Lee, H.S., Moon, G.I., Hong, J.H., Hwang, I.G., 2016. Environ. Res. 146, 245.
Kannan, K., Reiner, J.L., Yun, S.H., Perrotta, E.E., Tao, L., Restrepo, B.J., Rodan, B.D., 2005. Chemosphere 61, 693.
Kasper-Sonnenberg, M., Koch, H.M., Wittsiepe, J., Bruning, T., Wilhelm, M., 2014. Int. J. Hyg. Environ. Health 217, 830.
Kawamura, Y., Ogawa, Y., Nishimura, T., Kikuchi, Y., Nishikawa, J.I., Nishihara, T., Tanamoto, K., 2003. J. Health Sci. 49, 205.
Kay, V.R., Bloom, M.S., Foster, W.G., 2014. Crit. Rev. Toxicol. 44, 467.
Kim, S., Choi, K., 2014. Environ. Int. 70, 143.
Kim, K., Park, H., Yang, W., Lee, J.H., 2011. Environ. Res. 111, 1280.
Kim, S., Kang, S., Lee, G., Lee, S., Jo, A., Kwak, K., Kim, D., Koh, D., Kho, Y.L., Kim, S., Choi, K., 2014. Sci. Total Environ. 472, 49.
Klimova, Z., Hojerova, J., Berankova, M., 2015. Food Chem. Toxicol. 83, 237.
Koch, H.M., Rossbach, B., Drexler, H., Angerer, J., 2003. Environ. Res. 93, 177.
Koch, H.M., Wittassek, M., Bruning, T., Angerer, J., Heudorf, U., 2011. Int. J. Hyg. Environ. Health 214, 188.
Koniecki, D., Wang, R., Moody, R.P., Zhu, J., 2011. Environ. Res. 111, 329.
Kunisue, T., Chen, Z., Louis, G.M.B., Sundaram, R., Hediger, M.L., Sun, L., Kannan, K., 2012. Environ. Sci. Technol. 46, 4624.
Lankester, J., Patel, C., Cullen, M.R., Ley, C., Parsonnet, J., 2013. PLoS One 8, e80057.
Larsson, K., Bjorklund, K.L., Palm, B., Wennberg, M., Kaj, L., Lindh, C.H., Jonsson, B.A.G., Berglund, M., 2014. Environ. Int. 73, 323.
Latini, G., 2005. Clin. Chim. Acta 361, 20.
Lee, S.Y., Son, E., Kang, J.Y., Lee, H.S., Shin, M.K., Nam, H.S., Kim, S.Y., Jang, Y.M., Rhee, G.S., 2013. Bull. Korean Chem. Soc. 34, 1131.
Lee, S., Kim, S., Park, J., Kim, H.J., Lee, J.J., Choi, G., Choi, S., Kim, S., Kim, S.Y., Choi, K., Kim, S., Moon, H.B., 2015. Environ. Res. 140, 466.
Leon, Z., Chisvert, A., Tarazona, I., Salvador, A., 2010. Anal. Bioanal. Chem. 398, 831.
Li, X., Ying, G.G., Zhao, J.L., Chen, Z.F., Lai, H.J., Su, H.C., 2013. Environ. Int. 52, 81.
Liao, C., Kannan, K., 2014a. Environ. Sci. Technol. 48, 4103.
Liao, C., Kannan, K., 2014b. Sci. Total Environ. 475, 8.
Liao, C., Kannan, K., 2014c. Arch. Environ. Contam. Toxicol. 67, 50.
Liao, C., Liu, F., Kannan, K., 2013. Environ. Sci. Technol. 47, 3918.
Liebl, B., Ehrenstorfer, S., 1993. Chemosphere 27, 2253.
Lignell, S., Darnerud, P.O., Aune, M., Cnattingius, S., Hajslova, J., Setkova, L., Glynn, A., 2008. Environ. Sci. Technol. 42, 6743.
Lin, S., Ku, H.Y., Su, P.H., Chen, J.W., Huang, P.C., Angerer, J., Wang, S.L., 2011a. Chemosphere 82, 947.
Lin, L.C., Wang, S.L., Chang, Y.C., Huang, P.C., Cheng, J.T., Su, P.H., Liao, P.C., 2011b. Chemosphere 83, 1192.
Lorz, P.M., Towae, F.K., Enke, W., Jäckh, R., Bhargava, N., 2002. Phthalic Acid and Derivatives. Ullmann's Encyclopedia of Industrial Chemistry. Wiley-VCH, Weinheim.
Mackintosh, C.E., Maldonado, J.A., Ikonomou, M.G., Gobas, F.A.P.C., 2006. Environ. Sci. Technol. 40, 3481.
Main, K.M., Mortensen, G.K., Kaleva, M.M., Boisen, K.A., Damgaard, I.N., Chellakooty, M., Schmidt, I.M., Suomi, A.M., Virtanen, H.E., Petersen, J.H., Andersson, A.M., Toppari, J., Skakkebaek, N.E., 2006. Environ. Health Perspect. 114, 270.
Meeker, J.D., Cantonwine, D.E., Gonzalez, L.O.R., Ferguson, K.K., Mukherjee, B., Calafat, A.M., Ye, X., Toro, L.V.A.D., Hernandez, N.C., Velez, B.J., Alshawabkeh, A.N., Cordero, J.F., 2013. Environ. Sci. Technol. 47, 3439.

Moos, R.K., Angerer, J., Wittsiepe, J., Wilherm, M., Bruning, T., Koch, H.M., 2014. Int. J. Hyg. Environ. Health 217, 845.
Moreta, C., Tena, M.T., Kannan, K., 2015. Environ. Res. 142, 452.
Myridakis, A., Fthenou, E., Balaska, E., Vakinti, M., Kogevinas, M., Stephanou, E.G., 2015. Environ. Int. 83, 1.
Olesen, L.D., Cederberg, T., Pedersen, K.H., Hojgard, A., 2005. Chemosphere 61, 422.
Philippat, C., Mortamais, M., Chevrier, C., Petit, C., Calafat, A.M., Ye, X., Silva, M.J., Brambilla, C., Pin, I., Charles, M.A., Cordier, S., Slama, R., 2012. Environ. Health Perspect. 120, 464.
Pycke, B.F.G., Geer, L.A., Dalloul, M., Abulafia, O., Jenck, A.M., Halden, R.U., 2014. Environ. Sci. Technol. 48, 8831.
Pollack, A.Z., Louis, G.M.B., Chen, Z., Sun, L., Trabert, B., Guo, Y., Kannan, K., 2015. Environ. Res. 137, 101.
Raab, U., Preiss, U., Albrecht, M., Shahim, N., Parlar, H., Fromme, H., 2008. Chemosphere 72, 87.
Rastogi, S.C., Schouten, A., Kruijf, N.D., Weijland, J.W., 1995. Contact Dermat. 32, 28.
Reddy, B.S., Rozati, R., Reddy, S., Kodampur, S., Reddy, P., Reddy, R., 2006. Fertil. Steril. 85, 775.
Reiner, J.L., Kannan, K., 2011. Water Air Soil Pollut. 214, 335.
Reiner, J.L., Wong, C.M., Arcaro, K.F., Kannan, K., 2007. Environ. Sci. Technol. 41, 3815.
Riedel, J., Birner, G., Dorp, C.V., Neumann, H.G., Dekant, W., 1999. Xenobiotica 29, 573.
Rimkus, G.G., 2004. Synthetic Musk Fragrances in the Environment. Springer.
SCCP Scientific Committee on Consumer Products, 2010. Opinion on Triclosan, pp. 1–136. http//ec.europa.eu/health/ph_risk/committees/04_sccp/docs/sccp_o_166.pdf.
Schettler, T., 2006. Int. J. Androl, 29, 134.
Schlumpf, M., Kypke, K., Wittassek, M., Angerer, J., Mascher, H., Mascher, D., Vokt, C., Birchler, M., Lichtensteiger, W., 2010. Chemosphere 81, 1171.
Sexton, K., Needham, L.L., Pirkle, J.L., 2004. Am. Sci. 92, 38.
Shen, Q., Shi, H., Zhang, Y., Cao, Y., 2015. Arch. Public Health 73, 5.
Silva, M.J., Malek, N.A., Hodge, C.C., Reidy, J.A., Kato, K., Barr, D.B., Needham, L.L., Brock, J.W., 2003. J. Chromatogr. B 789, 393.
Silva, M.J., Samandar, E., Preau, J.J.L., Reidy, J.A., Needham, L.L., Calafat, A.M., 2007. J. Chromatogr. B 860, 106.
Singh, S., Li, S.S.L., 2011. Genomics 97, 148.
Slanina, P., 2004. Risk evaluation of dietary and dermal exposure to musk fragrances. In: The Handbook of Environmental Chemistry, vol. 3. Springer-Verlag, Berlin, p. 281. Part X.
Soni, M.G., Carabin, I.G., Burdock, G.A., 2005. Food Chem. Toxicol. 43, 985.
Soria, F.V., Ballesteros, O., Gomez, A.Z., Navalon, A., 2014. Talanta 121, 97.
Specht, I.O., Bonde, J.P., Toft, G., Lindh, C.H., Jonsson, B.A.G., Jorgensen, K.T., 2015. PLoS ONE. 10, e0120070. http://dx.doi.org/10.1371/journal.pone.0120070.
Swan, S.H., Main, K.M., Liu, F., Stewart, S.L., Kruse, R.L., Calafat, A.M., 2005. Environ. Health Perspect. 113, 1056.
Tanabe, S., 2005. Mar. Pollut. Bull. 50, 1025.
Tarazona, I., Chisvert, A., Salvador, A., 2013. Talanta 116, 388.
Tayama, S., Nakagawa, Y., Tayama, K., 2008. Mutat. Res. 649, 114.
Taylor, K.M., Weisskopf, M., Shine, J., 2014. Environ. Health 13, 14.
Teitelbaum, S.L., Britton, J.A., Calafat, A.M., Ye, X., Silva, M.J., Reidy, J.A., Galvez, M.P., Brenner, B.L., Wolff, M.S., 2008. Environ. Res. 106, 257.
Terasaki, M., Makino, M., Tatarazako, N., 2008. J. Appl. Toxicol. 29, 242.
Ueno, D., Moribe, M., Inoue, K., Someya, T., Ryuda, N., Ichiba, M., Miyajima, T., Kunisue, T., In, H., Maruo, K., Nakata, H., 2009. Interdisciplinary studies on environmental chemistry — environmental research in asia. In: Obayashi, Y., Isobe, T., Subramanian, A., Suzuki, S., Tanabe, S. (Eds.), Synthetic Musk Fragrances in Human Breast Milk and Adipose Tissue from Japan. Terrapub, pp. 247–252.
Wang, L., Kannan, K., 2013. Environ. Int. 59, 27.
Wang, H., Zhang, J., Gao, F., Yang, Y., Duan, H., Wu, Y., Berset, J.D., Shao, B., 2011. J. Chromatogr. B 879, 1861.

Wang, L., Wu, Y., Zhang, W., Kannan, K., 2013. Environ. Sci. Technol. 47, 2069.
Wang, L., Asimakopoulos, A.G., Kannan, K., 2015. Environ. Int. 78, 45.
Weiss, L., Arbuckle, T.E., Fisher, M., Ramsay, T., Mallick, R., Hauser, R., LeBlanc, A., Walker, M., Dumas, P., Lang, C., 2015. Int. J. Hyg. Environ. Health 218, 507.
Wittassek, M., Wiesmuller, G.A., Koch, H.M., Eckard, R., Dobler, L., Muller, J., Angerer, J., Schluter, C., 2007. Int. J. Hyg. Environ. Health 210, 319.
Xue, J., Wu, Q., Sakthivel, S., Pavithran, P.V., Vasukutty, J.R., Kannan, K., 2015. Environ. Res. 137, 120.
Yamagishi, T., Miyazaki, T., Horii, S., Kaneko, S., 1981. Bull. Environ. Contam. Toxicol. 26, 656.
Ye, X., Bishop, A.M., Reidy, J.A., Needham, L.L., Calafat, A.M., 2006. Environ. Health Perspect. 114, 1843.
Ye, X., Bishop, A.M., Needham, L.L., Calafat, A.M., 2008a. Anal. Chim. Acta 622, 150.
Ye, X., Tao, L.J., Needham, L.L., Calafat, A.M., 2008b. Talanta 76, 865.
Ye, X., Pierik, F.H., Hauser, R., Duty, S., Angerer, J., Park, M.M., Burdorf, A., Hofman, A., Jaddoe, V.W.V., Mackenbach, J.P., Steegers, E.A.P., Tiemeier, H., Longnecker, M.P., 2008c. Environ. Res. 108, 260.
Ye, X., Pierik, F.H., Angerer, J., Meltzer, H.M., Jaddoe, V.W.V., Tiemeier, H., Hoppin, J.A., Longnecker, M.P., 2009. Int. J. Hyg. Environ. Health 212, 481.
Yin, J., Wang, H., Zhang, J., Zhou, N., Gao, F., Wu, Y., Xiang, J., Shao, B., 2012. Chemosphere 87, 1018.
Yin, J., Wei, L., Shi, Y., Zhang, J., Wu, Q., Shao, B., 2016. Environ. Geochem. Health 38, 1125.
Zhang, X., Liang, G., Zeng, X., Zhou, J., Sheng, G., Fu, J., 2011. J. Environ. Sci. 23, 983.
Zhang, T., Sun, H., Qin, X., Wu, Q., Zhang, Y., Ma, J., Kannan, K., 2013. Sci. Total Environ. 461–462, 49.
Zhang, X., Jing, Y., Ma, L., Zhou, J., Fang, X., Zhang, X., Yu, Y., 2015. Int. J. Hyg. Environ. Health 218, 99.
Zhou, J., Zeng, X., Zheng, K., Zhu, X., Ma, L., Xu, Q., Zhang, X., Yu, Y., Sheng, G., Fu, J., 2012. Ecotoxicol. Environ. Saf. 84, 325.

CHAPTER 16

Environmental Monitoring of Cosmetic Ingredients

Alberto Chisvert[1], Dimosthenis Giokas[2], Juan L. Benedé[1], Amparo Salvador[1]
[1]University of Valencia, Valencia, Spain; [2]University of Ioannina, Ioannina, Greece

INTRODUCTION

Cosmetic ingredients have been at the forefront of industrial production for many decades owing to their high public acceptance and widespread use. Until the 1980s, however, the release of cosmetic ingredients into the environment was disregarded and these compounds were categorized as non-target contaminants. Although the first reports on the environmental release and bioaccumulation of cosmetic ingredients were published in the early 1980s by Yamagishi et al. (1981, 1983), it took almost 16 years until Rimkus et al. (1997) published the first overview on the occurrence of cosmetic ingredients (i.e., musks) in the environment. It is noteworthy that the first report on the environmental occurrence of UV filters also appeared in 1997 (Nagtegaal et al., 1997), and it concerned the accumulation of UV filters in fish from lakes used for recreational purposes. However, this study was published in German and it remained in the background of the international scientific literature until 1999, when Daughton and Ternes reviewed the state of the art of pharmaceuticals and cosmetics products in the environment, in one of the most popular review articles in the field (Daughton and Ternes, 1999). Since then, a spark has been lighted and a substantial amount of work has been done to evaluate the concentration levels and potential effects of cosmetic ingredients in the environment.

The first challenge in the environmental monitoring of cosmetic ingredients was the development of appropriate analytical methods that would be able to detect their presence at concentrations from nanograms to low micrograms per litre. By that time, almost all available analytical methods for the determination of cosmetic ingredients were designed for applications related to the compliance of commercial products with the regulatory limits (Salvador and Chisvert, 2007). Therefore, these analytical methods did not offer adequate sensitivity and did not take into consideration potential issues from environmental matrix effects. Today, this situation has changed dramatically. A wide range of analytical methods and (micro)extraction techniques have been put forth aiming at the determination of cosmetic ingredients in a variety of environmental samples such as bathing water (sea, river, lake, swimming pool), wastewater (influent and effluent wastewater from wastewater treatment plants (WWTPs)) and drinking water (groundwater, tap water and bottled water). Moreover, analytical methodologies to determine cosmetic ingredients in solid environmental samples, such as soil, sediments, beach sand and sewage sludge, and in indoor dust have been developed. The potential bioaccumulation of cosmetic ingredients in living organisms (i.e., biota) has also instigated the development of analytical methods for their determination in fish, mussels and even birds (Giokas et al., 2007; Peck, 2006; Pedrouzo et al., 2011; Wille et al., 2012).

Another challenge is related to the complexity of the matrix and the variety of interfering compounds that may be present in environmental samples. For example, influent and effluent wastewaters are usually burdened with high concentrations of organic matter; seawater has high ionic strength (i.e., salt content), whilst swimming pool water typically contains relatively high concentrations of chlorine that is used for disinfection. Therefore, extraction methods should be able to afford not only analyte pre-concentration but also sample clean-up. This requirement becomes even more demanding in the analysis of solid samples (e.g., wastewater sludge, sediments, etc.) and samples with a high fat content, such as biological tissues (e.g., fishes).

The last challenge is related to the large number and different natures of these compounds. Typically, research studies focus their attention on specific families of cosmetic ingredients, such as UV filters, preservatives, musk fragrances or insect repellents. Nevertheless, multi-residue methods are necessary to obtain a realistic representation of the occurrence and distribution of cosmetic ingredients in environmental samples. Therefore, the monitoring of a large number of compounds that usually belong to different chemical categories and have different physico-chemical properties is required.

Considering these challenges, this chapter aims to compile the latest developments and discuss the fundamental principles of sample preparation methodologies developed for the extraction of cosmetic ingredients from environmental samples. In this regard, a comprehensive outline of the literature dealing with the development and validation of analytical methods for the determination of cosmetic ingredients in the environment is provided. Alternative extraction techniques based on passive samplers, such as semipermeable membrane devices (Balmer et al., 2005; Sultana et al., 2016) and polar organic chemical integrative samplers (Zenker et al., 2008; Fent et al., 2010; Sultana et al., 2016; Iparraguirre et al., 2017), are not considered herein because these devices do not aim at sample preparation; instead, they primarily aim to mimic natural bioaccumulation in aquatic organisms and estimate the exposure of aquatic organisms to cosmetic ingredients.

UV FILTERS

UV filters, of either inorganic or organic nature, have the capacity to attenuate the deleterious UV radiation reaching the skin surface, either by reflecting and scattering sunlight or by absorbing it, respectively (see Chapter 5). There are many studies that provide evidence that UV filters can negatively affect both the flora and the fauna in the aquatic environment (Brausch and Rand, 2011; Tovar-Sánchez et al., 2013; Molins-Delgado et al., 2016). For this reason, they have been recently labelled as emerging contaminants and have attracted the attention of the scientific community. A variety of analytical methods have therefore been developed for the determination of UV filters in water samples (Table 16.1), solid samples (Table 16.2) and living organisms (Table 16.3).

Sample Preparation

The low levels of UV filters in environmental samples and the potential interference from co-existing matrix components necessitate the application of an appropriate sample preparation procedure prior to analysis. A plethora of analytical methodologies have been utilized, involving both liquid and solid-phase extraction and microextraction methods. Each of these methods has its own advantages and disadvantages and the selection of the most appropriate method depends on many factors, such as the complexity of the matrix, the detection technique and the target compounds.

Water Samples

Solid-phase extraction (SPE), in either offline or online mode, is the most widely applied method for the extraction and enrichment of UV filters from water samples. As shown in Table 16.1, large sample volumes (e.g., 500–1000 mL) are usually percolated throughout SPE cartridges or discs, and UV filters are eluted with appropriate organic solvents. The eluant solvents are evaporated to dryness and redissolved in a low volume of solvent that is compatible with the subsequent analytical technique. The wide acceptance of SPE as an analytical method is due to the availability of a plethora of commercially available sorbents that can be used to tailor the analytical method to the needs of the analysis. The most popular sorbents used so far are those enabling the simultaneous determination of both polar and non-polar UV filters, such as polyvinylpyrrolidone-divinylbenzene co-polymer (PVP-DVB) (Cuderman and Heath, 2007; Rodil et al., 2008, 2009a,b; Negreira et al., 2009b; Wick et al., 2010; Vosough and Mojdehi, 2011; Bratkovics and Sapozhnikova, 2011; Gracia-Lor et al., 2012; Kotnik et al., 2014; Emnet et al., 2015) or the polystyrene-divinylbenzene co-polymer modified with either pyrrolidone (PS-DVB/MP) (Pietrogrande et al., 2009; Matamoros et al., 2010; Da Silva et al., 2013; Purrà et al., 2014; Da Silva et al., 2015) or hydroxyl groups (Pedrouzo et al., 2009). When only non-polar UV filters are of interest, octadecyl functionalized silica (C_{18}) (Giokas et al., 2004; Li et al., 2007; Díaz-Cruz et al., 2012; Caldas et al., 2013) or polystyrene divinylbenzene co-polymer (PS-DVB) (Vosough and Mojdehi, 2011; Gago-Ferrero et al., 2013a) is well suited. For more polar compounds (e.g., BZ4, PBS, etc.) the PVP-DVB co-polymer modified with a cation exchanger (PVP-DVB/MCX) (Kasprzyk-Hordern et al., 2008; Maijó et al., 2013) provides enhanced efficiency because polar species are strongly retained by the ion-exchange moieties, whilst non-polar species are still effectively

Table 16.1 Published papers on UV filter determination in water samples (in chronological order)

Authors	Target compounds[a]	Sample[b]	Extraction technique[c]	Analytical technique[d]	MLOD (ng/L)[e]	Recovery (%)
Lambropoulou et al. (2002)	BZ3, EHDP	SP, SW	**(DI)SPME** (5-mL sample (pH 2, 1% NaCl); extracted with PDMS-coated fibre; 25°C, 45 min; thermal desorption (220°C, 8 min))	(TD)GC–FID	360–890	94–99 (SP) 82–99 (SW)
			(HS)SPME (5-mL sample (pH 2, 3% NaCl); extracted with PDMS-coated fibre; 90°C, 45 min; thermal desorption (220°C, 8 min))		220–1340	89–97 (SP) 91–98 (SW)
Giokas et al. (2004)	BMDM, BZ3, EHMC, MBC	SP, SW	**SPE** (500-mL sample (pH 3, 10% KCl, 1% methanol); extracted with C_{18} discs; eluted with 2 × 5 mL ethyl acetate/dichloromethane, evaporated to dryness and redissolved in 50 μL methanol (for LC) or 10 μL hexane (for GC))	LC–UV (for BMDM) GC–MS(EI+) (for the rest)	7.3 0.21–0.42	88 (SP) 87 (SW) 96–99 (SP) 93–96 (SW)
Poiger et al. (2004) Balmer et al. (2005)	BMDM, BZ3, EHMC, MBC, OC	EW, IW, LK, RV	**SPE** (100- to 1000-mL sample; extracted with PS-DVB cartridge at 10 mL/min; eluted with 5 mL methanol + 2 × 10 mL dichloromethane; then cleaned up with silica gel; evaporated to 0.1–1.5 mL)	GC–MS(EI+)	2–20	78–129 (EW) 42–90 (LK)

Reference	Analytes	Matrix	Extraction/Sample Preparation	Determination	Concentration	Recovery (%)
Giokas et al. (2005)	BMDM, BZ3, EHMC, MBC, PBSA	SW	**CPE** (50- to 100-mL sample (pH 3, 0.2 M NaCl); extracted with 0.1% Triton X-114; 60°C, 15 min; centrifuged, and surfactant-rich phase mixed with 100 μL methanol (for LC) or back-extracted in 200 μL hexane (for GC))	LC–UV (for BMDM and PBSA) GC–MS(EI$^+$) (for the rest)	300–1270 2.2–30.0	95–99 97–102
Parisis et al. (2005)	PBS, BZ3, MBC, EMC	SP	**DSPE** (250-mL sample (pH 8.2, 0.2 M NaBr, 0.7 mM didodecyldimethylammonium bromide); extracted with silica, filtered and eluted with 0.5 mL methanol)	LC–UV	180–1100	97.4–99.6 (DW)
Kawaguchi et al. (2006)	BZ, BZ3, BZ10	RV	**SBSE** (10-mL sample; extracted with PDMS-coated stir bar; 25°C, 120 min; thermal desorption (250°C, 5 min))	(TD)GC–MS(EI$^+$)	0.5–1	98–115
Kupper et al. (2006)	EHMC, EHT, MBC, OC and other compounds	EW, IW	**LLE + SPE** (700-mL sample (+50 g NaCl); extracted with 60 mL pentane + 60 mL diethyl ether + 100 mL diethyl ether (2 min each); dried with Na$_2$SO$_4$, evaporated to dryness and redissolved in 1 mL hexane; then SPE with silica and 50–70 mL hexane/diethyl ether; evaporated to dryness and redissolved in ethyl acetate)	GC–MS(EI$^+$)	3–34	74–91

Continued

Table 16.1 Published papers on UV filter determination in water samples (in chronological order)—cont'd

Authors	Target compounds[a]	Sample[b]	Extraction technique[c]	Analytical technique[d]	MLOD (ng/L)[e]	Recovery (%)
Jeon et al. (2006)	BZ, BZ1, BZ3, BZ8 and other benzophenones	LK, RV	**LLE** (100-mL sample (+10 g NaCl); extracted with 50 mL ethyl acetate (20 min); then washed with 50 mL 5% NaCl solution, dried with Na_2SO_4, evaporated to dryness and derivatized with 50 μL of MSTFA (80°C, 30 min))	GC–MS(EI+)	5–10	62–114
Cuderman and Heath (2007)	BMDM, BZ3, EHMC, HMS, MBC, OC and other compounds	LK, RV, SP, SW	**SPE** (500-mL sample (pH 3); extracted with PS-DVB cartridge at 5 mL/min; eluted with 3 × 0.5 mL ethyl acetate/dichloromethane, evaporated to dryness, redissolved with 0.4 mL toluene and derivatized with 100 μL MSTFA (60°C, 60 min))	GC–MS(EI+)	17–194 (LK) 26–181 (RV) 27–266 (SP) 13–129 (SW)	50–93 (LK) 65–97 (RV) 60–95 (SP) 75–93 (SW)
Li et al. (2007)	BZ3, EHMC, MBC, OC	EW, IW	**SPE** (1-L sample (pH 3); extracted with C_{18} cartridge at 10 mL/min; eluted with 2 × 5 mL ethyl acetate/dichloromethane, evaporated to dryness and redissolved in 1 mL hexane)	GC–MS(EI+)	10	67–118

Reference	Compounds	Matrix	Sample preparation	Determination	Concentration	Recovery (%)
Kasprzyk-Hordern et al. (2008)	BZ1, BZ2, BZ3, BZ4 and other compounds	EW, IW, RV	**SPE** (250- to 1000-mL sample (pH 3, 500 mg EDTA); extracted with PVP-DVB/MCX cartridge at 4 mL/min; eluted with 2 mL methanol + 2 mL 5% NH$_4$OH/methanol; evaporated to dryness and redissolved in 0.5 mL ammonium acetate/methanol)	LC–MS/MS(ESI$^-$)	0.5–25 (EW) 1–30 (IW) 0.1–5 (RV)	24–118 (EW) 17–50 (IW) 67–117 (RV)
Okanouchi et al. (2008)	BZ, BZ3, BZ10	RV	**(DI)SDME** (2-mL sample; extracted with 3 µL toluene; 25°C, 15 min)	GC–MS(EI$^+$)	10	93–101
Kawaguchi et al. (2008a)	BZ1, BZ3, BZ10 and others	RV	**SBSE** (10-mL sample (0.1 M K$_2$CO$_3$) simultaneously derivatized with 100 µL acetic anhydride and extracted with PDMS-coated stir bar; 25°C, 120 min; thermal desorption (250°C, 5 min))	(TD)GC–MS(EI$^+$)	0.5–2	102–128
Rodil and Moeder (2008a)	BMDM, BZ3, EHDP, EHMC, EHS, HMS, IMC, MBC, OC	EW, LK, RV	**SBSE** (20-mL sample (pH 2, 10% methanol); extracted with PDMS-coated stir bar; 25°C, 3 h; thermal desorption (250°C, 15 min))	(TD)GC–MS(EI$^+$)	0.2–63	75–115 (EW) 78–109 (LK) 77–116 (RV)
Rodil et al. (2008)	BMDM, BZ3, BZ4, EHDP, IMC, MBC, OC, PBSA, PDTA	EW, IW, RV, SW	**SPE** (200-mL sample (pH 4.5, 2% methanol, 50 mM tributylamine); extracted with PVP-DVB cartridge; eluted with 3 × 10 mL methanol, evaporated to 0.2 mL and diluted to 1 mL with methanol/water)	LC–MS/MS(ESI$^{+/-}$)	7–46	55–108 (EW) 29–93 (IW) 74–102 (RV) 66–91 (SW)

Continued

Table 16.1 Published papers on UV filter determination in water samples (in chronological order)—cont'd

Authors	Target compounds[a]	Sample[b]	Extraction technique[c]	Analytical technique[d]	MLOD (ng/L)[e]	Recovery (%)
Rodil et al. (2009a)	BZ3, BZ4, EHDP, EHMC, IMC, MBC, OC, PBSA and other compounds	EW, IW, TW, SW	**SPE** (200- to 500-mL sample (pH 7); extracted with PVP-DVB cartridge; eluted with 3 × 10 mL methanol, evaporated to 0.2 mL and diluted to 1 mL with methanol/water)	LC–MS/MS(ESI$^{+/-}$)	0.8–30	56–132(EW) 69–104 (IW) 77–128 (SW) 66–115 (TW)
Rodil et al. (2009b)	BMDM, BZ3, BZ4, EHDP, EHMC, EHS, HMS, IMC, MBC, OC, PBSA	EW, IW	**SPE** (200- to 500-mL sample (pH 7); extracted with PVP-DVB cartridge; eluted with 3 × 10 mL methanol, evaporated to 0.2 mL and diluted to 1 mL with methanol/water)	LC–MS/MS(ESI$^{+/-}$) LC–MS/MS(APPI$^{+/-}$)	0.7–84.3 0.5–59.5	29–106 (EW) 15–70 (IW) 45–113 (EW) 18–85 (IW)
Negreira et al. (2009a)	BZ1, BZ3, BZ8, EHS, HMS	EW, IW, RV	**(DI)SPME** (10-mL sample (pH 3); extracted with PDMS–DVB-coated fibre; 20°C, 30 min; then exposed to MSTFA (45°C, 10 min); thermal desorption (270°C, 3 min))	(TD)GC–MS/MS(EI$^+$)	0.15–3	48–93 (EW) 89–115 (IW) 97–106 (RV)
Rodil et al. (2009c)	BMDM, BZ3, EHDP, EHMC, EHS, HMS, IMC, MBC, OC	EW, IW, LK	**MALLE** (15-mL sample (+1.5 mL methanol); extracted with LDPE bags filled with 100 μL propanol; 40°C, 120 min)	LC–MS/MS(APPI$^{+/-}$)	0.4–16	52–114 (EW) 35–86 (IW) 66–106 (LK)

Reference	Compounds	Matrix	Sample preparation	Determination	LOD/LOQ (ng/L)	Recovery (%)
Pedrouzo et al. (2009)	BZ1, BZ8, BZ3, OC, EHDP and other compounds	EW, IW, RV	**SPE** (100- to 500-mL sample; extracted with PVP-DVB or PS-DVB/MH cartridge at 10 mL/min; eluted with 5 mL methanol + 5 mL dichloromethane, evaporated to 3–4 mL and diluted to 5 mL with water)	LC–MS/MS(ESI$^{+/-}$)	1–4 (RV) 3–10 (IW, EW)	20–71 (EW) 27–86 (IW) 46–97 (RV)
Negreira et al. (2009b)	BZ1, BZ2, BZ3, BZ4, BZ6, BZ8	EW, IW, RV	**SPE** (200- to 500-mL sample (pH 3); extracted with PVP-DVB cartridge at 10 mL/min; eluted with 3 mL methanol/NH$_4$OH; evaporated to dryness and redissolved in 1 mL methanol/NH$_4$OH)	LC–MS/MS(ESI$^{+/-}$)	0–4.2 (EW) 0.3–9.7 (IW) 0.1–2.4 (RV)	91–104 (EW) 83–101 (IW) 84–105 (RV)
Gómez et al. (2009)	BZ3, EHMC, MBC, OC and other compounds	EW, RV	**LLE** (500-mL sample (pH 3, 1% NaCl); extracted with 100 mL (3 min) + 50 mL (2 min) hexane; dried with Na$_2$SO$_4$; evaporated to 0.4–3 mL)	(LVI)GC–MS(EI$^+$)	10–30 (EW) 4–12 (RV)	124–187 (EW) 117–188 (RV)
Pietrogrande et al. (2009)	BZ, OC and other compounds	EW, TW	**SPE** (200-mL sample; extracted with PS-DVB/MP cartridge; eluted with 15 mL ethyl acetate; evaporated to dryness and redissolved in 0.2 mL methanol)	GC–MS(EI$^+$)	5–10	90–96
Haunschmidt et al. (2010)	BZ3, EHDP, EHS, HMS, MBC, OC	LK	**SBSE** (250-mL sample; extracted with PDMS-coated stir bar; 20 h; then DART)	(DART)MS	0.28–4.3	
Vidal et al. (2010)	BZ3, EHDP, EHMC IMC, MBC, OC	RV, SW	**IL-(DI)SDME** (20-mL sample (pH 2, 1% ethanol); extracted with 10 μL [C$_6$MIN][PF$_6$]; 37 min)	LC–UV	60–3000	96–115 (RV) 92–107 (SW)

Continued

Table 16.1 Published papers on UV filter determination in water samples (in chronological order)—cont'd

Authors	Target compounds[a]	Sample[b]	Extraction technique[c]	Analytical technique[d]	MLOD (ng/L)[e]	Recovery (%)
Wick et al. (2010)	BZ1, BZ2, BZ3, BZ4, PBSA and other compounds	EW, IW, RV	**SPE** (100- to 1000-mL sample (pH 6); extracted with PVP-DVB cartridge at 5 mL/min; eluted with 4 × 2 mL methanol/acetone; evaporated to 0.5 mL and diluted to 1 mL with aqueous formic acid)	LC–MS/MS(ESI$^{+/-}$)	0.15–1.5 (RV) 0.75–15 (EW) 1.5–15 (IW)	53–130 (RV) 66–105 (EW) 96–180 (IW)
Moeder et al. (2010)	BZ3, MBC, OC, EHMC and other compounds	EW, LK	**MEPS** (800-μL sample; extracted with C$_8$ sorbent; eluted with 2 × 25 μL ethyl acetate)	(LVI)GC–MS(EI$^+$)	35–87	60–100
Pedrouzo et al. (2010)	BZ3, BZ8, EHDP, OC and other compounds	EW, IW, RV	**SBSE** (50-mL sample (pH 5, 5% methanol); extracted with PDMS-coated stir bar; 180 min; eluted with 1 mL acetonitrile (30°C, 15 min); evaporated to dryness and redissolved in acetonitrile/water)	LC–MS/MS(ESI$^{+/-}$)	5–10 (IW, EW) 2.5 (RV)	28–89 (EW) 25–84 (IW) 31–87 (RV)
Oliveira et al. (2010)	BMDM, BZ3, EHMC, HMS	SP, SW	**(Online) SPE** (9-mL sample; extracted with PVP-DVB bead sorbent; eluted with 0.6 mL methanol/water; diluted to 0.9 mL with water)	LC–UV	450–3200	
Tarazona et al. (2010)	BZ1, BZ3, BZ8 and other	SW	**DLLME** (5-mL sample (pH 4, 10% NaCl); extracted with 60 μL chloroform (+1 mL acetone as disperser solvent); evaporated to dryness and redissolved in 60 μL BSTFA (75°C, 30 min))	GC–MS(EI$^+$)	32–33	65–169

Environmental Monitoring of Cosmetic Ingredients 445

Negreira et al. (2010)	BZ3, EHDP, EHMC, EHS, HMS, IMC, MBC, OC	EW, IW, RV, SP	**DLLME** (10-mL sample; extracted with 60 µL chlorobenzene (+1 mL acetone as disperser solvent))	GC–MS(EI⁺)	0.6–4.2	87–109 (EW, RV, SP) 80–117 (IW)
Liu et al. (2010)	BZ3, EHS, MBC, OC and other compounds	RV	**(DI)SPME** (3-mL sample; extracted with PDMS-coated fibre, 24°C, 90 min; thermal desorption (280°C, 7 min))	(TD)GC–MS(EI⁺)	0.2–2.0	64–117
Matamoros et al. (2010)	BZ3 and other compounds	RV	**SPE** (200-mL sample; extracted with PS-DVB/MP cartridge at 10 mL/min; eluted with 5 × 2 mL ethyl acetate; evaporated to 100 µL)	GC × GC–MS(EI⁺)	40	94
Zhang et al. (2011a)	BZ1, BZ3 and others	LK	**MSA-DLLME** (20-mL sample; extracted with 40 µL octanol; 20 min; diluted to 80 µL with methanol)	LC–UV	200–800	91–97
Kameda et al. (2011)	BZ3, EHDP, EHMC, EHS, HMS, MBC, OC and others	RV	**SPE** (1000-mL sample; extracted with C₁₈ + Ph cartridges in series at 10 mL/min; eluted with 10 mL dichloromethane; evaporated to 0.2 mL)	GC–MS(EI⁺)	0.1–0.3	80–113
Negreira et al. (2011a)	BZ3, EHDP, EHMC, EHS, HMS, IMC, MBC, OC	EW, IW, RV, SP	**(DI)SME** (100-mL sample (10% methanol); extracted with silicone discs; 14 h; eluted with 0.2 mL ethyl acetate (30 min))	(LVI)GC–MS(EI⁺)	1–12	75–93 (EW) 49–108 (IW) 90–104 (RV) 76–93 (SP)
Román et al. (2011)	BZ3, EHDP, EHMC, EHS, HMS, IMC, MBC, OC	RV, SW, TW	**DSPE** (75-mL sample (pH 3, 30% NaCl); extracted with CoFe₂O₄@oleic acid MNPs; 4 min; eluted with 2 × 1.5 mL hexane (4 min each); evaporated to dryness and redissolved in 50 µL BSTFA)	GC–MS(EI⁺)	0.2–6.0	74–119 (RV) 73–125 (SW) 63–110 (TW)

Continued

Table 16.1 Published papers on UV filter determination in water samples (in chronological order)—cont'd

Authors	Target compounds[a]	Sample[b]	Extraction technique[c]	Analytical technique[d]	MLOD (ng/L)[e]	Recovery (%)
Vosough and Mojdehi (2011)	BMDM, BZ3, EHMC, EHS, HMS, MBC, OC	EW	**SPE** (500-mL sample (5% NaCl); extracted with PS-DVB cartridge; 6 mL/min; eluted with 3 mL dichloromethane + 5 mL methanol; evaporated to dryness and redissolved in 1 mL methanol/acetonitrile)	LC–UV		76–130
Nguyen et al. (2011)	BZ3, EHDP, EHMC, EHS, HMS, OC	SW	**SBSE** (10-mL sample (pH 6, 5% methanol); extracted with PDMS-coated stir bar; 180 min; eluted with 1 mL methanol (30 min); 0.4 mL diluted with 0.1 mL water)	LC–MS/MS(APCI$^{+/-}$)	8–1200	71–100
Gómez et al. (2011)	BZ3, EHMC, MBC and other compounds	EW, RV	**SBSE** (25- to 100-mL sample (20 g NaCl, 10% methanol); extracted with PDMS-coated stir bar; 14 h; thermal desorption (295°C, 7 min))	(TD) GC×GC–MS(EI$^+$)	0.02–0.18	81–141 (EW) 109–146 (RV)
Bratkovics and Sapozhnikova (2011)	BMDM, BZ3, BZ4, BZ8, EHDP, EHMC, OC	SW, TW	**SPE** (200-mL sample (pH 2); extracted with PVP-DVB cartridges; 10 mL/min; eluted with 36 mL methanol/acetone; evaporated to dryness and redissolved in 0.5 mL methanol)	LC–MS/MS(ESI$^+$)	0.5–25	71–111 (SW) 74–109 (TW)
Díaz-Cruz et al. (2012)	BZ3, EHDP, EHMC, MBC, OC	BW, TW	**SPE** (200-mL sample; extracted with C$_{18}$ cartridge; 1 mL/min; eluted with 4×2.5 mL dichloromethane/ethyl acetate; evaporated to dryness and redissolved in 0.5 mL hexane)	GC–MS(EI$^+$)	0.14–7.4	74–111

Environmental Monitoring of Cosmetic Ingredients 447

Ge and Lee (2012a)	BZ, BZ1, BZ3, MBC	RV, TW	**IL-(DI)HFLPME** (10-mL sample (pH 3, 20% NaCl); extracted with 7 μL [C$_6$MIM][FAP]; 50 min)	LC–UV	200–500	95–105 (RV) 83–106 (TW)
Basaglia and Pietrogrande (2012)	BZ, BZ3 and other compounds	TW	**(HS)SPME** (40-mL sample; extracted with PA-coated fibre; 40°C, 125 min; then exposed to BSTFA (35°C, 30.5 min); thermal desorption (300°C, 2 min))	(TD)GC–MS(EI$^+$)	2–9	75–110
Zhang and Lee (2012a)	BZ, BZ3, EHS, HMS, MBC	RV	**(DI)SPME** (7-mL sample (pH 5); extracted with graphene-coated fibre; 25°C, 40 min; then exposed to MSTFA (45°C, 15 min); thermal desorption (280°C, 1 min))	(TD)GC–MS(EI$^+$)	0.5–6.8	99–114
Zhang and Lee (2012b)	BZ, BZ3, EHS, HMS	SP, RV, TW	**IL-USA-DLLME** (10-mL sample (pH 4); extracted with 20 μL [C$_6$MIM][FAP] (+100 μL methanol as disperser solvent); 25°C, 3 min)	LC–UV	200–5000	71–117 (SP) 81–118 (RV) 81–117 (TW)
Zhang and Lee (2012c)	BZ, BZ1, BZ3, EHS, HMS and other compounds	RV	**VA-DLLME** (10-mL sample (pH 4); extracted with 40 μL tetrachloroethene; 3 min; evaporated to dryness and derivatized with 30 μL BSTFA (75°C, 30 min))	GC–MS(EI$^+$)	8–30	76–120
Ge and Lee (2012b)	BZ, BZ1, BZ3, MBC	RV, TW	**IL-USAEME** (1.5-mL sample (pH 3); extracted with 100 μL [C$_6$MIM][FAP]; 24°C, 12 min)	LC–UV	500–1000	96–107 (RV) 96–105 (TW)

Continued

Table 16.1 Published papers on UV filter determination in water samples (in chronological order)—cont'd

Authors	Target compounds[a]	Sample[b]	Extraction technique[c]	Analytical technique[d]	MLOD (ng/L)[e]	Recovery (%)
Magi et al. (2012, 2013)	BZ3, EHDP, EHMC, EHS, HMS, OC	EW, IW, RV, SW	SBSE (50-mL sample; PDMS-coated stir bar; 5 h; eluted with 1 mL methanol (30 min); 0.4 mL diluted with 0.1 mL water)	LC–MS/MS(APCI+)	0.6–114	64–85
Giokas et al. (2012)	BMDM, BZ3, EHMC, MBC	LK, RV	CPE (50-mL sample (pH 4); 0.1% Triton X-114; 40°C, 5 min; addition of 10 mg PDMS-coated Fe$_2$O$_3$@C MNPs; eluted with 2 × 250 μL dichloromethane)	LC–UV	1.43–7.5	85–94 (LK) 88–97 (RV)
Gracia-Lor et al. (2012)	BZ, BZ1, BZ2, BZ3, BZ4 and other compounds	EW, LK, RV, SW	SPE (100-mL sample; extracted with PVP-DVB cartridge; eluted with 5 mL methanol; evaporated to dryness and redissolved in 1 mL methanol/water)	LC–MS/MS(ESI+/−)	2–12 (EW) 0.1–1 (RV, SW, LK)	70–120
Pintado-Herrera et al. (2013)	BZ3, OC and other compounds	EW, IW, SW	SBSE (100-mL sample (pH 2, 10% methanol); extracted with PDMS-coated stir bar; 25°C, 8 h; eluted with 200 μL ethyl acetate (30 min) and derivatized with 10 μL MTBSTFA)	GC–MS(EI+)	0.6–2	28–60
Zhang and Lee (2013a)	BZ, BZ3, EHS, HMS	SP, TW	IL-TC-DLLME (10-mL sample; extracted with 20 μL [C$_6$MIM][FAP]; 50°C; cooled to 0°C (20 min))	LC–UV	300–5000	88–116 (SP) 91–104 (TW)
Maijó et al. (2013)	BZ1, BZ3, BZ8, PBSA	RV	(Online)SPE (sample (pH 3); extracted with PVP-DVB/MCX-filled capillary at 0.93 bar for 15 min; eluted with acetonitrile at 0.05 bar for 30 s)	CE–MS(ESI−)	10–50	

Zhang and Lee (2013b)	BZ, BZ1, BZ3	SP	(DI)KWLPME (20-mL sample (pH 4); extracted with 8 μL perchloroethylene/octanol held in polyester wool; 30 min; eluted with 30 μL acetonitrile)	LC–UV	15–20	77–103
Ku et al. (2013)	BZ, BZ3, BZ8	EW, LK, SP	IL-DLLME (5-mL sample; extracted with 40 μL [C$_6$MIM][PF$_6$] (+200 μL methanol as disperser solvent); 3 min; extract diluted with 20 μL methanol)	LC–UV	200–1300	96–120 (EW) 93–109 (LK) 92–112 (SP)
Gago-Ferrero et al. (2013a)	BZ1, BZ2, BZ3, BZ4, BZ8, MBC and other compounds	EW, GW, IW, RV	(Online)SPE (5-mL sample; extracted with PS-DVB cartridge at 1 mL/min; eluted with aqueous formic acid/acetonitrile for ESI$^+$ or aqueous ammonium acetate/acetonitrile for ESI$^-$)	LC–MS/MS(ESI$^{+/-}$)	1–4 (EW) 0.3–3 (GW) 5–110 (IW) 0.5–3.5 (RV)	37–63 (EW) 86–101 (GW) 18–40 (IW) 65–89 (RV)
Li et al. (2013)	BZ, BZ3 and other compounds	RV	(DI)SPME (10-mL sample (20% NaCl); extracted with C$_{12}$-coated silver wire; 60 min; eluted in 0.2 mL methanol (10 min))	LC–UV	580–1860	70–102
Wu et al. (2013a)	BZ1, BZ3, BZ8, EHS, HMS	EW, RV	USA-DLLME (10-mL sample (pH 7, 5% NaCl); simultaneously extracted with 15 μL tetrachloroethylene (+750 μL acetone as disperser solvent) and derivatized with 20 μL BSTFA; 2 min)	GC–MS(EI$^+$)	1–2	73–91 (EW) 70–93 (RV)

Continued

Table 16.1 Published papers on UV filter determination in water samples (in chronological order)—cont'd

Authors	Target compounds[a]	Sample[b]	Extraction technique[c]	Analytical technique[d]	MLOD (ng/L)[e]	Recovery (%)
Xue et al. (2013)	BZ3, EHDP, EHMC, EHS	IW, RV, RW	IL–USA-DLLME (10-mL sample (pH 7); extracted with 30 µL [C$_8$MIM][PF$_6$] (+100 µL methanol as disperser solvent); 25°C; 5 min; extract diluted with 30 µL methanol)	LC–UV	60–160	93–114
Almeida et al. (2013)	BZ, BZ1, BZ3 and other compound	EW, SW	BAµE (25-mL sample (pH 5.5); extracted with modified pyrrolidone sorbent; 4 h; eluted with 1.5-mL methanol/acetonitrile; evaporated to dryness and redissolved in 200 µL methanol (15 min))	LC–UV	300–500	
Da Silva et al. (2013)	BZ3, EHMC, EHS, OC	EW, RV	SPE (500-mL sample (pH 3); extracted with PS-DVB/MP cartridge at 10 mL/min; eluted with 3 × 2 mL ethyl acetate; evaporated to 1 mL)	GC–MS/MS(EI$^+$)	2	62–107
Caldas et al. (2013)	BMDM, BZ4, MBC and other compounds	EW, RV	SPE (1000-mL sample (pH 3); extracted with C$_{18}$ cartridge at 10 mL/min; eluted with 2 × 1 mL methanol)	LC–MS/MS(ESI$^{+/-}$)	12	59–93
Gilart et al. (2013)	BZ1, BZ3, BZ8 and other compounds	EW, IW	SBSE (50-mL sample (pH 5); extracted with EGS-coated stir bar; 4 h; 25°C; eluted with 1 mL methanol (15 min); diluted with 1 mL water)	LC–MS/MS(ESI$^{+/-}$)	5–10	96–132
Benedé et al. (2014a)	BZ3, EHDP, EHMC, EHS, HMS, IMC, MBC, OC	SW	DLLME (5-mL sample (pH 2.5); extracted with 50 µL chloroform (+250 µL acetone as disperser solvent))	GC–MS(EI$^+$)	10–30	82–117

Kotnik et al. (2014)	BZ1, BZ3, BZ8 and other compounds	LK, RV, SW, TW	**SPE** (800-mL sample (pH 2–3); extracted with PVP-DVB cartridge at 2 mL/min; eluted with 4 × 450 μL ethyl acetate; evaporated to dryness and redissolved in 600 μL ethyl acetate; derivatized with 30 μL MSTFA (60°C, 1 h))	GC–MS(EI$^+$)	0.1–1.6 (LK) 0.2–1.9 (RV) 0.2–1.7 (SW) 0.4–1.3 (TW)	99–105
Pintado-Herrera et al. (2014)	BZ3, BZ10, EHS, EHDP, EHMC, HMS, MBC and other compounds	EW, RV, SW	**SBSE** (100-mL sample (1% Na$_2$CO$_3$, 10% NaCl); simultaneously derivatized with 500 μL acetic anhydride and extracted with PDMS-coated stir bar; 25°C; 5 h; eluted with 200 μL ethyl acetate (30 min))	GC–MS	0.01–12.4	
Li et al. (2014a)	BZ, BZ1, BZ3	LK, TW	**SPME** (70-mL sample; extracted with poly(MAA-EDMA) monolith; 20 min; eluted with 200 μL methanol (8 min))	LC–UV	300–800	99.3–99.5 (LK) 95.4–96.4 (TW)
Capriotti et al. (2014)	BMDM, BZ1, BZ2, BZ3, BZ4, DHHB, EHDP, EHMC, MBC, OC, PABA, PBSA	LK, TW	**SPE** (200-mL sample; extracted with graphite carbon black cartridge at 15–20 mL/min; eluted with 15 mL methanol/dichloromethane (10 mM ammonium formate); evaporated to dryness and redissolved in 500 μL water/methanol (0.1% formic acid))	LC–MS/MS(ESI$^{+/-}$)	0.7–3.5	49–113 (LK) 51–120 (TW)

Continued

Environmental Monitoring of Cosmetic Ingredients 451

Table 16.1 Published papers on UV filter determination in water samples (in chronological order)—cont'd

Authors	Target compounds[a]	Sample[b]	Extraction technique[c]	Analytical technique[d]	MLOD (ng/L)[e]	Recovery (%)
Purrà et al. (2014)	BZ1, BZ2, BZ3, BZ6, BZ8 and other benzophenones	BW, RV, TW	**SPE** (500-mL sample (pH 3); extracted with PS-DVB/MP cartridge at 2–3 mL/min; eluted with 3 mL methanol + 3 mL dichloromethane; evaporated to dryness and redissolved in 1 mL tetraborate buffer)	CZE–UV	72–415	87–118
Benedé et al. (2014b)	BZ3, EHDP, EHMC, EHS, HMS, IMC, MBC, OC	SW	**SBSDME** (25-mL sample; extracted with CoFe$_2$O$_4$@oleic acid MNPs; 20 min; eluted with 2.5 mL ethanol; evaporated to dryness and redissolved in 100 μL ethanol/aqueous acetic acid)	LC–UV	2400–30,600	79–120
Li et al. (2014b)	BZ3, EHDP, EHMC, EHS	IW, RV	**(DI)SPME** (15-mL sample; extracted with ZrO$_2$ NP-coated Ti–TiO$_2$ wire; 35°C, 30 min; liquid desorption into the injection port)	LC–UV	32–82	82–113 (RV) 81–116 (EW)
Da Silva et al. (2015)	BZ3, EHMC, EHS, OC	EW, IW	**SPE** (500-mL sample (pH 3); extracted with PS-DVB/MP cartridge at 10 mL/min; eluted with 3 × 2 mL ethyl acetate; evaporated to 1 mL)	GC–MS/MS(EI$^+$)	7.1–23.5	62–107

Environmental Monitoring of Cosmetic Ingredients 453

Emmet et al. (2015)	BZ1, BZ3, EHMC, MBC and other compounds	EW, SW	**SPE** (1- to 4-L sample (pH 2); extracted with PVP-DVB + Na$_2$SO$_4$ + Florisil cartridges in series; eluted with 6 × 5 mL dichloromethane/methanol; evaporated to dryness; derivatized with 200 µL BSTFA (80°C, 60 min))	GC-MS(EI$^+$)	0.1–0.5	68–177 (EW) 35–172 (SW)
Benedé et al. (2015)	BZ3, EHDP, EHMC, EHS, HMS, IMC, MBC, OC	LK, RV, SW, TW	**iSAME** (20-mL sample (1% NaCl); addition of SSA + CTAB and in situ formation of aggregate (10 min); aggregate filtered and dissolved in 2 mL isopropanol)	LC-UV	300–1700	80–112
Chung et al. (2015)	BZ1, BZ3, BZ8 and other compounds	EW, RV	**DSPE** (10-mL sample; extracted with PVP-DVB sorbent; 1 min; sorbent injection-port dried (122°C, 3.5 min); injection-port derivatized with 20 µL BSTFA (70°C, 2.5 min); then thermal desorption (340°C, 5.7 min))	(TD)GC-MS(EI$^+$)	0.5–1.0	87–94 (RV) 85–96 (EW)
Cunha et al. (2015)	BMDM, BZ1, BZ3, BZ8, DHHB, EHDP, EHMC, EHS, HMS, IMC, MBC, OC	EW, IW	**DLLME** (10-mL sample (pH 3); extracted with 50 µL trichloroethane (+1 mL acetone as disperser solvent); evaporated to dryness and derivatized with 40 µL BSTFA (microwave assisted, 5 min))	GC-MS(EI$^+$)	2–26	

Continued

Table 16.1 Published papers on UV filter determination in water samples (in chronological order)—cont'd

Authors	Target compounds[a]	Sample[b]	Extraction technique[c]	Analytical technique[d]	MLOD (ng/L)[e]	Recovery (%)
Benedé et al. (2016a)	BZ3, EHDP, EHMC, EHS, HMS, IMC, MBC, OC	RV, SP, SW	**SBSDME** (25-mL sample (pH 4, 5% NaCl); extracted with CoFe$_2$O$_4$@oleic acid MNPs; 30 min; thermal desorption (250°C, 10 min))	(TD)GC–MS(EI$^+$)	13–148	86–115 (RV) 88–112 (SW) 80–116 (SP)
Vila et al. (2016)	EHDP, EHMC, EHS, HMS, IMC, MA, MBC, OC	RV, SP, SW	**USAEME** (10-mL sample (20% NaCl); extracted with 100 μL chloroform (25°C, 5 min))	GC–MS/MS(EI$^+$)	0.08–1.5	65–105 (RV) 76–108 (SW) 60–106 (SP)
Suárez et al. (2016)	BZ3, EHS, HMS, MBC, OC	SP, SW	**(Online)MSA-IL-DLLME** (3.5-mL sample; extracted with 190 μL [C$_6$MIM][PF$_6$] (+200 μL acetonitrile as disperser solvent); 2.5 min; extract diluted with acetonitrile)	LC–UV	80–12000	89–114 (SW) 86–107 (SP)
Clavijo et al. (2016)	BZ3, EHS, EHMC, HMS, MBC, OC	SP, SW	**(Online)MSA-DLLME** (4-mL sample; simultaneously extracted with 250 μL trichloroethylene and derivatized with 150 μL BSTFA; 2.6 min)	GC–MS(EI$^+$)	19–160	82–122 (SW) 86–112 (SP)
Benedé et al. (2016b)	BZ4, PBSA, PDTA, TDSA	RV, SP, SW	**SBSDME** (25-mL sample (pH 2); extracted with CoFe$_2$O$_4$@SiO$_2$-nylon 6 composite; 30 min; eluted with 2.5 mL HCl 1M; evaporated to dryness and redissolved in 100 μL ethanol/aqueous acetate buffer)	LC–UV	1600–2900	92–99 (RV) 91–115 (SW) 90–101 (SP)

Vila et al. (2017)	BZ1, BZ3, BZ4, BZ8, EHS, EHDP, EHMC, HMS, IMC, MA, MBC, OC	RV, SP, SW	**(DI)SPME** (10-mL sample (0.1 g K$_2$CO$_3$); derivatized with 200 μL acetic anhydride, 100°C, 15 min; then extracted with DVB/CAR/PDMS-coated fibre; 30 min; thermal desorption (260°C, 5 min))	(TD)GC-MS/MS(EI$^+$)	0.045–8.2	80–106
Bu et al. (2017)	BZ1, BZ3	IW, RV	**IT-SPME** (40-mL sample (pH 7, 0.5% methanol); extracted with PANI functionalized basalt fibre at 1 mL/min; eluted with 600 μL acetonitrile/water)	LC-UV	20–50	97–114 (RV) 82–115 (IW)

[a]See Table 5.1 for abbreviation key, except BZ, benzophenone; BZ10, benzophenone-10.
[b]BW, bottled water; EW, effluent wastewater; GW, groundwater; IW, influent wastewater; LK, lake water; RV, river water; RW, rainwater; SP, swimming pool water; SW, seawater; TW, tap water.
[c]BAμE, bar adsorptive microextraction; BSTFA, N,O-bis(trimethylsilyl)trifluoroacetamide; CAR, carboxen; CPE, cloud point extraction; DI, direct immersion; DLLME, dispersive liquid–liquid microextraction; DSPE, dispersive solid-phase extraction; EGS, ethylene glycol-modified silicone; HFLPME, hollow-fibre liquid-phase microextraction; HS, headspace; IL, ionic liquid; iSAME, in situ suspended aggregate microextraction; IT, in-tube; KWLPME, knitting wool supported liquid-phase microextraction; LDPE, low-density polyethylene; LLE, liquid–liquid extraction; MALLE, membrane-assisted liquid–liquid extraction; MEPS, microextraction by packed sorbent; MNPs, magnetic nanoparticles; MSA, magnetic stirring-assisted; MSTFA, N-methyl-N-(trimethylsilyl)trifluoroacetamide; MTBSTFA, N-(tert-butyldimethylsilyl)-N-methyltrifluoroacetamide; NPs, nanoparticles; PA, polyacrylate; PDMS, polydimethylsiloxane; PDMS-DVB, polydimethylsiloxane–divinylbenzene; PS-DVB, polystyrene divinylbenzene co-polymer; PS-DVB/MH, polystyrene divinylbenzene co-polymer modified with hydroxyl groups; PS-DVB/MP, polystyrene divinylbenzene co-polymer modified with pyrrolidone; PVP-DVB, polyvinylpyrrolidone–divinylbenzene co-polymer; PVP-DVB/MCX, polyvinylpyrrolidone–divinylbenzene co-polymer modified with a cation exchanger; SBSDME, stir bar sorptive dispersive microextraction; SBSE, stir bar sorptive extraction; SDME, single-drop microextraction; SME, sorptive microextraction; SPE, solid-phase extraction; SPME, solid-phase microextraction; TC, temperature controlled; USA, ultrasound assisted; USAEME, ultrasound-assisted emulsification microextraction; VA, vortex-assisted.
[d]APCI, atmospheric pressure chemical ionization; APPI, atmospheric pressure photoionization; CE, capillary electrophoresis; CZE, capillary zone electrophoresis; DART, direct analysis in real time; EI, electron ionization; ESI, electrospray ionization; FID, flame ionization detector; GC, gas chromatography; GC × GC, multidimensional gas chromatography; LC, liquid chromatography; LVI, large volume injection; MS, mass spectrometry; MS/MS, tandem mass spectrometry; TD, thermal desorption; UV, ultraviolet spectrometry.
[e]MLOD, method limit of detection.

Table 16.2 Published papers on UV filter determination in solid samples (in chronological order)

Authors	Target compounds[a]	Sample[b]	Extraction technique[c]	Analytical technique[d]	MLOD (ng/g)[e]	Recovery (%)
Plagellat et al. (2006)	EHMC, EHT, MBC, OC	SS	**SLE + SPE** (60-g dry sample (+3 g NaCl); extracted with 20 mL pentane/acetone + 20 mL pentane/diethyl ether + 20 mL diethyl ether/dichloromethane (30 min each); dried with Na$_2$SO$_4$, evaporated to dryness and redissolved in 1 mL hexane; SPE with silica and 50–70 mL hexane/diethyl ether; then evaporated to dryness and redissolved in ethyl acetate (for GC) or ethanol (for LC))	LC–UV (for EHT) GC–MS(EI$^+$) (for the rest)	57 3–6	75 88–101
Jeon et al. (2006)	BZ, BZ1, BZ3, BZ8 and other benzophenones	SE	**SLE + LLE** (10-g sample (+10 g Na$_2$SO$_4$); extracted with 20 mL methanol (20 min); evaporated to 3 mL, mixed with 1 mL 5% NaCl and extracted with 5 mL ethyl acetate; then evaporated to dryness and derivatized with 50 µL MSTFA (80°C, 30 min))	GC–MS(EI$^+$)	0.1	60–125
Rodil and Moeder (2008b)	BZ3, EHDP, EHMC, EHS, HMS, IMC, MBC, OC	SE	**PLE** (4- to 5-g sample (+1 g Na$_2$SO$_4$): extracted with 4 × 5 min ethyl acetate/hexane; 160°C; 100 bar; evaporated to 0.5 mL and derivatized with 50 µL BSTFA (25°C, 60 min))	GC–MS(EI$^+$)	2–6	73–128

Environmental Monitoring of Cosmetic Ingredients 457

Nieto et al. (2009)	BZ3, EHDP, OC, PBSA and other compounds	SS	**PLE** (1-g dried sample; extracted with 2 × 5 min methanol + 2 × 5 min water (pH 7)/methanol; 100°C; 140 bar)	LC–MS/MS(ESI$^{+/-}$)	1.5–3.5	79–108
Negreira et al. (2009c)	EHS, HMS, IMC, EHMC, MBC, OC	ID	**MSPD** (0.5-g sample (+0.5 g Na$_2$SO$_4$); extracted with C$_{18}$/silica sorbent; eluted with 4 mL acetonitrile; evaporated to 1 mL)	GC–MS/MS(ESI$^{+/-}$)	3–12	77–99
Rodil et al. (2009d)	BMDM, BZ3, DHBT, EHDP, EHMC, EHS, EHT, HMS, IMC, MBC, OC	SS	**PMALE** (0.5-g dried sample (+1 mL ethyl acetate/hexane) into LDPE bags; extracted with 4 × 5 min ethyl acetate/hexane; 70°C; 10 mPa; evaporated to dryness and redissolved in 0.5 mL methanol/water)	LC–MS/MS(APPI$^{+/-}$)	0.3–25	95–124
Wick et al. (2010)	BZ1, BZ2, BZ3, BZ4, PBSA and other compounds	SS	**PLE + SPE** (0.2-g dried sample; extracted with 4 × 10 min water/methanol; 80°C; 30 mL final volume; diluted to 800 mL with water; then SPE as described for water samples)	LC–MS/MS(ESI$^{+/-}$)	0.75–7.5	74–118
Negreira et al. (2011b)	BZ3, EHDP, EHMC, EHS, HMS, IMC, MBC, OC	SS	**PLE + SPE** (0.5-g sample (+2 g diatomaceous earth); extracted with 1 × 5 min hexane/dichloromethane; 75°C; 103.4 bar; evaporated to 1 mL; subjected to SPE with PSA cartridge and 5 mL hexane/ether; then evaporated to dryness and redissolved in 1 mL isooctane)	GC–MS(EI$^+$)	5.2–18.5	73–112

Continued

Table 16.2 Published papers on UV filter determination in solid samples (in chronological order)—cont'd

Authors	Target compounds[a]	Sample[b]	Extraction technique[c]	Analytical technique[d]	MLOD (ng/g)[e]	Recovery (%)
Kameda et al. (2011)	BZ3, EHDP, EHMC, EHS, HMS, MBC, OC and others	SE, SS	**SLE + SPE** (4-g dried sample; extracted with 2×10 mL dichloromethane + 2×10 mL acetone (10 min each); SPE with MgSiO$_3$ cartridge and 2×40 mL acetone/hexane; SPE with graphite cartridge and 10 mL toluene/acetone; SPE with NH2 cartridge and 20 mL hexane + 7 mL acetone/hexane; then evaporated to 200 μL)	GC–MS(EI$^+$)	0.05–2.0	70–125
Zhang et al. (2011b)	BZ1, BZ2, BZ3, BZ8 and others	SE, SS	**SLE + SPE** (0.1- to 1-g dried sample; extracted with 3×5 mL methanol (30 min each); evaporated to 0.5 mL, diluted with water; SPE with PVP-DVB cartridge (1 mL/min) and 6 mL methanol/ethyl acetate; then evaporated to 1 mL)	LC–MS/MS(ESI$^-$)	0.041–0.67	38–116
Gago-Ferrero et al. (2011a,b)	BZ1, BZ3, EHDP, EHMC, MBC, OC and others	SE, SS	**PLE** (1-g dried sample (+1 g Al$_2$O$_3$); extracted with 2×5 min methanol + 2×5 min methanol/water; 100°C; 100 bar; 20 mL final volume; then diluted to 25 mL with methanol; 2 mL evaporated to dryness and redissolved in 250 μL acetonitrile)	LC–MS/MS(ESI$^+$)	0.5–15 (SE) 0.2–60 (SS)	58–125 (SE) 30–102 (SS)

Reference	Compounds	Sample	Extraction/Cleanup	Determination	LOQ (ng/g)	Recovery (%)
Sánchez-Brunete et al. (2011)	BZ1, BZ3, BZ6, BZ8, EHS, HMS	SE, SO	**MSPD** (2-g dried sample + 1 g C$_{18}$ + 1 g Na$_2$SO$_4$; extracted with 2 × 8 mL ethyl acetate/methanol (15 min each); evaporated to 0.5 mL and diluted to 1 mL; 100 μL derivatized with BSTFA (60°C, 10 min))	GC–MS(EI$^+$)	0.07–0.28	88–105
Albero et al. (2012a)	BZ3, EHS, HMS and other compounds	SO	**PLE** (1-g sample (+7 g diatomaceous earth); extracted with 2 × 10 min ethyl acetate/methanol; 80°C; 120 bar; evaporated to dryness, redissolved in 1 mL ethyl acetate and derivatized with BSTFA in the GC inlet)	GC–MS/MS(EI$^+$)	0.5–1.7	65–109
Amine et al. (2012)	EHDP, EHMC, OC	SE	**MAE** (5-g sample; extracted with 30 mL acetone/heptane; 115°C, 15 min; evaporated to dryness and redissolved in 1 mL heptane)	GC–MS/MS	1.5–2	97–115
Pintado-Herrera et al. (2013)	BZ3, OC and other compounds	SE	**PHWE + SBSE** (2g-sample (+18 g siliceous earth); extracted with 3 × 5 min water (10% methanol); 100°C; 103.4 bar; acidified to pH 2; then SBSE as described for water samples)	GC–MS(EI$^+$)	0.07–0.3	13–22
Barón et al. (2013)	BP3, OC, OD-PABA, BP1, 4HB, 4DHB, 4MBC, EHMC	SE	**PLE** (1-g sample (+1 g Al$_2$O$_3$); 25 mL methanol; 100°C; evaporated to dryness, redissolved in 250 μL acetonitrile)	LC–MS/MS(ESI$^+$)	0.4–9.9	58–125

Continued

Table 16.2 Published papers on UV filter determination in solid samples (in chronological order)—cont'd

Authors	Target compounds[a]	Sample[b]	Extraction technique[c]	Analytical technique[d]	MLOD (ng/g)[e]	Recovery (%)
Kotnik et al. (2014)	BZ1, BZ3, BZ8 + other compounds	SE	MAE (4-g sample; extracted with acetone/methanol (5% formic acid); 150°C; 30 min; extract evaporated to dryness and diluted to 100 mL with water; then SPE as described for water samples)	GC–MS(EI⁺)	01–1.4	101–106
Tarazona et al. (2014)	BZ3, EHDP, EHMC, EHS, HMS, IMC, MBC, OC	SE	SLE + DLLME (10-g sample; extracted with 1 × 5 mL and 2 × 1 mL acetone; extract diluted to 5 mL with acetone; then 2 mL subjected to DLLME with 60 μL chloroform and 5 mL water (pH 4))	GC–MS(EI⁺)	0.018–0.053	80–106
Li et al. (2016a)	BZ3, MBC, OC and other compounds	SS	MSPD (0.1-g sample; extracted with C₁₈ sorbent; eluted with 6 mL methanol + 10 mL acetonitrile/aqueous oxalic acid; evaporated to dryness and redissolved in 1 mL acetonitrile/water)	LC–MS/MS(ESI⁺/⁻)	0.856–1.60	54–81

[a] See Table 5.1 for abbreviation key, except BZ, benzophenone.
[b] ID, indoor dust; SE, sediment; SO, soil; SS, sewage sludge.
[c] BSTFA, N,O-bis(trimethylsilyl)trifluoroacetamide; DLLME, dispersive liquid–liquid microextraction; LDPE, low-density polyethylene; LLE, liquid–liquid extraction; MAE, microwave-assisted extraction; MSPD, matrix solid-phase dispersive; MSTFA, N-methyl-N-(trimethylsilyl)trifluoroacetamide; PHWE, pressurized hot water extraction; PLE, pressurized liquid extraction; PMALE, pressurized membrane-assisted liquid extraction; PSA, primary–secondary amine; PVP-DVB, polyvinylpyrrolidone–divinylbenzene co-polymer; SBSE, stir bar sorptive extraction; SLE, solid-liquid extraction; SPE, solid-phase extraction.
[d] APPI, atmospheric pressure photoionization; EI, electron ionization; ESI, electrospray ionization; GC, gas chromatography; LC, liquid chromatography; MS, mass spectrometry; MS/MS, tandem mass spectrometry; UV, ultraviolet spectrometry.
[e] MLOD, method limit of detection.

Table 16.3 Published papers on UV filter determination in biota samples (in chronological order)

Authors	Target compounds[a]	Sample[b]	Extraction technique[c]	Analytical technique[d]	MLOD (ng/g)[e]	Recovery (%)
Poiger et al. (2004) Balmer et al. (2005)	BMDM, BZ3, EHMC, MBC, OC	FI	**SLE + GPC** (20-g sample (+100 g Na$_2$SO$_4$); extracted with 150 mL cyclohexane/dichloromethane; GPC with dichloromethane/cyclohexane; then cleaned up with silica gel; evaporated to dryness and redissolved in 50–200 μL ethyl acetate)	GC–MS(EI$^+$)	3–53 (lipid weight)	93–115
Meinerling and Daniels (2006)	BZ3, EHMC, MBC, OC	FI	**Soxhlet + GPC + SPE** (10-g blended sample (+25 g Na$_2$SO$_4$); extracted with 200 mL hexane/acetone (reflux, 3 h); evaporated to dryness and redissolved in hexane/acetone; GPC with cyclohexane/ethyl acetate; then the fraction of interest evaporated to 1 mL; SPE with Florisil and elution with 3 × 5 mL hexane/acetone; evaporated to 0.5 mL and diluted to 2 mL with acetonitrile)	LC–MS/MS(ESI$^+$)	2.4	86–108

Continued

Table 16.3 Published papers on UV filter determination in biota samples (in chronological order)—cont'd

Authors	Target compounds[a]	Sample[b]	Extraction technique[c]	Analytical technique[d]	MLOD (ng/g)[e]	Recovery (%)
Buser et al. (2006)	MBC, OC	FI	**SLE + GPC + SPE** (10- to 25-g blended sample (+100 mL water); extracted with 2 mL aqueous oxalate + 100 mL ethanol + 50 mL diethyl ether + 70 mL pentane (1 min each); GPC with cyclohexane/ethyl acetate; then the fraction of interest evaporated to 1 mL, diluted with isooctane and hexane and evaporated to 2 mL; SPE with silica gel column and 40 mL hexane/dichloromethane + 40 mL dichloromethane; evaporated to 100 μL)	GC–MS(EI[+])	5–20 (lipid weight)	
Zenker et al. (2008)	3BC, BZ1, BZ2, BZ3, BZ4, EHMC, MBC and others	FI	**SLE + LC** (4-g cut sample (+4 mL water); extracted with ethyl acetate/heptane/water (10 min); evaporated to dryness and redissolved in 500 μL ethanol; LC with C$_{18}$ and methanol/water; then the fraction of interest evaporated to dryness and redissolved in 50 μL ethanol)	LC–MS/MS(ESI[+/−]) (rest) GC–MS/MS(EI[+]) (EHMC, MBC, BZ3, 3BC)	86–205 11–36	76–99
Fent et al. (2010)			**SLE** (1-g cut sample; extracted with 10 mL methanol/acetonitrile (10 min); evaporated to dryness and redissolved in 1 mL ethanol, filtered and diluted to 2 mL with ethanol; evaporated to dryness and redissolved in 50 μL ethanol)	LC–MS/MS(ESI[+/−]) (rest)	0.002–0.005 (lipid weight)	

Mottaleb et al. (2009)	BZ, MBC, OC and other compounds	FI	**SLE+SPE** (1-g blended sample; extracted with 10 mL acetone (25–35°C, 5–15 min); evaporated to dryness and redissolved in 200 μL hexane/acetone; SPE with silica gel column and hexane/acetone; evaporated to 50 μL and, after derivatization step for compounds other than UV filters, evaporated to 20 μL and diluted with 180 μL hexane)	GC–MS(EI⁺)	5.3–17	98–101
			SLE+SPE+GPC (1-g blended sample; extracted with 10 mL acetone (25–35°C, 5–15 min); evaporated to dryness and redissolved in 200 μL hexane/acetone; SPE with silica gel column and hexane/acetone; evaporated to dryness, redissolved in 700 μL dichloromethane; GPC; fraction of interest evaporated to 50 μL and, after derivatization step for compounds other than UV filters, evaporated to 20 μL and diluted with 180 μL hexane)	GC–MS/MS(EI⁺)	16–120	57–87
Bachelot et al. (2012)	EHDP, EHMC, OC	MU	**MAE+LC** (3-g freeze-dried and powdered sample; extracted with 25 mL acetone/heptane; 110°C, 15 min; evaporated to dryness and redissolved in 1.5 mL ethanol; then LC with C$_{18}$ and methanol/water; fraction of interest evaporated to dryness and redissolved in 1 mL heptane)	GC–MS/MS(EI⁺)	2	89–116

Continued

Table 16.3 Published papers on UV filter determination in biota samples (in chronological order)—cont'd

Authors	Target compounds[a]	Sample[b]	Extraction technique[c]	Analytical technique[d]	MLOD (ng/g)[e]	Recovery (%)
Gago-Ferrero et al. (2013b, 2015)	BZ1, BZ3, EHDP, EHMC, MBC, OC	FI	**PLE + SPE** (1-g blend and dried sample (+1 g Florisil); extracted with 2 × 5 min ethyl acetate/dichloromethane; 100°C; 103.4 bar; extract diluted to 250 mL; SPE with C_{18} cartridge and 7 mL ethyl acetate/dichloromethane + 2 mL dichloromethane; then evaporated to dryness and redissolved with 1 mL acetonitrile)	LC–MS/MS(ESI⁺)	0.1–6	36–112
Tsai et al. (2014)	BZ1, BZ3, BZ8, EHS, HMS	FI	**MSPD + SPE** (0.5-g blend and dried sample (+0.5 g Na_2SO_4); extracted with Florisil sorbent; charged in C_{18} cartridges; eluted with 7 mL acetonitrile; evaporated to dryness and redissolved in 100 μL dichloromethane; then injection-port derivatized with MSTFA (70°C, 2.5 min))	GC–MS/MS(EI⁺)	0.02–0.3	71–102
Picot-Groz et al. (2014)	EHDP, EHMC, OC + other	MU	**SLE** (2-g blend and dried sample (+10 mL water + 10 mL acetonitrile + 4 g Na_2SO_4 + 1 g NaCl + 1.5 g citrate); 1 min; acetonitrile layer cleaned up with C_{18}/PSA sorbent; 1 min; 1 mL extract evaporated to dryness and redissolved in heptane)	GC–MS/MS(EI⁺)	2.5–5	90–126

Emmet et al. (2015)	BZ1, BZ3, EHMC, MBC and other compounds	CL, FI	**PLE + SPE + GPC** (8-g blend and dried sample; extracted with 2 × 10 min water/isopropanol, 120–180°C, 100 bar; filtered and diluted with 50 mL phosphate buffer; SPE as above; evaporated to dryness; redissolved with 1.5 mL dichloromethane/methanol; GPC with the same solvent; then the fraction of interest evaporated to dryness and redissolved in 1 mL dichloromethane/methanol)	GC–MS(EI+)	0.6–2	52–67
Peng et al. (2015)	BMDM, BZ3, EHDP, EHMC, MBC, OC and others	FI, HT, SQ	**SLE + GPC + SPE** (4-g blend and dried sample; extracted with 3 × 20 mL methanol (15 min each); evaporated to dryness; redissolved in 1 mL ethyl acetate/cyclohexane; GPC with ethyl acetate/cyclohexane; fraction of interest evaporated, redissolved in hexane and subjected to SPE with silica gel column and 15 mL dichloromethane/ethyl acetate; evaporated to dryness and redissolved in 1 mL methanol)	LC–MS/MS(APCI+)	0.0015–3	41–116

[a]See Table 5.1 for abbreviation key, except *BZ*, benzophenone.
[b]*CL*, clams; *FI*, fish; *HT*, hairtails; *MU*, mussels; *SQ*, squids.
[c]*GPC*, gel-permeation chromatography; *MAE*, microwave-assisted extraction; *MSPD*, matrix solid-phase dispersion; *MSTFA*, N-methyl-N-(trimethylsilyl)trifluoroacetamide; *PLE*, pressurized liquid extraction; *PSA*, primary–secondary amine; *SLE*, solid–liquid extraction; *SPE*, solid-phase extraction.
[d]*APCI*, atmospheric pressure chemical ionization; *EI*, electron ionization; *ESI*, electrospray ionization; *GC*, gas chromatography; *LC*, liquid chromatography; *MS*, mass spectrometry; *MS/MS*, tandem mass spectrometry.
[e]*MLOD*, method limit of detection.

attracted by the less polar PVP-DVB skeleton. In the arsenal of SPE sorbents, graphite carbon black (Capriotti et al., 2014) has also been employed for the multi-residue determination of both polar and non-polar UV filters.

Online SPE for the automated extraction of UV filters has also been reported. Oliveira et al. (2010) used a PVP-DVB SPE cartridge adjusted into a multi-syringe-lab-on-valve apparatus, whilst Gago-Ferrero et al. (2013a) employed a commercial online SPE device with PS-DVB as the sorbent for the determination of UV filters prior to liquid chromatography (LC) analysis. Other automated methods are based on inline SPE with a PVP-DVB/MCX sorbent that was coupled to capillary electrophoresis (Maijó et al., 2013).

In addition to classic SPE, microextraction methods such as solid-phase microextraction (SPME), stir bar sorptive extraction (SBSE) and dispersive SPE (DSPE) have found many applications in the analysis of UV filters in environmental water samples. SPME is usually employed in direct immersion (DI) mode, because UV filters are generally non-volatile compounds, although some applications of headspace (HS) extraction mode have been also reported. SPME methods have relied on polydimethylsiloxane (PDMS), PDMS-DVB and polyacrylate (PA) coatings, which are well suited to the extraction of UV filters of low to medium polarity (Lambropoulou et al., 2002; Negreira et al., 2009a; Liu et al., 2010; Basaglia and Pietrogrande, 2012; Vila et al., 2017). Lab-made sorbents such as graphene-based sol-gel (Zhang and Lee, 2012a), basalt fibres coated with polyaniline (Bu et al., 2017), dodecyl-modified silver wire (Li et al., 2013), ZrO_2 nanoparticles (Li et al., 2014b) and a magnetic poly(methacrylic acid-co-ethylene dimethacrylate) monolith (Li et al., 2014a) have been proposed as alternative coatings to commercial sorbents with good analytical features. Nevertheless, SPME methods have not been applied to the extraction of the most polar UV filters (i.e., such as BZ4, PBS, TDS, etc.) possibly owing to the lack of appropriate fibre coatings.

In analogy to SPME, SBSE is also usually applied to non-polar compounds and typically with PDMS coatings (Kawaguchi et al., 2006, 2008a; Rodil and Moeder, 2008a; Haunschmidt et al., 2010; Pedrouzo et al., 2010; Nguyen et al., 2011; Gómez et al., 2011; Magi et al., 2012, 2013; Pintado-Herrera et al., 2013; Pintado-Herrera et al., 2014). The commercialization of PA-polyethyleneglycol and ethyleneglycol-modified silicone (EGS) coatings has enabled the application of the method to more polar UV filter compounds whilst affording good extraction efficiencies for non-polar species as well (Gilart et al., 2013). An innovative solution to the lack of appropriate coatings was conceived by Nogueira et al. (Almeida et al., 2013), who proposed the pasting of the solid sorbent on a polyethylene cylindrical tube with the aid of an adhesive tape. This method, called bar adsorptive microextraction (BAµE), extends the range of sorbents that can be employed in SBSE, thus enhancing its versatility in the multi-residue analysis of UV filters with a wide polarity. On the grounds of this principle, sorbents such as PS-DVB, modified pyrrolidone, cyano derivatives, and five activated carbons of different surface areas were evaluated (Almeida et al., 2013).

Another approach to the analysis of UV filters in water samples is DSPE using either magnetic or non-magnetic sorbents. By exploiting these advantages, Parisis et al. (2005) used silica as a sorbent for collecting vesicles previously used for extracting lipophilic UV filters, and later, Chung et al. (2015) used PVP-DVB for DSPE of lipophilic UV filters. Regarding magnetic sorbents, they are usually nanometre-sized particles which bear a magnetic core and an appropriate coating with various properties. The high contact areas, the ability to select different coatings and the facile collection of the dispersed sorbent (through magnetism) are some of the main advantages of this approach compared to the use of non-magnetic sorbent microparticles. On the basis of this principle, Román et al. (2011) presented the first study focusing on the use of magnetic nanomaterials for the extraction of UV filters from water samples by employing oleic acid-coated cobalt ferrite ($CoFe_2O_4$@oleic acid) magnetic nanoparticles (MNPs) with very satisfactory results. This application provided the springboard for alternative methods such as stir bar sorptive dispersive microextraction (Benedé et al., 2014b, 2016a,b). This method combines the principles of SBSE and DSPE by using a stirring bar physically coated with a magnetic sorbent. At low stirring speed, the magnetic sorbent acts as a coating material on the stir bar, thus acting in analogy to SBSE, whilst at high stirring speed the nano-sorbent is dispersed into the aqueous medium, in analogy to DSPE. The main advantage of this approach is that it significantly facilitates the extraction and the post-extraction (i.e., elution) steps with minimal external intervention. Most importantly, it enables the utilization of various coatings such as oleic acid (Benedé et al., 2014b, 2016a) and nylon-6 (Benedé et al., 2016b) for the extraction of hydrophobic and hydrophilic UV filters, respectively.

Finally, other sorbent-based microextraction approaches such as sorptive microextraction (SME) and microextraction by packed sorbent (MEPS) have also been used for the determination of UV filters. In SME different silicone discs are stirred within the sample solution for long periods of time (i.e., 14h) (Negreira et al., 2011a), whilst in MEPS the sample is aspirated into a barrel containing a packed non-polar sorbent (Moeder et al., 2010).

In parallel to the development of sample preparation methods based on the use of solid sorbents, liquid-phase extraction and microextraction methods have been successfully developed. Although the use of liquid–liquid extraction (LLE) is gradually phasing out, methods using *n*-hexane and ethyl acetate as extraction solvents have been reported (Jeon et al., 2006; Gómez et al., 2009). Cloud point extraction, using non-ionic surfactants of the Triton series, was one of the first methods utilized as a green alternative to LLE for the extraction of UV filters, employing centrifugation to separate the micellar phase containing the target compounds (Giokas et al., 2005). Later, highly hydrophobic MNPs (Fe_2O_3@C) were used to simplify and accelerate the procedure by retrieving the micellar phase on their surface; the MNPs were then collected by the force of a magnetic field, thus accomplishing extraction (Giokas et al., 2012). On both occasions, non-polar organic solvents were used to back-extract UV filters from the surfactant phase or the surfactant/MNP surface to minimize the co-extraction of surfactants in the final extract, which is delivered to the analytical detector. A simplification to micelle-mediated extraction is in situ suspended aggregate microextraction, which is based on the extraction of UV filters in a supramolecular aggregate phase, which is formed in situ in the sample through ion association between a cationic surfactant and an organic anion. The aggregate phase is easily separated by filtration and disrupted by polar organic solvents to release the target analytes. Therefore, there is no need for centrifugation or magnetic sorbents, thus significantly facilitating the overall sample preparation procedure (Benedé et al., 2015).

Beyond surfactant-based methods, liquid-phase microextraction (LPME) methods using non-polar organic solvents and ionic liquids (ILs) have gained increased popularity in the analysis of UV filters in water samples. Although various LPME methods such as DI–single-drop microextraction (SDME) (Okanouchi et al., 2008), knitting wool–supported LPME (Zhang and Lee, 2013b) and membrane-assisted LLE (MALLE) (Rodil et al., 2009c) have been successfully applied, the large number of analytical methods relying on the principles of classic dispersive liquid–liquid microextraction (DLLME) bear testimony to the fact that it is an efficient and expedient analytical tool for the environmental surveillance of UV filters. Initial efforts with DLLME were based on the use of organochlorine solvents and acetone as extracting and disperser solvents, respectively (Tarazona et al., 2010; Negreira et al., 2010; Benedé et al., 2014a; Cunha et al., 2015). Although the analytical figures of the merit of these methods were satisfactory, further improvement was pursued either by increasing the dispersion of the extracting solvent in the aqueous phase or by avoiding the use of a disperser solvent at all, because sometimes the use of a disperser solvent has been found to decrease the partition coefficient of some of the target analytes into the extracting solvent. On the basis of this reasoning, Wu et al. (2013a) used ultrasound to improve the dispersion of the extracting solvent (tetrachloroethylene) in the presence of acetone as a dispersion solvent and produced finer extracting droplets in the so-called ultrasound-assisted DLLME (USA-DLLME). Other studies have proposed the use of magnetic agitation (magnetic stirring–assisted DLLME, MSA-DLLME) (Zhang et al., 2011a; Clavijo et al., 2016), vortex mixing (vortex-assisted DLLME) (Zhang and Lee, 2012c) and USA emulsification (USAEME) (Vila et al., 2016), for the dispersion of the extracting solvent, alleviating the need for a disperser solvent.

The replacement of non-polar organic solvents with ILs has further advanced the analytical utility of DLLME in UV filter analysis. Owing to the high viscosity of ILs, external energy has to be introduced into the system (i.e., ultrasound, magnetic or mechanical agitation, temperature control) to accomplish the effective dispersion of the IL into the water sample. Typical examples are IL-USA-DLLME (Zhang and Lee, 2012b; Xue et al., 2013; Ge and Lee, 2012b), IL-MSA-DLLME (Suárez et al., 2016), temperature-controlled IL-DLLME (Zhang and Lee, 2013a) and end-over-end shaking (Ku et al., 2013). The only exceptions are IL-SDME and IL–hollow fibre LPME, where the IL is not dispersed into the solution. Instead, agitation is used to enhance the contact between the sample and the immersed IL droplet (Vidal et al., 2010) or the hollow fibre bearing the IL, respectively (Ge and Lee, 2012a).

Solid Samples

Owing to the high octanol-to-water partition coefficients of most of the UV filters (log K_{ow} >3), soils, sediments and sewage sludge have been shown to be an effective matrix for their accumulation. In contrast to the analysis of water samples, for which methods based on LLE are scant, solvent extraction is the main method of choice for the analysis of solid samples (see Table 16.2). Typically, UV filters are extracted with mixtures of non-polar and polar organic solvents using ultrasound or mechanical agitation (i.e., solid–liquid extraction (SLE)). Depending on the matrix and the solvent mixture this procedure is repeated (typically two or three times) and, after centrifugation or filtration, the extract is usually evaporated to a small volume to enhance the pre-concentration factor and facilitate its further treatment and clean-up. Because of the complexity of the matrix the extracts are cleaned up to relieve the subsequent chromatographic analysis from co-eluting and interfering matrix components. This has been usually accomplished by means of SPE using silica gel or C_{18} sorbents (Plagellat et al., 2006; Sánchez-Brunete et al., 2011; Zhang et al., 2011b; Kameda et al., 2011; Barón et al., 2013) as well as DLLME (Tarazona et al., 2014). Matrix solid-phase dispersion (MSPD) (Negreira et al., 2009c) and automated extraction methods have been also successfully utilized to accomplish faster, more efficient extraction and with less solvent consumption. Nevertheless, to accelerate the process, pressurized liquid extraction (PLE) is preferably used for the extraction of UV filters from solid samples (Rodil and Moeder, 2008b; Rodil et al., 2009d; Nieto et al., 2009; Wick et al., 2010; Negreira et al., 2011b; Gago-Ferrero et al., 2011a; Gago-Ferrero et al., 2011b), whilst other methods such as microwave-assisted extraction (MAE) coupled to an additional clean-up step have also been reported (Amine et al., 2012; Kotnik et al., 2014).

Living Organisms

Because of the high lipophilicity of most UV filters, their bioaccumulation in living organisms was the first issue to be examined (Nagtegaal et al., 1997). Most studies concerning the bioaccumulation of UV filters have been performed on fish, and only sparse results are available regarding other organisms such as macro-zoobenthos, mussels and birds (Gago-Ferrero et al., 2012). To accomplish the extraction of UV filters from biological tissues, methods similar to those employed for the analysis of solid samples were utilized (see Table 16.3). SLE (Poiger et al., 2004; Balmer et al., 2005; Buser et al., 2006;

Zenker et al., 2008; Mottaleb et al., 2009; Fent et al., 2010) and Soxhlet extraction (Nagtegaal et al., 1997; Meinerling and Daniels, 2006; Picot-Groz et al., 2014; Peng et al., 2015) were the first methods employed for the extraction of UV filters from the tissues of living organisms, whilst the availability of commercial apparatuses has facilitated the employment of more advanced extraction methods such as PLE (Gago-Ferrero et al., 2013b, 2015; Emnet et al., 2015) and microwave digestion (Bachelot et al., 2012). MSPD methods have also been reported (Tsai et al., 2014). None of these methods, however, can avoid the co-extraction of lipids, which are removed by gel-permeation chromatography (GPC), whilst additional clean-up is often applied to remove polar impurities using silica or Florisil columns. Enhanced clean-up can also be pursued with LC (Zenker et al., 2008; Bachelot et al., 2012), especially when UV filters with different polarities are of concern.

Analytical Techniques

The analytical techniques that are used for the determination of UV filters in environmental samples share many things in common with those developed for sunscreen analysis in commercial products (Salvador and Chisvert, 2007; Giokas et al., 2007). In that sense, liquid and gas chromatographic techniques are mainly used because they enable not only the separation of mixtures of UV filters but also the isolation of potential interfering compounds that may evade sample preparation and injection into the analytical system.

Owing to the low volatility of most UV filters, LC-based methods are particularity suitable for their analysis. They are also a good choice when organic solvents with low volatility are used for extraction (e.g., octanol, ILs, etc.) and are directly injected into the LC system for analysis (Vidal et al., 2010; Zhang et al., 2011a; Ge and Lee, 2012a; Xue et al., 2013; Ku et al., 2013; Cunha et al., 2015). In the overwhelming majority of applications reversed-phase C_{18} stationary phases are employed, whilst less hydrophobic materials (such as C_8, C_{12}, phenylhexyl) have also been applied with satisfactory results (Rodil et al., 2009a,b,c). In any case, the selection of the most appropriate mobile phase and elution conditions (i.e., isocratic or gradient) plays a crucial role in the efficient resolution of UV-absorbing chemicals from one another as well as from other matrix constituents. Both isocratic and gradient elution mobile phases (with or without phase modifiers such as weak organic acids) have been employed, depending mostly on the properties of the UV filters, the number of compounds to be separated, the intended application (i.e., determination of the parental compounds and/or byproducts) and the complexity of the matrix (water, soil, fish tissue). Isocratic mobile phases, typically consisting of hydro-organic mixtures, are usually preferred in simple matrices like water samples, whilst gradient elution often leads to better separation in more intricate matrices like wastewater, sewage sludge or fish tissues. It is interesting to note that many studies employ methanol/water as the mobile phase during gradient elution (Kupper et al., 2006; Plagellat et al., 2006; Rodil et al., 2008; Kasprzyk-Hordern et al., 2008; Rodil et al., 2009a,b,c,d; Nieto et al., 2009; Negreira et al., 2009b; Pedrouzo et al., 2009; Pedrouzo et al., 2010; Fent et al., 2010). With regard to detection, UV spectrometry and mass-selective detectors are applied. UV spectrometry is a convenient solution to the determination of UV filters because they exhibit intense absorbance in the UV region. However, owing to sensitivity and selectivity barriers they are usually selected when a limited number of analytes are of interest and in relatively simple matrices such as water (Table 16.1). On the other hand, mass spectrometry (MS) detectors are predominately used for realistic monitoring surveys in a wider variety of sample matrices because of the significantly higher sensitivity (and concurrently lower detection limits at the ng/L level), the improved analyte resolution in the presence of interference, as well as the feasibility of detecting degradation byproducts and isomers. From the literature gathered in Tables 16.1–16.3 it is made clear that electrospray ionization (ESI), either positive or negative, is predominately utilized for the analysis of UV filters. Atmospheric pressure chemical ionization (Nguyen et al., 2011; Magi et al., 2012, 2013; Peng et al., 2015) and atmospheric pressure photoionization (Rodil et al., 2009b,c,d) have also been used, mainly to reduce interference from co-eluting species but at the expense of sensitivity.

Gas chromatography (GC)–MS is another very popular analytical technique for the sensitive determination of UV filters at nanogram per litre levels. Simple quadrupole analysers with electron ionization (EI) in selected-ion monitoring mode are used on all occasions, whilst chemical ionization has never been reported for the determination of UV filters in environmental samples (Tables 16.1–16.3). More advanced analysers such as ion traps (Negreira et al., 2009a; Liu et al., 2010; Da Silva et al., 2013) and time-of-flight (Matamoros et al., 2010; Gómez et al., 2011) have been reported only scarcely. Separation of UV filters is accomplished primarily in 5% diphenyl–95% dimethylpolysiloxane or phenyl arylene polymer columns, which are non-polar, have a high temperature limit and have low bleed, thus offering improved signal-to-noise ratio for better sensitivity and mass spectral integrity.

The application of GC–MS to the analysis of UV filters is sometimes accompanied by derivatization reactions that improve the sensitivity and increase the volatility of UV filters. Silylation is the most common derivatization reaction using *N*-methyl-*N*-(trimethylsilyl)trifluoroacetamide (Jeon et al., 2006; Cuderman and Heath, 2007; Negreira et al., 2009a; Kotnik et al., 2014), *N,O*-bis(trimethylsilyl)trifluoroacetamide (Tarazona et al., 2010; Román et al., 2011; Basaglia and Pietrogrande, 2012; Zhang and Lee, 2012a,c; Wu et al., 2013a) or *N*-(*tert*-butyldimethylsilyl)-*N*-methyltrifluoroacetamide (MTBSTFA) (Pintado-Herrera et al., 2013). Other derivatization reactions such as acetylation with acetic anhydride (Kawaguchi et al., 2008a) and oxime formation with *O*-(2,3,4,5,6-pentafluorobenzyl)hydroxylamine (Kotnik et al., 2014) have been used only for the derivatization of benzophenone-type UV filters. Typically, derivatization is carried out after the extraction, by adding the derivatizing agent to the extract. However, in situ derivatization during the extraction step has been reported to enhance the extraction yields because some derivatives are more easily extracted than the parental compounds. This approach has been successfully demonstrated in combination with SPME (Negreira et al., 2009a; Basaglia and Pietrogrande, 2012; Zhang and Lee, 2012a), DLLME (Wu et al., 2013a) and SBSE (Kawaguchi et al., 2008a).

SYNTHETIC MUSKS

Synthetic musks are fragrance chemicals that are used not only in cosmetic products, but also in other household products like detergents, softeners, etc. They were created to replace natural musks that were extracted from natural (animal) sources, owing to economical but mainly ethical reasons. Although synthetic musks are completely different chemical compounds compared to their natural analogues, they possess very similar odour properties. Depending on their chemical structure, synthetic musks are divided into three subgroups: nitro musks, polycyclic musks and macrocyclic musks (see Chapter 10). There are different studies discussing the risks associated with the release of these compounds into the environment because they are not completely removed by WWTPs, thus appearing at trace levels in natural waters. From these studies it can be inferred that the environmental effects of synthetic musks are compound-dependent and can vary depending on the environmental compartment (i.e., water, soil, air, etc.) (Homem et al., 2015; Ramaswamy, 2015). In this regard, different analytical methods have been developed, allowing the determination of synthetic musks in different environmental compartments, such as water samples (Table 16.4) solid samples (Table 16.5) and in living organisms (Table 16.6).

Sample Preparation

The low levels of synthetic musks in environmental samples and the potential interference from co-existing matrix components require the application of appropriate sample preparation procedures. These procedures rely mainly on (micro)extraction techniques and depend on the nature of the target compounds and the complexity of the samples. Because synthetic musks are lipophilic compounds (log K_{ow} values within 2.5–4.9 for nitro musks, 3.4–5.0 for polycyclic musks and 2.7–6.0 for macrocyclic musks), extraction with non-polar media is the main choice. These procedures are presented in Tables 16.4–16.6 and are discussed below.

Water Samples

Traditional extraction techniques like LLE and SPE have been widely used for the extraction of synthetic musks from water samples (Table 16.4). In LLE, conventional organic solvents such as methanol, hexane, cyclohexane, toluene or dichloromethane are

Environmental Monitoring of Cosmetic Ingredients 473

Table 16.4 Published papers on musk fragrance determination in environmental water samples (in chronological order)

Authors	Target compounds[a]	Sample[b]	Extraction technique[c]	Analytical technique[d]	MLOD (ng/L)[e]	Recovery (%)
Winkler et al. (2000)	ADBI, AHTN, HHCB, MK	RV	**(DI)SPME** (3.5-mL sample; extracted with PDMS-DVB-coated fibre; 30°C, 45 min; thermal desorption (260°C, 5 min))	(TD)GC–MS (EI⁺)	14–22	
Simonich et al. (2000)	AHTN, HHCB, MK, MX and other compounds	EW, IW	**SPE** (0.5- to 1-L sample; extracted with C₁₈ disc at 5–25 mL/min; eluted with 20 mL dichloromethane; evaporated 0.5 mL)	GC–MS(EI⁺)	0.21–0.91	77–144 (EW) 79–116 (IW)
Fromme et al. (2001)	ADBI, AHMI, AHTN, ATII, HHCB	RV	**LLE** (600-mL sample (1.7% NaCl); extracted with 5 mL cyclohexane; dried with Na₂SO₄; evaporated to 0.5 mL)	GC–MS(EI⁺)	5–20	82–101
Osemwengie and Steinberg (2001)	ADBI, AETT, AHMI, AHTN, ATII, DPMI, HHCB, MA, MK, MM, MT, MX	EW	**(On-site)SPE** (60-L sample; extracted with PMA-DVB cartridge at 267 mL/min; eluted with 20 mL hexane + 20 mL ethyl acetate; then solvent exchanged to methylene chloride, evaporated to 1 mL)	GC–MS(EI⁺)	0.02–0.30	80–97 (EW)
García-Jares et al. (2002)	ADBI, AHMI, AHTN, ATII, DPMI, HHCB	EW, IW	**(HS)SPME** (100-mL sample; extracted with CAR–PDMS or PDMS-DVB-coated fibre; 100°C, 25 min; thermal desorption (250°C, 2 min))	(TD)GC–MS (EI)	0.1–9	

Continued

Table 16.4 Published papers on musk fragrance determination in environmental water samples (in chronological order)—cont'd

Authors	Target compounds[a]	Sample[b]	Extraction technique[c]	Analytical technique[d]	MLOD (ng/L)[e]	Recovery (%)
Einsle et al. (2006)	AHTN, HHCB and other compounds	EW, RV	MALLE (50-mL sample (pH 7, 25% NaCl); extracted with LDPE bag with 500 μL methanol; 50°C, 60 min)	GC–MS(EI+)	20	54–59 (EW)
Polo et al. (2007)	MK, MM, MT, MX	EW, IW, TW	(HS)SPME (10-mL sample; extracted with CAR–PDMS-coated fibre; 100°C, 25 min; thermal desorption (300°C, 2 min))	(TD)GC–ECD	0.25–3.60	96–108 (EW) 92–102 (IW) 84–106 (TW)
Ligon et al. (2008)	AHTN, HHCB, MK, MX and other compounds	EW, IW	LLE (1-L sample (pH <2); extracted with 20 mL toluene; diluted with 175 mL water; phase separation; evaporated to 1 mL)	GC–MS(EI+)	0.1–0.3	75–98 (EW) 66–94 (IW)
Regueiro et al. (2008)	ADBI, AHMI, AHTN, ATII, DPMI, HHCB, MK, MM, MX and other compounds	EW, IW, SP, SW, TW	USAEME (10-mL sample; extracted with 100 μL chloroform (25°C, 10 min))	GC–MS(EI+)	6–29	88–114 (EW) 85–113 (IW) 86–103 (SP) 80–91 (SW)
Zhou et al. (2009)	AHTN, HHCB	EW, IW	SPE (500-mL sample; extracted with PVP-DVB cartridge at <10 mL/min; eluted with ethyl acetate; evaporated to dryness and redissolved in 100 μL hexane)	GC–MS(EI+)	0.4	81–83

Moldovan et al. (2009)	AHTN, HHCB and other compounds	EW, IW, RV	**SPE** (500-mL sample; extracted with PVP-DVB cartridge; eluted with acetonitrile/dichloromethane; evaporated to dryness and redissolved in isooctane)	GC–MS(EI$^+$)		55–110
Wang and Ding (2009)	ADBI, AHMI, AHTN, ATII, DPMI, HHCB	EW	**(HS)SPME** (20-mL sample (pH 2–3); extracted with PDMS-DVB-coated fibre; 90°C, 25 min; thermal desorption (270°C, 2 min))	GC–MS(EI$^+$)	0.05–0.1	64–89
Panagiotou et al. (2009)	ADBI, AHMI, AHTN, ATII, HHCB and other compounds	LK, RV, SW	**DLLME** (5-mL sample; extracted with 250 μL carbon tetrachloride (+0.62 mL methanol as disperser solvent); evaporated to 20 μL)	GC–MS(EI$^+$)	7–69 (LK) 7–64 (RV) 7–60 (SW)	60–84 (LK) 61–89 (RV) 65–92 (SW)
Gómez et al. (2009)	ADBI, AHMI, AHTN, ATII, HHCB, MK, MX and other compounds	EW, RV	**LLE** (500-mL sample (pH 3, 1% NaCl); extracted with 100 mL (3 min) + 50 mL (2 min) hexane; dried with Na$_2$SO$_4$; evaporated to 0.4–3 mL)	(LVI)GC–MS(EI$^+$)	1–21 (EW) 0.4–11 (RV)	108–125 (EW) 104–139 (RV)
Lv et al. (2009)	AHTN, HHCB, MK, MX	EW, GW, IW, LK, TW	**SPE** (1-L sample (pH 7); extracted with C$_{18}$ cartridge at 5–10 mL/min; eluted with 5 mL hexane + 2 mL hexane/dichloromethane; evaporated to 1 mL)	GC–MS(EI$^+$)	0.09–0.18	86–107
Silva and Nogueira (2010)	ADBI, AHTN, HHCB, MK	EW, IW, SW, TW	**SBSE** (30-mL sample (pH 7); extracted with PDMS-coated stir bar; 25°C, 4 h; eluted with 200 μL hexane (25°C, 30 min))	(LVI)GC–MS(EI$^+$)	12–19	84–108

Continued

Table 16.4 Published papers on musk fragrance determination in environmental water samples (in chronological order)—cont'd

Authors	Target compounds[a]	Sample[b]	Extraction technique[c]	Analytical technique[d]	MLOD (ng/L)[e]	Recovery (%)
Moeder et al. (2010)	AHTN, HHCB and other compounds	EW, LK	**MEPS** (800-μL sample; extracted with C_8 sorbent; eluted with 2×25 μL ethyl acetate)	GC–MS(EI$^+$)	45–54	57–78
Liu et al. (2010)	ADBI, AHMI, AHTN, ATTI, HHCB and other compounds	RV	**(DI)SPME** (3-mL sample; extracted with PDMS-coated fibre; 24°C, 90 min; thermal desorption (280°C, 7 min))	(TD)GC–MS (EI$^+$)	0.4–9.6	64–117
Matamoros et al. (2010)	ADBI, AHTM, DPMI, HHCB and other compounds	RV	**SPE** (200-mL sample; extracted with PS-DVB/MP cartridge at 10 mL/min; eluted with 5×2 mL ethyl acetate; evaporated to 100 μL)	GC×GC–MS(EI$^+$)	2–51	41–96
Clara et al. (2011)	ADBI, AHMI, AHTN, ATII, DPMI, HHCB	EW, IW	**LLE + SPE** (100-mL sample; extracted with 3×5 mL hexane; evaporated to 5 mL; then SPE with aluminium oxide column; eluted with 35 mL hexane/ethyl acetate; evaporated to 700 μL)	GC–MS(EI$^+$)	0.10–0.51	79–115
Ramírez et al. (2011)	ADBI, AHMI, AHTN, ATII, DPMI, HHCB, MK, MM, MX	EW, IW, RV	**SBSE** (100-mL sample (pH 7); extracted with PDMS-coated stir bar; 25°C, 4 h; thermal desorption (300°C, 15 min))	(TD)GC–MS(EI$^+$)	0.02–0.30	82–95

Arbulu et al. (2011)	ADBI, AHMI, AHTN, ATII, DPMI, HHCB, MA, MK, MM, MX Ambrettolide, ethylene brassilate, globalide, helvetolide, muscone, thibetolide, tibetene	EW, GW, IW, RV	**SBSE** (30-mL sample (10% NaCl; extracted with PDMS-coated stir bar; 30°C, 240 min; thermal desorption (290°C, 5 min))	(TD)GC–MS(EI$^+$)	2–24	
López-Nogueroles et al. (2011)	MA, MK, MM, MT, MX	EW, IW, RV, SW	**DLLME** (5-mL sample; extracted with 50 µL chloroform (+1 mL acetone as disperser solvent))	GC–MS(EI$^+$)	4–33	93–116 (EW) 98–109 (IW) 92–105 (RV) 87–93 (SW)
Hu et al. (2011)	ADBI, AHMI, AHTN, ATII, HHCB, MK, MX	RV	**SPE** (500-L sample; extracted with C$_{18}$ cartridge; eluted with 25 mL hexane and 15 mL hexane/dichloromethane; evaporated to 1 mL)	GC–MS(EI$^+$)	1–1.2	79–106
Gómez et al. (2011)	AHMI, AHTN, ATII, HHCB, MK, MX	EW, RV	**SBSE** (25- to 100-mL sample (20 g NaCl, 10% methanol); extracted with PDMS-coated stir bar; 14 h; thermal desorption (295°C, 7 min))	(TD) GC × GC–MS(EI$^+$)	0.02–2.54 (EW) 0.04–1.86 (RV)	101–161 (EW) 123–153 (RV)

Continued

Table 16.4 Published papers on musk fragrance determination in environmental water samples (in chronological order)—cont'd

Authors	Target compounds[a]	Sample[b]	Extraction technique[c]	Analytical technique[d]	MLOD (ng/L)[e]	Recovery (%)
Yang and Ding (2012)	ADBI, AHMI, AHTN, ATII, DPMI, HHCB	EW, RV	USA-DLLME (10-mL sample (5% NaCl); extracted with 10 μL carbon tetrachloride (+1 mL isopropyl alcohol as disperser solvent); 1 min	GC–MS(EI+)	0.2	75–90 (EW) 70–95 (RV)
Chase et al. (2012)	ADBI, AHMI, AHTN, ATII, DPMI, HHCB, MK, MX	EW, IW, LK, RV	SPE (4-L sample; extracted with C$_{18}$ disc; eluted with acetone/hexane; evaporated to 2 mL) SBSE (30-mL sample; extracted with PDMS-coated stir bar; 96 h; eluted with 2 mL acetone/hexane (1 h))	GC–MS(EI+)	4 (EW, IW) 1 (LK, RV) 66.7	
Villa et al. (2012)	ADBI, AHTN, HHCB	RV	SPE (0.5- to 1-L sample; extracted with PVP-DVB cartridge at 10 mL/min; eluted with 10 mL hexane and 5 mL ethyl acetate)	GC–MS(EI+)	0.05–0.25	50–95
Basaglia and Pietrogrande (2012)	HHCB, MK and other compounds	TW	(HS)SPME (40-mL sample; extracted with PA-coated fibre; 40°C, 125 min; then exposed to BSTFA (35°C, 30.5 min); thermal desorption (300°C, 2 min))	(TD)GC–MS(EI+)	0.8–9	85–103

Ramírez et al. (2012)	ADBI, AHMI, AHTN, ATII, DPMI, HHCB, MK, MM, MX and other compounds	EW, IW, RV	SBSE (100-mL sample (100 μL acetic anhydride, 0.5% disodium hydrogen phosphate); extracted with PDMS-coated stir bar; 20°C, 4 h; thermal desorption (300°C, 15 min))	(TD)GC–MS(EI$^+$)	0.02–0.3	85–99
Vallecillos et al. (2012a)	ADBI, AHMI, AHTN, ATII, DPMI, HHCB, MA, MK, MM, MX	EW, IW	IL–(HS)SDME (10-mL sample (1:2) (3% NaCl); extracted with 1 μL [OMIM][PF$_6$]; 60°C, 45 min)	GC–MS/MS(EI$^+$)	10–30	
Vallecillos et al. (2012b)	Ambrettolide, civetone, exaltolide, exaltone, habanolide, muscone, Musk MC4, Musk-NN	EW, IW	MEPS (4-mL sample, extracted with C$_{18}$ sorbent; eluted with 50 μL ethyl acetate)	(LVI)GC–MS(EI$^+$)	5–10	54–95 (EW) 52–92 (IW)
Posada-Ureta et al. (2012)	ADBI, AHMI, AHTN, ATII, DPMI, HHCB, MA, MK, MM, MT, MX	EW, ES, IW	MALLE (150-mL sample; extracted with LDPE bag with 200 μL hexane; 25°C, 240 min)	(LVI)GC–MS(EI$^+$)	4–25	50–126 (EW) 64–138 (ES) 47–124 (IW)
Klaschka et al. (2013)	HHCB and other compounds	EW, IW	SBSE (100-mL sample; extracted with PDMS-coated stir bar; 3 h; thermal desorption)	(TD)GC–MS(EI$^+$)		

Continued

Table 16.4 Published papers on musk fragrance determination in environmental water samples (in chronological order)—cont'd

Authors	Target compounds[a]	Sample[b]	Extraction technique[c]	Analytical technique[d]	MLOD (ng/L)[e]	Recovery (%)
He et al. (2013)	AHTN, HHCB	EW, IW	**SPE** (500-mL sample; extracted with PVP-DVB cartridge at 10 mL/min; eluted with 5 mL hexane and 5 mL hexane/methylene chloride; evaporated to 1 mL)	GC–MS(EI$^+$)	0.4–0.5	99–111
López-Nogueroles et al. (2013)	MA, MK, MM, MT, MX	EW, RV, SW	**SPE** (200-mL sample (5% methanol); extracted with MIS cartridge; eluted with 1 mL acetonitrile; evaporated to dryness and redissolved in 200 μL acetonitrile)	GC–MS(EI$^+$)	1.5–2.7	52–87 (EW) 63–92 (RV) 57–87 (SW)
Cavalheiro et al. (2013)	ADBI, AHMI, AHTN, ATII, DPMI, HHCB, MA, MK, MM	EW, ES, IW	**MEPS** (5.5-mL sample; extracted with C$_{18}$ sorbent; eluted with 2 × 25 μL ethyl acetate/hexane)	(LVI)GC–MS(EI$^+$)	7–39 (EW) 5–25 (IW)	75–133 (EW) 76–135 (ES) 76–135 (IW)
Wang et al. (2013a)	AHMI, AHTN, ATII, DPMI, HHCB	EW, LK, TW	**SPE** (1-L sample; extracted with PVP-DVB cartridge at 5 mL/min; eluted with 10 mL ethyl acetate; evaporated to 1 mL (35°C, 30 min))	GC–MS/MS(EI$^+$)	1.01–2.01 (EW) 1.01–2.04 (LK) 1.04–1.56 (TW)	96–113 (EW) 93–116 (LK) 87–112 (TW)
Chung et al. (2013)	ADBI, AHMI, AHTN, ATTI, HHCB	RV	**DSPE** (10-mL sample; extracted with C$_{18}$ sorbent; 1 min; thermal desorption (337°C, 3.8 min))	(TD)GC–MS(EI$^+$)	0.5–1	80–93

Reference	Analytes	Matrix	Sample Treatment	Technique	LOD/LOQ (ng/L)	Recovery (%)
Caballero-Díaz et al. (2013)	MK	RV	**MEPS** (500-μL sample, extracted with C$_{18}$ sorbent; eluted with 10 μL methanol)	SERS	2×10^{-5}	47–63
Vallecillos et al. (2013a)	Ambrettolide, civetone, exaltolide, exaltone, habanolide, muscone, Musk-MC4, Musk-NN	EW, IW	**(HS)SPME** (10-mL sample; extracted with PDMS-DVB-coated fibre; 100°C, 45 min; thermal desorption (250°C, 3 min))	(TD)GC–MS(EI$^+$)	0.75–5 (EW) 1–5 (IW)	
Kwon and Rodriguez (2014)	HHCB and other compounds	EW, IW	**LLE** (100- to 500-mL sample; extracted with 2 × 30–50 mL hexane (1 min); dried with Na$_2$SO$_4$; evaporated to dryness and redissolved in 0.5 mL hexane)	GC–MS(EI$^+$)		72–105
Sun et al. (2014)	AHTN, HHCB	EW	**LLE** (10-mL sample; extracted with 10 mL methanol)	LC–MS/MS(EI$^+$)	0.02–0.08	91
Chung et al. (2014)	ADBI, AHMI, AHTN, ATII, HHCB, MK, MX	EW, RV	**DSPE** (50-mL sample; extracted with 10.1 mg C$_{18}$ sorbent; 10.4 min; eluted with 200 μL hexane (0.5 min))	(LVI)GC–MS(EI$^+$)	0.5–2	73–90 (EW) 70–97 (RV)
Wang et al. (2014)	ADBI, AHMI, AHTN, ATII, HHCB	ES, SW	**(DI)SDME** (6-μL sample (pH 7); extracted with 1 μL hexane; 25 min)	GC–MS(EI$^+$)	3.4–11.0	85–120

Continued

Table 16.4 Published papers on musk fragrance determination in environmental water samples (in chronological order)—cont'd

Authors	Target compounds[a]	Sample[b]	Extraction technique[c]	Analytical technique[d]	MLOD (ng/L)[e]	Recovery (%)
Pintado-Herrera et al. (2014)	ADBI, AHMI, AHTN, ATII, HHCB, MA, MK, MM, MT, MX Exaltenone, habanolide, muscenone, muscone, Musk R1 and other compounds	EW, GW, RV, SW	**SBSE** (100-mL sample (1% Na$_2$CO$_3$, 10% NaCl, 500 μL acetic anhydride); extracted with PDMS-coated stir bar; 25°C; 5 h; eluted with 200 μL ethyl acetate (30 min)]	GC–MS(TOF)	0.02–1.4	
Vallecillos et al. (2014a)	ADBI, AHMI, AHTN, ATII, DPMI, HHCB, MK, MM, MX Ambrettolide, civetone, exaltolide, exaltone, muscone, Musk-MC4, Musk-NN	EW, IW	**(Online)SPE** (10-mL sample (50% methanol); extracted with PVP-DVB at 2 mL/min; eluted with 100 μL ethyl acetate)	GC–MS(EI$^+$)	1–25 (EW) 1–30 (IW)	
Lange et al. (2015)	AHTN, HHCB	RV	**LLE** (2-L sample; extracted with 2 × 80 mL dichloromethane; evaporated to 2 mL and dried with Na$_2$SO$_4$; evaporated to 100 μL)	GC–MS(EI$^+$)		

Environmental Monitoring of Cosmetic Ingredients   483

Godayol et al. (2015)	AHTN, HHCB and other compounds	EW	**(HS)SPME** (10-mL sample (24% NaCl); extracted with PDMS–DVB-coated fibre; 90°C, 45 min; thermal desorption (220°C, 2 min))	(TD)GC–MS(EI⁺)	40	30–84
Vallecillos et al. (2015a)	ADBI, AHMI, AHTN, ATII, DPMI, HHCB, MM, MX	EW, IW	**(HS)MEPS** (10-mL sample; 500 μL headspace extracted with C₁₈ sorbent set at the needle; 60°C, 15 min; thermal desorption (230°C, 3 min))	(TD)GC–MS/MS(EI⁺)	2.5–10 (EW) 2.5–12 (IW)	
Lu et al. (2015)	ADBI, AHMI, AHTN, HHCB, MK, MX	RV	**SPE** (1-L sample; extracted with C₁₈ cartridge; eluted with 10 mL hexane and 5 mL hexane/dichloromethane; evaporated to 1 mL)	GC–MS(EI⁺)	0.3–0.6	
Cavalheiro et al. (2015)	ADBI, AHMI, AHTN, ATII, HHCB, MA, MK, MM, MX	ES, EW, IW	**(DI)SPME** (100-mL sample (10% NaCl); extracted with polyethersulphone fibre; 15 h; eluted with 300 μL ethyl acetate (8 min))	(LVI)GC–MS(EI⁺)	0.05–60	64–100 (ES) 76–100 (EW) 42–119 (IW)
Zhang et al. (2015a)	ADBI, AHMI, AHTN, ATII, DPMI, HHCB, MA, MK, MM, MT, MX and other compounds	SW	**SPE** (2.2- to 2.6-L sample; extracted with C₁₈ discs at 100 mL/min; eluted with 35 mL dichloromethane; dried with Na₂SO₄; solvent exchanged to isooctane and evaporated to 50 μL)	GC–MS/MS(EI⁺)	3.2–51.1	69–165
Maidatsi et al. (2015)	ADBI, AHMI, AHTN, DPMI, HHCB	EW, LK	**DSPE** (100-mL sample (2% Na₂SO₄); extracted with Fe₃O₄@graphene-C₈ MNPs; 20°C, 15 min; eluted with 2 × 100 μL ethyl acetate/methyl-*tert*-butyl ether (2 min); evaporated to 100 μL)	GC–MS(EI⁺)	0.64–1.40	83–105

Continued

Table 16.4 Published papers on musk fragrance determination in environmental water samples (in chronological order)—cont'd

Authors	Target compounds[a]	Sample[b]	Extraction technique[c]	Analytical technique[d]	MLOD (ng/L)[e]	Recovery (%)
Homem et al. (2016)	ADBI, AHMI, AHTN, DPMI, HHCB, MA, MK, MM, MT, MX Exaltolide, ethylene brassylate	EW, IW, RV, SW, TW	USA-DLLME (6-mL sample (3.5% NaCl); extracted with 80 μL chloroform (+880 μL acetonitrile as disperser solvent); 2 min)	GC–MS(EI+)	0.004–54	71–115 (EW) 71–118 (IW) 79–107 (RV) 71–108 (SW) 75–106 (TW)
Li et al. (2016b)	AHTN, HHCB, MK, MX Muscone	RV,TW	(DI)SPME (10-mL sample; extracted with amino-modified graphene fibre; 60°C, 60 min; thermal desorption (250°C, 5 min))	(TD)GC–MS(EI+)	0.46–5.8	82–110 (RV) 96–112 (TW)
Guo et al. (2016)	MA, MK, MM, MT, MX	RV	(HS)BID-SDME (1-mL sample (10% NaCl); extracted with 1 μL 1-octanol; 40°C, 20 min)	GC–MS(EI+)	12–42	92–105

[a]ADBI, celestolide; AHMI, phantolide; AHTN, tonalide; ATII, traseolide; DPMI, cashmeran; HHCB, galaxolide; MA, musk ambrette; MK, musk ketone; MM, musk moskene; MT, musk tibetene; MX, musk xylene.
[b]ES, estuarine; EW, effluent wastewater; GW, groundwater; IW, influent wastewater; LK, lake water; RV, river water; SP, swimming-pool water; SW, seawater; TW, tap water.
[c]BID, bubble-in-drop; BSTFA, N,O-bis(trimethylsilyl)trifluoroacetamide; CAR, carboxen; DI, direct immersion; DLLME, dispersive liquid–liquid microextraction; DSPE, dispersive solid-phase extraction; HS, headspace; IL, ionic liquid; LDPE, low-density polyethylene; LLE, liquid–liquid extraction; MALLE, membrane-assisted liquid–liquid extraction; MEPS, microextraction by packed sorbent; MIS, molecularly imprinted silica; MNPs, magnetic nanoparticles; PA, polyacrylate; PDMS, polydimethylsiloxane; PDMS-DVB, polydimethylsiloxane–divinylbenzene; PMA-DVB, polymethylmethacrylate–divinylbenzene; PS-DVB/MP, polystyrene divinylbenzene co-polymer modified with pyrrolidone; PVP-DVB, polyvinylpyrrolidone–divinylbenzene co-polymer; SBSE, stir bar sorptive extraction; SDME, single-drop microextraction; SME, sorptive microextraction; SPE, solid-phase extraction; SPME, solid-phase microextraction; USA, ultrasound-assisted; USAEME, ultrasound-assisted emulsification microextraction.
[d]ECD, electronic capture detector; EI, electron ionization; GC, gas chromatography; GC×GC, multidimensional gas chromatography; LC, liquid chromatography; LVI, large-volume injection; MS, mass spectrometry; MS/MS, tandem mass spectrometry; SERS, surface-enhanced Raman spectroscopy; TD, thermal desorption; TOF, time-of-flight.
[e]MLOD, method limit of detection.

Table 16.5 Published papers on musk fragrance determination in environmental solid samples (in chronological order)

Authors	Target compounds[a]	Sample[b]	Extraction technique[c]	Analytical technique[d]	MLOD (ng/g)[e]	Recovery (%)
Simonich et al. (2000)	AHTN, HHCB, MK, MX and other compounds	SS	PLE + SPE (25-g sample (+3 g silica); extracted with 2 × 15 min dichloromethane; 60°C, 2000 psi; evaporated to 0.5 mL and solvent exchanged to hexane; then SPE in silica gel column and eluted with 15 mL dichloromethane; concentrated as above)	GC–MS(EI⁺)		68–91
Fromme et al. (2001)	ADBI, AHMI, AHTN, ATII, HHCB	SE	SLE (10-g dried sample (+10 g NaCl + 600 mL water); extracted with 5 mL cyclohexane; dried with Na₂SO₄; evaporated to 0.5 mL)	GC–MS(EI⁺)	4–30	78–96
Llompart et al. (2003)	ADBI, AHMI, AHTN, ATII, HHCB, MK, MM, MT, MK	SS	(HS)SPME (0.16-g dried sample + 3 mL water; extracted with PDMS-DVB-coated fibre; 100°C, 15 min; thermal desorption (250°C, 3 min))	(TD)GC–MS(EI⁺)	0.028–0.611	
Burkhardt et al. (2005)	AHTN, HHCB and other compounds	SE	PLE + SPE (15- to 30-g wet sample; extracted with 3 × 10 min water/isopropanol; 200°C; diluted with phosphate buffer (pH 7); then SPE in PS-DVB cartridge at 20 mL/min; eluted with 3 × 10 mL dichloromethane/diethyl ether; evaporated to 1 mL)	GC–MS(EI⁺)	12.5–16.5	77–79

Continued

Table 16.5 Published papers on musk fragrance determination in environmental solid samples (in chronological order)—cont'd

Authors	Target compounds[a]	Sample[b]	Extraction technique[c]	Analytical technique[d]	MLOD (ng/g)[e]	Recovery (%)
Ternes et al. (2005) Carballa et al. (2008)	AHTN, HHCB and other compounds	SS	**SLE + SPE** (0.2-g dried sample; extracted with 6 mL methanol + 3 × 2 mL acetone (5 min each); evaporated to 200 μL and redissolved in 150 mL water; then SPE in C$_{18}$ cartridge and eluted with 1 mL methanol at 20 mL/min; evaporated to dryness and redissolved in 200 μL hexane)	GC–MS(EI$^+$)	76	64–109
Rice and Mitra (2007)	MK and other compounds	SE, SO	**MAE + GPC** (3-g sample; extracted with 3 × 10 mL (115°C, 800W, 15 min); evaporated to 1 mL hexane; then GPC with methylene chloride/acetone; evaporated to dryness and redissolved in 1 mL hexane)	GC–MS(EI$^+$)	0.26	90
Zeng et al. (2008)	ADBI, AHMI, AHTN, ATII, DPMI, HHCB	SE	**Soxhlet + SPE** (20-g dried sample; extracted with dichloromethane (reflux, 72h); evaporated to 1 mL, solvent exchanged to hexane and evaporated to 0.5 mL; then SPE in silica/alumina column and eluted with 50 mL dichloromethane; evaporated to 0.5 mL)	GC–MS(EI$^+$)	0.3–0.67	56–109

Ligon et al. (2008)	AHTN, HHCB, MK, MX and other compounds	SS	**PLE** (0.5-g dried sample; extracted with 2 × 45 min ethyl acetate; 130°C; diluted with 2 mL toluene; evaporated to 1 mL)	GC–MS(EI$^+$)	0.2–1.1	91–127
Zhou et al. (2009)	AHTN, HHCB	SS	**SLE + SPE** (0.02-g dried sample; extracted with 5 mL methanol/water and 3 × 5 mL hexane/acetone (10 min each); evaporated to 200 µL and diluted with 500 mL water; then SPE in PVP-DVB cartridge at <10 mL/min; eluted with ethyl acetate; evaporated to dryness and redissolved in 100 µL hexane)	GC–MS(EI$^+$)	180–670	61–75
Chen and Bester (2009)	AHTN, HHCB, MK, MX and other compounds	SS	**PLE + SPE + GPC** (40-g sample (+10 g diatomaceous earth); extracted with ethyl acetate; 80°C, 70 mbar; SPE in silica cartridge and eluted with 12 mL ethyl acetate; then GPC with cyclohexane/ethyl acetate: the fraction of interest evaporated to 1 mL)	GC–MS(EI$^+$)	3–10	36–77

Continued

Table 16.5 Published papers on musk fragrance determination in environmental solid samples (in chronological order)—cont'd

Authors	Target compounds[a]	Sample[b]	Extraction technique[c]	Analytical technique[d]	MLOD (ng/g)[e]	Recovery (%)
Sumner et al. (2010)	ADBI, AHMI, AHTN, HHCB, MK, MX	SE	**Soxhlet + SPE** (3-g sample; extracted with 200 mL hexane/dichloromethane (reflux, 12 h); dried with Na$_2$SO$_4$; evaporated to 1 mL; then SPE in alumina cartridge and eluted with 5 mL ethyl acetate; evaporated to 300 μL)	GC–MS(EI$^+$)	1.1–8	
Guo et al. (2010)	ADBI, AHMI, AHTN, ATII, DPMI, HHCB, MA, MK, MM, MT, MX	SS	**PLE + GPC** (1-g dried sample; extracted with 2 × 10 min acetone/hexane; 100°C; evaporated to 2 mL and diluted with 10 mL hexane; then GPC with 100 mL acetone/hexane; evaporated to dryness and redissolved in dichloromethane)	GC–MS(EI$^+$)	3–10	71–85
Wu and Ding (2010)	ADBI, AHMI, AHTN, ATII, DPMI, HHCB	SE, SS	**MA-HS(SPME)** (5-g sample + 20 mL water (+3 g NaCl, pH 1); extracted with PDMS–DVB-coated fibre; 80 W, 5 min; thermal desorption (270°C, 2 min))	(TD)GC–MS(EI$^+$)	0.04–0.1	67–88 (SE) 68–89 (SS)
Hu et al. (2011)	ADBI, AHMI, AHTN, ATII, HHCB, MK, MX	SE	**PLE + SPE** (0.1-g sample (+28 g Na$_2$SO$_4$ + 4 g activated silica); extracted with 2 × 15 min hexane/dichloromethane; 60°C, 1500 psi; then SPE in silica column and eluted with hexane/dichloromethane; evaporated to 1 mL)	GC–MS(EI$^+$)	0.25–0.33	84–105

Chase et al. (2012)	ADBI, AHMI, AHTN, ATII, DPMI, HHCB, MK, MX	SE, SO	**SLE** (1-g dried sample; extracted with 120 mL acetone/hexane (2h); half of the extract evaporated to 2 mL)	GC–MS(EI⁺)	0.3	
Villa et al. (2012)	ADBI, AHTN, HHCB	SO	**Soxhlet + SPE** (<0.07-g sample; extracted with hexane (reflux, 24 h); evaporated to 2 mL; then SPE in graphitized carbon cartridge; eluted with hexane, hexane/dichloromethane and ethyl acetate; evaporated to 0.5 mL)	GC–MS(EI⁺)	0.07–0.35	80–108
Matamoros et al. (2012)	ADBI, AHTN, DPMI, HHCB	SS	**SLE** (0.5-g dried sample; extracted with 3 × 5 mL hexane (15 min each); evaporated to 300 µL)	GC–MS(EI⁺)	2–10	
Vallecillos et al. (2012c)	ADBI, AHMI, AHTN, ATII, DPMI, HHCB, MK, MM, MX	SS	**PLE + IL–(HS)SDME** (1-g dried sample (+1 g diatomaceous earth); extracted with 2 × 5 min water/methanol; 80°C; 1500 psi; evaporated with methanol and diluted up to 10 mL with water; then subjected to SDME, extracted with 1 µL [OMIM][PF₆]; 60°C, 45 min)	GC–MS/MS(EI⁺)	1–3	72–98

Continued

Table 16.5 Published papers on musk fragrance determination in environmental solid samples (in chronological order)—cont'd

Authors	Target compounds[a]	Sample[b]	Extraction technique[c]	Analytical technique[d]	MLOD (ng/g)[e]	Recovery (%)
Chamorro et al. (2013)	AHTN, HHCB and other compounds	SE	**PLE + SPE** (5-g dried sample; extracted with 3 × 15 min hexane/acetone; 110°C, 140 bar; then SPE in Florisil column and eluted with hexane/ethyl acetate; evaporated to dryness and redissolved in ethyl acetate)	GC × GC–MS(TOF)		
Vallecillos et al. (2013b)	Ambrettolide, civetone, exaltolide, exaltone, habanolide, muscone, Musk MC4, Musk-NN	SS	**(HS)SPME** (0.25-g dried sample + 0.5 mL water; extracted with PDMS-DVB-coated fibre; 80°C, 45 min; thermal desorption (250°C, 3 min))	(TD)GC–MS(EI⁺)	0.005–0.025	
Wang et al. (2013b)	AHTN, HHCB	SO	**PLE + SPE** (10-g sample; extracted with 3 × 5 min hexane/dichloromethane; 100°C; evaporated to 5 mL; then SPE in PVP-DVB cartridge and eluted with ethyl acetate)	GC–MS(EI⁺)	0.06–0.09	70–132
Aguirre et al. (2014)	ADBI, AHMI, AHTN, ATII, DPMI, HHCB, MA, MK	AS	**SBSE** (0.5-g sample + 9 mL water/methanol; extracted with PDMS-coated stir bar; 40°C, 3 h; thermal desorption (300°C, 10 min))	(TD)GC–MS(EI⁺)	0.01–1.1	74–126

Vallecillos et al. (2014b)	ADBI, AHMI, AHTN, ATII, DPMI, HHCB, MK, MM, MX Ambrettolide, exaltolide, exaltone	SS	(HS)SBSE (100-mg sample + 0.2 mL water, extracted with PDMS-coated stir bar; 80°C, 45 min; thermal desorption (300°C, 15 min))	(TD)GC–MS(EI$^+$)	5–30
Sun et al. (2014)	AHTN, HHCB	SS	SLE (1-g sample; extracted with 3 × 20 mL methanol (30 min); diluted with methanol)	LC–MS/MS(EI$^+$)	0.03–0.15	111–112
Guo et al. (2014)	ADBI, AHMI, AHTN, ATII, HHCB, MK, MX and other compounds	SE	PLE + GPC (5-g dried sample; extracted with 3 × 5 min hexane/acetone; 100°C; solvent exchanged to hexane; then GPC with dichloromethane; solvent exchanged to hexane, 2 mL final volume)	GC–MS/MS(EI$^+$)	0.0127–0.0560	102–110
Zhang et al. (2015a)	ADBI, AHMI, AHTN, ATII, DPMI, HHCB, MA, MK, MM, MT, MX and other compounds	SE	SLE + SPE (1-g dried sample; extracted with 3 × 15 mL hexane/dichloromethane (15 min each); evaporated to 1 mL; then SPE in Florisil column and eluted with 15 mL hexane/dichloromethane; solvent exchanged to isooctane and evaporated to 50 µL)	GC–MS/MS(EI$^+$)	0.0006–0.0439	72–111
Lu et al. (2015)	ADBI, AHMI, AHTN, HHCB, MK, MX	SE	Soxhlet + SPE (1-g sample; extracted with dichloromethane (reflux, 72 h); then SPE in silica/alumina column; evaporated to 1 mL)	GC–MS(EI$^+$)	0.06–0.15

Continued

Table 16.5 Published papers on musk fragrance determination in environmental solid samples (in chronological order)—cont'd

Authors	Target compounds[a]	Sample[b]	Extraction technique[c]	Analytical technique[d]	MLOD (ng/g)[e]	Recovery (%)
Zhang et al. (2015b)	ADBI, AHMI, AHTN, ATII, HHCB, MA, MK, MM, MT, MX and other compounds	SE	**SLE + SPE** (5- to 10-g sample + 50 g Na$_2$SO$_4$; extracted with 3 × 50 mL hexane/dichloromethane (15 min each); evaporated to 2 mL; then SPE in Florisil column and eluted with 160 mL hexane/dichloromethane; evaporated to dryness and redissolved in isooctane)	GC–MS/MS(EI$^+$)	0.0022–0.042	77–120
Pintado-Herrera et al. (2016)	ADBI, AHMI, AHTN, ATII, HHCB, MA, MK, MM, MT, MX Helvetolide, Musk-R1 and other compounds	SE	**PLE** (2-g sample + 1 g activated alumina + 1.5 g diatomaceous earth; extracted with 3 × 5 min dichloromethane; 100°C, 1500 psi; evaporated to dryness and redissolved in 500 μL ethyl acetate + 10 μL MTBSTFA (as derivatizing agent))	GC–MS/MS(EI$^+$)	0.013–2.400	51–144

[a]*ADBI*, celestolide; *AHMI*, phantolide; *AHTN*, tonalide; *ATII*, traseolide; *DPMI*, cashmeran; *HHCB*, galaxolide; *MA*, musk ambrette; *MK*, musk ketone; *MM*, musk moskene; *MT*, musk tibetene; *MX*, musk xylene.
[b]*AS*, amended soil; *SE*, sediment; *SO*, soil; *SS*, sewage sludge.
[c]*GPC*, gel-permeation chromatography; *HS*, headspace; *IL*, ionic liquid; *MAE*, microwave-assisted extraction; *MTBSTFA*, N-(*tert*-butyldimethylsilyl)-N-methyltrifluoroacetamide; *PDMS*, polydimethylsiloxane; *PDMS-DVB*, polydimethylsiloxane–divinylbenzene; *PLE*, pressurized liquid extraction; *PS-DVB*, polystyrene divinylbenzene co-polymer; *PVP-DVB*, polyvinylpyrrolidone–divinylbenzene co-polymer; *SBSE*, stir bar sorptive extraction; *SDME*, single-drop microextraction; *SLE*, solid-liquid extraction; *SPE*, solid-phase extraction; *SPME*, solid-phase microextraction.
[d]*EI*, electron ionization; *GC*, gas chromatography; *GC × GC*, multidimensional gas chromatography; *LC*, liquid chromatography; *MS*, mass spectrometry; *MS/MS*, tandem mass spectrometry; *TD*, thermal desorption; *TOF*, time-of-flight.
[e]*MLOD*, method limit of detection.

Table 16.6 Published papers on musk fragrance determination in biota samples (in chronological order)

Authors	Target compounds[a]	Sample[b]	Extraction technique[c]	Analytical technique[d]	MLOD (ng/g)[e]	Recovery (%)
Fromme et al. (2001)	ADBI, AHMI, AHTN, ATII, HHCB	FI	**Soxhlet + GPC + SPE** (10-g sample (+15–20 g Na$_2$SO$_4$); extracted with cyclohexane/ethyl acetate (reflux, 10 h); evaporated to 20 mL; GPC with cyclohexane/ethyl acetate; then SPE in silica column and eluted with toluene/acetone)	GC–MS(EI$^+$)	4–30	78–95
Osemwengie and Steinberg (2003)	ADBI, AHMI, AHTN, ATII, DPMI, HHCB, MA, MK, MM, MT, MX	FI	**PLE + GPC + SPE** (2.3-g sample (+5 g diatomaceous earth); extracted with 2 × 5 min ethyl acetate/hexane; 80°C, 2000 psi; evaporated to 300 µL, diluted with 2 mL methylene chloride and evaporated to 1 mL; GPC with methylene chloride; evaporated to 1 mL hexane; then SPE in amino cartridge and eluted with 8 mL methylene chloride; evaporated to 100 µL)	GC–MS(EI$^+$)	0.45–2.0	31–56
Nakata (2005) Nakata et al. (2007)	AHTN, HHCB, MA, MK, MX	MM	**Soxhlet + GPC + SPE** (1- to 4-g sample; extracted with dichloromethane/hexane (reflux, 7 h); GPC with hexane/dichloromethane; then SPE with silica gel cartridge and eluted with 60 mL dichloromethane/hexane; evaporated to 200 µL)	GC–MS(EI$^+$)	1.9–9.1	92–108

Continued

Table 16.6 Published papers on musk fragrance determination in biota samples (in chronological order)—cont'd

Authors	Target compounds[a]	Sample[b]	Extraction technique[c]	Analytical technique[d]	MLOD (ng/g)[e]	Recovery (%)
Duedahl-Olesen et al. (2005)	ADBI, AHMI, AHTN, ATII, HHCB, MA, MK, MM, MT, MX	FI	**SLE + GPC + SPE** (10-g sample; extracted with 2 × 100 mL acetone/pentane; evaporated to dryness and redissolved in ethyl acetate/cyclohexane; GPC with ethyl acetate/cyclohexane; then SPE in Florisil column and eluted with 20 mL ethyl acetate/pentane; evaporated and redissolved in 1 mL isooctane)	GC–MS(EI$^+$)	0.03–0.6	70–97
Kannan et al. (2005)	AHTN, HHCB	FI, MM	**Soxhlet + GPC + SPE** (1- to 5-g sample; extracted with dichloromethane/hexane (reflux, 12 h); evaporated to 11 mL; GPC with hexane/dichloromethane; then SPE with silica gel cartridge; evaporated to 200 μL)	GC–MS(EI$^+$)	0.3	85–98
Rüdel et al. (2006)	ADBI, AHMI, AHTN, ATII, DPMI, HHCB, MK, MX	FI, MU	**PLE + GPC + SPE** (1- to 5-g sample; extracted with 2 × 10 min hexane; 80°C; filled up to 50 mL; evaporated to 0.2 mL; GPC with dichloromethane/cyclohexane; then SPE with silica gel cartridge and eluted with 15 mL of the same solvent; evaporated to 100 μL)	GC–MS(EI$^+$)	0.03–0.15	83–135

Wan et al. (2007)	ADBI, AHTN, HHCB, MA, MK, MT, MX and other compounds	FI	**Soxhlet + LLE + SPE** (0.5- to 10-g sample (+20 g Na$_2$SO$_4$); extracted with 250 mL dichloromethane/methanol (reflux, 24 h); evaporated to dryness, redissolved in hexane and extracted with acetonitrile; transferred to 500 mL of 5% NaCl solution and extracted with 50 mL hexane; then SPE with alumina column and eluted with hexane and hexane/dichloromethane; evaporated to dryness and redissolved in 0.2 mL hexane)	GC–MS(EI$^+$)	0.2–1.3	74–111
Mottaleb et al. (2009)	ADBI, AHTN, HHCB, MK, MX and other compounds	FI	**SLE + SPE** (1-g blended sample; extracted with 10 mL acetone (25–35°C, 5–15 min); evaporated to dryness and redissolved in 200 μL hexane/acetone; SPE with silica gel column and hexane/acetone; evaporated to 50 μL and, after derivatization step with 100 μL MSTFA, evaporated to 20 μL and diluted with 180 μL hexane)	GC–MS(EI$^+$)	4–17	87–105
			SLE + SPE + GPC (1-g blended sample; extracted with 10 mL acetone (25–35°C, 5–15 min); evaporated to dryness and redissolved in 200 μL hexane/acetone; SPE with silica gel column and hexane/acetone; evaporated to dryness, redissolved in 700 μL dichloromethane; GPC; fraction of interest evaporated to 50 μL and, after derivatization step with 100 μL MSTFA, evaporated to 20 μL and diluted with 180 μL hexane)	GC–MS/MS(EI$^+$)	12–397	67–107

Continued

Table 16.6 Published papers on musk fragrance determination in biota samples (in chronological order)—cont'd

Authors	Target compounds[a]	Sample[b]	Extraction technique[c]	Analytical technique[d]	MLOD (ng/g)[e]	Recovery (%)
Subedi et al. (2011)	AHTN, HHCB, MK, MX and other compounds	FI	PLE + GPC (2.5-g sample (+Na$_2$SO$_4$); extracted with 2 × 5 min dichloromethane/ethyl acetate; 80°C, 1500 psi; evaporated to 0.5 mL and redissolved in 700 μL dichloromethane; GPC, the fraction of interest evaporated to 200 μL, solvent exchanged to hexane and, after derivatization step with 100 μL MSTFA (70°C, 1 h), evaporated to 200 μL hexane)	GC–MS/MS(EI$^+$)	1.5–38	57–76
Hu et al. (2011)	ADBI, AHMI, AHTN, ATII, HHCB, MK, MX	FI	PLE + GPC + SPE (1-g sample (+35 g Na$_2$SO$_4$); extracted with 2 × 15 min hexane; 100°C, 1500 psi; evaporated to dryness and redissolved in 10 mL; GPC with cyclohexane/ethyl acetate; then SPE in alumina column and eluted with hexane/dichloromethane; evaporated to 1 mL)	GC–MS(EI$^+$)	1–1.2	90–110
Wu et al. (2012)	AHTN, HHCB	OY	MA–HS(SPME) (5-g sample (+10 mL water + 3 g NaCl, pH 1); extracted with PDMS-DVB-coated fibre; 80W, 5 min; thermal desorption (270°C, 10 min))	(TD) GC–MS(EI$^+$)	0.04	80–89
Zhang et al. (2013)	ADBI, AHMI, AHTN, ATII, HHCB, MK, MX	FI	Soxhlet + GPC + SPE (4-g sample; extracted with hexane/acetone; GPC with hexane/dichloromethane; then SPE in silica/alumina column and eluted with hexane/dichloromethane)	GC–MS(EI$^+$)	0.02–0.04 0.4–1.0 (lipid weight)	72

Wu et al. (2013b)	AHTN, HHCB, MK, MX	FI	**MA–HS(SPME)** (2-g sample (+4 mL methanol + 15 mL water + 4 g NaCl, pH 2); extracted with PDMS-DVB-coated fibre; 80 W, 5 min; thermal desorption (270°C, 2 min))	(TD) GC–MS(EI⁺)	0.15–0.50	80–92
Klaschka et al. (2013)	HHCB and other compounds	FI	**PLE + GPC** (1-g sample (+10 g Na₂SO₄); extracted with cyclohexane/ ethyl acetate; then GPC with the same solvent; evaporated to 1 mL)	GC–MS/ MS(EI⁺)	10	
Foltz et al. (2014)	AHTN, HHCB, MK	FI	**SLE + SPE** (1-g sample; extracted with 10 mL acetone (25°C, 15 min); evaporated to dryness and redissolved in 200 µL hexane/acetone; then SPE with silica gel column and hexane/ acetone; evaporated to 20 µL and diluted with 120 µL hexane/acetone)	GC–MS(EI⁺)		
Picot-Groz et al. (2014)	ADBI, DPMI, HHCB, MK and other compounds	MU	**SLE** (2-g blended and dried sample (+10 mL water + 10 mL acetonitrile + 4 g Na₂SO₄ + 1 g NaCl + 1.5 g citrate); 1 min; acetonitrile layer cleaned up with C₁₈/ PSA sorbent; 1 min; 1 mL extract evaporated to dryness and redissolved in heptane)	GC–MS/ MS(EI⁺)	0.5–50	81–122
Lange et al. (2015)	AHTN, HHCB	FI	**Soxhlet + SPE** (2-g dried sample; extracted with 100 mL hexane (reflux, 6 h); evaporated to 10 mL; then SPE in silica column and eluted with hexane, hexane/dichloromethane and acetone; evaporated to 100 µL)	GC–MS(EI⁺)		

Continued

Table 16.6 Published papers on musk fragrance determination in biota samples (in chronological order)—cont'd

Authors	Target compounds[a]	Sample[b]	Extraction technique[c]	Analytical technique[d]	MLOD (ng/g)[e]	Recovery (%)
Vallecillos et al. (2015b)	ADBI, AHMI, AHTN, ATII, DPMI, HHCB, MK, MM, MX	FI, MU	**PLE** (0.5-g dried sample (+1 g Florisil + 1 g diatomaceous earth); extracted with 5 min dicholoromethane; 60°C, 1500 psi; evaporated to dryness and redissolved in 2 mL ethyl acetate) **MSPD + DSPE** (0.5-g dried sample + 10 mL water; extracted with a salt packet (4 g MgSO$_4$, 1 g NaCl, 0.5 g sodium citrate dibasic, 1 g sodium citrate dihydrate); 3 min; eluted with 10 mL acetonitrile; DSPE with 1 g Florisil, 3 min; evaporated to 1 mL and redissolved to 2 mL ethyl acetate)	(LVI) GC–MS/MS(EI$^+$)	0.25–5 (FI) 0.5–5 (MU) 0.25–10 (FI) 0.5–7.5 (MU)	61–109 (FI) 45–91 (MU) 41–110 (FI) 24–110 (MU)
Ziarrusta et al. (2015)	AHTN, HHCB and other compounds	MU	**MSPD + SPE** (0.3-g dried sample; extracted with 0.30 g Florisil sorbent; then charged in C$_{18}$ cartridge; eluted with 25 mL dichloromethane; evaporated to dryness and redissolved in 140 µL hexane)	GC–MS(EI$^+$) GC–MS/MS(EI$^+$)	14–29 4.1–6.3	45–116 45–123
Ros et al. (2015)	AHTN, HHCB and other compounds	FI (bile)	**SPE + SPE** (100-µL sample (+1.5 mL phosphate buffer); extracted with PCX cartridge; eluted with 4 mL ethyl acetate; evaporated to 1 mL; then SPE in Florisil cartridge and eluted with 3 mL hexane/ethyl acetate; evaporated to dryness and redissolved in 175 µL hexane)	GC–MS(EI$^+$)	0.028–0.089	75–90

Zhang et al. (2015b)	ADBI, AHMI, AHTN, ATII, HHCB, MA, MK, MM, MT, MX and other compounds	FI	**SLE + GPC + SPE** (5- to 10-g sample + 50 g Na$_2$SO$_4$; extracted with 3 × 50 mL hexane/dichloromethane (15 min each); evaporated to 2 mL; GPC with the same solvent; then the fraction of interest to SPE in Florisil column with 160 mL hexane/dichloromethane; evaporated to dryness and redissolved in isooctane)	GC–MS/MS(EI$^+$)	0.0003–0.022	79–128
Yao et al. (2016)	AHTN, HHCB, MK, MX and other compounds	FI	**MSPD** (0.5-g sample; extracted with C$_{18}$ sorbent; eluted with acetonitrile; evaporated to dryness and redissolved in 200 μL methanol; separated in two aliquots, evaporated to dryness and redissolved in 100 μL dichloromethane)	GC–MS(EI$^+$)	0.10–3.66	52–145

[a]*ADBI*, celestolide; *AHMI*, phantolide; *AHTN*, tonalide; *ATII*, traseolide; *DPMI*, cashmeran; *HHCB*, galaxolide; *MA*, musk ambrette; *MK*, musk ketone; *MM*, musk moskene; *MT*, musk tibetene; *MX*, musk xylene.
[b]*FI*, fish; *MM*, marine mammal; *MU*, mussel; *OY*, oyster.
[c]*DSPE*, dispersive solid-phase extraction; *GPC*, gel-permeation chromatography; *HS*, headspace; *MA*, microwave assisted; *MSPD*, matrix solid-phase dispersion; *MSTFA*, N-methyl-N-(trimethylsilyl)trifluoroacetamide; *MTBSTFA*, N-((*tert*-butyldimethylsilyl))-N-methyltrifluoroacetamide; *PCX*, polymeric cation exchange; *PDMS-DVB*, polydimethylsiloxane-divinylbenzene; *PLE*, pressurized liquid extraction; *PSA*, primary-secondary amine; *SLE*, solid-liquid extraction; *SPE*, solid-phase extraction; *SPME*, solid-phase microextraction.
[d]*EI*, electron ionization; *GC*, gas chromatography; *LVI*, large-volume injection; *MS*, mass spectrometry; *MS/MS*, tandem mass spectrometry; *TD*, thermal desorption.
[e]*MLOD*, method limit of detection.

employed (Fromme et al., 2001; Ligon et al., 2008; Gómez et al., 2009; Kwon and Rodriguez, 2014; Sun et al., 2014; Lange et al., 2015), whilst in the case of SPE, C_{18} (Simonich et al., 2000; Lv et al., 2009; Hu et al., 2011; Chase et al., 2012; Lu et al., 2015; Zhang et al., 2015a) and PVP-DVB (Zhou et al., 2009; Moldovan et al., 2009; Villa et al., 2012; He et al., 2013; Wang et al., 2013a; Vallecillos et al., 2014a) are the preferred options, although the use of other sorbents such as poly(methyl methacrylate)-divinylbenzene (Osemwengie and Steinberg, 2001) and PS-DVB/MP (Matamoros et al., 2010) has also been reported. In addition to commercial sorbents (such as those described earlier), the use of tailor-made sorbents based on molecularly imprinted silica was also proposed by López-Nogueroles et al. (2013). Clara et al. (2011) used LLE first, and the extract was further cleaned up by SPE using alumina.

Other than LLE and SPE, a variety of microextraction techniques, either in solid- or in liquid-phase mode, have been used for the determination of synthetic musks. SPME either in DI (Winkler et al., 2000; Liu et al., 2010; Cavalheiro et al., 2015) or in HS (García-Jares et al., 2002; Polo et al., 2007; Wang and Ding, 2009; Basaglia and Pietrogrande, 2012; Vallecillos et al., 2013a; Godayol et al., 2015; Li et al., 2016b) mode has been successfully applied using commercial coatings such as PDMS, PDMS-DVB, carboxen-PDMS (CAR-PDMS) and PA, as well as synthesized polyethersulphone (Cavalheiro et al., 2015) or amino-modified graphene (Li et al., 2016b).

Other microextraction methods that utilize solid sorbents for the extraction of synthetic musks are SBSE (Silva and Nogueira, 2010; Ramírez et al., 2011; Arbulu et al., 2011; Gómez et al., 2011; Chase et al., 2012; Ramírez et al., 2012; Klaschka et al., 2013; Pintado-Herrera et al., 2014) employing the typical PDMS-coated stir bar, DSPE using C_{18} microparticles (Chung et al., 2013, 2014) or Fe_3O_4@graphene-C_8 MNPs (Maidatsi et al., 2015) and MEPS using conventional C_8 (Moeder et al., 2010) or C_{18} sorbents (Vallecillos et al., 2012b; Cavalheiro et al., 2013; Caballero-Díaz et al., 2013). It is worth mentioning the paper published by Vallecillos et al. (2015a), in which a miniaturized MEPS was accomplished by setting the C_{18} sorbent inside the needle of a micro-syringe.

Liquid-phase-based microextraction techniques have also been used for the extraction and enrichment of musks, yet to a lesser extent compared to SPME techniques. Techniques like DLLME (Panagiotou et al., 2009; López-Nogueroles et al., 2011), USA-DLLME (Yang and Ding, 2012; Homem et al., 2016) USAEME (Regueiro et al., 2008) and SDME, using either conventional solvents (Wang et al., 2014) or ILs (Vallecillos et al., 2012a), have been reported. A modification to SDME was described by Guo et al. (2016), who intentionally inserted an air bubble within the drop during the SDME procedure (i.e., bubble-in-drop SDME), to increase the surface area of the extraction droplet whilst maintaining the same volume of liquid. MALLE is another LPME technique that has been sparsely reported for the extraction of synthetic musks (Einsle et al., 2006; Posada-Ureta et al., 2012).

Solid Samples

Similar to lipophilic UV filters, the high octanol-to-water partition coefficients of synthetic musks (log K_{ow} > 2.5) favour their accumulation in soils, sediments and sewage sludge. Solvent extraction is the main method of choice for the analysis of solid samples (see Table 16.5), by applying either ultrasound or mechanical agitation (i.e., SLE) to accelerate the extraction kinetics (Fromme et al., 2001; Ternes et al., 2005; Carballa et al., 2008; Zhou et al., 2009; Chase et al., 2012; Matamoros et al., 2012; Sun et al., 2014; Zhang et al., 2015a,b), or by applying heat by means of microwave irradiation (i.e., MAE) (Rice and Mitra, 2007), or by using pressurized solvents (i.e., PLE) (Simonich et al., 2000; Burkhardt et al., 2005; Ligon et al., 2008; Chen and Bester, 2009; Guo et al., 2010; Hu et al., 2011; Vallecillos et al., 2012c; Chamorro et al., 2013; Wang et al., 2013b; Guo et al., 2014; Pintado-Herrera et al., 2016). Finally, Soxhlet extraction has been also used for the extraction of musks from solid samples (Zeng et al., 2008; Sumner et al., 2010; Villa et al., 2012; Lu et al., 2015). In all these methods, the solvent collected after the extraction is evaporated to a smaller volume to enhance the pre-concentration factor and facilitate further treatment and clean-up. Clean-up is customarily accomplished by means of SPE using silica or alumina (Simonich et al., 2000; Zeng et al., 2008; Sumner et al., 2010; Hu et al., 2011; Lu et al., 2015), C_{18} (Ternes et al., 2005; Carballa et al., 2008), PS-DVB (Burkhardt et al., 2005), PVP-DVB (Zhou et al., 2009; Wang et al., 2013b), graphitized carbon (Villa et al., 2012) or Florisil sorbents (Chamorro et al., 2013; Zhang et al., 2015a,b). On a few occasions, the clean-up step has been carried out by means of GPC (Rice and Mitra, 2007; Guo et al., 2010, 2014), or by combining SPE and GPC (Chen and Bester, 2009). It is worth commenting on the paper published by Vallecillos et al. (2012c), who extracted musks from solid samples by PLE, evaporated the extract, reconstituted it in water and then subjected it to (HS)SDME by employing an IL as the extraction phase.

Clean-up procedures based on microextraction techniques have also been applied, although to a minor extent. Specifically, HS(SPME) has been used by Llompart et al. (2003), Wu and Ding (2010) and Vallecillos et al. (2013b), whilst SBSE was used by Aguirre et al. (2014) by suspending the solid sample in water/methanol. Vallecillos et al. (2014b) also used SBSE in the HS mode, by holding a stir bar in the HS of a vial containing the solid sample suspended in water.

Living Organisms

To achieve the extraction of synthetic musks from biological tissues, methods similar to those employed for the analysis of solid samples were utilized: SLE, PLE, Soxhlet extraction (see Table 16.6) and, on some occasions, MSPD (Vallecillos et al., 2015b; Ziarrusta et al., 2015; Yao et al., 2016). However, owing to the complexity of the biological tissues, e.g., fishes, analysis is not feasible; therefore, it is common to resort to a clean-up step to remove the co-extracted lipids. GPC (Subedi et al., 2011; Klaschka et al., 2013), SPE

(Mottaleb et al., 2009; Foltz et al., 2014; Lange et al., 2015; Ziarrusta et al., 2015) and GPC followed by SPE (Fromme et al., 2001; Osemwengie and Steinberg, 2003; Nakata, 2005; Nakata et al., 2007; Duedahl-Olesen et al., 2005; Kannan et al., 2005; Rüdel et al., 2006; Mottaleb et al., 2009; Hu et al., 2011; Zhang et al., 2013, 2015b) are the most popular clean-up methods. Amongst the arsenal of clean-up procedures, (HS)SPME has also been reported, by Wu et al. (2012, 2013b).

Analytical Techniques

Owing to the high volatility of the synthetic musks, GC is the most frequently applied analytical technique for their sensitive determination in environmental samples (see Tables 16.4–16.6). Analysis of musks by GC requires no derivatization; however, derivatization facilitates the analysis of other organic pollutants; therefore, in multi-residue studies, analysis of musks, along with other pollutants, has been performed after derivatization (Ramírez et al., 2012; Pintado-Herrera et al., 2014; Basaglia and Pietrogrande, 2012; Pintado-Herrera et al., 2016; Mottaleb et al., 2009; Subedi et al., 2011). As in the case of most organic pollutants, GC is coupled to MS to achieve high selectivity and sensitivity. Simple-quadrupole, triple-quadrupole and ion traps have been used in EI-positive (EI$^+$) mode. In only one case, an electronic capture detector (ECD) was used for the determination of nitro musks (Polo et al., 2007), because this detector is sensitive to the nitro moieties.

In contrast to other cosmetic ingredients, LC has seldom been used for the analysis of synthetic musks. In fact, LC–MS/MS has been used only once for the determination of synthetic musks in wastewater and sewage sludge (Sun et al., 2014).

PRESERVATIVES

Preservatives are chemical compounds belonging to very different chemical families that are added to a wide variety of products such as food, pharmaceuticals, household products, paints and cosmetic formulations to inhibit the development of microorganisms in the products themselves. The compounds with antimicrobial activity that are allowed to be used in cosmetic products are listed in the European Regulation on Cosmetic Products, and a detailed discussion can be found in Chapter 9.

Preservatives, like UV filters and synthetic musks, reach the aquatic environment mainly because their removal rate by WWTPs is low, and to a lesser extent they come from direct sources such as bathing activities. Their antimicrobial actions, mainly related to the disturbance of metabolic pathways or transport mechanisms through the cellular membranes of microorganisms, can cause various harmful effects to living organisms (Brausch and Rand, 2011; Carbajo et al., 2015; Montaseri and Forbes, 2016; Evans et al., 2016; Dambal et al., 2017). In this regard, several analytical methods have been developed for the environmental surveillance of preservatives in water samples (Table 16.7), solid samples (Table 16.8) and living organisms (Table 16.9).

Environmental Monitoring of Cosmetic Ingredients 503

Table 16.7 Published papers on preservative determination in environmental water samples (in chronological order)

Authors	Target compounds[a]	Sample[b]	Extraction technique[c]	Analytical technique[d]	MLOD (ng/L)[e]	Recovery (%)
Singer et al. (2002)	TCS	EW, LK, RV	**SPE** (1-L sample (pH 3); extracted with PVP-DVB cartridge at 15 mL/min; eluted with 6 mL ethyl acetate/acetone; evaporated to 150 µL; derivatized with 800 µL diazomethane (30 min); evaporated to 150–200 µL)	GC–MS(EI⁺)	1.5 (EW) 0.3 (LK)	105 (EW, LK)
Agüera et al. (2003)	TCS	EW, IW	**SPE** (50-mL sample (pH 4); extracted with C₁₈ cartridge at 10 mL/min; eluted with 8 mL methanol and 10 mL acetone; evaporated to dryness and redissolved in 1 mL ethyl acetate)	GC–MS(CI⁻)	20	84
Canosa et al. (2005)	TCS and other compounds	EW, IW, RV	**(DI)SPME** (22-mL sample (pH 4.5); extracted with PA-coated fibre; 20°C, 30 min; on-fibre derivatized with 20 µL MTBSTFA (20°C, 10 min); thermal desorption (280°C, 3 min))	(TD)GC–MS(EI⁺)	4–14 (EW, IW) 2–7 (RV)	95–105
Canosa et al. (2006)	BP, BzP, EP, MP, PP	EW, IW, RV	**(DI)SPME** (20-mL sample (pH 6, 0.15% NaCl); extracted with PA-coated fibre; 20°C, 40 min; on-fibre derivatized with 20 µL MTBSTFA (20°C, 10 min); thermal desorption (260°C, 3 min))	(TD)GC–MS/MS(EI⁺)	0.3–8	87–96 (EW) 92–104 (IW) 98–114 (RV)

Continued

Table 16.7 Published papers on preservative determination in environmental water samples (in chronological order)—cont'd

Authors	Target compounds[a]	Sample[b]	Extraction technique[c]	Analytical technique[d]	MLOD (ng/L)[e]	Recovery (%)
Ying and Kookana (2007)	TCS	EW, RV	**SPE** (1-L sample; extracted with C_{18} cartridge; eluted with 2×4 mL ethyl acetate; evaporated to dryness and redissolved in 1 mL acetonitrile)	GC–MS(EI$^+$)	0.9	89–96 (EW) 101–104 (RV)
Gatidou et al. (2007)	TCS and other compounds	EW, IW	**SPE** (100-mL sample; extracted with C_{18} cartridge at 10 mL/min; eluted with 4×2 mL dichloromethane/hexane; evaporated to dryness and redissolved in 50 μL BSTFA + 50 μL pyridine (65°C, 20 min))	GC–MS(EI$^+$)	130	72–87
Wu et al. (2007)	TCS	EW, IW, RV, SW	**SPE** (1-L sample (pH 2); extracted with C_{18} cartridge at 3 mL/min; eluted with 3×2 mL; dried with Na_2SO_4; evaporated to dryness and redissolved in 200 μL ethyl acetate)	GC–MS/MS(EI$^+$)	0.25	83–110
Zhao et al. (2007)	TCS	TW	**(DI)HFLPME** (10-mL sample (0.1 M NaOH) + 200 μL acetic anhydride; extracted with 5 μL n-dodecane; 20 min)	GC–MS(EI$^+$)	20	84–114
Cuderman and Heath (2007)	CLP, TCS and other compounds	LK, RV, SP, SW	**SPE** (500-mL sample (pH 3); extracted with PS-DVB cartridge at 5 mL/min; eluted with 3×0.5 mL ethyl acetate/dichloromethane, evaporated to dryness, redissolved with 0.4 mL toluene and derivatized with 100 μL MSTFA (60°C, 60 min))	GC–MS(EI$^+$)	10–28 (LK) 17–18 (RV) 113–178 (SP) 143–163 (SW)	79–96 (LK) 82–97 (RV) 87–98 (SP) 88–95 (SW)

Rafoth et al. (2007)	CMI, MI and other compounds	EW, IW, RV, TW	**SPE** (500-mL sample (pH 5); extracted with C$_{18}$/PS-DVP/MP cartridge; eluted with 6 mL acetone; evaporated to 200 μL)	GC–MS(EI$^+$)	3–85	10–103
Ligon et al. (2008)	TCS and other compounds	EW, IW	**LLE** (1-L sample (pH <2); extracted with 20 mL toluene; diluted to 175 mL with water; evaporated to 1 mL)	GC–MS(EI$^+$)	0.3	68 (EW) 67 (IW)
Kasprzyk-Hordern et al. (2008)	BzPh, CLP, CXL, TBC, TCS, BP, EP, MP, PP and other compounds	EW, IW, RV	**SPE** (250- to 1000-mL sample (pH 3, 500 mg EDTA); extracted with PVP-DVB/MCX cartridge at 4 mL/min; eluted with 2 mL methanol + 2 mL 5% NH$_4$OH/methanol; evaporated to dryness and redissolved in 0.5 mL ammonium acetate/methanol)	LC–MS/MS(ESI$^-$)	1–26 (EW) 0.6–31 (IW) 0.05–5 (RV)	8–186 (EW) 6–139 (IW) 40–140 (RV)
Silva and Nogueira (2008)	TCS	EW, IW	**SBSE** (25-mL sample (pH 7); extracted with PDMS-coated stir bar; 25°C, 2 h; eluted with 1.5 mL acetonitrile (25°C, 60 min))	LC–UV	100	79
Kawaguchi et al. (2008b)	TCS	RV	**SBSE** (10-mL sample; extracted with PDMS-coated stir bar; 120 min; thermal desorption (240°C, 5 min))	(TD)GC–MS(EI$^+$)	5	92–108
Blanco et al. (2008, 2009)	BP, BzP, EP, MP, PP	EW, IW, RV, TW	**SPE** (500-mL sample (pH 2.5); extracted with PS-DVB cartridge; eluted with 2 mL methanol)	(LVSS)CE–UV	25–31	90–111 (EW) 62–94 (IW) 99–106 (RV) 97–104 (TW)

Continued

Table 16.7 Published papers on preservative determination in environmental water samples (in chronological order)—cont'd

Authors	Target compounds[a]	Sample[b]	Extraction technique[c]	Analytical technique[d]	MLOD (ng/L)[e]	Recovery (%)
Saraji and Mirmahdieh (2009)	BP, EP, IPP, MP, PP,	RV	(DI)SDME (3-mL sample; extracted with 3 μL hexyl acetate; 47°C, 20 min; derivatized with 0.4 μL BSA (2 min))	GC–MS(EI$^+$)	1–15	72–99
Rodil et al. (2009a)	TCS and other compounds	EW, IW, SW, TW	SPE (200- to 500-mL sample (pH 7); extracted with PVP-DVB cartridge; eluted with 3 × 10 mL methanol; evaporated to 0.2 mL and diluted to 1 mL with methanol/water)	LC–MS/MS(ESI$^{+/-}$)	6	105 (EW) 57 (IW) 79 (SW) 82 (TW)
Montes et al. (2009)	TCS	EW, IW, RV, TW	DLLME (10-mL sample; extracted with 40 μL trichloroethane (+1 mL methanol as disperser solvent and 40 μL MTBSTFA as derivatization reagent))	GC–MS/MS(EI$^+$)	0.6–1.5	96 (EW) 93 (IW) 103 (RV) 103 (TW)
Guo et al. (2009)	TCC, TCS	EW, IC, RV, TW	DLLME (5-mL sample; extracted with 15 μL dichlorobenzene (+1 mL tetrahydrofuran as disperser solvent; evaporated to dryness and redissolved in 35 μL methanol)	LC–UV	42.1–134	77–81 (EW) 64–85 (IC) 71–96 (RV) 81–106 (TW)
González-Mariño et al. (2009)	TCC, TCS, BP, BzP, EP, MP, PP	EW, IW, LK, RV	SPE (200- to 500-mL sample; extracted with PVP-DVB cartridge at 10 mL/min; eluted with 4 mL methanol; evaporated to 0.5 mL and diluted to 1 mL with water)	LC–MS/MS(ESI$^-$)	0.02–50 (IW) 0.008–20 (RV)	69–123 (EW) 62–137 (IW) 69–118 (RV)

Reference	Compounds	Matrix	Sample preparation	Technique	LOD/LOQ (ng/L)	Recovery (%)
Regueiro et al. (2009a)	TCS, BP, EP, MP, PP	EW, IW, RV, SP	**(HS)SPME** (10-mL sample (35% NaCl + 100 µL acetic anhydride); extracted with DVB-CAR-PDMS-coated fibre; 100°C, 15 min; thermal desorption (240°C, 2 min))	GC-MS/MS(EI⁺)	4–17	82–102
Regueiro et al. (2009b)	TCS, BP, EP, MP, PP	EW, IW, RV, SP	**USAEME** (10-mL sample (+200 µL acetic anhydride); extracted with 100 µL trichloroethane; 25°C, 5 min)	GC-MS/MS(EI⁺)	3.90–27.5	85–95
Pedrouzo et al. (2009)	TCC, TCS, BzP, EP, MP, PP and other compounds	EW, IW, RV	**SPE** (100- to 500-mL sample (pH 3); extracted with PVP-DVB or PS-DVB/MH cartridge at 10–15 mL/min; eluted with 5 mL methanol and 5 mL dichloromethane; evaporated to 3–4 mL and diluted to 5 mL with water)	LC-MS/MS(ESI⁻)	3–10 (EW, IW) 1–3 (RV)	20–92 (EW) 27–85 (IW) 69–101 (RV)
Gómez et al. (2009)	TCS and other compounds	EW, RV	**LLE** (500-mL sample (pH 3, 1% NaCl); extracted with 100 mL (3 min) + 50 mL (2 min) hexane; dried with Na₂SO₄; evaporated to 0.4–3 mL)	(LVI)GC-MS(EI⁺)	44 (EW) 18 (RV)	149 (EW) 142 (RV)
Jonkers et al. (2010)	BzP, BP, EP, MP, PP and other compounds	EW, LK, RV, SW	**SPE** (150- to 250-mL sample (5% methanol); extracted with PVP-DVB cartridge; eluted with 3 mL methyl-*tert*-butyl ether/propanol and 3 mL methanol; evaporated to dryness and redissolved in 250 µL methanol/water)	LC-MS/MS(ESI⁺/⁻)	0.06–0.6	107–121

Continued

Table 16.7 Published papers on preservative determination in environmental water samples (in chronological order)—cont'd

Authors	Target compounds[a]	Sample[b]	Extraction technique[c]	Analytical technique[d]	MLOD (ng/L)[e]	Recovery (%)
Wick et al. (2010)	CLP, IPBC, TCC, TCS and other compounds	EW, IW, RV	**SPE** (100- to 1000-mL sample (pH 6); extracted with PVP-DVB cartridge at 5 mL/min; eluted with 4 × 2 mL methanol/acetone; evaporated to 0.5 mL and diluted to 1 mL with aqueous formic acid)	LC–MS/MS(ESI$^{+/-}$)	0.75–7.5 (EW) 1.5–15 (IW) 0.15–0.6 (RV)	90–100 (EW) 95–108 (IW) 97–120 (RV)
Pedrouzo et al. (2010)	TCC, TCS and other compounds	EW, IW, RV	**SBSE** (50-mL sample (pH 5, 5% methanol); extracted with PDMS-coated stir bar; 180 min; eluted with 1 mL acetonitrile (30°C, 15 min); evaporated to dryness and redissolved in acetonitrile/water)	LC–MS/MS(ESI$^{+/-}$)	5 (EW, IW) 2.5 (RV)	44–84 (EW) 46–89 (IW) 50–87 (RV)
Klein et al. (2010)	TCC	EW, GW	**SBSE** (10-mL sample; extracted with PDMS-coated stir bar; 22 h; eluted with 1.5 mL methanol)	LC–MS/MS(ESI$^-$)	1	92–96 (EW) 93 (GW)
Zhao et al. (2010)	TCC, TCS	EW, TW	**IL-DLLME** (5-mL sample (pH 6–8); extracted with 60 µL [C$_6$MIM][PF$_6$] (+0.5 mL methanol as disperser solvent); diluted to 100 µL with methanol)	LC–MS/MS(ESI$^-$)	40–580	70–98 (EW) 72–104 (TW)
Villaverde-de-Sáa et al. (2010)	TCS, BP, BzP, EP, MP, PP	EW, IW, TW	**MALLE** (18-mL sample (+0.2 g K$_2$HPO$_4$ + 200 µL acetic anhydride); extracted with LDPE bag with 400 µL chloroform; 35°C, 90 min)	(LVI)GC–MS/MS(EI$^+$)	0.1–14	83–104

Environmental Monitoring of Cosmetic Ingredients 509

Speksnijder et al. (2010)	CMI, DCMI, MI	EW, RV, TW	Direct injection	(LVI)LC–MS/MS(APCI⁺)	30–110	10–95 (EW) 21–99 (RV) 82–109 (TW)
López-Darias et al. (2010)	BP, BzP and other compounds	BW, SP	(DI)SPME (20-mL sample; extracted with poly(VBHDIm⁺NTf₂⁻)-coated fused silica fibre; 60 min; thermal desorption (250°C, 5 min))	(TD)GC–FID	1500–3400	83.5–95.7
Matamoros et al. (2010)	TCS and other compounds	RV	SPE (200-mL sample (pH 2); extracted with PS-DVB/MP cartridge at 10 mL/min; eluted with 5 × 2 mL ethyl acetate; evaporated to 100 μL)	GC × GC–MS(EI⁺)	3	93
Gómez et al. (2011)	CLP, TCS and other compounds	EW, RV	SBSE (25- to 100-mL sample (20 g NaCl, 10% methanol); extracted with PDMS-coated stir bar; 14 h; thermal desorption (295°C, 7 min))	(TD) GC × GC–MS(EI⁺)	0.08–0.12 (EW) 0.06–0.55 (RV)	121–147 (EW) 132–144 (RV)
Cheng et al. (2011)	TCS	EW, IW, RV	SPE (100-mL sample (pH 3); extracted with C₁₈ cartridge at 3–5 mL/min; eluted with 4 mL dichloromethane; evaporated to dryness and redissolved in 100 μL dichloromethane)	(LVI)GC–MS(EI⁺)	0.4	73–99 (EW) 60–71 (IW) 78–110 (RV)
Casas-Ferreira et al. (2011)	TCS, BP, BzP, IPP, MP, PP	EW, IW, TW	SBSE (5-mL sample (+5 μL acetic anhydride); extracted with PDMS-coated stir bar; 25°C, 60 min; thermal desorption (275°C, 6 min))	(TD)GC–MS(EI⁺)	0.54–4.12	48–105

Continued

Table 16.7 Published papers on preservative determination in environmental water samples (in chronological order)—cont'd

Authors	Target compounds[a]	Sample[b]	Extraction technique[c]	Analytical technique[d]	MLOD (ng/L)[e]	Recovery (%)
González-Mariño et al. (2011)	TCS, BP, EP, IPP, MP, PP	EW, IW	MEPS (2-mL sample (pH 3); extracted with C$_{18}$ sorbent; eluted with 2×25 μL ethyl acetate)	(LVI)GC–MS(EI$^+$)	10–240 (EW) 20–590 (IW)	87–118 (EW) 87–121 (IW)
Zheng et al. (2011)	TCS	LK, RV, TW	DLLME-SFO (5-mL sample (pH 6); extracted with 12 μL 1-dodecanol (+300 μL acetonitrile as disperser solvent))	LC–MS/MS(ESI$^-$)	2	84–116 (LK) 93–11 (RV) 87–105 (TW)
Prichodko et al. (2012)	BP, EP, MP, PP	RV, TW	DLLME (8-mL sample (+8 μL acetic anhydride); extracted with 20 μL chlorobenzene (+0.28 mL acetone as disperser solvent))	GC–FID	2500–22,000	
Basaglia and Pietrogrande (2012)	CLP, CLX, TCS and other compounds	TW	(HS)SPME (40-mL sample; extracted with PA-coated fibre; 40°C, 125 min; then exposed to BSTFA (35°C, 30.5 min); thermal desorption (300°C, 2 min))	(TD)GC–MS(EI$^+$)	2.5–7	85–103
Tahmasebi et al. (2012)	EP, MP, PP	EW	DSPE (70-mL sample (pH 8); extracted with Fe$_3$O$_4$@polyaniline MNPs sorbent; 2 min; eluted with 100 μL acetonitrile)	LC–UV	300–400	88–109
Gracia-Lor et al. (2012)	BP, EP, MP, PP and other compounds	EW, LK, RV, SW	SPE (100-mL sample; extracted with PVP-DVB cartridge; eluted with 5 mL methanol; evaporated to dryness and redissolved in 1 mL methanol/water)	LC–MS/MS(ESI$^{+/-}$)	4–10 (EW) 0.3–1 (LK, RV, SW)	70–120

Reference	Analytes	Matrix	Method	Technique	LOD/LOQ	Recovery (%)
Ramírez et al. (2012)	BP, EP, IPP, MP, PP and other compounds	EW, IW, RV	**SBSE** (100-mL sample (+100 μL acetic anhydride, 0.5% disodium hydrogen phosphate); extracted with PDMS-coated stir bar; 20°C, 4 h; thermal desorption (300°C, 15 min))	(TD)GC–MS(EI⁺)	0.03–0.3	47–102
Çabuk et al. (2012)	BP, EP, MP, PP	TW	**DLLME** (5-mL sample; extracted with 50 μL decanol (+1 mL acetonitrile as disperser solvent))	LC–UV	21–46	61–108
Shaaban and Górecki (2012)	EP, MP, PP and other compounds	IW, LK, RV	**SPE** (150- to 500-mL sample; extracted with PVP-DVB cartridge at 4 mL/min; eluted with 5 mL ammonia/methanol; evaporated to dryness and redissolved in 0.5 mL acetonitrile/water)	LC–UV	7500	73–81
Chen et al. (2012)	TCC, TCS, BP, EP, MP, PP and other compounds	EW, IW, TW	**SPE** (1-L sample (pH 3); extracted with PVP-DVB cartridge at 5–10 mL/min; eluted with 3 × 4 mL ethyl acetate; evaporated to dryness and redissolved in 1 mL methanol)	LC–MS/MS(ESI⁺/⁻)	0.01–1.17 (EW) 0.02–1.50 (IW) 0.01–0.09 (TW)	39–149 (EW) 20–156 (IW) 59–148 (TW)
Pintado-Herrera et al. (2013)	TCS and other compounds	EW, IW, SW	**SBSE** (100-mL sample (pH 2, 10% methanol); extracted with PDMS-coated stir bar; 25°C, 8 h; eluted with 200 μL ethyl acetate (30 min) and derivatized with 10 μL MTBSTFA)	GC–MS(EI⁺)	0.2	24

Continued

Table 16.7 Published papers on preservative determination in environmental water samples (in chronological order)—cont'd

Authors	Target compounds[a]	Sample[b]	Extraction technique[c]	Analytical technique[d]	MLOD (ng/L)[e]	Recovery (%)
Caldas et al. (2013)	TCC, TCS, PP and other compounds	EW, RV	**SPE** (1000-mL sample (pH 3); extracted with C$_{18}$ cartridge at 10 mL/min; eluted with 2 × 1 mL methanol)	LC–MS/MS(ESI$^{+/-}$)	0.2–2	76–115
Gilart et al. (2013)	TCC, TCS, BzP, MP, PP	EW, IW	**SBSE** (50-mL sample (pH 5); extracted with EGS-coated stir bar; 25°C, 4 h; eluted with 1 mL methanol (15 min); diluted with 1 mL water)	LC–MS/MS(ESI$^{+/-}$)	5–10	
Abbasghorbani et al. (2013)	BP, EP, IPP, MP, PP	RV	**DSPE** (10-mL sample; extracted with Fe$_3$O$_4$@aminopropyl MNP sorbent; 5 min; eluted with 2 × 25 μL acetonitrile (2 min each); derivatized with 10 μL acetic anhydride and 10 μL pyridine (5 min); evaporated to dryness and redissolved in 5 μL acetonitrile)	GC–PID	50–300	95–103
Shih et al. (2013)	TCS	LK, RV	**USAEME** (5-mL sample (2.5% NaCl); extracted with 20 μL 1-octanol; 0.5 min)	GC–ECD	4	91–97
Gorga et al. (2013)	TCC, TCS, BzP, EP, MP, PP and other compounds	EW, IW, RV	**(Online)SPE** (2- to 5-mL sample; extracted with a C$_{18}$ column; eluted with 2 × 5 mL methanol)	LC–MS/MS(ESI$^{+/-}$)	0.12–1.5 (EW) 0.18–2.1 (IW) 0.021–0.27 (RV)	59–125 (EW) 59–126 (IW) 62–116 (RV)

Lorenzo et al. (2013)	BP	RV	**SPE** (100-mL sample; extracted with MIP cartridge; eluted with 4 mL acetonitrile; evaporated to dryness and redissolved in 10 mL electrolyte)	AdSV	16,000	79–154
Alcudia-León et al. (2013)	BP, EP, MP, PP	SP, SW	**MCNPME** (30-mL sample; extracted with 10 mg Fe$_3$O$_4$@SiO$_2$@C$_{18}$ MNPs sorbent; 20 min; eluted with 100 µL ethyl acetate (2 min))	GC–MS(EI$^+$)	23–86	96–102 (SP) 99–106 (SW)
Kotowska et al. (2014)	TCS and other compounds	EW, IW	**USAEME** (5-mL sample (+300 µL acetic anhydride); extracted with 40 µL carbon tetrachloride; 25°C, 5 min)	GC–MS(EI$^+$)	3 (EW, IW)	102 (EW) 97 (IW)
Mudiam et al. (2014)	BP, EP, MP, PP and other compounds	EW, IW	**USA-DLLME** (5-mL sample; extracted with 100 µL dichloromethane (+200 µL acetonitrile as disperser solvent); 1.5 min)	GC–MS/MS(EI$^+$)	8–230	94–98 (EW) 86–95 (IW)
Martín et al. (2014)	EP, MP, PP	EW, RV	**SPE** (250-mL sample (pH 2); extracted with PVP-DVB cartridge at 10 mL/min; eluted with 4 × 1 mL methanol; evaporated to dryness and redissolved in 1 mL methanol)	LC–MS/MS(ESI$^-$)	1.81–3.95 (EW) 1.98–3.94 (RV)	98–113 (EW) 99–106 (RV)
Carmona et al. (2014)	TCC, TCS, BP, EP, MP, PP and other compounds	EW, IW, RV, TW	**SPE** (0.25- to 1-L sample; extracted with PS-DVB cartridge; eluted with 6 mL methanol; evaporated to dryness and redissolved in 1 mL methanol/water)	LC–MS/MS(ESI$^{+/-}$)	0.15–0.45 (EW) 0.3–1.5 (IW) 0.06–0.90 (RV) 0.03–0.30 (TW)	89–99 (EW) 78–90 (IW) 93–110 (RV) 98–109 (TW)

Continued

Table 16.7 Published papers on preservative determination in environmental water samples (in chronological order)—cont'd

Authors	Target compounds[a]	Sample[b]	Extraction technique[c]	Analytical technique[d]	MLOD (ng/L)[e]	Recovery (%)
Almeida and Nogueira (2014)	BP, EP, MP, PP	EW, ES, SP, TW	BAμE (25-mL sample (pH 5.5); extracted with activated carbon sorbent; 25°C, 16 h; eluted with 200 μL methanol/acetonitrile (45 min))	LC–UV	100	85–101
Celano et al. (2014)	TCC, TCS, BP, BzP, EP, MP, PP and other compounds	EW, IW, RV, SW, TW	SPE + DLLME (250- to 500-mL sample (pH 2); extracted with PVP-DVB cartridge at 5 mL/min; eluted with 5 mL methanol/acetonitrile/dichloromethane; then 2 mL injected into 10 mL water (5% NaCl); evaporated to dryness and redissolved in 200 μL acetonitrile/water)	LC–MS/MS(ESI−)	0.2–1.8 (EW, IW) 0.1–0.4 (RV) 0.1–0.6 (SW) 0.1–0.7 (TW)	54–94 (EW, IW) 83–98 (RV) 55–95 (SW) 49–95 (TW)
Azzouz and Ballesteros (2014)	TCS, BP, BzP, EP, IPP, MP, PP and other compounds	EW, RV, SP, TW	SPE (100-mL sample (pH 4); extracted with EVB-DVB cartridge at 4 mL/min; eluted with 400 μL acetonitrile; evaporated to 25 μL and derivatized with 70 μL BSTFA (3 min))	GC–MS(EI+)	0.01–0.08	92–101 (EW) 90–101 (RV) 91–101 (SP, TW)
Hashemi et al. (2015)	EP, MP, PP	TW	SPE + DLLME-SFO (100-mL sample (pH 6); extracted with C₁₈ cartridge at 10 mL/min; eluted with 1.5 mL acetone; evaporated to 250 μL; then DLLME; extracted with 20 μL 1-undecanol)	LC–UV	300–1700	96–112

González-Hernández et al. (2015)	TCS, BP, BzP, EP, IPP, MP, PP	EW, SP, SW, TW	**VAEME** (8-mL sample (pH 5, 15% NaCl); extracted with 200 µL trichloromethane (3 min))	LC–UV	30–1650	99–113
Li et al. (2015)	TCC, TCS	EW, IW	**SPE** (300-mL sample (pH 6–8); extracted with SiO₂/PS cartridge at 10 mL/min; eluted with 3 × 5 mL methanol/acetone; evaporated to dryness and redissolved in 0.5 mL methanol)	LC–UV	28–40	95–103 (EW) 90–104 (IW)
Rocío-Bautista et al. (2015)	BP, BzP, EP, IPP, MP, PP	SP, TW	**VA-DSPE** (20-mL sample; extracted with 150 mg HKUST-1 MOF; 5 min; eluted with 2 mL methanol (5 min); evaporated to dryness and redissolved in 500 µL acetonitrile/water)	LC–UV	1500–2600	61–98 (SP)
Salvatierra-Stamp et al. (2015)	TCS and other compounds	RV	**SPE** (500-mL sample (pH 6); extracted with C₁₈ cartridge at 8 mL/min; eluted with 3 mL methanol; evaporated to dryness and redissolved in 225 µL methanol/water)	LC–UV	673.6	101–105
Martín et al. (2015)	EP, MP, PP and other compounds	RV, TW	**USA-DLLME-SFO** (10-mL sample (pH 1); extracted with 80 µL 1-undecanol (+500 µL methanol as disperser solvent; 5 min)	LC–MS/MS(ESI⁻)	62–296 (RV) 75–343 (TW)	32–81 (RV) 29–75 (TW)

Continued

Table 16.7 Published papers on preservative determination in environmental water samples (in chronological order)—cont'd

Authors	Target compounds[a]	Sample[b]	Extraction technique[c]	Analytical technique[d]	MLOD (ng/L)[e]	Recovery (%)
Dias et al. (2015)	TCC, EP, MP and other compounds	LK	BAμE (150-mL sample (pH 5.5, 25% NaCl); extracted with cork sorbent; 1.5h; eluted with 200μL methanol/acetonitrile (30min))	LC–UV	500	65–110
Caldas et al. (2016)	TCC, TCS, MP, PP and other compounds	RV	SD-DLLME (10-mL sample (pH 2, 1% Mg$_2$SO$_4$); extracted with 120μL 1-octanol (+0.75mL acetone as disperser solvent); 0.75mL acetone as demulsification solvent; diluted to 250μL with methanol)	LC–MS/MS(ESI$^-$)	8–379	73–120
Kapelewska et al. (2016)	BP, EP, MP, PP and other compounds	GW	USAEME (5-mL sample (pH 2); extracted with 70μL chloroform and derivatized with 50μL acetic anhydride; 5min)	GC–MS(EI$^+$)	1	91–119
Yi et al. (2016)	EP, MP, PP and other compounds	EW, RV, TW	(DI)HFLPME (10-mL sample (pH 5); extracted with 1-octanol; 25°C, 30min)	CE–AD	64–740	82–109 (EW) 91–97 (RV) 86–117 (TW)
Casado-Carmona et al. (2016)	BP, EP, MP, PP and other compounds	RV, SP, SW	DSPE (100-mL sample (pH 8, 30% NaCl); extracted with 100mg Fe$_3$O$_4$@SiO$_2$@MIM-PF$_6$; 20min; eluted with 500μL methanol (6min))	LC–MS/MS(ESI$^-$)	260–1350	97–99 (RV) 88–98 (SP) 87–99 (SW)

Chen et al. (2016)	TCC, TCS, BP, BzP, EP, MP, PP and other compounds	EW, IW, RV	SPE (500-mL sample (pH 2.5); extracted with PVP-DVB cartridge at 3 mL/min; eluted with 12 mL acetonitrile/ethyl acetate; evaporated to dryness and redissolved in 100 μL water/methanol)	LC–MS/MS(ESI⁻)	0.47–1.93 (EW) 0.53–2.87 (IW) 0.33–2.14 (RV)	50–130 (EW) 72–148 (IW) 66–112 (RV)
Wluka et al. (2016)	TCS and other compounds	EW, IW	SPE (500- to 1000-mL sample (pH 9); extracted with PVP-DVB cartridge at 5 mL/min; eluted 2×4 mL acetone and 2×4 mL methanol; evaporated to 50 μL)	GC–MS(EI⁺)	4.5	104
Pérez et al. (2016)	TCS and other compounds	RV, TW	DLLME + DSPE (25-mL sample (pH 3); extracted with 60 μL 1-octanol (+4 mL methanol as disperser solvent); then DSPE: extracted with 50 mg sodium oleate-coated Fe_3O_4 MNPs as sorbent; 1 min; eluted with 2×2 mL ethyl acetate (1 min each); evaporated to dryness and redissolved in 200 μL ethyl acetate)	GC–MS/MS(EI⁺)	22 (RV) 54 (TW)	89–90 (RV) 99–100 (TW)

Continued

Table 16.7 Published papers on preservative determination in environmental water samples (in chronological order)—cont'd

Authors	Target compounds[a]	Sample[b]	Extraction technique[c]	Analytical technique[d]	MLOD (ng/L)[e]	Recovery (%)
Lopes et al. (2017)	TCC, MP and other compounds	GW	**HF-DLLME** (20-mL sample (pH 4, 30% NaCl); extracted with 50 μL hexane/acetone as extraction/disperser solvent mixture; eluted with 100 μL acetonitrile (10 min))	LC–UV	900–3000	90–135
Ramos-Payan et al. (2017)	BP, EP, PP	LK, RV	**DF-μLPME** (50-μL sample (pH 3.5, 5% NaCl) at 10 μL/min; 1 μL dihexyl ether as SLM; 5.6 mM NaOH aqueous solution (pH 11.75) as acceptor phase; 1 μL/min)	LC–UV	1800–3500	82–86 (LK) 75–97 (RV)
Fumes and Lanças (2017)	BP, BzP, EP, MP, PP	IW, LK, SP, TW	**MEPS** (1-mL sample (pH 4–7, 30% NaCl); extracted with Si-G sorbent; eluted with 6 × 50 μL acetonitrile/methanol)	LC–MS/MS(ESI⁻)	60–90	82–119
Aparicio et al. (2017)	EP, MP, PP and other compounds	RV, TW	**SBSE** (100-mL sample (pH 3, 38% NaCl); extracted with EG-silicone-coated stir bar, 24 h; eluted with 0.5 mL methanol (15 min); evaporated to dryness and redissolved in 100 μL methanol/water)	LC–MS/MS(ESI⁻)	3.6–6.3 (RV) 3.3–5.4 (TW)	83–117 (RV) 84–103 (TW)

| Wei et al. (2017) | TCS, BP, BzP, EP, IPP, MP, PP and other compounds | RV | **USA-DLLME** (100-μL sample (pH 9.5); extracted with 90 μL 4-bromoanisole (+200 μL acetonitrile as disperser solvent and carbonyl chloride rosamine as derivatization reagent); 1 min; diluted to 50 μL with acetonitrile) | LC–MS/MS(ESI⁺) | 0.02–0.05 | 93–113 |

[a]*BP*, butylparaben; *BzP*, benzylparaben; *CLP*, chlorophene; *CMI*, chloromethylisothiazolinone; *CXL*, chloroxylenol; *DCMI*, dichloromethylisothiazolinone; *EP*, ethylparaben; *IPBC*, iodopropynyl butylcarbamate; *IPP*, isopropylparaben; *MP*, methylparaben; *MI*, methylisothiazolinone; *PP*, propylparaben; *TBC*, tetrabromcresol; *TCC*, triclocarban; *TCS*, triclosan.
[b]*BW*, bottled water; *EW*, effluent wastewater; *GW*, groundwater; *IC*, irrigation channel; *IW*, influent wastewater; *LK*, lake water; *RV*, river water; *SP*, swimming-pool water; *SW*, seawater; *TW*, tap water.
[c]*BAμE*, bar adsorptive microextraction; *BSA*, N,O-bis(trimethylsilyl)acetamide; *BSTFA*, N,O-bis(trimethylsilyl)trifluoroacetamide; *CAR*, carboxen; *DF-μLPME*, double-flow microfluidic device–based liquid-phase microextraction; *DI*, direct immersion; *DLLME*, dispersive liquid–liquid microextraction; *DSPE*, dispersive solid-phase extraction; *DVB*, divinylbenzene; *EDTA*, ethylenediaminetetraacetic acid; *EGS*, ethyleneglycol silicone; *EVB*, ethylvinylbenzene; *HFLPME*, hollow-fibre liquid-phase microextraction; *HS*, headspace; *IL*, ionic liquid; *LDPE*, low-density polyethylene; *LLE*, liquid–liquid extraction; *MALLE*, membrane-assisted liquid–liquid extraction; *MCNPME*, magnetically confined nanoparticle microextraction; *MEPS*, microextraction by packed sorbent; *MIP*, molecularly imprinted polymer; *MNPs*, magnetic nanoparticles; *MOF*, metal organic framework; *MSTFA*, N-methyl-N-(trimethylsilyl)trifluoroacetamide; *MTBSTFA*, N-(tert-butyldimethylsilyl)-N-methyltrifluoroacetamide; *PA*, polyacrylate; *PDMS*, polydimethylsiloxane; *PS-DVB*, polystyrene divinylbenzene co-polymer; *PS-DVB/MH*, polystyrene divinylbenzene co-polymer modified with hydroxyl groups; *PS-DVB/MP*, polystyrene divinylbenzene co-polymer modified with pyrrolidone; *PVP-DVB*, polyvinylpyrrolidone-divinylbenzene co-polymer; *PVP-DVB/MCX*, polyvinylpyrrolidone-divinylbenzene co-polymer modified with cationic exchange groups; *SBSE*, stir bar sorptive extraction; *SD*, solvent demulsification; *SDME*, single-drop microextraction; *SFO*, solidification of a floating organic drop; *SPE*, solid-phase extraction; *SPME*, solid-phase microextraction; *USA*, ultrasound-assisted; *USAEME*, ultrasound-assisted emulsification microextraction; *VAEME*, vortex-assisted emulsification microextraction.
[d]*AD*, amperometric detector; *AdSV*, adsorptive stripping voltammetry; *APCI*, atmospheric pressure chemical ionization; *CE*, capillary electrophoresis; *CI*, chemical ionization; *ECD*, electronic capture detector; *EI*, electron ionization; *ESI*, electrospray ionization; *FID*, flame ionization detector; *GC*, gas chromatography; *GC × GC*, multidimensional gas chromatography; *LC*, liquid chromatography; *LVI*, large-volume injection; *LVSS*, large-volume sample stacking; *MS*, mass spectrometry; *MS/MS*, tandem mass spectrometry; *PID*, photoionization detector; *TD*, thermal desorption; *UV*, ultraviolet spectrometry.
[e]*MLOD*, method limit of detection.

Table 16.8 Published papers on preservative determination in environmental solid samples (in chronological order)

Authors	Target compounds[a]	Sample[b]	Extraction technique[c]	Analytical technique[d]	MLOD (ng/g)[e]	Recovery (%)
Singer et al. (2002)	TCS	SE, SS	**PLE + SPE** (1-g dried sample; extracted with 3 × 5 min dichloromethane; 100°C, 103 bar; evaporated to 2 mL; then SPE in silica column and eluted with 2 mL dichloromethane; evaporated to 150 µL; derivatized with 800 µL diazomethane (30 min); evaporated to 150–200 µL)	GC–MS(EI⁺)	1.5	100
Agüera et al. (2003)	TCS	SE	**PLE + SPE** (10-g dried sample; extracted with 1 × 5 min dichloromethane; 100°C, 1500 psi; evaporated to 5 mL; then SPE in silica cartridge and eluted with 5 mL acetone; evaporated to dryness and redissolved in 1 mL ethyl acetate)	GC–MS(CI⁻)	0.09	100
Bester (2003)	TCS	SS	**Soxhlet + SPE + GPC** (10-g sample; extracted with ethyl acetate (reflux, 6 h); solvent exchanged to toluene; SPE in silica cartridge and eluted with ethyl acetate; then GPC with cyclohexane/ethyl acetate)	(PTV) GC–MS(EI⁺)	1.2	94
Morales et al. (2005)	TCS and other compounds	SE, SS	**MA-SLE + SPE** (0.5- to 1-g sample; extracted with 30 mL acetone/methanol (130°C, 20 min); diluted with 100 mL NaOH 0.2 M and 30 mL hexane; then SPE in PVP-DVB cartridge and eluted with 5 mL ethyl acetate; evaporated to 2 mL and derivatized with 50 µL MTBSTFA)	GC–MS/MS(EI⁺)	0.06	99–100 (SE) 82–97 (SS)

Morales-Muñoz et al. (2005)	TCS and other compounds	SE	**SHLE + LLE** (1-g sample; extracted with dichloromethane (120°C, 50 bar, 25 min) and water (200°C, 50 bar, 10 min); then LLE to water extracts with 20 mL hexane; evaporated to dryness and redissolved in 200 μL ethyl acetate)	GC–MS(EI⁺)	0.2	102
Gatidou et al. (2007)	TCS and other compounds	SS	**USA-SLE + SPE** (20-mg sample; extracted with 8 mL methanol/water (50°C, 30 min); diluted to 100 mL with water; then SPE in C$_{18}$ cartridge and eluted with 4 × 2 mL dichloromethane/hexane; evaporated to dryness and redissolved in 50 μL BSTFA + 50 μL pyridine (65°C, 20 min))	GC–MS(EI⁺)	0.15	71–86
Heidler and Halden (2007)	TCS	SS	**PLE** (samples extracted with 1 × 5 min acetone; 100°C, 1500 psi; evaporated to dryness and redissolved in mobile phase)	LC–MS(ESI⁻)	1000	78
Ying and Kookana (2007)	TCS	SS	**USA-SLE + SPE** (0.5-g dried sample; extracted with 3 × 4 mL ethyl acetate (15 min each); evaporated to dryness and redissolved in 1 mL methanol; diluted to 10 mL with water; then SPE in C$_{18}$ cartridge and eluted with ethyl acetate; evaporated to dryness and redissolved in 1 mL acetonitrile)	GC–MS(EI⁺)	1.5	74
Rice and Mitra (2007)	TCS and other compounds	SE, SO	**MA-SLE + GPC** (3-g sample; extracted with 3 × 10 mL (115°C, 800 W, 15 min); evaporated to 1 mL hexane; then GPC with methylene chloride/acetone; evaporated to dryness and redissolved in 1 mL hexane)	GC–MS(EI⁺)	0.10	90

Continued

Table 16.8 Published papers on preservative determination in environmental solid samples (in chronological order)—cont'd

Authors	Target compounds[a]	Sample[b]	Extraction technique[c]	Analytical technique[d]	MLOD (ng/g)[e]	Recovery (%)
Chu and Metcalfe (2007)	TCC, TCS	SS	**PLE + SPE** (0.1- to 0.2-g wet sample; extracted with 3 × 5 min dichloromethane; 60°C, 1500 psi; evaporated to 2 mL; solvent exchanged into hexane and evaporated to 0.2 mL; SPE in PVP-DVB cartridge and eluted with 3 × 3 mL methanol/acetone; evaporated to dryness and redissolved in 200 μL methanol)	LC–MS/MS(ESI[−])	0.2–1.5	90–103
Ligon et al. (2008)	TCS and other compounds	SS	**SLE + LLE** (0.5-g dried sample; extracted with 30 mL NaOH 1M (30 min); adjusted to pH 2; extracted with 20 mL toluene (30 min); evaporated to 1 mL; derivatized with 100 μL trimethylsulphonium hydroxide and 20 μL triethylamine (70°C, 1 h))	GC–MS(EI[+])	1.4	116
Núñez et al. (2008)	BP, BzP, EP, IPP, MP, PP	SE, SO	**USA-SLE** (10-g sample; extracted with 2 and 7 mL acetonitrile (15 min each); evaporated to 1 mL)	LC–MS/MS(ESI[−])	0.06–0.14	83–110
Cha and Cupples (2009)	TCC, TCS	SO	**PLE** (5-g dried sample; extracted with 1 × 5 min acetone; 100°C, 1500 psi; evaporated to dryness and redissolved in 200 μL acetonitrile)	LC–MS/MS(ESI[−])	0.05–0.58	96–87
Nieto et al. (2009)	TCC, TCS, BzP, EP, MP, PP and other compounds	SS	**PLE** (1-g sample; extracted with 2 × 5 min methanol and 2 × 5 min methanol/water (pH 7); 100°C, 140 bar)	LC–MS/MS(ESI[+/−])	1.25–8	72–106

Chen and Bester (2009)	TCS and other compounds	SS	**PLE + SPE + GPC** (40-g sample (+10 g diatomaceous earth); extracted with ethyl acetate; 80°C, 70 mbar; SPE in silica cartridge and eluted with 12 mL ethyl acetate; then GPC with cyclohexane/ethyl acetate; the fraction of interest evaporated to 1 mL)	GC–MS(EI⁺)	30	114
Durán-Alvarez et al. (2009)	TCS and other compounds	SO	**PLE + SPE** (10-g sample (+2 g diatomaceous earth); extracted with 2 × 5 min acetone/hexane/acetic acid; evaporated to 3–4 mL; SPE in PVP-DVB cartridge and eluted with 6 mL acetone/dichloromethane; evaporated to dryness and redissolved in 25 µL pyridine; derivatized with 50 µL BSTFA (60°C, 30 min))	GC–MS(EI⁺)	1	88–128
Wick et al. (2010)	CLP, IPBC, TCC, TCS and other compounds	SS	**PLE** (200-mg dry sample; extracted with 4 × 10 min water/methanol; 80°C; diluted to 800 mL with water; SPE in PVP-DVB cartridge and eluted with 4 × 2 mL methanol/acetone; evaporated to 0.5 mL and diluted to 1 mL with aqueous formic acid)	LC–MS/MS(ESI⁺/⁻)	0.8–3.2	96–108
Núñez et al. (2010)	BP, BzP, EP, IPP, MP, PP	SE, SO	**USA-SLE + SPE** (15-g sample; extracted with 2 × 8 mL acetonitrile (15 min each); SPE in MIP cartridge and eluted with 1 mL acetonitrile, 1 mL methanol and 5 × 1 mL methanol/acetic acid; evaporated to dryness and redissolved in 0.5 mL acetonitrile/water)	LC–UV	0.17–0.33	80–91

Continued

Table 16.8 Published papers on preservative determination in environmental solid samples (in chronological order)—cont'd

Authors	Target compounds[a]	Sample[b]	Extraction technique[c]	Analytical technique[d]	MLOD (ng/g)[e]	Recovery (%)
Sánchez-Brunete et al. (2010)	TCS	SO, SS	**MSPD** (1- to 2-g dry sample; extracted with 2 g C_{18} sorbent; eluted with 13 mL acetonitrile; evaporated to 1 mL; aliquot of 100 μL derivatized with 50 μL MTBSTFA (60°C, 10 min))	GC–MS(EI⁺)	0.05 (SO) 0.08 (SS)	99 (SO) 97–101 (SS)
González-Mariño et al. (2010)	TCS	SE, SS	**MSPD** (0.5-g sample (+1 g diatomaceous earth); extracted with 2 g silica sorbent; eluted with 10 mL dichloromethane; evaporated to dryness and redissolved in 1 mL ethyl acetate; derivatized with 0.1 mL MTBSTFA	GC–MS(EI⁺)	2.1	94
Yu et al. (2011)	TCC, TCS, BP, EP, MP, PP and other compounds	SE, SS	**USA-SLE + SPE** (0.1- to 1-g sample; extracted with 3 × 8 mL acetonitrile/water (15 min each); SPE in PVP-DVB cartridge and eluted with 3 × 2 mL methanol; evaporated to dryness and redissolved in 1 mL acetonitrile)	LC–MS/MS(ESI⁺/⁻)	0.009–0.03 (SE) 0.06–0.21 (SS)	81–108 (SE) 80–126 (SS)
Ferreira et al. (2011)	TCS, BP, BzP, IPP, MP, PP	SE, SO, SS	**SBSE** (0.5-g sample (+5 mL $NaHCO_3$ 0.4 M); extracted with PDMS-coated stir bar; +400 μL acetic anhydride; 60 min; thermal desorption (275°C, 6 min))	(TD) GC–MS(EI⁺)	0.08–1.06	94–110
Chen et al. (2012)	TCC, TCS, BP, EP, MP, PP and other compounds	SE, SO, SS	**USA-SLE + SPE** (0.5- to 2-g sample; extracted with 3 × 10 mL methanol (15 min each); diluted to 300 mL with water; SPE in PVP-DVB cartridge and eluted with 3 × 4 mL ethyl acetate; evaporated to dryness and redissolved in 1 mL methanol)	LC–MS/MS(ESI⁺/⁻)	0.01–0.04 (SE) 0.01–0.02 (SO) 0.01–0.07 (SS)	82–128 (SE) 65–131 (SO) 44–147 (SS)

Albero et al. (2012a)	TCS, EP, MP, PP and other compounds	SO	**PLE** (1-g sample; extracted with 2×10 min ethyl acetate/methanol; 80°C, 120 bar; evaporated to 1 mL)	GC–MS/MS(EI$^+$)	0.1–1.1	77–100
Albero et al. (2012b)	BP, BzP, EP, IPP, MP, PP	SS	**MSPD + SPE** (1-g sample; extracted with 2 g C$_{18}$ sorbent; eluted with 10 mL ethyl acetate/methanol; aliquot of 100 μL derivatized with 50 μL MTBSTFA (60°C, 12 min))	GC–MS/MS(EI$^+$)	0.3–1.7	85–1250
Lozano et al. (2013)	TCC, TCS	SS	**PLE + SPE** (0.3- to 0.5-g sample; extracted 3×10 min water/isopropyl alcohol; 120°C, 2000 psi; SPE in PVP-DVB cartridge and eluted with 3×10 mL dichloromethane/diethyl ether; evaporated to dryness and redissolved in 1.5 mL methanol)	LC–MS(ESI$^-$)	7.9–13.9	89–92
Pintado-Herrera et al. (2013)	TCS and other compounds	SE	**PHWE + SBSE** (2-g dry sample (+18 g siliceous earth), extracted with 3×5 min methanol/water; 100°C, 1500 psi; SBSE with PDMS–coated stir bar; 25°C, 8h; eluted with 200 μL ethyl acetate (30 min) and derivatized with 10 μL MTBSTFA)	GC–MS(EI$^+$)	0.03	62
Liao et al. (2013)	BzP, BP, EP, MP, PP	SE, SS	**SLE + SPE** (100- to 500-mg sample; extracted with 2×5 mL methanol/water (60 min); evaporated to 4 mL and diluted to 10 mL with formic acid in water; SPE in cationic exchanged cartridge and eluted with 5 mL methanol; evaporated to 1 mL)	LC–MS/MS(ESI$^-$)	0.015–0.30	81–119

Continued

Table 16.8 Published papers on preservative determination in environmental solid samples (in chronological order)—cont'd

Authors	Target compounds[a]	Sample[b]	Extraction technique[c]	Analytical technique[d]	MLOD (ng/g)[e]	Recovery (%)
Carmona et al. (2014)	TCC, TCS, BP, EP, MP, PP and other compounds	SE	DSPE (1-g sample (+7.5 mL water and 10 mL acetonitrile); 1 mL of supernatant extracted with 50 mg PSA, 150 mg MgSO$_4$ and 50 mg C$_{18}$ sorbent (2 min); evaporated to dryness and redissolved in 1 mL methanol/water)	LC–MS(ESI$^{+/-}$)	0.3–8.9	57–87
Cerqueira et al. (2014)	TCS, MP, PP and other compounds	SS	SLE + DSPE (10-g sample; extracted with 10 mL acetonitrile (+100 μL acetic acid); 1 min; 2 mL of extract mixed with 300 mg MgSO$_4$ and 125 mg PSA as clean-up step (1 min))	LC–MS(ESI$^-$)	0.15–1.5	73–93
Souchier et al. (2015)	TCC, TCS and other compounds	SE	PLE (0.1-g sample; extracted with 2 × 5 min acetone/methanol; 80°C, 1500 psi; evaporated to dryness and redissolved in 5 mL methanol/water)	LC–MS(ESI$^-$)	0.01–0.12	115–119
Mijangos et al. (2015)	TCS and other compounds	SO	USA-SLE + DSPE (0.5-g dried sample; extracted with 10 mL acetone/hexane (5 min); evaporated to dryness and redissolved in 1.5 mL acetonitrile; DSPE with 75 mg graphitized carbon as sorbent (40 s); evaporated to dryness and redissolved in 250 μL methanol)	LC–MS(ESI$^{+/-}$)	2.6	58–80
Zhang et al. (2015b)	TCS and other compounds	SE	USA-SLE + SPE (5- to 10-g sample + 50 g Na$_2$SO$_4$; extracted with 3 × 50 mL hexane/dichloromethane (15 min each); evaporated to 2 mL; SPE in Florisil column and eluted with 160 mL hexane/dichloromethane; evaporated to dryness and redissolved in isooctane)	GC–MS(EI$^+$)	0.0062	117

DeSousa et al. (2015)	TCS and other compounds	SE	**USA-SLE + SPE** (2-g sample; extracted with 5 mL methanol (10 min); 5 mL methanol/water and 2 mL acetone; evaporated to dryness and redissolved in 1.5 mL methanol; diluted to 250 mL with water; SPE in PVP–DVB cartridge and eluted 2 × 3 mL methanol and 3 mL acetone)	LC–MS/MS(ESI$^{+/-}$)	0.013	85–92
Azzouz and Ballesteros (2016)	TCS, BzP, BP, EP, IPP, MP, PP and other compounds	SE, SO, SS	**MA-SLE + SPE** (0.5- to 2-g sample; extracted with 10 mL methanol; 3 min; evaporated to 100 µL and diluted to 10 mL with water (pH 4); SPE in amino-modified cartridge and eluted with ethyl acetate; evaporated to 25 µL and derivatized with 70 µL BSTFA (350 W, 3 min))	GC–MS(EI$^+$)	0.0005–0.0045	92–102
Camino-Sánchez et al. (2016)	TCC, TCS, BP, EP, MP, PP and other compounds	SO	**USA-SLE** (0.5-g sample; extracted with 15 mL methanol (25 min); evaporated to dryness and redissolved in 500 µL water/methanol)	LC–MS/MS(ESI$^{+/-}$)	0.03–0.20	80–98
Wluka et al. (2016)	TCS and other compounds	SS	**PLE + GPC** (10-g sample; extracted with 30 mL acetone/hexane; 100°C; evaporated to 0.5 mL; GPC with dichloromethane and methanol; evaporated to 50 µL)	GC–MS(EI$^+$)		77
Vakondios et al. (2016)	TCS and other compounds	SS	**USA-SLE + SPE + (DI)SPME** (1-g sample; extracted with 6 mL methanol (10 min); SPE in silica column and eluted with 10 mL methanol; evaporated to 50 µL; diluted to 10 mL with water; SPME in PA-coated fibre; 50°C, 1 h)	GC–MS(EI$^+$)	7.5	

Continued

Table 16.8 Published papers on preservative determination in environmental solid samples (in chronological order)—cont'd

Authors	Target compounds[a]	Sample[b]	Extraction technique[c]	Analytical technique[d]	MLOD (ng/g)[e]	Recovery (%)
Li et al. (2016a)	TCC, TCS, BzP, MP, PP and other compounds	SS	**MSPD** (0.1-g sample; extracted with 0.4 g C$_{18}$ sorbent; eluted with 6 mL methanol and 10 mL acetonitrile/oxalic acid; evaporated to dryness and redissolved in 1 mL acetonitrile/water)	LC–MS/MS(ESI$^{+/-}$)	0.07–1.45	50–107
Martín et al. (2017)	TCC, TCS, EP, MP, PP and other compounds	SE	**SLE + DSPE** (0.5-g sample; extracted with 2 × 5 mL acetonitrile (2 min each); DSPE with 800 mg C$_{18}$ sorbent (2 min); evaporated to dryness and redissolved in 0.25 mL methanol/water)	LC–MS/MS(ESI$^{+/-}$)	0.01	80–103
Chen et al. (2017)	BzP, BP, EP, MP, PP	SS	**USA-SLE + SPE** (0.1-g sample; extracted with 2 × 5 mL methanol/acetone (60 min); evaporated to 0.5 mL and diluted to 5 mL with acetic acid; SPE in cationic exchange cartridge and eluted with 4 mL methanol; evaporated to dryness and redissolved in 1 mL methanol)	LC–MS/MS(ESI$^{+/-}$)	0.28–0.97	78–113

[a]*BP*, butylparaben; *BzP*, benzylparaben; *CLP*, chlorophene; *EP*, ethylparaben; *IPBC*, iodopropynyl butylcarbamate; *IPP*, isopropylparaben; *MP*, methylparaben; *PP*, propylparaben; *TCC*, triclocarban; *TCS*, triclosan.
[b]*SE*, sediment; *SO*, soil; *SS*, sewage sludge.
[c]*BSTFA*, N,O-bis(trimethylsilyl)trifluoroacetamide; *DI*, direct immersion; *DSPE*, dispersive solid-phase extraction; *GPC*, gel-permeation chromatography; *LLE*, liquid–liquid extraction; *MA*, microwave assisted; *MIP*, molecularly imprinted polymer; *MSPD*, matrix solid-phase dispersion; *MTBSTFA*, N-(*tert*-butyldimethylsilyl)-N-methyltrifluoroacetamide; *PA*, polyacrylate; *PDMS*, polydimethylsiloxane; *PHWE*, pressurized hot water extraction; *PLE*, pressurized liquid extraction; *PSA*, primary–secondary amine; *PVP-DVB*, polyvinylpyrrolidone–divinylbenzene co-polymer; *SBSE*, stir bar sorptive extraction; *SHLE*, superheated liquid extraction; *SLE*, solid-liquid extraction; *SPE*, solid-phase extraction; *SPME*, solid-phase microextraction; *USA*, ultrasound-assisted.
[d]*CI*, chemical ionization; *EI*, electron ionization; *ESI*, electrospray ionization; *GC*, gas chromatography; *LC*, liquid chromatography; *MS*, mass spectrometry; *MS/MS*, tandem mass spectrometry; *PTV*, programmed temperature vaporizer inlet; *TD*, thermal desorption; *UV*, ultraviolet detector.
[e]*MLOD*, method limit of detection.

Table 16.9 Published papers on preservative determination in biota samples (in chronological order)

Authors	Target compounds[a]	Sample[b]	Extraction technique[c]	Analytical technique[d]	MLOD (ng/g)[e]	Recovery (%)
Adolfsson-Erici et al. (2002)	TCS	FI (bile)	LLE (200-mg sample + 2 mL water; extracted with 2 × 3 mL hexane/methyl-*tert*-butyl ether; evaporated to dryness and redissolved in 100 μL ethyl acetate)	GC–MS(ESI[+])		
Canosa et al. (2008)	TCS	FI	MSPD (0.5-g sample; extracted with 3 g of silica impregnated with sulphuric acid; eluted with 10 mL dichloromethane; evaporated to dryness and redissolved in 1 mL ethyl acetate)	GC–MS/MS(EI[+])		79–112
Ramírez et al. (2009)	TCS and other compounds	FI	USA-SLE (0.5- to 1-g sample; extracted with 8 mL methanol/water (15 min); evaporated to dryness and redissolved in 0.5–1 mL mobile phase)	LC–MS/MS(ESI[−])	38	
Mottaleb et al. (2009)	TCS and other compounds	FI	USA-LLE + SPE (1-g blended sample; extracted with 10 mL acetone (25–35°C, 5–15 min); evaporated to dryness and redissolved in 200 μL hexane/acetone; SPE with silica gel column and hexane/acetone; evaporated to 50 μL and, after derivatization step with 100 μL MSTFA, evaporated to 20 μL and diluted with 180 μL hexane)	GC–MS(EI[+])	5.5	98
			USA-LLE + SPE + GPC (1-g blended sample; extracted with 10 mL acetone (25–35°C, 5–15 min); evaporated to dryness and redissolved in 200 μL hexane/acetone; SPE with silica gel column and hexane/acetone; evaporated to dryness, redissolved in 700 μL dichloromethane; GPC; fraction of interest evaporated to 50 μL and, after derivatization step with 100 μL MSTFA, evaporated to 20 μL and diluted with 180 μL hexane)	GC–MS/MS(EI[+])	38	93

Continued

Table 16.9 Published papers on preservative determination in biota samples (in chronological order)—cont'd

Authors	Target compounds[a]	Sample[b]	Extraction technique[c]	Analytical technique[d]	MLOD (ng/g)[e]	Recovery (%)
Fair et al. (2009)	TCS	MM (plasma)	**LLE** (2- to 4-g sample; extracted with methyl-*tert*-butyl ether/hexane; derivatized with diazomethane; evaporated to 100 μL)	GC–(EI+)	0.033	51
Kim et al. (2011) Ramaswamy et al. (2011)	TCC, TCS, BP, EP, MP, PP and other compounds	FI	**SLE + SPE** (5-g dried sample; extracted with hexane/acetone; 30°C, 30 min; evaporated to dryness and redissolved in 10 mL hexane; SPE in silica column and eluted with 100 mL dichloromethane; evaporated to dryness and redissolved in 1 mL methanol)	LC–MS/MS(ESI+/−)	0.001–0.015	79–89
Subedi et al. (2011)	TCS and other compounds	FI	**PLE + GPC** (2.5-g sample (+Na₂SO₄); extracted with 2 × 5 min dichloromethane/ethyl acetate; 80°C, 1500 psi; evaporated to 0.5 mL and redissolved in 700 μL dichloromethane; GPC, the fraction of interest evaporated to 200 μL, solvent exchanged to hexane and, after derivatization step with 100 μL MSTFA (70°C, 1 h), evaporated to 200 μL hexane)	GC–MS/MS(EI+)	3.4	66
Zarate et al. (2012)	TCC, TCS	PL	**VA-SLE** (500-mg sample (+7.5 mL water); extracted with 30 mL hexane/ethyl acetate (2 min); evaporated to dryness and redissolved in 100 μL acetonitrile; derivatized with 50 μL MSTFA (60°C, 2 h); evaporated to dryness and redissolved in 70 μL dichloromethane and 10 μL MSTFA)	LC–MS(ESI−)	6–11	103–142

Rüdel et al. (2013)	TCS	FI	**PLE + GPC** (1- to 2.5-g sample; extracted with cyclohexane; 100°C, 14 MPa; GPC with dichloromethane/cyclohexane; SPE in silica column and eluted with hexane/acetone; evaporated to dryness and redissolved in 2 mL acetone; derivatized with 10 µL pentafluorobenzylbromide (60°C, 1 h))	GC–MS(CI⁻)	0.20	101
Jakimska et al. (2013)	TCS, BzP, EP, MP, PP	FI	**SLE + DSPE** (0.5-g dried sample; extracted with acetonitrile; DSPE with 900 mg MgSO₄, 150 mg PSA and 150 mg C₁₈ sorbent (1 min); evaporated to dryness and redissolved in 0.5 mL methanol/water)	LC–MS/MS(ESI⁺/⁻)	0.002–0.3	32–113
Escarrone et al. (2014)	TCS	FI	**VA-MSPD** (0.3-g sample; extracted with 0.5 g C₁₈ sorbent; eluted with 5 mL acetonitrile)	LC–MS/MS(ESI⁻)	16	79–108
Miao et al. (2014)	TCC	PL	**VA-SLE + SPE** (2-g sample; extracted with 20 mL acetone/dichloromethane (2 min); evaporated to dryness and redissolved in 2 mL hexane; SPE in silica cartridge and eluted with hexane/dichloromethane; evaporated to dryness and redissolved in 1 mL methanol)	LC–MS/MS(ESI⁻)	0.05	91–106
Emmet et al. (2015)	TCS, BP, EP, MP, PP and other compounds	CL, FI	**PLE + SPE + GPC** (8-g blended and dried sample; extracted with 2 × 10 min water/isopropanol, 120–180°C, 100 bar; filtered and diluted with 50 mL phosphate buffer; SPE in PVP-DVB cartridge and eluted with 3 × 5 mL dichloromethane/methanol; evaporated to dryness and redissolved in 1.5 mL dichloromethane/methanol; GPC with the same solvent; the fraction of interest evaporated to dryness and redissolved in 1 mL dichloromethane/methanol)	GC–MS(EI⁺)	0.3–0.6	

Continued

Table 16.9 Published papers on preservative determination in biota samples (in chronological order)—cont'd

Authors	Target compounds[a]	Sample[b]	Extraction technique[c]	Analytical technique[d]	MLOD (ng/g)[e]	Recovery (%)
Zhang et al. (2015b)	TCS and other compounds	FI	**USA-SLE + GPC + SPE** (5- to 10-g sample + 50 g Na$_2$SO$_4$; extracted with 3 × 50 mL hexane/dichloromethane (15 min each); evaporated to 2 mL; GPC with the same solvent; the fraction of interest to SPE in Florisil column with 160 mL hexane/dichloromethane; evaporated to dryness and redissolved in isooctane)	GC–MS/MS(EI$^+$)	0.0023	115–117
Villaverde-de-Sáa et al. (2016)	BzP, BP, EP, IPP, MP, PP	MU	**MSPD** (0.5-g dried sample; extracted with 3 g C$_{18}$ sorbent; eluted with 10 mL acetonitrile; evaporated to dryness and redissolved in 100 μL methanol)	LC–MS/MS(ESI$^-$)	0.04–0.43	71–117
Aznar et al. (2017)	TCS, MP, PP and other compounds	PL	**USA-MSPD + SPE** (1-g sample; extracted with 4 g Florisil (15 min); eluted with 8 mL ethyl acetate and 5 mL acetonitrile; evaporated to 1 mL; SPE in C$_{18}$ column and eluted with 5 mL acetonitrile; evaporated to dryness and redissolved in 0.5 mL acetonitrile)	GC–MS(EI$^+$)	0.3–1.1	70–99

[a]*BP*, butylparaben; *BzP*, benzylparaben; *EP*, ethylparaben; *IPP*, isopropylparaben; *MP*, methylparaben; *PP*, propylparaben; *TCC*, triclocarban; *TCS*, triclosan.
[b]*CL*, clams; *FI*, fish; *MM*, marine mammal; *MU*, mussels; *PL*, plants.
[c]*DSPE*, dispersive solid-phase extraction; *GPC*, gel-permeation chromatography; *LLE*, liquid–liquid extraction; *MSPD*, matrix solid-phase dispersion; *MSTFA*, N-methyl-N-(trimethylsilyl)trifluoroacetamide; *PLE*, pressurized liquid extraction; *PSA*, primary–secondary amine; *PVP-DVB*, polyvinylpyrrolidone–divinylbenzene co-polymer; *SLE*, solid–liquid extraction; *SPE*, solid-phase extraction; *USA*, ultrasound-assisted; *VA*, vortex-assisted.
[d]*CI*, chemical ionization; *EI*, electron ionization; *ESI*, electrospray ionization; *GC*, gas chromatography; *LC*, liquid chromatography; *MS*, mass spectrometry; *MS/MS*, tandem mass spectrometry.
[e]*MLOD*, method limit of detection.

Sample Preparation
Water Samples
Sample preparation for the determination of preservatives in water samples is primarily accomplished through extraction on a solid sorbent material. These include SPE (using C$_{18}$, PVP-DVB, PS-DVB, PVP-DVB/MCX, molecularly imprinted polymer, amongst other sorbents), SPME (mainly in PA-coated fibres, although DVB-CAR-PDMS or polymeric ILs like poly(VBHDIm$^+$NTf$_2^-$) have been also used as coating materials), SBSE (with PDMS- or EGS-coated stir bars) and BAµE (using activated carbon and cork sorbents). Methods that are based on the dispersion of the sorbent material in the aqueous phase, such as DSPE, usually using MNPs coated with different sorbents, such as Fe$_3$O$_4$@polyaniline, Fe$_3$O$_4$@aminopropyl, Fe$_3$O$_4$@SiO$_2$@C$_{18}$ and Fe$_3$O$_4$@SiO$_2$@MIM-PF$_6$ and, less frequently, other materials like the HKUST-1 metal organic framework, have also attracted attention. A great deal of effort has been devoted to the development of liquid-phase extraction methods. DLLME and its various modifications (such as with solidification of the organic floating drop (DLLME-SFO), IL-DLLME, SD-DLLME) and USAEME are the most frequently reported methods for the liquid-phase microextraction of preservatives from water samples. Combined liquid- and solid-phase-based extraction methods such as SPE + DLLME, SPE + DLLME-SFO and DLLME + DSPE have also been developed, enabling enhanced sample clean-up and the handling of large sample volumes, thus offering improved detection limits. Detailed descriptions of the methods developed for the determination of preservatives in water samples are given in Table 16.7.

Solid Samples
The extraction of preservatives from solid samples, such as soils, sediments or sewage sludge, is performed through the classic two-step extraction process. First, the analytes are released from the solid matrix through an appropriate extraction technique such as PLE, LLE, SLE or Soxhlet extraction and then the extracts are purified with a solid sorbent using SPE, GPC or DSPE. Direct extraction of preservatives from solid samples with MSPD has also been successfully accomplished. A detailed list of the methods for the extraction of preservatives from solid samples are given in Table 16.8.

Living Organisms
Monitoring of preservative accumulation in living organisms is accomplished through a one- or two-step process (Table 16.9). One-step methods involve the use of LLE or MSPD, whilst two-step methods involve the combination of an extraction step (usually PLE or SLE) and a clean-up step employing GPC, SPE or a combination of them.

Analytical Techniques
The overwhelming majority of methods developed for the determination of preservatives employ mass selective detectors and more specifically GC–MS/MS in EI mode or

LC–MS/MS in ESI mode. LC–UV has also been proven as a useful tool in the monitoring of preservatives in various environmental samples, whilst other detectors such as GC–ECD, GC with flame ionization detector and GC with photoionization detector have been reported less frequently. In several studies reporting on the use of GC a derivatization step precedes analysis. Silylating agents and acetic anhydride are usually employed to increase the volatility of the target compounds, whilst other derivatization agents (for example, trimethylsulphonium hydroxide and triethylamine, diazomethane and pentafluorobenzylbromide) have been successfully utilized, expanding the available options depending on the needs of the analysis, especially in complex sample matrices and multi-residue methods.

INSECT REPELLENTS

Insect repellents are cosmetic products that are applied on the skin surface to deter insects, especially mosquitoes, and minimize the risk of a sting. These cosmetic products contain active ingredients that interfere with the sensory perception of insects, thus avoiding or minimizing their attraction by the host. Diethyltoluamide (DEET) is the most common active ingredient in this type of product, whilst hydroxyethyl isobutyl piperidine carboxylate, also known as icaridin, piperonyl butoxide and permethrin are occasionally used.

These compounds can cause harmful effects to the aquatic environment due to their capacity to interact with enzymes, mainly cholinesterase, involved in the nervous system of vertebrates and insects (Costanzo et al., 2007; Roy et al., 2017). In this regard, different analytical methods have been developed with the aim of monitoring insect repellents in water samples (Table 16.10), solid samples (Table 16.11) and living organisms (Table 16.12).

Sample Preparation
Water Samples
The extraction of insect repellents from water samples is usually accomplished by means of solid-phase extraction methods. SPE (using PVP-DVB, C_{18} and PS-DVB sorbents) is once more the most frequently applied method for sample preparation, followed by SBSE with PDMS-coated stir bars. Surprisingly, LLE is the most popular liquid-phase extraction method for the pre-concentration and isolation of insect repellents from water samples, whilst liquid-phase-based microextraction methods such as USAEME and DLLME are seldom used. Detailed descriptions of the methods developed for the determination of insect repellents in water samples are given in Table 16.10.

Solid Samples
The analytical methods for the determination of insect repellents in solid samples are still at an early stage of development. PLE and SLE have been successfully used for the extraction

Table 16.10 Published papers on insect repellent determination in environmental water samples (in chronological order)

Authors	Target compounds[a]	Sample[b]	Extraction technique[c]	Analytical technique[d]	MLOD (ng/L)[e]	Recovery (%)
Kolpin et al. (2002)	DEET and other compounds	IW	**LLE** (1-L sample; extracted with dichloromethane; evaporated to 1 mL)	GC–MS(EI⁺)		81
Weigel et al. (2004)	DEET and other compounds	LK, RV	**SPE** (1-L sample (pH 7); extracted with PVP-DVB cartridge at 15 mL/min; eluted with 5 mL hexane, 5 mL ethyl acetate and 14 mL methanol; evaporated to 50 µL)	LC–UV	0.05	82
Knepper (2004)	ICA	IW, RV	**SPE** (0.5- to 1-L sample; extracted with PS-DVB/C₁₈ cartridge; eluted with 3 × 1.5 mL acetone/ethyl acetate; evaporated to 100 µL)	GC–MS(EI⁺)	50 (IW) 10 (RV)	98 (RV)
Standler et al. (2004)	ICA	RV, SP	**SPE** (500-mL sample; extracted with C₈ cartridge; eluted with 2 mL ethyl acetate; evaporated to 0.3 mL) **SBSE** (250-mL sample; extracted with PDMS-coated stir bar; 14 h; thermal desorption (300°C, 10 min))	(TD) GC–MS(EI⁺)	25 25	105
Sandstrom et al. (2005)	DEET	RV	**LLE** (1-L sample; extracted with methylene chloride; evaporated to 1 mL)	GC–MS(EI⁺)	20	74
Schwarzbauer and Heim (2005)	DEET and other compounds	RV	**LLE** (1-L sample (pH 2); extracted with 50 mL pentane and dichloromethane; evaporated to 25 µL)	GC–MS(EI⁺)		
Rodil and Moeder (2008c)	DEET, ICA, PBO, PER and other compounds	EW, IW, LK, RV	**SBSE** (20-mL sample (20% NaCl); extracted with PDMS-coated stir bar; 180 min; thermal desorption (250°C, 5 min))	(TD) GC–MS(EI⁺)	0.5–150	12–99 (EW) 3–94 (IW) 82–102 (RV, LK)

Continued

Table 16.10 Published papers on insect repellent determination in environmental water samples (in chronological order)—cont'd

Authors	Target compounds[a]	Sample[b]	Extraction technique[c]	Analytical technique[d]	MLOD (ng/L)[e]	Recovery (%)
Rodil et al. (2009a,b,c,d)	DEET, ICA, PBO and other compounds	EW, IW, SW, TW	**SPE** (200- to 500-mL sample (pH 7); extracted with PVP-DVB cartridge; eluted with 3 × 10 mL methanol, evaporated to 0.2 mL and diluted to 1 mL with methanol/water)	LC–MS/MS(ESI$^{+/-}$)	0.6–3.7	64–107 (EW) 72–109 (IW) 64–124 (SW) 72–117 (TW)
Quednow and Püttmann (2009)	DEET and other compounds	RV	**SPE** (2.5-L sample; extracted with C$_{18}$ cartridge; eluted with 1 mL methanol/acetonitrile; evaporated to dryness and redissolved in 100 µL acetonitrile)	GC–MS(EI$^+$)	8	
Sui et al. (2010)	DEET and other compounds	EW, IW	**SPE** (500-mL sample (pH 7); extracted with PVP-DVB cartridge at 5–10 mL/min; eluted with 5 mL methanol; evaporated to 0.4 mL)	LC–MS/MS(ESI$^-$)	0.4 (EW) 1.0 (IW)	99 (EW) 118 (IW)
Loos et al. (2010)	DEET and other compounds	GW	**SPE** (1-L sample; extracted with PVP-DVB cartridge at 5 mL/min; eluted with 6 mL methanol; evaporated to 500 µL)	LC–MS/MS(ESI$^{+/-}$)	0.4	
Chen et al. (2012)	DEET, ICA and other compounds	EW, IW, TW	**SPE** (1-L sample (pH 3); extracted with PVP-DVB cartridge at 5–10 mL/min; eluted with 3 × 4 mL ethyl acetate; evaporated to dryness and redissolved in 1 mL methanol)	LC–MS/MS(ESI$^{+/-}$)	0.01 (EW, TW) 0.02 (IW)	74–106 (EW) 65–108 (IW) 80–137 (TW)
Pintado-Herrera et al. (2013)	DEET and other compounds	EW, IW, SW	**SBSE** (100-mL sample (pH 2, 10% methanol); extracted with PDMS-coated stir bar; 25°C, 8 h; eluted with 200 µL ethyl acetate (30 min) and derivatized with 10 µL MTBSTFA)	GC–MS(EI$^+$)	74	12

Almeida et al. (2014)	DEET, PER	GW, RV, SP, SW, TW	**BAµE** (25 mL sample (pH 2), extracted with activated carbon sorbent; 16 h; eluted with 200 µL acetonitrile (15 min))	(LVI) GC–MS(EI+)	8–20	74–96
Celano et al. (2014)	DEET and other compounds	EW, IW, RV, SW, TW	**SPE + DLLME** (250- to 500-mL sample (pH 2); extracted with PVP-DVB cartridge at 5 mL/min; eluted with 5 mL methanol/acetonitrile/dichloromethane; then 2 mL injected into 10 mL water (5% NaCl); evaporated to dryness and redissolved in 200 µL acetonitrile/water)	LC–MS/MS(ESI−)	0.2 (EW, IW) 0.1 (RV, SW, TW)	84 (EW, IW) 91 (RV) 85 (SW) 74 (TW)
Šatínský et al. (2015)	PER and other compounds	LK, RV	**(Online)SPE** (1.5-L sample; extracted with C$_{18}$ column; eluted with acetonitrile/water)	LC–UV		
Kapelewska et al. (2016)	DEET and other compounds	GW	**USAEME** (5-mL sample (pH 2); extracted with 70 µL chloroform and derivatized with 50 µL acetic anhydride; 5 min)	GC–MS(EI+)	1	108–114

[a]*DEET*, N,N-diethyl-m-toluamide; *ICA*, icaridin; *PBO*, piperonyl butoxide; *PER*, permethrin.
[b]*EW*, effluent wastewater; *GW*, groundwater; *IW*, influent wastewater; *LK*, lake water; *RV*, river water; *SP*, swimming-pool water; *SW*, seawater; *TW*, tap water.
[c]*BAµE*, bar adsorptive microextraction; *DLLME*, dispersive liquid–liquid microextraction; *LLE*, liquid–liquid extraction; *MTBSTFA*, N-(tert-butyldimethylsilyl)-N-methyltrifluoroacetamide; *PDMS*, polydimethylsiloxane; *PS-DVB*, polystyrene divinylbenzene co-polymer; *PVP-DVB*, polyvinylpyrrolidone–divinylbenzene co-polymer; *SBSE*, stir bar sorptive extraction; *SPE*, solid-phase extraction; *USAEME*, ultrasound-assisted emulsification microextraction.
[d]*EI*, electron ionization; *ESI*, electrospray ionization; *GC*, gas chromatography; *LC*, liquid chromatography; *LVI*, large-volume injection; *MS*, mass spectrometry; *MS/MS*, tandem mass spectrometry; *TD*, thermal desorption; *UV*, ultraviolet spectrometry.
[e]*MLOD*, method limit of detection.

Table 16.11 Published papers on insect repellent determination in environmental solid samples (in chronological order)

Authors	Target compounds[a]	Sample[b]	Extraction technique[c]	Analytical technique[d]	MLOD (ng/g)[e]	Recovery (%)
Chen et al. (2012)	DEET, ICA and other compounds	SE, SO, SS	**USA-SLE + SPE** (0.5- to 2-g sample; extracted with 3 × 10 mL methanol (15 min each); diluted to 300 mL with water; SPE in PVP-DVB cartridge and eluted with 3 × 4 mL ethyl acetate; evaporated to dryness and redissolved in 1 mL methanol)	LC–MS/MS(ESI+/−)	0.01 (SE, SO) 0.01–0.03 (SS)	92–133 (SE) 115–137 (SO) 143–193 (SS)
Prestes et al. (2012)	PBO and other compounds	SO	**DSPE** (5-g sample + 2.5 mL water; extracted with a salt packet (4 g MgSO4, 4 g NaCl, 0.5 g sodium citrate dibasic, 1 g sodium citrate dihydrate); 5 min; eluted with 5 mL acetonitrile)	LC–MS/MS(ESI+)	1	91–101
Pintado-Herrera et al. (2013)	DEET and other compounds	SE	**PHWE + SBSE** (2-g dry sample (+18 g siliceous earth), extracted with 3 × 5 min methanol/water; 100°C, 1500 psi; SBSE with PDMS-coated stir bar; 25°C, 8 h; eluted with 200 µL ethyl acetate (30 min) and derivatized with 10 µL MTBSTFA)	GC–MS/MS(EI+)	2	5
Chen et al. (2013)	DEET and other compounds	SS	**PLE + SPE** (0.5-g sample; extracted with 3 × 5 min methanol; 100°C, 10 MPa; SPE in PVP-DVB cartridge and eluted with 2 mL methanol; evaporated to less than 1 mL)	LC–MS/MS(ESI+/−)	0.4	93–98
Pintado-Herrera et al. (2016)	DEET and other compounds	SE	**PLE** (2-g sample + 1 g activated alumina +1.5 g diatomaceous earth; extracted with 3 × 5 min dichloromethane; 100°C, 1500 psi; evaporated to dryness and redissolved in 500 µL ethyl acetate + 10 µL MTBSTFA (as derivatizing agent))	GC–MS/MS(EI+)	0.082	62–86

[a]DEET, N,N-diethyl-m-toluamide; ICA, icaridin; PBO, piperonyl butoxide.
[b]SE, sediment; SO, soil; SS, sewage sludge.
[c]DSPE, dispersive solid-phase extraction; MTBSTFA, N-(tert-butyldimethylsilyl)–N-methyltrifluoroacetamide; PDMS, polydimethylsiloxane; PHWE, pressurized hot water extraction; PLE, pressurized liquid extraction; PVP-DVB, polyvinylpyrrolidone-divinylbenzene co-polymer; SBSE, stir bar sorptive extraction; SLE, solid-liquid extraction; SPE, solid-phase extraction; USA, ultrasound-assisted.
[d]EI, electron ionization; ESI, electrospray ionization; GC, gas chromatography; LC, liquid chromatography; MS, mass spectrometry; MS/MS, tandem mass spectrometry.
[e]MLOD, method limit of detection.

Table 16.12 Published papers on insect repellent determination in biota samples (in chronological order)

Authors	Target compounds[a]	Sample[b]	Extraction technique[c]	Analytical technique[d]	MLOD (ng/g)[e]	Recovery (%)
Mottaleb et al. (2009)	DEET and other compounds	FI	**USA-LLE + SPE** (1-g blended sample; extracted with 10 mL acetone (25–35°C, 5–15 min); evaporated to dryness and redissolved in 200 μL hexane/acetone; SPE with silica gel column and hexane/acetone; evaporated to 50 μL and, after derivatization step with 100 μL MSTFA, evaporated to 20 μL and diluted with 180 μL hexane)	GC–MS(EI+)	3.5	110
			USA-LLE + SPE + GPC (1-g blended sample; extracted with 10 mL acetone (25–35°C, 5–15 min); evaporated to dryness and redissolved in 200 μL hexane/acetone; SPE with silica gel column and hexane/acetone; evaporated to dryness, redissolved in 700 μL dichloromethane; GPC; fraction of interest evaporated to 50 μL and, after derivatization step with 100 μL MSTFA, evaporated to 20 μL and diluted with 180 μL hexane)	GC–MS/MS(EI+)	5.1	102
Tamoue et al. (2014)	DEET and other compounds	BI, FI	**USA-SLE + LLE + SPE** (0.5-g sample; extracted with 5 mL acetonitrile; evaporated to 5 mL and diluted to 50 mL with 5% NaCl; extracted with methyl-*tert*-butyl ether; SPE with PVP-DVB cartridge and eluted with methanol/methyl-*tert*-butyl ether)	LC–MS/MS(ESI+)	3.2 (BI) 1.6 (FI)	92–117 (BI) 88–107 (FI)

[a] *DEET*, *N,N*-diethyl-*m*-toluamide.
[b] *BI*, birds; *FI*, fish.
[c] *GPC*, gel-permeation chromatography; *LLE*, liquid–liquid extraction; *MSTFA*, *N*-methyl-*N*-(trimethylsilyl)trifluoroacetamide; *PVP-DVB*, polyvinylpyrrolidone-divinylbenzene co-polymer; *SLE*, solid–liquid extraction; *SPE*, solid-phase extraction; *USA*, ultrasound-assisted.
[d] *EI*, electron ionization; *ESI*, electrospray ionization; *GC*, gas chromatography; *LC*, liquid chromatography; *MS*, mass spectrometry; *MS/MS*, tandem mass spectrometry.
[e] *MLOD*, method limit of detection.

of these compounds from solid samples, whilst solid-phase extraction methods such as DSPE and SBSE have also been reported. Detailed descriptions of the methods developed for the determination of insect repellents in solid samples are given in Table 16.11.

Living Organisms

With regard to the determination of insect repellents in living organisms, only two experimental procedures have been published as of this writing (see Table 16.12). These methods share many things in common with those used for the extraction of other cosmetic ingredients from living organisms and rely on the combination of LLE or SLE with SPE and/or GPC clean-up of the extracts.

Analytical Techniques

The determination of insect repellents is almost exclusively performed with mass spectrometric detectors such as GC–MS in EI$^+$ and LC–MS/MS in ESI$^-$ mode. In most studies the GC separation and detection of these compounds are accomplished without derivatization reactions. However, the successful derivatization with MTBSTFA and acetic anhydride has been reported. Last, but not least, a limited number of studies have shown the possibility of using LC–UV detectors combined with SPE extraction to obtain low detection limits.

REFERENCES

Abbasghorbani, M., Attaran, A., Payehghadr, M., 2013. J. Sep. Sci. 36, 311.
Adolfsson-Erici, M., Petterson, M., Parkkonen, J., Sturve, J., 2002. Chemosphere 46, 1485.
Agüera, A., Fernández-Alba, A.R., Piedra, L., Mézcua, M., Gómez, M.J., 2003. Anal. Chim. Acta 480, 193.
Aguirre, J., Bizkarguenaga, E., Iparraguirre, A., Fernández, L.A., Zuloaga, O., Prieto, A., 2014. Anal. Chim. Acta 812, 74.
Albero, B., Sánchez-Brunete, C., Miguel, E., Perez, R.A., Tadeo, J.L., 2012a. J. Chromatogr. A 1248, 9.
Albero, B., Pérez, R.A., Sánchez-Brunete, C., Tadeo, J.L., 2012b. J. Hazard. Mater. 239–240, 48.
Alcudia-León, M.C., Lucena, R., Cárdenas, S., Valcárcel, M., 2013. Microchem. J. 110, 643.
Almeida, C., Nogueira, J.M.F., 2014. J. Chromatogr. A 1348, 17.
Almeida, C., Stepkowska, A., Alegre, A., Nogueira, J.M.F., 2013. J. Chromatogr. A 1311, 1.
Almeida, C., Strzelczyk, R., Nogueira, J.M.F., 2014. Talanta 120, 126.
Amine, H., Gomez, E., Halwani, J., Casellas, C., Fenet, H., 2012. Mar. Pollut. Bull. 64, 2435.
Aparicio, I., Martín, J., Santos, J.L., Malvar, J.L., Alonso, E., 2017. J. Chromatogr. A 1500, 43.
Arbulu, M., Sampedro, M.C., Unceta, N., Gómez-Caballero, A., Goicolea, M.A., Barrio, R.J., 2011. J. Chromatogr. A 1218, 3048.
Aznar, R., Albero, B., Sánchez-Brunete, C., Miguel, E., Martín-Girela, I., Tadeo, J.L., 2017. Environ. Sci. Pollut. Res. 24, 7911.
Azzouz, A., Ballesteros, E., 2014. J. Chromatogr. A 1360, 248.
Azzouz, A., Ballesteros, E., 2016. Anal. Bioanal. Chem. 408, 231.
Bachelot, M., Li, Z., Munaron, D., Le Gall, P., Casellas, C., Fenet, H., Gomez, E., 2012. Sci. Total Environ. 420, 273.
Balmer, M.E., Buser, H.R., Müller, M.D., Poiger, T., 2005. Environ. Sci. Technol. 39, 953.
Barón, E., Gago-Ferrero, P., Gorga, M., Rudolph, I., Mendoza, G., Zapata, A.M., Díaz-Cruz, S., Barra, R., Ocampo-Duque, W., Páez, M., Darbra, R.M., Eljarrat, E., Barceló, D., 2013. Chemosphere 92, 309.
Basaglia, G., Pietrogrande, M.C., 2012. Chromatographia 75, 361.
Benedé, J.L., Chisvert, A., Salvador, A., Sánchez-Quiles, D., Tovar-Sánchez, A., 2014a. Anal. Chim. Acta 812, 50.

Benedé, J.L., Chisvert, A., Giokas, D.L., Salvador, A., 2014b. J. Chromatogr. A 1362, 25.
Benedé, J.L., Giokas, D.L., Chisvert, A., Salvador, A., 2015. J. Chromatogr. A 1408, 63.
Benedé, J.L., Chisvert, A., Giokas, D.L., Salvador, A., 2016a. Talanta 147, 246.
Benedé, J.L., Chisvert, A., Giokas, D.L., Salvador, A., 2016b. Anal. Chim. Acta 926, 63.
Bester, K., 2003. Water Res. 37, 3891.
Blanco, E., Casais, M.C., Mejuto, M.C., Cela, R., 2008. Electrophoresis 29, 3229.
Blanco, E., Casais, M.C., Mejuto, M.C., Cela, R., 2009. Anal. Chim. Acta 647, 104.
Bratkovics, S., Sapozhnikova, Y., 2011. Anal. Methods 3, 2943.
Brausch, J.M., Rand, G.M., 2011. Chemosphere 82, 1518.
Bu, Y., Feng, J., Wang, X., Tian, Y., Sun, M., Luo, C., 2017. J. Chromatogr. A 1483, 48.
Burkhardt, M.R., ReVello, R.C., Smith, S.G., Zaugg, S.D., 2005. Anal. Chim. Acta 534, 89.
Buser, H.-R., Balmer, M.E., Schmid, P., Kohler, M., 2006. Environ. Sci. Technol. 40, 1427.
Caballero-Díaz, E., Simonet, B.M., Valcárcel, M., 2013. Anal. Bioanal. Chem. 405, 7251.
Çabuk, H., Akyüz, M., Ata, Ş., 2012. J. Sep. Sci. 35, 2645.
Caldas, S.S., Bolzan, C.M., Guilherme, J.R., Silveira, M.A.K., Escarrone, A.L.V., Primel, E.G., 2013. Environ. Sci. Pollut. Res. 20, 5855.
Caldas, S.S., Rombaldi, C., Arias, J.L.O., Marube, L.C., Primel, E.G., 2016. Talanta 146, 676.
Camino-Sánchez, F.J., Zafra-Gómez, A., Dorival-García, N., Juárez-Jiménez, B., Vílchez, J.L., 2016. Talanta 150, 415.
Canosa, P., Rodríguez, I., Rubí, E., Cela, R., 2005. J. Chromatogr. A 1072, 107.
Canosa, P., Rodríguez, I., Rubí, E., Bollaín, M.H., Cela, R., 2006. J. Chromatogr. A 1124, 3.
Canosa, P., Rodríguez, I., Rubí, E., Ramil, M., Cela, R., 2008. J. Chromatogr. A 1188, 132.
Capriotti, A.L., Cavaliere, C., Piovesana, S., Samperi, R., Stampachiacchiere, S., Ventura, S., Laganà, A., 2014. J. Sep. Sci. 37, 2882.
Carbajo, J.B., Perdigón-Melón, J.A., Petre, A.L., Rosal, R., Letón, P., Garcia-Calvo, E., 2015. Water Res. 72, 174.
Carballa, M., Fink, G., Omil, F., Lema, J.M., Ternes, T., 2008. Water Res. 42, 287.
Carmona, E., Andreu, V., Picó, Y., 2014. Sci. Total Environ. 484, 53.
Casado-Carmona, F.A., Alcudia-León, M.C., Lucena, R., Cárdenas, S., Valcárcel, M., 2016. Microchem. J. 128, 347.
Casas-Ferreira, A.M., Möder, M., Fernández-Laespada, M.E., 2011. Anal. Bioanal. Chem. 399, 945.
Cavalheiro, J., Prieto, A., Monperrus, M., Etxebarria, N., Zuloaga, O., 2013. Anal. Chim. Acta 773, 68.
Cavalheiro, J., Prieto, A., Zuloaga, O., Preudhomme, H., Amouroux, D., Monperrus, M., 2015. J. Sep. Sci. 38, 2298.
Celano, R., Piccinelli, A.L., Campone, L., Rastrelli, L., 2014. J. Chromatogr. A 1355, 26.
Cerqueira, M.B.R., Guilherme, J.R., Caldas, S.S., Martins, M.L., Zanella, R., Primel, E.G., 2014. Chemosphere 107, 74.
Cha, J., Cupples, A.M., 2009. Water Res. 43, 2522.
Chamorro, S., Hernández, V., Matamoros, V., Domínguez, C., Becerra, J., Vidal, G., Piña, B., Bayona, J.M., 2013. Chemosphere 90, 611.
Chase, D.A., Karnjanapiboonwong, A., Fang, Y., Cobb, G.P., Morse, A.N., Anderson, T.A., 2012. Sci. Total Environ. 416, 253.
Chen, X., Bester, K., 2009. Anal. Bioanal. Chem. 395, 1877.
Chen, Z.-F., Ying, G.-G., Lai, H.-J., Chen, F., Su, H.-C., Liu, Y.-S., Peng, F.-Q., Zhao, J.-L., 2012. Anal. Bioanal. Chem. 404, 3175.
Chen, Y., Cao, Q., Deng, S., Huang, J., Wang, B., Yu, G., 2013. Int. J. Environ. Anal. Chem. 93, 1159.
Chen, W., Huang, H., Chen, C.-E., Qi, S., Price, O.R., Zhang, H., Jones, K.C., Sweetman, A.J., 2016. Chemosphere 163, 99.
Chen, J., Pycke, B.F.G., Brownawell, B.J., Kinney, C.A., Furlong, E.T., Kolpin, D.W., Halden, R.U., 2017. Sci. Total Environ. 593–594, 368.
Cheng, C.-Y., Wang, Y.-C., Ding, W.-H., 2011. Anal. Sci. 27, 197.
Chu, S., Metcalfe, C.D., 2007. J. Chromatogr. A 1164, 212.
Chung, W.-H., Tzing, S.-H., Ding, W.-H., 2013. J. Chromatogr. A 1307, 34.
Chung, W.-H., Tzing, S.-H., Huang, M.-C., Ding, W.-H., 2014. J. Chin. Chem. Soc. 61, 1031.

Chung, W.-H., Tzing, S.-H., Ding, W.-H., 2015. J. Chromatogr. A 1411, 17.
Clara, M., Gans, O., Windhofer, G., Krenn, U., Hartl, W., Braun, K., Scharf, S., Scheffknecht, C., 2011. Chemosphere 82, 1116.
Clavijo, S., Avivar, J., Suárez, R., Cerdà, V., 2016. J. Chromatogr. A 1443, 26.
Costanzo, S.D., Watkinson, A.J., Murby, E.J., Kolpin, D.W., Sandstrom, M.W., 2007. Sci. Total Environ. 384, 214.
Cuderman, P., Heath, E., 2007. Anal. Bioanal. Chem. 387, 1343.
Cunha, S.C., Pena, A., Fernandes, J.O., 2015. J. Chromatogr. A 1414, 10.
Da Silva, C.P., Emídio, E.S., de Marchi, M.R.R., 2013. J. Braz. Chem. Soc. 24, 1433.
Da Silva, C.P., Emídio, E.S., de Marchi, M.R.R., 2015. Talanta 131, 221.
Dambal, V.Y., Selvan, K.P., Lite, C., Barathi, S., Santosh, W., 2017. Ecotoxicol. Environ. Saf. 141, 113.
Daughton, C.G., Ternes, T.A., 1999. Environ. Health Perspect. 107, 907.
DeSousa, D.N.R., Grosseli, G.M., Mozeto, A.A., Carneiro, R.L., Fadini, P.S., 2015. J. Sep. Sci. 38, 3454.
Dias, A.N., da Silva, A.C., Simão, V., Merib, J., Carasek, E., 2015. Anal. Chim. Acta 888, 59.
Díaz-Cruz, M.S., Gago-Ferrero, P., Llorca, M., Barceló, D., 2012. Anal. Bioanal. Chem. 402, 2325.
Duedahl-Olesen, L., Cederberg, T., Pedersen, K.H., Hojgard, A., 2005. Chemosphere 61, 422.
Durán-Alvarez, J.C., Becerril-Bravo, E., Castro, V.S., Jiménez, B., Gibson, R., 2009. Talanta 78, 1159.
Einsle, T., Paschke, H., Bruns, K., Schrader, S., Popp, P., Moeder, M., 2006. J. Chromatogr. A 1124, 196.
Emnet, P., Gaw, S., Northcott, G., Storey, B., Graham, L., 2015. Environ. Res. 136, 331.
Escarrone, A.L.V., Caldas, S.S., Soares, B.M., Martins, S.E., Primel, E.G., Nery, L.E.M., 2014. Anal. Methods 6, 8306.
Evans, W.A., Davies, P.J., McRae, C., 2016. Environ. Sci. Water Res. Technol. 2, 733.
Fair, P.A., Lee, H.-B., Adams, J., Darling, C., Pacepavicius, G., Alaee, M., Bossart, G.D., Henry, N., Muir, D., 2009. Environ. Pollut. 157, 2248.
Fent, K., Zenker, A., Rapp, M., 2010. Environ. Pollut. 158, 1817.
Ferreira, A.M.C., Möder, M., Laespada, M.E.F., 2011. J. Chromatogr. A 1218, 3837.
Foltz, J., Mottaleb, M.A., Mezini, M.J., Islam, M.R., 2014. Chemosphere 107, 187.
Fromme, H., Otoo, T., Pilz, K., 2001. Water Res. 35, 121.
Fumes, B.H., Lanças, F.M., 2017. J. Chromatogr. A 1487, 64.
Gago-Ferrero, P., Díaz-Cruz, M.S., Barceló, D., 2011a. Anal. Bioanal. Chem. 400, 2195.
Gago-Ferrero, P., Díaz-Cruz, M.S., Barceló, D., 2011b. Chemosphere 84, 1158.
Gago-Ferrero, P., Díaz-Cruz, M.S., Barceló, D., 2012. Anal. Bioanal. Chem. 404, 2597.
Gago-Ferrero, P., Mastroianni, N., Díaz-Cruz, M.S., Barceló, D., 2013a. J. Chromatogr. A 1294, 106.
Gago-Ferrero, P., Díaz-Cruz, M.S., Barceló, D., 2013b. J. Chromatogr. A 1286, 93.
Gago-Ferrero, P., Díaz-Cruz, M.S., Barceló, D., 2015. Sci. Total Environ. 518–519, 518.
García-Jares, C., Llompart, M., Polo, M., Salgado, C., Macías, S., Cela, R., 2002. J. Chromatogr. A 963, 277.
Gatidou, G., Thomaidis, M.S., Stasinakis, A.S., Lekkas, T.D., 2007. J. Chromatogr. A 1138, 32.
Ge, D., Lee, H.K., 2012a. J. Chromatogr. A 1229, 1.
Ge, D., Lee, H.K., 2012b. J. Chromatogr. A 1251, 27.
Gilart, N., Miralles, N., Marcé, R.M., Borrull, F., Fontanals, N., 2013. Anal. Chim. Acta 774, 51.
Giokas, D.L., Sakkas, V.A., Albanis, T.A., 2004. J. Chromatogr. A 1026, 289.
Giokas, D.L., Sakkas, V.A., Albanis, T.A., Lampropoulou, D.A., 2005. J. Chromatogr. A 1077, 19.
Giokas, D.L., Salvador, A., Chisvert, A., 2007. Trends Anal. Chem. 26, 360.
Giokas, D.L., Zhu, Q., Pan, Q., Chisvert, A., 2012. J. Chromatogr. A 1251, 33.
Godayol, A., Besalú, E., Anticó, E., Sánchez, J.M., 2015. Chemosphere 119, 363.
Gómez, M.J., Gómez-Ramos, M.M., Agüera, A., Mezcua, M., Herrera, S., Fernández-Alba, A.R., 2009. J. Chromatogr. A 1216, 4071.
Gómez, M.J., Herrera, S., Solé, D., García-Calvo, E., Fernández-Alba, A.R., 2011. Anal. Chem. 83, 2638.
González-Hernández, P., Pino, V., Ayala, J.H., Afonso, A.M., 2015. Anal. Methods 7, 1825.
González-Mariño, I., Quintana, J.B., Rodríguez, I., Cela, R., 2009. Rapid Commun. Mass Spectrom. 23, 1756.
González-Mariño, I., Rodríguez, I., Quintana, J.B., Cela, R., 2010. Anal. Bioanal. Chem. 398, 2289.
González-Mariño, I., Quintana, J.B., Rodríguez, I., Schrader, S., Moeder, M., 2011. Anal. Chim. Acta 684, 59.
Gorga, M., Petrovic, M., Barceló, D., 2013. J. Chromatogr. A 1295, 57.
Gracia-Lor, E., Martínez, M., Sancho, J.V., Peñuela, G., Hernández, F., 2012. Talanta 99, 1011.
Guo, J.-H., Li, X.-H., Cao, X.-L., Li, Y., Wang, X.-Z., Xu, X.-B., 2009. J. Chromatogr. A 1216, 3038.

Guo, R., Lee, I.-S., Kim, U.-J., Oh, J.-E., 2010. Sci. Total Environ. 408, 1634.
Guo, J., Li, Z., Sandy, A.L., Li, A., 2014. J. Chromatogr. A 1370, 1.
Guo, L., Nawi, N.B., Lee, H.K., 2016. Anal. Chem. 88, 8409.
Hashemi, B., Shamsipur, M., Fattahi, N., 2015. J. Chromatogr. Sci. 53, 1414.
Haunschmidt, M., Klampfl, C.W., Buchberger, W., Hertsens, R., 2010. Anal. Bioanal. Chem. 397, 269.
He, Y.-J., Chen, W., Zheng, X.-Y., Wang, X.-N., Huang, X., 2013. Sci. Total Environ. 447, 248.
Heidler, J., Halden, R.U., 2007. Chemosphere 66, 362.
Homem, V., Silva, J.A., Ratola, N., Santos, L., Alves, A., 2015. J. Chem. Technol. Biotechnol. 90, 1619.
Homem, V., Alves, A., Alves, A., Santos, L., 2016. Talanta 148, 84.
Hu, Z., Shi, Y., Cai, Y., 2011. Chemosphere 84, 1630.
Iparraguirre, A., Prieto, A., Vallejo, A., Moeder, M., Zuloaga, O., Etxebarria, N., Paschke, A., 2017. Talanta 164, 314.
Jakimska, A., Huerta, B., Barganska, Z., Kot-Wasik, A., Rodríguez-Mozas, S., Barceló, D., 2013. J. Chromatogr. A 1306, 44.
Jeon, H.-K., Chung, Y., Ryu, J.-C., 2006. J. Chromatogr. A 1131, 192.
Jonkers, N., Sousa, A., Galante-Oliveira, S., Barroso, C.M., Kohler, H.-P.E., Giger, W., 2010. Environ. Sci. Pollut. Res 17, 834.
Kameda, Y., Kimura, K., Miyazaki, M., 2011. Environ. Pollut. 159, 1570.
Kannan, K., Reiner, J.L., Yun, S.H., Perrotta, E.E., Tao, L., Johnson-Restrepo, B., Rodan, B.D., 2005. Chemosphere 61, 693.
Kapelewska, J., Kotowska, U., Wisniewska, K., 2016. Environ. Sci. Pollut. Res. 23, 1642.
Kasprzyk-Hordern, B., Dinsdale, R.M., Guwy, A.J., 2008. Anal. Bioanal. Chem. 391, 1293.
Kawaguchi, M., Ito, R., Endo, N., Sakui, N., Okanouchi, N., Saito, K., Sato, N., Shiozaki, T., Nakazawa, H., 2006. Anal. Chim. Acta 557, 272.
Kawaguchi, M., Ito, R., Honda, H., Endo, N., Okanouchi, N., Saito, K., Seto, Y., Nakazawa, H., 2008a. J. Chromatogr. A 1200, 260.
Kawaguchi, M., Ito, R., Honda, H., Endo, N., Okanouchi, N., Saito, K., Seto, Y., Nakazawa, H., 2008b. J. Chromatogr. A 1206, 196.
Kim, J.-W., Ramaswamy, B.R., Chang, K.-H., Isobe, T., Tanabe, S., 2011. J. Chromatogr. A 1218, 3511.
Klaschka, U., von der Ohe, P.C., Bschorer, A., Krezmer, S., Sengl, M., Letzel, M., 2013. Environ. Sci. Pollut. Res. 20, 2456.
Klein, D.R., Flannelly, D.F., Schultz, M.M., 2010. J. Chromatogr. A 1217, 1742.
Knepper, T.P., 2004. J. Chromatogr. A 1046, 159.
Kolpin, D.W., Furlong, E.T., Meyer, M.T., Thurman, E.M., Zaugg, S.D., Barber, L.B., Buxton, H.T., 2002. Environ. Sci. Technol. 36, 1202.
Kotnik, K., Kosjek, T., Krajnc, U., Heath, E., 2014. Anal. Bioanal. Chem. 406, 3179.
Kotowska, U., Kapelewska, J., Sturgulewska, J., 2014. Environ. Sci. Pollut. Res. 21, 660.
Ku, Y.-C., Leong, M.-I., Wang, W.-T., Huang, S.-D., 2013. J. Sep. Sci. 36, 1470.
Kupper, T., Plagellat, C., Brändli, R.C., de Alencastro, L.F., Grandjean, D., Tarradellas, J., 2006. Water Res. 40, 2603.
Kwon, J.-W., Rodriguez, J.M., 2014. Arch. Environ. Contam. Toxicol. 66, 538.
Lambropoulou, D.A., Giokas, D.L., Sakkas, V.A., Albanis, T.A., Karayannis, M.I., 2002. J. Chromatogr. A 967, 243.
Lange, C., Kuch, B., Metzger, J.W., 2015. J. Hazard. Mater. 282, 34.
Li, P., Liu, X., Wang, X., 2015. Acta Chromatogr. 27, 255.
Li, W., Ma, Y., Guo, C., Hu, W., Liu, K., Wang, Y., Zhu, T., 2007. Water Res. 41, 3506.
Li, J., Ma, L., Tang, M., Xu, L., 2013. J. Chromatogr. A 1298, 1.
Li, J., Xu, L., Yu, Q.-W., Shi, Z.-G., Zhang, T., Liu, Y., 2014a. J. Sep. Sci. 37, 2732.
Li, Y., Yang, Y., Liu, H., Wang, X., Du, X., 2014b. Anal. Methods 6, 8519.
Li, M., Sun, Q., Li, Y., Lv, M., Lin, L., Wu, Y., Ashfaq, M., Yu, C.-P., 2016a. Anal. Bioanal. Chem. 408, 4953.
Li, S., Zhu, F., Jiang, R., Ouyang, G., 2016b. J. Chromatogr. A 1429, 1.
Liao, C., Lee, S., Moon, H.-B., Yamashita, N., Kannan, K., 2013. Environ. Sci. Technol. 47, 10895.
Ligon, A.P., Zuehlke, S., Spiteller, M., 2008. J. Sep. Sci. 31, 143.
Liu, H., Liu, L., Xiong, Y., Yang, X., Luan, T., 2010. J. Chromatogr. A 1217, 6747.
Llompart, M., García-Jares, C., Salgado, C., Polo, M., Cela, R., 2003. J. Chromatogr. A 999, 185.

Loos, R., Locoro, G., Comero, S., Contini, S., Schwesig, D., Werres, F., Balsaa, P., Gans, O., Weiss, S., Blaha, L., Bolchi, M., Gawlik, B.M., 2010. Water Res. 44, 4115.
Lopes, D., Dias, A.N., Simão, V., Carasek, E., 2017. Microchem. J. 130, 371.
López-Darias, J., Pino, V., Meng, Y., Anderson, J.L., Afonso, A.M., 2010. J. Chromatogr. A 1217, 7189.
López-Nogueroles, M., Chisvert, A., Salvador, A., Carretero, A., 2011. Talanta 85, 1990.
López-Nogueroles, M., Lordel-Madeleine, S., Chisvert, A., Salvador, A., Pichon, V., 2013. Talanta 110, 128.
Lorenzo, M.A., Sánchez-Arribas, A., Moreno, M., Bermejo, E., Chicharro, M., Zapardiel, A., 2013. Microchem. J. 110, 510.
Lozano, N., Rice, C.P., Ramirez, M., Torrents, A., 2013. Water Res. 47, 4519.
Lu, B., Feng, Y., Gao, P., Zhang, Z., Lin, N., 2015. Environ. Sci. Pollut. Res. 22, 9090.
Lv, Y., Yuan, T., Hu, J., Wang, W., 2009. Anal. Sci. 25, 1125.
Magi, E., Di Carro, M., Scapolla, C., Nguyen, K.T.N., 2012. Chromatographia 75, 973.
Magi, E., Scapolla, C., Di Carro, M., Rivaro, P., Nguyen, K.T.N., 2013. Anal. Methods 5, 428.
Maidatsi, K.V., Chatzimitakos, T.G., Sakkas, V.A., Stalikas, C.D., 2015. J. Sep. Sci. 38, 3758.
Maijó, I., Fontanals, N., Borrull, F., Neusü, C., Calull, M., Aguilar, C., 2013. Electrophoresis 34, 374.
Martín, J., Camacho-Muñoz, D., Santos, J.L., Aparicio, I., Alonso, E., 2014. Anal. Bioanal. Chem. 406, 3709.
Martín, J., Santos, J.L., Aparicio, I., Alonso, E., 2015. Talanta 143, 335.
Martín, J., Zafra-Gómez, A., Hidalgo, F., Ibáñez-Yuste, A.J., Alonso, E., Vilchez, J.L., 2017. Talanta 166, 336.
Matamoros, V., Jover, E., Bayona, J.M., 2010. Anal. Chem. 82, 699.
Matamoros, V., Nguyen, L.X., Arias, C.A., Nielsen, S., Laugen, M.M., Brix, H., 2012. Water Res. 46, 3889.
Meinerling, M., Daniels, M., 2006. Anal. Bioanal. Chem. 386, 1465.
Miao, H.-H., Wang, Y.-N., Zhao, R.-S., Guo, W.-L., Wang, X., Shen, T.-T., Wang, C., Wang, X.-K., 2014. Anal. Methods 6, 2227.
Mijangos, L., Bizkarguenaga, E., Prieto, A., Fernández, L.A., Zuloaga, O., 2015. J. Chromatogr. A 1389, 8.
Moeder, M., Schrader, S., Winkler, U., Rodil, R., 2010. J. Chromatogr. A 1217, 2925.
Moldovan, Z., Chira, R., Alder, A.C., 2009. Environ. Sci. Pollut. Res. 16, S46.
Molins-Delgado, D., Gago-Ferrero, P., Diaz-Cruz, M.S., Barceló, D., 2016. Environ. Res. 145, 126.
Montaseri, H., Forbes, P.B.C., 2016. Trends Anal. Chem. 85, 221.
Montes, R., Rodríguez, I., Rubí, E., Cela, R., 2009. J. Chromatogr. A 1216, 205.
Morales, S., Canosa, P., Rodríguez, I., Rubí, E., Cela, R., 2005. J. Chromatogr. A 1082, 128.
Morales-Muñoz, S., Luque-García, J.L., Ramos, M.J., Fernández-Alba, A., Luque de Castro, M.D., 2005. Anal. Chim. Acta 552, 50.
Mottaleb, M.A., Usenko, S., O'Donnell, J.G., Ramirez, A.J., Brooks, B.W., Chambliss, C.K., 2009. J. Chromatogr. A 1216, 815.
Mudiam, M.K.R., Jain, R., Singh, R., 2014. Anal. Methods 6, 1802.
Nagtegaal, M., Ternes, T.A., Baumann, W., Nagel, R., 1997. Umweltchem. Ökotox. 9, 79.
Nakata, H., 2005. Environ. Sci. Technol. 39, 3430.
Nakata, H., Sasaki, H., Takemura, A., Yoshioka, M., Tanabe, S., Kannan, K., 2007. Environ. Sci. Technol. 41, 2216.
Negreira, N., Rodríguez, I., Ramil, M., Rubí, E., Cela, R., 2009a. Anal. Chim. Acta 638, 36.
Negreira, N., Rodríguez, I., Ramil, M., Rubí, E., Cela, R., 2009b. Anal. Chim. Acta 654, 162.
Negreira, N., Rodríguez, I., Rubí, E., Cela, R., 2009c. J. Chromatogr. A 1216, 5895.
Negreira, N., Rodríguez, I., Rubí, E., Cela, R., 2010. Anal. Bioanal. Chem. 398, 995.
Negreira, N., Rodriguez, I., Rubi, E., Cela, R., 2011a. Anal. Bioanal. Chem. 400, 603.
Negreira, N., Rodríguez, I., Rubí, E., Cela, R., 2011b. J. Chromatogr. A 1218, 211.
Nguyen, K.T.N., Scapolla, C., Di Carro, M., Magi, E., 2011. Talanta 85, 2375.
Nieto, A., Borrull, F., Marcé, R.M., Pocurull, E., 2009. J. Chromatogr. A 1216, 5619.
Núñez, L., Tadeo, J.L., García-Valcárcel, A.I., Turiel, E., 2008. J. Chromatogr. A 1214, 178.
Núñez, L., Turiel, E., Martin-Esteban, A., Tadeo, J.L., 2010. Talanta 80, 1782.
Okanouchi, N., Honda, H., Ito, R., Kawaguchi, M., Saito, K., Nakazawa, H., 2008. Anal. Sci. 24, 627.
Oliveira, H.M., Segundo, M.A., Lima, J.L.F.C., Miró, M., Cerdà, V., 2010. J. Chromatogr. A 1217, 3575.
Osemwengie, L.I., Steinberg, S., 2001. J. Chromatogr. A 932, 107.
Osemwengie, L.I., Steinberg, S., 2003. J. Chromatogr. A 993, 1.
Panagiotou, A.N., Sakkas, V.A., Albanis, T.A., 2009. Anal. Chim. Acta 649, 135.

Parisis, N.A., Giokas, D.L., Vlessidis, A.G., Evmiridis, N.P., 2005. J. Chromatogr. A 1097, 17.
Peck, A.M., 2006. Anal. Bioanal. Chem. 386, 907.
Pedrouzo, M., Borrull, F., Marcé, R.M., Pocurull, E., 2009. J. Chromatogr. A 1216, 6994.
Pedrouzo, M., Borrull, F., Marcé, R.M., Pocurull, E., 2010. Anal. Bioanal. Chem. 397, 2833.
Pedrouzo, M., Borrull, F., Marcé, R.M., Pocurull, E., 2011. Trends Anal. Chem. 30, 749.
Peng, X., Jin, J., Wang, C., Ou, W., Tang, C., 2015. J. Chromatogr. A 1384, 97.
Pérez, R.A., Albero, B., Tadeo, J.L., Sánchez-Brunete, C., 2016. Anal. Bioanal. Chem. 408, 8013.
Picot-Groz, M., Martinez-Bueno, M.J., Rosain, D., Fenet, H., Casellas, C., Pereira, C., Maria, V., Bebianno, M.J., Gomez, E., 2014. Sci. Total Environ. 493, 162.
Pietrogrande, M.C., Basaglia, G., Dondi, F., 2009. J. Sep. Sci. 32, 1249.
Pintado-Herrera, M.G., González-Mazo, E., Lara-Martín, P.A., 2013. Anal. Bioanal. Chem. 405, 401.
Pintado-Herrera, M.G., González-Mazo, E., Lara-Martín, P.A., 2014. Anal. Chim. Acta 851, 1.
Pintado-Herrera, M.G., González-Mazo, E., Lara-Martín, P.A., 2016. J. Chromatogr. A 1429, 107.
Plagellat, C., Kupper, T., Furrer, R., de Alencastro, L.F., Grandjean, D., Tarradellas, J., 2006. Chemosphere 62, 915.
Poiger, T., Buser, H.-R., Balmer, M.E., Bergqvist, P.-A., Müller, M.D., 2004. Chemosphere 55, 951.
Polo, M., García-Jares, C., Llompart, M., Cela, R., 2007. Anal. Bioanal. Chem. 388, 1789.
Posada-Ureta, O., Olivares, M., Navarro, P., Vallejo, A., Zuloaga, O., Etxebarria, N., 2012. J. Chromatogr. A 1227, 38.
Prestes, O.D., Padilla-Sánchez, J.A., Romero-González, R., Grio, S.L., Frenich, A.G., Martínez-Vidal, J.L., 2012. J. Sep. Sci. 35, 861.
Prichodko, A., Janenaite, E., Smitiene, V., Vickackaite, V., 2012. Acta Chromatogr. 24, 589.
Purrà, M., Cinca, R., Legaz, J., Núñez, O., 2014. Anal. Bioanal. Chem. 406, 6189.
Quednow, K., Püttmann, W., 2009. Environ. Sci. Pollut. Res. 16, 630.
Rafoth, A., Gabriel, S., Sacher, F., Brauch, H.-J., 2007. J. Chromatogr. A 1164, 74.
Ramaswamy, B.R., 2015. In: Díaz-Cruz, M.S., Barceló, D. (Eds.), Environmental Risk Assessment of Personal Care Products. Springer.
Ramaswamy, B.R., Kim, J.-W., Isobe, T., Chang, K.-H., Amano, A., Miller, T.W., Siringan, F.P., Tanabe, S., 2011. J. Hazard. Mater. 192, 1739.
Ramírez, A.J., Brain, R.A., Usenko, S., Mottaleb, M.A., O'Donnell, J.G., Stahl, L.L., Wathen, J.B., Snyder, B.D., Pitt, J.L., Perez-Hurtado, P., Dobbins, L.L., Brooks, B.W., Chambliss, C.K., 2009. Environ. Toxicol. Chem. 28, 2587.
Ramírez, N., Marcé, R.M., Borrull, F., 2011. J. Chromatogr. A 1218, 156.
Ramírez, N., Borrull, F., Marcé, R.M., 2012. J. Sep. Sci. 35, 580.
Ramos-Payan, M., Maspoch, S., Llobera, A., 2017. Talanta 165, 496.
Regueiro, J., Llompart, M., Garcia-Jares, C., Garcia-Monteagudo, J.C., Cela, R., 2008. J. Chromatogr. A 1190, 27.
Regueiro, J., Becerril, E., Garcia-Jares, C., Llompart, M., 2009a. J. Chromatogr. A 1216, 4693.
Regueiro, J., Llompart, M., Psillakis, E., Garcia-Monteagudo, J.C., Garcia-Jares, C., 2009b. Talanta 79, 1387.
Rice, S.L., Mitra, S., 2007. Anal. Chim. Acta 589, 125.
Rimkus, G.G., Buttle, W., Geyer, H.J., 1997. Chemosphere 35, 1497.
Rocío-Bautista, P., Martínez-Benito, C., Pino, V., Pasán, J., Ayala, J.H., Ruiz-Pérez, C., Afonso, A.M., 2015. Talanta 139, 13.
Rodil, R., Moeder, M., 2008a. J. Chromatogr. A 1179, 81.
Rodil, R., Moeder, M., 2008b. Anal. Chim. Acta 612, 152.
Rodil, R., Moeder, M., 2008c. J. Chromatogr. A 1178, 9.
Rodil, R., Quintana, J.B., López-Mahía, P., Muniategui-Lorenzo, S., Prada-Rodríguez, D., 2008. Anal. Chem. 80, 1307.
Rodil, R., Quintana, J.B., López-Mahía, P., Muniategui-Lorenzo, S., Prada-Rodríguez, D., 2009a. J. Chromatogr. A 1216, 2958.
Rodil, R., Schrader, S., Moeder, M., 2009b. Rapid Commun. Mass Spectrom. 23, 580.
Rodil, R., Schrader, S., Moeder, M., 2009c. J. Chromatogr. A 1216, 4887.
Rodil, R., Schrader, S., Moeder, M., 2009d. J. Chromatogr. A 1216, 8851.
Román, I.P., Chisvert, A., Canals, A., 2011. J. Chromatogr. A 1218, 2467.

Ros, O., Izaguirre, J.K., Olivares, M., Bizarro, C., Ortiz-Zarragoitia, M., Cajaraville, M.P., Etxebarria, N., Prieto, A., Vallejo, A., 2015. Sci. Total Environ. 536, 261.
Roy, D.N., Goswami, R., Pal, A., 2017. Environ. Toxicol. Pharmacol. 50, 91.
Rüdel, H., Böhmer, W., Schröter-Kermani, C., 2006. J. Environ. Monit. 8, 812.
Rüdel, H., Böhmer, W., Müller, M., Fliedner, A., Ricking, M., Teubner, D., Schröter-Kermani, C., 2013. Chemosphere 91, 1517.
Salvador, A., Chisvert, A., 2007. Analysis of Cosmetic Products. Elsevier.
Salvatierra-Stamp, V.C., Ceballos-Magaña, S.G., Gonzalez, J., Jurado, J.M., Muñiz-Valencia, R., 2015. Anal. Bioanal. Chem. 407, 4661.
Sánchez-Brunete, C., Miguel, E., Albero, B., Tadeo, J.L., 2010. J. Sep. Sci. 33, 2768.
Sánchez-Brunete, C., Miguel, E., Albero, B., Tadeo, J.L., 2011. J. Chromatogr. A 1218, 4291.
Sandstrom, M.W., Kolpin, D.W., Thurman, E.M., Zaugg, S.D., 2005. Environ. Toxicol. Chem. 24, 1029.
Saraji, M., Mirmahdieh, S., 2009. J. Sep. Sci. 32, 988.
Šatínský, D., Naibrtová, L., Fernández-Ramos, C., Solich, P., 2015. Talanta 124.
Schwarzbauer, J., Heim, S., 2005. Water Res. 39, 4735.
Shaaban, H., Górecki, T., 2012. Talanta 100, 80.
Shih, H.-K., Lin, C.-W., Ponnusamy, V.K., Ramkumar, A., Jen, J.-F., 2013. Anal. Methods 5, 2352.
Silva, A.R.M., Nogueira, J.M.F., 2008. Talanta 74, 1498.
Silva, A.R.M., Nogueira, J.M.F., 2010. Anal. Bioanal. Chem. 396, 1853.
Simonich, S.L., Begley, W.M., Debaere, G., Eckhoff, W.S., 2000. Environ. Sci. Technol. 34, 959.
Singer, H., Müller, S., Tixier, C., Pillonel, L., 2002. Environ. Sci. Technol. 36, 4998.
Souchier, M., Benali-Raclot, D., Benanou, D., Boireau, V., Gomez, E., Casellas, C., Chiron, S., 2015. Sci. Total Environ. 502, 199.
Speksnijder, P., Van Ravestijn, J., de Voogt, P., 2010. J. Chromatogr. A 1217, 5184.
Standler, A., Schatzl, A., Klampfl, C.W., Buchberger, W., 2004. Microchim. Acta 148, 151.
Suárez, R., Clavijo, S., Avivar, J., Cerdà, V., 2016. Talanta 148, 589.
Subedi, B., Mottaleb, M.A., Chambliss, C.K., Usenko, S., 2011. J. Chromatogr. A 1218, 6278.
Sui, Q., Huang, J., Deng, S., Yu, G., Fan, Q., 2010. Water Res. 44, 417.
Sultana, T., Murray, C., Ehsanul-Hoque, M., Metcalfe, C.D., 2016. Environ. Monit. Assess. 189, 1.
Sumner, N.R., Guitart, C., Fuentes, G., Readman, J.W., 2010. Environ. Pollut. 158, 215.
Sun, P., Casteel, K., Dai, H., Wehmeyer, K.R., Kiel, B., Federle, T., 2014. Sci. Total Environ. 493, 1073.
Tahmasebi, E., Yamini, Y., Mehdinia, A., Rouhi, F., 2012. J. Sep. Sci. 35, 2256.
Tanoue, R., Nomiyama, K., Nakamura, H., Hayashi, T., Kim, J.-W., Isobe, T., Shinohara, R., Tanabe, S., 2014. J. Chromatogr. A 1355, 193.
Tarazona, I., Chisvert, A., León, Z., Salvador, A., 2010. J. Chromatogr. A 1217, 4771.
Tarazona, I., Chisvert, A., Salvador, A., 2014. Anal. Methods 6, 7772.
Ternes, T.A., Bonerz, M., Herrmann, N., Löffler, D., Keller, E., Lacida, B.B., Alder, A.C., 2005. J. Chromatogr. A 1067, 213.
Tovar-Sánchez, A., Sánchez-Quiles, D., Basterretxea, G., Benedé, J.L., Chisvert, A., Salvador, A., Moreno-Garrido, I., Blasco, J., 2013. PLoS One 8, e65451.
Tsai, D.-Y., Chen, C.-L., Ding, W.-H., 2014. Food Chem. 154, 211.
Vakondios, N., Mazioti, A.A., Koukouraki, E.E., Diamadopoulos, E., 2016. J. Environ. Chem. Eng. 4, 1910.
Vallecillos, L., Pocurull, E., Borrull, F., 2012a. Talanta 99, 824.
Vallecillos, L., Pocurull, E., Borrull, F., 2012b. J. Chromatogr. A 1264, 87.
Vallecillos, L., Borrull, F., Pocurull, E., 2012c. J. Sep. Sci. 35, 2735.
Vallecillos, L., Borrull, F., Pocurull, E., 2013a. Anal. Bioanal. Chem. 405, 9547.
Vallecillos, L., Pocurull, E., Borrull, F., 2013b. J. Chromatogr. A 1314, 38.
Vallecillos, L., Borrull, F., Pocurull, E., 2014a. J. Chromatogr. A 1364, 1.
Vallecillos, L., Pedrouzo, M., Pocurull, E., Borrull, F., 2014b. J. Sep. Sci. 37, 1322.
Vallecillos, L., Borrull, F., Sanchez, J.M., Pocurrull, E., 2015a. Talanta 132, 548.
Vallecillos, L., Pocurull, E., Borrull, F., 2015b. Talanta 134, 690.
Vidal, L., Chisvert, A., Canals, A., Salvador, A., 2010. Talanta 81, 549.
Vila, M., Lamas, J.P., Garcia-Jares, C., Dagnac, T., Llompart, M., 2016. Microchem. J. 124, 530.
Vila, M., Celeiro, M., Lamas, J.P., Garcia-Jares, C., Dagnac, T., Llompart, M., 2017. J. Hazard. Mater. 323, 45.

Villa, S., Assi, L., Ippolito, A., Bonfanti, P., Finizio, A., 2012. Sci. Total Environ. 416, 137.
Villaverde-de-Sáa, E., González-Mariño, I., Quintana, J.B., Rodil, R., Rodríguez, I., Cela, R., 2010. Anal. Bioanal. Chem. 397, 2559.
Villaverde-de-Sáa, E., Rodil, R., Quintana, J.B., Cela, R., 2016. J. Chromatogr. A 1459, 57.
Vosough, M., Mojdehi, N.R., 2011. Talanta 85, 2175.
Wan, Y., Wei, Q., Hu, J., Jin, X., Zhang, Z., Zhen, H., Liu, J., 2007. Environ. Sci. Technol. 41, 424.
Wang, Y.-C., Ding, W.-H., 2009. J. Chromatogr. A 1216, 6858.
Wang, L., McDonald, J.A., Khan, S.J., 2013a. J. Chromatogr. A 1303, 66.
Wang, M., Peng, C., Chen, W., Markert, B., 2013b. Ecotoxicol. Environ. Saf. 97, 242.
Wang, X., Yuan, K., Liu, H., Lin, L., Luan, T., 2014. J. Sep. Sci. 37, 1842.
Wei, N., Zheng, Z., Wang, Y., Tao, Y., Shao, Y., Zhu, S., You, J., Zhao, X.-E., 2017. Rapid Commun. Mass Spectrom. 31, 937.
Weigel, S., Kallenborn, R., Hühnerfuss, H., 2004. J. Chromatogr. A 1023, 183.
Wick, A., Fink, G., Ternes, T.A., 2010. J. Chromatogr. A 1217, 2088.
Wille, K., De Brabander, H.F., Vanhaecke, L., De Wulf, E., Van Caeter, P., Janssen, C.R., 2012. Trends Anal. Chem. 35, 87.
Winkler, M., Headley, J.V., Peru, K.M., 2000. J. Chromatogr. A 903, 203.
Wluka, A.-K., Rüdel, H., Pohl, K., Schwarzbauer, J., 2016. Environ. Sci. Pollut. Res. 23, 21894.
Wu, S.-F., Ding, W.-H., 2010. J. Chromatogr. A 1217, 2776.
Wu, J.-L., Lam, N.P., Martens, D., Kettrup, A., Cai, Z., 2007. Talanta 72, 1650.
Wu, S.-F., Liu, L.-L., Ding, W.-H., 2012. Food Chem. 133, 513.
Wu, J.-W., Chen, H.-C., Ding, W.-H., 2013a. J. Chromatogr. A 1302, 20.
Wu, M.-W., Yeh, P.-C., Chen, H.-C., Liu, L.-L., Ding, W.-H., 2013b. J. Chin. Chem. Soc. 60, 1169.
Xue, L.-K., Ma, W.-W., Zhang, D.-X., Du, X.-Z., 2013. Anal. Methods 5, 4213.
Yamagishi, T., Miyazaki, T., Horii, S., Kaneko, S., 1981. Bull. Environ. Contam. Toxicol. 26, 656.
Yamagishi, T., Miyazaki, T., Horii, S., Akiyama, K., 1983. Arch. Environ. Contam. Toxicol. 12, 83.
Yang, C.-Y., Ding, W.-H., 2012. Anal. Bioanal. Chem. 402, 1723.
Yao, L., Zhao, J.-L., Liu, Y.-S., Yang, Y.-Y., Liu, W.-R., Ying, G.-G., 2016. Anal. Bioanal. Chem. 408, 8177.
Yi, F., Zheng, Y., Wang, T., Liu, L., Yu, Q., Xu, S., Ma, H., Cheng, R., Ye, J., Chu, Q., 2016. Chromatographia 79, 619.
Ying, G.-G., Kookana, R.S., 2007. Environ. Int. 33, 199.
Yu, Y., Huang, Q., Cui, J., Zhang, K., Tang, C., Peng, X., 2011. Anal. Bioanal. Chem. 399, 891.
Zarate, F.M., Schulwitz, S.E., Stevens, K.J., Venables, B.J., 2012. Chemosphere 88, 323.
Zeng, X., Mai, B., Sheng, G., Luo, X., Shao, W., An, T., Fu, J., 2008. Environ. Toxicol. Chem. 27, 18.
Zenker, A., Schmutz, H., Fent, K., 2008. J. Chromatogr. A 1202, 64.
Zhang, H., Lee, H.K., 2012a. Anal. Chim. Acta 742, 67.
Zhang, Y., Lee, H.K., 2012b. Anal. Chim. Acta 750, 120.
Zhang, Y., Lee, H.K., 2012c. J. Chromatogr. A 1249, 25.
Zhang, Y., Lee, H.K., 2013a. J. Chromatogr. A 1271, 56.
Zhang, Y., Lee, H.K., 2013b. J. Chromatogr. A 1273, 12.
Zhang, P.-P., Shi, Z.G., Yu, Q.-W., Feng, Y.-Q., 2011a. Talanta 83, 1711.
Zhang, Z., Ren, N., Li, Y.-F., Kunisue, T., Gao, D., Kannan, K., 2011b. Environ. Sci. Technol. 45, 3909.
Zhang, X., Xu, Q., Man, S., Zeng, X., Yu, Y., Pang, Y., Sheng, G., Fu, J., 2013. Environ. Sci. Pollut. Res. 20, 311.
Zhang, H., Bayen, S., Kelly, B.C., 2015a. Sci. Total Environ. 523, 219.
Zhang, H., Bayen, S., Kelly, B.C., 2015b. Talanta 143, 7.
Zhao, R.-S., Yuan, J.-P., Li, H.-F., Wang, X., Jiang, T., Lin, J.-M., 2007. Anal. Bioanal. Chem. 387, 2911.
Zhao, R.-S., Wang, X., Sun, J., Wang, S.-S., Yuan, J.-P., Wang, X.-K., 2010. Anal. Bioanal. Chem. 397, 1627.
Zheng, C., Zhao, J., Bao, P., Gao, J., He, J., 2011. J. Chromatogr. A 1218, 3830.
Zhou, H., Huang, X., Gao, M., Wang, X., Wen, X., 2009. J. Environ. Sci. 21, 561.
Ziarrusta, H., Olivares, M., Delgado, A., Posada-Ureta, O., Zuloaga, O., Etxebarria, N., 2015. J. Chromatogr. A 1391, 18.

PART IV

Safety Evaluation of Cosmetic Products

CHAPTER 17

Alternative Methods to Animal Testing in Safety Evaluation of Cosmetic Products

Octavio Díez-Sales, Amparo Nácher, Matilde Merino, Virginia Merino
University of Valencia, Valencia, Spain

INTRODUCTION

The Cosmetics Directive of the EU (76/768/EEC) demands that '*A cosmetic product put on the market within the Community must not cause damage to human health when applied under normal or reasonably foreseeable conditions of use…*'. Thus the responsibility for the product's safety lies with the cosmetic manufacturer or the person placing a cosmetic product on the Community market, who must be able to demonstrate that this product is safe for consumer use. Concerning assessment of the safety for human health of the finished product, Article 7a (d) states '*To that end the manufacturer shall take into consideration the general toxicological profile of the ingredients, their chemical structure and their level of exposure*'.

On the other hand, the 7th Amendment to the Cosmetics Directive (2003/15/EC) introduces new provisions related to non-animal testing of finished cosmetic products and ingredients. In particular, it bans the testing of finished cosmetic products and cosmetic ingredients on animals and prohibits marketing such products in the European Community.

The new EU regulation (1223/2009 on cosmetic products, in force since 11 July 2013). The regulation replaces Directive 76/768/EC, which was adopted in 1976 and had been substantially revised on numerous occasions. It provides a robust, internationally recognized regime, which reinforces product safety while taking into consideration the latest technological developments, including the possible use of nanomaterials. The previous rules on the ban on animal testing have not been modified.

For this reason recent years have witnessed important efforts being made to develop new alternative methods to ensure the safety of cosmetic preparations without using laboratory animals. These alternative methods involve new techniques to determine a toxicological endpoint that results in reduction (of the number of animals per test), refinement (of the methodologies by reducing the pain and distress of the animals) or replacement (of the animals by non-sentient material). The 'Three Rs' provide a strategy for a rational and stepwise approach to minimizing animal use without compromising the quality of the scientific work being done, while having, as the ultimate aim, total replacement of animal models with non-animal alternatives (Russell et al., 1959).

In Europe, the validation of alternative methods is officially coordinated by the European Centre for the Validation of Alternative Methods (ECVAM), while the US counterpart is the Interagency Coordinating Committee on the Validation of Alternative Methods (ICCVAM). The tendency for close collaboration between the two organizations exists and joint reports are generated. When an alternative method has been formally validated and accepted by the EU, its use becomes mandatory. Consequently, this implies that animal testing for the same endpoint is prohibited (Herráez Domínguez and Díez-Sales, 2007).

Nevertheless, it is important to point out that the laboratories where the different validation methods are carried out should be certified in the execution of good laboratory practice. These principles of good laboratory practice are a group of rules, procedures and established practices developed by the OECD (Organization for Economic Cooperation and Development). This normative has been assumed by the EU in Directive 87/18/CEE and 99/11/CE: *'The purpose of these principles of good laboratory practice (GLP) is to promote the development of quality test data. Comparable quality of test data forms the basis for the mutual acceptance of data among countries. If individual countries can confidently rely on test data developed in other countries, duplicative testing can be avoided, thereby saving time and resources. The application of these principles should help to avoid the creation of technical barriers to trade, and further improve the protection of human health and the environment'*.

This chapter reviews the methods recommended for the different areas listed as follows, which are included in the EU legislation (Regulation CE No. 440/2008) and in the 'OECD Guidelines for the Testing of Chemicals' (OECD), as well as other alternative methods that, given the current state of the art, should be validated in the near future. The areas in which important advances have taken place in the use of new alternative methods are:

- acute toxicity
- skin corrosion/irritation
- eye irritation
- skin sensitization
- skin absorption
- repeated dose toxicity
- genotoxicity/mutagenicity
- UV-induced toxic effects (phototoxicity, photogenotoxicity, photoallergy)

- carcinogenicity
- reproductive and developmental toxicity
- toxicokinetics studies

ACUTE TOXICITY

The main objective is basically to classify chemicals according to their intrinsic toxicity as required by the EC directive on classification, packaging and labelling of dangerous substances (Regulation EC No. 1272/2008). This requirement aims to protect public health by regulating exposure to potentially dangerous materials.

Chemicals are classified on the basis of the median lethal dose (LD_{50}) value, defined as 'the statistically derived single dose of a substance that can be expected to cause death in 50% of the animals in an experimental group'. The LD_{50} test procedure has been modified in various ways to reduce the number of animals required and to reduce the suffering caused to any animal used. These modifications to the classical LD_{50} test include:

- the fixed-dose method (EC B1 bis; OECD TG 420), which is a useful refinement to EC B1/OECD TG 401;
- the acute oral toxicity method (EC B1 tris; OECD TG 423), which does not aim to calculate a precise LD_{50} value, but allows the determination of a range of exposure dosages for which lethality is expected, offering a significant reduction in the number of animals tested;
- the up and down procedure (OECD TG 425);
- the acute inhalation toxicity method (EC B2; OECD TG 403). After this method, a reduction and refinement that describes the acute toxic class method by the inhalation route is described in the following method (EC B52; OECD TG 436);
- the acute inhalation toxicity–fixed-dose procedure (OECD TG 433), as an alternative to OECD TG 403;
- the acute dermal toxicity method, which provides information on health hazards likely to arise from a short-term exposure to a solid or liquid test substance by the dermal route (OECD TG 402).

These tests cannot be predictive for acute toxicity as single methods, but they become good alternatives if integrated into a tiered approach and/or in a test battery. As of this writing there are no validated alternative methods able to completely replace the use of animals in the field of acute toxicity. Validation of an alternative model for acute oral toxicity is very complex and the time estimated to achieve complete animal replacement for acute toxicity is not clearly defined.

Finally, there are a number of software packages [quantitative structure–activity relationship (QSAR) models] to predict effects on human health and related toxicities. These systems are used because they are user friendly and fast and they can also predict toxicity directly from chemical structure (Cronin et al., 2003; Lapenna et al., 2010).

SKIN CORROSION/IRRITATION

The skin is often exposed, either intentionally or unintentionally, to cosmetic products. It is clear that the potential for a particular product/ingredient to cause skin irritation or corrosion needs to be carefully evaluated as part of the overall safety assessment process. In Table 17.1 are listed the alternative tests for skin corrosion.

Skin Corrosion

Skin corrosion is defined as the production of irreversible tissue damage to the skin, visible necrosis through the epidermis and into the dermis, following the application of a test substance for up to 4 h. Corrosivity is not a feature one expects to find with cosmetics, but can occasionally occur after a manufacturing error or misuse by the consumer. On the other hand, a cosmetic ingredient that has an intrinsic corrosive property is not necessarily excluded for use in cosmetics. It very much depends on its final concentration in the cosmetic product, as well as the presence of 'neutralizing' substances, the excipients used, the exposure route, the conditions of use, etc.

Table 17.1 Summary of validated alternative methods for skin corrosion, accepted by regulatory authorities in EU legislation (Regulation EC No. 440/2008) and in the OECD testing guidelines for chemicals

Alternative test	Test system	Endpoint	Status (EU/OECD TG)
Rat skin transcutaneous electrical resistance assay	Rat skin	Stratum corneum integrity and barrier function	B.40/430
EpiSkin human skin model (commercial system)	RHE	Cell viability (MTT test)	B.40bis/431
EpiDerm human skin model (commercial system)	–	–	B.40bis/431
SkinEthic RHE (commercial system)	–	–	B.40bis/431
EST1000 (commercial system)	–	–	B.40bis/431
Corrositex (commercial system)	Biobarrier membrane	Visually detectable change	Not included in EU regulation/435

EU, European Union; *MTT*, 3-(4,5-dimethyldiazol-2-yl)-2,5 diphenyltetrazolium bromide; *OECD*, Organization for Economic Cooperation and Development; *RHE*, reconstructed human epidermis; *TG*, testing guideline.

In the past, skin corrosion was assessed using animal studies (EC B.4; OECD TG 404) but as of this writing five validated in vitro replacement alternatives have been taken up in Regulation (EC) No. 440/2008:

- TER test (rat skin transcutaneous electrical resistance test) (EC B.40; OECD TG 430)
- EpiSkin (EC B.40bis; OECD TG 431)
- EpiDerm (EC B.40bis; OECD TG 431)
- SkinEthic (EC B.40bis; OECD TG 431)
- EST1000 (epidermal skin test 1000) (EC B.40bis; OECD TG 431)

On the other hand, another corrosivity test is the Corrositex test (OECD TG 435), which uses penetration of test substances through a collagen matrix (biobarrier) and supporting filter membrane. Although the Corrositex test passed the ECVAM Scientific Advisory Committee (ESAC), it has not been taken up in the EU legislation.

Rat Skin Transcutaneous Electrical Resistance Test

This ex vivo test is used to assess the skin corrosivity of a test substance following topical application to the epidermal surface of skin discs taken from a single rat. The contact period between the test substance and the skin is 24 h. Corrosive materials are identified by their ability to produce a loss of normal stratum corneum integrity and barrier function, which is measured as a reduction in the TER below a threshold level of 5 kΩ. The mean TER results are accepted on condition that concurrent positive and negative control values fall within the acceptable ranges for the method.

If the TER values of test substances, which are either surfactants or neutral organics, are less than or equal to 5 kΩ, an assessment of sulphorhodamine B dye penetration can be carried out on the tissues (skin disc) to reduce false positives obtained specifically with these types of chemicals (Fentem et al., 1998).

Human Skin Model Assay

The reconstructed human skin models are three-dimensional models generated by growing keratinocyte cultures at the air–liquid interface on various substrates and that enable the topical application of either neat or diluted test materials.

Corrosive materials are identified by their ability to produce a decrease in cell viability. The test material is applied topically for up to 4 h to a three-dimensional human skin model comprising a reconstructed epidermis with a functional stratum corneum (Fig. 17.1). For liquid (a minimum of 25 μL/cm^2) and solid materials, sufficient test substance must be applied to cover the skin surface.

The principle of the assay is in accordance with the hypothesis that corrosive chemicals are those that are able to penetrate the stratum corneum (by diffusion or erosion) and are sufficiently cytotoxic to cause cell death in the underlying cell layers.

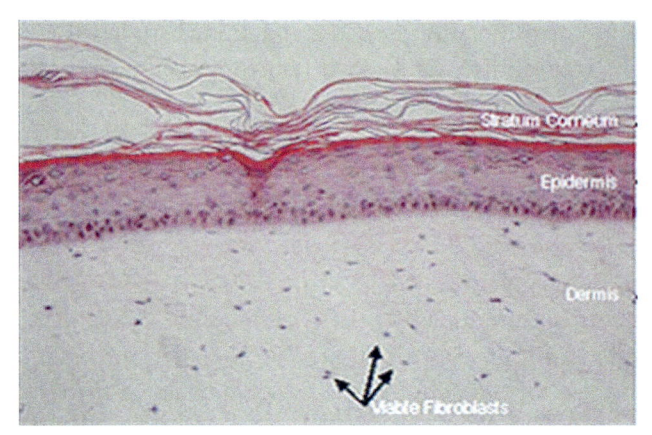

Figure 17.1 Full-thickness skin model (EpiDerm, MatTek Corporation, Ashland, MA, USA).

The viability of the living cells in the model must be high enough to discriminate properly between the positive and the negative control substances. Cell viability is measured by the amount of MTT [3-(4,5-dimethyldiazol-2-yl)-2,5 diphenyltetrazolium bromide] dye reduction: the water-soluble yellow dye is converted to insoluble purple-coloured formazan within cells. This technique has been shown to give accurate and reproducible results in various laboratories. The skin disc, after treatment with the test material, is placed in an MTT solution of 0.3 mg/mL at 20°C–28°C for 3 h. The precipitated blue formazan product is then extracted (solvent extraction) and the concentration of the formazan is measured with a wavelength of between 545 and 595 nm.

Corrositex Assay

The Corrositex test (InVitro International, Irvine, CA, USA) is a standardized, quantitative in vitro test for skin corrosivity, based upon determination of the time required for a test material to pass through a biobarrier membrane (a reconstituted collagen matrix, constructed to have physico-chemical properties similar to those of rat skin) and produce a visually detectable change. The time required for this change to occur (the breakthrough time) is reported to be inversely proportional to the degree of corrosivity of the test material.

Corrositex can be used to assess the corrosivity of seven categories of chemicals: acids, acid derivatives, acyl halides, alkylamines/polyalkylamines, bases, chlorosilanes, metal halides and oxyhalides, according to the US regulations.

Skin Irritation

Dermal irritation is defined as the production of 'reversible damage of the skin following the application of a test substance for up to 4 hours'. It is generally assessed by the potential of a certain substance to cause erythema/eschar and/or oedema after a single topical application on rabbit skin and is based on the Draize score (OECD TG 404). The main overall objective is to identify those in vitro alternatives capable of discriminating skin irritants from non-irritants.

In vitro alternatives in the field vary from simpler models such as keratinocyte cultures to more complex reconstituted human skin models. A major advantage of these in vitro models is that the test substance can be applied directly (topically) to the culture surface (stratum corneum) at the air interface, thereby closely mimicking dermal exposure in humans. Therefore, the models are particularly appropriate for irritancy testing of products intended for topical exposure in humans, such as treatments for skin conditions, cosmetics, wound dressings, transdermal delivery systems and medical devices (Botham et al., 1998).

A number of in vitro skin irritation tests have been officially validated (EC B.46):
- EpiSkin
- modified EpiDerm skin irritation test
- SkinEthic reconstructed human epidermis (RHE)

  Thus the in vitro test methods, based on RHE, EpiSkin, modified EpiDerm and SkinEthic RHE, are included in OECD TG 439 and were endorsed by ESAC. EC B.46 [Regulation (EC) No. 640/2012] is the counterpart of OECD 439. Depending on the regulatory framework and the classification system in use, OECD 439 may be used to determine the skin irritancy of chemicals as a stand-alone replacement test for in vivo skin irritation testing or as a partial replacement test within a tiered testing strategy (Table 17.2).
- The EpiSkin model (EPISKIN, Lyon, France) is another three-dimensional human skin model (Tinois et al., 1991). Its use for skin irritation testing involves topical application of test materials to the surface of the skin, and the subsequent assessment of their effects on cell viability by using the MTT assay. The endpoint used to distinguish between potential skin irritants and non-irritants is the percentage of cell viability (Roguet et al., 1998).
- The modified EpiDerm model (MatTek) is a reconstituted human skin model, in which human skin-derived keratinocytes (Fig. 17.2) are grown on specially prepared Millicell cell culture inserts, forming multiple layers. The epidermal tissue model is used to assess the skin irritation potential of test substances following a timed exposure of the tissues to the test substance. Irritant substances are identified by their ability to produce a decrease in cell viability (as determined by using the MTT reduction assay). The endpoint of the assay is the ET_{50}, effective time of exposure to reduce tissue viability to 50% in the treated tissues compared to untreated controls (Zuang et al., 2002).
- Other models exist in which the keratinocyte cultures grown at the air–liquid interface are cultured in different substrates, e.g., the SkinEthic model.

Consequently, the human skin model assays appear to be the most promising in vitro methods for skin irritation testing. However, there is a need to develop new endpoints that are more predictive of skin irritation than simple cytotoxicity determinations.

Table 17.2 Summary of some alternative methods for skin irritation (Regulation EC No. 640/2012) and in the OECD testing guidelines

Alternative test	Test system	Endpoint	Status (EC/OECD TG)
EpiSkin human skin model (commercial system)	Reconstructed human epidermis	Cell viability (MTT test)	B 46/439
Modified EpiDerm human skin model (commercial system)	Reconstructed human epidermis	Cell viability (MTT test)	B 46/439
SkinEthic human skin model (commercial system)	Reconstructed human epidermis	Cell viability (MTT test)	B 46/439

EC, European Commission; *MTT*, 3-(4,5-dimethyldiazol-2-yl)-2,5 diphenyltetrazolium bromide; *OECD*, Organization for Economic Cooperation and Development; *TG*, testing guideline.

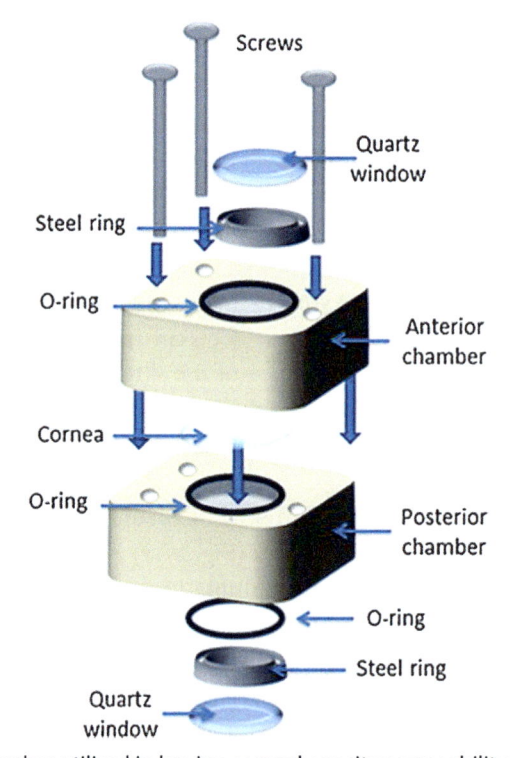

Figure 17.2 Testing chamber utilized in bovine corneal opacity permeability test (Wilson et al., 2015).

EYE IRRITATION TESTS

The eye can be exposed to cosmetic products and their ingredients (e.g., mascaras, eye creams) or through accidental exposure (e.g., shampoos). Therefore, the evaluation of eye irritation potential for a cosmetic product and its ingredients is essential to provide reassurance that a product is safe for consumers.

As of this writing there are no fully validated alternative methods replacing the classical Draize in vivo eye irritation test. The conventional test for the irritant and corrosive potential of chemicals is the rabbit eye test (Draize et al., 1944), and it has become the international standard assay for acute eye irritation and corrosion (EC B.5, OECD TG 405). The test material is applied to the conjunctival sac of the animal's eye and subsequent grading of the ocular lesion is established: cornea opacity, iris lesion, redness of the conjunctiva and oedema of the conjunctiva (chemosis).

Major validation studies took place in the 1990s to replace the Draize test for eye irritation testing (Balls et al., 1995; Gettings et al., 1996; Bradlaw et al., 1997; Ohno et al., 1991). Good reproducibility and reliability of the most valuable alternative methods have been demonstrated, but so far it has not been possible to identify a single method able to replace the Draize rabbit eye test. This is due to different factors including the limited quality of the existing in vivo data, limitations of the animal test method, and the fact that the range of criteria for injury and inflammation covered by the Draize rabbit eye test is unlikely to be replaced by a single in vitro test.

This section includes some of the most promising methods that may replace animals in the eye irritation test (Table 17.3). The alternative methods in this field comprise isolated organs, chorioallantoic membrane methods, tissue and cell culture systems and physico-chemical tests (Christian and Diener, 1996; Chamberlain et al., 1997).

The available alternative methods for eye irritation/corrosion consist of a screening battery of two assays, namely the BCOP (bovine corneal opacity permeability) (OECD 437; EC B.47) and the ICE (isolated chicken eye) (OECD 438; EC B.48). They can be used in the process of hazard identification (not risk assessment) and allow the elimination of severe eye irritants, but fail to detect mild irritants. Two other screening tests, namely the IRE (isolated rabbit eye) and the HET–CAM (hen's egg test–chorioallantoic membrane), also provide supportive evidence for cosmetic ingredient safety assessment.

- The BCOP test uses freshly isolated corneas from slaughterhouse material. The cornea is mounted horizontally on a holder, which is placed inside a specially modified opacitometer with controlled temperature. This cornea divides the test chamber into two compartments (Fig. 17.2). The test compound is added to the compartment enclosing the epithelial surface of the cornea. After the opacity is measured, a fluorescein-containing solution is added to the epithelial side (i.e., the upper compartment) to determine corneal permeability by assessing the optical density of the medium in the lower compartment. The measured numerical values for opacity and permeability can be used to calculate a so-called in vitro score (Gautheron et al., 1994; Sina et al., 1995). Test materials can be classified according to this score. Better prediction of certain chemical classes can be obtained by including a histological evaluation of the cornea (Curren et al., 1999). The BCOP test is well suited to classifying moderate, severe and very severe eye irritants because the depth of injury can

Table 17.3 In vitro eye irritation alternative methods for achieving animal replacement (Commission Regulation EU No. 1152/2010 and OECD)

Alternative test	Test system	Endpoint	Status (EU/OECD TG)
Bovine corneal opacity and permeability test	Excised cornea from the bovine eye	Opacity and permeability of the cornea	B47/437
Isolated chicken eye	Isolated chicken eye corneal swelling	Corneal opacity and fluorescein retention	B48/438
Isolated rabbit eye test	Isolated rabbit eye corneal swelling	Corneal opacity and fluorescein retention	Under validation
Hen's egg test–chorioallantoic membrane	Hen's egg chorioallantoic membrane	Damage to chicken chorioallantoic membrane	Under validation
Cytosensor microphysiometer test	Monolayer of adherent cells (mouse L929 fibroblasts)	Decrease in metabolic rate (glucose utilization rate)	Validated (ESAC)
Fluorescein leakage	Monolayer of MDCK CB997 cells	Permeability	Validated (ESAC)/460

ESAC, European Centre for the Validation of Alternative Methods Scientific Advisory Committee; *EU*, European Union; *OECD*, Organization for Economic Cooperation and Development; *TG*, testing guideline.

be measured in this ex vivo tissue model. However, the resolution of mild to very mild levels of irritancy would be better suited to an epithelial tissue construct or cytotoxicity assay. For these reasons, the BCOP assay in tandem has been put forward with assays measuring other endpoints such as cytotoxicity (Swanson et al., 1995) or as part of a test battery (Sina and Gautheron, 1994).

- The ICE test assesses the ocular irritation potential of a test substance following topical application onto the surface of the cornea of an enucleated chicken eye (obtained from freshly slaughtered chickens). The eyes (pre- and post-treatment) are maintained in a superfusion chamber at 32°C. The direct effect of the test substance on the cornea can be assessed using multi-endpoint analysis following evaluation of changes in corneal thickness, changes in corneal opacity, alteration of the corneal epithelium and fluorescein uptake up to 240 min following treatment. The ICE test has been shown to reliably differentiate between negligible and moderate to severe irritants (Prinsen, 1996).

The following methods are under validation (ECVAM) as of this writing.

- The IRE test (Burton et al., 1999) determines the opacification of the cornea and the increase in corneal thickness (corneal swelling) after exposure to irritant substances. Whole eyeballs obtained by immediate dissection from humanely killed laboratory rabbits with healthy eyes are mounted and maintained in a vertical position in a superfusion chamber, with controlled temperature and humidity. This ensures that the eyes remain viable throughout the duration of the test. Pre-warmed saline solution is applied drop by drop directly onto the cornea at regular intervals to keep it moist. Prior to treatment with the test sample in question, the eye is checked visually for opacity together with an evaluation of fluorescein penetration and cornea swelling, and damaged tissues are excluded. The eyeball is then either taken out of the chamber or left in situ (depending on the type of test material) and exposed to the test chemical; for example, 10 s for identification of severe irritants and 1 min (or longer) for the ranking of less severely damaging materials (Whittle et al., 1992; York et al., 1994). After removal of the chemical, the eye is repositioned in the chamber and the cornea is examined for evidence of opacification and corneal thickness is measured. Further assessments are made at different times (30 min; 1, 2, 3 and 4 h) after dosing. A check on fluorescein penetration is carried out 4 h after treatment. Scores for corneal opacity (similar to Draize scores) and fluorescein penetration are recorded (qualitative assessments). For each test sample the mean percentage of the corneal swelling of three eyes is calculated and compared to that of an untreated control eye. The preparation and examination of histological sections of the treated corneas can be used to confirm the level and depth of corneal damage. Overall damage is assessed by means of combining the different parameters scored, depending on the nature of the effects observed, and in-house classification systems may vary (Whittle et al., 1992). Chemical substances causing the cornea to swell by more than 15% were considered to have the potential to cause severe eye irritation in vivo; however, more recently a more complex classification model combining opacity, corneal swelling and histological observations of the corneal epithelium has been published. The IRE test is considered to be well suited to identifying substances that are 'severely irritating to the eye' according to EU classification R 41. It is considered to be a valid test for the (pre-)screening of severely irritating materials/formulations (Chamberlain et al., 1997). The assay is useful for in-house comparative testing of materials/formulations to obviate the need for any animal testing during product development.

- The CAM of the fertilized chicken egg is considered a suitable model for establishing what effects substances have on the conjunctival tissues of the eye (Christian and Diener, 1996).
  - The HET–CAM enables irritant reactions to be identified that are similar to those which occur in the eye using the standard Draize rabbit eye test. In the HET–CAM system (Fig. 17.4), three reactions are determined, namely, haemorrhage, lysis and coagulation (sometimes hyperaemia is also used as a parameter) of the chorioallantoic membrane on the ninth day of embryonation, when nerve tissue and pain perception have not yet developed. After the test sample is placed

directly onto the CAM, the aforementioned parameters are evaluated over a 5-min observation period. The most widely used approach is the reaction time method, which determines the time taken for each of the three endpoints to appear. Another approach is the irritation threshold method, which determines the test material concentration at which effects on these parameters are first observed. While these approaches are mainly used for transparent test materials, a third approach for non-transparent insoluble and solid materials can be used by exposing the CAM to test samples for a fixed time (e.g., 30 s or 5 min) and examining the membrane after careful rinsing to remove the sample (Fig. 17.3). The majority of the validation studies carried out showed a useful correlation between the HET–CAM and the Draize rabbit eye test for the assessment of raw materials and cosmetic products. These in vivo versus in vitro correlations revealed good results in the area of mild and non-irritating test materials as well as for surfactants and surfactant-based formulations (Spielmann et al., 1993). As a consequence, the HET–CAM can be regarded as an alternative, which is already leading to a reduction in animal experiments.

For the cytosensor microphysiometer (CM) and the fluorescein leakage (FL) test methods, which have been validated by ECVAM in 2009, a draft OECD guideline is in progress.

- The CM test is a cytotoxicity and cell function–based in vitro assay that is performed on a sub-confluent monolayer of adherent cells (mouse L929 fibroblasts) cultured on a transwell polycarbonate insert with a porous membrane, which functions as electrode, and a light-addressable potentiometric sensor detecting changes in pH (acidity). Mechanistically, the CM test is intended to model the cytotoxic action of an irritant chemical on the cell membranes of the corneal and conjunctival epithelium where the test chemical would reside in an in vivo exposure. The CM estimates the metabolic rate (glucose utilization rate) of a population of cells maintained in

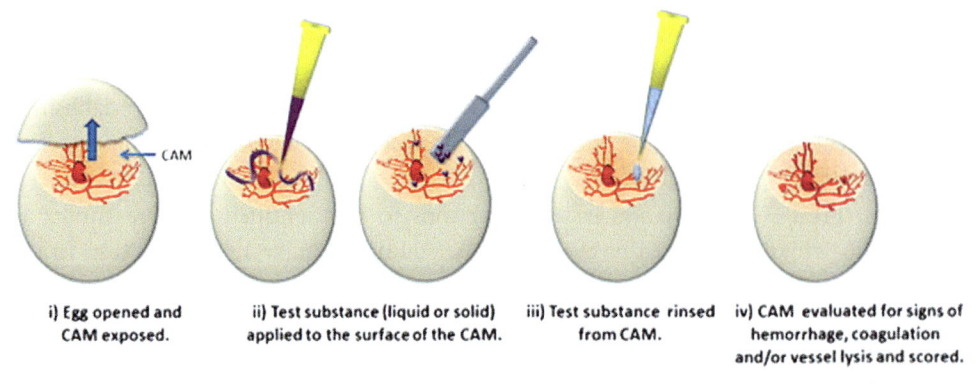

Figure 17.3 Hen's egg test–chorioallantoic membrane (*CAM*) (Wilson et al., 2015).

low-volume flow through chambers by measuring the rate of excretion of acid byproducts and the resulting decrease in pH of the surrounding medium. The metabolic rate is determined indirectly by the number of protons excreted into the low-buffer medium (change in pH) per unit time (OECD, 2010).

- The FL test is an in vitro test that can be used under certain circumstances and with specific limitations to classify chemicals (substances and mixtures) as ocular corrosives and severe irritants. The FL has been adopted by the OECD and is recommended as part of a tiered testing strategy for regulatory classification and labelling but only for limited types of chemicals (i.e., water-soluble substances and mixtures) (OECD TG 460). The FL test is a cytotoxicity and cell function–based in vitro assay that is performed on a confluent monolayer of MDCK CB997 tubular epithelial cells that are grown on semipermeable inserts and model the non-proliferating state of the in vivo corneal epithelium. The MDCK cell line is well established and forms tight junctions and desmosomal junctions similar to those found on the apical side of conjunctival and corneal epithelia. Tight and desmosomal junctions in vivo prevent solutes and foreign materials from penetrating the corneal epithelium. Loss of trans-epithelial impermeability, due to damaged tight junctions and desmosomal junctions, is one of the early events in chemically induced ocular irritation. Cells are grown on inserts so that, when confluent, the cell layer separates the medium into two compartments. The cells are exposed to the test substance in the upper compartment for a set period. Then the sample solution is removed and, after a washing step, the cells are incubated with a sodium fluorescein solution (usually for 30 min). Disruption of the cell layer permits the diffusion of fluorescein into the lower compartment, which is measured spectrophotometrically.

Two tests, including human reconstructed tissue models, have been validated (ESAC).

- The SkinEthic in vitro reconstituted human corneal epithelium consists of immortalized human corneal epithelial cells (HCE cell line, LSU Eye Center, New Orleans, LA, USA) that are cultivated at the air–liquid interface in a chemically defined medium on a polycarbonate substrate and form an air–epithelial tissue, devoid of stratum corneum, morphologically resembling the corneal mucosa of the human eye.

    Triplicate in vitro reconstituted human corneal epithelial tissues (size 0.5 cm^2) are topically dosed with a small amount of test agent for different times:
    - For finished products, tissues are dosed for 10 min and 1, 3 and 24 h.
    - For chemical raw materials, tissues are dosed for 10, 20, 30 and 60 min.

    A negative control (phosphate-buffered saline solution), as well as positive controls (sodium dodecyl sulphate 0.5% and 1%) are run in parallel. At each time point, duplicate tissues are assessed for tissue viability (MTT assay), and one culture is fixed in a balanced 10% formalin solution for histological analysis, which is performed when the MTT assay data show no tissue toxicity. Meanwhile, the culture medium underneath the tissues is stored at −20°C for pro-inflammatory mediator analysis.

Possible endpoint measurements are tissue viability using the MTT assay or lactate dehydrogenase release, histology and quantification of cytokine release (e.g., IL-1α, IL-6, IL-8, PGE$_2$). Until this test has passed formal validation, it is suitable as a pre-screen for eye irritation of chemicals and finished products (Nguyen et al., 2003).

- The EpiOcular corneal model involves culturing normal, human-derived epidermal keratinocytes to form a stratified, squamous epithelium similar to that found in the cornea. The epidermal cells, which are cultured on specially prepared cell culture inserts using serum-free medium, differentiate to form a multilayered structure which closely parallels the corneal epithelium. EpiOcular is mitotically and metabolically active and releases many of the pro-inflammatory agents (cytokines) known to be important in ocular irritation and inflammation. Comparison with in vivo animal data has been carried out, using the ET$_{50}$ value determined by MTT assay. Using the variable of time rather than dose allows ingredients and formulations to be tested without dilution in medium. Thus both hydrophobic and hydrophilic materials may be tested.

As a stratified epithelium, the EpiOcular system is intended to model damage to the corneal epithelium and conjunctiva (with its very thin epithelium). Therefore, it can be used to resolve degrees of irritancy potential (cellular damage) in the moderate to very mild irritancy range (Stern et al., 1998). The model is also capable of identifying high, moderate and severe irritants by their very short ET$_{50}$ values. However, based on the stromal changes associated with severe irritation, an epithelial construct would not be expected to provide the degree of resolution in the severe range that a full-thickness cornea (e.g., ex vivo cornea) would provide (Blazka et al., 2003). The EpiOcular system can be used to differentiate between mild and moderate irritants and identify potentially severe irritants.

On the other hand, a number of cytotoxicity/cell function–based assays for water-soluble substances (the CM test, the FI test and the neutral red release, FI and red blood cell haemolysis test) have undergone retrospective validation and peer review by ESAC. These tests, however, are only screening assays and are not suitable for determining the potency of eye irritancy.

Finally, eye irritation is a difficult endpoint to model in silico because of the complexity of the biological mechanisms that may be involved. Therefore, most approaches have modelled eye irritation resulting from physical effects, such as ocular penetration or corrosion (Cronin et al., 2003).

SKIN SENSITIZATION

A skin sensitizer is an agent that is able to cause an allergic response in susceptible individuals. The consequence of this is that following subsequent exposure via the skin, the characteristic adverse health effects of allergic contact dermatitis may be provoked. As of this writing, there is not a validated in vitro test method accepted for skin sensitization.

Since the mid-1960s, skin-sensitization potential assessment has been of paramount importance for ensuring the safety of cosmetic products. Different human sensitization tests, such as the Schwartz–Peck test, human repeated-insult patch tests (Marzulli and Maibach, 1973; Griffith and Buehler, 1976) and the human maximization test (Kligman and Epstein, 1975), have been used.

On the other hand, skin sensitization is a very complex biological process, involving a series of interrelated events, many of which are either not understood or only partially understood. Therefore, it is not surprising that at the time of writing, there are no validated non-animal tests for assessment of skin-sensitization potential.

There are three common in vivo laboratory animal test methods to evaluate the potential of a substance to cause skin sensitization:

- The local lymph node assay (LLNA) has already been validated and adopted for regulatory use in the EU (B.42 Regulation EC No. 440/2008) and is equivalent to the OECD TG 429 (OECD, 2002). The LLNA has been accepted by the ICCVAM in the United States as a stand-alone alternative to the current guinea pig tests. This is a method to assess the skin-sensitization potential of chemicals in animals (Kimber et al., 1994, 1998). Therefore, the LLNA is an in vivo method (Kimber et al., 2002) and, consequently, will not eliminate the use of animals in assessing contact sensitizing activity. It can, however, potentially reduce the number of animals required for this purpose. The basic principle underlying the LLNA is that sensitizers induce a primary proliferation of lymphocytes in the lymph node draining the site of chemical application. This proliferation is proportional to the dose applied (and to the potency of the allergen) and provides a quantitative measurement of sensitization. The LLNA assesses this proliferation as a dose–response relationship in which the proliferation in test groups is compared to that in vehicle-treated controls. The method is based on the use of radioactive labelling ($[^3H]$methyl thymidine) to measure cell proliferation. The ratio of the proliferation in treated groups to that in vehicular controls, the stimulation index (SI), is determined. The test is positive when the SI ≥ 3.

  LLNA methods using a non-radioactive methodology have been adopted by the European Commission and OECD:
  - The LLNA–DA is a modified LLNA method using adenosine triphosphate (ATP) as an endpoint. The mice are exposed four times instead of three times and the ATP content is used as a measure of the proliferation of the lymph node cells (EC B.50; OECD 442A).
  - The LLNA–BrdU–ELISA is a cell proliferation ELISA (enzyme-linked immunosorbent assay) using 5-bromo-2-deoxyuridine, which is a second-generation ELISA with calorimetric or chemiluminescence detection which quantifies the DNA synthesis within the lymph node cells (NICEATM–ICCVAM website, test method evaluation report LLNA: BrdU–ELISA1) (EC B.51; OECD 442B).

A reduced LLNA (rLLNA) was adopted by ESAC after a retrospective analysis of published data. However, as the rLLNA uses only the negative control group and the equivalent of the high-dose group of the original LLNA, no determination of the sensitizing potency is possible. Therefore the rLLNA is suitable only for screening purposes to distinguish between sensitizers and non-sensitizers (SCCS/1294/10).

- The Magnusson–Kligman guinea pig maximization test (GPMT) and Buehler guinea pig test are available as No. B.06 (Directive 96/54/EC, 1996) and OECD 406. These methods are accepted for hazard identification of skin-sensitizing substances. The GPMT is a highly sensitive method using Freund's complete adjuvant (FCA) as an immune enhancer (Magnusson and Kligman, 1970), but the Buehler test is less sensitive and may underestimate the sensitization potential of a substance. The limitation of the tests is that the evaluation is based on visual inspection of erythema and personal judgement. For these reasons, evaluation of coloured test substances, e.g., pigments and dye stuffs, is often impossible owing to staining of the skin by the test substance.
- The mouse ear swelling test is a useful model for identifying strong contact sensitizers (Asherson and Ptak, 1968). Several weeks prior to and during the test period mice are fed a diet enriched in vitamin A for enhancement contact sensitization. The induction phase comprises clipping off the fur on the belly region and removal of the outer layers of the epidermis by tape stripping. FCA is injected intradermally before the test substance is applied topically, in vehicle (test mice), or vehicle alone (control mice). The ear thickness of test and control mice is measured under anaesthesia with a micrometer 24 and 48 h after the test substance is applied. Ear swelling is expressed as the difference between test and control ears in percentage.

On the other hand, various alternative in vitro models are being developed to assess the skin-sensitizing potential of chemicals and products. These methods are based on cell systems which have been shown to be capable of distinguishing between sensitizers and non-sensitizers, depending upon the profile of cytokine release. However, they are still at a basic research level.

SKIN ABSORPTION STUDIES

Human exposure to cosmetic substances occurs mainly via the skin. To reach the circulation (blood and lymph vessels) cosmetic ingredients must cross a number of cell layers of the skin, of which the rate-determining layer is considered to be the stratum corneum (SC). A number of factors play a key role in this process, including the lipophilicity of the compounds, the thickness and composition of the SC (which depends on the body site), the duration of exposure and the amount of topically applied product (Schaefer and Redelmeir, 1996).

Skin absorption testing of chemicals is normally not required by legislation governing dangerous chemical substances. However, cosmetic companies still require skin absorption data from ingredient suppliers, in particular in the case of 'actives', e.g., colourants, preservatives, UV filters and substances with restrictions on concentration or site of application. For this purpose, knowledge of dermal absorption studies is essential. The methods for measuring dermal absorption and dermal delivery can be divided into two categories: in vivo and in vitro. Both in vivo and in vitro testing protocols form part of the lists of official EU and OECD test methods (EC B.44, 45; Regulation No. 440/2008; OECD TG 427, 428).

In Vitro Dermal Penetration

The purpose of in vitro dermal absorption studies of cosmetic substances is to obtain qualitative and/or quantitative information on the compounds that may enter, under in-use conditions, into the systemic compartment of the human body. The quantities can then be taken into consideration to calculate the margin of safety using the no-observed-adverse-effect level of an appropriate repeated-dose toxicity study with the respective substance.

In vitro testing is carried out on excised pig or human skin, according to OECD Guideline 428 (OECD, 2004). In general animal skin (e.g., rat skin) is more permeable and therefore may overestimate human percutaneous absorption. The use of in vitro dermal absorption studies on isolated skin is based on the fact that the epidermis, in particular the SC, forms the principal in vivo barrier of the skin against the penetration and uptake of xenobiotics in the body.

The test substance is applied in an appropriate formulation on the skin sample, which is usually placed in a diffusion cell (Fig. 17.4). The diffusion cell consists of an upper donor and a lower receptor chamber, separated by the skin preparation. The cells are

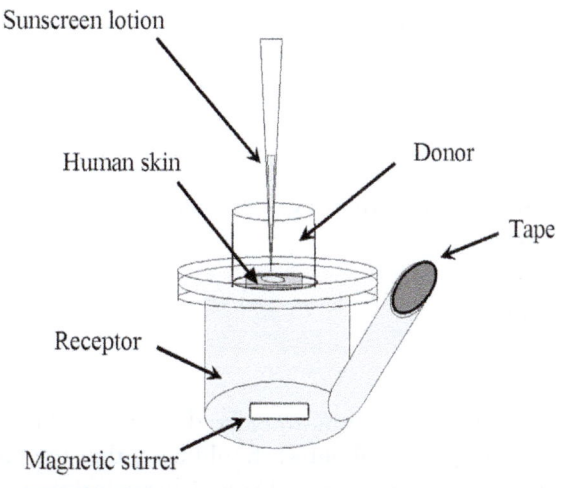

Figure 17.4 The Franz diffusion cell consists of an upper donor and a lower receptor chamber, separated by a skin preparation (Herráez Domínguez and Díez-Sales, 2007).

made preferably from an inert non-adsorbing material. Temperature control of the receptor fluid is crucial throughout the experiment. The skin surface temperature in the diffusion cell should be kept at the in vivo skin temperature of 32°C.

The receptor fluid is well mixed throughout the experiment. The composition of the receptor fluid is chosen so that it does not limit the extent of diffusion of the test substance, i.e., the solubility and stability in the receptor fluid of the chemical under assay must be guaranteed. Saline or buffered saline solutions are commonly used for hydrophilic compounds. For lipophilic molecules, serum albumin or appropriate solubilizers/emulsifiers are added in amounts which do not interfere with membrane integrity. One must ensure that the amount of penetrant in the receptor fluid is less than 10% of its saturation level at any time. The substance must remain stable in the receptor fluid for the duration of the in vitro test and the subsequent analysis (SCCNFP/0750/03).

Both the dose and the contact time (exposure) with the skin are chosen to mimic intended use conditions. The amount of the formulation to be applied is between 2 and 5 mg/cm^2 for solids and semi-solid preparations, and up to 10 μL/cm^2 for liquids. The volume of formulation used should be enough to spread the sample homogeneously over the skin surface. This strongly depends on the viscosity of the formulation.

The exposure time and sampling period should be defined in the protocol. The normal exposure time is 24 h. Longer duration may result in membrane deterioration and requires membrane integrity to be carefully checked. Barrier integrity should be checked using a suitable method. This is achieved either by measuring the penetration of a marker molecule, e.g., tritiated water, caffeine or sucrose, or by physical methods like transepidermal water loss or TER measurements. Data obtained should be reported.

Sampling frequency depends on the rate/extent of dermal absorption. Appropriate analytical techniques, e.g., scintillation counting, high-performance liquid chromatography or gas chromatography, should be used. Their validity, sensitivity and detection limits should be documented in the report. When an increase in sensitivity is needed, the test substance should, whenever possible, be radiolabelled. Qualitative or semi-quantitative methods such as microautoradiography can be useful tools for skin distribution assessments.

Amounts of the test compound must be determined in (SCCNFP/0750/03):
- the surplus on the skin
- the SC (e.g., adhesive tape strips)
- the epidermis without SC
- the dermis
- the receptor fluid

The mass balance of the applied dose must be determined. The overall recovery of the test substance (including its metabolites) should be within the range of 85%–115%. If lower recoveries of the test substance are obtained, the reasons need to be investigated and explained. One should check for substance adsorption in the equipment.

The results of dermal absorption studies should be expressed as an absolute amount (μg/cm² of skin surface) and as a percentage of the amount of test substance contained in the intended dose applied per square centimetre of skin surface.

The amounts of penetrated substance found in the receptor fluid are considered to be systemically available. The epidermis (except for the SC) and dermis are considered a sink, therefore the amounts found in these tissues are equally considered absorbed and are added to those found in the receptor fluid. The amounts that are retained by the SC at the time of sampling are not considered dermally absorbed and, thus, they do not contribute to the systemic dose. The absorption rate and mass balance should be calculated separately for each diffusion cell. Only then can the mean ± SD be calculated.

In Vivo Studies

This type of study has some advantages over in vitro methods, which include the generation of systemic kinetic and metabolic information (Schaefer and Redelmeier, 1996). The disadvantages are the use of live animals, the need for radiolabelled material to facilitate reliable results, difficulties in determining the early absorption phase and the differences in permeability of the preferred species (rat) and human skin (Sanco/222/2000).

Animal testing should be carried out in accordance with OECD TG 427 (OECD, 2004). The rat is the most commonly used species for tests in vivo. In rats, the application site should be about 10 cm² and should be defined by a device, which is secured on the skin surface. The test preparation is applied to the skin surface, and remains there for a specified period of time, relating to potential human exposure. During exposure, animals are housed individually in metabolic cages from which excreta are collected. At the end of the study, the amount found in the skin, the carcass and the excreta is determined. These data give an estimate of the total recovery of the test substance. The skin absorption of the test substance can be expressed as the percentage of dose absorbed per unit time or in terms of an average absorption rate per unit area of skin (e.g., μg/cm²/h).

REPEATED-DOSE TOXICITY

Chronic toxicity is a consequence of the persistent or progressively deteriorating dysfunction of cells, organs or multiple organ systems, resulting from long-term exposure to a chemical.

The following in vivo repeated-dose toxicity tests are available:
- repeated-dose (28 days) toxicity (oral) (EC B.7; OECD 407)
- repeated-dose (21–28 days) toxicity (dermal) (EC B.9; OECD 410)
- repeated-dose (28 days) toxicity (inhalation) (EC B.8; OECD 412)
- subchronic oral toxicity test: repeated-dose 90-day oral toxicity study in rodents (EC B.26; OECD 408)

- subchronic oral toxicity test: repeated-dose 90-day oral toxicity study in non-rodents (EC B.27; OECD 409)
- subchronic dermal toxicity study: repeated-dose 90-day dermal toxicity study using rodent species (EC B.28; OECD 411)
- subchronic inhalation toxicity study: repeated-dose 90-day inhalation toxicity study using rodent species (EC B.29; OECD 413)

The 28- and 90-day oral toxicity tests in rodents are the most commonly used repeated-dose toxicity tests and often give a good indication on target organs and type of systemic toxicity. Preferably studies of 90 days or more should be used in safety assessments. For repeated-dose toxicity testing, no validated or generally accepted alternative method is available at the time of writing to replace animal testing.

In the case of developing cosmetic ingredients with specific biological properties and which will come into contact with human skin for long periods of time, evaluation of the systemic risk is a key element in evaluating the safety of these new ingredients. Therefore the Scientific Committee on Consumer Safety (SCCS) considers that in certain cases the use of long-term animal experiments to study one or more potential toxic effects remains a scientific necessity. It is self-evident that animal use should be limited to a minimum, but never at the expense of consumer safety.

On the other hand, more efforts are needed to further evaluate the usefulness of QSAR approaches to predict toxicity. In particular, the statistically based system TOPKAT has been proposed as a model for rat chronic lowest observed adverse effect level (Cronin et al., 2003).

GENOTOXICITY/MUTAGENICITY

These areas are an important part of the hazard assessment of chemicals for regulatory purposes. Genotoxicity is a broader term that refers to the ability to interact with DNA and/or the cellular apparatus that regulates the fidelity of the genome, such as the spindle apparatus and topoisomerase enzymes (2006/1907/EC). Mutagenicity refers to the induction of permanent transmissible changes in the structure of the genetic material of cells or organisms. These changes (mutations) may involve a single gene or a block of genes (SCCNFP/0755/03).

Several well-established in vitro mutagenicity/genotoxicity tests are available, described in the OECD guidelines and/or in Regulation (EC) No. 440/2008. In principle, the SCCS recommends, for the base level testing of cosmetic substances, three assays, represented by the following test systems:
- tests for gene mutation
  - bacterial reverse mutation test (EC B.13/14; OECD 471)
  - in vitro mammalian cell gene mutation test (EC B.17; OECD 476)
- tests for clastogenicity and aneugenicity

- in vitro micronucleus test (EC B.49; OECD 487) or
- in vitro mammalian chromosome aberration test (EC B.10; OECD 473)

Several in vitro tests are routinely used and accepted by regulatory authorities, but they present crucial limitations which affect the usefulness of the assays to predict mutagenicity/genotoxicity potential of a substance in vivo in mammals and especially in humans. Because of these limitations, no single in vitro test can fully replace an existing in vivo animal test yet.

ULTRAVIOLET-INDUCED TOXIC EFFECTS

Cosmetic ingredients and mixtures of ingredients absorbing UV light, in particular UV filter chemicals used, e.g., to ensure light stability of cosmetics, or used in sun protection products, need to be tested for acute phototoxicity and photogenotoxicity potential. Testing for photosensitization (immunological photoallergy) is not specifically required; nevertheless it is often performed.

Acute phototoxicity is defined as a toxic response that is elicited after the first exposure of skin to certain chemicals and subsequent exposure to light, or that is induced similarly by skin irradiation after systemic administration of a chemical. In this area there are three possible methods to use.

- The in vitro 3T3 neutral red uptake phototoxicity test (3T3 NRU PT) is used with a basic screen to identify acute phototoxic potential (EC B41; OECD 432). It is based on a comparison of the cytotoxicity of a chemical when tested in the presence and in the absence of exposure to a non-cytotoxic dose of UVA/VIS light. Cytotoxicity in this test is expressed as a concentration-dependent reduction in the uptake of the vital dye neutral red 24h after treatment with the test chemical and irradiation.

  A permanent mouse fibroblast cell line, Balb/c 3T3, is maintained in culture for 24h for the formation of monolayers. Two 96-well plates per test chemical are then preincubated with eight different concentrations of the chemical for 1h. Thereafter one of the two plates is exposed to a non-cytotoxic UVA/VIS light dose of 5 J/cm^2 UVA, whereas the other plate is kept in the dark. In both plates, the treatment medium is then replaced by culture medium and followed with 24h of incubation; cell viability is determined by neutral red uptake for 3h (Fig. 17.5). The neutral red penetrates cell membranes and accumulates intracellularly in lysosomes. Alterations in the cell surface or sensitive lysosomal membranes lead to a decreased uptake of the neutral red. To discriminate between photoirritant and non-photoirritant chemicals, the photoirritation factor (PIF) was defined as the ratio of the IC$_{50}$ values, determined in the absence of UVA and the presence of UVA (Eq. 17.1):

$$\text{PIF} = \text{IC}_{50}(-\text{UV})/\text{IC}_{50}(+\text{UV}) \tag{17.1}$$

```
┌─────────────────────────────┐
│ A permanent mouse fibroblast│
│    cell line (Balb/c 3T3)   │
└──────────────┬──────────────┘
               │
┌──────────────┴──────────────┐
│     Pre-incubation (24 h)   │
└──────────────┬──────────────┘
               │
┌──────────────┴──────────────────────────┐
│          Treatment (1 h)                │
│ [eight different concentrations of the  │
│              chemical]                  │
└────────┬───────────────────────┬────────┘
         │                       │
┌────────┴────────┐      ┌───────┴────────┐
│  +UV experiment │      │ -UV experiment │
└────────┬────────┘      └───────┬────────┘
         └───────────┬───────────┘
                     │
         ┌───────────┴───────────┐
         │  Neutral red uptake   │
         └───────────┬───────────┘
                     │
         ┌───────────┴─────────────────┐
         │ Treatment of cellular       │
         │      disintegration         │
         └───────────┬─────────────────┘
                     │
         ┌───────────┴───────────┐
         │   Determination of NR │
         │    ($\lambda$ = 540 nm)│
         └───────────────────────┘
```

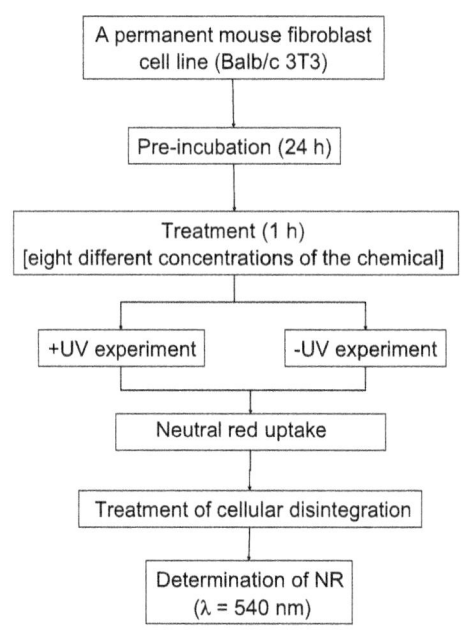

Figure 17.5 3T3 neutral red (NR) uptake phototoxicity test is used with a basic screen to identify acute phototoxic potential.

- Two additional tests, the red blood cell phototoxicity test (RBC PT) (Pape et al., 1994) and the human 3-D skin model in vitro phototoxicity test (H3D PT) (Roguet et al., 1994; Bernard et al., 1999; Liebsch et al., 1999; Jones et al., 2003), are regarded as useful and important adjunct tests to overcome some limitations of the 3T3 NRU PT, like the fairly low UVB tolerance of 3T3 fibroblasts. Moreover, the RBC PT enables evaluation of the phototoxic mechanisms involved (Okamoto et al., 1999) and the H3D PT model is qualified as an adjunct test to further investigate chemicals with (probably false) positive outcomes in the 3T3 NRU PT.

In conclusion, in vitro tests are regarded as sufficiently covering identification of acute phototoxic hazards; therefore, at present animal testing with this objective for that endpoint can be replaced 100%.

In the area of *photochemical genotoxicity*, almost the whole battery of in vitro genetic toxicity tests has been converted into test protocols of photogenotoxicity tests, and several in vitro tests can be used:

- Photo Ames test (P-AMES): Prokaryotic, bacterial mutation tests are easier and cheaper to perform than any photogenotoxicity tests with eukaryotic mammalian cells (Chételat et al., 1993). Therefore, the P-AMES test was the first in vitro photogenotoxicity test adapted to the parallel use of light (Jose, 1979) and later proposed for safety testing of cosmetics (Loprieno, 1999). Nevertheless, the P-AMES test has never been validated formally. The critical point (Brendler-Schwaab et al., 2004) of all P-AMES test protocols is the UV sensitivity of the *Salmonella typhimurium* or

Escherichia coli strain used. Chemicals often have to be pre-irradiated in the absence of the bacteria to achieve doses of light necessary to activate the photogenotoxins.
- The photo chromosome aberration test (P-CAT) is used to detect photochemically induced clastogenicity, the most relevant in vitro endpoint for the assessment of the photocarcinogenic hazard potential of substances (Gocke et al., 2000; Brendler-Schwaab et al., 2004). The purpose of the in vitro P-CAT is to identify agents that cause structural chromosomal aberrations in cultured mammalian cells in the presence of a non-clastogenic UV/VIS radiation. Structural aberrations fall into two categories, chromosome or chromatid. With the majority of chemical mutagens, induced aberrations correspond to the chromatid type, but chromosome-type aberrations also occur. The in vitro P-CAT may employ cultures of established cell lines (mostly Chinese hamster lines CHO, V79, CHL) or primary cell cultures (e.g., human lymphocytes). The P-CAT has not been formally validated so far. Despite this, the P-CAT is recommended and accepted by the Scientific Committee on Cosmetic Products and Non-food Products for safety testing of cosmetic ingredients.
- The photo micronucleus test (P-MNT) is an in vitro mutagenicity test system to detect chemicals which induce the formation of small membrane-bound DNA fragments, i.e., micronuclei in the cytoplasm of interphase cells. These micronuclei may originate from chromosome fragments lacking a centromere or whole chromosomes which are unable to migrate with the rest of the chromosomes during the anaphase of cell division. Thus, the micronucleus assay is in principle suitable to detect clastogenic and aneugenic effects (the latter is not relevant to photogenotoxicity testing).

The method employing Chinese hamster V79 cells (Kalweit et al., 1999) was successfully adapted to photogenotoxicity testing (Kersten et al., 1999, 2002). The method is based on concentration-response experiments performed with and without irradiation with a UV/VIS sunlight simulation.
- The photo comet test (P-COMET) is a method for electrophoretic measurement of DNA strand breaks. The P-COMET is applied to detect DNA strand breaks induced by a substance and subsequent or concurrent UV/VIS irradiation, using a small number of eukaryotic cells. P-COMET is an extremely sensitive test, but it has produced negative results with some compounds that provide positive results in other photogenotoxicity tests, like the P-MNT. Such chemicals are, for example, nalidixic acid and 8-methoxypsoralen, of which only the latter result is understood as DNA–DNA cross-linking under UV light (Brendler-Schwaab et al., 2004).

CARCINOGENICITY

Substances are defined as carcinogenic if they induce tumours (benign or malignant), increase tumour incidence or malignancy, or shorten the time of tumour occurrence when they are inhaled, ingested, dermally applied or injected. The process of carcinogenesis is now recognized to be caused by the transition of normal cells into cancer cells, via a sequence of stages and complex biological interactions. It is generally accepted that carcinogenesis is a multistep process that is strongly influenced by factors such as age, diet, environment, hormonal balance, etc. The complexity of this process makes it not only difficult to extrapolate findings in animals to humans, but also difficult to develop in vitro alternative test models.

Because the induction of cancer involves genetic alterations, which can be induced directly or indirectly, carcinogens have conventionally been divided into two categories according to their presumed mode of action: genotoxic carcinogens (=initiators) and non-genotoxic carcinogens (epigenetic carcinogens = promoters). Most potent mutagens are also carcinogens in animal experiments. Information about the genotoxic carcinogenic potential of a substance may be obtained from mutagenicity/genotoxicity studies.

Two in vitro tests, cell transformation assay and gap junction intercellular communication (GJIC), have been proposed as tests that may provide information on possible non-genotoxic as well as genotoxic carcinogens. In cases in which the in vitro short-term tests suggest possible carcinogenic potential, carcinogenicity bioassays in animals are useful to determine potency and target organs.

- The cell transformation assay uses mammalian cell culture systems to detect phenotypic changes in vitro induced by chemical substances associated with malignant transformation in vivo.
- GJIC is the intercellular exchange of low-molecular-weight molecules (less than 1000–1500 Da) through gap junction channels between adjacent cells, and has been found to play an important role in the regulation of cell growth and differentiation. Several methods exist to determine GJIC in different cell types (Rosenkranz et al., 2000). Dysfunction in this type of communication has been observed to result in abnormal cell growth and behaviour, and is associated with several pathological conditions in humans. Structure–activity studies have shown a relationship between the ability of substances to inhibit GJIC and their ability to induce tumours in rodents, but not between inhibition of GJIC and genotoxic activity. This suggests that GJIC inhibition is involved in non-genotoxic cancer induction, and is a candidate endpoint in screening assays for the identification of non-genotoxic carcinogens and tumour promoters, which is not detected by conventional genetic toxicology tests.

Finally, a large number of systems and models dedicated to predicting carcinogenicity have been developed (QSARs and SARs).

REPRODUCTIVE AND DEVELOPMENTAL TOXICITY

Reproductive toxicity refers to the adverse effects of a substance on any aspect of the reproductive cycle, including the impairment of reproductive function and the induction of adverse effects on the embryo, such as growth retardation, malformation and death. Owing to the complexity of the mammalian reproductive cycle it is not possible to model the whole cycle on one in vitro system to detect chemical effects on mammalian reproduction. However, the cycle can be broken down into its biological components, which can then be studied individually or in combination. This has the enormous advantage that the target tissue/organ of an agent can be identified.

Three embryotoxicity tests have been formally validated (ECVAM) to replace OECD TG 414 (developmental toxicity testing):

- The embryonic stem (ES) cell test is composed of two procedures, a cytotoxicity test, which is conducted with the mouse ES cell line D3 and cells of the differentiated mouse fibroblast cell line 3T3, and a differentiation assay using D3 cells (Spielmann et al., 1997; Genschow et al., 2000). Three toxicological endpoints were identified to classify the embryotoxic potential of chemicals:
    - the inhibition of differentiation of ES cells into cardiomyocytes (ID_{50})
    - the decrease in viability of adult 3T3 cells ($IC_{50}3T3$)
    - ES cells ($IC_{50}D3$) in an MTT cytotoxicity test. The validation of the ES cell test provides an overall accuracy of 78% (Genschow et al., 2002).
- The whole embryo culture derived from mice, rats or rabbits is used to detect developmental toxicants (Bechter and Schmid, 1987; Piersma et al., 1996). Head-fold or early somite-stage embryos are dissected free from the maternal tissue, parietal yolk sac and Reichert's membrane, leaving the visceral yolk sac and ectoplacental cone intact. The conceptus is cultured in medium under defined conditions for 24–48 h. Medium containing a high proportion of serum is usually used and test compounds are added to the cultures for appropriate periods of time. A defined number of endpoints are analysed after the incubation period, such as dysmorphogenic effects, embryonic growth, differentiation, yolk sac circulation and vascularization, effects on the haematopoiesis, etc. (Genschow et al., 2002).
- Micromass culture makes use of cell cultures of the limb bud and/or neuronal cells. The cells are isolated from the limb or the cephalic tissues of mid-organogenesis embryos (Whittaker and Faustman, 1994). After a single-cell solution is prepared, the cells are seeded at high density and undergo differentiation into chondrocytes and neurons without additional stimulation. The differentiation after exposure to test chemicals is analysed by using defined toxicological endpoints (Genschow et al., 2002).

These tests can be used only in a test strategy that covers the main manifestation of developmental toxicity.

TOXICOKINETICS STUDIES

The term 'toxicokinetics studies' is, in the context of chemical substances such as cosmetic ingredients, used to describe the time-dependent fate of a substance within the body. This includes absorption, distribution, biotransformation and/or excretion. The term 'toxicodynamics' means the process of interactions of chemical substances with target sites and the subsequent reactions leading to adverse effects (EC B.36; OECD 417).

In the context of the EU cosmetics legislation, a review of the actual status of alternatives to animal toxicokinetics studies was carried out in this field (Coecke et al., 2005).

CONCLUSIONS

Concerning acute oral toxicity, validation of an alternative model is very complex and the time estimated to achieve complete animal replacement for acute toxicity testing is not clearly defined.

Alternative methods for skin corrosion have been validated and accepted for regulatory use in the EU and the OECD member countries. Furthermore, the human skin model assays (e.g., EpiDerm and EpiSkin) appear to be the most promising in vitro methods for skin irritation testing. However, there is a need to develop new endpoints that are more predictive of skin irritation than simply being cytotoxicity determinations.

In the case of eye irritation tests, no single method is able to replace the Draize rabbit eye test. This is due to different factors including the limited quality of the existing in vivo data, limitations of the animal testing method, and the fact that the range of criteria for injury and inflammation covered by the Draize rabbit eye test is unlikely to be replaced by a single in vitro test.

For predicting skin sensitization, methods that imply refinement and reduction have been developed and are widely used. Nevertheless, there is no single test method which will adequately identify all substances with a potential for sensitizing human skin and which is relevant for all substances.

The in vitro assessment of dermal absorption of cosmetic ingredients can be seen as a full alternative to the in vivo test. However, the quality of submitted in vitro data is still not always satisfying with respect to documentation and technical aspects, which have been published.

No generally accepted alternative methods are available to replace the usual repeated-dose toxicity in vivo assays.

To determine the genotoxic and mutagenic profile of a compound, no single validated test can provide information on gene mutations. As a consequence, a battery of in vitro tests is needed.

In the field of UV-induced toxic effects, the identification of acute phototoxic hazards is regarded as sufficiently covered by in vitro tests; consequently, animal testing for that endpoint can be 100% replaced. On the other hand, the in vitro genetic toxicity tests have been converted into test protocols of photogenotoxicity tests, and for in vivo photoallergy testing (photosensitization) no standard testing protocol exists.

Two in vitro tests, cell transformation assay and GJIC, have been proposed as tests that may provide information on possible carcinogenicity effects.

In the reproductive and developmental toxicity field three embryotoxicity tests have been formally validated (ECVAM) to replace OECD TG 414.

Finally, complete replacement of animal usage will represent an enormous scientific and technical challenge.

REFERENCES

Introduction

Commission Directive 1999/11/EC of 8 March 1999 Adapting to Technical Progress the Principles of Good Laboratory Practice as Specified in Council Directive 87/18/EEC on the Harmonisation of Laws, Regulations and Administrative Provisions Relating to the Application of the Principles of Good Laboratory Practice and the Verification of Their Applications for Tests on Chemical Substances. Official Journal L 77.

Commission Regulation (EU) No 1152/2010 of 8 December 2010 amending, for the purpose of its adaptation to technical progress, Regulation (EC) No 440/2008 laying down test methods pursuant to Regulation (EC) No 1907/2006 of the European Parliament and of the Council on the Registration, Evaluation, Authorisation and Restriction of Chemicals (REACH) Text with EEA relevance.

Official Journal L 66. Council Directive 2003/15/EC of the European Parliament and of the Council of 27 February 2003 Amending Council Directive 76/768/EEC on the Approximation of the Laws of the Member States Relating to Cosmetic Products, 11 March, 2003, p. 26.

Official Journal L 262. Council Directive 76/768/EEC of 27 July 1976 on the Approximation of the Laws of the Member States Relating to Cosmetic Products, 27 September, 1976, p. 169.

Council Directive 87/18/CEE of 18 December 1986 on the Harmonization of Laws, Regulations and Administrative Provisions Relating to the Application of the Principles of Good Laboratory Practice and the Verification of Their Applications for Tests on Chemical Substances. Official Journal L 15.

Herráez Domínguez, M., Díez-Sales, O., 2007. Safety evaluation. In: Salvador, A., Chisvert, A. (Eds.), Analysis of Cosmetic Products. Elsevier, Amsterdam, pp. 423–463.

Official Journal L 342. Regulation (EC) No 1223/2009 of the European Parliament and of the Council of 30 November 2009 on Cosmetic Products, 22 December, 2009, p. 59.

Official Journal L 142. Regulation (EC) No 440/2008 de la Commission Directive, 30 May, 2008, p. 141.

Russell, B., Russell, W.M.S., Burch, R.L., 1959. The Principles of Humane Experimental Technique. Methuen and Co Ltd., London, UK(Reprinted by the Universities Federation for Animal Welfare UFAW, 1992, Potters Bar, Herts).

Acute Toxicity

Cronin, M.T.D., Jaworska, J.S., Walker, J.D., Comber, M.H.I., Watts, C.D., Worth, A.P., 2003. Environ. Health Perspect. 111 (10), 139.

EC B.1 bis – Acute Oral Toxicity. Fixed Dose Procedure, 31 May, 2008. Council Regulation (EC) No 440/2008 of 30 May 2008 Laying Down Test Methods Pursuant to Regulation (EC) No 1907/2006 of the European Parliament and of the Council on the Registration, Evaluation, Authorisation and Restriction of Chemicals (REACH). Official Journal L 142, p. 145.

EC B.1 tris – Acute oral toxicity. Acute Toxic Class Method, 31 May, 2008. Council Regulation (EC) No 440/2008 of 30 May 2008 Laying Down Test Methods Pursuant to Regulation (EC) No 1907/2006 of the European Parliament and of the Council on the Registration, Evaluation, Authorisation and Restriction of Chemicals (REACH). Official Journal L 142, p. 158.

EC B.1-Acute Toxicity (Oral), 29 December, 1992. Commission Directive 92/69/EEC of 31 July 1992 Adapting to Technical Progress for the Seventeenth Time Council Directive 67/548/EEC on the Approximation of Laws, Regulations and Administrative Provisions Relating to the Classification, Packaging and Labelling of Dangerous Substances. Official Journal L 383A, p. 110.

EC B.2 – Acute Toxicity (Inhalation), 31 May, 2008. Council Regulation (EC) No 440/2008 of 30 May 2008 Laying Down Test Methods Pursuant to Regulation (EC) No 1907/2006 of the European Parliament and of the Council on the Registration, Evaluation, Authorisation and Restriction of Chemicals (REACH). Official Journal L 142, p. 174.

EC B.3-Acute Toxicity (Dermal), 31 May, 2008. Council Regulation (EC) No 440/2008 of 30 May 2008 Laying Down Test Methods Pursuant to Regulation (EC) No 1907/2006 of the European Parliament and of the Council on the Registration, Evaluation, Authorisation and Restriction of Chemicals (REACH). Official Journal L 142, p. 178.

EC B.52-Acute Inhalation Toxicity – Acute Toxic Class Method.

Lapenna, S., Fuart-Gatnik, M., Worth, A., 2010. Review of QSAR Models and Software Tools for Predicting Acute and Chronic Systemic Toxicity. Joint Research Centre–Institute for Health and Consumer Protection. Publications Office of the European Union, Luxembourg. 26 pp.

OECD 401, 1981. OECD Guideline for Testing of Chemicals – Guideline 401: Acute Oral Toxicity. Organization for Economic Cooperation and Development, Paris. Adopted 12 May, 1981, last updated 24 February, 1997 and deleted 17 December, 2002.

OECD 402, 1997. OECD Guideline for Testing of Chemicals – Guideline 402: Acute Dermal Toxicity. Organization for Economic Cooperation and Development, Paris. Adopted 24 February, 1997.

OECD 403-OECD Guideline for Testing of Chemicals – Guideline 403, 7 September, 2009. Acute Inhalation Toxicity Organization for Economic Cooperation and Development. Paris.

OECD 420, 1992. OECD Guideline for Testing of Chemicals – Guideline 420: Acute Oral Toxicity – Fixed Dose Procedure. Organization for Economic Cooperation and Development, Paris. Adopted 17 July, 1992, last updated 17 December, 2001.

OECD 423, 1996. OECD Guideline for Testing of Chemicals – Guideline 423: Acute Oral Toxicity – Acute Toxic Class Method. Organization for Economic Cooperation and Development, Paris. Adopted 22 March, 1996, last updated 17 December, 2001.

OECD 425, 2008. OECD Guideline for Testing of Chemicals – Guideline 425: Acute Oral Toxicity – Upand-Down-Procedure. Organization for Economic Cooperation and Development, Paris. Adopted 3 October, 2008.

OECD 433, 8 June, 2004. OECD Guideline for Testing of Chemicals – Draft Proposal for a New Guideline 433: Acute Inhalation Toxicity – Fixed Concentration Procedure Organization for Economic Cooperation and Development. Paris, second version.

OECD 436, 2009. OECD Guideline for Testing of Chemicals – Guideline 436: Acute Inhalation Toxicity – Acute Toxic Class (ATC) Method. Organization for Economic Cooperation and Development, Paris. Adopted 7 September, 2009.

Official Journal L 353. Regulation (EC) No 1272/2008 of the European Parliament and the Council of 16 December 2008 on Classification, Labelling and Packaging of Substances and Mixtures, Amending and Repealing Directives 67/548/EEC and 1999/45/EC, and Amending Regulation (EC) No 1907/2006, 31 December, 2008, p. 1.

Skin Corrosion/Irritation

B.40. In Vitro Skin Corrosion, 31 May, 2008. Transcutaneous Electrical Resistance Test (TER). Council Regulation (EC) No 440/2008 of 30 May 2008 Laying Down Test Methods Pursuant to Regulation (EC) No 1907/2006 of the European Parliament and of the Council on the Registration, Evaluation, Authorisation and Restriction of Chemicals (REACH). Official Journal L 142, p. 381.

Botham, P.A., Earl, L.K., Fentem, J.H., Roguet, R., van de Sandt, J.J.M., 1998. ATLA 26, 195.

EC B.40bis – In Vitro Skin Corrosion: Human Skin Model Test, 31 May, 2008. Council Regulation (EC) No 440/2008 of 30 May 2008 Laying Down Test Methods Pursuant to Regulation (EC) No 1907/2006 of the European Parliament and of the Council on the Registration, Evaluation, Authorisation and Restriction of Chemicals (REACH). Official Journal L 142, p. 394.

EC B.46 – In Vitro Skin Irritation: Reconstructed Human Epidermis Test Method, 20 July, 2012. Commission Regulation (EU) No 640/2012 of 6 July 2012 Amending, for the Purpose of Its Adaptation to Technical Progress, Regulation (EC) No 440/2008 Laying Down Test Methods Pursuant to Regulation (EC) No 1907/2006 of the European Parliament and of the Council on the Registration, Evaluation, Authorisation and Restriction of Chemicals (REACH). Official Journal L 193, p. 17.

EC B.4-Acute Toxicity: Dermal Irritation/Corrosion, 31 May, 2008. Council Regulation (EC) No 440/2008 of 30 May 2008 Laying Down Test Methods Pursuant to Regulation (EC) No 1907/2006 of the European Parliament and of the Council on the Registration, Evaluation, Authorisation and Restriction of Chemicals (REACH). Official Journal L 142, p. 182.

Fentem, J.H., Archer, G.E.B., Balls, M., Botham, P.A., Curren, R.D., Earl, L.K., Esdaile, D.J., Holzhutter, H.G., Liebsch, M., 1998. Toxicol. Vitro 12, 483.

OECD 404, 1981. OECD Guideline for Testing of Chemicals – Guideline 404: Acute Dermal Irritation/Corrosion. Organization for Economic Cooperation and Development, Paris. Adopted 12 May, 1981, last updated 24 April, 2002.

OECD 430, 2004. OECD Guideline for Testing of Chemicals – Guideline 430: In Vitro Skin Corrosion: Transcutaneous Electrical Resistance Test (TER). Organization for Economic Cooperation and Development, Paris. Adopted 13 April, 2004.

OECD 431, 2004. OECD Guideline for Testing of Chemicals – Guideline 431: In Vitro Skin Corrosion: Human Skin Model Test. Organization for Economic Cooperation and Development, Paris. Adopted 13 April, 2004.

OECD 435, 2006. OECD Guideline for Testing of Chemicals – Guideline 435: In Vitro Membrane Barrier Test Method for Skin Corrosion. Organization for Economic Cooperation and Development, Paris. Adopted 19 July, 2006.

OECD 439, 2010. OECD Guideline for Testing of Chemicals – Guideline 439: In Vitro Skin Irritation: Reconstructed Human Epidermis Test Method. Organization for Economic Cooperation and Development, Paris. Adopted 22 July, 2010.

Roguet, R., Cohen, C., Robles, C., Courtellemont, P., Tolle, M., Guillot, J.P., Pouradier Duteil, X., 1998. Toxicol. Vitro 12, 295.

Tinois, E., Tillier, J., Gaucherand, M., Dumas, H., Tardy, M., Thivolet, J., 1991. Exp. Cell Res. 193, 310.

Zuang, V., Balls, M., Botham, P.A., Coquette, A., Corsini, E., Curren, R.D., Elliott, G.R., Fentem, J.H., Heylings, J.R., Liebsch, M., Medina, J., Roguet, R., van de Sandt, J.J.M., Wiemann, C., Worth, A.P., 2002. ATLA 30, 109.

Eye Irritation

Balls, M., Botham, P.A., Bruner, L.H., Spielmann, H., 1995. Toxicol. vitro 9 (6), 871.

Blazka, M., Harbell, J.W., Klausner, M., Merrill, J.C., Kubilus, J., Kloss, C., Bagley, D.M., 2003. Toxicologist 72, 221.

Bradlaw, J., Gupta, K., Green, S., Hill, R., Wilcox, N., 1997. Food Chem. Toxicol. 35, 75.

Burton, A.B.G., York, M., Lawrence, R.S., 1999. Food Cosmet. Toxicol. 19, 471.

Chamberlain, M., Gad, S.C., Gautheron, P., Prinsen, M.K., 1997. Food Chem. Toxicol. 35 (1), 23.

Christian, M.S., Diener, R.M., 1996. J. Am. Coll. Toxicol. 15 (1), 1.

Curren, R., Evans, M., Raabe, H., Dobson, T., Harbell, J., 1999. ATLA 27, 344.

Draize, J.H., Woodard, G., Calvery, H.O., 1944. J. Pharmacol. Exp. Ther. 82, 377.

EC B.47 – Bovine Corneal Opacity and Permeability Test Method for Identifying Ocular Corrosives and Severe Irritants, 09 December, 2010. Commission Regulation (EC) No 761/2009 of 23 July 2009 Amending, for the Purpose of Its Adaptation to Technical Progress, Regulation (EC) No 440/2008 Laying Down Test Methods Pursuant to Regulation (EC) No 1907/2006 of the European Parliament and of the Council on the Registration, Evaluation, Authorisation and Restriction of Chemicals (REACH). Official Journal L 324, p. 14.

EC B.48 – Isolated Chicken Eye Test Method for Identifying Ocular Corrosives and Severe irritants, 09 December, 2010. Commission Regulation (EC) No 761/2009 of 23 July 2009 Amending, for the Purpose of Its Adaptation to Technical Progress, Regulation (EC) No 440/2008 Laying Down Test Methods Pursuant to Regulation (EC) No 1907/2006 of the European Parliament and of the Council on the Registration, Evaluation, Authorisation and Restriction of Chemicals (REACH). Official Journal L 324, p. 14.

EC B.5-Acute Toxicity: Eye Irritation/Corrosion, 31 May, 2008. Council Regulation (EC) No 440/2008 of 30 May 2008 Laying Down Test Methods Pursuant to Regulation (EC) No 1907/2006 of the European Parliament and of the Council on the Registration, Evaluation, Authorisation and Restriction of Chemicals (REACH). Official Journal L 142, p. 191.

Gautheron, P., Giroux, J., Cottin, M., Audegond, L., Morilla, A., Mayordomo-Blanco, L., Tortajada, A., Haynes, G., Vericat, J.A., 1994. Toxicol. vitro 8 (3), 381.

Gettings, S.D., Lordo, R.A., Hintze, K.L., Bagley, D.M., Casterton, P.L., Chudkowski, M., Curren, R.D., Demetrulias, J.L., DiPasquale, L.C., Earl, L.K., Feder, P.I., Galli, C.L., Glaza, S.M., Gordon, V.C., Janus, J., Kurtz, P.J., Marenus, K.D., Moral, J., Pape, W.J.W., Renskers, K.J., Rheins, L.A., Roddy, M.T., Rozen, M.G., Tedeschi, J.P., Zyracki, J., 1996. Food Chem. Toxicol. 34 (1), 79.

Nguyen, D.H., Beuerman, R.W., De Wever, B., Rosdy, M., 2003. Alternative methods for the New Millenium (Chapter 14). In: Three-Dimensional Construct of the Human Corneal Epithelium for In Vitro Toxicology. CRC Press.

OECD 2010-Draft OECD Guideline for the Testing of Chemicals – The Cytosensor Microphysiometer Test Method, 20 December, 2010. An In Vitro Method for Identifying Chemicals Not Classified as Irritant, as Well as Ocular Corrosive and Severe Irritant Chemicals. Organization for Economic Cooperation and Development, Paris.

OECD 405, 1981. OECD Guideline for Testing of Chemicals – Guideline 405: Acute Eye Irritation/Corrosion. Organization for Economic Cooperation and Development, Paris. Adopted 12 May, 1981, last updated 2 October, 2012.

OECD 437, 2009. OECD Guideline for Testing of Chemicals – Guideline 437: Bovine Corneal Opacityand Permeability Test Method for Identifying Ocular Corrosives and Severe Irritants. Organization for Economic Cooperation and Development, Paris. Adopted 7 September, 2009.

OECD 438, 2009. OECD Guideline for Testing of Chemicals – Guideline 438: Isolated Chicken Eye Test Method for Identifying Ocular Corrosives and Severe Irritants. Organization for Economic Cooperation and Development, Paris. Adopted 7 September, 2009.

OECD 460, 2 October, 2012. OECD Guideline for Testing of Chemicals – Guideline 460: Fluorescein Leakage Test Method for Identifying Ocular Corrosives and Severe Irritants. Organization for Economic Cooperation and Development, Paris.

Ohno, Y., Kaneko, T., Inoue, T., Morikawa, Y., Yoshida, T., Fuji, A., Masuda, M., Ohno, T., Pape, W.J.W., Hoppe, U., 1991. Skin Pharmacol. 4, 205.

Prinsen, M.K., 1996. Food Chem. Toxicol. 34 (3), 291.

Sina, J.F., Gautheron, P.D., 1994. Toxicol. Methods 4 (1), 41.

Sina, J.F., Galer, D.M., Sussman, R.G., Gautheron, P.D., Sargent, E.V., Leong, B., Shah, P.V., Curren, R.D., Miller, K., 1995. Fundam. Appl. Toxicol. 26, 20.

Spielmann, H., Kalweit, S., Liebsch, M., Wirnsberger, T., Gerner, I., Bertram-Neiss, E., Krauser, K., Kreiling, R., Miltenburger, H.G., Pape, W., Steiling, W., 1993. Toxicol. Vitro 7 (4), 505.

Stern, M., Klausner, M., Alvarado, R., Reskers, K., Dickens, M., 1998. Toxicol. Vitro 12, 455.

Swanson, J.E., Lake, L.K., Donnelly, T.A., Harbell, J.W., Huggins, J., 1995. J. Cutan. Ocul. Toxicol. 14 (3), 179.

Whittle, E., Basketter, D., York, M., Kelly, L., McCall, J., Botham, P., Esdaile, D., Gardner, J., 1992. Toxicol. Methods 2, 30.

Wilson, S.L., Ahearne, M., Hopkinson, A., 2015. Toxicology 327, 32.

York, M., Wilson, A.P., Newsome, C.S., 1994. Toxicol. Vitro 8 (6), 1265.

Skin Sensitisation

Asherson, G.L., Ptak, W., 1968. Immunology 15, 405.

EC B.42-Skin Sensitisation: Local Lymph Node Assay, 20 July, 2012. Commission Regulation (EU) No 640/2012 of 6 July 2012 Amending, for the Purpose of Its Adaptation to Technical Progress, Regulation (EC) No 440/2008 Laying Down Test Methods Pursuant to Regulation (EC) No 1907/2006 of the European Parliament and of the Council on the Registration, Evaluation, Authorisation and Restriction of Chemicals (REACH). Official Journal L 193, p. 3.

EC B.50 – Skin Sensitisation: Local Lymph Node Assay: DA, 20 July, 2012. Commission Regulation (EU) No 640/2012 of 6 July 2012 Amending, for the Purpose of Its Adaptation to Technical Progress, Regulation (EC) No 440/2008 Laying Down Test Methods Pursuant to Regulation (EC) No 1907/2006 of the European Parliament and of the Council on the Registration, Evaluation, Authorisation and Restriction of Chemicals (REACH). Official Journal L 193, p. 46.

EC B.51 – Skin Sensitisation: Local Lymph Node Assay: BrdU-ELISA, 20 July, 2012. Commission Regulation (EU) No 640/2012 of 6 July 2012 Amending, for the Purpose of Its Adaptation to Technical Progress, Regulation (EC) No 440/2008 Laying Down Test Methods Pursuant to Regulation (EC) No 1907/2006 of the European Parliament and of the Council on the Registration, Evaluation, Authorisation and Restriction of Chemicals (REACH). Official Journal L 193, p. 56.

Griffith, J.F., Buehler, E., 1976. In: Drill, V.P., Lazer, P. (Eds.), Cutaneous Toxicity: Prediction of Skin Irritancy and Sensitization Potential by Testing with Animals and Man. Academic Press, New York.

Kimber, I., Derman, R.J., Scholes, E.W., Basketter, D.A., 1994. Toxicology 93, 13.

Kimber, I., Hilton, J., Dearman, R.J., Gerberick., G.F., Ryan, C.A., Basketter, D.A., Lea, L., House, R.V., Ladies, G.S., Loveless, S.E., Hastings, K.L., 1998. J. Toxicol. Environ. Health 53, 563.

Kimber, I., Dearman, R.J., Basketter, D.A., Ryan, C.A., Gerberick, G.F., 2002. Contact Dermat. 47, 315.

Kligman, A.M., Epstein, W., 1975. Contact Dermat. 1, 231.

Magnusson, B., Kligman, A.M., 1970. In: Charles, C. (Ed.), Allergic Contact Dermatitis in the Guinea Pig. Thomas, Springfield, Illinois.

Marzulli, F.N., Maibach, H.I., 1973. J. Soc. Cosmet. Chem. 24, 399.

OECD 406, 1992. OECD Guideline for Testing of Chemicals – Guideline 406: Skin Sensitisation. Organization for Economic Cooperation and Development, Paris. Adopted 17 July, 1992.

OECD 429, 2002. OECD Guideline for Testing of Chemicals – Guideline 429. Skin Sensitisation: Local Lymph Node Assay. Organization for Economic Cooperation and Development, Paris. Adopted 24 April, 2002.

OECD 442A, 2010. OECD Guideline for Testing of Chemicals – Guideline 442A: Skin Sensitization: Local Lymph Node Assay: DA. Organization for Economic Cooperation and Development, Paris. Adopted 22 July, 2010.

OECD 442B, 2010. OECD Guideline for Testing of Chemicals – Guideline 442B: Skin Sensitization: Local Lymph Node Assay: BrdU-ELISA. Organization for Economic Cooperation and Development, Paris. Adopted 22 July, 2010.

Skin Absorption

EC B.44-Skin Absorption: In Vivo Method, 31 May, 2008. Council Regulation (EC) No 440/2008 of 30 May 2008 Laying Down Test Methods Pursuant to Regulation (EC) No 1907/2006 of the European Parliament and of the Council on the Registration, Evaluation, Authorisation and Restriction of Chemicals (REACH). Official Journal L 142, p. 432.

EC B.45-Skin Absorption: In Vitro Method, 31 May, 2008. Council Regulation (EC) No 440/2008 of 30 May 2008 Laying Down Test Methods Pursuant to Regulation (EC) No 1907/2006 of the European Parliament and of the Council on the Registration, Evaluation, Authorisation and Restriction of Chemicals (REACH). Official Journal L 142, p. 438.

OECD 427, 2004. OECD Guideline for Testing of Chemicals – Guideline 427: Skin Absorption: In Vivo Method. Organization for Economic Cooperation and Development, Paris. Adopted 13 April, 2004.

OECD 428, 2004. OECD Guideline for Testing of Chemicals – Guideline 428: Skin Absorption: In Vitro Method. Organization for Economic Cooperation and Development, Paris. Adopted 13 April, 2004.

Sanco/222/2000, 2004. Guidance Document on Dermal Absorption. European Commission, Health and Consumer Protection Directorate-General. Doc. Sanco/222/2000 rev. 7, of 19 March, 2004.

SCCNFP/0750/03, 2003. Basic Criteria for the In Vitro Assessment of Dermal Absorption of Cosmetic Ingredients. Adopted by the SCCNFP during the 25th plenary meeting of 20 October, 2003.

Schaefer, H., Redelmeier, T.E. (Eds.), 1996. Skin Barrier, Principles of Percutaneous Absorption. Karger, Basel.

Repeated Dose Toxicity

Cronin, M.T.D., Jaworska, J.S., Walker, J.D., Comber, M.H.I., Watts, C.D., Worth, A.P., 2003. Environ. Health Perspect. 111, 1391.

EC B.26-Sub-Chronic Oral Toxicity Test: Repeated Dose 90-day Oral Toxicity Study in Rodents, 31 May, 2008. Council Regulation (EC) No 440/2008 of 30 May 2008 Laying Down Test Methods Pursuant to Regulation (EC) No 1907/2006 of the European Parliament and of the Council on the Registration, Evaluation, Authorisation and Restriction of Chemicals (REACH). Official Journal L 142, p. 302.

EC B.27-Sub-Chronic Oral Toxicity Test: Repeated Dose 90-day Oral Toxicity Study in Non-Rodents, 31 May, 2008. Council Regulation (EC) No 440/2008 of 30 May 2008 Laying Down Test Methods Pursuant to Regulation (EC) No 1907/2006 of the European Parliament and of the Council on the Registration, Evaluation, Authorisation and Restriction of Chemicals (REACH). Official Journal L 142, p. 308.

EC B.28-Sub-Chronic Dermal Toxicity Study: 90-day Repeated Dermal Dose Study Using Rodent Species, 31 May, 2008. Council Regulation (EC) No 440/2008 of 30 May 2008 Laying Down Test Methods Pursuant to Regulation (EC) No 1907/2006 of the European Parliament and of the Council on the Registration, Evaluation, Authorisation and Restriction of Chemicals (REACH). Official Journal L 142, p. 314.

EC B.29-Sub-Chronic Inhalation Toxicity Study: 90-day Repeated Inhalation Dose Study Using Rodent Species, 31 May, 2008. Council Regulation (EC) No 440/2008 of 30 May 2008 Laying Down Test Methods Pursuant to Regulation (EC) No 1907/2006 of the European Parliament and of the Council on the Registration, Evaluation, Authorisation and Restriction of Chemicals (REACH). Official Journal L 142, p. 318.

EC B.30-Chronic Toxicity Test, 31 May, 2008. Council Regulation (EC) No 440/2008 of 30 May 2008 Laying Down Test Methods Pursuant to Regulation (EC) No 1907/2006 of the European Parliament and of the Council on the Registration, Evaluation, Authorisation and Restriction of Chemicals (REACH). Official Journal L 142, p. 323.

EC B.7-Repeated Dose (28 days) Toxicity (Oral), 31 May, 2008. Council Regulation (EC) No 440/2008 of 30 May 2008 Laying Down Test Methods Pursuant to Regulation (EC) No 1907/2006 of the European Parliament and of the Council on the Registration, Evaluation, Authorisation and Restriction of Chemicals (REACH). Official Journal L 142, p. 210.

EC B.8-Repeated Dose (28 days) Toxicity (Inhalation), 31 May, 2008. Council Regulation (EC) No 440/2008 of 30 May 2008 Laying Down Test Methods Pursuant to Regulation (EC) No 1907/2006 of the European Parliament and of the Council on the Registration, Evaluation, Authorisation and Restriction of Chemicals (REACH). Official Journal L 142, p. 216.

EC B.9-Repeated Dose (28 days) Toxicity (Dermal), 31 May, 2008. Council Regulation (EC) No 440/2008 of 30 May 2008 Laying Down Test Methods Pursuant to Regulation (EC) No 1907/2006 of the European Parliament and of the Council on the Registration, Evaluation, Authorisation and Restriction of Chemicals (REACH). Official Journal L 142, p. 221.

OECD 407, 2008. OECD Guideline for Testing of Chemicals – Guideline 407: Repeated Dose 28-Day. Oral Toxicity Study in Rodents. Organization for Economic Cooperation and Development, Paris. Adopted 3 October, 2008.

OECD 408, 1998. OECD Guideline for Testing of Chemicals – Guideline 408: Repeated Dose 90-Day. Oral Toxicity Study in Rodents. Organization for Economic Cooperation and Development, Paris. Adopted 21 September, 1998.

OECD 409, 1998. OECD Guideline for Testing of Chemicals – Guideline 409: Repeated Dose 90-Day. Oral Toxicity Study in Non-Rodents. Organization for Economic Cooperation and Development, Paris. Adopted 21 September, 1998.

OECD 410, 1981. OECD Guideline for Testing of Chemicals – Guideline 410: Repeated Dose Dermal. Toxicity: 21/28-Day Study. Organization for Economic Cooperation and Development, Paris. Adopted 12 May, 1981.

OECD 411, 1981. OECD Guideline for Testing of Chemicals – Guideline 411: Subchronic Dermal. Toxicity: 90-Day Study. Organization for Economic Cooperation and Development, Paris. Adopted 12 May, 1981.

OECD 412, 2009. OECD Guideline for Testing of Chemicals – Guideline 412: Subacute Inhalation. Toxicity: 28-Day Study. Organization for Economic Cooperation and Development, Paris. Adopted 7 September, 2009.

OECD 413, 2009. OECD Guideline for Testing of Chemicals – Guideline 413: Subchronic Inhalation. Toxicity: 90-Day Study. Organization for Economic Cooperation and Development, Paris. Adopted 7 September, 2009.

Genotoxicity/Mutagenicity

EC B.10-Mutagenicity – In Vitro Mammalian Chromosome Aberration Test, 31 May, 2008. Council Regulation (EC) No 440/2008 of 30 May 2008 Laying Down Test Methods Pursuant to Regulation (EC) No 1907/2006 of the European Parliament and of the Council on the Registration, Evaluation, Authorisation and Restriction of Chemicals (REACH). Official Journal L 142, p. 225.

EC B.13/14-Mutagenicity – Reverse Mutation Test Using Bacteria, 31 May, 2008. Council Regulation (EC) No 440/2008 of 30 May 2008 Laying Down Test Methods Pursuant to Regulation (EC) No 1907/2006 of the European Parliament and of the Council on the Registration, Evaluation, Authorisation and Restriction of Chemicals (REACH). Official Journal L 142, p. 248.

EC B.17-Mutagenicity – In Vitro Mammalian Cell Gene Mutation Test, 31 May, 2008. Council Regulation (EC) No 440/2008 of 30 May 2008 Laying Down Test Methods Pursuant to Regulation (EC) No 1907/2006 of the European Parliament and of the Council on the Registration, Evaluation, Authorisation and Restriction of Chemicals (REACH). Official Journal L 142, p. 262.

EC B.49 – In Vitro Mammalian Cell Micronucleus Test, 20 July, 2012. Commission Regulation (EU) No 640/2012 of 6 July 2012 Amending, for the Purpose of Its Adaptation to Technical Progress, Regulation (EC) No 440/2008 Laying Down Test Methods Pursuant to Regulation (EC) No 1907/2006 of the European Parliament and of the Council on the Registration, Evaluation, Authorisation and Restriction of Chemicals (REACH). Official Journal L 193, p. 30.

OECD 471, 1983. OECD Guideline for Testing of Chemicals – Guideline 471: Bacterial Reverse Mutation Test. Organization for Economic Cooperation and Development, Paris. Adopted 26 May, 1983, last updated 21 July, 1997.

OECD 473, 1997. OECD Guideline for Testing of Chemicals – Guideline 473: In Vitro Mammalian Chromosomal Aberration Test. Organization for Economic Cooperation and Development, Paris. Adopted 21 July, 1997.

OECD 476, 1997. OECD Guideline for Testing of Chemicals – Guideline 476: In Vitro Mammalian Cell Gene Mutation Test. Organization for Economic Cooperation and Development, Paris. Adopted 21 July, 1997.

OECD 487, 2010. OECD Guideline for Testing of Chemicals – Guideline 487: In Vitro Mammalian Cell Micronucleus Test (MNvit). Organization for Economic Cooperation and Development, Paris. Adopted 22 July, 2010.

SCCNFP/0755/03. SCCNFP Mutagenicity/Genotoxicity Tests Recommended for the Safety Testing of Cosmetics Ingredients to be Included in the Annexes to Council Directive 76/768/EEC.

UV-Induced Toxic Effects

Bernard, F.X., Barrault, C., Deguery, A., de Wever, B., Rosdy, M., 1999. Alternatives to animal testing II. In: Clark, D., Lisansky, S., Macmillan, R. (Eds.), Proceedings of the Second International Scientific Conference Organised by the European Cosmetic Industry Brussels, Belgium.

Brendler-Schwaab, S., Czich, A., Epe, B., Gocke, E., Kaina, B., Müller, L., Pollet, D., Utesch, D., 2004. Mutat. Res. 566, 65.

Chételat, A., Albertini, S., Dresp, J.H., Strobel, R., Gocke, E., 1993. Mutat. Res. 292, 241.

EC B.41-In vitro 3T3 NRU phototoxicity test, 31 May, 2008. Council Regulation (EC) No 440/2008 of 30 May 2008 Laying Down Test Methods Pursuant to Regulation (EC) No 1907/2006 of the European Parliament and of the Council on the Registration, Evaluation, Authorisation and Restriction of Chemicals (REACH). Official Journal L 142, p. 400.

Gocke, E., Müller, L., Guzzie, P.J., Brendler-Schwaab, S., Bulera, S., Chignell, C.F., Henderson, L.M., Jacobs, A., Murli, H., Snyder, R.D., Tanaka, N., 2000. Environ. Mol. Mutagen. 35, 173.

Jones, P.A., King, A.V., Earl, L.K., Lawrence, R.S., 2003. Toxicol. Vitro 17, 471.

Jose, J.G., 1979. Proc. Natl. Acad. Sci. U.S.A. 76, 469.

Kalweit, S., et al., 1999. Mutat. Res. 439, 183.

Kersten, B., Zhang, J., Brendler-Schwaab, S.Y., Kasper, P., Müller, L., 1999. Mutat. Res. 445 (1), 55.

Kersten, B., Kasper, P., Brendler-Schwaab, S.Y., Müller, L., 2002. Mutat. Res. 519, 49.

Liebsch, M., Traue, D., Barrabas, C., Spielmann, H., Gerberick, G.F., Cruse, L., Diembeck, W., Pfannenbecker, U., Spieker, J., Holzhütter, H.G., Brantom, P., Aspin, P., Southee, J., 1999. Alternatives to animal testing II. In: Clark, D., Lisansky, S., Macmillan, R. (Eds.), Proceedings of the Second International Scientific Conference Organised by the European Cosmetic Industry: Prevalidation of the EpiDerm Phototoxicity Test Brussels, Belgium.

Loprieno, N., 1991. Mutagenesis 6, 331.

OECD 432, 2004. OECD Guideline for Testing of Chemicals – Guideline 432: In Vitro 3T3 NRU Phototoxicity Test. Organization for Economic Cooperation and Development, Paris. Adopted 13 April, 2004.

OECD, 2002. Test Guideline 432: In Vitro 3T3 NRU Phototoxicity Test.
Okamoto, Y., Ryu, A., Ohkoshi, K., 1999. ATLA 27, 639.
Pape, W.J.W., Brandt, M., Pfannenbecker, U., 1994. Toxicol. Vitro 8, 755.
Roguet, R., Cohen, C., Rougier, A., 1994. Alternative methods in toxicology. In: Rougier, A., Goldberg, A., Maibach, H. (Eds.), 10: In Vitro Skin Toxicology – Irritation, Phototoxicity, Sensitization: A Reconstituted Human Epidermis to Assess Cutaneous Irritation, Photoirritation and Photoprotection In Vitro. Mary Ann Libert Publ., New York.

Carcinogenicity
Rosenkranz, H.S., Pollack, N., Cunningham, A.R., 2000. Carcinogenesis 21 (5), 1007.

Reproductive and Development Toxicity
Bechter, R., Schmid, B.P., 1987. Toxicol. Vitro 1, 11.
Genschow, E., Scholz, G., Brown, N., Piersma, A., Brady, M., Clemann, N., Huuskonen, H., Paillard, F., Bremer, S., Becker, K., Spielmann, H., 2000. Vitro Mol. Toxicol. 13, 51.
Genschow, E., Spielmann, H., Scholz, G., Seiler, A., Brown, N.A., Piersma, A., Brady, M., Clemann, N., Huuskonen, H., Paillard, F., Bremer, B., Becker, K., 2002. ATLA 30, 151.
OECD, 2001. Test Guideline 414: Prenatal Developmental Toxicity Study.
Piersma, A.H., Bechter, R., Krafft, N., Schmid, B.P., Stadler, J., Verhoef, A., Verseil, C., Zijlstra, J., 1996. ATLA 24, 201.
Spielmann, H., Pohl, I., Döring, B., Liebsch, M., Moldenhauer, F., 1997. Vitro Toxicol. 10, 119.
Whittaker, S.G., Faustman, E.M., 1994. In vitro toxicology. In: Cox Gad, S. (Ed.), In Vitro Assays for Developmental Toxicity. Raven Press, New York.

Toxicokinetics
Coecke, S., Blaauboer, B.J., Elaut, G., Freeman, S., Freidig, A., Gensmantel, N., Hoet, P., Kapoulas, V.M., Ladstetter, B., Langley, G., Leahy, D., Mannens, G., Meneguz, A., Monshouwer, M., Nemery, B., Pelkonen, O., Pfaller, W., Prieto, P., Proctor, N., Rogiers, V., Rostami-Hodjegan, A., Sabbioni, E., Steiling, W., van de Sandt, J.J., 2005. Toxicokinetics and metabolism. Altern. Lab. Anim. 33, 147.
EC B.36 – Toxicokinetics, 31 May, 2008. Council Regulation (EC) No 440/2008 of 30 May 2008 Laying Down Test Methods Pursuant to Regulation (EC) No 1907/2006 of the European Parliament and of the Council on the Registration, Evaluation, Authorisation and Restriction of Chemicals (REACH), p. 365 Official Journal L 142.
OECD 417, 1984. OECD Guideline for Testing of Chemicals – Guideline 417: Toxicokinetics. Organization for Economic Cooperation and Development, Paris. Adopted 4 April, 1984, last updated 22 July, 2010.

CHAPTER 18

Microbiological Quality in Cosmetics

Gabriel A. March[1], Maria C. Garcia-Loygorri[2], José M. Eiros[3], Miguel A. Bratos[1,3], Raúl Ortiz de Lejarazu[1,3]

[1]University Clinic Hospital of Valladolid, Valladolid, Spain; [2]Medina del Campo Hospital, Medina del Campo, Spain; [3]University of Valladolid, Valladolid, Spain

INTRODUCTION

Most cosmetic products are considered to be appropriate growth media for microorganisms because they incorporate water and nutrients, such as lipids, polysaccharides, alcohol, proteins, amino acids, glucosides, steroids, peptides and vitamins. Furthermore, the conditions of oxygenation, pH, temperature and osmotic pressure and some perfume ingredients favour microbial multiplication (Herrera, 2004). Therefore, development of several bacteria, yeasts or moulds is highly likely for many reasons. For example, during manufacturing, raw materials can contribute to a significant level of microbial contamination to the finished product. Moreover, conditions such as insufficient hygiene of both manufacturing areas and personnel can significantly contribute to bacterial contamination. During their use, cosmetics are exposed to microorganisms from the environment and the consumer's hands and body; in this case, purity after opening depends on the preservative ability of the product. In addition, consumers should be responsible for the proper use of the cosmetic product; fingers dipped into the product and spillage of water into the product should be avoided, and label integrity on cosmetic products must be respected.

Microbial growth in cosmetic products causes organoleptic alterations, such as offensive odours and changes in viscosity and colour. In addition, cosmetics contaminated by microorganisms are considered potentially hazardous to human health. For these reasons, these products are recalled from the market. Hence, cosmetic microbiology is a matter of great importance for the industry because microbial contamination can become a major cause of economic losses (Orus and Leranoz, 2005).

The aim of this chapter is to provide insight into the field of cosmetic microbiology and expand knowledge of the general demands on microbiological analysis of cosmetic products. Data about legislation in the European Union and the United States of America and the published harmonized standards for microbiological control of cosmetic products before, during and at the end of use are presented. Moreover, information about the recall of cosmetics contaminated by microorganisms from the market and bacterial resistance to cosmetic preservatives is given.

IMPACT ON HUMAN HEALTH OF COSMETIC PRODUCTS CONTAMINATED BY MICROORGANISMS

Cases of blindness as a consequence of bacterial contamination by *Pseudomonas* spp. in mascaras in the latter half of the 1970s are a milestone for the microbiological analysis of cosmetic products (Reid and Wood, 1979; Wilson and Ahearn, 1997). Since then, incidences of human injury due to contaminated cosmetics have been reported over time. Although individual infections due to contaminated cosmetics are unlikely to be discovered or documented, there are some reported outbreaks of infectious diseases in hospitals. Generally, the outbreaks affect immunocompromised patients, who are at higher risk for infectious diseases than immunocompetent patients. Stephenson et al. (1985) described an outbreak caused by mouthwash contaminated with *Pseudomonas aeruginosa*; seven immunocompromised patients with haematological malignancy developed sepsis and one of them died. Becks and Lorenzoni (1995) described an outbreak of *P. aeruginosa* in which contaminated hand lotion applied to clean hands of health care workers led to infection of six patients. Since 1999, four outbreaks due to *Burkholderia cepacia* complex strains have been reported. In three of them the cosmetic product involved was alcohol-free mouthwash, containing hexetidine 0.1% in one case (Molina-Cabrillana et al., 2006) and cetylpyridinium chloride in the two others (Matrician et al., 1999; Kutty et al., 2007). In these three outbreaks, patients were injured by the mouthwash used for prevention of health care–associated pneumonia; 24 hospitals were affected in the United States and one in Spain. Globally, 211 patients were involved and 40 of them died. Finally, the fourth outbreak was reported in Spain in 2008 (Alvarez-Lerma et al., 2008). In this last, the contaminated cosmetic product was moisturizing milk; five patients were affected and three of them died. It is worth mentioning that, in the four outbreaks due to *B. cepacia*, unopened bottles of the same batch of the cosmetic product involved in the outbreak showed bacterial contamination and, in most cases, deaths could not be directly attributed to the infection.

LEGISLATION ON COSMETIC MICROBIOLOGY

The European Union (EU) (European Parliament and the Council, 2009) and United States (US Food and Drug Administration, 2014) legislations state that companies and individuals who market cosmetics are responsible for their safety under normal or reasonably foreseeable conditions of use. This means that a commercialized cosmetic product must be free from microbial contamination, not only at the moment of reaching the market but also throughout the product life. To achieve this goal, exigent legislation is needed.

The processes of manufacturing and analytical controls are continually being improved. Since the late 1980s, the implementation of good manufacturing practices (GMP) has been the backbone for improving industrial quality control analysis. For cosmetic products, GMP providing guidelines for production, control, storage and shipment were published in 2007 (International Organization for Standardization, 2007). Moreover, according to the EU Cosmetics Regulation (European Parliament and the Council, 2009), each cosmetic product placed on the market must have its own Product Information File (PIF). The PIF is a leaflet that contains a great deal of information about the cosmetic product, including the physico-chemical stability, traces, impurities, packaging interactions and microbiological quality of both raw materials (substances or mixtures) and the cosmetic product.

EU (European Parliament and the Council, 2009) and US (U.S. Food and Drug Administration, 2016) legislations have specific demands on the microbial load of finished products as well as on the products' ability to withstand contamination. Regarding the EU, by means of the Scientific Committee on Consumer Safety (SCCS) (2015), in the section entitled 'Guidelines on microbiological quality of the finished cosmetic product' of the Notes on Guidance for Testing of Cosmetic Ingredients and Their Safety Evaluation, cosmetics are divided into two different categories, depending on microbiological requirements:

- Category 1: products specifically intended for children under 3 years or to be used in the eye area and on mucous membranes. For these products, total viable counts for microorganisms must not exceed 1×10^2 colony-forming units (CFU) in 1 g or 1 mL of the product. Microorganisms screened belong to aerobic mesophilic bacteria (bacteria that grow between 20 and 45°C), yeasts (mainly represented by *Candida* spp.) and mould (mainly represented by *Aspergillus* spp.). Furthermore, the pathogenic bacteria *Escherichia coli*, *P. aeruginosa* and *Staphylococcus aureus* and the yeast *Candida albicans* must not be present in 1 g or 1 mL of the product.
- Category 2: other products. For these products, total viable counts of microorganisms must not exceed 1×10^3 CFU/g or 1×10^3 CFU/mL, and the aforementioned pathogens must not be present in 1 g or 1 mL of the analysed product.

According to US legislation (U.S. Food and Drug Administration, 2016), cosmetic products are also divided into two different categories: eye area products and non–eye area products. Microbial counts below 500 CFU/g or 500 CFU/mL for eye area products and 1×10^3 CFU/g or 1×10^3 CFU/mL for non–eye area products are accepted. However, a precise limit for specific pathogenic bacteria is not regulated in the US legislation; as a consequence, the presence of particular pathogenic bacteria should be investigated if the total viable count is near the threshold of acceptance.

MICROBIOLOGICAL ANALYSIS OF COSMETIC PRODUCTS

Manufacturers should follow the GMP described in International Organization for Standardization (ISO) 22716 (International Organization for Standardization, 2007) and adopt adequate precautions to limit the introduction of microorganisms, not only from raw materials, but also during processing and packaging of the cosmetic product. Testing of raw materials before use is important, especially those of natural origin. The specifications of the raw materials must include microbiological purity. Water is the most common ingredient among raw materials and must be continuously tested for microbial growth.

In the United States, the US Food and Drug Administration (FDA) (2016) has provided detailed procedures for cosmetic microbiology. Moreover, the ISO and the European Committee for Standardization (CEN), through their technical committees for cosmetics (ISO/TC 217 and CEN/TC 392), have presented a number of international standards for the standardization of microbiological analysis of cosmetic products. They are summarized in Table 18.1.

It is strongly recommended that they be incorporated into the microbiological testing of cosmetic products. Some cosmetic products considered to have low microbiological risk may not need to be subjected to routine microbiological testing. Thus, manufacturers can decide not to test these products. They can consult ISO 29621 (under review at the time of writing, as is indicated in Table 18.1).

Microbiological testing can be performed according to two general standards: (1) Technical Report 19838, which provides general guidelines about the use of ISO cosmetic microbiological standards depending on the objective (in-market control, product development, etc.) and the product to be tested, and (2) ISO 21148, which is intended for the protection of the health of laboratory personnel and to achieve homogeneous results among laboratories.

For microbial detection, ISO 18415 is based on the subculture of microorganisms present in the cosmetic product in a non-selective liquid medium (enrichment broth) to increase their number, followed by the isolation of the microorganisms in a non-selective solid agar medium. Growing media must have the capacity to partially inactivate preservative systems commonly found in cosmetic products, which could inhibit bacterial growth. Finally, once colonies are grown in culture plates, microbial identification is performed by routine microbiological methods, which will be reviewed later. In addition, for the detection of some microorganisms, such as *E. coli*, *P. aeruginosa*, *S. aureus* and *C. albicans*, ISO standards 21150, 22717, 22718 and 18416, respectively, have been published.

Regarding the quantitative detection of microorganisms, ISO 17516 is applicable for all cosmetics and assists interested parties in the assessment of the microbiological quality of the products. More specifically, ISO 21149 is intended for aerobic mesophilic bacteria and ISO 16212 for yeasts and moulds. Both ISO standards are based on culturing a known amount of cosmetic product on agar medium; after several days of incubation, colonies are counted and microorganisms are identified. The application of both standards allows detection of the number of colonies per unit of volume (mL) or weight (g) in the final product, and results are expressed in CFU/mL or CFU/g. These values have

Table 18.1 International standards for microbiological analysis of cosmetic products

Standard on microbiology in cosmetics	International Organization for Standardization/TC 217		European Committee for Standardization/TC 392	
Cosmetics—Microbiology—General instructions for microbiological examination	ISO 21148	2005 2006 corrected	EN ISO 21148	2009
Cosmetics—Microbiology—Enumeration and detection of aerobic mesophilic bacteria	ISO 21149	2006	EN ISO 21149	2009
Cosmetics—Microbiology—Detection of specified and non-specified microorganisms	ISO 18415	2007	EN ISO 18415	2011
Cosmetics—Microbiology—Enumeration of yeast and mould	ISO 16212	2008	EN ISO 16212	2011
Cosmetics—Microbiology—Guidelines for the risk assessment and identification of microbiologically low-risk products	ISO 29621 ISO/DIS 29621	2010 Under development	EN ISO 29621 prEN ISO 29621	2011 Under development
Cosmetics—Microbiology—Evaluation of the antimicrobial protection of a cosmetic product	ISO 11930	2012	EN ISO 11930	2012
Cosmetics—Microbiology—Microbiological limits	ISO 17516	2014	EN ISO 17516	2014
Cosmetics—Microbiology—Detection of *Candida albicans*	ISO 18416	2015	EN ISO 18416	2015
Cosmetics—Microbiology—Detection of *Escherichia coli*	ISO 21150	2015	EN ISO 21150	2015
Cosmetics—Microbiology—Detection of *Pseudomonas aeruginosa*	ISO 22717	2015	EN ISO 22717	2015
Cosmetics—Microbiology—Detection of *Staphylococcus aureus*	ISO 22718	2015	EN ISO 22718	2015
Microbiology—Cosmetics—Guidelines for the application of ISO standards on cosmetic microbiology	ISO/TR 19838	2016	FprCE N ISO/TR 19838	2016

Technical Committee for Cosmetics (TC)

limits clearly stated by EU (Scientific Committee on Consumer Safety, 2015) and US (U.S. Food and Drug Administration, 2016) legislations.

In addition to the maximum number of microorganisms allowed in the finished product, it is also important that the cosmetic product, once opened, is not damaged by in-use bacterial contamination. For this reason, during manufacture, preservatives are added into cosmetic products. In this way, preservatives prevent microbial spoilage, extending the shelf life of the product and protecting the consumer from an infection. The capacity of preservatives to avoid microbial contamination during the use of the cosmetic product is investigated by a test called the challenge test, preservative effectiveness test, or antimicrobial effectiveness test. Briefly, this test shows the ability of the cosmetic product to reduce the counts of microorganisms after a microbial contamination. This assay was advised by various pharmacopoeias and cosmetic organizations, such as the Cosmetic, Toiletry and Fragrance Association. Additionally, different in-house test protocols have been established (Briteksoz et al., 2013; Campana et al., 2006; Na'was and Alkofahi, 1994; Okeke and Lamikanra, 2001). To meet the needs of challenge test standardization, ISO 11930, devoted to the evaluation of antimicrobial protection in cosmetic products, was published in April 2012. According to the procedure described in this standard, the cosmetic product is intentionally and heavily contaminated with known inocula of *P. aeruginosa* ATCC 9027, *S. aureus* ATCC 6538, *E. coli* ATCC 8739, *C. albicans* ATCC 10231 and *Aspergillus brasiliensis* ATCC 16404. After 7, 14 and 28 days, samples of the previously contaminated cosmetics are removed and serially diluted for counting viable microorganisms by means of colony counts after incubation of subculture agar for the appropriate periods of time at specified temperatures. Thus, it is possible to calculate the reduction from the value of log CFU/mL at time zero to that value at a given time. Depending on the reduction of the starting inocula, the cosmetic product will meet two different criteria. According to Section 3.3 of the guidelines (European Parliament and the Council, 2013) in Annex I of the EU Cosmetic Regulation (European Parliament and the Council, 2009) and the aforementioned guidance of the SCCS (Scientific Committee on Consumer Safety, 2015), the challenge test is mandatory for all cosmetic products that under normal conditions of storage and use may deteriorate or be a risk for consumers.

Microbial Identification

The last step of the majority of microbiological analyses consists in the identification of the microbial organisms obtained from colonies grown in culture plates. For performing bacterial and yeast identification there are two major systems: those that rely on genotypic characteristics and those that rely on phenotypic characteristics. The first step of genotypic identification involves DNA extraction. Subsequently, molecular methods such as the polymerase chain reaction (PCR) are applied. PCR is based on the ability of DNA polymerase to synthesize a new strand of DNA. For this step, a segment of DNA complementary to a given DNA sequence is required to initiate replication by DNA polymerase; this segment is called the primer and allows amplification of a specific region of a DNA target. For performing bacterial identification, the amplified genes are those which codify the smaller subunit of the ribosomes because these genes are highly conserved within the same microbial species, and are different between microorganisms of other genera and species. At the end of the PCR, target DNA will be accumulated in billions of copies (amplicons). Finally, the amplicons are sequenced and compared with known databases to identify the microorganism (Tang et al., 1998; Woo et al., 2008). Genotypic identification is considered the definitive identification of the microorganism. Moreover, this identification can be performed not only from colonies grown in culture plates, but also directly from the enrichment broth in which bacteria present in the cosmetic product were incubated for several hours (Jimenez et al., 2000; Jimenez, 2001).

Classical phenotypic identification is based on observable physical or metabolic characteristics after bacterial incubation of a colony previously grown on culture plates in the presence of selected substrates for at least 17 h. Thus, phenotypic identification depends on characteristics such as requirement of specific nutrients to grow, ability of the microorganism to produce one or more specific bioproducts, production of specific enzymes, fermentation of one or more specific sugars or susceptibility or resistance to one or more specific antibacterial agents (Bou et al., 2011). There are some commercialized systems for achieving a phenotypic identification. These systems allow the identification of microorganisms with a high degree of accuracy, not only of the most frequent isolates in clinical microbiology laboratories but also of microorganisms grown in cosmetic products.

More recently, within the phenotypic methods, protein analysis using matrix-assisted laser desorption/ionization time-of-flight (MALDI–TOF) mass spectrometry has led to considerable improvement in microbial identification. One major improvement is a significant savings in the time required for achieving microbial identification, which can be obtained in less than 5 min. MALDI–TOF identification requires a certain amount of microorganisms previously grown in culture plates to be mixed with a chemical matrix on a metal sample plate surface. After crystallization of the matrix by evaporation of the solvent at room temperature, the metal sample plate is bombarded in the mass spectrometer with brief laser pulses from a nitrogen laser. The matrix absorbs energy from the laser leading to the sublimation and ionization of microbial proteins, which are ejected through a metal flight tube until they reach a detector. The time of flight of the ionized proteins is proportional to their mass (m)/charge (z) relation. In this way, microbial proteins generate a mass spectrum that is characterized by both the m/z and the intensity of the ions, which represents the number of ions of a particular m/z that struck the detector. Then, the obtained protein mass spectrum, also called the peptide mass fingerprint, which is unique to each microorganism, is compared to the fingerprints of known microorganisms in the database in such a way that the problem fingerprint may be linked to the most similar fingerprint, enabling the microorganism to be identified (Croxatto et al., 2012; March-Rosselló et al., 2013).

When identification results obtained by MALDI–TOF were compared to traditional phenotypic identification (morphology, differential fermentation test, etc.), some discrepancies were observed. In studies carried out with bacteria and yeasts, such discrepancies were solved by microbial DNA sequencing, and in most cases, the sequencing results confirmed the results obtained by MALDI–TOF (Benagli et al., 2011; Sendid et al., 2013). For this reason, microbial identification by MALDI–TOF is more reliable and accurate than traditional phenotypic identification for most microorganisms.

Regarding moulds, they are identified by both macroscopic and microscopic observation of colony morphology grown in culture plates. For performing microscopic examination, a lactophenol blue stain should be prepared. The microscopic examination is difficult to perform and is often supported by textbook descriptions. Hence, this approach requires personnel appropriately trained, and identification at the species level can be problematic in some cases. To solve these difficulties, publications of mould identifications by MALDI–TOF and PCR are emerging (Boch et al., 2015; Cassagne et al., 2011; Schulthess et al., 2014).

RECALL OF COSMETICS CONTAMINATED BY MICROORGANISMS FROM THE MARKET

Despite legislation and microbiological controls, contaminated cosmetics appear on the market. For this reason, microbiological analysis must also be done once the product is placed on the market. Thus, it will be possible to detect cosmetics contaminated by microorganisms and recall them from the market. In more detail, from October 1993 to September 1998, the FDA recalled 56 cosmetics contaminated by microorganisms, and *P. aeruginosa*, the main contaminant, was isolated from more than 45% of the contaminated products (Wong et al., 2000). The EU has an intergovernmental rapid alert system (RAPEX) (European Commission, 2004) that operates when unsafe consumer products appear on the market. RAPEX issued 24 recalls of microbiological-contaminated cosmetic products from 2005 until week 17 of 2008. Microbial counts ranged from 600 to 8×10^6 CFU/g and high levels of contamination ($>1 \times 10^5$ CFU/g) were found in more than 60% of cosmetic products placed on the market. The most frequently isolated microorganism was also *P. aeruginosa*, but in this case the percentage of isolation was 40% (Lundov and Zachariae, 2008), slightly lower than that reported by the FDA (Wong et al., 2000). Birteksoz Tan et al. (2013) evaluated 93 commercial cosmetic products. Fourteen cosmetic products were found to be contaminated by microorganisms. Bacterial counts ranged between 1.5×10^2 and 5.5×10^5 CFU/mL. The most often isolated bacteria were *S. aureus* and *B. cepacia*, which were identified from six and four products, respectively. On the basis of this work, it can also be observed that bacterial contamination of cosmetic products with high water content (creams, gels and shampoos) is more frequent than that of cosmetics with low water content (lotions). This is because water is an essential element for microbial growth. For this reason, microbiological analysis of cosmetic products with high water content should be done with special attention.

To assess the efficacy of preservatives added into cosmetics to avoid microbial contamination during their use, Okeke and Lamikanra (2001) evaluated the microbiological safety of 49 commercial moisturizing products at time of purchase and after being used daily for 14 days. Eight products provided viable bacterial counts that exceeded 10^3 CFU/mL or CFU/g at time of purchase. The most commonly recovered bacteria were *Staphylococcus* spp., *E. coli*, *Pseudomonas* spp. and *Bacillus* spp., which were isolated in nine, eight, seven and six cases, respectively. When the bacterial contamination was studied along the product life, an increase in the isolation of *E. coli* and other gram-negative microorganisms was obtained. Later, Campana et al. (2006) evaluated the bacterial load from 91 cosmetics in three different states of use: intact, in-use and ending product. The intact products showed no bacterial contamination. However, in five ending products, and in one case of in-use product as well, bacterial contamination was observed. Bacterial counts ranged between 1×10^2 and 1×10^4 CFU/mL. The species isolated were *Staphylococcus warneri* and *Staphylococcus epidermidis* in two cases, and *Pseudomonas putida* in one case.

BACTERIAL RESISTANCE IN COSMETICS

The modes of action of preservatives are much broader than those of antibiotics used in clinical settings. The mode of action of antibiotics is known at the molecular level because they act via specific biochemical reactions. However, preservatives act in a more general way by denaturing cellular proteins or by affecting membrane permeability; in this way, transport of nutrients and energy generation are blocked (Brannan, 1995).

There are some methods for determining the effectiveness of antimicrobial preservatives in cosmetics. One of them is the challenge test. However, to select adequate preservatives to add into cosmetic products, the susceptibility of bacteria commonly isolated from cosmetic products and manufacturing plants to preservatives should be previously determined. Despite the importance of these studies, as of this writing no standard method for evaluating the susceptibility of bacteria to preservatives can be found. Nevertheless, the minimum inhibitory concentration (MIC) method, described for antimicrobial drugs in clinical microbiology, can be considered as a useful approach for studying the phenotypic behaviour of bacteria when incubated in the presence of preservatives (Mokhtari, 2008). In this case, MIC is defined as the lowest concentration of preservative at which the growth of a tested microorganism is completely inhibited. The minimum bactericidal concentration (MBC) method, rarely used in clinical microbiology, can also be used to evaluate the susceptibility of bacteria to antibiotics. The MBC is defined as the lowest concentration of antibacterial agent that reduces the viability of the initial bacterial inoculum by ≥99.9% (French, 2006). The application of these methods would enable one to select the most suitable preservatives for a specific cosmetic product and, moreover, it could be possible to approximate the preservative concentration suitable for avoiding microbial growth.

The problem of bacterial resistance to preservatives in cosmetic products should be focused on the group non-fermenting gram-negative rods, which include the species *P. aeruginosa* and *B. cepacia* complex. *P. aeruginosa* is the most frequently isolated bacterium from recalled contaminated cosmetic products in the EU and the United States (Lundov and Zachariae, 2008; Wong et al., 2000) because of its resistance to some of the preservatives commonly used in cosmetics. Ferrarese et al. (2003) noted that, against the preservatives imidazolidinil urea, dimethylol dimethyl hydantoin, methylisothiazolinone and parabens mix in phenoxyethanol, the strains of *P. aeruginosa* isolated from contaminated cosmetic products and industrial plants had enhanced resistance compared to collection bacterial strains, which have not been in contact with preservatives. Many publications have demonstrated that *Pseudomonas* spp., via progressive subculture in the presence of sublethal concentrations of the preservatives benzalkonium chloride (Mc Cay et al., 2010), phenoxyethanol (Abdel Malek and Badran, 2010) and isothiazolinone biocides (Brozel and Cloete, 1994; Winder et al., 2000), are able to develop resistance to them. Moreover, Mokhtari (2008) calculated the MICs of parabens and isothiazolinones in *S. aureus*, *E. coli*, *P. aeruginosa*, *C. albicans* and *Aspergillus niger* and noted that among the microorganisms tested, *P. aeruginosa* provided the highest MICs. Regarding *B. cepacia* complex, Rushton et al. (2013) calculated the MIC and MBC of eight commercially available preservatives (benzisothiazolinone, benzethonium chloride, dimethylol dimethyl hydantoin, methylisothiazolinone, methylisothiazolinone/chloromethylisothiazolinone at the proportion 3:1, methylparaben, phenoxyethanol and sodium benzoate) in 83 strains. They found that, for six of the eight preservatives evaluated, MIC and/or MBC values were above the maximum level permitted for use in personal care products. They also observed that maximum permitted levels of dimethylol dimethyl hydantoin (0.3%) and phenoxyethanol (1%) had the greatest activity against *B. cepacia* complex, inhibiting and killing all strains tested. Rose et al. (2009) identified two strains belonging to *B. cepacia* complex with chlorhexidine MBCs of 1000 mg/L and 31 *B. cepacia* complex strains with cetylpyridinium chloride MBCs in excess of 1000 mg/L. It is important to note that chlorhexidine and cetylpyridinium chloride are used in a variety of commercial products in concentrations ranging from 0.1% to 4% (1000–40,000 mg/L). Hence, the lowest concentration of chlorhexidine and cetylpyridinium used in cosmetic products may not be enough to kill strains of *B. cepacia* complex.

Antibacterial agents and antibiotics share the same resistance problem: resistance will certainly increase as the drug persists, especially at low levels for long periods of time.

In clinical settings, owing to misuse and overuse of antibiotics, bacteria resistant to all standard antimicrobial agents have emerged (Sopirala et al., 2010). Preservatives should be used at optimal concentrations, not only to kill bacteria and avoid the emergence of bacterial strains resistant to preservatives, but also to exclude the occurrence of adverse reactions such as skin irritation. Moreover, preservatives should be compatible with all ingredients of a cosmetic product and its packaging and should be active in the complete formulation and stable over the range of pH values (Mokhtari, 2008). Thus, a major challenge for cosmetics microbiology is to ensure that preservatives are able to kill all bacteria that might be present in the cosmetic and do not lead to adaptive antimicrobial resistance. For a better understanding of bacterial resistance mechanisms to preservatives, some papers about phenotypic changes in *Enterobacter gergoviae* and *B. cepacia* in the adaptive response to preservatives have appeared in the literature (Periame et al., 2015a,b; Rushton et al., 2013). It is important to make progress in this field and discover the molecular mechanism of how bacteria become resistant to preservatives.

CONCLUSIONS

In conclusion it can be stated that microbial contamination of cosmetic products occurs sometimes. To avoid microbial contamination, strong legislation and regular and effective microbiological checking are required. Furthermore, more thorough surveillance of microbiological contamination of cosmetic products used in adults predisposed to infection should be mandatory. Microbiological analyses must be performed in accordance with published harmonized methods, which are based on identification and enumeration of microorganisms grown in culture plates. Depending on the accuracy of microbial identification from colonies grown in culture plates, genotypic or phenotypic methods can be applied. Regarding preservatives added to cosmetic products to avoid bacterial contamination during their use, an ISO standard to calculate their MIC or MBC is urgently required. By means of these calculations it could be possible to use preservatives at their optimal concentration, which could delay the emergence of strains resistant to preservatives.

REFERENCES

Abdel Malek, S.M., Badran, Y.R., 2010. Folia Microbiol. (Praha) 55, 588.
Alvarez-Lerma, F., Maull, E., Terradas, R., Segura, C., Planells, I., Coll, P., Knobel, H., Vázquez, A., 2008. Crit. Care 12, R10.
Becks, V.E., Lorenzoni, N.M., 1995. Am. J. Infect. Control 23, 396.
Benagli, C., Rossi, V., Dolina, M., Tonolla, M., Petrini, O., 2011. PLoS One 6, e16424.
Birteksoz Tan, A.S., Tuysuz, M., Otuk, G., 2013. Pak. J. Pharm. Sci. 26, 153.
Boch, T., Reinwald, M., Postina, P., Cornely, O.A., Vehreschild, J.J., Heußel, C.P., Heinz, W.J., Hoenigl, M., Eigl, S., Lehrnbecher, T., Hahn, J., Claus, B., Lauten, M., Egerer, G., Müller, M.C., Will, S., Merker, N., Hofmann, W.K., Buchheidt, D., Spiess, B., 2015. Mycoses 58, 735.
Bou, G., Fernandez-Olmos, A., Garcia, C., Saez-Nieto, J.A., Valdezate, S., 2011. Enferm. Infecc. Microbiol. Clin. 29, 601.

Brannan, D.K., 1995. J. Soc. Cosmet. Chem. 46, 199.
Brozel, V.S., Cloete, T.E., 1994. J. Appl. Bacteriol. 76, 576.
Campana, R., Scesa, C., Patrone, V., Vittoria, E., Baffone, W., 2006. Lett. Appl. Microbiol. 43, 301.
Cassagne, C., Ranque, S., Normand, A.C., Fourquet, P., Thiebault, S., Planard, C., Hendrickx, M., Piarroux, R., 2011. PLoS One 6, e28425.
Croxatto, A., Prod'hom, G., Greub, G., 2012. FEMS Microbiol. Rev. 36, 380.
European Commission, 2004. Keeping European Consumers Safe, Rapid Alert System for Non-Food Dangerous Products. http://ec.europa.eu/consumers/consumers_safety/safety_products/rapex/index_en.htm.
European Parliament and the Council, 2013. Commission Implementing Decision of 25 November 2013 on Guidelines on Annex I to Regulation (EC) No 1223/2009 of the European Parliament and of the Council on Cosmetic Products.
European Parliament, the Council, 2009. Regulation (EC) No 1223/2009 of the European Parliament and of the Council of 30 November 2009 on Cosmetic Products. http://eur-lex.europa.eu/LexUriServ/LexUriServ.do?uri=OJ: L:2009:342:0059:0209:en:PDF.
Ferrarese, L., Paglia, P., Ghirardini, A., 2003. Ann. Microbiol. 53, 477.
French, G.L., 2006. J. Antimicrob. Chemother. 58, 1107.
Herrera, A.G., 2004. Methods Mol. Biol. 268, 293.
International Organization for Standardization, Cosmetics – Good Manufacturing Practices (GMP) – Guidelines on Good Manufacturing Practices. ISO 22716:2007.
Jimenez, L., 2001. J. AOAC Int. 84, 671.
Jimenez, L., Smalls, S., Jimenez, L., Smalls, S., 2000. J. AOAC Int. 83, 963.
Kutty, P.K., Moody, B., Gullion, J.S., Zervos, M., Ajluni, M., Washburn, R., Sanderson, R., Kainer, M.A., Powell, T.A., Clarke, C.F., Powell, R.J., Pascoe, N., Shams, A., LiPuma, J.J., Jensen, B., Noble-Wang, J., Arduino, M.J., McDonald, L.C., 2007. Chest 132, 1825.
Lundov, M.D., Zachariae, C., 2008. Int. J. Cosmet. Sci. 30, 471.
March-Rossello, G.A., Munoz-Moreno, M.F., Garcia-Loygorri-Jordan de Urries, M.C., Bratos-Perez, M.A., 2013. Eur. J. Clin. Microbiol. Infect. Dis. 32, 699.
Matrician, L., Ange, G., Burns, S., Fanning, L., Kioski, C., Cage, G., Harter, G., Reese, D., McFall, D., Komatsu, K., Englund, R., 1999. JAMA 281, 318.
Mc Cay, P.H., Ocampo-Sosa, A.A., Fleming, G.T., 2010. Microbiology 156, 30.
Mokhtari, F., 2008. Res. J. Biol. Sci. 3, 984.
Molina-Cabrillana, J., Bolanos-Rivero, M., Alvarez-Leon, E.E., Martín Sánchez, A.M., Sánchez-Palacios, M., Alvarez, D., Sáez-Nieto, J.A., 2006. Infect. Control Hosp. Epidemiol. 27, 1281.
Na'was, T., Alkofahi, A., 1994. J. Clin. Pharm. Ther. 19, 41.
Okeke, I.N., Lamikanra, A., 2001. J. Appl. Microbiol. 91, 922.
Orus, P., Leranoz, S., 2005. Int. Microbiol. 8, 77.
Periame, M., Pages, P.M., Davin-Regli, A., 2015a. J. Appl. Microbiol. 118, 49.
Periame, M., Philippe, N., Condell, O., Fanning, S., Pages, J.M., Davin-Regli, A., 2015b. Lett. Appl. Microbiol. 61, 121.
Reid, F.R., Wood, T.O., 1979. Arch. Ophthalmol. 97, 1640.
Rose, H., Baldwin, A., Dowson, C.G., Mahenthiralingam, E., 2009. J. Antimicrob. Chemother. 63, 502.
Rushton, L., Sass, A., Baldwin, A., Dowson, C.G., Donoghue, D., Mahenthiralingam, E., 2013. Antimicrob. Agents Chemother. 57, 2972.
Schulthess, B., Ledermann, R., Mouttet, F., Zbinden, A., Bloemberg, G.V., Böttger, E.C., Hombach, M., 2014. J. Clin. Microbiol. 52, 2797.
Scientific Committee on Consumer Safety (SCCS), SCCS/1564/15. The SCCS notes of guidance for the testing of cosmetic ingredients and their safety evaluation. 9th revision. Adopted by the SCCS at its 11th Plenary Meeting of 29 September, 2015 http://ec.europa.eu/health/scientific_committees/consumer_safety/docs/sccs_o_190.pdf.
Sendid, B., Ducoroy, P., Francois, N., Lucchi, G., Spinali, S., Vagner, O., Damiens, S., Bonnin, A., Poulain, D., Dalle, F., 2013. Med. Mycol. 51, 25.
Sopirala, M.M., Mangino, J.E., Gebreyes, W.A., Biller, B., Bannerman, T., Balada-Llasat, J.M., Pancholi, P., 2010. Antimicrob. Agents Chemother. 54, 4678.
Stephenson, J.R., Heard, S.R., Richards, M.A., Tabaqchali, S., 1985. J. Hosp. Infect. 6, 369.

Tang, Y.W., Ellis, N.M., Hopkins, M.K., Smith, D.H., Dodge, D.E., Persing, D.H., 1998. J. Clin. Microbiol. 36, 3674.
U.S. Food, Drug Administration, 2014. Cosmetics Safety Q&A: Contaminants. http://www.fda.gov/Cosmetics/ResourcesForYou/Consumers/ucm290083.htm.
U.S. Food, Drug Administration, Anthony, D., Tony, T., James, E., 2016. Bacteriological Analytical Manual, Chapter 23: Microbial Methods for Cosmetics. http://www.fda.gov/Food/FoodScienceResearch/LaboratoryMethods/ucm073598.htm.
Wilson, L.A., Ahearn, D.G., 1997. Am. J. Ophthalmol. 84, 112.
Winder, C.L., Al-Adham, I.S., Abdel Malek, S.M., Buultjens, T.E., Horrocks, A.J., Collier, P.J., 2000. J. Appl. Microbiol. 89, 289.
Wong, S., Street, D., Delgado, S.I., Klontz, K.C., 2000. J. Food Prot. 63, 1113.
Woo, P.C., Lau, S.K., Teng, J.L., Tse, H., Yuen, K.Y., 2008. Clin. Microbiol. Infect. 14, 908.

INDEX

Note: Page numbers followed by "f" indicate figures and "t" indicate tables.
Due to the large number of substances mentioned in the different chapters (ingredients, contaminants, etc.), they are not cited in this Index. Readers can easily find them in the related chapters.

A

Acute toxicity, 553–554, 576
Aftershave, 85, 151–153, 199t–200t, 234t–241t, 244–245, 361t–362t
After-sun, 22, 108, 151, 199t–200t, 234t–241t
Agência Nacional de Vigilância Sanitária (ANVISA), 24–25
Allergen, allergy, allergic, allergically, 19–20, 42, 45–46, 76t, 77, 183–184, 229–231, 233–245, 234t–241t, 323, 333t–334t, 370, 565
Allowed substance/s, allowed ingredient/s, 31–32, 47, 164–165, 184–185, 336
Alternative method/s, 12, 147, 356, 467, 549–584
Amperometry, amperometric, 119, 166–171, 213, 366t–368t
Animal testing, 6t, 8, 12, 19–20, 24, 34, 184, 549–584
Anionic surfactant/s, 249, 253–262, 253t, 255t–256t, 265, 267–274
Anti-ageing, 22, 85, 290–291, 340t–341t
Antidandruff, 13, 22, 24–26, 47, 51, 69t–70t, 76t, 77–78, 181, 340t–341t, 345t–348t, 372t–373t
Antimicrobial, 40–41, 77–78, 175–183, 266–267, 290, 314–315, 339–341, 340t–341t, 391, 401, 405–408, 502, 589t, 590, 593–595
Antioxidant/s, 167t–170t, 171–172, 175, 182–183, 213, 217–219, 290–291, 318–319, 327, 340t–341t, 388
Antiperspirant/s, 13, 24–26, 28–29, 47, 69t–70t, 186t–197t, 221, 234t–241t, 243, 340t–343t
Anti-wrinkle, 22, 24, 30, 199t–200t
Association of Official Analytical Chemists (AOAC), 338
Association of Southeast Asia Nations (ASEAN), 3–4, 20–22, 89t–90t
Australia, 22–24, 395t–397t, 398, 401, 402t–404t, 405, 406t–407t
Australian Competition and Consumer Commission (ACCC), 23–24
Australian Register of Therapeutic Goods (ARTG), 23

B

Banned substance/s, banned ingredient/s, 9, 13–14, 17–18, 21, 27, 30, 33, 57, 282
Batch certified, 47, 124–132, 134, 142, 165
Bath foam/s, bath salt/s, 15–16, 151–153, 199t–200t, 255t–256t, 275t–276t, 282, 361t–362t
Bibliographic resource/s, 55–66
Biomonitoring, 385–434
Biota, 435–436, 461t–465t, 493t–499t, 529t–532t, 539t
Bleaching, 24, 31, 76t, 80, 107, 112, 366t–368t
Blood, 390, 392, 400, 402t–404t, 408–410, 411t–414t, 564, 566, 572
Blush, 123, 147, 148t, 150, 152, 341–344
Body, 7, 12–17, 22, 24–25, 27–32, 43, 48t–49t, 72–73, 85, 87, 124, 136, 151–152, 164, 186t–197t, 199t–200t, 221, 227, 234t–241t, 243–244, 268–270, 269t, 275t–276t, 291–293, 297, 300, 323, 336, 339, 345t–348t, 353t–355t, 364, 370, 372t–373t, 375, 387, 390, 393, 400–401, 410, 566–567, 575–576, 585
Body fluid/s, 390–391
Borderline, 7, 33
Brazil, 3, 19–20, 19t, 24–25, 336, 395t–397t
Breast milk, 390, 392, 401–405, 406t–407t, 409–410, 411t–414t, 415, 419t–429t
British Pharmacopoeia, 259–260

C

Canada, 19, 25–27, 88, 134, 185, 336, 338, 341–344, 405–408, 406t–407t
Carcinogenicity, 48t–49t, 333t–334t, 399–400, 573–574
Carcinogenic, mutagenic or toxic for reproduction (CMR), 6, 9, 78
Cationic surfactant/s, 249–257, 253t, 255t–256t, 259–260, 262–265, 267–268, 270, 282, 467–468
Centre for Drug Evaluation and Research (CDER), 14

599

Centre for Food Safety and Applied Nutrition (CFSAN), 14
Certification-exempt, 124–132, 129t–131t, 134, 136
Certified Reference Materials (CRMs), 344, 377
Chemiluminescence (CL), 119, 171, 198–213, 215, 353t–355t, 356, 565
Chinese Food and Drug Administration (CFDA), 28–29, 339
Chromatography, chromatographic, 59, 63, 93–102, 113–119, 138–140, 166–171, 198, 214, 219, 232–233, 242, 263–272, 277–282, 360–363, 361t–362t, 374, 469
Coal-tar colours, coal-tar colors, 18, 47, 124
Code of Federal Regulations (CFR), 12–13, 46–47, 48t–49t, 112, 124–132, 135, 138, 142, 146, 165
Cologne/s, 17, 153, 228t
Colour additive/s, color additive/s, 12, 15, 47, 112, 123–132, 125t–131t, 134–135, 138, 146–147, 149f, 165, 341, 342t–343t
Colourant/s, colorant/s, 5, 7, 43, 47, 51–52, 67, 132–133, 135–136, 164, 342t–343t, 566–567
Colour Index (CI), 45, 133–135
Colouring agent/s, coloring agent/s, 42–47, 50, 52, 61, 123–154, 159–162, 165, 246, 340t–341t, 388
Contaminant/s, 57–58, 75, 79, 87, 137–138, 142, 231–232, 246, 272, 274, 331–384, 389–390, 435, 592–593
Cosmeceutical/s, 30, 289, 326
Cosmetic Ingredients Database (CosIng), 44, 227, 337, 374–375
Cosmetic Ingredients Review (CIR), 50, 185, 186t–197t, 252, 332, 370
Cosmetic Products Notification Portal (CPNP), 6, 9, 252
Cosmetics Europe, 3, 12, 42–43, 252, 337
Cosmetic, Toiletries and Fragrance Association (CTFA), 42
Cosmetovigilance, 25
Cream/s, 22, 24, 112, 152–153, 186t–197t, 199t–200t, 218, 221, 234t–241t, 243–245, 261t, 263, 275t–276t, 349, 353t–355t, 361t–362t, 366t–368t, 370–374, 372t–373t, 558, 592–593
Customs Union, 31

D

Databases, 14, 44, 61–65, 79, 113, 166, 231–233, 590–592
Decorative, 45, 133, 388
Demi-permanent, 159, 161, 163
Densitometry, densitometric, 150–151, 171
Deodorant/s, 13, 15–17, 24, 28–29, 51, 153, 181, 186t–197t, 234t–241t, 243–244, 290, 335, 340t–341t, 387, 401, 405–408
Depilatory/ies, 17, 25, 28–30, 51
Dermal, 22, 80–81, 245, 292, 391, 401, 408–410, 556–557, 566–569
Dermatology, dermatologic, dermatologically, 40, 43, 109–110, 136
Drug Identification Number (DIN), 26
Dye/s, 123, 129t, 135–138, 140f, 142, 148t, 152–154, 159–160, 161f, 162, 164–165, 171, 262–263, 341–344, 387–388, 556, 571

E

Eco-friendly, 303–316
Efficacy, 3–4, 8, 11–16, 20–34, 39, 72–73, 81, 86–88, 108, 110, 112, 163, 176, 182, 263–264, 290–291, 365, 593
Electroanalysis, electroanalytical, 119, 254–260, 255t–256t
Electrodes, 94t–100t, 114t–118t, 166–171, 167t–170t, 213, 215, 254–260, 255t–256t, 266–267, 369, 371–374, 562–563
Electrophoresis, electrophoretic, 119, 154, 166–171, 185–198, 214–215, 242–243, 245, 263, 369, 466, 573
Emergent, 87
Emerging, 295, 387, 436–437, 592
Emulsification, 171–172, 220, 243–244, 345t–348t, 350, 468
Emulsion, 77, 82, 160–162, 182, 199t–200t, 254–257, 279, 335, 345t–348t, 350, 357–358, 361t–362t, 365–369, 371, 374
Endocrine disrupting, 183, 389
Environmental monitoring, 435–548
Environmental Working Group (EWG), 332, 387–388
Essential oil/s, 27, 182–183, 225–228, 234t–241t, 243, 319–321, 321t–322t, 323, 324f
Estimated Daily Intake (EDI), 392, 400, 415–416
EU Cosmetics Directive, EU Directive, 3–5, 7, 11, 24, 28–29, 87–88, 124t, 132–133, 229, 337, 375

EU Cosmetics Regulation, EU Regulation, 3–6, 6t, 8–9, 11–12, 27–29, 31–34, 43–46, 45t, 67–68, 71–72, 81, 87–88, 111, 132–133, 164, 175, 180, 186t–197t, 198, 229, 230t, 231, 233, 242–243, 249–252, 250t–251t, 295–296, 298–300, 335–337, 341, 364, 370–371, 551, 560t, 586–587

European Centre for the Validation of Alternative Methods (ECVAM), 552, 555, 560–562, 560t, 575, 577

European Commission (EC), 5–6, 11, 43–44, 67–69, 71, 72t, 73, 75, 77–78, 80–81, 87–88, 94t–100t, 111, 113, 123, 132–133, 148, 164–166, 175, 185, 186t–197t, 198, 227, 229–231, 233, 244–245, 250t–251t, 298, 303, 315, 323, 337, 339–341, 342t–343t, 344, 352, 365, 370–371, 392, 551, 553, 554t, 555, 557, 558t, 565–566, 569–571, 575–576, 592–593

European Committee for Standardization (CEN), 67–69, 72–73, 74t, 75, 76t, 77, 80–82, 252, 588, 589t

European Union (EU), 3–13, 6t, 19–21, 19t, 24–28, 30–35, 39, 42–47, 45t, 50–52, 67–68, 71–72, 71t, 76t, 81–82, 87–88, 89t–90t, 91f–92f, 93, 102, 111–113, 123, 124t, 132–136, 152–153, 164–166, 175–176, 180, 184–185, 186t–197t, 198, 229, 230t, 231, 233–245, 249–252, 250t–251t, 289–291, 290f, 293, 295–300, 335–338, 341–344, 352, 364, 370–371, 375, 392, 399–400, 551–552, 554t, 555, 560t, 565–567, 576, 585–587, 590, 592–594

Extract/s, 42–43, 71t–72t, 74t, 77–80, 94t–100t, 103, 109–110, 112–113, 114t–118t, 124, 129t–131t, 145, 149–154, 167t–170t, 183, 199t–200t, 216–218, 220, 225–226, 228–229, 231–242, 234t–241t, 244–245, 263–264, 273–274, 278, 280, 317–318, 321t–322t, 325–327, 353t–355t, 356, 361t–362t, 365, 366t–368t, 371–374, 376, 438t–465t, 467–469, 471–500, 473t–484t, 493t–499t, 503t–532t, 533, 535t–537t, 539t, 540, 556

Extraction, 75, 77, 80, 94t–100t, 103, 112, 145, 148, 151–153, 166, 167t–170t, 171–172, 185–198, 216–221, 226, 234t–241t, 243–246, 253t, 254–257, 262–263, 268, 272, 277–282, 303, 317–327, 325f–326f, 331–332, 345t–348t, 350–351, 353t–355t, 361t–362t, 363–364, 376–377, 391, 395t–397t, 402t–404t, 406t–407t, 411t–414t, 435–437, 438t–465t, 466–501, 473t–499t, 503t–532t, 533–534, 535t–539t, 540, 556, 590–591

Eye irritation, 184, 558–564, 576

Eye shadow/s, 25, 123, 147, 148t, 150–151, 153, 341–344, 345t–348t

Eyelash/es, 165, 340t–343t

Eyeliner/s, 123, 147, 341–344, 345t–348t

F

Face, 22, 30, 147, 151–153, 186t–197t, 199t–200t, 234t–241t, 242, 261t, 275t, 341–344, 342t–343t, 364, 370, 372t–373t

Federal Food, Drugs, and Cosmetics Act (FD&C Act), 12–15, 48t–49t, 123, 132, 134, 136, 337–338, 341–344

Food and Drug Administration (FDA), 12–16, 29, 42, 46–47, 48t–49t, 50, 89t–90t, 112, 123–132, 124t–126t, 128t–129t, 134, 138, 142, 165, 176, 185, 186t–197t, 229, 231, 252, 273–277, 293–295, 332, 335–338, 341–344, 342t–343t, 351–352, 359, 364, 370–371, 375, 588, 592–593

Forensic, 135–136, 154

Formaldehyde releasers, 177t, 180, 184, 370–371

Fragrance/s, 6–7, 11, 13–14, 18, 23–24, 32–33, 42, 76t, 77, 148t, 176, 182–183, 225–246, 228t, 234t–241t, 309–311, 323, 331, 335, 350, 472, 473t–499t, 590

Functional cosmetic/s, 30, 289, 323

G

Gas chromatography (GC), 63–64, 71t, 76t, 77, 79–80, 93–102, 112, 114t–118t, 119, 142–146, 171, 198, 199t–200t, 213–214, 217–219, 221, 232–246, 234t–241t, 253t, 268, 269t, 271–274, 271f, 273t, 275t–276t, 278–282, 319–321, 321t–322t, 353t–355t, 356–358, 359f, 360, 361t–362t, 363–364, 372t–373t, 374, 391, 406t–407t, 411t–414t, 438t–465t, 471, 473t–499t, 502, 503t–532t, 533–534, 535t–538t, 540, 568

Gel/s, 22, 77–78, 150–151, 153, 159–160, 180–181, 186t–197t, 199t–200t, 234t–241t, 243–244, 254–259, 255t–256t, 261t, 269t, 270–271, 271f, 308, 353t–355t, 361t–362t, 366t–368t, 371–374, 372t–373t, 411t–414t, 438t–455t, 461t–465t, 466, 469, 485t–499t, 529t–532t, 539t, 592–593
Genital organs, 7
Genotoxicity, 183, 552, 570–572, 574, 582
Good Manufacture Practices (GMP), 4, 8, 14–15, 17, 20–21, 31–33, 67, 70, 335–337, 341–344, 376, 586–588
Good Quality Practices (GQP), 17
Good Vigilance Practises (GVP), 17
Green chemistry, 61, 303–305, 312, 317
Green ingredient/s, 303–330

H

Hair, 7, 13, 15–18, 22, 24–25, 28–31, 68–69, 69t–70t, 85, 111, 132–133, 159–165, 163f, 180–181, 199t–200t, 228, 234t–241t, 243, 260–264, 261t, 264t, 267–268, 269t, 275t–276t, 280, 323, 342t–343t, 353t–355t, 370–374, 372t–373t, 387–388, 390, 398–399, 401, 410
Hair dye/s, 15–17, 24–26, 28–31, 33, 42, 47, 51–52, 61, 69t–70t, 111, 132, 134, 159–174, 199t–200t, 275t–276t, 278–279, 339–341, 345t–348t, 356, 364
Harmonization, harmonized, 5, 8–9, 12, 43, 58, 67–68, 71t–72t, 72, 74t, 134, 335–337, 585, 595
Head-space, 219, 277–278
Health Canada, 25–27, 338, 341–344, 364, 370–371, 375
Human exposure, 389–390, 393–394, 400–401, 408–409, 416–418, 566

I

Impurity/ies, 43, 48t–49t, 60, 75, 136–138, 138f, 142–147, 274–277, 281–282, 294–295, 331–341, 333t–334t, 351–352, 364, 376–377, 469–470, 586–587
In vitro, 108, 295, 304, 306, 320–321, 323, 327, 555–557, 559–564, 560t, 566–577
In vivo, 295, 399–400, 557, 559–569, 571, 574, 576
Ingredient/s, 3–6, 9–10, 13–16, 18, 19t, 21, 23–30, 32–34, 39–54, 57–58, 61, 67–68, 73, 75, 80, 85, 87, 102, 108, 111–112, 123, 132–134, 159–166, 160f, 171–172, 176, 180–185, 186t–197t, 198, 218, 225–246, 234t–241t, 249–252, 250t–251t, 265, 267, 271, 281, 290–291, 294, 298, 300, 303–330, 332–341, 333t–334t, 340t–343t, 344, 351–352, 361t–362t, 364–365, 374–376, 385–548, 551, 554, 558, 564, 570–571, 573, 575–576, 585, 587–588, 594–595
In-market control, 4–6, 9, 588
Insect repellent/s, 17, 436, 534–540, 535t–539t
Institute for Reference Materials and Measurements (IRMM), 58, 166
Interagency Coordinating Committee on the Validation of Alternative Methods (ICCVAM), 552, 565
Intermediate dye/s, 161–162
Intermediate/s, 7, 58–60, 124, 137–138, 142–146, 161–162, 338, 393
International Cooperation on Cosmetics Regulation (ICCR), 19–20, 331–332, 336, 360, 376
International Dialogue for the Evaluation of Allergens (IDEA), 231
International Fragrance Association (IFRA), 50, 227, 231, 242
International Nomenclature of Cosmetic Ingredients (INCI), 4, 9–10, 15–16, 18, 19t, 21, 23, 25–26, 28–29, 42–43, 51–52, 89t–90t, 109t, 133, 332
International Non-proprietary Name (INN), 44
International Organization for Standardization (ISO), 18–20, 32–33, 41, 67–70, 71t, 73, 75, 76t, 77, 80, 82, 86, 252–254, 253t, 272–274, 273t, 278–279, 335–336, 344, 352, 358, 359f, 587–590, 589t

J

Japan, 3, 7, 16–19, 19t, 39, 42, 50–52, 88, 89t–90t, 112, 123, 124t, 133, 135–136, 148, 150–151, 165, 176, 185, 229, 245, 336, 339, 342t–343t, 399–400, 408, 411t–414t, 417–418, 419t–429t
Japanese Cosmetic Industry Association (JCIA), 17–18
Joint Research Centre (JRC), 68–69, 71, 73–75

K

Kingdom of Saudi Arabia, 27–28
Kovats index, 232–233

L

Label, 13–16, 20–21, 25–26, 29, 32–34, 39–40, 43, 46, 51, 132, 136, 176, 229, 272, 298, 370–371, 585
Labelling, 4, 8, 10–11, 14, 16, 18, 20–21, 23, 26, 28, 32–34, 42, 45–46, 77, 124–132, 134, 292, 298, 331, 335, 370, 553, 563, 565
Lake/s, 87, 132, 435–436
Lip gloss/es, 153
Lipstick/s, 13, 25, 85, 136, 147–154, 148t, 275t–276t, 335–336, 341–344, 345t–348t, 349–351, 353t–355t, 388, 398–399
Liquid chromatography (LC), 75, 77–82, 93, 112, 137–138, 139f, 141f, 143f–144f, 151–153, 166, 198–213, 233, 242, 244–245, 263, 265–268, 266t, 269t, 273t, 278–280, 308, 350–351, 359f, 391, 395t–397t, 402t–404t, 406t–407t, 419t–429t, 438t–460t, 466, 502, 539t, 568
Liquid-phase, 220–221, 374, 468, 500, 533–534
Lotion/s, 24, 30, 69t–70t, 82, 147, 150–153, 159–161, 186t–197t, 199t–200t, 216, 234t–241t, 243–245, 266t, 267, 275t–276t, 345t–348t, 353t–355t, 359, 361t–362t, 364–369, 366t–368t, 374, 387, 392, 394, 401, 405–408, 410, 586, 592–593

M

Make-up, 13, 17, 24, 28–29, 32, 73, 148, 148t, 149f, 150–152, 199t–200t, 275t–276t, 341–344, 345t–348t, 353t–355t, 387, 394
Mask/s, 18, 85, 199t–200t, 228, 234t–241t, 270, 366t–368t, 372t–373t
Mass spectrometry, mass spectrometric (MS), 76t, 77–80, 82, 94t–100t, 102, 142, 171, 232–233, 242–246, 263–265, 264t, 266t, 267, 271, 271f, 275t–276t, 277–279, 319–321, 321t–322t, 344, 345t–348t, 350, 352, 353t–355t, 358, 359f, 361t–362t, 363, 374, 376, 391, 395t–397t, 402t–404t, 411t–414t, 419t–429t, 438t–465t, 473t–499t, 503t–532t, 538t–539t, 591–592
Mercosur, 3–4, 24, 89t–90t, 165
Microbiological, 19–20, 25, 29, 32, 41, 51, 585–598
Microextraction, 103, 146, 171–172, 185–198, 218–221, 243–246, 277–278, 345t–348t, 358, 361t–362t, 364, 374, 437, 466–468, 500, 533–534
Microwave/s, 82, 147, 215, 220, 303–304, 308f, 316t–317t, 319–327, 320t–322t, 324f–326f, 327t, 345t–348t, 349, 469, 501
Milk/s, 48t–49t, 79, 199t–200t, 218, 234t–241t, 243–244, 345t–348t, 390, 392, 395t–397t, 401–405, 402t–404t, 406t–407t, 409–410, 411t–414t, 415, 419t–429t, 586
Minimum bactericidal concentration, 94t–100t, 102, 438t–465t, 593–595
Minimum inhibitory concentration, 593–595
Ministry of Food and Drug Safety (MFDS), 30
Ministry of Health, 51, 133, 165, 250t–251t
Ministry of Health, Labour and Welfare (MHLW), 16–18, 50–51, 123, 133, 339
Mouth, 22, 340t–341t
Mouthwash/es, 22, 30, 150, 182, 184, 186t–197t, 199t–200t, 216, 252, 255t–256t, 259–260, 262–263, 586
Musk/s, 226, 231, 234t–241t, 245, 335, 389f, 408–410, 411t–414t, 435, 472–502, 473t–499t
Mutagenicity, 570–571, 573

N

Nail polish/es, 25, 123, 147, 148t, 150–153, 199t–200t, 234t–241t, 371–374, 372t–373t, 388
Nail/s, 7, 18, 22, 24, 81, 111, 186t–197t, 364, 370–371, 374, 390, 405–408, 410
Nanomaterial/s, 6–7, 9–10, 15, 19–20, 31–32, 103, 289–302, 467, 551
Natural moss extract/s, 233–242, 244–245
Natural perfume/s, 225–226
Natural water/s, 472
Nitrosamine/s, 15, 50, 76t, 77, 183, 186t–197t, 250t–251t, 272, 274, 278–280, 331, 333t–334t, 335, 339, 351–358, 353t–355t, 359f
Non-ionic surfactant/s, 103–104, 249, 263–268, 266t, 270–272, 277–278, 361t–362t, 364, 467–468
Non-oxidative hair dye/s, 159–162, 164, 166–171

O

Oil/s, 22, 27, 108, 147, 148t, 150, 176, 182–183, 199t–200t, 216, 225–227, 234t–241t, 243, 254–257, 255t–256t, 265, 266t, 305, 311, 313, 318–321, 321t–322t, 323, 333t–334t, 376, 388
Over-the-counter (OTC), 13–16, 19t, 47, 52
Oxidative hair dye/s, 111, 161–164, 166–171

P

Paraben/s, 175–176, 180, 183, 186t–197t, 198, 213–221, 315, 387–388, 390–398, 594
People's Republic of China, 3, 19t, 20, 28–30
Perfume/s, 17, 22, 24–25, 27–28, 31, 42–43, 45–46, 61, 77, 79, 153, 199t–200t, 225–248, 312, 319, 323, 335, 356, 401, 408, 410, 585
Period After Opening (PaO), 10, 28, 30, 34
Permanent, 12, 17, 24–25, 44, 51, 73, 109–110, 133, 159–163, 570–571
Personal care, 3, 245, 265–266, 291–292, 300, 303–304
Personal Care Products Council (PCPC), 14, 42–43, 47, 50
Pharmaceutical Affairs Law (PAL), 16–18
Photoallergy, 552, 571, 576
Photogenotoxicity, 552, 571–573, 576
Phototoxicity, 306, 552, 571, 572f
Phthalate/s, 76t, 79, 335, 388, 390–391, 410–418
Physical UV filter/s, 85–86
Pigment/s, 69t–70t, 107, 123–124, 125t, 129t, 135, 147–148, 148t, 150–151, 153–154, 318–319, 340t–341t, 341–344
Plasticizer/s, 79, 255t–256t, 257–260, 388, 410
Platform of European Market Surveillance (PEMSAC), 68–69, 335–336
Potentially allergens substance/s (PAS), 229–230, 233–243, 234t–241t
Pre-market approval, 3–4, 14, 16, 19t, 26, 29, 31, 51–52, 124–132, 337–338
Preparative/s, 137f–138f, 138–140, 140f, 153
Preservative/s, 3–5, 7, 9, 18, 19t, 21, 30, 33, 40–41, 43, 48t–49t, 51–52, 61, 63–64, 67, 77–78, 110–111, 113–119, 175–224, 246, 249–252, 250t–251t, 278, 280, 333t–334t, 340t–341t, 341, 370, 388, 502–534, 585, 588, 590, 593–594
Principal display panel (PDP), 15–16, 26
Product information, 6–8, 10, 20–21, 28, 34, 296
Prohibited substance/s, prohibited ingredient/s, 42–43, 45, 47, 50–52, 57, 67, 72, 75, 164–165, 231, 234t–241t, 249–252, 337, 339, 364
Purification, 137–140, 216–217, 416
Purity, 47, 137–138, 186t–197t, 314, 318–319, 332, 335, 376–377, 585, 587–588

Q

Quality, 16–18, 24, 28, 39, 41, 67–70, 136–137, 226, 231–232, 295, 317–318, 332, 344, 376, 551–552, 559, 576, 585–598
Quality control, 17, 30–31, 39–54, 57, 67, 70, 93, 102, 113, 136–137, 154, 185–198, 231–232, 263–265, 268–270, 336, 360, 586–587
Quasi-drugs, 17–18, 30, 51–52, 133, 165

R

Raw material/s, 15, 27, 41, 43, 45–46, 51–52, 79–80, 182, 225–229, 231–232, 234t–241t, 252–272, 274, 275t–276t, 277–280, 304, 313, 315, 331–335, 337, 339–376, 387, 561–563, 585–588
Reference material/s, 58, 138–142, 145, 166
Repeated dose toxicity, 12, 567, 569–570
Reproductive and developmental toxicity, 574–575
Republic of Korea, 30–31
Research Institute for Fragrance Materials (RIFM), 231
Restricted substances, restricted ingredients, 5, 9, 14–15, 19t, 21, 29–33, 43, 48t–49t, 50–52, 67, 73, 77, 164–165, 243, 249–252, 342t–343t, 364
Russian Federation, 31–32

S

Safety, 3–4, 6t, 8–10, 12–34, 19t, 39–40, 40f, 44, 50–52, 61, 67, 86–88, 110, 112, 132–133, 163–165, 175, 183–184, 226, 231, 290f, 294–295, 297, 299, 303–305, 317–318, 323, 327, 331–332, 335–339, 344, 370, 375, 388, 549–584, 586, 593
Saudi Food and Drug Authority (SFDA), 27
Scientific Committee on Consumer Products (SCCP), 164–165, 290, 405
Scientific Committee on Consumer Safety (SCCS), 9–10, 44, 88, 163–165, 184, 227, 229, 243, 252, 278, 290, 293, 295–297, 300, 337, 351–352, 360, 364, 370, 375, 570, 587, 590
Scientific Committee on Cosmetic Products and Non-Food Products (SCCNFP), 164–165, 227, 229, 231, 568, 570
Semi-permanent, 159–161
Separation, 40, 77–78, 93–102, 113–119, 137f–138f, 140f–141f, 143f–144f, 146–148, 150–152, 154, 166–171, 198, 214–216, 220, 226, 232–233, 245, 263–264, 266–271, 278–280, 282, 345t–348t, 350–351, 358, 365, 369, 377, 470–471

Serious Undesirable Effects (SUE), 12
Serum, 366t–368t, 394, 395t–397t, 401–405, 402t–404t, 411t–414t, 415, 417–418, 419t–429t, 564, 568, 575
Shampoo/s, 13, 24–26, 51, 69t–70t, 77–78, 85, 147, 148t, 151–153, 159–160, 180–181, 199t–200t, 234t–241t, 243, 254–257, 255t–256t, 259–260, 261t, 262–267, 264t, 270–271, 275t–276t, 278, 342t–343t, 345t–348t, 359, 361t–362t, 364, 370–374, 372t–373t, 387, 392, 398–399, 410, 558, 592–593
Shaving, 13, 18, 24, 199t–200t, 387, 392
Skin, 13, 17–18, 22, 24–25, 28–30, 33, 39, 48t–49t, 76t, 78, 81, 85–88, 107–108, 110–119, 165, 175, 180–184, 199t–200t, 218, 228–229, 228t, 234t–241t, 252, 262, 265–266, 275t–276t, 278, 281, 289–292, 300, 306, 316, 333t–334t, 345t–348t, 351–352, 359–360, 365, 366t–368t, 387–388, 391, 398–399, 401, 554–557, 566–569
Skin absorption, 359, 566–569
Skin corrosion, 554–557
Skin irritation, 48t–49t, 165, 183, 229, 401, 554–557, 594–595
Skin sensitisation, 580
Soap/s, 13, 17, 24–25, 28–29, 81, 114t–118t, 151–152, 181, 186t–197t, 234t–241t, 243, 253t, 257–258, 261t, 275t–276t, 341–344, 361t–362t, 370, 392, 401, 405–408, 410
Solid-phase, 75, 103, 146, 153, 171–172, 216–219, 243–245, 270–271, 350–351, 358, 363, 391, 437–466, 533–534
South Africa, 20, 32–33, 37, 88
Spanish Association for Standardization and Certification (AENOR), 252, 344
Spectrometry, spectrometric, 77–80, 82, 93, 102–104, 113–119, 142, 146–147, 166, 171, 232–246, 258, 261t, 262–265, 271f, 319–320, 321t–322t, 344, 349–351, 359f, 391, 470–471, 540, 591–592
Spectrophotometry, spectrophotometric, 75, 147–148, 150–153, 260–263, 261t, 344, 366t–368t, 369–371, 372t–373t, 374, 563
Spectroscopy, spectroscopic, 103, 113–119, 136–137, 213, 232, 242, 260–263, 377

Stability, 25, 41, 85, 88, 136, 259–260, 277–278, 291, 335, 391, 568, 571, 586–587
Starting material/s, 137–138, 142, 146
Subsidiary colour/s, 137–140, 142, 145–146
Sun protection factor (SPF), 18, 22, 79, 81, 86
Sunscreen/s, 12–13, 17–18, 22–26, 28–30, 33, 40–41, 47, 48t–49t, 73, 79, 85–88, 110, 199t–200t, 221, 234t–241t, 275t–276t, 281f, 290, 306, 315, 340t–341t, 357, 361t–362t, 374, 392, 398–401, 470
Surfactant/s, 42, 61, 76t, 77–78, 94t–101t, 103–104, 154, 214, 249–288, 291, 340t–341t, 350, 359–364, 361t–362t, 370, 388, 438t–455t, 467–468, 555, 561–562
Switzerland, 33–34, 395t–397t
Synthetic perfume/s, 225–226

T

Tanning, 13, 22, 42, 85, 107–122
Temporary, 12, 159–161
Tissue/s, 87, 183, 319, 325–326, 387, 390–392, 395t–397t, 398, 401–405, 406t–407t, 408, 411t–414t, 436, 469–471, 501–502, 554–555, 557, 559–563, 569, 574–575
Toilet, 24, 42, 79, 199t–200t, 218, 234t–241t, 274–277, 275t–276t, 356, 387, 590
Tonic, 153, 199t–200t, 365–369, 366t–368t, 371–374, 372t–373t
Tooth/teeth, 7, 18, 22, 24–25, 76t, 80, 234t–241t
Toothpaste/s, 13, 15, 18, 24–26, 30, 47, 51–52, 68–69, 69t–70t, 150, 152, 181, 184, 186t–197t, 199t–200t, 216, 228, 228t, 234t–241t, 243–244, 255t–256t, 258, 261t, 268–270, 290, 344, 345t–348t, 349, 359, 371–374, 372t–373t
Toxicity, 12, 138, 184, 281–282, 316, 331, 364, 376, 393–394, 418, 553–554, 569–570, 574–575
Toxicokinetics, 12, 575–576
Trace/s, 9, 19–20, 68, 73, 75, 76t, 79, 82, 145–146, 226, 331–332, 333t–334t, 335–337, 339, 341, 345t–348t, 350–352, 360, 376, 472, 586–587

U

Ultrasound-assisted, 171–172, 220, 243–244, 280, 318, 350, 468
United States Adopted Names (USAN), 47
United States Pharmacopeia (USP), 252, 273–274

United States (USA), 19t, 89t–90t, 393–394, 395t–397t, 468, 556, 556f, 563–564
Urine, 390–391, 393–398, 394f, 395t–397t, 400–410, 402t–404t, 406t–407t, 415–417, 419t–429t
UV filter/s, 3–5, 9, 18–21, 19t, 27, 30, 33, 42–44, 45t, 51–52, 61, 76t, 78–79, 81–82, 85–88, 93–104, 111, 164, 215, 306–309, 314–315, 342t–343t, 387–388, 399–400, 399f, 402t–404t, 436–472, 502, 566–567, 571

V

Validation, 19–20, 58–62, 71, 71t–72t, 74, 166, 242, 264–265, 278–279, 344, 356–357, 377, 436, 552–553, 559, 561–562
Volatile organic compounds (VOC), 16, 34, 234t–241t

Voltammetry, voltammetric, 102, 166–171, 215, 369
Vortex-assisted, 220, 468

W

Waving, 17, 24, 28–29, 31, 51–52, 69t–70t
Whitening, 22, 31, 42, 61, 76t, 80–81, 107–122, 275t–276t, 345t–348t, 365–369, 366t–368t, 374

X

X-ray fluorescence (XRF), 103, 147